Principles of Environmental Sciences

Jan J. Boersema • Lucas Reijnders
Editors

Principles of Environmental Sciences

With Editorial Assistance by
Joeri Bertels and Henk Bezemer

Springer

Editors
Jan J. Boersema
Institute for Environmental Studies
Vrije Universiteit
De Boelelaan 1087
1081 HV Amsterdam
The Netherlands
jan.boersema@ivm.vu.nl

Lucas Reijnders
IBED/EDCO
University Amsterdam
Nieuwe Achtergracht 166
1018 WV Amsterdam
The Netherlands
L.Reijnders@uva.nl

ISBN 978-1-4020-9157-5   e-ISBN 978-1-4020-9158-2

Library of Congress Control Number: 2008936827

© 2009 Springer Science+Business Media B.V.
No part of this work may be reproduced, stored in a retrieval system, or transmitted in any form or by any means, electronic, mechanical, photocopying, microfilming, recording or otherwise, without written permission from the Publisher, with the exception of any material supplied specifically for the purpose of being entered and executed on a computer system, for exclusive use by the purchaser of the work.

Chapters 23, 24, 25 © Edward Elgar Publishing, Aldershot, United Kingdom

The Editors acknowledge that there are instances where they were unable to trace or contact the copyright holder for permission to reproduce selected material in this volume. The Editors have included complete source references for all such material and take full responsibility for these matters. If notified, the Publisher will be pleased to rectify any errors or omissions at the earliest opportunity.

Cover illustration: Landscape near Emden (Germany) Photo: Jan J. Boersema

Printed on acid-free paper

springer.com

Preface

Why this book?

This academic textbook is meant to be *complementary* to the many existing textbooks on environmental science. It distinguishes itself for two main reasons:

- Environmental problems, the object of the environmental sciences, are seen and described as resulting from observed phenomena in our natural environment on one hand, and the societal awareness and evaluation of these phenomena on the other. A combination of the two causes a phenomenon to be considered 'an environmental problem'. Therefore, students must learn that right from the beginning there is a *cultural and historical dimension* when it comes to defining and analysing environmental problems. That is why we pay attention to environmental history and to the variations in both perception and the implementation of solutions. In more philosophical terms: this book tries to avoid the 'Scylla of positivism' (as natural scientists, we know what the problems are) and the 'Charybdis of constructivism' (problems only exist if we see them as problems).
- This book provides a comprehensive picture of the various principles, concepts, and methods applicable to environmental problems, and relates these methods to underlying guiding principles and the adjacent policy measures. The focus is on multi- and interdisciplinary methods, although most of the methods originate from a specific discipline and many have a limited domain. Wherever possible, examples of application of a method in practice are given, as well as evaluations in terms of gains of a particular method over other methods. This *focus on methodology* distinguishes this book from other textbooks. To give just two examples: The LCA-approach (Life Cycle Analysis/Assessment) is given only a few lines in most basic textbooks, although this methodology has become widely and officially accepted by professionals and scientists (as an ISO-standard). Modelling is not explicitly dealt with in most textbooks either, although many different kinds of models are common practice in environmental studies.

Throughout this book the term sciences is meant to include the social sciences and even disciplines of the humanities.

Aim

The aim of this academic textbook is threefold:

- To describe environmental problems in their historical context
- To delineate how complex environmental problems can be analysed and tackled by using various (inter)disciplinary concepts, methods and tools and
- To illustrate how solutions work out in their social context

Readership

The book is intended to be a course text for students who take environmental science as a major or as a minor. So, the book is primarily meant for:

- Undergraduate and graduate students of multi- and interdisciplinary courses in environmental studies/sciences and courses focusing on methodology
- Graduate students specialising in environmental topics of their discipline

To a lesser degree, the book or chapters of the book may be useful as a reference to students of some post-academic course or 'lifetime learning' course for professionals in the environmental field.

Assumed background: an introductory course in environmental science and/or some years of disciplinary training.

Outline

In line with the three aims, the book is subdivided into three parts.

Part I, *Stating the Problem* (Chapters 1–6)

This part introduces the environmental sciences and gives an overview of the historical context. This is done on a large timescale, including geological and human history. It concludes with a concise description of recent developments and trends.

Part II, *Principles and Methods* (Chapters 7–18)

This is the core of the book. It starts with two chapters on the guiding principles, followed by seven chapters in which disciplinary and multidisciplinary methods are described and explained at length. The text will include many practical examples, including evaluations of the pros and cons of each example. This part concludes with three chapters on integrative methods. Special emphasis is given to the concept of integration, modelling (both as a learning and research tool) and integrated assessment.

Part III, *Context and Perspectives* (Chapters 19–28)

The last part is designed to illustrate the way solutions work in a specific societal context. The first chapter introduces the topic, which is followed by three case studies on different spatial scales. Solutions need to be implemented in and/or accepted by a given society. The same (technical/practical) solution of a more or less similar problem may provoke quite different reactions in different societies. The case studies will be used to illustrate this point. Finally the book offers perspectives on economic growth and on major societal sectors and the most likely course they will take in the future.

Although it is acknowledged that (by definition) there is no such phenomenon as an objective description of environmental problems, the book tries to avoid a too outspoken standpoint.

Learning Objectives

We expect students to learn some specific skills, e.g. the essentials of building a model or applying LCA, but our main objective is to improve their ability to analyse and conceptualise environmental problems in context, to make students aware of the value and scope of different methods and to teach them the results and insights of previous work in this field.

Amsterdam, August 2008   Jan J. Boersema & Lucas Reijnders,

Acknowledgements

As editors we are ultimately responsible for the book as it stands, although authors remain responsible for the content of their chapters.

The Ministry of Housing, Spatial Planning and the Environment is greatly acknowledged for their generous financial support. The Institute of Environmental Sciences, Leiden University is acknowledged for hosting the 'book project' in its initial period.

Many people have provided help, encouragement, and advice over the years. We are grateful to all of them in many ways:

To: Paul Roos and Betty van Herk of Springer Publisher for their enduring support and belief in the project.
To: Nachshon Rodrigues Pereira and Marleen Rengers for their technical assistance and to Edith de Roos for her bibliographical contributions.
To: The following colleagues for their constructive criticism or useful comments:

René Benders, University of Groningen
Joop de Boer, Vrije Universiteit Amsterdam
Jeroen van den Bergh, Autonomous University of Barcelona
William C. Clark, Harvard University Cambridge
Michiel van Drunen, Vrije Universiteit Amsterdam
Jodi de Greef, Netherlands Environmental Assessment Agency (MNP) Bilthoven
Wander Jager, University of Groningen
D. Jimenez-Beltran, European Environment Agency, Copenhagen
Andrew Johnson, White Horse Press, Strond UK
Paul de Jongh, Ministry of Housing, Spatial Planning and the Environment, The Hague
Henk Moll, University of Groningen
Roland Scholz, Eidgenössische Technische Hochschule, Zürich
Jan van der Straaten, Saxifraga Foundation Tilburg
Egbert Tellegen, Utrecht University

Amsterdam, August 2008   Jan J. Boersema and Lucas Reijnders

Contents

Preface	v
Acknowledgements	ix
Contributors	xv

Part I Stating the Problem

1 **Environmental Sciences, Sustainability, and Quality**   3
   Jan J. Boersema

2 **Biogeochemical Cycles** 15
   Lucas Reijnders

3 **Reconstructing Environmental Changes over
   the Last 3 Million Years** 19
   A. M. Mannion

4 **Environmental History: Object of Study and Methodology** 25
   Petra J. E. M. van Dam and S. Wybren Verstegen

5 **Human Environmental History Since the Origin of Agriculture** 33
   Clive Ponting

6 **Recent Developments and Trends** 49
   Lucas Reijnders, Jan Bakkes, and Klaas van Egmond

Part II Principles and Methods

7 **General Principles** 75
   Craig Dilworth

8 **Specific Principles** 85
   Lucas Reijnders

9 **Social Science and Environmental Behaviour** 97
   Linda Steg and Charles Vlek

10 **The Natural Environment** 143
   Piet H. Nienhuis with contributions of Egbert Boeker
   (Sections 10.2 and 10.3)

| 11 | **Analytical Tools for the Environment-Economy Interaction**........ 177
Paul Ekins |

| 12 | **Analysis of Physical Interactions Between the Economy and the Environment**.. 207
Helias A. Udo de Haes and Reinout Heijungs |

| 13 | **Environmental Policy Instruments**........................... 239
Gjalt Huppes and Udo E. Simonis |

| 14 | **Environmental Institutions and Learning: Perspectives from the Policy Sciences**......................... 281
Matthijs Hisschemöller, Jan Eberg, Anita Engels, and Konrad von Moltke[†] |

| 15 | **Technology for Environmental Problems**..................... 305
Karel Mulder |

| 16 | **Integration**.. 335
Jan J. Boersema |

| 17 | **Environmental Modelling**.................................. 345
H.J.M. (Bert) de Vries |

| 17A | **An Illustration of the LCA Technique**........................ 375
Reinout Heijungs |

| 18 | **Integrated Assessment**..................................... 385
Jean-Paul Hettelingh, Bert J. M. de Vries, and Leen Hordijk |

Part III   Context and Perspectives

| 19 | **Environmental Policies in Their Cultural and Historical Contexts** .. 423
Maurie J. Cohen |

| 20 | **National Policy Styles and Waste Management in The Netherlands and Bavaria**............................. 433
Jan Eberg |

| 21 | **Land Use in Zimbabwe and Neighbouring Southern African Countries**................................. 445
Ignas M. A. Heitkönig and Herbert H. T. Prins |

| 22 | **Climate Change Policy of Germany, UK and USA**............... 459
Richard van der Wurff |

| 23 | **Technical Progress, Finite Resources and Intergenerational Justice**................................. 473
Wilfred Beckerman |

| 24 | **Sustainability Is an Objective Value**.......................... 483
Herman E. Daly |

| 25 | **A Rejoinder to Wilfred Beckerman and Herman Daly**............ 491
Roefie Hueting |

26	**Transitions to Sustainability as Societal Innovations**	503
	Anna J. Wieczorek and Frans Berkhout	
27	**Agriculture and Food Problems**	513
	David Pimentel	
28	**Tracing the Sustainable Development of Nations with Integrated Indicators**	517
	Bastiaan Zoeteman	

Author Index 529

Subject Index 537

Contributors

Jan Bakkes, M.Sc. is senior scientist at the Netherlands Environmental Assessment Agency (MNP), Bilthoven, The Netherlands. jan.bakkes@mnp.nl

Dr. Wilfred Beckerman is emeritus fellow at Balliol College, Oxford, UK. wilfred.beckerman@econs.ox.uk

Dr. Frans Berkhout is professor of Innovation and Sustainability, and Director of The Institute for Environmental Studies (IVM) at the VU University Amsterdam, The Netherlands. frans.berkhout@ivm.vu.nl

Dr. Egbert Boeker is professor emeritus of Theoretical Physics at the VU University Amsterdam, The Netherlands. egbertb@xs4all.nl

Dr. Jan J. Boersema is professor of Environmental Science and Worldviews at the Institute for Environmental Studies (IVM) at the VU University Amsterdam, The Netherlands. jan.boersema@ivm.vu.nl

Dr. Maurie Cohen is associate professor of Environmental Policy and Sustainability at the New Jersey Institute of Technology, Newark, USA. maurie.cohen@njit.edu

Dr. Herman E. Daly is an ecological economist and professor at the School of Public Policy of University of Maryland, College Park, USA. hdaly@umd.edu

Dr. H. J. M. (Bert) de Vries is senior scientist at the Netherlands Environmental Assessment Agency (MNP) Bilthoven and professor of Global Change and Energy at the Copernicus Institute Utrecht University, The Netherlands. bert.devries@pbl.nl

Dr. Craig Dilworth is reader in theoretical philosophy at Uppsala University, Sweden. Craig.Dilworth@comhem.se

Dr. Jan Eberg is a lecturer of Politics and Social Sciences at the Institute of Safety and Security Management, *Hogeschool* of Utrecht, The Netherlands. jan.eberg@casema.nl

Dr. Paul Ekins is professor of Sustainable Development at the University of Westminster and head of the environmental group at the Policy Studies Institute, London, UK. p.ekins@psi.org.uk

Dr. Anita Engels is professor of Sociology at the Centrum für Globalisierung und Governance Fakultät für Wirtschafts- und Sozialwissenschaften Universität Hamburg, Germany. sofv001@sozialwiss.uni-hamburg.de

Dr. Reinout Heijungs is an assistant professor at the Institute of Environmental Sciences at Leiden University (CML), The Netherlands. heijungs@cml.leidenuniv.nl

Dr. Ignas M. A. Heitkönig is lecturer at the Resource Ecology Group of Wageningen University, The Netherlands. ignas.heitkonig@wur.nl

Dr. Jean-Paul Hettelingh directs the Coordination Centre for Effects at the Netherlands Environmental Assessment Agency (MNP), Bilthoven, The Netherlands. jean-paul.hettelingh@pbl.nl

Dr. Matthijs Hisschemöller is assistant professor Environmental Policy Analysis at the Institute for Environmental Studies (IVM) at the VU University Amsterdam, The Netherlands. Matthijs.hisschemöller@ivm.vu.nl

Dr. Leen Hordijk is director of the International Institute for Applied Systems Analysis (IIASA) at Laxenburg, Austria. hordijk@iiasa.ac.at

Dr. Roefie Hueting is an economist and retired as head of the Department of Environmental Statistics at the Central Bureau of Statistics (CBS), The Hague, The Netherlands. rhig@hetnet.nl

Dr. Gjalt Huppes is head of the Department of Industrial Ecology, Institute of Environmental Sciences at Leiden University (CML), The Netherlands. huppes@cml.leidenuniv.nl

Dr. A. M. Mannion is a senior lecturer in Geography at the University of Reading, UK. a.m.mannion@reading.ac.uk

Dr.ir. Karel Mulder is senior lecturer in Technology Dynamics & Sustainable Development at the Faculty of Technology, Policy & Management of Delft University of Technology, The Netherlands. k.f.mulder@tudelft.nl

Dr. P. H. (Piet) Nienhuis is professor emeritus of Environmental Science at the Department of Environmental Science, Institute for Wetland and Water Research, Radboud University Nijmegen, P.O. Box 9010, 6500 GL Nijmegen, The Netherlands. p.nienhuis@science.ru.nl

Dr. David Pimentel is professor emeritus of Ecology and Agriculture at Cornell University, Ithaca, USA. dp18@cornell.edu

Clive Ponting served as a reader in the Department of Politics and International Relations at the University of Wales, Swansea, until his retirement in 2004. He now lives in Greece.

Dr. Herbert H. T. Prins is professor and chair of Resource Ecology at Wageningen University, The Netherlands. herbert.prins@wur.nl

Dr. Lucas Reijnders is professor of Environmental Science(s) at the University of Ámsterdam and the Open University of the Netherlands, The Netherlands. l.reijnders@uva.nl

Dr. Udo E. Simonis is professor emeritus of Environmental Policy at the Wissenschaftscentrum für Sozialforschung, Berlin, Germany. simonis@wzb.eu

Dr. Linda Steg is a senior lecturer in Environmental Psychology at the Faculty of Behavioural and Social Sciences, University of Groningen, The Netherlands. E.M.Steg@rug.nl

Dr. Helias A. Udo de Haes is professor emeritus of Environmental Sciences at the Institute of Environmental Sciences at Leiden University (CML), The Netherlands. udodehaes@cml.leidenuniv.nl

Dr. Petra J. E. M. van Dam is a senior investigator and lecturer in Environmental History at the Amsterdam Institute for Heritage and Society at the VU University Amsterdam, The Netherlands. pjem.van.dam@let.vu.nl

Dr. Richard van der Wurff is associate professor at the Department of Communication and the Amsterdam School of Communications Research *ASCoR*, University of Amsterdam, The Netherlands. R.J.W.vanderWurff@uva.nl

Ir. Klaas van Egmond is professor of Earth and Sustainability at the faculty of Geoscience Utrecht University and was director of the Netherlands Environmental Assessment Agency (MNP), Bilthoven, The Netherlands. nd.van.egmond@wxs.nl

Dr. S. W. (Wybren) Verstegen is lecturer in Economic, Social and Environmental History at the VU University Amsterdam, The Netherlands. sw.verstegen@let.vu.nl

Dr. Charles Vlek is professor emeritus of Environmental Psychology and Behavioural Decision Research at the University of Groningen, The Netherlands. c.a.j.vlek@rug.nl

Dr. Konrad von Moltke (†) was director of the Institute for European Environmental Policy in Bonn, Germany and Senior Fellow at the International Institute for Sustainable Development in Canada. He passed away in 2005.

Anna J. Wieczorek, M.Sc. is an executive officer of the Industrial Transformation project of the International Human Dimensions Programme on Global Environmental Change hosted by the Institute for Environmental Studies (IVM), VU University Amsterdam, The Netherlands. anna.wieczorek@ivm.vu.nl

Dr. Bastiaan Zoeteman is professor of Sustainable Policies in International Perspective at the Faculty for Economics and Business Administration at Tilburg University and active at Telos, Brabants Centre for Sustainability Issues, Tilburg, The Netherlands. zoeteman@uvt.nl

Part I
Stating the Problem

Chapter 1
Environmental Sciences, Sustainability, and Quality

Jan J. Boersema

Contents

1.1	Introduction	3
1.2	**Concepts, Definitions, and Delineation**	3
1.2.1	Environment	3
1.2.2	Environmental Quality	4
1.2.3	Environmental Sciences	4
1.3	**Environmental Problems and Solutions in Context**	5
1.3.1	Whose Problems?	5
1.3.2	Science and Society	5
1.3.3	Solutions in Context	6
1.3.4	Two Positions for Environmental Scientists	8
1.4	**Measuring Human Impact on the Environment**	9
1.4.1	Measuring and Indicators	9
1.4.2	The IPAT-Equation	9
1.4.3	Ecological Footprint and MIPS	11
1.4.4	Measuring Sustainable Development	13
References		13

1.1 Introduction

This chapter examines the contribution of environmental sciences and scientists to the finding to solutions to environmental problems. It defines and describes important concepts, highlights methods used to analyse human impacts on the environment, and it discusses the ways in which sustainability can be measured. The chapter is subdivided into three sections:

J.J. Boersema (✉)
Environmental Sciences and Worldviews at the Institute for Environmental Studies (IVM) VU University Amsterdam, The Netherlands.
e-mail: jan.boersema@ivm.vu.nl

1.2 Concepts, definitions, and delineation
1.3 Environmental problems and solutions in context and
1.4 Measuring human impact on the environment

1.2 Concepts, Definitions, and Delineation

1.2.1 Environment

The term *environment* in environmental sciences is derived from the science of ecology. The term ecology or *oekologie* was coined by the German biologist Ernst Haeckel in 1866, when he defined it as 'the comprehensive science of the relationship of the organism to the environment'. In the environmental sciences these organisms are humans. This explains why the term *human ecology* is used sometimes as a synonym for environmental sciences. By using the latter term we want to avoid that humans are only seen as biological beings and to emphasise that we consider them primarily as social beings and as members of a society. A further restriction is placed on the use of *environment*: the social environment is excluded as an object for study. The focus is on the physical (living and not living) environment: air, water, land, and all the biota that grows and live therein. Environmental scientists are not concerned with angry neighbours, although they may well be interested in noisy traffic, the fate of cod and smokestacks (at least nowadays).

Therefore, the environment is to be defined as: *the physical, non-living and living, surrounding of a society with which it has a reciprocal relationship.*

In this definition, the living world is included and the relationship with society explicitly mentioned, contrary to the more narrow definitions of 'environment'. An example of such a narrow definition is for instance, the definition of environment used in the UK Environmental Protection Act 1990: 'consists of all, or any, of the following media, namely, the air, water and land' (Porteous 2000: 217). In the narrow sense, the term environment can also be equivalent to the term 'nature', which is defined as comprising all biota. Combinations were later made, like *the natural environment*, as opposed to the social environment. Use of the term 'environment' in the broad sense, as is done in this textbook, reflects the growing understanding of the interrelationship between both the non-living and the living world.

1.2.2 Environmental Quality

To properly assess and value the actual state of the environment, we need to have an idea of what that state could or should be; it requires the setting of a norm or adopting a reference. The difference between actual state and reference points to the concept of quality, being somehow a valuation of the gap between 'is' and 'ought'. Following the German philosopher Schopenhauer (1848) we consider quality to be a *relational* as well as *relative* concept. The quality of a specific entity always depends on the needs of a 'user'. Surface water with very low oxygen content may be of poor quality for a pike while it is up to the mark for an anaerobic. Besides being relational, quality is also relative, since it has no absolute size. Something can only have (more or less) quality with respect to a chosen or given reference. For instance, the counterpart of quality in a different field would not be 'income' but rather 'prosperity'. One's annual income can, as a rule, be measured objectively and amounts to a certain number of euros or dollars per year. Whether this can be seen as prosperous is dependent on the outcome of the comparison with a subjectively chosen reference, like the average of a given country or your neighbour's income, or what you think you deserve.

A definition which would do justice to what has been postulated above, may be the following: *quality is the level at which a quantity satisfies the function which it is expected to satisfy*. Applied to the environment and to users of the environment this has led to the listing of a whole catalogue of environmental functions, or parts thereof (Groot 1992). Once the function is known, it is easier to set a standard. A standard is defined as the state of (parts of) the environment in which or by which the function is e.g. fully or sufficiently satisfied. The more accurate the user and the function are defined and the more we understand about causal relations and other relevant mechanisms, the more precise the standards that can be set.

Quality can then be assessed by comparing the actual (or expected) state of the environment with the standard. It is noted that the actual state of (parts of) the environment can change due to either human interventions or to 'natural' fluctuations, and the emphasis here is on the human induced or enhanced changes. If people in a given society view the differences or the expected changes as negative effects, then we are referring to environmental problems. As we will see in Section 1.3 science plays an important role in this process of awareness and valuation. Environmental problems vary largely in scale and gravity (Chapter 6). On a higher level, as environmental problems can be considered a deterioration of the relationship between a society and the environment, the relationship is deemed to be unsustainable (this will be elaborated in Section 1.3.3).

1.2.3 Environmental Sciences

Environmental science(s) can now be defined as *the study of man-made environmental problems*. In the title of this book we use the plural *sciences* to acknowledge the fact that many sciences take part in these studies, ranging from natural sciences and the social sciences to the humanities. All have their distinctive language, methods and approaches (set out in the first chapters of Part II of this book). This is not to deny or undervalue the need to employ multidisciplinary and interdisciplinary approaches to analyse and solve the often intricate and complex environmental problems: it is to emphasise the interdependences and complementarities of the scientific efforts in this field. Interdisciplinary builds on disciplinary.

It is important to recognise the limits of science and scientists while trying to solve problems that are ultimately societal problems. When it comes to analysing causes, science may be argued to have a virtual monopoly but efforts required to solve environmental problems may involve factors beyond science; like funding, political will, or the cooperation of stakeholders. Most complex problems require a thorough scientific analysis to understand their root causes and underlying mechanisms but this knowledge does not always translate easily into action.

1.3 Environmental Problems and Solutions in Context

1.3.1 Whose Problems?

Environmental problems, the main subject of the environmental sciences, are currently important for society. Surveys have shown that they have now been on the public and political agenda for nearly 40 years (Dunlap 1991, 2002). At the same time, surveys do not always provide a clear idea of what people understand by environmental 'problems' or how important environmental problems are considered to be. People are generally asked what they consider to be major social issues, or they are asked to assign priorities to a number of issues specified in no further detail ('criminality', 'unemployment', 'the' environment, etc.). In either case, the result is a hierarchical listing of the issues as perceived by society at that particular instance in time. When asked to characterise environmental problems in greater detail, it makes a world of difference whether people are questioned about matters confronting them in their own everyday environment or about environmental problems in general. Such a discrepancy is to be expected, because not every general environmental problem is experienced as a problem in one's own living environment. Being aware of problems is not the same as experiencing them.

1.3.2 Science and Society

There may also be differences between the general problems cited by the public and the issues discussed in academic textbooks, journals, and reference works, or in government documents. This can be explained in a variety of ways. A given environmental problem may be quite familiar to a broad section of the public, but still only rarely cited spontaneously in surveys (e.g. the CFC refrigeration fluids causing the well known 'hole' in the ozone layer). It may also be the case that although a problem is deemed of vital importance, as well as topical interest, by policy-makers and public alike, it has little appeal as a scientific problem, i.e. worthy of research (like dog excrement in the public domain). Yet other problems may not have fully permeated the public consciousness, even though scientists and policy-makers may already have been wrestling with them for some time, an example being the worldwide loss of biological diversity (now usually termed biodiversity). However, due caution should be exercised here, as there may simply be a lack of public familiarity with the specific terminology employed; i.e. the public are familiar with the loss of certain species, like the Panda, but are less aware of the more general problem of biodiversity decline. Finally, there are also problems that are recognized by some sections of the scientific community but are yet to be acknowledged by other scientists, policy-makers and the general public (like the release of methane out of methane clathrates resulting from the warming of the permafrost or oceans). Whether such recognition indeed follows depends partly on the robustness of the data brought forward as evidence, and partly on how the problem is picked up by policy-makers and society at large. A relatively recent example of an issue becoming a recognized problem is 'hormonal pollution' by endocrine disrupters (substances acting like hormones and adversely affecting animals and humans) described by Colborn et al. (1996). If such recognition is not forthcoming, then by definition the issue at hand does not constitute an environmental problem in the sense of something requiring public attention, for public recognition, and acceptance of the problematical nature of the issue is a *sine qua non* in this respect. Of course, facts exist whether or not they are acknowledged – social constructivists seem to miss this point – but in order to turn facts into an environmental problem there needs to be some recognition on the societal level. The question, though, is this: when does a change in the environment materialise into an environmental problem? With many environmental problems, in retrospect we frequently see a shifting of public concern with time.

The social scientist Anthony Downs (1972) introduced the compelling idea of an 'issue attention cycle', later reflected in the 'policy cycle' approach of the former Dutch minister of the environment Pieter Winsemius. They identified a general pattern whereby every environmental problem goes through successive policy phases, each associated with a different degree of attention on the part of the various actors involved. The approach taken by Downs and Winsemius is a good reflection of how environmental problems always have both a factual and a perceptual side. These two aspects stand in complex relation to one another. Problems may well 'drop out of the picture', as it were, while still remaining just as topical as real-world phenomena. Facts can also be interpreted differently. Certain issues may draw massive public attention even though 'objectively speaking' there is little reason for such a sharp rise in interest. The 'oil crisis' of 1973 is a case in point. At the time there was little if any physical scarcity, and the economic scarcity (resulting from the price rise) was modest, when compared with later rises at the end of the 1970s and in the early 1980s. However the modest scale of the problem was not reflected in the massive attention it received from the politicians and the public.

Considered over a longer time span, too, there may be major changes in both public and political awareness of the issues at hand. In just a few decades the smokestack underwent a metamorphosis from a symbol of progress and reconstruction to one of environmental pollution; an emission source to be controlled. The writer and former president of the Czech Republic Václav Havel (1989) goes further, and holds the smokestack that 'fouls the heavens' to be 'a symbol of a civilization that renounces the absolute, denies the natural world and despises her imperatives'.

We also see national and cultural differences in how environmental problems are perceived and described. Such differences may be due to material circumstances, but cultural perception is also generally involved. The sentiment voiced by Havel is not as likely to be heard in contemporary China, say, and even less likely to be voiced in the same way by China's political leaders. This cannot be explained from the physical conditions, for in some regions of contemporary China the number of smokestacks is considerably greater than in the Czech Republic and the heavens at least as heavily fouled.

What can be said of environmental problems holds also true of solutions: the context is important. The case studies described in the Chapters 20–22 reveal differences between countries with respect to the way they solve their waste problems, make use of their natural resources or try to reach agreement on measures to be taken to curb emissions of greenhouse gases. In these cases there were no major differences in problem perception; nevertheless, the chosen solutions and the approach were different and these differences were – to a considerable extent – culturally determined.

There is every reason, then, to include the cultural-historical background of environmental problems and their perception within the domain of environmental research. In the next paragraph, we will do so in order to explain the emergence of the concepts of sustainability and sustainable development.

1.3.3 Solutions in Context

Human beings live in an intensive relationship with their natural surroundings. This relationship, described so aptly as 'metabolism' by Marx, forms the basis of every human society. From time immemorial, humans have made use of the natural environment to satisfy their basic needs for food, clothing, shelter, warmth, security and transport. In this respect, human beings are in principle no different from other creatures. In addition, though, humanity makes claims to the natural environment to satisfy what Maslow terms 'social needs' and 'the need for self-actualisation' (Maslow 1954).

This process of metabolism has had a severe impact; human beings have in fact radically altered the face of the earth (Marsh 1864; Thomas 1956; Simmons 1990). Even in their early stages of development, human societies were confronted with the consequences of their actions, in the form of soil salinisation and exhaustion, erosion and desertification. The extinction of plant and animal species also has a long history, going back to the Pleistocene (>10,000 BCE) according to some scholars. At a later date, but still centuries old, is local pollution of the soil, water and air with toxic substances. Clive Ponting in Chapter 5 provides a concise overview of this environmental history.

The fact that we can nevertheless speak of environmental degradation as a modern problem, despite its ultimately long history, is partly due to the global scale that humanity's environmental impacts have assumed in our era and also to our vastly expanded knowledge of the nature of that impact. But that is not all. Just as important, if not more so, is that it is now also increasingly seen as a structural problem of societies. This somewhat remarkable reversal of attitude cannot be explained by the fact that people were previously blind to the negative impacts of human action: for the most evident forms of environmental pollution, protest is almost as old as the pollution itself (Brimblecombe and Pfister 1990; Simmons 1993). The change in thinking is due mainly to the fact that today, far more so than in former times, we have become aware of the inter-relatedness, scale and scope of environmental problems, no longer categorising them as being nasty but unavoidable side-effects of our social evolution.

This process has occurred gradually over the past few decades. In the early stages, publications like *Silent Spring* by Rachel Carson (1962) and *The Limits to Growth*, the first report to the Club of Rome (Meadows et al. 1972), played a major role. Carson's work described graphically how persistent toxic chemicals were being transported through food webs around the globe, wreaking havoc with animal populations often far distant from pollution sources. *The Limits to Growth* focused minds on the inescapably finite nature of non-renewable resources and the well-nigh-impossible marriage of exponential growth (of resource use and population, for example) and sustainability in a finite world. With the United Nations' conference on the human environment in Stockholm in 1972, for the first time the environment issue was placed squarely on the international political agenda.

The growing concern worldwide in the 1970s about the global character of the emerging environmental problems did not result in any coherent strategy or 'solution' at the global level, however. Although an important initial step had been taken, the issue was by no means universally recognised and neither was there any common analysis of the problems involved. For example, in the centrally planned economies of Central and Eastern Europe authorities focused on the toxicity of substances in the workplace, while virtually ignoring the pollution outside the factory gates or the issue of natural resource depletion (Komarov 1980). In the industrialised western nations, technological cleanup was seen as the ultimate solution. At the same time, the core message of *The Limits to Growth* also met with resistance from certain quarters. For most developing nations, as well as for dominant liberal and socialist political currents in the developed world, the Club of Rome's report exuded an 'anti-growth' ideology. There was major apprehension that this would hold back growth of gross domestic product (GDP), which was deemed absolutely vital.

At the regional level in most developed countries, though, this period saw a growing focus on tackling concrete environmental problems, using an increasingly sophisticated institutional and legislative toolkit. By the end of the 1970s, most developed countries had national environmental legislation in place to control soil, water and air pollution. 'Environmental affairs' were situated within the apparatus of government, with new ministries and policy departments being established worldwide. Standards were set for many toxic substances, and filters and other technologies were installed to reduce both indoor and outdoor pollution. Economic instruments were also introduced, based on the *polluter pays principle*. Some of these measures proved very effective. By introducing levies on emissions into the atmosphere and surface waters, a shift was effectuated from end-of-the-pipe measures to process-integrated strategies. Frequently, the latter approach benefits not only the environment but also business results. Coal-burning by households was largely phased out. The results gradually became apparent within the natural environment: in rivers like the Thames and the Rhine fish stocks began recovering and in many major Western cities the 'London smog' became a thing of the past.

In some areas, though, progress was not quite as straightforward. It became increasingly apparent that e.g. over-fishing, and landscape fragmentation were leading to an overall decline in the quality of the world's ecosystems. These problems, it was realised, are related very intimately to the functioning of many of the essential sectors of today's economy: industry, transport, agriculture, fisheries and households. To control the environmental impacts of these sectors requires an integrated strategy designed for each specific sector, employing such instruments as the Environmental Impact Statement, the Environmental Audit and Life Cycle Assessment (see Chapters 11–13 and 17A). The compartmentalised strategy of the 1970s was no longer effective, and the first long-term cross-sectoral and cross-media environmental masterplans were drawn up. At the end of the 1980s, the result is paradoxical. With the environment now a fully-fledged 'issue', understood in far greater detail and depth, a realisation begins to dawn of the fundamentally unsustainable nature of much of modern humanity's interaction with the natural environment. Climate change, land degradation through erosion, declining fish catches through over-harvesting and biological extinction on a massive scale through habitat destruction are the most convincing and best-documented examples of man's unsustainable use of the natural environment. Many countries adopt the objective of reversing these broader trends and *sustainability* becomes not only a policy goal but is also seen as an important precondition for the long term viability of socio-economic and socio-cultural development (IUCN 1980; Clark and Munn 1986).

It is in this climate that the UN Commission on Environment and Development, chaired by Mrs. Gro Harlem Brundtland, embarked on its mission. The commission acknowledged and further substantiated the gravity of the world's environmental problems. Humanity's treatment of its natural environment was described as threatening not only the environment itself but also the legitimate economic and social needs of the present and, above all, future generations. The fact that many countries still faced poverty and hardship and viewed strong, sustained economic growth as their only way forward, formed an important motive for the commission to identify development as an essential objective. Moreover, to a certain extent poverty and environmental degradation prove to be positively correlated.

This explains why, in its final report, the commission recommended that the goal of sustainable development be adopted worldwide. In many countries, the ideas embodied in the Brundtland Report (WCED 1987) were subsequently adopted as a basis for government policy. For the first time, the relationship between the environment and the economy and the implications for future generations featured prominently on the international agenda.

Acceptance of sustainable development as an overall guiding framework (a solution in context) led to an explosion of studies and publications in which the concept was critically analysed and fleshed out in increasing detail. Once the aforementioned tension between sustainability and development is taken seriously, the key question becomes how prosperity can be increased while at the same time reducing environmental pressure (see Chapters 23–26).

1.3.4 Two Positions for Environmental Scientists

The considerations discussed in the foregoing section lead to two basic positions. Both relate to academic environmental research and environmental scientists.

First, environmental scientists should not align themselves too much with short-term perceptions of problems in society, nor with the problem definitions currently in sway in policy circles. The environmental scientist should seek to lay bare the underlying, more fundamental problems and unravel the relationships between them. This quest may (temporarily) lead the researcher away from the current, often whimsical, public debate, but it is essential: not only because it is the only avenue by which to arrive at more readily practicable answers to familiar problems, but above all because it holds out greater prospects for identifying as yet unknown problems and solutions. Science has a role in the disinterested definition and analysis of problems. Scientists can endeavour to present a clear picture of the situation, what positions have been and might be adopted, and where – scientifically speaking – the important and researchable problems lie. This may obviously include pronouncements on the nature, scale and relative gravity of the problems concerned.

Secondly, environmental scientists should perhaps be less pretentious about their ability to resolve social issues. Two arguments can be given for greater modesty in this respect. One, in many fields there is too little scientific knowledge available to expect 100% reliable recommendations for solutions from this quarter. Indeed, in some fields (human behaviour; complex ecosystems; the effects of climate change) this will probably remain the case in the foreseeable future. Scientists should not lay claim to unassailable knowledge. Two, although scientists may well succeed in carefully dissecting the problem at hand, as well as all the associated dilemmas, they are rarely able to resolve those dilemmas on purely scientific grounds. Conscientious scientists return the ball of these dilemmas into the court of society, or to the party commissioning the research. Less scrupulous scientists – when challenged or otherwise – may champion their own personal choices as scientific solutions. Although the first approach may leave the researcher dissatisfied, the latter also has its drawbacks. It involves a twofold risk: the solutions pursued may prove erroneous, for the scientist's expertise is not generally in the realm where the ultimate decision is to be taken; and democratic transparency may be lost, with those responsible for decisions hiding behind science or scientists. Scientists should clearly indicate when they are speaking as professionals and when they are participating in a debate as engaged citizens.

This position thus advocates a more social debate on how to resolve environmental problems and a more transparent role for science and scientists and scientific books and journals in that debate. This line of reasoning underpins the approach chosen in this textbook.

1.4 Measuring Human Impact on the Environment

1.4.1 *Measuring and Indicators*

Measuring is at the heart of science and the environmental sciences are no exception to that rule. Over the last decades a wide range of metrics and indicators has been developed (Adriaanse 1993; OECD 1998). Indicators can be used at national and international levels in state-of-the-environment reporting, measurement of environmental performance and in reporting on progress towards sustainable development. Once a set of environmental indicators has been established and measured there will be a tendency to integrate these measurements into one overarching indicator (see Chapter 16). To do this quantitatively requires a common denominator like hectares (the ecological footprint) or kilograms (the MIPS). Qualitative comparison requires an overall reference system (AMOEBA). In Section 1.4.3 three of these integrative indicators will be discussed. Since the Brundtland Report we have also seen many attempts to develop indicators for sustainable development (SD) (Kuik and Verbruggen 1991; Bell and Morse 2003), and there is a growing consensus on what can be considered a suitable indicator. The concept of SD links economic, environmental, and some social aspects and an important question is how far these essential dimensions can be integrated. It seems logical to develop indicators for each dimension, with further integration being envisaged in a subsequent phase (see Chapter 28). In the last section we will discuss some of the methods used to track progress towards a better environmental quality and to SD, but we will start with an analytical framework for analysing the human impact.

1.4.2 *The IPAT-Equation*

In the early 1970s, Paul Ehrlich and John Holdren were fighting a full-scale academic war with Barry Commoner over the question of what contributes most to environmental problems. Ehrlich and Holdren pointed to (over) population as being the worst for the planet, while Commoner argued that technology is the dominant reason for environmental degradation in modern societies (Ehrlich and Holdren 1971, 1972; Commoner 1971, 1972).

Ehrlich was not the first to blame population size and growth. The idea that population growth affects natural resources and human welfare is perhaps as old as written history itself. The Greek historian Herodotus (484–ca. 425 BCE) already noted how the population of the Lydians had outpaced (food) production, which led to a prolonged famine (*The Histories* Book I) and the Latin author Seneca the Younger (ca. 4 BCE–AD 65), living in Rome, noted a connection between population and pollution in his *Naturales Quaestiones*. Of a more recent date is Malthus' famous *Essay on the Principle of Population* (1798) in which he posed the question: 'what effect does population growth have on the availability of resources needed for human welfare'. Malthus' answer was that 'geometric' growth (exponential growth as we call it now) would eventually outstrip 'arithmetic' (linear) growth in the means of subsistence. He concluded that population growth has to be controlled. If not, the inevitable outcome will be misery and poverty. This Malthusian message was echoed clearly in Paul Ehrlich's bestseller *The Population Bomb* published in 1968 and repeated in many publications that follow thereafter (1971, 1972, 1990). The first formula presented by Ehrlich and his collaborators was intended to refute the notion that population was a minor contributor to the environmental crisis. It reads:

$$I = P \times F$$

I is the total Impact, P the population size and F the impact per capita.

Commoner had fewer predecessors to follow. In his popular book *The Closing Circle* (1971) he was concerned with measuring the amount of pollution resulting from economic growth in the United States. However, just measuring economic growth was not enough. 'The fact that the economy has grown – that GNP has increased – tells us very little about the possible environmental consequences. For that we need to know *how* the economy has grown' (Commoner 1971: 129). The economy had followed a 'counter-ecological pattern of growth' in which productive technologies with intense impacts on the environment have displaced less destructive ones' (Commoner 1971: 175). For that reason he felt that we have to include a specific factor to measure the impact per unit of economic production. 'Impact' he defines as pollution. The equation he published in 1972 reflected this view:

I = Population × (Economic Good/Population) × (Pollutant/Economic Good)

Population was used to express the size of the (US) population in a given year or the change in population over a defined period. Economic good (referred to as Affluence) was used to express the amount of a particular good produced or consumed during a given year or the change over a defined period. Pollutant refers to the amount of a specific pollutant released per economic good and reflecting the nature of the productive technology.

After cancelling out the identical factors what remains is: I = Pollutant. By defining the factors more rigorously Commoner became the first to apply the IPAT concept in a quantitative way.

Both combatants try to prove themselves as correct by adapting and applying 'their' equation to specific processes and/or products. Over the years of their dialogue the equation grew into the following form:

$$I = P \times A \times T$$

With P being the population or population growth, A being a measure of individual or collective welfare (GDP, goods or services per capita) and T the environmental impact per unit of A, reflecting the technological performance.

The use of the IPAT equation has been met with criticism. The criticism concerned especially the supposed independence of the three factors and the truncation of Technology. Is it possible to consider the technology as being fully independent or is there a mutual dependence? Most likely there is a reciprocal relationship and it is quite possible that Boserup (1981) is partially right in stating that it is precisely the population growth which is the driving force for new technological development and innovations, thereby increasing the affluence. Maybe Julian Simon, who believes that increasing population and wealth together evokes new technologies, deserves some credit (Simon 1980, 1981). And finally we have the so called Kuznets curve, a nuanced relationship between A and I, such that an environmental emission might rise as income increases until a particular level is reached, at which point emission levels begin to fall (Arrow et al. 1995).

The value of the equation in providing conclusive answers to the question raised by Ehrlich and Commoner may be overstated. In this book, we will not elaborate on this but instead refer the reader to the literature: for instance, to the thoughtful reviews made about the history and various interpretations of the IPAT equation by Dietz and Rosa (1994) and by Chertow (2000).

Now we would like to illustrate the use of the equation as an analytic framework with an example taken from Amory Lovins. It concerns the environmental impact of a coffee mug in connection with its use of energy (see Fig. 1.1).

Itemising the separate factors A and T clarifies where each environmental pressure is created and what the possibilities of reducing it are at that detailed level, what the estimated size of it is and what time scales we have to keep in mind in that connection. The figure shows in 'Window-like scrolls', as it were, what is hidden behind the aggregated factors P, A and T. This creates a useful framework for research strategies and policy measures. According to Chertow this kind of application has proven to be the most valuable. He writes 'the use of the IPAT equation in research related to climate change, specifically energy-related carbon emission studies, may be the most enduring legacy of IPAT' (Chertow 2000: 19; see also Chapter 6.1 for a similar use of IPAT).

Whereas Commoner introduced the factor T because he considered technology to be the largest contributor to pollution, since a few years we notice that this is reversed, which leads to a remarkable optimism whereby technological improvements are regarded as an essential part of the *solution* (Heaton et al. 1991). This technological optimism is also apparent in approaches that use the IPAT concept for future oriented programs, setting goals for research and policy alike. Based on the same IPAT concept the emphasis now has been placed on the need to substantially reduce

Energy-related environmental impact =

population × affluence × impact =

population × [stock/person × throughput/stock] × [energy/throughput × impact/energy]

Example

population × cups/person × (plastic/y, ceramics/y,...)/(cup maintained) × (GJ, kW-h,...)/(kg material/y) × (CO_2, NO_x, land-use,...)/(GJ, kW-h,...)

Applicable tools

family planning	values	product longevity	end-use efficiency	benign sources
female literacy	prices	materials choice	conversion efficiency	scale
social welfare	full costing	minimum-materials design	distribution efficiency	siting
role of women	what do we want?	near-net-shape mfg.	system integration	technical mitigation
land tenure	how much is enough?	recycle, reuse, remfg.	process redesign	offsets
		scrap recovery		

Approximate scope for long-term change

~ 2x   ?   ~ 3-10x   ~ 5-10x   ~ 10^2-10^3x

Timescale of major change

~ 50-100 y   ~ 0-50 y   ~ 0-20 y   ~ 0-30 y   ~ 0-50 y

Fig. 1.1 IPAT applied to the energy related impact of a coffee mug
Source: After Amory Lovins personal communication 1996

global material flows or on completely re-usable materials (McDonough and Braungart 2002). The Factor 10 club for instance has advocated that the current productivity of resources must be increased by an average of a factor ten during the next decades (Schmidt-Bleek 1998). Von Weizsäcker et al. (1997) state in their book Factor Four that the amount of wealth extracted from a unit of natural resources can quadruple by doubling the A while halving the T factor. They define technological progress overall as ecoefficiency, a gain in productivity of resources. For an overview of this 'Factor X debate' see Reijnders (1998).

1.4.3 Ecological Footprint and MIPS

In order to determine the total human impact on the environment the separate impacts have to be joined together. To this end, in principle two courses are available. The first course is a quantitative one. It involves reducing the impacts to a common denominator which then may be added up. The best known and most developed methods are the ecological footprint and the Material Intensity per Service unit (MIPS).

The *ecological footprint* became known in particular by the work of Wackernagel and Rees (1996) and since then has continuously been developed further. It is defined by the authors as follows: 'Ecological footprint analysis is an accounting tool that enables us to estimate the resource consumption and waste assimilation requirements of a defined human population or economy in terms of a corresponding productive land area' (Wackernagel and Rees 1996: 20) Recently, internationally there has been an attempt at attaining a generally accepted methodology (www.footprintnetwork.org). For the purpose of calculating the footprint, the human consumptive activities are being converted to the use of land which is required for making those activities possible. Our food requires agricultural and pastureland, our houses require land for building etc. For energy consumption another conversion takes place. At the end, a footprint per person, per city or country may be determined (see Box 1.1).

> **Box 1.1** The ecological footprint: tracking human demand on nature
>
> Ecological Footprint comparisons of human demand on nature with nature's regenerative capacity are updated each year. Recent calculations, available on the website of the European Environment Agency, show that the average Canadian required, in 2002, over 7.5 global hectares (or 18.5 *acres*) to provide for his or her consumption. The average Italian lived on a footprint almost half that size (4.0 global hectares or 10 *acres*). The average Mexican occupied 2.4 global hectares (6 *acres*), the average Indian lived on 0.7 global hectares (1.7 *acres*). Average demand globally was 2.2 ha per person (5.4 *acres*).
>
> In contrast, globally there were 1.8 ha (4.5 *acres*) of biologically productive land and sea area available per person in 2002. Maintenance of biodiversity also depends on this area.
>
> Comparison of supply and demand shows that humanity's Ecological Footprint exceeded the Earth's biocapacity by over 20% (2.2/1.8 ha = 1.2). In other words, it took 1 year and more than 2 months to regenerate the resources humanity consumed in 2002.
>
> The Ecological Footprint can be applied at scales from single products to households, organizations, cities, regions, nations, and humanity as a whole.

Humanity's Ecological Footprint Exceeds Earth's Biological Capacity

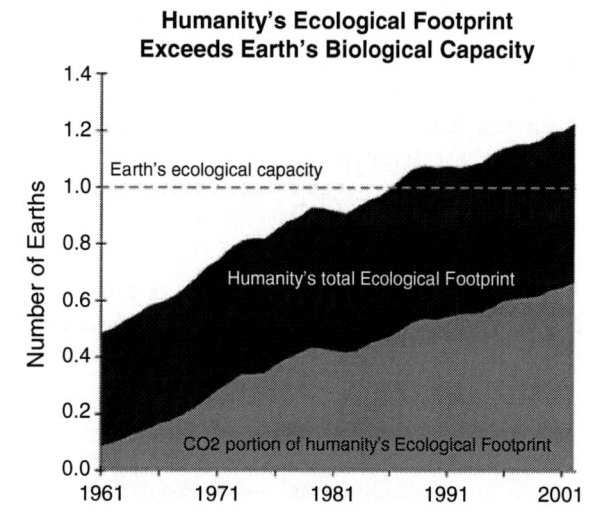

Such calculations show large differences between countries which eat up space and countries which are within or even below the 'norm'. In this way, it has been calculated for the earth as a whole that actually the hectares which are needed are 1.2 times the earth if all our consumptive needs are to be met. From the global perspective we are living beyond our means.

The ecological footprint has turned out to be a powerful means of communication with which to express that most likely with our production and consumption we are exceeding the earth's supporting power. Some researchers (Bergh and Verbruggen 1999) have expressed criticism at the reliability and especially the use of lower-scale levels. The method discriminates, for instance, too little with regard to differences in quality of the land use, whereas the impact of man on the biodiversity cannot be included properly in the calculations either.

The metaphor of *ecological backpack* was introduced by Schmidt-Bleek (1994) in order to illustrate the concept of material intensity of a product or a service per unit (*MIPS*). This MIPS is the amount of material which is required for the production and use of certain goods and products. The calculation is carried out on the whole life cycle of a good, including all the emissions and flows of waste. Goods are supposed to provide services. A car, for instance, provides 'transportation kilometers'. In order to make a comparison with other modalities of transportation the backpack may be calculated per 'unit of service' i.e. transportation kilometer. That way the effect of recycling and a lengthening of the life span becomes visible as well. For the production of a golden ring, for instance, which may weigh some 20 g, several tons of minerals have been transmuted. Lignite, too, has a backpack which is ten times as heavy as the weight of lignite itself.

Although the ecological backpack may be considered to be an appealing metaphor, the conversion into kilograms has not become a widespread method with which to determine integrally the pressure on the environment.

In addition, there are integration methods which do require quantification, but where the indicators are not reduced to a common denominator and added. The so-called *AMOEBA* approach is in this category. It is a visualisation in which the actual situation of a certain system such as for instance the North Sea, is compared with a reference after having been measured in a number of indicators (for an example see Chapter 10, Fig. 10.3). It is a *distance to target* (*DTT*) method that can

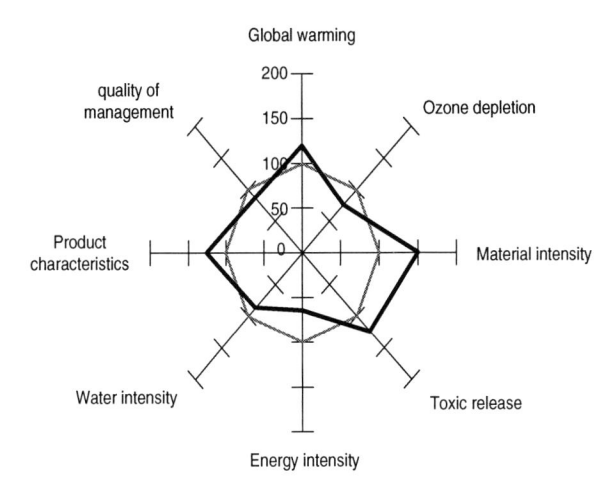

Fig. 1.2 AMOEBA applied to a firm

be applied to many systems. The reference values are put to 100% and the observed values are represented as a percentage of the reference. The general idea is made visible in Fig. 1.2.

The advantage of this approach is that at a single glance the differences in the diverging surface become clear. The method has taken its name from the often fanciful patterns which may form, for the organism amoebe may take many shapes as well. Another advantage is the wide applicability. The system and the indicators may be chosen freely and may be absorbed in any dimension in the same figure. If the crucial parameters of a certain system are known, they may be represented as indicators in relation to the reference. This reference may be the desired ideal as well. In addition, some points of criticism may be mentioned. First of all, the method is very sensitive to the choice of reference and this may not always be made objectively. The method gives little insight into the meaning of the visualized differences. The indicators included in the model may be very heterogeneous and therefore measured in totally different metrics. It is a first approach after which further analysis needs to be done.

1.4.4 Measuring Sustainable Development

As explained above, the concept of SD links economic and environmental aspects. However most indicators discussed so far point to the environment and are therefore sometimes referred to as indicators of 'environmental sustainability' (see Chapter 28). If the quality of the environment is to be integrated into the economic indicators we have to look at the most dominant indicator for economic growth: Gross Domestic Product (GDP). This gave rise to the so called 'greening of GDP'. By incorporating external effects into the determinants of growth this figure is corrected in order to better reflect the 'real' growth (Darmstadter 2000). However, it is unclear to what extent this aim can be duly achieved and how far this process will bring us towards sustainability. New alleys were explored and within the economic domain one of the most promising as well as far reaching proposals of greening growth is the Index of Sustainable Welfare (Daly 1996).

Inspired by John Elkingtons book *Cannibals with Forks* (1998) – written for the business community – it became fashionable to define sustainability as having three dimensions of equal importance. He included the social dimension as a separate third pillar or corner in a triangle. 'Sustainability is the principle of ensuring that our actions do not limit the range of economic, social and environmental options open to future generations' (Elkington 1998: 20). Following this definition sustainability might turn into a mixture of three types of sustainabilities, a trade-of in a triangle. Viewing sustainability as an important precondition for the long term viability of socio-economic and socio-cultural development (as is done in this chapter) emphasise the environmental sustainability to be a prerequisite. Which social indicators should be integrated and how this could be done is by and large unknown.

Zoeteman, in Chapter 28, is navigating such uncharted waters when he calculates the sustainability of nations by combining indicators taken from three domains. For the social domain he uses indicators from the human development index (UNDP 2001).

References

Adriaanse, A. (1993). *Environmental policy performance indicators: A study on the development of indicators for environmental policy in the Netherlands*. The Hague: SDU.

Arrow, K., Bolin, B., Costanza, R., Dasgupta, P., Folke, C., Holling, C.S., Jansson, B-O., Levin, S., Mäler, K-G., Perrings, Ch., & Pimentel, D. (1995). Economic growth, carrying capacity, and the environment. *Science, 268*, 520–521.

Bell, S., & Morse, S. (2003). *Measuring sustainability. Learning by doing.* London: Earthscan.

Bergh, J. C. J. M. van den, Boserup, E. (1981). *Population and technological change. A study of long-term trends.* Chicago: University of Chicago Press.

Brimblecombe, P., & Pfister, C. (Eds.). (1990). *The silent countdown: Essays in European environmental history.* Berlin: Springer.

Carson, R. (1962). *Silent spring.* New York: Houghton Mifflin.

Chertow, M. R. (2000). The IPAT equation and its variants: Changing views of technology and environmental impact. *Journal of Industrial Ecology, 4*(4), 13–29.

Clark, W. C., & Munn, R. E. (Eds.). (1986). *Sustainable development of the biosphere.* International Institute for Applied Systems Analysis (IIASA). Cambridge: Austria/Cambridge University Press.

Colborn, T., Dumanoski, D., & Peterson Myers, J. (1996). *Our stolen future.* New York: E.P. Dutton.

Commoner, B. (1971). *The closing circle. Nature, man and technology.* New York: Bantam.

Commoner, B. (1972). A bulletin dialogue on the 'The Closing Circle': Response. *Bulletin of Atomic Scientist, 28*(5), 42–56.

Daly, H. E. (1996). Beyond Growth. Boston: Beacon Press.

Darmstadter, J. (2000). *Greening the GDP: Is it desirable, is it feasible?* Washington, DC: Resources for the Future.

Dietz, Th., & Rosa, E. A. (1994). Rethinking the environmental impacts of population, affluence and technology. *Human Ecology Review, 1*, 277–300.

Downs, A. (1972). Up and down with ecology – The issue attention cycle. *The Public Interest, 2*, 38–50.

Dunlap, R. E. (1991). Trends in public opinion towards environmental issues: 1965–1990. *Society and Natural Resources, 4*, 285–312.

Dunlap, R. E. (2002). An enduring concern. *Public Perspective*, September/October, 10–14.

Ehrlich, P. R. (1968). *The population bomb.* New York: Ballantine.

Ehrlich, P. R., & Holdren, J. (1971). Impact of population growth. *Science, 171*, 1212–1217.

Ehrlich, P. R., & Holdren, J. (1972). A bulletin dialogue on 'The Closing Circle': Critique: One dimensional ecology. *Bulletin of Atomic Scientists, 28*(5), 16–27.

Ehrlich, P. R., & Ehrlich, A. (1990). *The population explosion.* New York: Simon & Schuster.

Elkington, J. (1998). *Cannibals with forks. The triple bottom line of 21st century business.* Gabriola Island, BC: New Society.

Groot, R. S. de (1992). *Functions of nature. Evaluation of nature in environmental planning, management and decisionmaking.* Groningen: Wolters-Noordhoff.

Havel, V. (1989). *Do R°uzných Stran.* Prague: Scheinfeld-Schwarzenberg.

Heaton, G., Repetto, R., & Sobin, R. (1991). *Transforming technology: An agenda for environmentally sustainable growth in the 21st century.* Washington, DC: World Resources Institute.

Herodotus *The Histories.* With an English Translation by A.D. Godley. 4 Vols. Loeb Classical Library. Cambridge, MA: Harvard University Press.

IUCN (1980). *World conservation strategy: Living resource conservation for sustainable development.* Gland: International Union for the Conservation of Nature.

Komarov, B. (1980). *The destruction of nature in the Soviet Union.* Translation from Russian by M. E. Sharpe. New York: White Plains.

Kuik, O., & Verbruggen, H. (Eds.). (1991). *In search of indicators for sustainable development.* Dordrecht/Boston: Kluwer.

Malthus, R. T. (1798). *An essay on the principle of population as it affects the future improvement of society, with remarks on the speculations of Mr. Godwin, M. Condorcet, and Other Writers.* London: J. Johnson.

Marsh, G. P. (1864/1965). *Man and nature. Or physical geography as modified by human action.* New York: Charles Scribner (1864). Reprinted in 1965 by Harvard University Press, Cambridge, MA.

Maslow, A. (1954). *Motivation and personality.* New York: Harper.

McDonough, W., & Braungart, M. (2002). *Cradle to cradle.* New York: North Point Press.

Meadows, D. H., Meadows, D. L., Randers, J., & Behrens III, W. W. (1972). *The limits to growth. A report for the club of Rome's project on the predicament of mankind.* New York: Potomac Associates.

OECD (1998). *Towards sustainable development: Environmental indicators.* Paris: Organisation for Economic Cooperation and Development.

Porteous, A. (2000). *Dictionary of environmental science and technology* (3rd ed.). Chichester: Wiley.

Reijnders, L. (1998). The Factor X Debate: setting targets for eco-efficiency. *Journal of Industrial Ecology, 2*(1), 13–22.

Schmidt-Bleek, F. (1994). *Wieviel Umwelt braucht der Mensch? MIPS – Maß für ökologisches Wirtschaften.* Basel/Berlin: Birkhäuser.

Schmidt-Bleek, F. (1998). *Das MIPS Konzept. Weniger Naturverbrauch – mehr Lebensqualität durch Faktor 10.* München: Droemer-Knaur.

Schopenhauer, A. (1848/1961). *Parerga und Paralipomena: kleine philosophische Schriften* (Zweiter Band, Par. 177 Ueber das Christentum). Wiesbaden: Sämtliche Werke, Brockhaus.

Seneca *Naturales Quaestiones.* (1971). With an English Translation by T.H. Corcoran, 2 Vols. Loeb Classical Library, Cambridge, MA: Harvard University Press.

Simmons, I. G. (1990). *Changing the face of the earth. Culture, environment, history.* Oxford: Basil Blackwell.

Simmons, I. G. (1993). *Environmental history. A concise introduction.* Oxford: Basil Blackwell.

Simon, J. (1980). Resource, population, environment: An oversupply of false bad news. *Science, 208*, 1431–1437.

Simon, J. (1981). Environmental disruption or environmental improvement? *Social Science Quarterly, 62*(1), 30–43.

Thomas, W. L. Jr. (Ed.). (1956). *Man's role in changing the face of the earth.* University of Chicago: Chicago Press.

UNDP (2001). *Human development report.* Oxford: Oxford University Press.

Wackernagel, M., & Rees, W. (1996). *Our ecological footprint. Reducing human impact on the earth.* Gabriola Island BC: New Society Publishers.

WCED (World Commission on Environment and Development) (1987). *Our common future.* Oxford: Oxford University Press.

Weizsäcker, E. U. von, Lovins, A. B., & Hunter Lovins, L. (1997). *Factor four: Doubling wealth, halving resource use.* (German edition published in 1995.) London: Earthscan.

Chapter 2
Biogeochemical Cycles

Lucas Reijnders

Contents

References 17

It is now often assumed that life first appeared on planet Earth about 3,500 million years ago. Since then 'our' Sun has changed considerably. While the flux of solar energy to the Earth has increased by about 30% over this period, though, this has not led to a corresponding increase in the Earth's temperature or the amount of ultraviolet radiation reaching the planet's surface.

The main reason for the absence of any major change in the Earth's temperature over this extended period is that the concentrations of so-called greenhouse gases – i.e. gases transparent to visible light but absorbing infrared radiation – such as carbon dioxide (CO_2) and methane (CH_4) have fallen dramatically. Ultraviolet irradiation of the Earth's surface has in all probability declined substantially since life's first origins, a crucial development because DNA and other vital cell components are easily damaged by ultraviolet radiation. The decrease in the UV radiation striking the Earth's surface is due to the presence of an 'ozone layer' in the stratosphere, the section of the atmosphere 15–50 km above the Earth's surface containing about 90% of atmospheric ozone. The ozone in this layer is a strong absorber of UV radiation. This long-term decline in atmospheric levels of greenhouse gases and the formation of the ozone layer are intimately linked to the development of life on Earth.

The decrease in concentrations of CO_2 and CH_4 is due largely to the biogeochemical 'carbon cycle'. This cycle, involving both biotic and abiotic processes,

L. Reijnders (✉)
Environmental Science(s) University of Amsterdam and the Open University of the Netherlands, The Netherlands
e-mail: l.reijnders@uva.nl

transfers carbon within and between four major reservoirs: the lithosphere (the solid outer crust), the hydrosphere (the aqueous envelope, i.e. water bodies), the atmosphere and the biosphere. The carbon cycle is not and has never been a perfect cycle. It has led, rather, to burial in the lithosphere of large amounts of carbon originally present in the atmosphere. The White Cliffs of Dover, the oil reserves of the Middle East and coalmines of China are all places where carbon was buried in the remote past. Figure 2.1 shows the carbon cycle as it is at present. One of the notable aspects of the situation today is the man-made transfer of carbon from the lithosphere to the atmosphere, increasing the temperature of the lowest part, the troposphere.

The genesis of the Earth's ozone layer is closely bound up with the emergence and development of photosynthesis: the conversion of atmospheric CO_2 into organic matter by plants, a process known as carbon fixation and driven by the energy provided by sunlight. Photosynthesis is accompanied by the emission of oxygen (O_2), which can in turn be converted to ozone (O_3), a process occurring mainly in the stratosphere and driven energetically by ultraviolet radiation. Through its capacity to absorb damaging UV, the ozone layer vastly increased the capacity of life-forms to colonise the land and the upper layer of the hydrosphere.

More generally, photosynthetic production of oxygen has increased the atmospheric concentration of this gas from mere traces to its current level of 21%. This permitted development of relatively complex and warm-blooded animals such as mammals, which need a large amount of energy to maintain their bodily processes; with insufficient atmospheric oxygen, the energy generated by the metabolic conversion of food is inadequate.

On the other hand, there are also limits to the amount by which the oxygen concentration of the atmosphere can safely rise. Under an 'over-oxygenated' atmo-

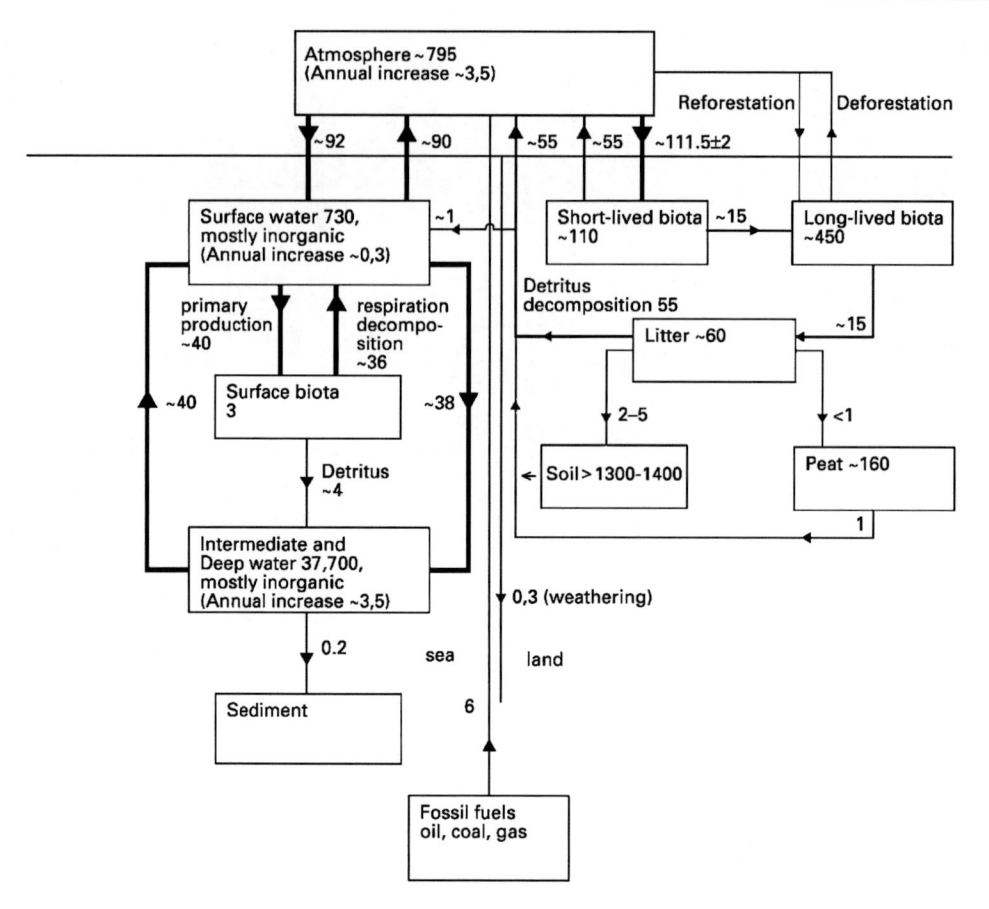

Fig. 2.1 Reservoirs and fluxes, in units of 10^{12} kg C, for the part of the current global carbon cycle that has a turnover time of less than 1,000 years
Source: Adapted from Bolin and Cook (1983)

sphere, biomass would be more often subject to fire. As fire converts biomass carbon to CO_2, however, oxygen levels would be lowered once more.

Current atmospheric levels of oxygen and carbon dioxide are just two of the aspects of our environment that have been shaped by biogeochemical cycles. In fact, many elements undergo transferral between the atmosphere, hydrosphere, lithosphere and biosphere as a result of biotic and abiotic processes. Several of these biogeochemical cycles, including the chlorine, sulphur and nitrogen cycles, which are e.g. important determinants of the concentrations of atmospheric trace gases such as N_2O, CH_3Cl and dimethyl sulfide, will be discussed in Chapter 6.

The remarkable long-term stability of the Earth's surface temperature and the decrease in ultraviolet irradiation of the biosphere were also noted several decades ago by the British scientist James Lovelock and were instrumental in development of his 'Gaia theory'. This theory (Lovelock 1989), named after the Greek Earth goddess Gaia, suggests that the planet is essentially a 'super-organism', characterised by homeostasis: the tendency for organisms to maintain a fairly constant internal environment, as in the case of temperature control in the human body, which is likewise regulated by means of 'negative feedback'. As we have already glimpsed in the case of spontaneous biomass combustion in an over-oxygenated atmosphere and its subsequent 'correction', our planet is clearly susceptible to such feedback mechanisms. Another example is the intensification of photosynthesis with rising atmospheric levels of CO_2 (in the absence of other limiting factors), with an attendant increase in carbon fixation and oxygen production.

However, there are also cases of *positive* feedback that tend to accelerate processes of environmental change. All else remaining equal, a rise in atmospheric CO_2 levels will be mirrored in a temperature rise, promoting microbial respiration of the carbon present in soils, in turn for instance leading to elevated soil emissions of CO_2 from arable soils (Ogle et al. 2005). Other examples of positive feedback are encountered in the context of the Ice Ages of the past 3 million years, discussed in more detail in the next chapter. These Ice Ages were triggered by the so-called Milankovitch cycles associated with peculiarities of the Earth's movement around the Sun (see Fig. 3.1). During recent Ice Ages, however, following the initial cooling brought on by this cycle, photosynthesis in the oceans increased, thereby reinforcing the cooling trend. When the Milankovitch cycle triggers atmospheric warming, there is probably also positive feedback, as the huge, frozen reservoirs of methane 'ice' accumulating in tundra soils and in oceans in the course of the previous Ice Age begin to melt, releasing gaseous methane into the atmosphere. As methane is a greenhouse gas, this reinforces global warming. Thus, the Earth is characterised by both positive and negative feedbacks involving the biosphere in many ways. This means that it is not certain that perturbations will have a homeostatic outcome.

References

Bolin, B., & Cook, R. B. (Eds.). (1983). *The major biogeochemical cycles and their interactions*. Chichester: Wiley.

Lovelock, J. E. (1989). *The ages of gaia*. Oxford: Oxford University Press.

Ogle S. M., Breidt F. J., Paustian, K. (2005). Agricultural management impacts on soil organic carbon storage under moist and dry climatic conditions of temperate and tropical regions. *Biogeochemistry, 72*, 87–121.

Chapter 3
Reconstructing Environmental Changes over the Last 3 Million Years

A.M. Mannion

Contents

3.1	Background	19
3.2	Past Environmental Change	21
References		24

3.1 Background

Ever since the Earth's creation, some 5 billion years ago, environmental change has been a defining characteristic of our planet. At first these changes were purely inorganic in nature: weathering and erosion of the Earth's surface, and tectonic processes beneath the crust. As life forms began to develop, though, a new, organic influence came to be exerted on the planetary environment. These abiotic and biotic influences continue to this day and are reciprocally related through the various biogeochemical cycles that transport chemical elements within and between the atmosphere, hydrosphere, lithosphere and biosphere. In addition to these 'internal' planetary characteristics and mechanisms, external factors also exert a degree of control over processes of environmental change, the most important of which is the periodicity of the Earth's movement around the Sun, defined by so-called Milankovitch cycles, as shown in Fig. 3.1.

Over the last 200 years our understanding of the Earth's dynamics and evolution has improved enormously. One major early contributor was undoubtedly James Hutton, who in the late 1700s first proposed the theory that parts of the Earth had in the distant past experienced an extended period of glaciation. Other important landmarks include Charles Darwin's work in the mid-1800s, *The Geographical Cycle*, published by William Morris Davis in 1899, and Alfred Wegener's ideas on Continental Drift, published in 1915. More recently, further conceptual elements of a dynamic Earth were introduced by Sir Arthur Tansley's ecosystem theory of the 1930s, the plate tectonics proposed by J. Tuzo Wilson in the 1960s, a time when the systems approach was being adopted in the Earth and Environmental Sciences, and the 'Gaia hypothesis' advanced by James Lovelock in the early 1970s. Equally important is the role of ice and sedimentary archives whose gas and fossil content (pollen, insect remains etc. see below) respectively has facilitated environmental reconstruction and drawn attention to the significant environmental impact of humans during the Holocene. This issue has recently been revisited by Ruddiman (2003) who has suggested, controversially, that human impact on the atmosphere can be detected as far back as 8,000 years ago due to carbon dioxide release following deforestation for expanding agriculture. His opinion is based on a comparison of interglacial and Holocene ice-core carbon dioxide and methane trends though others (e.g. Claussen et al. 2005) indicate that anomalous trends might be expected due to a non-linear response of the carbon cycle to external factors such as insolation. Thus Ruddiman's observations may be due to natural variation rather than land-cover change by humans.

It is also important to signal the role of contemporary environmental issues in bringing the reality of a rapidly changing environment to the attention of the general public. Increasing travel and tourism and especially the rapid growth in media focus and access (notably television) have highlighted such issues as deforestation, loss of biodiversity, acidification, stratospheric ozone depletion and global warming. This has not only made individuals aware of their role in

A.M. Mannion (✉)
University of Reading, U.K
e.mail: a.m.mannion@reading.ac.uk

The astronomical forcing factors involved in the Milankovitch theory (astronomical theory) of climatic change

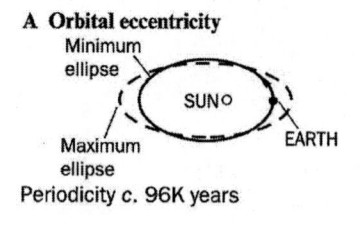

A Orbital eccentricity

Periodicity c. 96K years

The Earth's orbit around the Sun varies, and is elliptical rather than circular. When the Earth is furthest from the Sun, cooling occurs. The periodicity of orbital eccentricity is considered to be a major factor in the waxing and waning of ice ages.

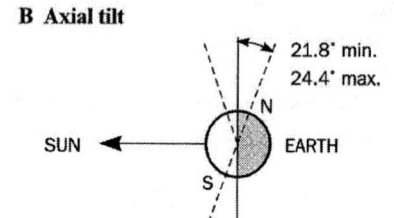

B Axial tilt

Periodicity c. 42K years

The tilt of the axis around which the Earth rotates causes its seasonality. It also determines the intensity of incident radiation. When the angle of tilt is at its minimum, 21.8°, incident radiation in the Northern Hemisphere is c. 15 per cent less than when the angle of tilt is at its maximum, 24.4°. Periods of minimum tilt therefore relate to cooling.

C Precession of the equinoxes

Periodicity c. 21K years

This occurs due to the wobble of the Earth's axis. It controls the amount of solar radiation received at the Earth's surface by influencing the season in which the Northern Hemisphere is closest to the Sun. In particular, an ice age is likely to develop when the Northern Hemisphere is furthest from the Sun in summer.

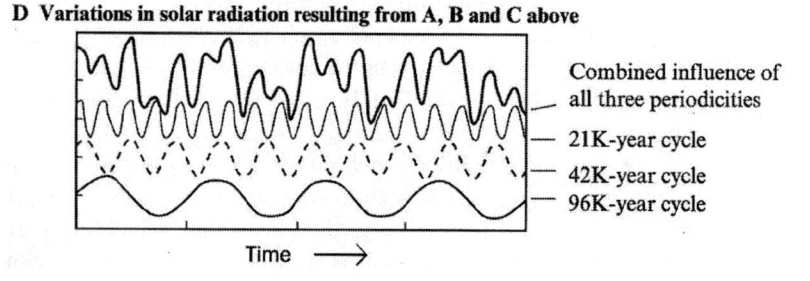

D Variations in solar radiation resulting from A, B and C above

- Combined influence of all three periodicities
- 21K-year cycle
- 42K-year cycle
- 96K-year cycle

Fig. 3.1 The Milankovitch cycles
Source: After Mannion (1999b)

environmental change but has also demonstrated the interdependence of people around the globe. Moreover, concerns at the local, national and international level about the potential socio-economic impacts of environmental change have renewed interest in research into environmental change (and its funding) and prompted greater international cooperation.

These new initiatives fall into three main categories, the foci of which are past, present and future environmental change. In relation to the past, the premise is that former environmental change provides information on both naturally and culturally driven change today, opening up possibilities for identifying thresholds in environmental systems and the responses of individual plant and animal species to changing environmental conditions. Research on the earth's present surface characteristics and processes has greatly illuminated the physics, chemistry and biology of environmental systems and the myriad ways in which the biotic and abiotic components are related. Observations of today's environmental systems provide not only analogues for interpreting past conditions and processes but also information that can be usefully applied for the purposes of environmental management. Cases in point include various forms of environmental modelling and prediction, which are often based on data relating to the past and the impacts of climatic change, for example, on agriculture and sea level rise. Such models can be validated and uncertainties assessed, moreover, through reference to past and/or present circumstances.

3.2 Past Environmental Change

In all likelihood environmental change played a significant role in hominid evolution and the later spread of *Homo erectus* and subsequently *Homo sapiens* out of Africa. Similarly, a combination of environmental change and socio-cultural circumstance may have been what triggered the emergence of human agriculture as the last Ice Age drew to a close, about 10,000 years ago (reviewed in Mannion 1999a).

The period of hominid presence on Earth provides a good illustration of the planet's remarkable dynamism. Since the acceptance of Hutton's glaciation theory in the mid-19th century an enormous body of information has accrued on the environmental changes that have occurred over the last 3 million years, i.e. the latter part of the Tertiary period and the whole of the Quaternary. In official geological terms the latter comprises the last 1.8 million years, though there are in fact arguments for placing the Tertiary-Quaternary boundary much earlier, at about 2.6 million years before present (BP), based on evidence of climatic cooling. (For reviews of this debate, see Mannion 1997 and Lowe and Walker 1997.) Even though this is the most recent period of Earth's history, the environmental record is difficult to interpret. This is despite the enormous scientific strides made in the last century, with physics, chemistry, biology, mathematics, geography, geology and archaeology all contributing to enhancing our understanding of the distant past. In common with most avenues of research, however, an answer to one particular problem often raises numerous questions about related topics. The research is thus as dynamic as the glacial era it purports to investigate.

Early research, which was focused on terrestrial geological deposits, led to the original glaciation theory being abandoned and replaced by a theory of multiple glaciations. Study of glacial and periglacial deposits and other evidence led to ice-sheet reconstructions, a better understanding of directions of movement, more sophisticated theories of climatic change and consequently increasingly successful attempts at local, regional and intercontinental correlation. In the early 1900s three major glacial advances were recognised. In the 1950s, though, ocean drilling provided a new source of evidence, revealing the pattern of change to be far more complex than indicated by the terrestrial record. Marine sediment cores have revealed that there have been numerous glacial advances, or 'cold stages', separated by warm periods, termed interglacials. Today, as many as 50 climatic cycles are recognised, each comprising a cold and warm stage. The evidence for this derives from the so-called oxygen isotope record contained in marine fossils such as foraminifera (see Wilson et al. 2000 for details). In particular, the ratio of $\delta^{18}O$ to $\delta^{16}O$ correlates with the prevailing temperature and its signature, preserved within foraminiferal calcium carbonate, thus provides a means of determining past temperatures.

Cores obtained from all the world's oceans provide a record of oxygen isotope changes that facilitates correlation as well as temperature reconstructions. Marine oxygen isotope signatures have also revealed that cold stages, or 'stadials', are not uniform but are often punctuated by relatively short-lived warm stages known as interstadials. Age estimates of the various oxygen isotope stages (OIS) have brought to light other features of the last 3 million years, notably that cold stages are much longer than warm stages, lasting about 100,000 and 20,000 years, respectively. Thus, the 'average' condition of Planet Earth is cold rather than warm. Other evidence for environmental change gleaned from ocean sediments includes ice-rafted debris, sediment whose composition indicating the source and direction of transport, and other fossil indicators such as coccolithophores and diatoms (see details in Mannion 1999b).

Over the last three decades new, deep archives of information have been discovered in the continental masses. These include the loess sequences of central Asia, China and North America, some of which provide an unbroken sedimentary record of the last 3 million years rivalling that of the marine cores. The marine and continental records have now been correlated, using a variety of techniques, and the synchronicity of environmental change worldwide has now been firmly established. Similarly, lakes contain a number of deep 'lacustrine' sequences providing records of many or all sections of the last 3 million years. Examples include Lake Biwa in Japan, the former lake at Funza on the High Plain of Bogota, Colombia, Lake George in Australia and the Hula basin in Israel. The biological remains in these sediments provide a record of environmental change in terms of aridity, humidity, forest cover, tree-line shifts and vegetation composition. They provide valuable pieces of the jigsaw of past global change and have reinforced the consensus regarding the pivotal mechanism underpinning that change: the (astronomically driven) Milankovitch cycle referred to above.

During the 20th century further major advances in palaeoenvironmental research were made by investigating plant and animal remains preserved in a wide variety of archives, including lakes, sediments, peat bogs, cave sediments and ice cores. Pollen analysis and the study of plant macrofossils, in particular, have made a major contribution to our understanding the environmental changes of the last 3 million years. The waterlogged and low pH of lake sediments and peats make them good media for preserving both plant macrofossils and pollen and spores. As the layers of sediment and peat accumulate, the plant fossils become anaerobically incarcerated making them available for extraction in layered cores. The plant material can then be treated, using a variety of chemical preparation techniques, and identified using detailed keys and so-called 'type collections' (see Moore et al. 1991 for details of pollen analysis). The resulting data can thus provide a record of vegetational change since the sediment began to accumulate, permitting researchers to infer other aspects of environmental change, including climatic change. Although pollen analysis has been employed throughout the world, from high-latitude ice cores to tropical peats, it is in the temperate zones that it has been used most widely to identify the dynamics of environmental change.

Many other types of biological remains have been employed in palaeoenvironmental research; the most frequently used are listed in Table 3.1. Details of the techniques involved in reading these archives and examples of their application can be found in Lowe and Walker (1997), Mannion (1997, 1999b) and Williams et al. (1998). One archive category meriting specific mention are insect remains, and notably those of coleoptera (beetles), because of the contribution studies based on them have made to current understanding of climatic change. This is because, in contrast to plant pollen and macrofossils, many species of beetle respond rapidly to climatic change, especially those not ecologically tied to plants and vegetation, such as carnivores and detrivores. Most of these studies, which were pioneered by Russell Coope in the 1960s, have focused on the mid-latitudes and many specifically on the end of the last Ice Age. This work has revealed that a rapid rise in temperature occurred during the last glacial-interglacial transition. In Britain, for instance, July and January temperatures increased by about 13°C and 20°C, respectively, between 13,000 and 12,000 years BP (see Lowe and Walker 1997). These findings parallel those from marine sediment cores (see above) and ice cores (below), where a similarly rapid change in oxygen isotope ratios has also been found. In contrast with most other palaeoenvironmental techniques, analysis of beetle remains facilitates relatively precise temperature reconstructions.

At the other end of the spectrum of animal fossils are bone remains. These also merit special mention, for historical reasons, as it was the preservation of large fossilised bones in cliff sections and quarries that first betrayed the existence of an ancient fauna with no modern parallels. These fossils inspired the inauguration of museum collections and encouraged natural scientists in their pursuit of understanding the past. Today, the bone assemblages studied by palaeontologists range all the way down to the tiny bones and teeth of small mammals, and have made a solid contribution

Table 3.1 The various archives and indicators that can provide information on palaeoenvironments

Archives	Biological indicators
Glacial sediments	Pollen
Periglacial sediments	Plant macrofossils
Lake sediments	Diatoms
Fluviatile sediments	Other algae, e.g. Chrysophytes
Peats	Fungal remains
Loess	Foraminifera
Cave soils	Radiolavia
Speleothems	Coccolithophores
Archaeological sediments	Non-marine molluscs
Marine sediments	Marine molluscs
Palaeosols	Ostracods
Estuarine muds	Beetles
Ice cores	Vertebrate remains
	Cledocera
	Corals
	Testate amoebae
	Chironomids

to elucidating the environments of the last several million years. One major result has been to demonstrate that the interglacial periods were characterised by many large animals that are now extinct. A famous case is that of Trafalgar Square in central London, where excavations associated with the construction of New Zealand House in the 1950s revealed deposits from the last interglacial, i.e. about 140,000 to 116,000 years BP. Contained within them were bones of hippopotamus and straight-tusked elephant, both grazers of the Thames floodplain, and since found in many other parts of Britain and Europe. Animal bone assemblages indicate that past interglacials were characterised by quite different species than the present one, while plant and pollen remains point to more similarities than differences.

Recently, a new focus of attention has been the discovery of a well-preserved mammoth carcass in the Siberian permafrost. While such finds are not uncommon, the excellent state of preservation has opened up the possibility of extracting mammoth DNA and, inevitably, the scope for replication and cloning. This may be a far-fetched idea, but the recovery of DNA will at least facilitate genetic comparison with other species. DNA evidence from Neanderthal remains (Noonan et al. 2006) also allows insights into Neanderthal-Modern Human relationships. For example, it indicates the divergence of Neanderthals from an ancestor shared with modern humans some 370,000 years ago. This was well before modern humans emerged as a species. If it proves possible to obtain DNA from a range of hominid fossils it will contribute significantly to the study of human evolution.

In terms of exploiting archives of information on the last 10 million years, major advances have been made since the 1970s. An important step was the extraction of the first ice cores from the polar caps. Although the results of initial studies yielded few secrets about past environments, an important breakthrough came in the mid-1980s with the extraction of the 'Vostok' core from Antarctica. Extending to a depth of some 2.5 km, this core stretches back 160,000 years or so, thus including the whole of the last glacial-interglacial cycle. Since the mid-1980s deep cores have been extracted from the Greenland ice cap, too, and an even deeper core, extending 3.35 km and representing the past 450,000 years (the last four glacial-interglacial cycles), has been obtained from Vostok. A great deal of palaeoenvironmental information has been obtained from these ice cores, contributing enormously to further refine the framework of global environmental change already provided by marine sediment cores and long terrestrial sequences such as loess deposits in Asia (see above).

The reciprocal relationship between ice cores and marine sediments is reflected in oxygen isotope ratios, those of marine sediments being a mirror-image of those in ice cores. Thus, in glacial periods the oceans and marine organisms become enriched with the heavier isotope δ^{18}O and the polar ice caps with the lighter δ^{16}O, with the reverse obtaining during warm (interglacial) periods. Temperature reconstructions based on the isotope ratios of marine sediments mirror those from the ice cores, which contain a second temperature indicator in the form of deuterium, i.e. heavy hydrogen. Temperature reconstructions from ice core data confirm the general trends deduced from marine sediment data, reflecting a degree of global synchronicity.

However, perhaps the most important aspect of ice-core research involves the air bubbles trapped within the ice. These bubbles contain a replica of the atmosphere in which they were formed and of particular significance is the composition of this air: in warm interglacials there is about 25% more carbon dioxide and 100% more methane in the atmosphere than in cold, glacial periods. This provides quantitative linkage between greenhouse gas levels and natural warming and cooling of the Earth's climate and thus baseline data for assessing the impact of anthropogenic emissions on climatic change. Other aspects of ice-core research yield data on volcanic eruptions, inputs of sea salts and soil erosion regimes on landside continental masses.

Studies of environmental change have benefited not only from the mining of various archives for physical, chemical and biological data (see Table 3.1) but also from developments in techniques of age estimation. De Geer's work on lake sediments in the 1890s was the first attempt at tackling the question as to when, and at what rate, past environmental changes have occurred. This was the first of several 'incremental' methods of age estimation; others include measurement of annual ice layers, lichens (lichenometry) and annual tree-rings (dendrochronology). In the case of sediment, peat and ice, stratigraphic 'markers' like ash resulting from a precisely dated volcanic eruption may provide a useful proxy for age. This is known as tectochronology and is often used in conjunction with radiometric methods, discussed in the following paragraph. Given the long-term changes that have occurred in the Earth's magnetic field, the palaeomagnetism of sediments can be exploited similarly to oxygen-isotope stratigraphy.

Major advances in age estimation have also been made by exploiting the decay of radioactive isotopes, and notably carbon-14, an approach pioneered by Willard Libby in the 1950s. The 'radiocarbon method', which measures radiocarbon levels in ancient organic matter and atmospheres, relies on the fact that radioactive elements decay at a constant rate, expressed in terms of the isotope's half-life. With this method, materials can be dated with a fair degree of accuracy back to about 60,000 years BP. Its use in the 1950s and 1960s made a major contribution to palaeoenvironmental studies and today it continues to be the most widely applied technique. Recent advances in calibration (Balter 2006) are improving the reliability of the technique, especially in the 25,000 to 50,000 year range. Several other techniques based on radioactive decay have also been developed during the last 30 years; these include the potassium-argon method, and techniques exploiting the decay of the uranium series, lead-210 and caesium-137. The importance of age estimation is also reflected in many other recent developments in the field of radiometric techniques e.g. fission track dating, thermoluminescence, optically stimulated luminescence and electron spin resonance. Another approach to age estimation focuses on chemical changes, such as amino-acid racemisation, obsidian hydration, soil formation, rock-surface weathering and various changes in bone chemistry.

Not only have many new archives of geophysical information been exploited during the last few decades; the geographical range of investigations has also increased. As well as facing the rigours of the oceans and polar regions, researchers are exploring the hot deserts and the remaining wildernesses of the tropical forests. In addition, increasing political openness has facilitated creation of international research teams and a historically unparalleled exchange of information between East and West – cooperation that is essential in a rapidly changing world that needs to prepare itself for environmental change in this new millennium.

References

Balter, M. (2006). Radiocarbon dating's final frontier. *Science, 313*, 1560–1563.

Claussen, M., Brovkin, V., Calov, R., Ganopoliski, A., & Kubatzki, C. (2005). Did humankind prevent a Holocene glaciation? Comment on Ruddiman's hypothesis of a pre-historic anthropocene. *Climatic Change, 69*, 409–417.

Lowe, J. J., & Walker, M. J. C. (1997). *Reconstructing quaternary environments* (2nd ed.). Harlow: Longman.

Mannion, A. M. (1997). *Global environmental change. A natural and cultural environmental history* (2nd ed.). Harlow: Longman.

Mannion, A. M. (1999a). Domestication and the origins of agriculture: An appraisal. *Progress in Physical Geography, 23*, 37–56.

Mannion, A. M. (1999b). *Natural environmental change, the last 3 million years*. London: Routledge.

Moore, P. D., Webb, J. A., & Collinson, M. D. (1991). *Pollen analysis* (2nd ed.). Oxford: Blackwell.

Noonan, J. P. et al. (2006). Sequencing and analysis of Neanderthal genomic DNA. *Science, 314*, 1113–1118.

Ruddiman, W. F. (2003). The anthropogenic greenhouse era began thousands of years ago. *Climatic Change, 61*, 261–293.

Williams, M., Dunkerley, D., De Deckker, P., Kershaw, P., & Chappell, J. (1998). *Quaternary environments* (2nd ed.). London: Arnold.

Wilson, R. C. L., Drury, S. A., & Chapman, J. L. (2000). *The great ice age*. London/New York: Routledge and the Open University.

Chapter 4
Environmental History: Object of Study and Methodology

Petra J.E.M. van Dam and S. Wybren Verstegen

Contents

4.1 Introduction 25
4.2 Object of Study and Approaches 25
4.3 Periodisation 26
4.4 Sources and Methodology 28
4.5 Conclusion 30
References 30

4.1 Introduction

Environmental history (in Europe also commonly named ecological history) emerged as a separate sub-discipline of environmental science in the 1980s. Although there had been several earlier historical studies on the state of the environment, many spurred by the first report to the Club of Rome and the first oil crisis of the 1970s (Van Zon 1993), it was in the 1980s that environmental history first began to develop as a dynamic field of research in its own right, pulling ever more themes into its orbit as the scope of the environmental debate broadened (Winniwarter et al. 2004).

While early environmental history studies were infused mainly by concerns about pollution and resource depletion, today even climate history falls within its province. One of the consequences of this dynamic, in a sense 'cyclonic', development is that a variety of longer-standing fields of study are now subsumed under the umbrella term 'environmental history'. Such

Petra.J. E. M. van Dam (✉)
Environmental History Amsterdam Institute for Heritage and Society VU University Amsterdam, The Netherlands
e-mail: pjem.van.dam@let.vu.nl

is the case with historical studies of health and sanitation, for example, subjects traditionally studied as part of social history (Corbain 1982). Similarly, some aspects of economic history and the history of technology are today studied within the context of environmental history. In European historiography much has been written about whaling, (over)fishing, deforestation, erosion and overgrazing, for example. The contemporary environmental debate has given these older studies new political relevance; as they are re-read from an environmental rather than economic or geographic perspective.

English historians down the ages have written about deforestation, for example, about the transition from charcoal to coal as a fuel, the smog it caused and the bronchitis (or 'British disease') it gave rise to, and the cultural changes brought on by the new coal era (Wilkinson, in Worster 1988). In Germany, the Age of Romance had its influence on forest and landscape studies. Mining, pollution and the country's industrialisation in the 19th century provide examples of themes of long-standing academic interest to historians that were given a new impulse with the coming of environmental history. In a similar vein, the renowned French *Annales* school had been studying climate and landscape history for many decades, and it was no accident that Emmanuel Le Roy Ladurie became one of the driving forces behind the *Studies in Environmental History* series published by Cambridge University Press.

4.2 Object of Study and Approaches

At first environmental historians described their object of study simply as the history of environmental themes such as pollution, natural resource depletion and natural beauty (Clapp 1994). They hoped to gain a better insight into such problems, as the past provides examples of both sustainable and unsustainable behaviour. More recently, environmental history seems to have taken one step back from its object, becoming the story of environmental change, its human causes and consequences (McNeill 2000). Through their studies, historians have become more conscious of a certain 'relativity' affecting perceptions of environmental problems over time. To evaluate environmental changes as good or bad, one must first define the criteria being used: good or bad for whom or what, for which people or species, and so on?

Studies in environmental history can be classified in different ways. One type of study traces the history of particular environmental problems like urban and industrial pollution, following one of two basic approaches. Either one starts in the present and seeks to identify the problem's origins in the past (Poulussen 1987). Or, conversely, one studies past environmental problems and investigates whether or not they persisted later, and if so their circumstances and dynamics. Towns, for example, have long witnessed periods of heavy pollution and had in most cases found adequate solutions, often applied for many centuries. Today's pollution problems are therefore not the direct result of their historical antecedents, but caused by relatively new changes. Other examples of historical environmental problems include the impact of the salt industry's massive fuel needs and overgrazing by sheep, which created treeless heath land (Jäger 1994). With time, this came to be seen in terms of fuel shortage and dangerous erosion. Today, however, the heather-clad landscapes are highly valued nature conservation areas. More generally, in contrast to today people in the past regarded 'wild' nature as unpleasant, of little use and even dangerous, and preferably to be conquered, reclaimed and 'civilised.' It is thus a society's value system that determines which environmental changes are perceived as problems and which are not. This first category of environmental history studies also includes those that are concerned with historical aspects of solutions to environmental problems. There has thus been a recent spate of research into how what we now call 'sustainability' was conceived by societies in the past (e.g. Van Zon 2002).

Environmental history studies can also be classified according to how authors conceive of the relationship between nature, the material world and ideas. The bottom-up approach starts from the premise that the natural world establishes the conditions for human life, and can be termed the *ecocentric* approach. It is based largely on the methods developed by the French school of historians grouped around the journal *Annales*. It examines the role of nature in human life at three levels: nature (landscape), human modes of production and power, and mentality (Worster 1988). Such historians discern three levels of time. At the level of the history of nature, changes in nature tend to be so slow that commonly humans barely perceive those as changes. Nature, thus, seems a constant in human histories. At the second level developments are very slow: the secular trend of prices and wages, for instance, has a cycle expressed in centuries. At the third level, that of ideas and political history, things change very quickly, with a pace of years or even days. However, new research findings from archaeology and other palaeo-sciences point to the fact that even so-called undisturbed nature, at the first level, was often influenced by humans at a very early stage reaching back into prehistory. One example here is the impact of early American Indians setting fire to the landscape to create open forests more conducive to hunting (Budiansky 1995). In other environments early changes have been even more profound, as in the case of wetlands, where human occupation only became feasible after drainage (Van Dam 2002a).

As a result a top-down or *anthropocentric* approach, starting at the level of ideas, has been gaining ground in recent years (Van Zanden and Verstegen 1993). Research within this category includes the study of changing conceptions of nature since the Middle Ages, for example. Studying the history of ideas has brought to light the roots of environmental thinking and the links with (the protection of) nature and romanticism (Bramwell 1989). Historians have, for example, rediscovered the influence of the 'founding fathers' of modern environmental thinking. These include, in the United States, such people as George Perkins Marsh, John Muir and Henry Thoreau, and, in Europe, Ernst Haeckel, Rudolf Steiner and Frederik van Eeden.

4.3 Periodisation

In endeavouring to structure complex historical processes, historians generally divide history into bounded periods. It is a means to differentiate between continuity and change. The name of the period often reflects the essential new change, such as the period of the Industrial Revolution, of the Enlightenment, or of Ecological Imperialism. One of the key issues in this process of 'periodisation' is which historical developments are to be deemed more important: political, economic, social or cultural changes. Also essential is the question whether the change is to be seen as a quantitative one, including numbers of people affected, kilojoules used or pollution units emitted, say, or qualitative, as with the impact of spiritual or political leaders, innovations or introduction of new methods of payment in the economy (first money instead of goods, now the internet). The period boundaries taken will depend on how the object of study is defined. Environmental historians examining the history of energy, for example, will not want to restrict their studies to a history of energy systems focusing solely on types of energy resource and energy converters and their respective efficiencies. This would lead to an overly simplistic periodisation based on calculations of the number of joules available to a society. The approach taken in environmental history therefore includes the particular structures of a society at a specified moment in time, within which conversion technologies are created and developed and in which energy is appropriated for particular ends. This implies that social, technological, political, economic, cultural and cognitive dimensions must all be taken into account in the historical narrative (Sieferle 2001).

In Europe, the diffusion of large-scale coal burning was a key step in the transition from the solar to the fossil era. This is a major item in Chapter 5. Previously, the total amount of energy available in a given landscape was defined to a large extent by total surface area. Wood, wind and water – all forms of energy deriving ultimately from the sun – are limited in their availability per unit time, although ultimately infinite. As fossil fuels became widely available, energy production and consumption rose dramatically. An important period in this respect is the European Industrial Revolution of the 18th century that started in England. By 1800, the capacity of the first steam engines was already equivalent to 200 labourers, and by 1900 they had become 30 times more powerful still (McNeill 2000). Although agriculture, industry and transportation profited tremendously from the coal-based energy systems, there were also other socio-economic developments of great influence on the Industrial Revolution. The commercialisation of agriculture, starting in Flanders in Late Medieval times, and the specialisation of agriculture, in particular the transition from grain growing to wool production for the textile industry in England from the 17th century onwards, was the result of agricultural intensification combined with a new international division of labour. It freed up huge numbers of now landless labourers for the new coal-based industry. So the 1800 boundary is not just about the increasing energy potential of society, but also about large-scale social and economic transformations, including the rise of industries and large cities, new crops and division of labour in the countryside and peasant migration (Van Dam 2002b).

Yet future historians may see the period from 1800 to 1950 as merely the first, introductory phase of the fossil fuel era in Europe. The transition to oil after the Second World War may prove to be even more important. It made enormous amounts of energy available at a very cheap price. Today's global air and water pollution are to a large extent due to the environmental dispersion of myriad synthetic substances produced from or with the aid of oil. The mass production of consumer items made of novel materials and fabrics, the explosive rise of the use of packaging and the emissions of traffic and the manufacturing industry are among the many petroleum-related developments that over the past five decades have had such a far-reaching impact on the natural environment (Pfister 1995). Once more, though, these effects cannot be seen in isolation from a number of major social and economic changes, including the rise of new technology, inspired by space and arms technology. According to McNeill (2000), it was nation-states that played a key incentive role in this respect.

The history of the introduction of foreign species of plants and animals is also of interest in the context of environmental periodisation. One of the key questions facing historians here is whether the rising number of new species introductions is in itself significant (quantitative change), or whether it is more important to identify specific plants and animals or assemblages that have caused particular changes to ecosystems, including the role of human action therein (structural, qualitative change).

The botanical historian traditionally takes the year 1500 as marking the beginning of a new era. Prior to that date a few plants had been introduced one by one from Southern Europe, most of them originating from Asia. In many cases monastery gardens and castle estates had served as reserves for these so-called archaeophytes. From excavations as well as the first botanical treatises of the 11th and 12th centuries we know of newly introduced fruit-bearing plants and medicinal herbs. With the scientific expeditions to the Americas and to East Asia that began around 1500, however, many hundreds of new plants were introduced to Europe. These so-called neophytes became established in the botanical gardens of universities and, from the 17th century onwards, diffused across the new estates of the nobility and urban elites. They include many decorative species still popular in homes and gardens today. Other, economically important crop species introduced during this period include tomatoes, maize and potatoes (Dirkx 2001; Van Dam 2002b).

Ecological historians frame all cases of species introduction tightly into human history and they take a broad perspective that focuses on the complex web of interactive relationships among humans, animals, plants and micro-biota (Beinart and Middleton 2004). For as long as Europeans have explored, conquered and migrated between continents, they have exchanged, wittingly or unwittingly, both beneficial and obnoxious organisms. Crosby (1986) has classified the period between 900 and 1900, from the time of the Vikings and the Crusaders through to the modern industrial era, as an era of 'ecological imperialism'. One of the most devastating combinations of natural organisms and human practices will be familiar to all: the introduction into the indigenous population of Latin America of hitherto unknown childhood diseases like measles and smallpox in the 16th century, which in decimating their numbers by about 75% (Mc Neill 1976; Slicher van Bath 1992) cleared the way for later conquest and colonisation by European settlers.

A second example is the introduction of European cattle and their fodder plants to the New World. The American pampas and prairie grasses could not withstand the trampling of the new breeds of cattle and were forced into retreat. Meanwhile, the seeds of aggressive European grasses and weeds, such as white clover (*Trifolium retens*), was carried by cattle and trading horses in their pelts and transported in the animals' hooves and soon spread like wildfire. By 1877, so archives tell us, no less than 153 European plants were found in the province of Buenos Aires alone, the list headed by the ubiquitous white clover (Crosby 1986).

Although 1500 may remain an important boundary for the introduction of new species, because of the massive new influx of species and the large-scale changes effected, the late 20th century clearly represents a new era of bio-invasion. Structural changes in society, such as globalised mass transportation of passengers (chartered flights) and goods (container ships) have created hitherto unknown opportunities for biological exchange. Thus, water recycled from the ballast tanks of ocean-going vessels has brought about a partial homogenisation of the world's coastal, harbour and estuarial species, each tank displacing some 3,000–4,000 water organisms each time it is evacuated (McNeill 2000).

In summary, then, establishing historical watersheds depends on the object of the study and the perspective taken, with different research questions requiring different periodisations.

4.4 Sources and Methodology

The classic sources of information for the historian, environmental or otherwise, are texts. As the content of a text is generally determined by its origins, the historian prefers to consult different types of source, resembling a detective hearing a series of witnesses. The administrative documents of public authorities have survived in large numbers and wide variety. 'Normative' sources, i.e. documents setting some form of standards, originate at central, regional and local levels of governance and include laws, legal decisions and police ordinances. Another type of administrative sources are documents arising during 'day-to-day routine', such as minutes of meetings, surveying reports, the annual accounts of landowners and pollution concessions. The latter sources are often serial in character: the form of the document remains constant, making the type of data predictable, but with contents (mainly figures) changing over time, reflecting economic cycles or changing attitudes. Since many serial sources come in large quantities going back hundreds of years, they require a statistical approach (Bayerl et al. 1997; Hahn and Reith 2001; Verbruggen 2002).

The loss of forests after the Middle Ages is a good example of how sources from various origins sustain different arguments. From 1500 onwards, all over Europe early modern states issued forest ordinances in order to protect timber resources and game. Forests were closed off and a hierarchical system of control and management was set up under supervision of the authorities. Henceforth, peasants were to pay for use of the forest and its products, with severe penalties imposed on those transgressing the new rules. However, studies of local institutional arrangements reveal that in the Middle Ages peasant communities managed forests as common pool resources quite well as an integral part of their farming economy. Examination of administrative documents shows that in the 16th century states began to protect and promote new fuel-intensive industries such as ore smelting and salt distillation. The taxes levied on these new industrial products were very welcome in the process of (military and bureaucratic) state formation. Research into the minutes of the 'Holthinge' or 'holthink', the meetings of the German forest communities, shows that local institutional arrangements started to lose their effectiveness quite independently of state action. A struggle between the old, locally rooted peasants and new, market-oriented agricultural entrepreneurs explains the breakdown of local systems controlling the use of the forests. Analysis of the pollen from particular forests shows that trees favoured by cattle disappeared first, while contemporary paintings reveal that these forests lacked their natural undergrowth (Dirkx 1998; Sonnlechner and Winiwarter 1999; Radkau 2002; De Moor et al. 2002; Ward 2003).

Summarising, then, the historical phenomenon of forest destruction can be attributed to several sets of explanations, depending on the approach and the type of source consulted. The political historian will focus on state formation and see forest degradation as an example of failed resource policy, while the economic historian will see forest degradation as an effect of the rise of the market, the commercialisation of agricultural production and demographic growth. The environmental historian will investigate the biological aspects of forest destruction: which species first declined, and was fuelwood gathering or overgrazing the more important factor of destruction? Then he or she will assess the extent to which political, economical, social, technological and cultural developments contributed to either fuelwood consumption or (over)grazing.

Besides the various types of administrative sources described, a very different group of historical sources are diaries, logbooks, letters and other sources of a personal or individual nature. Such sources are excellent if the aim is to write a history of environmental ideas and awareness. Many preachers, and later doctors and scientists, made daily observations of the stars, the weather and seasonal changes in flora and fauna, occasionally including earthquakes, storm tides and other natural disasters. Prior to our era, most scientists were in touch with an extensive network of correspondents, and many of their letters survive in private archives. One of the things these sources show us is how the medieval notion that the world was created by God, with nature a kind of 'picture book' to the Bible, started to compete with more empirical, 'mechanical' notions of nature, in particular in the Age of Scientific Revolution and the Enlightenment, 1650–1800. These discussions paved the way and opened minds for the great revolution of the 19th century: Darwin's new evolutionary paradigm. Evolution gave time a place in the life sciences and gave ecology a historical character (Thomas 1983; Kempe and Rohr 2003).

There is an enormous body of historical sources in existence. Written sources kept in archives are usually referred to in terms of shelf length: prior to 1800 in metres and subsequently in kilometres. Since historians can never do more than examine a finite sample of the material available to them, it makes no great difference that the number of written sources has expanded so tremendously since the Second World War, due in part to the use of the typewriter and photocopier. More important, though, are the major revolutions in source production that have taken place over the last decade or so. With the rise of electronic communications, commonplace and day-to-day sources such as letters and minutes, and draft versions of normative documents revealing discussions and changes in environmental awareness and changes in monitoring rules, often disappear before ever reaching the archives.

Written sources are not all that interest environmental historians, though. They may also consult pollen diagrams, for example, which provide valuable information on former landscapes based on the dispersion of fossilised pollen in geological sediments. Similarly, tree rings, ice cores and isotopes in peat bogs and other deposits can provide the researcher with detailed information on climate change (see Chapter 3). Climate history is in fact the first historical field to have attempted to cross the disciplinary boundary between the humanities and the natural sciences (Pfister 2001).

At the same time, though, the use of non-written sources poses major challenges. There are two options open to historians. They can simply rely on the reports of other historical researchers: palynologists (pollen experts), historical geographers, palaeozoologists, geologists or archaeologists. Alternatively, they may be an expert in more than one field and able to interpret specialist data themselves. As nobody is proficient in all disciplines, however, the basic problem remains. As a consequence, interdisciplinary cooperation is a key element of every ecological history research project. (A classic example of an interdisciplinary survey is Simmons 1989). It should also be noted that an important aspect of all such studies is the use of everyday language, to facilitate communication (Bayerl et al. 1997).

A successful example of interdisciplinary research concerns the environmental history of fish, fisheries and fish consumption. Lists of fish in the annals of abbeys and kitchen records of palaces, abbeys and urban hospitals reflect the assemblages of living species, destined for the pot or otherwise, as well as their relative rarity or abundance in a specific area. In some cases, identification of fish bones found in pits and kitchen floors in excavated settlements can yield confirmatory data on assemblages and distribution. Such archaeological data may also provide an indication of the size of individual fish as well as population structure and associated trends. It has, for example, been discovered that oxygen-loving fishes such as trout and salmon adapted to fast-flowing water had already become extinct in the upper reaches of the main European river systems by 1400. This was due to a progressive deterioration of their habitats, the result of overfishing and other human activities such as dam-building (for watermills) and canalisation of watercourses (Hoffmann 1996). The environmental history of fish is a fine example of how questions about the interaction between humans and nature lead to the use of several types of source, both written sources of various origin, and unwritten sources provided by various sciences.

4.5 Conclusion

Although environmental or ecological history is a relatively new discipline, it has developed into an important field for environmentalists of all persuasions. The story of environmental change, its human causes and its consequences, contributes to a broader and more profound understanding of environmental problems. It has shown, for example, that environmental problems change over time. They arise as a consequence of large-scale historical changes in society, be they political, social, economic, technological, cultural, or indeed a blend of these. Environmental history has demonstrated that current environmental problems and solutions are neither unique nor new. Problems are context-dependent and time-dependent. They require specific solutions, geared to the specific situation. Solutions from one period or culture cannot always be applied to others, because so many basic parameters vary with time, just as they vary from one place to another. Environmental history provides important examples of both sustainable and unsustainable behaviour over the ages and how such behaviour relates to the ideas of a culture. It demonstrates, in short, that for understanding and ultimately solving today's environmental problems, knowledge of both the material and the mental facets of the past is indispensable.

References

Bayerl, G., Fuchsloch, N., & Meyer, T. (1997). *Umweltgeschichte, Methoden, Themen, Potentiale*. Tagung des Hamburger Arbeitskreises für Umweltgeschichte Hamburg 1994, Münster.

Beinart, W., & Middleton, K. (2004). Plant transfers in historical perspective: A review article. *Environment and History, 10*(1), 3–29.

Bramwell, A. (1989). *Ecology in the 20th century. A history*. New Haven: Yale University Press.

Budiansky, S. (1995). *Nature's keepers. The new science of nature management*. London: Weidenfeld & Nicholson.

Clapp, B. W. (1994). *An environmental history of Britain since the Industrial Revolution*. London: Longman.

Corbain, A. (1982). Le miasme et la jonquille: l'odorat et l'imaginaire social, 18e-19e siècles. Aubier, Paris.

Crosby, A. W. (1986). *Ecological imperialism. The biological expansion of Europe, 900–1900*. Cambridge: Cambridge University Press.

De Moor, M., L. Shaw-Taylor, & Warde, P. (2002). *The management of common land in North West Europe, ca. 1500–1850*. Turnhout: Brepols.

Dirkx, G. P. H. (1998). Wood-pasture in Dutch common woodlands and the deforestation of the Dutch landscape. In: K. J. Kirby & C. Watkins (Eds.), *The ecological history of European forests* (pp. 53–62). New York: Oxon.

Dirkx, G. P. H. (Ed.). (2001). Vreemdelingen in de natuur. Special issue of: *Jaarboek voor Ecologische Geschiedenis 2000*. Ghent: Academia Press.

Hahn, S., & Reith, R. (2001). *Umwelt-Geschichte. Arbeitsfelder, Forschungsansätze, Perpektiven*. Munich: Oldenbourg.

Hoffmann, R. C. (1996). Economic development and aquatic ecosystems in medieval Europe. *American Historical Review, 101*(3), 631–669.

Jäger, H. (1994). *Einführung in die Umweltgeschichte.* Darmstadt: Wissenschaftliche Buchgesellschaft.

Kempe, M., & Chr. Rohr (Eds.). (2003). Coping with the unexpected – Natural disasters and their perceptions. Special issue of *Environment and History* 9, 2 May 2003.

McNeill, J. R. (2000). *Something new under the sun. An environmental history of the twentieth century.* London: Penguin.

McNeill, W. H. (1976). *Plagues and peoples.* New York: Doubleday.

Pfister, C. (1995). *Das 1950's Syndrom. Der Weg in die Konsumgesellschaft.* Bern: Haupt.

Pfister, C. (2001). Klimawandel in der Geschichte Europas. Zur Entwicklung und zum Potenzial der Historische Klimatologie. *Österreichische Zeitschrift für Geschichtswissenschaften, 12*(2), 7–43.

Poulussen, P. (1987). Van burenlast tot milieuhinder: het stedelijk leetmilieu 1500–1800. Kapellen (België): Pelckmans.

Radkau, J. (2002). *Natur und Macht. Eine Weltgeschichte der Umwelt.* Munich: C.H. Beck.

Sieferle, R.-P. (2001). *The subterranean forest. Energy systems and the industrial revolution.* Cambridge UK: White Horse Press.

Simmons, I. G. (1989). *Changing the face of the earth. Culture, environment, history.* Oxford: Basil Blackwell.

Slicher van Bath, B. H. (1992). *Indianen en Spanjaarden. Latijns Amerika 1500–1800. Een ontmoeting tussen twee werelden.* Amsterdam: Bert Bakker.

Sonnlechner, C., & V. Winiwarter (1999) Recht und Verwaltung in grundherrschaftlichen Waldordnungen Niederösterreichs und Salzburgs (16.-18. Jahrhundert). In: *Jahrbuch für europäische Verwaltungsgeschichte 11*, 57–85.

Thomas, K. (1983). *Man and the natural world. Changing attitudes in England 1500–1800.* London: Allan Lane.

Van Dam, P. J. E. M. (2002a). Het onderaardse bos. Chronologische afbakeningen in de ecologische geschiedenis. *Tijdschrift voor Sociale Geschiedenis, 28*(2), 176–202.

Van Dam, P. J. E. M. (2002b). New habitats for the rabbit in Northern Europe, 1300–1600. In: J. Howe & M. Wolfe (Eds.), *Inventing medieval landscapes. Senses of place in Western Europe* (pp. 57–69). Gainesville: University Press of Florida.

Van Zanden, J. L., & Verstegen, S.W. (1993). *Groene geschiedenis van Nederland.* Utrecht: Het Spectrum.

Van Zon, H. (1993). Milieugeschiedenis, een vak apart. *Economisch- en Sociaal Historisch Jaarboek, 56*, 1993.

Van Zon, H. (2002). *Geschiedenis en duurzame ontwikkeling.* Nijmegen: Netwerk Duurzaam Hoger Onderwijs.

Verbruggen, C. (2002). Nineteenth century reactions to industrial pollution in Ghent, the Manchester of the continent. The case of the chemical industry. In: C. Bernhardt & G. Massard-Guilbaud (Eds.), *Le demon moderne/The modern demon. Pollution in urban and industrial European societies* (pp. 377–391). Clermont-Ferrand: Presses Universitaires Blaise-Pascal.

Ward, P. (2003). Forests, energy and politics in the early modern German states. In: S. Cavaciocchi (Ed.), *Economia ed energia secc. XIII–XVIII.* Atti della trentaquattresima Settimana di Studi 15–19 aprile 2002. Firenze: Le Monnier.

Winiwarter, W. et al. (2004). Environmental history in Europe from 1994 to 2004, enthusiasm and consolidation. *Environment and History, 10*(4), 501–530.

Worster, D. (Ed.). (1988). The ends of the earth. Perspectives on modern environmental history. Cambridge: Cambridge University Press.

Chapter 5
Human Environmental History Since the Origin of Agriculture

Clive Ponting

Contents

5.1	Human History from an Environmental Perspective...............................	33
5.2	Population, Land and Food...............................	34
5.2.1	Gathering and Hunting...............................	34
5.2.2	Farming...............................	34
5.2.3	The Environmental Impact of Farming...............	35
5.2.4	Population...............................	35
5.2.5	Feeding the World's Population...............................	36
5.3	Resources and Energy...............................	38
5.4	Inequality...............................	40
5.5	Pollution...............................	41
5.5.1	Global Problems: CFCs...............................	42
5.5.2	Global Problems: Global Warming...............	43
5.6	Ideas and Actions...............................	44
5.7	Conclusion...............................	46
References...............................		47

5.1 Human History from an Environmental Perspective

This chapter examines human history from an environmental perspective and, at the broadest level, isolates those trends which have shaped current environmental problems. It argues that it is impossible to understand the complexity of modern environmental problems without considering their historical background. It is built around a number of key themes:

- Population, land and food
- Resources and energy
- Inequality
- Pollution
- Ideas and actions

Although these themes are considered separately they do of course run in parallel.

Before considering these central features of human history, our starting point is the fact that all living things on earth, including humans, form part of ecosystems – the complex webs of interdependence built around the various food chains which stretch from the photosynthesisers at the bottom through the herbivores to the carnivores at the top. Because of the decreasing energy efficiency of food chains, the number of creatures that can be supported at each level gets progressively smaller. When the first, distant ancestors of modern humans evolved in East Africa some 4 million years ago they operated mainly as herbivores, but they were also carnivores – scavenging dead animals and undertaking a small amount of hunting. Their numbers were therefore limited by the ability of these ecosystems to support animals operating towards the top of the food chain. Human history is, at its most fundamental level, the story of how these limitations have been circumvented and of the consequences for the environment of doing so.

C. Ponting (✉)
Department of Politics and Internatioal Relations University of Wales, Swansea

5.2 Population, Land and Food

5.2.1 Gathering and Hunting

The most important departure from basic ecological constraints has been the increase in human numbers far beyond the level that could be supported by natural ecosystems. The first step in this process, though it did not, in itself, imply any departure from normal ecological constraints, was the spread of humans from the area in which they evolved (East Africa) to cover the whole globe. This was achieved when humans were still gatherers and hunters and depended on a number of key developments such as:

- The development of speech and language and the resulting huge increase in social interaction and co-operation
- The development of increasingly complex technologies – not just much more effective stone tools but also use of new materials such as bone and the making of boats, bows and arrows, clothing and other items essential for life in more difficult ecosystems

For more than 90% of human history people subsisted by gathering and hunting for their food. They lived in small, mobile groups exploiting an extensive territory and a wide range of resources. They could obtain all the food they needed relatively easily, leaving plenty of time for leisure activities. Gathering and hunting was the longest-lived of all human adaptations and its impact on the environment was minimal. People had few possessions (they were a hindrance to mobility) and this mobility, the use of a wide range of resources and their relatively small numbers, produced minimal ecological damage. In the course of their evolution, humans had developed adaptations enabling them to live in the harshest of Ice Age environments in Europe and Siberia, and by 10,000 years ago humans had settled every major area of the world, including Australia and the Americas. However there is a longstanding debate about the possibility that some hunter-gatherers (e.g. on Madagascar) contributed substantially to loss of species. Also there is the possibility that hunter-gatherers that made use of fire may have had a significant impact on the landscape (Reijnders 2006). As the last Ice Age drew to a close there were only about 4 million people in the world.

5.2.2 Farming

Perhaps the most crucial of all the changes in human history was the creation of artificial ecosystems in which humans could grow crops and tend domesticated animals – what we call farming. It happened on at least four separate occasions across the world in the period between about 8000 and 4000 BCE – in south-west Asia, China, Mesoamerica and the Andes (Crops were also domesticated in the tropics but little evidence of how, when and where this was done has survived). The Near east is particularly important since it is the best documented and earliest example of the origins of food production (Byrd 2005). From these core areas farming gradually spread across nearly the whole world. Farming requires far more work than gathering and hunting and produces increased reliance on a small number of crops and animals. It was only adopted because slowly rising population pushed gathering and hunting groups into more marginal areas and climatic changes meant that in these areas more intensive techniques were required in order to obtain enough food. At this stage a ratchet effect came into operation. In return for greater effort, farming increased the amount of food available and therefore allowed the population to grow. Once this had happened it was impossible to revert to gathering and hunting.

The adoption of farming was the most fundamental transition in the whole of human history. First, humans slowly abandoned their mobile way of life and settled in one place. Second, an individual farmer can normally produce far more food than is required for the support of his family and this food surplus has formed the fundamental foundation of everything we call 'civilisation'. Over the course of a few thousand years after the adoption of farming, the food surplus was appropriated by the elite and used to support groups of non-producers such as priests, rulers, soldiers, bureaucrats and craftsmen. By about 3000 BCE in Mesopotamia, a little later in Egypt and slightly later again in China, cities and literate societies had developed. However, the productivity of agriculture was still low and almost nowhere until the last 200 years or so could more than about 20% of the population be supported in non-farming occupations and normally it was only about half this level.

5.2.3 The Environmental Impact of Farming

Farming had a fundamental impact on the environment by concentrating the human demand for resources and the impact of the pollution humans produced. Removing the natural vegetation and cultivating the soil relatively intensively could easily lead to deterioration including increased soil erosion. Even more important was the impact of the demand for wood – used particularly as the easiest source of fuel but also to build houses and in numerous other ways. Settled societies very rapidly led to extensive deforestation and subsequent soil erosion. In the Jordan valley within a few thousand years of the adoption of farming these effects were apparent and led to the abandonment of numerous villages. The long-term impact was even greater. In China the steady cutting down of the forests to create new agricultural land and provide fuel and building materials meant that by the 20th century little more than about 3% of the original forests still survived. The environmental impact has been catastrophic. It led to extensive flooding in valleys such as that of the Yellow river and in the 1850s it produced one of the greatest disasters ever to strike China. The Yellow river shifted its course so much that it now reaches the sea to the north of the Shantung peninsula rather than the south — a distance of over 200 km. In south-west Asia and the Mediterranean the long-term impact of farming has been equally devastating in removing the natural vegetation and forests and leaving behind bare hillsides where the soil has been eroded away leaving large areas of poor quality scrub.

Many of the early civilisations emerged in river valleys and depended on irrigation to provide enough food to support the non-producers. In Egypt the modifications to the natural flood of the Nile were relatively small scale and did not result in major environmental damage. This was not the case in Mesopotamia. Here the adoption of irrigation led to a major environmental disaster that eventually destroyed the foundations of the earliest civilisation in the world and its cities such as Uruk, Ur and Lagash, which had developed in the southern Mesopotamian area of Sumer. The twin rivers, the Tigris and Euphrates, were at their highest in the spring following the melting of the winter snows near their sources and at their lowest between August and October, the time when the newly planted crops needed the most water. Water storage and irrigation were the essential foundations for the development of civilisation in Sumer. At first the advantages in terms of higher crop yields would have been readily apparent. Only later did the problems become apparent. In summer, temperatures were high, often up to 40°C, which increased evaporation from the surface and therefore the amount of salt in the soil. Waterlogging was a major problem because the soils of the area had very low permeability and the rate of drainage was very slow because the land was very flat (and made more so by the amount of silt coming down the rivers following extensive deforestation upstream). As the land became waterlogged and the water table rose, more salt was brought to the surface where the high evaporation rates produced a thick layer of salt. The only way to cure the problem was to leave the land fallow and unwatered for a long period. The internal pressures within the early cities, the increasing competition between them, and the rising population made this impossible.

The deterioration of the environment can be clearly traced. About 3500 BCE, just before the full emergence of literacy and highly complex, stratified societies, roughly equal amounts of wheat and barley were grown. But barley can grow in soil with twice the level of salt that wheat can tolerate. By 2500 BCE wheat was only 15% of the total crop and within another 500 years it was no longer grown on any significant scale. Even more important were declining yields. Until about 2400 BCE yields were high – at least at the level of medieval Europe and possibly even higher. Salinisation of the soil reduced yields dramatically – by 1700 BCE they had fallen to a third of their earlier levels. The consequences were predictable – the number of non-producers that could be supported fell equally dramatically and as the number of soldiers declined the area fell to external conquerors. Soon the whole infrastructure of the area collapsed and the focus of civilisation in Mesopotamia shifted northwards. The societies of Sumer failed to recognise the problems they faced and did not take the action necessary to reverse the decline in the amount of food available.

5.2.4 Population

In general, the first settled societies were able to avoid the worst of the environmental disasters that farming could cause. As more areas were brought under cultivation, the population of the world rose (Table 5.1).

In the centuries after about 1000 BCE there was a major technological transformation in human societies with the widespread production of iron across the Eurasian continent. From a farming perspective iron tools were important in that they:

- Allowed much easier clearing of land from the forests
- Led to the introduction of heavier ploughs capable of cultivating heavy soils
- Made possible more effective hoes and digging tools

The result was a second wave of population growth as more land was brought into production. By the end of the first millennium BCE the world's population had more than tripled to about 170 million. However, at this point it reached a new plateau – little new land could be cultivated and technology was largely static. Over the next millennium to 900 CE the world's population only increased by just over 40% to about 240 million.

A major change came around the end of the first millennium CE. An increasing shift in the Chinese population to the intensive rice growing areas south of the Yangtze, together with a major expansion in the settled area across north-west Europe, produced a 50% rise in world population to reach 360 million by 1300. Famine and plague reduced numbers and growth was slow – in 1500 the European population had only recovered to the levels of 200 years earlier. World population reached 425 million in 1600 and then grew steadily to 610 million by 1700, almost three times the total only a thousand years earlier. Nevertheless this rate of growth was still very slow compared with the modern population history of the world.

Modern growth began in the 18th century and occurred across the world – indeed it began in China, not Europe. The great stability and prosperity of 18th century China saw population more than double from 160 million to 330 million. Overall world population rose to just over 900 million by 1800. It was the start of the phenomenal population growth of the last two centuries (Table 5.2). By the 1990s, the world population was increasing by about 90 million people every year.

5.2.5 Feeding the World's Population

The huge increase in world population has been fed by two processes:

1. By bringing even more land into cultivation
2. Through the development of high-intensity farming

Between 1860 and 1980, about 8 million square kilometres of new land was converted to agricultural use. Most of it was in the newly settled lands of the Great Plains of the United States, in South America and Australasia, followed by the then Soviet Union. In the 100 years after 1860 the area under crops in the United States increased two-and-a-half-fold, in the Soviet Union it quadrupled, in Canada it increased eight-fold and in Australia the rise was nearly 30-fold. From the mid-20th century, as population rose rapidly in Africa and parts of Latin America, huge areas of forest were cleared to provide more land for farming so that people could grow enough food to survive.

The problem, from an environmental point of view, was that much of this land was highly marginal and easily subject to degradation. The worst effects of extending cultivation were found on the Great Plains. Before the latter half of the 19th century these areas were avoided by settlers and left to the large herds of bison and cattle. Then, the development of heavy steel ploughs pulled by teams of up to twelve oxen made cereal cultivation possible. However, the area was highly marginal – the climate was semi-arid with only about 50 cm of rain a year and the thin topsoil was only held together by the roots of the grass. From the 1880s to the late 1920s nearly 200,000 km² land were

Table 5.1 World population between 10,000 and 1000 BCE

Date (BCE)	World population (millions)
10,000	4
4000	7
3000	14
1000	50

Table 5.2 Population growth in the last two centuries

World Population (billion)	Date	Time taken to add 1 billion to world population (years)
1	1825	2 million+
2	1925	100
3	1960	35
4	1975	15
5	1987	12
6	1999	11

brought under cultivation. Disaster struck in the 1930s during one of the periodic droughts that affect the area. Huge dust storms swept the across the Great Plains – by 1938, more than 40,000 km² of land had lost the top 12 cm of soil and nearly 56,000 km² the top 6 cm. Over 850 million tonnes of soil were being lost every year and 3.5 million people abandoned farms in the region. Another drought in the 1950s caused wind erosion over twice the area damaged in the 1930s and a further drought in the 1970s led to severe erosion across another 40,000 km². The situation was bad elsewhere in the United States too, as a survey in the 1970s showed. By then, a third of the topsoil of the country had been lost and nearly 800,000 km² of cropland either ruined or made highly marginal for cultivation. Nearly 3,000 km² of productive land were being lost to soil erosion every year. The Soviet Union replicated this disaster following the 'virgin land' programme of the 1950s – the attempt to cultivate the marginal grasslands of Kazakhstan. In the late 1950s, about 400,000 km² were ploughed up but practices such as deep ploughing and leaving the soil bare during fallow periods led to severe erosion following a major drought in 1963. In just 3 years, almost half the new land was lost, and thereafter, wind erosion in Kazakhstan led to the abandonment of over 4,000 km² of land every year.

The brunt of the destruction in the second half of the 20th century was borne by the tropical forests. Since 1950 about half of the world's tropical forests have been destroyed and about three-quarters of that loss was caused by land clearance for agriculture. About two-thirds of the total loss was in Africa, although from the 1970s the scale of the destruction in the Amazon was rising rapidly as more landless peasants moved into the area – they were landless because most of the land in the settled areas of Brazil was in the hands of the small number of people who owned vast estates. However, in much of Latin America forest destruction was carried out to provide pastureland to raise cattle for export to the United States. By the 1990s two-thirds of the available farmland in Costa Rica was devoted to cattle ranching for export. Amid widespread reports of deforestation, some nations have nevertheless experienced transitions from deforestation to forestation. Deforestation does continue in about half of the 50 nations with most forest. However, 36% of the 50 increased its forest area and 44% increased the biomass (Kauppi et al. 2006).

The large scale intensification of agriculture began in the early 19th century. Until then farms were almost entirely dependent for the maintenance of fertility on manure and compost produced on the farm itself. The most productive of all agricultural systems were the wet-rice paddy fields of Asia but the Europeans and the Americans were able to import increasing quantities of fertiliser. The first guano was brought from South America to Europe in the 1820s, later phosphates were mined around the world. New artificial fertilisers were also developed – superphosphates in the 1840s and nitrogenous fertilisers in the 1920s. It was through these artificial products that food output in the industrialised world doubled between 1950 and 1980. (British use of nitrogen fertilisers increased 50-fold between 1930 and 1980). In addition new crop varieties were developed – the so-called 'Green Revolution'. Although they increased output their social effects were disastrous. They needed large inputs of chemical fertilisers, pesticides and water and therefore their cost meant that they were mainly grown by large landlords. Overall world pesticide use increased dramatically in the second half of the 20th century (Table 5.3). Agriculture was also intensified through greater use of irrigation (Table 5.4).

Irrigation now accounts for almost three-quarters of the world's water use and most of it is used very inefficiently. In India and China almost two-thirds is lost through evaporation and seepage from the canals and in the United States the comparable figure is a half. However, the problems that affected Mesopotamia 5,000 years ago are now being repeated across the world – salinisation affects 80% of the irrigated land in

Table 5.3 Annual world pesticide use

Date	Thousands of tonnes
1945	70
1980	2,300
1995	2,600

Table 5.4 Amount of irrigated land in the world

Date	Thousand km²
1800	80
1900	400
1950	1,200
1980	2,000
1987	2,300
1997	2,700

the Punjab and overall the world is now losing irrigated land as fast as new schemes are introduced.

By far the greatest environmental disaster associated with irrigation has occurred in and around the Aral Sea – a large inland sea in southern Kazakhstan with two rivers draining into it but none flowing out (the sea was maintained in size by very high summer evaporation). In the early 1970s the Soviet Union devised an ambitious plan to divert the rivers to irrigate over 72,000 km^2 to grow rice and cotton in highly unsuitable soils and climates. This major diversion of water predictably caused the Aral Sea to shrink dramatically. By the late 1980s two-thirds of the sea had dried up, exposing the seabed over an area of 12,000 square miles. This caused major climatic changes as temperatures rose and humidity and rainfall decreased. The local fishing industry died out as nearly all the fish became extinct. The salinity of the sea tripled and as the water table fell the sewage system collapsed. By the early 1990s nine out of ten children in the area were diagnosed as being permanently ill, typhoid rates rose over 30-fold and there was an outbreak of bubonic plague.

5.3 Resources and Energy

Human history has, at one level, been one of steadily increasing technological sophistication. It began with the slow development of ever more complex and difficult to make stone tools, followed by tools from other materials, the development of pottery and the use of metals. In this perspective the so-called 'Industrial Revolution' of the last 200 years is no more than the latest stage of this process during which the pace of change has continued to increase exponentially. This revolution has enormously increased the use of the world's mineral resources, but from an environmental point of view, the far more important change has been the energy revolution of the last three centuries or so. Although the huge increase in production in the 20th century has involved parallel increases in resource and energy use, there has not been a shortage of either. New fields and deposits have been exploited with increasing efficiency. Recycling has been increased in many areas, in particular copper use. Despite the fears expressed by many environmentalists in the 1960s and 1970s, there is little likelihood of any shortage of important resources in the foreseeable future.

Until the last few centuries, all human societies were very short of energy. At first, all that was available was human and animal power and the first was by far the most important. The states and empires that depended on the agricultural surplus the farmers could create then conscripted and coerced the population into working on the great projects desired by the elite. The Pyramids and other great monuments that still survive were built almost entirely by human labour. In China the building of the Great Wall involved about 1 million workers and the construction of the Grand Canal, to bring food from the rice producing area in the south to capital and armies in the north, took about 5.5 million workers of whom about half died on the project. Everywhere humans were used as porters to carry loads. The only other source of energy was animals – and the societies in the Americas such as the Maya, Aztec and Inca lacked even those. It took human societies some time to learn how to harness animals (especially the horse) effectively. But they then provided a major source of energy – in agriculture for ploughing and more generally, for transport. The first major departure from these limitations was the development of water power. The first water mills were operating by about the last century BCE in both the Mediterranean region and China. Very slowly they were adapted to mill cereals and to power industry – this was particularly the case in the highly sophisticated Chinese iron industry where water-powered bellows were very important. Shortly after 1100 CE the first windmills were developed for use in areas where water power was difficult to harness or non-existent. They too were gradually adapted to a multitude of uses. Throughout this long period the major source of heating in both industry and the home was wood.

Until about 1600 therefore all human societies relied upon renewable sources of energy (wood is a renewable resource although it was rarely treated as such). The first of the non-renewable fossil fuels to be used was coal. Coal was a much more difficult source of energy to use – people preferred wood in the house and in industry the impurities in coal meant it could not be used in many processes such as iron production. The switch to fossil fuels only happened because of the increasing shortage of wood. This first became apparent in Europe in the 16th century, in particular in shipbuilding – from the mid-16th century Portugal had to build nearly all its ships in India and when Philip II of Spain fitted out the Armada to sail against England and the Dutch Republic he had to buy trees in Poland. Growing shortages of fuel affected salt production – England had to import two-thirds of its requirements from France where the sun could be used and in Poland works closed altogether because of the shortage. By the 18th century Britain was even running out of charcoal for iron production. In 1717 a newly constructed furnace in Wales could not begin production for 4 years until it had accumulated enough stocks of charcoal and even then it only had enough fuel to operate for 36 weeks before it was forced to close. Only slowly were techniques developed to enable coal to be used.

The 19th century was the age of coal in Europe (until the last few decades of the century the United States industrialised using water power and wood). In 1900 coal provided 90% of the world's energy consumption. It became the major domestic source of fuel and its major by-product, town gas, was used to light streets, factories and homes. It was also used to fuel power stations that produced electricity, the first easily distributed source of energy that could be transmitted over large distances. Electricity became a major driving force behind the huge increase in energy consumption and fundamental to later technological advances. (World electricity consumption increased twice as fast as overall energy consumption – in Europe its use rose 26-fold in the 50 years after 1920). In Europe, coal remained the primary source of energy until the 1950s and it was also central to the industrialisation of China in the late 20th century. In the last two centuries there has been a phenomenal increase in the world's consumption of coal (Table 5.5).

During the 20th century world coal use trebled, yet in the same period the share of coal in world energy consumption fell from 90% to just over 30%. It was replaced as the world's main source of energy by oil. This was first extracted on a commercial basis in 1859 and initially was mainly used as a lubricant and refined into kerosene (illuminating oil). Then fuel oil became the main product before being replaced in the 1930s by gasoline as the number of cars in the world began to rise substantially. By the 1980s oil provided over 40% of the world's consumption of energy. The growth of oil consumption was even more rapid than that of coal (Table 5.6).

The expansion of natural gas consumption in the 20th century was also large – in 1900 it provided about 1% of the world's energy but by the 1980s it was almost 20% of the total.

The shift to non-renewable fossil fuels to provide the world's energy needs was the second fundamental transition in human history. It lay behind the huge increases in industrial output from the early 19th century onwards. One of the primary industries in the early stages of industrialisation was the production of iron (and later steel); see Table 5.7.

Production of other metals also increased dramatically – copper output rose 70-fold in the century after 1880 and nickel by the same amount in the 60 years after 1920. The new metal of the 20th century was aluminium, which requires large amounts of electricity to produce (it takes six and a half times more energy to produce a tonne of aluminium than a tonne of steel); see Table 5.8.

The increase in metal use was made possible by technological changes enabling lower grades of ore to be exploited. In 1900 the lowest workable grade of copper ore was about 3% – by the end of the 20th century this had fallen to 0.35%. However, exploiting this ore involved breaking, transporting and milling over 300 t of rock and then removing it as waste in order

Table 5.5 Annual world coal consumption

Date	Millions of tonnes
1800	15
1860	132
1900	701
1950	1,454
1990	2,100
1997	2,255

Table 5.6 Annual world oil consumption

Date	Millions of tonnes
1890	10
1920	95
1940	294
1960	1,051
1990	3,000
1997	3,409

Table 5.7 Annual world iron and steel production

Date	Tonnes
1400	>100,000
1700	300,000
1850	12,000,000
1980	1,200,000,000

Table 5.8 Annual world aluminium production

Date	Tonnes
1895	223
1928	250,000
1947	7,000,000
1980	20,000,000
1995	27,250,000

Table 5.9 World car production and ownership

Date	Annual world car production	World car Ownership
1900	30,000	100,000
1950	8,000,000	53,000,000
1990	36,000,000	445,000,000
1999	39,000,000	520,000,000

Table 5.10 The rise of the city

Date	Percentage of world population living in cities	Number of people living in cities (millions)	Number of cities with population over 1 million
1800	2.5	20	9
1900	10	160	
1920			27
1980			230
1995	48	3,000	

to obtain a tonne of copper. The amounts of energy consumed in this process were immense – by the second half of the 20th century over a fifth of US energy consumption was used to extract and process metals.

In the 20th century the crucial industry for the world economy was vehicle production. The United States' economy was the first to be dominated by the car as the number of vehicles on the road rose from 80,000 in 1905 to 30 million by 1938. At this time vehicle production was taking up half of all steel production, four-fifths of rubber output, two-thirds of plate glass production and about a third of nickel and lead output. Mass car ownership did not begin in Europe until the 1950s and in Japan a decade later. The result was a profound transformation in economies and societies across the world (Table 5.9).

The overall result of all these trends was a phenomenal growth in the world economy in the 20th century. In the early stages of world industrialisation in the 19th century industrial output tripled. During the 20th century world industrial output rose 35-fold. The pace of change was such that world industrial output in the period 1953–1973 was equivalent to the total industrial output in the period 1800–1950. By the 1990s the annual growth in the world economy was such that every 2 years it added to total world output the equivalent of the total world output in 1900.

One of the other fundamental changes in human society in the last 200 years has been the rise of the city. Agricultural societies could only support a limited number of non-producers and until the early 19th century no society in the world had more than a fifth of its population living in cities, and even this proportion was exceptionally high and confined to a few areas such as China in the 11th and 12th centuries, The Netherlands in the 17th century and Britain in the late 18th century. Industrial societies are urban societies and by 1900 about four out of five people in countries such as Britain were living in towns. This growth continued throughout the 20th century so that by the end of the 1990s about half of the world's greatly increased population lived in cities (Table 5.10).

5.4 Inequality

Until the late 18th century the various societies in the world were roughly equal in terms of their wealth. For most of the time China and India were the richest and it was not until the exploitation of the wealth of the Americas after 1500 that the Europeans were slowly able to catch up with the long-established societies of Asia. Since about 1800 the world has become increasingly unequal and this has had a fundamental impact on environmental problems and their possible solutions. Rapid industrialisation in western Europe and North America in the 19th century and the rapid expansion of European control over large parts of the world created massive inequality. The colonial economies and societies were re-made through a variety of imperial policies:

- De-industrialisation so as to provide a market for domestic products
- The introduction of single crops grown for export (such as cotton, sugar, tobacco, coffee, cocoa and rubber)
- Concentration on the production of a few minerals (as in Northern Rhodesia – the present Zambia, and the Belgian Congo – the present Democratic Republic of the Congo, formerly known as Zaïre)
- Very limited infrastructure development (a few ports and railways to the interior to aid exports)

Similar effects were found in the 'informal' empires in Latin America, which concentrated on grain and meat exports to Europe or metal production (such as copper from Chile). Only a few countries such as Japan were able to avoid European control or domination and begin the process of industrialisation.

In the early 20th century about 90% of the world's industrial output came from the major economies of Western Europe (Britain, Germany, France and Belgium) together with the United States. That proportion was reduced by the industrialisation of the Soviet Union after 1930 and Japan and Eastern Europe after 1945, but it was only slightly reduced by the 'newly industrialising countries', such as Taiwan, South Korea and Brazil, after 1960. By the end of the 20th century half the world's industrial production took place in just three countries – the United States, Japan and Russia. Three-quarters of world production came from seven countries, those three plus China, Germany, France and Britain. Most of the rest came from the other industrialised countries in Europe (Sweden, Belgium, The Netherlands and Italy) – Brazil contributed less than 2% of world industrial output and for both Taiwan and South Korea the figure was less than 1%.

Between 1900 and 1980, on average everybody in the world's population became wealthier and experienced an improvement in their quality of life – particularly in terms of increased life expectancy and lower child mortality. However, overall, world inequalities grew greater, not smaller in the 20th century. In 1900, on average, a person living in one of the industrialised countries was about three times wealthier than someone living elsewhere in the world. By the late 1990s they were about seven times richer. By the 1990s income per head in the United States was, on average, 80 times greater than in the Congo. Overall the poorest 20% of the world's population (about 1.2 billion people) received little more than 1% of the world's income.

These inequalities were reflected in every aspect of life. Energy and resource consumption were highly unequal. By the end of the 20th century the United States, with about 5% of the world's population, used a third of all the energy and resources consumed in the world. The average American used twice as much energy as a European and 30 times more than a person living in India. Similarly, although in the late 20th century there was enough food in the world to feed even 6 billion people adequately, its consumption was highly unequal. Put simply, the people of the industrialised countries ate half the world's food although they constituted only a quarter of the world's population. More food was sent from the poorest countries to the richest than travelled in the opposite direction. A cat in the United States ate more meat than the average person living in Central America, where most of the meat produced was exported to the United States. In the 1980s about a quarter of the world's population (about 1.2 billion people) suffered from chronic malnutrition and at least as many again had a grossly inadequate diet. In the 1990s about 40 million people a year were dying from hunger and its related diseases.

5.5 Pollution

All human activities throughout history have produced pollution. One of the most intractable problems has been separating human wastes from drinking water and that was not accomplished on any scale until the 20th century. A notable exception might be ancient Rome and a few other cities in the Roman Empire that had aquaducts and sewers resulting in a substantial separation of human wastes from drinking water. Nineteenth century European and American cities (like many of their counterparts in the poorer countries of the world today) had grossly inadequate or non-existent sewage systems. In Paris in 1925 half the houses were not connected to a sewerage system and until the 1960s half the city's sewage went into the Seine untreated. As late as 1970 half the population of Japan did not have mains drainage. Conditions in the poorer countries were even worse – in the 1970s the Pasig river in Manila was, by volume, 70% raw sewage. Early industrial activities were equally damaging. Industries such as tanning, linen bleaching, starch making and cotton dyeing polluted rivers for miles downstream and early mining operations simply dumped their wastes into the rivers.

Large-scale industrialisation in the early 19th century soon had a major environmental impact. The burning of coal on a growing scale increased sulphur dioxide in the atmosphere and produced 'smogs' across most of the industrial areas, such as the Ruhr in Germany, the 'Black Country' in England, the Monongahela valley near Pittsburgh in the United States (which had 14,000 smokestacks belching fumes into the atmosphere) and in London where there were 3.5 million fireplaces, nearly all burning coal. The exact conditions in 19th century industrial areas are difficult to determine because so few records were kept, but an idea of what they were like can be gauged by conditions in Eastern Europe and China during their basic industrialisation in the second half of the 20th century. In the area around Most in Czechoslovakia, sulphur dioxide levels were twenty times WHO-recommended levels and schoolchildren had to carry portable respirators. Around Cracow in Poland almost two-thirds of the food produced was unfit for human consumption because of metal residues and nearly three-quarters of the water could not be drunk. The Vistula was unfit even for industrial use over two-thirds of its length because it was so corrosive and over 100,000 square kilometres of the Baltic was biologically dead from the wastes brought down by the rivers.

One of the first types of pollution to be recognised, from the early 1850s in English industrial towns, was acid rain, mainly caused by increasing coal consumption and the sulphur dioxide emissions it produced (Table 5.11). At first this problem was largely concentrated in the areas around industrial cities. However, increasing output and the misguided policy of building very tall chimneys to disperse the pollutants produced major regional problems downwind (especially in Canada and Scandinavia). By the late 1980s, the lakes in western Sweden were a hundred times more acidic than they had been 50 years earlier, and across Scandinavia a quarter of all the lakes no longer contained any life. This was almost entirely the result of pollution originating in Western Europe, including Britain. During the late 1980s and early 1990s some agreements were reached among the major industrial countries to limit emissions. However, little action was taken elsewhere and in China, where industrialisation was largely based on very high sulphur content coal, conditions were as bad as anywhere in the world.

The increasing number of cars in the world, together with the practice (adopted in the 1920s) of adding tetraethyl lead to petrol to enable lower octane rated fuel to be used (reducing costs to the oil companies), caused major pollution. This was first apparent in the United States, the first country to have large-scale car ownership. It was noticed in Los Angeles in 1943 where the natural inversion layer trapped the car exhaust fumes in the valley. By the late 1980s over a hundred American cities were similarly affected.

As car ownership increased in other countries, the same problems developed. Photochemical smog became serious in Tokyo in the late 1960s and also in Mexico City which, like Los Angeles, has a natural inversion layer. A few measures were taken to improve the situation in the industrialised countries, mainly through the fitting of catalytic converters – in the United States from the late 1970s and a decade later in the European Union.

5.5.1 Global Problems: CFCs

Until the last quarter of the 20th century pollution resulting from human activities was mainly found at a local and regional level. Then two problems emerged which threatened the complex global mechanisms which make life on earth possible. The first was the impact on the stratospheric ozone layer of chlorofluorocarbons (CFCs). CFCs were invented in the 1920s by Thomas Midgley (the same scientist who suggested tetraethyl lead as an additive in petrol) and at first they seemed a classic example of the beneficial aspects of technological progress. They were non-poisonous, did not burn and did not react with other substances. They were widely used as coolants in refrigerators and air-conditioning systems, to make expanded foam containers, clean electronic circuit boards and as propellants in spray cans. Nearly all the CFCs were used by the major industrialised states – by the 1970s US consumption was eight times the world average. Only very late in the century did states such as China make an impact – the number of people owning a fridge in Beijing rose from 3% of the population to over 60% in the decade after 1975. Emissions of CFCs into the atmosphere rose dramatically (Table 5.12).

The first scientific papers linking CFCs to ozone depletion were published in the early 1970s and some (industrialized) countries took action regarding spray cans, but many did not. Removing them as propellants in spray cans was painless because other gases could easily be substituted. However, the chemical industry

Table 5.11 Annual world sulphur dioxide emissions

Date	Millions of tonnes
1860	10
1910	50
1990	170

Table 5.12 Annual world CFC emissions

Date	Tonnes
1931	100
1950	35,000
1960	135,000
1970	500,000
1990	650,000

refused to accept the links or phase out production until substitutes (hydrochlorofluorocarbons – H(C)FCs) had been developed. Under pressure from the chemical companies, governments did not take effective action until H(C)FCs were available. Then, from the late 1980s, CFC production in the main industrial countries was restricted[1] and finally ended in 1996, with arrangements to transfer the new technology to the newly industrialising countries. Global consumption of CFCs fell from 1.1 million tonnes in 1986 to 160,000 t by 1996 (UNEP 2000). However, HCFCs also destroy the ozone layer, although they are not as long lasting. Moreover global warming of the troposphere (see below) is associated with a cooling of the stratosphere, which is conducive to the destruction of its ozone layer. Levels of destruction will presumably peak in 2015 and it will not be until the mid-21st century that ozone levels will return to the levels of the 1980s, even though alternatives without chlorine (such as hydrofluorocarbons, or HFCs, and hydrocarbons, or HCs (such as butane), are now widely available.

5.5.2 Global Problems: Global Warming

Ozone layer depletion was a relatively simple problem caused by a small group of chemicals, wrongly thought to be benign. It could be solved by the development of a substitute without involving any fundamental reappraisal of human activities. However, global warming is a very different problem. Its causes lie deep within the way in which human societies have evolved, in particular with the high energy consumption and use of fossil fuels in the last 250 years. The problem of finding a solution is made more difficult because it involves taking into account the massive inequalities in the world which have emerged over the same period.

The major gas responsible for global warming is carbon dioxide, which is produced whenever fossil fuels are burnt to provide energy (a fifth of global emissions now come from vehicles) and during the burning of forests (a reduction in their size also reduces the amount of carbon dioxide absorbed from the atmosphere). The huge increase in fossil fuel use in the last 250 years has been set out above. The consequence is that the amount of carbon dioxide in the atmosphere has risen by a third (Table 5.13). The slow increase in the 19th century reflects the relatively slow pace of industrialisation and increasing energy use. Half of the overall increase has occurred since the 1950s and reflects just how fast the world economy has grown in that period. Global carbon dioxide emissions more than tripled from 1.6 billion tonnes a year in 1950 to 5.4 billion tonnes in the mid-1980s and have continued to rise since.[2]

The other major gas involved in global warming is methane. Its increase is linked less to industrialisation than to rising population and the need to grow more food and also supply the increasing food trade to the richest countries in the world. Methane is produced in a number of ways:

- From the decaying matter at the bottom of paddy fields, which helps to keep the soil fertile. The need to feed an increasing population across Asia has led to a steady increase in the number of paddy fields in the world – by the 1980s the growth rate was about 0.7% a year.
- In the digestive tracts of cattle – the number of cattle in the world doubled between 1960 and 1980 and has continued to rise at about 1% a year.
- The digestive tracts of the billions of termites in the world – their number has increased enormously

Table 5.13 Atmospheric concentration of carbon dioxide

Date	Parts per million
1750	270
1850	280
1900	300
1958	316
1985	350
1997	363
1998	367

---

[1] In 1987, 27 nations signed a global environmental treaty, the Montreal Protocol to Reduce Substances that Deplete the Ozone Layer, that had a provision to reduce 1986 production levels of these compounds by 50% before the year 2000. More forceful amendments were approved in London in 1990, Copenhagen in 1992, Montreal in 1997 and Beijing in 1999. The manufacture of CFCs ended for the most part on January 1, 1996, with the exception of production within developing countries and for some exempted applications in medicine (i.e. asthma inhalers) and research. The Montreal Protocol included enforcement provisions by applying economic and trade penalties should a signatory country trade or produce these banned chemicals. 173 countries are now Parties to the Convention and 172 to the Protocol, of which about 130 are developing countries.

[2] In 1998, there was a slight decrease of 0.5% in global CO_2 emission, for the first time since the oil crisis in the early 1970s. Preliminary figures for 1999 indicate a second consecutive annual decline (Worldwatch Institute 2000).

because of the destruction of the tropical forests, which has provided large extra sources of their food.

Overall, the amount of methane in the atmosphere has increased by almost 140% since the 18th century. Other important gases causing global warming are CFCs, HFCs and nitrous oxide (the latter mainly from increased fertiliser production and use and from vehicle exhausts).

Global warming was first identified as a problem by the Swedish scientist Svante Arrhenius in 1896, but no effective action has yet been taken. The essence of the difficulty is that global warming is an international problem involving very many activities which cannot be solved by any one country or group of countries. A solution needs to take account of the different national interests of the 200 or so states in the world (there were less than 50 a century ago). The responsibility for the current problem lies overwhelmingly with the main industrialised countries, who have pumped most of the carbon dioxide into the atmosphere since 1750 through their industrialisation, high vehicle use and large energy consumption. Although there is clearly scope for major increases in energy efficiency (for example if American cars used the same amount of petrol as European cars) there seems little prospect of major cuts in energy use given the way in which it is built into the high consumption patterns of the major industrialised states. However, freezing emissions at current levels would simply entrench the current inequalities in the world by denying the majority of the world's people the ability to industrialise and improve their standard of living. During the negotiations that have continued since the UN Conference on Environment and Development (UNCED) in Rio de Janeiro, Brazil, in 1992, the industrialised countries have argued that they cannot make all the changes and that the newly industrialising world must play its part. That argument has considerable validity. Even if the major industrialised countries cut their carbon dioxide output by 50%, Chinese industrialisation over the next 2 decades would, on current plans involving heavy coal use, put an equivalent amount back into the atmosphere. Within the next decade, China will have recovered its historic position and once again become the largest economy in the world.

Given these acute difficulties stemming from the history of the last two centuries, it is hardly surprising that no effective international agreements have so far been implemented. After almost a decade of difficult negotiations, the first global agreement on greenhouse gas reduction targets was reached at the Kyoto conference at the end of 1997. Although the international scientific panel on climate change recommended reductions over 1990 levels of 50–60% in order to stabilise carbon dioxide levels, the conference agreed cuts of 5–8% among the industrialised countries. Although the conference fell well short of the action most environmentalists felt necessary, it was a start. There were a number of omissions – in particular the absence of any targets for developing countries, in particular China, Brazil and India. Although the Clinton administration signed the Kyoto agreement, there were always doubts whether the US senate would ever ratify the agreement, given the lack of action by developing countries. The conference in The Hague to devise monitoring procedures fell through, and in the spring of 2001 the new Bush administration withdrew from Kyoto. It is now unclear how far the process to limit (even marginally) greenhouse gas emissions can proceed.

5.6 Ideas and Actions

Over the last 30 years the world has become increasingly aware of a growing environmental crisis. A number of authors have suggested that the roots of the problem lie in the way in which people have come to see the world around them. In this context blame has usually been laid on the Judaeo-Christian heritage. Certainly a number of texts from the Bible can be quoted which endorse the human domination of nature and the idea that God has placed the natural world in human hands to be exploited as they see fit. However, this is a highly Eurocentric perspective. The majority of the world has never been Christian (let alone Jewish) and levels of environmental destruction have been as high in China and the rest of Asia as anywhere else in the world. Another culprit has been identified as modern economics, which makes the fundamental mistake of ignoring resource depletion and dealing only with the secondary problem of the distribution of resources between competing ends. The earth's resources are therefore treated as capital – a set of assets to be turned into profit – and their price is simply the cost of extracting them and turning them into marketable commodities. (Some such as air never enter a market mechanism). Resources are therefore used up at whatever rate current conditions dictate and it is assumed, in defiance of all the facts, that resources are inexhaustible, substitution of one material or source of energy for another can continue indefinitely and that prices need not (indeed cannot) reflect the fact that all resources are finite and will eventually run out. Marxist economics (as a derivative of classical economics) simply reflects these attitudes in even stronger form. A subsidiary argument has been the so-called 'problem of the commons' (a mistaken term since it is used to refer to resources that are not owned at all rather than those held in common). Many of the world's resources (for example fish and whales) are not owned – their price is simply the cost of obtaining them. In these circumstances the most logical course for any exploiter is to obtain as many fish or whales as quickly as possible – any self-denial will simply leave more resources for a rival. In the absence of any regulatory regime animals will be driven either to extinction or to the point where it is no longer worthwhile exploiting them. This is what has happened to the world's whale stocks (control by the International Whaling Commission was for a long time almost entirely ineffective) and the fish stocks in the coastal seas have been in decline over a long period of time. Paleoecological, archaeological, and historical data show that time lags of decades to centuries accrued between the onset of overfishing and consequent changes in ecological communities (Jackson et al. 2001).

It seems better to regard ideas about humans and their relation to the natural world as reflecting basic human needs and the way humans have acted, rather than as being responsible for the way in which humans have treated their environment. One of the basic factors has been the rise in population. For much of human history after the development of farming growth was slow because of the difficulty in increasing agricultural productivity and matching food output with human numbers. Nevertheless the rise from less than 10 million in about 5000 BCE (when agriculture was first widely adopted) to 1 billion by the 1820s was bound to place an increasing demand on the world's resources. As more land was needed to grow food, more natural ecosystems were destroyed. Timber was needed for buildings and wood for cooking and heating and so forests were gradually cleared. Metal ores were wanted to make both tools and luxury items. People had to be clothed and so land had to be cultivated for crops such as cotton, animals kept to provide wool and leather and wild animals trapped for their skins and fur. These basic demands are more than enough to explain the way humans have treated the world around them. What has happened in the last two centuries has been a phenomenal expansion of these basic demands. There has been a huge increase in population and an even greater increase in consumption and production (and therefore resource and energy use) as some members of the human community have increased their consumption to levels that would have been unimaginable even a hundred years ago.

Many technological changes have been reactions to these basic forces. In clothing, for example, the first humans were able to clothe themselves using the skins of the animals they killed or scavenged. As numbers rose this was no longer possible and textiles were made from natural fibres such as flax, cotton and wool. This required extra land for crops and grazing animals but together with fur trapping proved enough to clothe the world's population. However, with present agricultural production technology the current population of the world could not be clothed in this way – there is not enough land available. Only the development in the 20th century of ways of manufacturing artificial fibres from chemicals has enabled people to be clothed. But these more complex manufacturing processes involve the use of more resources and energy. Similar trends can be seen in writing materials. Until the 1st centuries CE these were natural products – animal skins (parchment and vellum) or plant products (such as papyrus reed). The total amount was therefore limited by the amount of land that could be devoted to raising animals – in Europe in particular this was very limited and many medieval monasteries had to stop the production of manuscripts until more parchment and vellum became available. The invention of paper by the Chinese broke from these constraints even though it was widely regarded (particularly in Europe) as an inferior product. Further changes from the 19th century that relied on using wood pulp and extensive chemical treatment produced an inferior product but one which was available on an even larger scale. Similarly coal was only adopted as a fuel when other materials were in short supply. Yet the use of coal made the industrial and technical advances of the 19th century possible. By 1900 the world's coal consumption was equivalent to destroying and transporting a forest three times the size of Britain every year. There was not enough wood in the world to sustain consumption on this level (let alone the levels of coal and oil consumption reached by the late 20th century) for very long and such huge quantities could never have been transported. Similarly in 1900 Western Europe and North America were reaching the maximum number of horses the agricultural system could support – numbers had risen phenomenally in the 19th century as a result of railway construction, which increased the amount of goods transported and the number of people travelling. In the early 20th century horse feed took up a quarter of the cropland in the United States and this was near the maximum the agricultural system could sustain. Only the development of vehicles with internal combustion engines provided an escape from this problem (and the pollution horses caused in cities) but they then rapidly created their own, even worse problems.

From one perspective this invention of new techniques and more complicated production processes and the utilisation of more resources can be viewed as progress – the increasing ability of human societies to control and modify the environment to meet their needs through sheer ingenuity and a capacity to respond to challenges. The idea of progress has become deeply imbedded in human societies in the last few centuries, particularly in Europe and North America. (Until then few societies believed in progress and most saw the world as being in a state of decline). It has become almost axiomatic in modern societies that increasing wealth and consumption, particularly as measured through that dubious indicator – Gross National Product – can be equated with progress and human development. From an ecological perspective these processes appear to be a succession of ever more complex and environmentally damaging ways of meeting the same basic human needs.

5.7 Conclusion

The aim of this chapter has been to indicate the extent to which current environmental problems, and human attitudes to them, lie deeply embedded in human history. As Abraham Lincoln put it: 'we cannot escape history'. Current global environmental problems can be divided into two kinds:

- Soil degradation, deforestation, desertification and salinisation, which are largely related to one of the basic forces of human history – rising population and the need to feed, clothe and provide fuel for the rising numbers. In the 20th century these problems have been mainly (but not entirely) found in the poorer countries of the world.
- Rapidly rising energy and resource consumption, the unparalleled increase in production and the pollution this causes, from acid rain, photochemical smog, toxic waste, ozone depletion and the output of global warming gases, is primarily (but again not entirely) the responsibility of the industrialised and rich countries of the world.

Both of these problems have to be tackled in an increasingly unequal world where a small minority of the world's population consume most of the resources and have most of the wealth and four-fifths of the world's people (about 4.5 billion people) have about a fifth of the world's wealth.

These problems also have to be tackled by nearly 200 sovereign entities called states which have very unequal power. The United States – the 5% of the world's population that uses about a third of the world's consumption of resources every year and causes an even bigger share of the pollution – has proved singularly unwilling to make any sacrifices that might imperil its standard of living in the short-term. The other industrialised countries have been little different. Neither are newly industrialising countries prepared to forego the expected benefits of higher levels of consumption. The fundamental problem is that if the current population of the world were to live at European (not American) levels of consumption it would require a huge increase in world production. It is unlikely that there are enough mineral and energy resources on earth to sustain such an increase and the consequences of doing so in terms of pollution would be catastrophic.

Another fundamental problem to have emerged in the 20th century is the increasing lack of control by governments over an increasingly globalised economy. Most governments in history have allowed trade and manufacturing a high degree of autonomy while seeking revenue from taxing such activities. The rise of transnational corporations (able to transfer production around the world to suit their own requirements) and globalised financial markets has meant that states and national economies have lost power. The world economic and financial system appears to have developed on autopilot and to be beyond control – indeed governments have largely become suppliants for the favours of the major transnational corporations. Other international institutions such as the World Trade Organisation and the World Bank/International Monetary Fund reflect these priorities. By the 1980s a third of all world trade took place as transactions within transnational corporations and a half of all American and Japanese imports and exports were similar transactions. By the 1990s of the hundred largest economic entities in the world half were states and half were corporations. The Exxon oil company had a turnover six times bigger than Morocco's Gross National Product and General Motors had a turnover twice the GNP of Egypt. This essentially autonomous world economy made it even more difficult for states to implement policies that might affect its continued expansion. In these circumstances the possibility of dealing effectively with the environmental legacy of 10,000 years of human history, and especially that of the last 200 years, seemed increasingly remote.

References

Byrd, B. F. (2005). Reassessing the emergence of village life in the Near East. *Journal of Archaeological Research*, *13*(3), 231–290.
Jackson, J. B. C. et al. (2001). Historical overfishing and the recent collapse of coastal ecosystems. *Science*, *293*, 629–638.
Kauppi, P. E., Ausubel, J. H., Jingyun Fang, Mather, A. S., Sedjo, R. A., & Waggoner, P. E. (2006). Returning forest analyzed with the forest identity. *PNAS*, *103* (46), 17574–17579.
Reijnders, L. (2006). Is increased energy utilization linked to greater cultural complexity? Energy utilization by Australian Aboriginals and traditional swidden agriculturalists. *Environmental Sciences*, *3*(3), 207–220.
UNEP (2002). *Global environment outlook 3*. London: UNEP/Earthscan.
WRI (1998). *World resources 1998*. Washington, DC: World Resources Institute.
Worldwatch Institute (2002). *Vital signs 2002–2003: The trends that are shaping our future*. London: Earthscan.

Suggestions for further reading

Connelly, J., & Smith, G. (1999). *Politics and the environment: From theory to practice*. London: Routledge.
Elliott, L. (1998). *The global politics of the environment*. London: Macmillan.
Garner, R. (2000). *Environmental politics: Britain, Europe and the global environment*. Contemporary Political Studies Series (2nd ed.). New York: St. Martin's Press.
Glacken, C. (1967). *Traces on the Rhodian shore: Nature and culture in western thought from ancient times to the end of the eighteenth century*. Berkeley: University of California Press.
Goudie, A. (2001). *The human impact on the natural environment* (5th ed.). Oxford: Blackwell.
Ponting, C. (1991). *A green history of the world*. London: Penguin Books.
Simmons, I. G. (1996). *Changing the face of the earth: Culture, environment, history* (2nd ed.). Oxford: Blackwell.
State of the World series from 1984 through 2002, an annual report on the progress toward a sustainable society; Lester Brown et al. Worldwatch Institute). UNEP (1999) *Global environment outlook 2000*. United Nations Environmental Programme. London: Earthscan.
Vital Signs series from 1992 through 2002, an annual report on the trends that are shaping our future; Lester Brown et al. Worldwatch Institute.
Wilkinson, R. G. (1973). *Poverty and progress: An ecological model of economic development*. London: Methuen.
Worster, D. (Ed.). (1988). *The ends of the earth: Perspectives on modern environmental history*. Cambridge: Cambridge University Press.

On-line data

Center for International Earth Science Information Network: www.ciesin.org
World Resources Institute: www.wri.org

Chapter 6
Recent Developments and Trends

Lucas Reijnders, Jan Bakkes, and Klaas van Egmond

Contents

6.1	Introduction	49
6.2	Natural Resources	52
6.2.1	General Description	52
6.2.2	Recent Trends and Current Status	52
6.3	Living Nature	55
6.3.1	General Description, Trends and Current Status	55
6.4	Pollution	58
6.4.1	General Description	58
6.4.2	Recent Trends and Current Status	64
6.5	Environmental Hazards	68
6.5.1	General Description	68
6.5.2	Trends	68
6.6	Allocation of Physical Space	68
6.6.1	General Description	68
6.6.2	Recent Trends	69
References		70

6.1 Introduction

Although we often speak of 'the' environmental problem, what we in fact have is a large family of problems affecting the geophysical system, ecosystems and human health and varying enormously in magnitude, nature and temporal and spatial scale. There is a similar diversity in how these problems are framed conceptually by individual societies, under the influence of myriad historical, cultural and social factors. It is therefore instructive to start out by considering a few historical examples of the way in which perceptions of environmental issues have changed over time.

In 19th-century Western Europe the main focus of what we today term environmental hygiene was on the agents of infectious disease. The classic anecdote concerns the London doctor John Snow, who in 1849 suspected that the prevalence of cholera in the vicinity of Broad Street was related to a mixing of sewerage and drinking water. His advice, correctly, was to take the handle off the Broad Street water pump. Since then there has been a thorough-going separation of sewerage and drinking water supply systems in industrial countries, where the incidence of (drinking) water-borne infections has consequently plummeted. Measured in so-called disability-adjusted lost years the burden of disease associated with drinking water, sanitary facilities and poor personal hygiene is now about 0.1% of what it once was. Indeed, in industrialised countries infectious diseases associated with drinking water now scarcely feature on the environmental agenda. The situation in the developing world is quite different, though. Here substandard drinking water and sewerage systems and poor standards of personal hygiene are responsible for 7.6% of the burden of disease (Murray and Lopez 1997; Prüss and Havelaar 2001).

Also telling in this broad, historical context is the fate of a hypothesis formulated in the late 19th century by the chemist Svante Arrhenius. He supposed, on the basis of the absorption characteristics of carbon dioxide (CO_2), that a rise in the atmospheric concentration of this gas would lead to a warming of global climate. It was not until the late 1980s, though, that widespread public and political interest in this theory began to emerge. Following a hot summer in the

L. Reijnders (✉)
Environmental science(s) University of Amsterdam and the Open University of the Netherlands, The Netherlands
e-mail: l.reijnders@uva.nl

United States, testimony was given at US congressional hearings documenting a possible link between higher summer temperatures and elevated atmospheric concentrations of CO_2 and other 'greenhouse gases'. This sparked a lively worldwide debate that is still raging today.

Similarly, society's interest in natural resources may sometimes be triggered by dramatic historical events. Wind erosion, which in the 1930s transformed the Great Plains of the USA into a 'dust bowl', sparked considerable interest in the long-term availability of fertile land. Likewise, the 1973 OPEC move to suspend oil exports to Western countries seen to be too supportive of Israel, focused minds on the future availability of mineral oil. This is not to say that dramatic events always have this power to focus, though, as is all too apparent in the case of European fish stocks. Though some of these stocks are in spectacular decline, society often seemed to be concerned mainly with the problems to the fishing industry caused by diminishing quota, with EU Fisheries ministers getting a better press the larger the quota taken home from negotiations in Brussels.

The way we perceive current environmental problems, and our assessment of their scale and urgency can thus only be understood within the wider economic and socio-cultural context. At the same time, though, we can certainly try and make some kind of objective assessment of the severity of the current environmental situation in terms of socio-economic parameters. In a very simplified way this is done by the 'IPAT' equation discussed in Chapter 1, which states that environmental impact (I) is a function of population (P), affluence (A) – as an approximation for production and consumption – and technology (T). Figure 6.1, showing European emissions of sulphur dioxide, illustrates the importance of these factors.

First we see an increase in emissions driven by population growth (P) and industrialisation, the latter associated with an increase in energy generation and more generally with affluence (A). More recently emissions have been decreasing and consequently become 'decoupled' from production growth. This is a result of technical measures (T) such as desulphurisation of fuels, flue gas treatment and energy-saving provisions. As production and population increase further, improved and probably costlier technology must be applied, for otherwise there would be a 'recoupling' of emissions and their impact (I) with trends in P × A. The latter quantity is more or less reflected in GDP (Gross Domestic Product).

More generally it can be observed that two of the three drivers of environmental impact, population size and affluence, have increased enormously. Since 1950 the world's population has more than doubled (Fig. 6.2), while energy consumption has more than quadrupled (Fig. 6.3). Over this period environmental impact (I) has been largely though not exclusively limited by applying environmentally more benign technologies and forms of production management (T).

There are several ways to structure the current family of environmental problems. In this chapter, in reviewing a number of key environmental problems we shall follow a commonly adopted scheme that distinguishes problems involving natural resources, living nature, pollution, environmental hazards and allocation of physical space (under growing pressure from changes in land use). Any particular activity may obviously have impacts in two or more categories. For instance, burning fossil fuels leads to pollution, reduces the natural resource base, may involve hazards such as oil spills and has an impact on living nature.

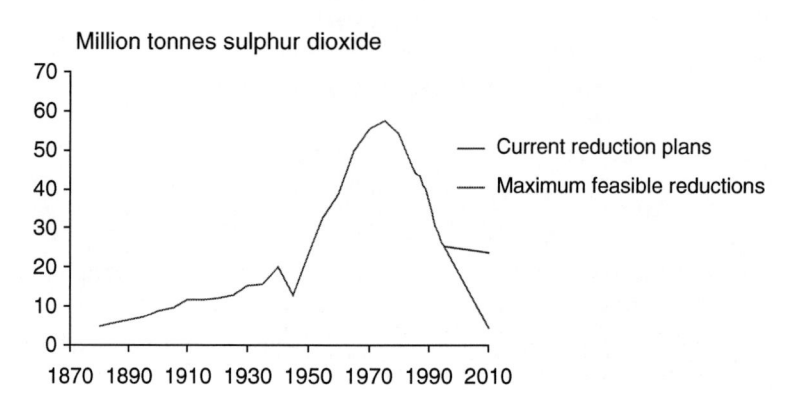

Fig. 6.1 Emission of sulphur dioxide from Europe West of the Urals
Source: Compiled by RIVM

Fig. 6.2 Population (A) and population growth (B)
Source: UN-FPA

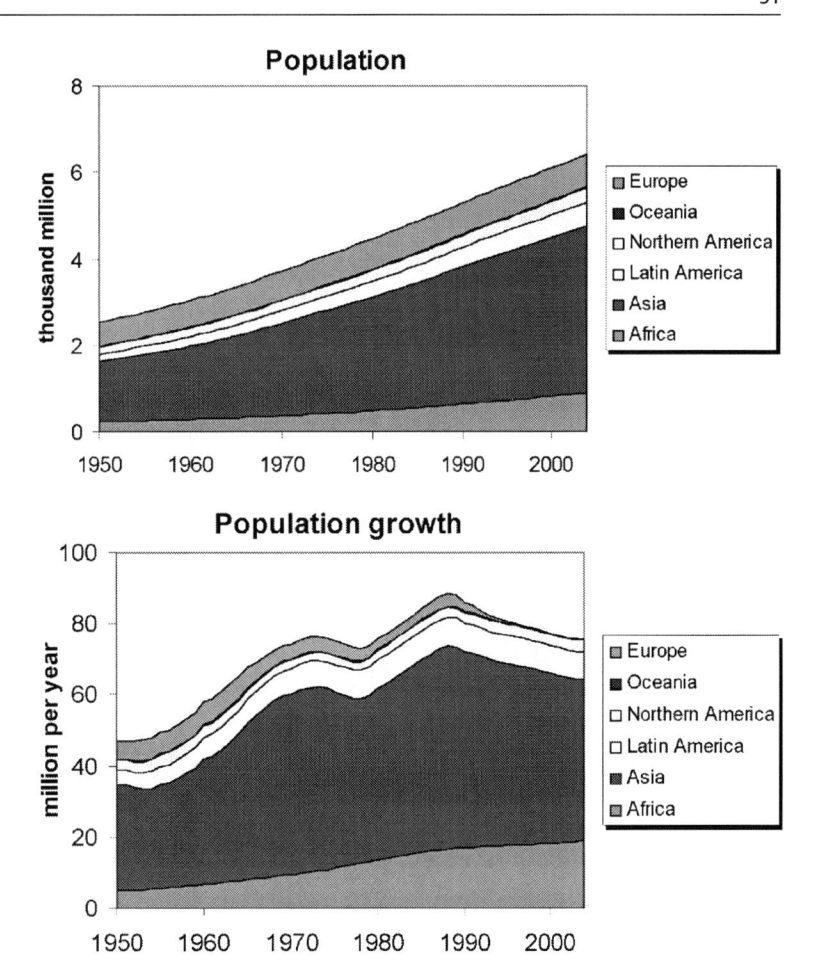

Fig. 6.3 World commercial fuel use
Source: Data before 1925: Etemad et al. (1991). Data for 1925–1965 period: Darmstadter (1971). Data after 1970: British Petroleum (2003). Shares per fuel type: IEA (1971–2001)

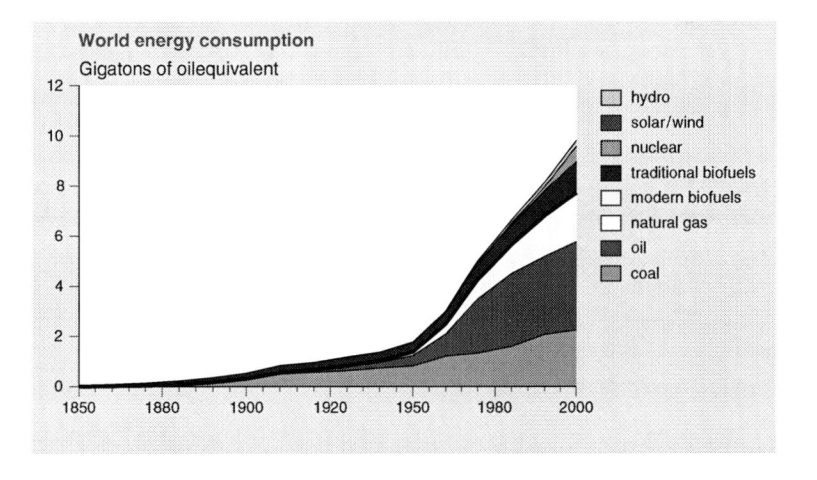

Below we describe the main defining features of each of these five categories of environmental problems, discussing the scope for further systematisation within each and reviewing recent trends as well as the current status of these problems. From the foregoing it will be clear that developments in recent decades have been shaped to no uncertain extent by trends in population, production and consumption ('affluence') and technology.

6.2 Natural Resources

6.2.1 General Description

Natural resources are what we take or 'borrow' from nature to meet our needs in life. Sunlight, fertile soil, groundwater, metal ores and fish from the sea are thus all natural resources. There are major differences in how these resources are generated, however. Some of them are in fact part of a flow, with the convenient property that today's use leaves future availability unaffected. Obvious cases are sunlight and wind used for power generation. The kilowatt-hours produced today by photovoltaic cells and windmills have zero impact on future availability of sunshine and wind. In the field of spatial allocation (see Section 6.6), however, use of such resources may run into problems.

The characteristic of present use not affecting future availability is a far from common property of natural resources. Many of the natural resources we currently use are formed in very slow geological processes. Fossil carbon compounds and metal ores are in this category. They are often called *virtually non-renewable* resources, indicating that their present rate of formation is extremely slow. Other natural resources generated in relatively short time spans, geologically speaking, are often called *renewable*. In this category are most, but not all, groundwater resources, river water, fuelwood and fish. Rain replenishes groundwater and rivers, while fish stocks are maintained by reproduction and fuelwood stocks by tree growth.

It holds for both virtually non-renewable and renewable resources that present-day use may affect future availability. For instance, dissipative use of copper (i.e. applications leading to dispersal during use) may eventually exhaust copper ores, resources of which are virtually non-renewable. Although copper is not destroyed during dissipative use, converting low-grade wastes with their origin in dissipative use into usable copper would require enormous inputs of energy and water and create massive amounts of new waste (Gordon et al. 1987). Under these circumstances copper may no longer be economically attractive in certain applications. In the same vein, a shift towards lower-grade metal ores as high-grade ores grow scarce may lead to quite substantial increases in energy use and environmental side-effects. Dissipative use of high-grade resources has important environmental and economic consequences. At the same time, though, it should be noted that different opinions have been voiced regarding the future of mineral reserves and the implications thereof (Van Vuuren et al. 1999). Similarly, if extraction, loss and/or harvest of renewable resources systematically exceed additional formation, depletion will occur. Thus, overexploitation of fish may lead to the collapse of fish stocks (Worm et al. 2006), erosion may degrade the fertility of farmland and extracting more groundwater than is replenished by rain will lead to a progressive lowering of the water table. There may be a decline in the quality of certain renewable resources, moreover, through pollution for example, limiting the functions they can usefully fulfil. Fertile soils used in irrigated agriculture may become degraded as a result of salinisation (i.e. accumulation of natural salts present in irrigation water). In some areas of the world, groundwater stocks have been rendered unfit for use as drinking water through contamination with microbial vectors of infectious disease and with compounds such as arsenic, nitrate and pesticides.

6.2.2 Recent Trends and Current Status

Worldwide, use of renewable and virtually non-renewable natural resources by *Homo sapiens* is still increasing in absolute terms, although in a number of industrial countries resource intensities (i.e. the input of resources per unit Gross Domestic Product) have decreased overall since 1970. In physical terms the world's richest stocks of virtually non-renewable resources are being depleted, but remaining stocks will generally be sufficient to meet foreseeable demand in the near future, though there may be bottlenecks for specific resources such as those necessary for the supply of bismuth, germanium, tin and platinum (Graedel 2002). Over the longer term there is likely to be a depletion of those natural resources that are scarce in geochemical terms and subject to large-scale dissipative use. Conventional mineral oil, i.e. petroleum, provides an important example here.

In 1956, M.K. Hubbert presented a simple model for predicting future oil production and ultimate oil recovery in the USA (Hubbert 1956). Since then his model has been made substantially more sophisticated (Pesaran and Samier 1995). Actual US oil production has closely followed Hubbert's predictions, colloquially known as the Hubbert curve. Maximum production levels were expected to occur around 1970 and so they did. After that a fall in US oil production was predicted and this indeed occurred (Pesaran and Samier 1995) and for this reason the USA has become structurally more dependent on imports of foreign oil. In the future other oil producers, too, will be confronted with their own variety of the Hubbert curve, moreover. In terms of energy, then, the era of conventional mineral oil as the dominant source of energy is likely to draw to a close within the present century. The same holds for conventional natural gas.

For all types of virtually non-renewable resources, an end to conventional means of extraction will become a fact of life. It will occur relatively soon in the case of high-grade deposits of intensively exploited and geochemically scarce resources and very much later for geochemically abundant resources such as aluminium and iron ores. Much work has been done on this topic and interestingly this has taught us that price signals may well be false indicators of scarcity in the physical sense. Reynolds (1999) has shown that when it comes to depletion of virtually non-renewable resources, prices may fall for many years, with true scarcity revealed only when exhaustion is imminent. For this reason, he suggests, it is conceivable that after a century of price decline and production growth a resource may suddenly increase 10- or 100-fold in price within the space of a year or 2, with a correspondingly sudden decline in production. For geochemically scarce metals like silver and tin such developments are well feasible during the present century.

Acuter today in physical terms, however, are a series of problems associated with renewable natural resources. As an example, Box 6.1 briefly reviews the main problems associated with supplies of freshwater, a vital human resource. There are many other problems in the realm of renewable resources, though. About one-third of the world's major fishing grounds are being

---

Box 6.1 Freshwater resources

On average there is no shortage of water on our planet. Of the estimated 1,386 million cubic kilometres of water, however, only 2.5% is fresh as opposed to salt; of this, some 70% is in the form of ice sheets and glaciers in remote areas, mostly in Antarctica and Greenland (UNESCO 2003). The freshwater resources practically available for human use total about 9,000–14,000 km^3 annually, about half of which is indeed currently used by our species (Raskin et al. 1997).

Irrigation accounts for the largest proportion of water use, roughly two-thirds of total consumption worldwide. Much agricultural water use is inefficient, with only about 40% of the water withdrawn effectively contributing to crop production (UNESCO 2003). In Africa and the Middle East, especially, competition for water is dominated by agricultural demand.

The amount of freshwater available per capita varies widely both between and within countries, making national annual averages relatively meaningless in large countries where water availability and demand may be geographically far apart, in different drainage basins. In addition, population growth, climate change and coping capacities are all spread unevenly. Current projections based on the water balances of individual drainage basins (i.e. more detailed than country statistics) show the number of people in areas with severe water stress increasing in all but the high-income regions, up from the current 300 million to 2,500 million by 2050 (UNEP 2002; World Water Council 2000; see also Fig. 6.4). Finally, on top of the issue of matching water availability and demand within each drainage basin comes the issue of fair and effective water supply. The urban poor are at particular risk here, as their drinking water supply and the price they must pay for it are highly dependent on local governance, including such factors as corruption, weak government, whether and how water services have been privatised and arrangements regarding investments in infrastructure. If the projected growth of 'megacities' like Mexico City is taken tentatively into account, the potential increase in the population at risk from water stress becomes even larger, especially in parts of Africa, Latin America and Asia (RIVM 2000).

There are, moreover, widespread problems with water quality. It has already been mentioned that 7.6% of the burden of disease in developing countries is associated with problems linked to unsafe drinking water and inadequate sanitation. Parts of the Indian subcontinent are confronted with high concentrations

(continued)

Box 6.1 (continued)

of arsenic in drinking and irrigation water, leading to a high incidence of skin lesions. All in all, it is estimated that over 35 million people in Bangladesh, West Bengal and Nepal will eventually suffer the effects of arsenic poisoning (Anawar et al. 2002). The increased incidence of childhood disease around the Aral Sea is probably also due in part to water pollution.

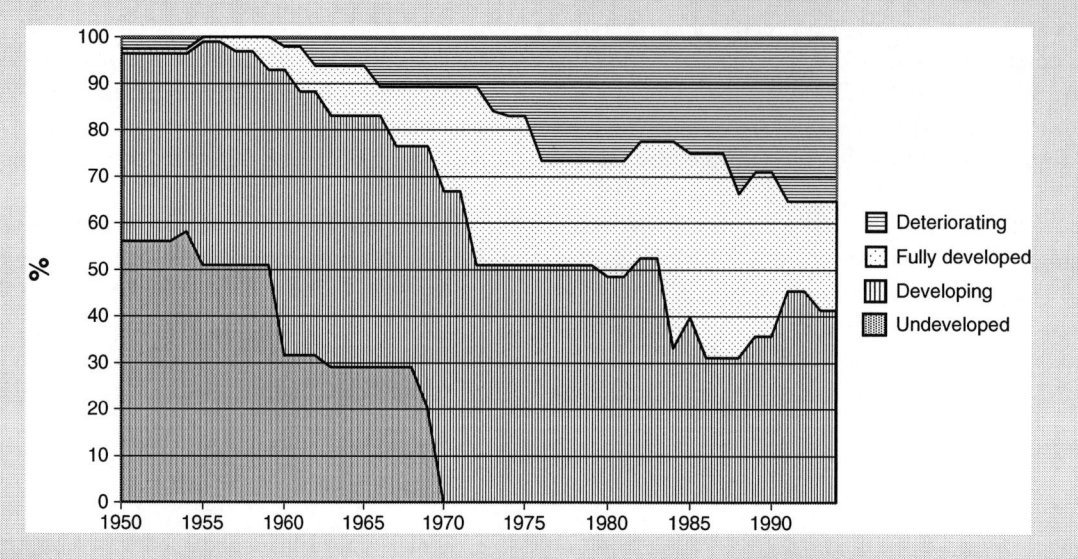

Fig. 6.4 Population living in areas with severe water stress. Note: based on analysis per drainage basin. When annually more than 40% of the renewable water resources of a river basin are being withdrawn for human use, the river basin is considered to be under severe water stress

Source: Center for Environmental Systems Research, University of Kassel (UNEP 2002)

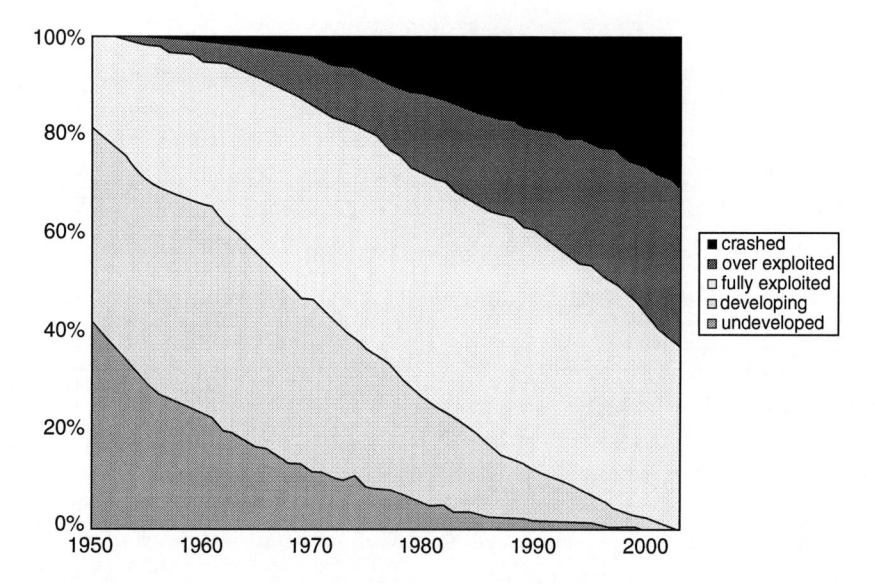

Fig. 6.5 Exploitation status of marine fish stocks, 1950–2002
Source: Pauly (2007)
Note: The exploitation status is defined with respect to the highest catch of each time series representing one stock, usually a species, within 1 of 18 FAO statistical areas covering the world oceans. Underdeveloped: 0–10%; developing: 10–50%; fully exploited: above 50% of maximum; overexploited: 50–10%; and crashed: 10–0% of maximum catch

overexploited (FAO 1997, 2002), leading to dwindling or even vanishing returns on fishing (see Fig. 6.5). Parts of Asia, Africa and Latin America are suffering a fuelwood crisis. And in many areas of the world soil fertility has declined through erosion, salinisation and reduced rainfall associated with the loss of natural vegetation.

Although the precise worldwide extent of these problems is subject to controversy, their impact locally may be very significant. Worst affected by land and soil degradation are Africa and West Asia. In these areas, unless due action is taken, the deterioration of fertile land and water shortages will necessitate a growing reliance on food imports (also called 'virtual water' in this context) and may well lead to decreased food security of the poor. Similarly, in the absence of remedial and mitigating action in Africa, financial losses relating to land degradation may be considerable, amounting to the equivalent of up to 15% of GDP in some countries. In both Africa and Asia there is also increasing migration due to land degradation. The collapse of codfish populations off the coast of Newfoundland (EEA 2001) has been a disaster for the fishermen concerned. In parts of Nepal, due to a fuelwood crisis, a family may spend up to 30 h a week gathering the fuel they need.

6.3 Living Nature

6.3.1 General Description, Trends and Current Status

Nature has been defined by some in terms of the absence of human intervention. In this sense, though, living nature has ceased to exist on our planet, for all the world's ecosystems have been affected to at least some extent by human activities. If one defines nature as 'not being under close human control', however, living nature is still there, though the total area it encompasses is certainly diminishing; and so is biodiversity, see Fig. 6.6.

According to the Convention on Biological Diversity (CBD), biodiversity encompasses the overall variety found in the living world and includes the variation in genes, populations, species and ecosystems. Several complementary indices are used within the CBD framework. In Fig. 6.6 biodiversity loss has been expressed for each biome in *the mean abundance of the original species* (MSA) compared to the natural or low-impacted state.

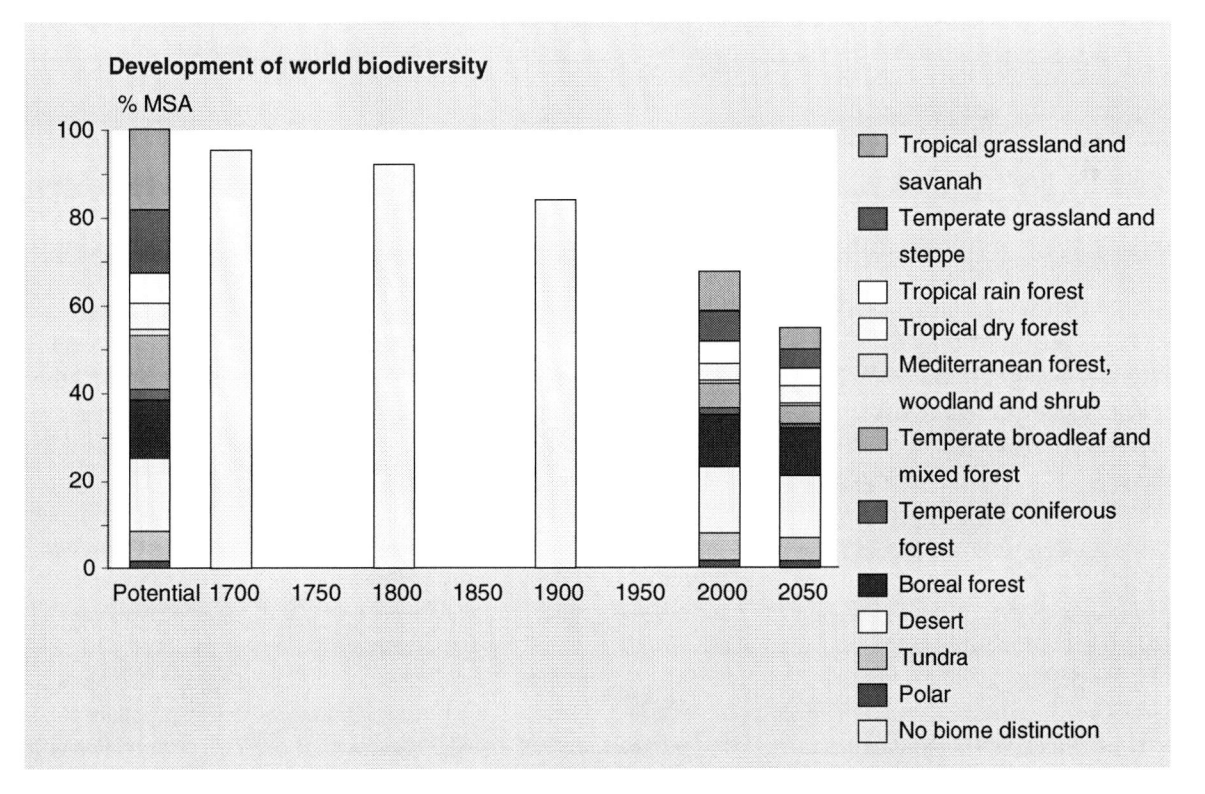

Fig. 6.6 Development of world biodiversity. For explanation: see text
Source: MNP, UNEP, WCMC and UNEP/GRID Arendal (2006)

The various nature types in the world, also called 'biomes' vary greatly in the number of species, their species composition and their species abundance. Obviously a tropical rainforest is entirely different from tundra or tidal mudflats. The *mean abundance of the original species* at the global and regional levels is the sum of the underlying biome values (in km^2) in which each biome is equally weighted. A key feature is that the indicator treats the biodiversity value of all ecosystems alike, whether tundra or tropical rainforest. This indicator can be interpreted as a measure of 'naturalness' or 'intactness', and can be disaggregated to regions, biomes and causes of loss. While not being an absolute measure of biodiversity and not intended to highlight individual species under threat, the indicator is very suitable for use in comparing different future projections.

It is not an absolute measure of biodiversity. If the indicator is 100%, the biodiversity is similar to the natural state. If the indicator is 50%, the average abundance of the original species is 50% of the natural state, and so on. The abundance of exotic or invasive species is not included in the indicator but their impact shows up by the decrease in the abundance of the original species they replace.

Future changes in mean species abundance are estimated using model results for pressures including climate change, nitrogen deposition, land-use change (agriculture, forestry, urban), infrastructure and human settlement, fragmentation and agricultural intensification on a 0.5 × 0.5 degree spatial grid. These projections are then weighted on the basis of literature-based dose-response relationships for their effect on biodiversity (MNP, UNEP, WCMC and UNEP/GRID Arendal 2006). It should be emphasized that Fig. 6.6 reflects rough estimates only and is based on pressures that are model outcomes in themselves.

In reality, impacts of the various pressures are probably not linear. Furthermore, effects related to climate change are long-term. This means that in ecosystems in the northern regions, which are already under significant pressure due to the current rate of climate change, the effects shown by the indicator have not yet materialized. Finally, the indicator may mask differences between regions or biomes by averaging.

The only biomes shown not to decrease significantly in biodiversity so far, when using the approach reflected in Fig. 6.6, are the relatively inhospitable biomes of deserts, boreal forests, tundra and polar areas. All in all, living nature is today subject to major changes, one of the chief elements of which is species extinction. Although extinction is itself a natural process, recent and current extinction rates suggest that the latter exceed by at least two orders of magnitude the expected rate without human influence (Groombridge and Jenkins 2000).

The first major cause of species extinction is human appropriation of primary biological production, i.e. plant biomass produced by photosynthesis, on which most of life as we know is ultimately based. Estimates suggest that about 70% of all primary production takes place on the continents and 30% in the seas and oceans. An ever greater proportion of this is being diverted for use by *Homo sapiens*. On land a substantial share of worldwide net primary production is currently appropriated by mankind. Estimates of the size of this share vary between about 20% and 50% (Haberl et al. 2007; Imhoff et al. 2004; Pimentel 2001; Vitousek et al. 1986). Though further research into this matter is desirable, it would seem that the relatively low estimates of about 20–24% (Haberl et al. 2007; Imhoff et al. 2004) are best supported by currently available empirical data. Imhoff et al. (2004) also estimate that in Western Europe and South Central Asia human appropriation of net primary production on land is over 70%. For coastal seas the appropriation of net primary production has been estimated at between 25% and 35% (Pauly and Christensen 1995). Today, agriculture is the main form in which terrestrial biomass is harvested. The extent to which agriculture replaces nature on land depends on a number of factors. The first is population size. The second is consumption of animal produce. To produce 1 kg of animal protein requires approximately 10 kg of plant protein. Thus, other things being equal, the more animal produce is eaten, the more biomass needs to be appropriated. The third key factor is technology, which is a major determinant of primary production per hectare of farmland. Technology-based agricultural 'intensification' may reduce claims and pressure on available land. On the other hand, this process of intensification has frequently led to greater emissions of nutrients and pesticides, which may have a negative impact on natural species. In Europe increased food production in recent decades has been achieved mainly through agricultural intensification, in Africa and Latin America by increasing farmland acreage.

The expansion of farm acreage in particular has led to a physical 'roll-back' of living nature and an overall decline in the number of natural organisms. The appropriation of biomass associated with surface waters, both inland and marine has also reduced species diversity (e.g. Wolff 2000). Another development affecting living nature is climate change, in particular the rate of change. This factor becomes increasingly important. In a study that covered about 20% of the Earth's terrestrial surface, Thomas et al. (2004) estimated that due to climate change 15–37% of species may become subject to regional extinction before 2050. These are not the only pressures on terrestrial biodiversity. Others include:

- Forms of human land use other than agriculture, often leading to 'soil sealing'
- Water management, including manipulation of groundwater tables
- Over-harvesting and over-hunting
- Logging and clear-cutting of forests
- Changes in the levels of plant nutrients such as nitrogen compounds
- Fragmentation of habitats by roads and other infrastructure

Offsetting some of the pressures towards species extinction, there has been growing funding of professionally managed nature reserves in recent decades. To the extent that nature conservation has been evaluated, the overall effect on species extinction has remained limited. In case of birds a further increase of the extinction rate has been prevented by conservation efforts (Rodrigues 2006). In addition, in regions with a relatively stable population and highly productive agriculture, farmland may be taken out of production and subsequently allowed to revegetate naturally. In terms of biodiversity, though, this reconverted land will generally remain of relatively poor quality for the first few decades.

A new factor in this context is the growing field application of genetically modified organisms (GMOs), the full influence of which – either directly on other species or through changes in agricultural practices and trade – has yet to become clear.

A rough but useful way of summarising the combined effect on biodiversity of changes in habitat area and human pressure is to express it as an index, as in Fig. 6.6. On balance, terrestrial biodiversity is projected to decline in every region of the world over the next few decades, but particularly in Africa. Some regions, such as North America, may see a spatial shift in species distribution (Groombridge and Jenkins 2002; UNEP 2002).

Finally, living nature is gradually becoming 'globalised' as a result of ever more international transportation. In Canada so-called 'exotics' now account for about 20% of recorded natural species (Muradian 2001) and North American coastal waters are now home to at least 289 exotic invertebrates brought there by ocean-going vessels (Ruiz et al. 2000). Exotics tend to displace local species. In parts of the southern United States the original vegetation has been wiped out entirely by the successful invader *Tamarix*, or salt cedar (Simberloff and Von Holle 1999). In South Africa 181 alien plant species have to date been counted; these have affected ecosystems on 10 million hectares of land and reduced the value of 'fijnbos' (a forest variety) by about US$11.5 billion (Van Wilgen et al. 2001). Many of the endemic cichlid fish species of Lake Victoria have been driven to extinction by human introduction of the Nile perch (Witte et al. 2000). A number of other dramatic extinctions relate to island species. Inadvertent introduction of a brown tree snake to the Pacific island of Guam in 1950, for example, led to the extinction of 12 endemic bird species and a number of reptiles and bats. By 1990 only three of Guam's original vertebrates still survived (Fritts and Rodda 1998).

Worldwide, species diversity is in general decline. In most of the familiar divisions of the animal kingdom between 5% and 20% of species have already been driven to extinction by human activity (Sala 2001). Specialised organisms, species living in rivers and large land animals have been hit relatively hard. In centuries past, extinctions were restricted mainly to island species. More recently, though, extinction rates on the continents are beginning to approach those known previously only from islands (Lomolino and Perrault 2001). Many of the species that have escaped extinction have been vastly reduced in number. A telling example in this respect is our closest natural relatives, the primates. Between 1983 and 2000 the number of chimpanzees and gorillas in equatorial Africa plummeted at a rate of 4.7% a year. Since then the reduction of chimpanzee and gorilla populations has continued relentlessly. Unless there is a change from 'business as usual', the next decade will likely see our closest relatives pushed to the very brink of extinction (Bermejo et al. 2006; Walsh et al. 2003).

As a result of species extinctions and sharp declines in numbers, some ecosystems have changed dramatically. The collapse of cichlid stocks in Lake Victoria has had a major knock-on effect in the form of increased turbidity and 'algal blooms' (Witte et al. 2000). The reduction in the number of oysters in Chesapeake Bay (USA) has greatly reduced water filtration in the bay and has altered its natural life dramatically. Similarly, the disappearance of the Pacific's kelp forests has been ascribed to overexploitation of fish stocks (Jackson et al. 2001), while the demise of Steller's sea cow has led to major changes in the Bering Strait ecosystem.

On a planetary scale, the very functioning of natural ecosystems has also changed. The roll back of nature and the loss of species from ecosystems leads to reduced ecosystem services (services freely provided to mankind by living nature) and changes in biogeochemical cycling. The widespread destruction of the earth's forests has led to a rise in atmospheric CO_2 levels and a decline in concentrations of the hydroxide radical, an important 'scavenging agent' that helps remove pollutants from the atmosphere. (Monson and Holland 2001). In some areas deforestation has increased water flux variability in river basins. Full-blown clearance of tropical rainforests has turned relatively wet areas into dry savannas (Silveira and Sternberg 2001). Impoverishment of plant communities is known to affect the functioning of ecosystems, for instance leading to a lower abundance of biomass (Cardinale et al. 2006) and an increase of the rate at which nutrients such as nitrate leak to the groundwater (Sala 2001). Finally, there is the hypothesis that the continuing roll-back of living nature may disturb or disrupt feedback systems that are currently stabilising our abiotic environment (Lovelock 1987).

6.4 Pollution

6.4.1 General Description

Pollution can be characterised as matter or energy (such as electromagnetic fields or noise) in the wrong place and in the wrong amount. There is a bewildering variety of pollutants in the environment today. In the course of human development there has been massive mobilisation of many of the constituents of the earth's crust, including heavy metals, arsenic, fluoride, selenium, asbestos, phosphate and other substances potentially harmful to organisms, including ourselves (Platt et al. 2001). Besides these natural elements there are also over 10,000 man-made chemicals in widespread use that may turn up as pollutants. Furthermore, a substantial number of pollutants are formed as by-products during industrial conversion processes such as combustion and chemical synthesis as well as during biological metabolism. High levels of noise, radioactivity and actual exposures to 50 Hz electromagnetic fields have been found to affect living organisms adversely. Biological organisms, natural or man-made, are increasingly being recognised as environmental contaminants, too. This follows a remarkable historical reversal in the global campaign to wipe out infectious disease. In 1969 the US Surgeon General W.H. Stewart felt the time had come to pronounce the book on such diseases closed. Since then, however, many new infectious agents have emerged, while other diseases that seemed to be on the wane have bounced right back. One of the factors contributing to this trend has been the excessive use of antibiotics, both medically and in livestock production, which has led to the emergence of antibiotic-resistant strains of a number of common bacteria. Another more recent development is field planting of genetically engineered crops, bringing with them the risk of weeds having multiple herbicide resistance.

Ever since humans built their first dwellings, pollution of the indoor environment has constituted a major problem. The indoor use of fire is traditionally important in this respect and in many places still is. In India, for example, indoor air pollution is currently estimated to be responsible for 4–6% of the national burden of disease (Smith 2000). Similarly, damp indoor conditions may lead to the growth of moulds that aggravate adult asthma, still an important problem in the European Union (Zock et al. 2002). Air polluted by combustion products and moulds are examples of 'old' concerns affecting the indoor environment, but there are also more recent ones. A new phenomenon in this context is the Legionella bacterium, which has been found to thrive in warm water systems at a temperature of 30–40°C. When that water is dispersed in the form of fine droplets (by air conditioners, cooling towers or whirlpools, for instance) the bacteria may lodge in the human respiratory system, causing pneumonia and even death.

In the wider, outdoor environment the dynamics of pollution can be extremely variable in both spatial and temporal terms. A pollutant like cadmium infiltrating a clayey soil will essentially have a local impact, for example, while atmospheric emissions of a gas like carbon dioxide have an impact on a global scale. Noise is short-lived and its effects are therefore highly localised, while atmospheric emissions causing smog are known to enter hemispheric circulation and to drift to the next continent. Emissions of other compounds such as the propellant gas carbon tetra fluoride CF_4 that is emitted by aluminium smelters may have a far longer effective lifetime, however, affecting atmospheric temperature for many thousands of years. Similar environmental lifetimes of thousands of years have been reported for metals discharged into the oceans.

How long-lived a particular pollutant is depends on a multitude of factors. Bacteria and other living pollutants may proliferate by means of reproduction, for example, while metals are 'conservative', remaining unchanged. Organic chemical compounds may degrade, though the extent to which they do so over a certain time span may depend very much on the precise context. In some cases effects may persist long after the pollutant itself has disappeared, as when the germ line of an organism undergoes mutation following exposure to a mutagenic chemical or ionising radiation, with effects sometimes turning up only after many generations.

Systematic categorisation of the pollutants found in the environment today is rendered all the more difficult by their sheer variety. In an attempt to bring at least some order, though, we can identify several useful categories based on pollutant lifetimes and other temporal dynamics in the overall environment and on relations of cause and effect. In doing so we shall focus on abiotic, i.e. non-living, pollutants in the outdoor environment. For the general public and professional environmental scientists alike, it has been abiotic pollutants that have attracted greatest attention over the past few decades. This does not mean that present knowledge of the effects of these substances is anything like perfect, however. Of the approximately 3,000 air pollutants thus far identified, for example, only about 200 have been investigated as to their impacts (Fenger 1999).

It is now nonetheless clear that two types of abiotic pollutants can be usefully distinguished: flux-type and sink-type, the former characterised by the pollutant's mobility in the environment, the latter by its accumulation in some part of the environment that serves as a 'sink'. This distinction has implications for the relationship between (human) discharge and (environmental) concentration. If a pollutant passes relatively rapidly through the environment, its behaviour can be characterised as 'fast-flux'. A case in point is river discharge of a soluble pollutant, which may be transported many kilometres downstream in a single day. If the discharge is voluminous, downstream concentrations will be high; if it is relatively small, concentrations will also be low. In both cases, though, concentrations will respond quickly to any change in the discharge and if and when the latter ceases, river clean up will also be rapid.

Pollutant movement is not always this fast, however. In groundwater, for instance, pollutants may travel at a rate of only several millimetres a day. This is termed slow-flux pollution. In the case of groundwater pollution, moreover, effects may only become apparent after the water is brought to the surface from a well, which may be long after the occurrence of the polluting activity. This also means that there may be a long time lag between cuts in discharges and cleaner wells.

In the case of sink-type pollution, discharge and concentration stand in a different relationship. If a soluble compound is discharged into a river at a constant rate, river-water concentrations will, as a first approximation, remain constant. If compounds with a long environmental lifetime are discharged at a constant rate into an environmental sink, in contrast, local concentrations may well increase. This is exemplified by Fig. 6.7A, reflecting the choices facing negotiators in 1986 during discussion of the Montreal Protocol to Protect the Ozone Layer. Their concern was to cut emissions of chlorofluorocarbons (CFCs) known to damage the ozone layer. Figure 6.7A shows that a 85% cut in emissions of CFC-12 was necessary to stabilise the atmospheric concentration of this compound and prevent further deterioration of the ozone layer. In deference to alleged economic interests, the Montreal negotiators finally settled for a 15% cut. As the figure shows, this limited reduction still leads to a major rise in atmospheric CFC concentrations. During renegotiation of the Montreal Protocol in the 1990s parties agreed to implement far larger cuts and by 1997 an 85% reduction in worldwide emissions of CFC-12 and other ozone-depleting substances had been agreed to Actual global consumption is shown in Fig. 6.7B. Remarkably, though, while use of certain HCFCs approved temporarily as less-depleting alternatives remained constant as of 1997, use of non-ozone layer depleting HFCs that act as 'greenhouse gases' has been allowed to grow by about 20% a year since the mid-1990s (AFEAS 2002; Olivier and Bakker 2000; Smythe 2000). At the same time, by reducing emissions of substances that both deplete the ozone layer and act as greenhouse gases, the Montreal Protocol has produced sizeable ancillary benefits in terms of preventing global warming (Velders et al. 2007).

There are essentially three major environmental sinks for longlived abiotic pollutants: the atmosphere, soils and oceans. The temporal dynamics of sink-type pollution makes it a relatively hard nut to crack. To complicate matters still further, however, long-lived pollutants can accumulate not only abiotically, as with fluorocarbons in the atmosphere, but also biotically, i.e. in living organisms. In some cases biological concentrations may rise to far in excess of ambient levels. An important example of a chemical with a propensity for such bioaccumulation is the insecticide DDT, which accumulates in fatty tissues of birds and mammals, where it may become toxically mobilised in times of stress. In such cases, moreover, the pollutant also accumulates along the food chain, with top predators such as otters and eagles receiving a relatively heavy burden.

There are many other substances that can accumulate biotically, usually in specific parts of the body. In humans, for instance, cadmium accumulates in the kidneys, methyl mercury in the nervous system, lead in bones and organobromine fire retardants in body fat.

Although a division of pollutants into flux-type and sink-type is useful as a first approximation, it provides an

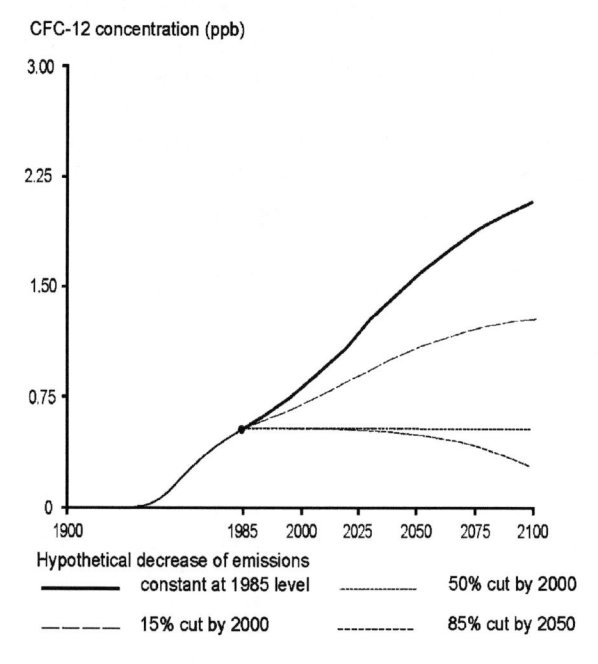

Fig. 6.7A Concentration of CFC-12 in the atmosphere since 1900 and concentration at the chosen scenarios
Source: RIVM (2000)

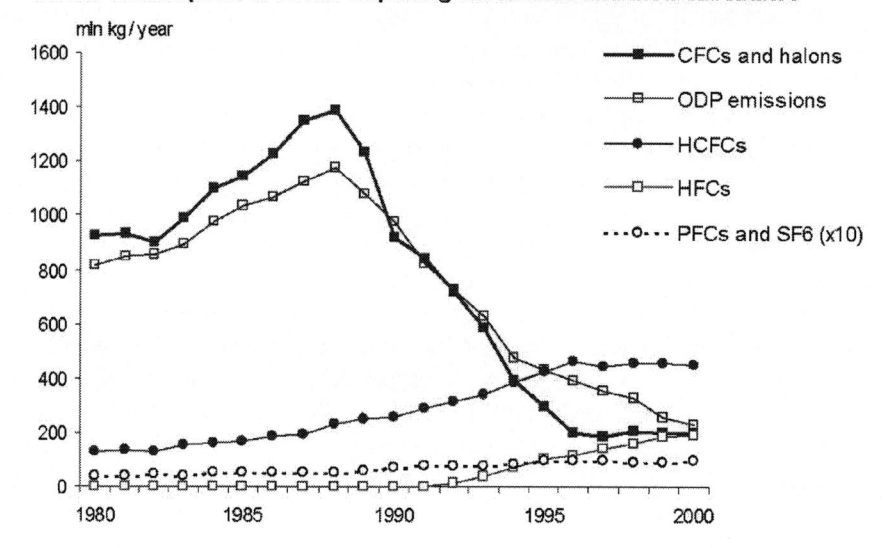

Fig. 6.7B Global consumption of ozone depleting substances and their substitutes
Source: AFEAS (2002), Olivier and Bakker (2000), Smythe (2000)

incomplete picture of what is actually happening, as is clearly exemplified when one and the same compound contributes to more than one environmental problem. For instance emissions of nitrogen oxides may first lead to smog (fast-flux) and later to acidification of soils (sink-type) and/or nitrate pollution of aquifers (slow-flux). This example is elaborated in Box 6.2, briefly reviewing the environmental problems involving nitrogen compounds.

Not only for nitrogen, though, but for quite a number of other pollutants too, it is useful to consider the

---

Box 6.2 Nitrogen compounds in the environment

The element nitrogen (N) is involved in regulating numerous essential ecological characteristics and biogeochemical processes including species composition, diversity, population growth and dynamics, productivity, decomposition and nutrient cycling in many terrestrial, freshwater and marine ecosystems.

The Earth's biogeochemical nitrogen cycle has been markedly altered by a number of human activities that have greatly increased the rate of *nitrogen fixation* compared to the pre-industrial era (Galloway et al. 1995; Arrigo 2005). Nitrogen fixation is the transformation of the highly abundant but biologically unavailable form dinitrogen (N_2) in the atmosphere to '*reactive*' forms, including ammonium (NH_4^+), nitrite (NO_2^-) and nitrate (NO_3^-) in the biosphere and ammonia (NH_3), the nitrogen oxides NO and NO_2, together denoted as NO_x, and nitrous oxide or dinitrogen oxide (N_2O), in the atmosphere.

Prior to the human alteration of the biosphere, the primary pathways for transfer from inert to available forms of nitrogen were *biological nitrogen fixation* by specialized bacteria and algae, accounting for about 100 Tg N year^{-1} in terrestrial ecosystems and 100–300 Tg N year^{-1} in marine systems, and chemical fixation by lightning, accounting for perhaps 10 Tg N year^{-1}.

Anthropogenic pathways of nitrogen fixation include industrial nitrogen fixation in the form of fertilizers, biological nitrogen fixation by nitrogen-fixing crops, and fixation during stationary and mobile fuel combustion including fossil fuels and biofuels (see reviews by Galloway et al. 1995; Krug and Winstanley 2002; Smil 1999; Vitousek et al. 1997). The main nitrogen-fixing crops are legumes such as beans, peas, soybeans and alfalfa, which live in symbiosis with N-fixing Rhizobium bacteria. The human input of reactive N into the global biosphere has increased to about 150 Tg N year^{-1}. This increase stems mainly from *food production* (about 125 Tg N year^{-1}) and energy production (about 30 Tg N year^{-1}) for the world's growing population.

It seems likely that human activities have increased not only the input of reactive nitrogen, but also its *mobility* within and between terrestrial and aquatic ecosystems, as a consequence of land clearing, biomass burning, wetland drainage and other processes (Vitousek et al. 1997; Krug and Winstanley 2002). Reactive nitrogen species can be transported over great distances in the atmosphere and in aquatic systems. Nitrogen is a highly mobile element, so that while deliberate applications occur at the local level, its influence becomes regional and global as it spreads through water and air across political and geographical boundaries.

Reactive nitrogen influences numerous essential processes in the atmosphere and the terrestrial and aquatic biosphere. In each of these spheres nitrogen has effects at different scales.

Local scale. Some consequences of increasing nitrogen inputs play out as local air and water pollution problems, as in the case of ammonia emissions, which can harm vegetation, and the leaching of nitrate, which can pollute adjacent groundwater, rivers and lakes. Such local problems are most widespread in industrialised regions (Europe, North America), but they may also occur in areas with intensive agricultural production, e.g. regions with wetland rice cultivation (South and East Asia) or intensive livestock production.

Regional scale. Other problems are reflected in regional changes in net primary production and in eutrophication. Bouwman and Van Vuuren (1999) have reported that long-range atmospheric transport of nitrogen may be causing eutrophication even in remote regions of Africa, South America and Asia (see Fig. 6.8).

(continued)

Box 6.2 (continued)

Global scale. Finally there are nitrogen-related problems that are global in scale, such as those due to changes in atmospheric chemistry and climate change.

Over the centuries human activity has caused atmospheric nitrogen deposition rates to increase to three to over ten times natural rates (Holland et al. 1999). One of the consequences of this is nitrogen eutrophication, which now affects substantial areas all over the world; see Fig. 6.8. In contrast to acidification, which is spreading globally but at the same time declining in the industrialised regions where it once started, eutrophication pressure by nitrogen compounds is increasing in all regions (see Fig. 6.9).

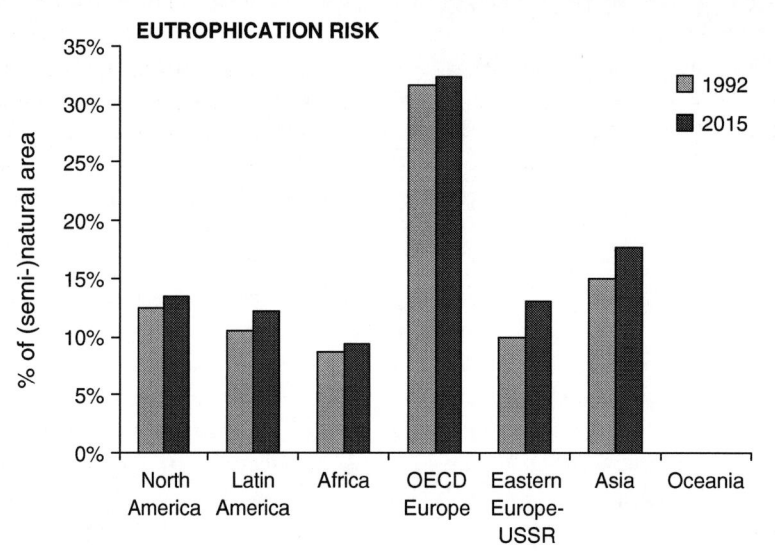

Fig. 6.8 Risk of nitrogen eutrophication in natural and seminatural ecosystems, based on 'current reduction plans' and medium estimate of critical loads
Source: UNEP/RIVM (Bouwman and Van Vuuren 1999)

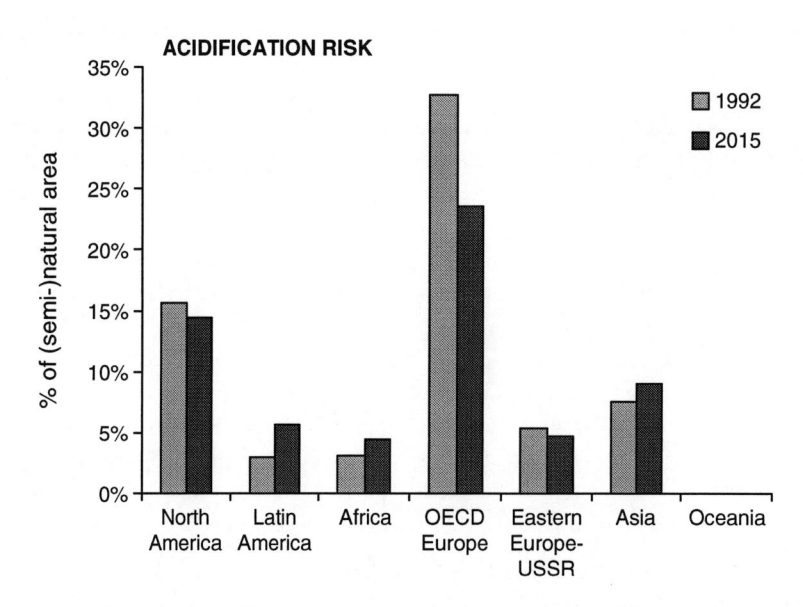

Fig. 6.9 Risk of acidification in natural and seminatural ecosystems, based on 'current reduction plans' and medium estimate of critical loads
Source: UNEP/RIVM (Bouwman and Van Vuuren 1999)

biogeochemical cycles in which they are involved. Thus, the effects of emissions of sulphur oxides are best studied against the background of the sulphur cycle: see Fig. 6.10, showing among other things that the sulphur flux from land to sea has doubled as a result of human activities. Similarly, studies of biogeochemical cycles have shown that the phosphorus flux from land to sea has roughly doubled in historical times. Around 1970 the flux of lead through the atmosphere was about 350 times the natural flux.

In general, then, conceptualising the fate of substances that occur naturally is well served by examining biogeochemical cycles. When dealing with man-made chemicals with no natural counterparts, in contrast, one should be well aware of the pitfalls of viewing them against the background of biogeochemical cycles. Good examples here are the chlorofluorocarbons already discussed. Natural fluxes of chlorine compounds are 2,000–3,000 times larger than those generated by mankind. Similarly, in 1993 the net addition of man-made chlorine to the stratosphere was 8.10^{10} g chlorine, while volcanic emissions were 700.10^{10} g (Graedel and Crutzen 1993). Nevertheless, because of the specific nature of the CFCs accumulating in the stratosphere these had a dramatic effect on the amount of ozone present, whereas volcanic emissions of HCl did not.

A second approach to categorising pollutants and helping us understand their behaviour proceeds from the severity of their impact on living organisms. Thus, pollutants can be classified in terms of dose-effect relationships, which may quite different for different compounds as well as for different organisms.

The first type of dose-effect relationship holds for nutrients and quite a number of endogenous (internally created) substances, such as hormones. In this case, dose-effect curves are U-shaped, or biphasic: see curve A in Fig. 6.11. Too little has a negative effect, as does too much, with the optimum in between. Where this optimum lies may vary quite substantially from species to species. Indeed, this is precisely what is driving

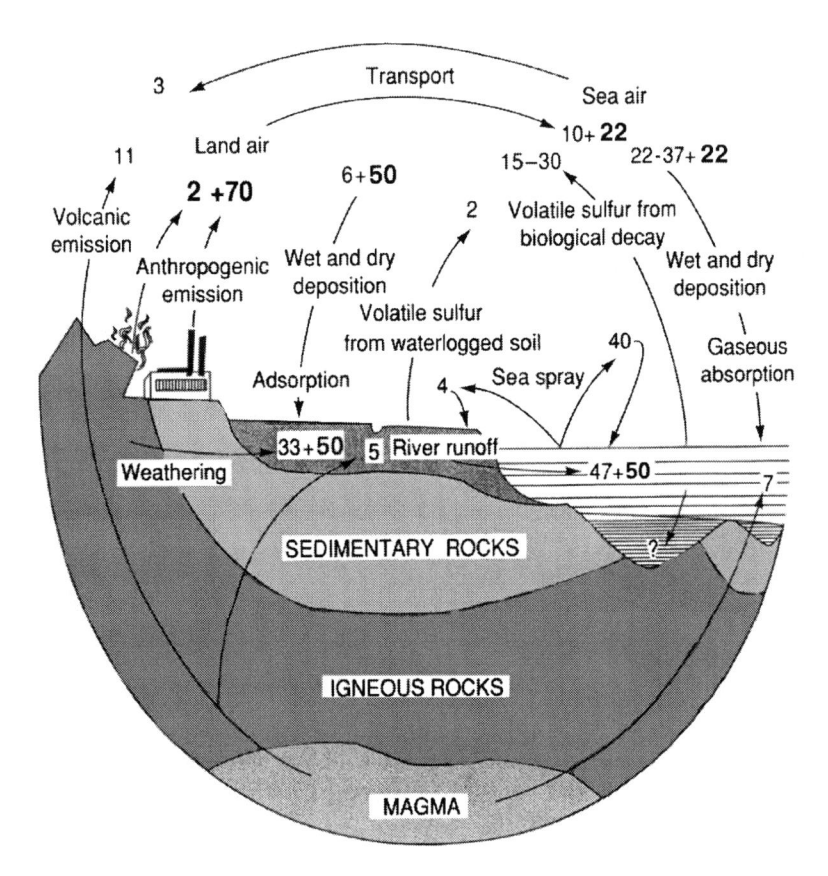

Fig. 6.10 Fluxes involved in the global sulphur cycle. Unit: Tg sulphur
Source: Adapted from Svensson and Söderlund (1976)
Note: Large bold numbers refer to anthropogenic sources, small numbers to sulphur fluxes of a natural origin

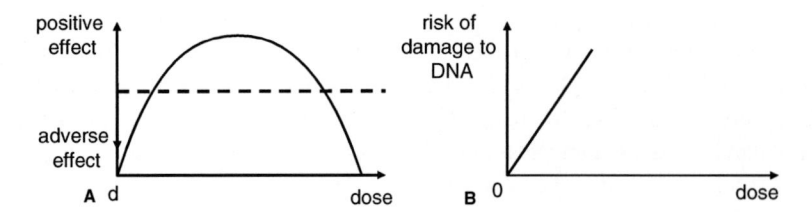

Fig. 6.11 (A) U-shaped or biphasic dose-effect relation; (B) dose-effect relation for agents that can damage DNA

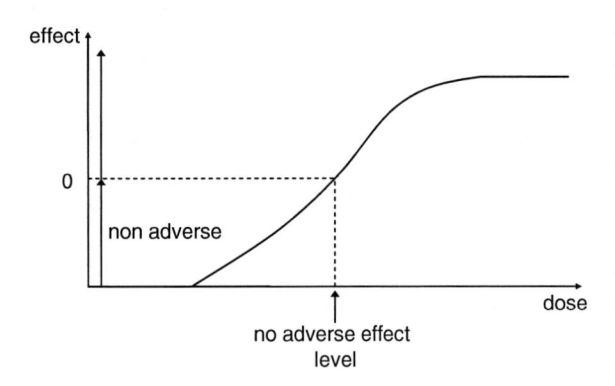

Fig. 6.12 Dose-effect relation involving a no-adverse effect level

the major shifts in ecosystems due to eutrophication by e.g. nitrogen-containing nutrients. As organisms thriving on low levels of such nutrients disappear, those that do well at high levels flourish. In most cases the net effect is a major reduction in biodiversity. Certain synthetic substances that mimic endogenous compounds also have a biphasic dose-effect relationship. One example is the phenoxy acids that mimic plant hormones and are used as herbicides. Applied in very small amounts these compounds stimulate plant growth, while in large doses they lead to plant death.

Many pollutants not belonging to the categories of nutrient or endogenous substance have different types of dose-effect relationship. The effects of some such substances are due to damage to (or mutations in) DNA, the hereditary material of the cell nucleus. Examples of such 'mutagens' include ionising radiation, emitted by radioactive substances, and compounds like benzpyrene and benzene that are common air pollutants. Genetic mutation may in turn lead to cancer or other health effects or, if the germ line is involved, hereditary defects in future generations. If it is assumed that an organism's DNA is already subject to substantial 'background' levels of damage, there can be no such thing as a 'safe' dose of any mutagen; exposure to growing amounts will therefore increase the risk of disease or mortality (as with mammals; see B in Fig. 6.11). The rate at which risk increases with dose may vary considerably, depending on the organism and the mutagenic agent involved.

For most pollutants, however, it is assumed that there is a 'no adverse effect level': a dose below which no harm is estimated to occur. This is illustrated in Fig. 6.12. For these pollutants a 'safe' level of exposure is assumed to exist for any given organism, though it may vary substantially depending on the organism involved. Exposing mammals to ever-greater organic solvent levels, for instance, will cause growing impairment of their nervous system. Exposure to noise levels over 80 dB(A) increases perceived stress among a considerable proportion of the human population. Sometimes the effects occurring at relatively low exposures may be the opposite of those at higher exposures. For instance, exposing wheat germ to organic solvents stimulates growth at low doses but inhibits it at high doses (Calabrese and Baldwin 2001).

6.4.2 Recent Trends and Current Status

Trends in environmental pollution and consequent damage vary widely. When it comes to fast-flux pollution of rivers and the troposphere, in industrialised western countries the overriding trend since the early 1970s has been one of improvement, though there have certainly been local exceptions. By and large, technical change has been the major driver of improvement.

In developing countries the situation has been quite different. There, fast-flux pollution has generally increased over the past few decades, though more recently it has begun to stabilise or even improve in a few places. Air quality in cities like New Delhi, Beijing and Mexico City is now much worse than in the cities of the industrialised countries of Western Europe and North America, leading to a much elevated risk of respiratory disorder and cancer (Aneja et al. 2001; Fenger 1999; Raga et al. 2001; Zhang et al. 2000). Though air pollution from stationary sources may now be declining in a number of mega-cities in the third world, any such gains tend to be offset or outweighed by traffic pollution (Fenger 1999; Han and Naeher 2006). In and around many larger third-world cities water pollution also tends to be rife, with some streams essentially consisting of virtually untreated wastewater from households and industry (Falkenmark 1998).

Among the cities of the industrialised north there is also wide variation in the severity of air pollution (Fenger 1999). Even in relatively clean cities in industrialised countries, though, air pollution is still leading to excess respiratory illness, cardiovascular disease and cancer (Fenger 1999; Han and Naeher 2006; Samet et al. 2000). Similarly, the decline in river pollution in the industrialised world does not mean that all is well. Eutrophication, causing major shifts in plant communities and reducing biodiversity, is still an important problem and pollutant levels also tend to restrict and influence river faunas. Several European rivers are noted for the sex changes in fish and amphibian fauna caused by pollutants that mimic natural sex hormones, often referred to as endocrine disruptors. The mutagenic activity of river water may also be high, a development associated largely with unregulated chemical emissions and discharges. In an evaluation of environmental quality in the Netherlands carried out in 2003 it was found that only a few bodies of surface water met basic water quality standards (OECD 2003).

Generally speaking, the volume of significantly polluted groundwater seems to be on the increase. The same holds for soil pollutant loads. In the latter case clean-up efforts in certain western industrialised countries have been limited in scale and scope and have probably not kept apace with the ongoing accumulation of pollutants in soils, which function largely as sinks. As a result of human activity the amount of lead and cadmium present in topsoil has on average increased about five to nine fold (Han et al. 2002) and this may have a negative impact on sections of the general population (Akesson et al. 2005; Laidlaw et al. 2006; Lamphear et al. 2000).

Though data are limited, the available evidence suggests that in developing countries there may well have been an overall increase in human exposure to bioaccumulating compounds since 1970. In the industrialised west, on the other hand, the situation seems to be improving in this respect. In many of these countries emissions of organochlorine compounds have fallen dramatically, benefiting top predators such as birds of prey, seals, otters and *Homo sapiens*. Even so, environmental levels of these compounds still remain well above the no-adverse effect level in some areas and countries. Similarly, emissions of lead, mercury and cadmium have generally declined in the industrial world. On the other hand emissions of readily bioaccumulating organobromine compounds have been steadily rising since 1970, with a variety of environmental impacts. Past discharges of organochlorines by the industrialised nations are now accumulating in the Northern Atlantic, moreover, posing a growing threat to the local marine fauna and polar bears. Because of the temperature gradient between the tropics and the pole, volatile organic chemicals like these emitted in temperate zones may be 'distilled of' on a planetary scale and deposited at higher latitudes. This is indeed an instructive example of how pollutants emitted in one part of the world may later specifically accumulate in other regions.

In the stratosphere concentrations of the main organohalogen gases impacting on the ozone layer are now stabilising or slowly decreasing. On the other hand, levels of greenhouse gases such as methane, carbon dioxide, dinitrogen oxide and fluorocarbons are still rising. This is leading to increased climate forcing (IPCC 2001; Lastovicka et al. 2006), i.e. higher temperatures in the troposphere and lower temperatures in the stratosphere. This latter process is slowing down recovery of the ozone layer, as nitrogen compounds that inhibit the breakdown of stratospheric ozone are increasingly 'frozen out'. The wider issue of climate change is discussed in more detail in Box 6.3.

Levels of radioactive contamination have been on the decrease since the atmospheric testing of nuclear weapons was officially banned in 1963, though locally there have been increases, some of them very significant. Discharges of radioactivity in the former Soviet Union in particular come to mind, including the fall-out from the 1986 Chernobyl nuclear disaster. In temperate regions exposure to ultraviolet radiation has been rising in recent decades, owing to changes in lifestyle (more outdoor leisure activities including sunbathing), and depletion of the ozone layer. Ambient exposure to electromagnetic fields has also grown. While the precise effects of this exposure remain unclear, reliable epidemiological studies consistently show a correlation between exposure to 50 Hz fields in excess of 0.2–0.4 microTesla and the risk of childhood leukaemia (Wartenberg 2001). Exposure to noise has also increased in many places throughout the industrialised world, with exposures over 80 dB(A) being linked to increased stress and negative impacts on health.

Box 6.3 Climate change

The Earth's energy balance is an important determinant of climate. About one-third of the solar radiation striking the Earth is reflected back to space, the remainder being absorbed and ultimately re-emitted as heat, i.e. infrared radiation. This energy balance is not stable but in constant flux, because of variations in incoming and reflected radiation and in the concentrations of water vapour and trace atmospheric gases that absorb infrared, such as carbon dioxide, methane, dinitrogen oxide (N_2O) and several halocarbons. The latter are called greenhouse gases (often abbreviated to GHGs). The natural presence of these gases gives rise to a natural 'greenhouse effect': without them, the Earth's average temperature would be about −18°C instead of the present +15°C.

Human activity has led to a marked increase in atmospheric levels of greenhouse gases. The concentration of carbon dioxide has increased by 31% between 1750 and 2000, for instance, to a level not previously attained in the past 420,000 years. The balance of evidence suggests that the increase in GHG concentrations is having a discernible warming influence on global climate (IPCC 2001). Halting the rise in atmospheric levels of these gases is far from easy, however: in the case of carbon dioxide it means a 60–80% reduction in worldwide emissions. Even then it may be thousand's of years before all the carbon dioxide emitted by human society is out of the atmosphere (Archer 2005).

Between 1970 and 2000 human emissions of the six most important GHGs (the 'Kyoto gases') increased worldwide by 50%. Over this period global emissions of carbon dioxide rose by approximately 60%, methane by 25% and dinitrogen oxide by 45%. This rise in methane emissions can be explained by the increase in emissions from fossil fuel use (+60%), from particular gaseous fuels, and from landfills and wastewater treatment (together +70%), while agricultural emissions remained on balance the same. The rise in emissions of dinitrogen oxide is mainly due to a major increase in the use of chemical fertilisers and manure and an expansion of production volumes in crop and cattle farming (IEA 2002; Olivier and Berdowski 2001; Olivier et al. 2001).

As to the effects of this significant increase in atmospheric GHG concentrations, recent years have turned out to be warm indeed; see Fig. 6.13.

Climate change refers to more than just a rise in the temperature, however, encompassing regional changes in both precipitation and temperature as well as changes in their variability and seasonal pattern. For example, winter precipitation in a

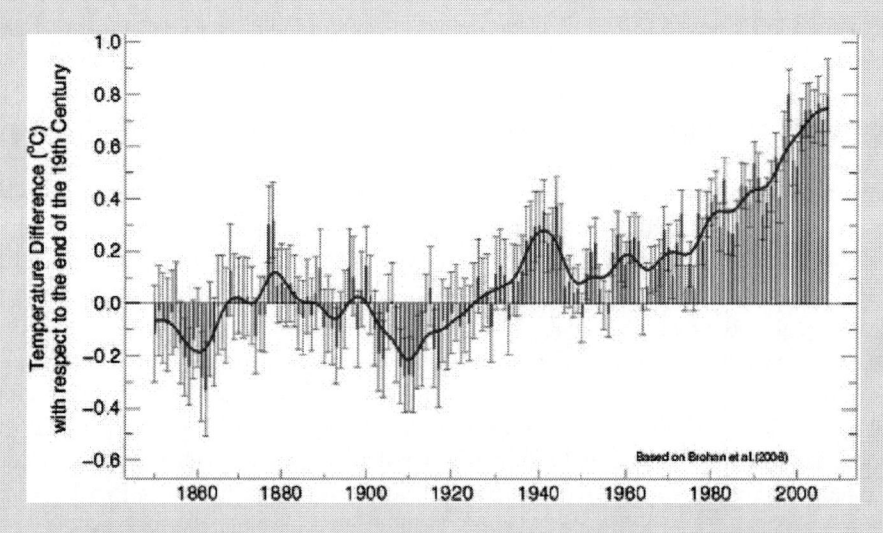

Fig. 6.13 Global average near-surface temperatures, temperature difference with respect to the end of the 20th century. Unit: °C Source: UK MetOffice, Hadley Center (2007)

certain region may no longer fall as snow, melting in spring, but fall straight away as rain. Also climate change may slow down the Gulf Stream that transports a large amount of heat from the Carribean Sea and the Gulf of Mexico into the North Atlantic Ocean and is a major determinant of temperatures in Western Europe (Rahmstorf 2003). Future global average changes are easier to estimate than regional changes, although the various computer models developed in recent years are relatively consistent when it comes to regional changes in the northern hemisphere (see for example IPCC 2001). Thus, changes in global mean temperature may not be telling the whole story.

Nevertheless, the outlook for the change in global mean temperature (see Fig. 6.14) by itself implies considerable change during the coming decades, with rates of change up to 0.3°C per decade. This is well over the value of 0.1°C per decade that has been suggested as the rate above which ecosystems are likely to suffer damage, being unable to cope with such a pace of change (Vellinga and Swart 1991). A comparison of the impacts of contrasting scenarios in Fig. 6.14 shows that the change in mean temperature over the next 30 years is largely unavoidable, as an inheritance of the past 30 years. However, the trend by the end of that period – and thus the changes in the decades thereafter – depends entirely on the scenario the world follows over the next 30 years (UNEP 2002).

Of the many factors affecting the complexities of climate, the effect of sulphate aerosols deserves particular mention. Aerosols formed by emissions of sulphur oxides have a cooling effect on the atmosphere. Although this effect is comparatively short-lived, on a regional scale it partly counteracts global warming. As sulphur emissions per unit energy must eventually be reduced, however, a certain amount of additional climate forcing will at some stage be 'unveiled' in those regions where these emissions are currently high. This effect is already accounted for in the global projections of Fig. 6.14 and is discussed in the Fourth Assessment Report of IPCC (IPCC 2007, in particular the reporting by Working Group II).

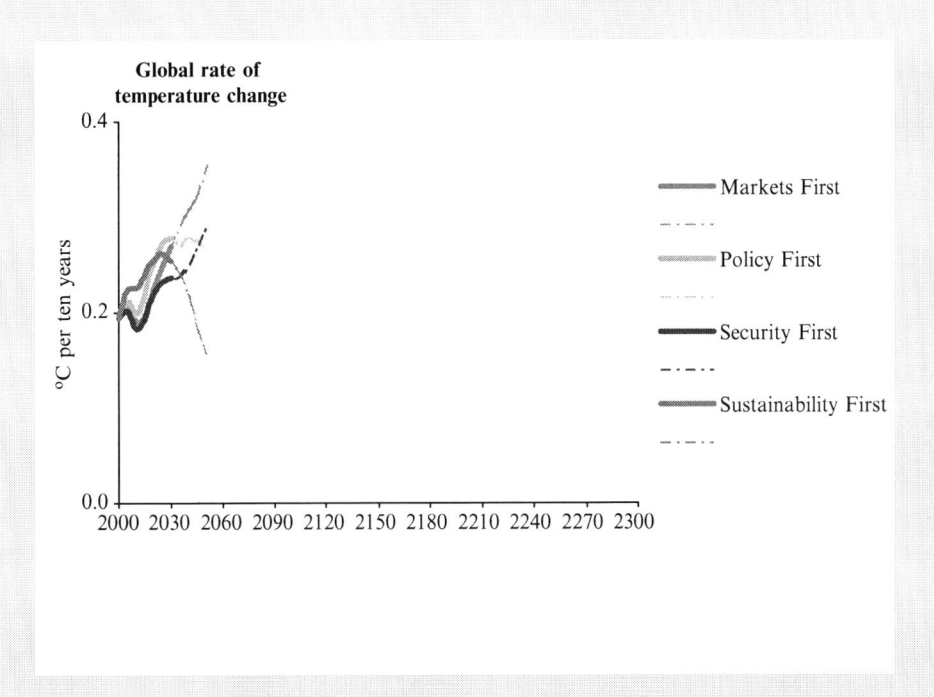

Fig. 6.14 Global rate of temperature change, based on the scenarios of the Third Global Environment Outlook (UNEP 2002) Source: RIVM (2000)

6.5 Environmental Hazards

6.5.1 General Description

'Hazards' have three defining features: they are associated with extreme events, have a less than 100% chance of occurrence and lead to material losses. In former times a distinction was often made between man-made and natural hazards, the latter sometimes being referred to as 'Acts of God'. The category of man-made hazards includes major accidents like the 1986 Chernobyl nuclear disaster or the SE Fireworks explosion in the Dutch city of Enschede in 2000. Natural hazards are associated with extreme geophysical events like violent storms, excessive rainfall, earthquakes and volcanism. Increasingly, though, these two hazard domains are beginning to merge, and responses to man-made and natural hazards often show major similarities (White et al. 2001). More importantly, it is now realised that human factors are involved in most natural disasters in one way or another. Anthropogenic climate change may well increase the likelihood of extreme weather (rains and storms) (Berz et al. 2001; Emanuel 2005), for example, and deforestation and river canalisation can exacerbate floodplain hazards (Disse and Engel 2001). Moreover, floods and volcanism would not pose so much of a hazard if human populations did not settle on floodplains and volcano slopes. Here there is an interesting interplay between natural resources and hazards, as hazard-prone floodplains and coastal and volcanic areas are often particularly appealing for human settlement and industry because of their relatively good supplies of natural resources.

6.5.2 Trends

Although the data available on these issues is not always that reliable (Hittelman et al. 2001) worldwide loss of life from extreme geophysical events has probably been declining in recent decades, owing to a variety of developments and responses, including more sophisticated rescue operations. Losses of property due to such events have, in contrast, probably been increasing worldwide (White et al. 2001). In part this reflects the cited 'appeal' of vulnerable locations. In developing countries, especially, an additional factor is probably that rapid social and cultural change has led to the collapse of traditional responses to limit property loss. A case in point are urban squatter settlements, which are often built on slopes prone to landslides during excessive rain. The same holds for floodplain occupation in several third-world mega-cities where river channels have become clogged by garbage (Falkenmark 1998). Moreover, a number of short-term responses to 'natural' hazards have been identified that may in fact prove counterproductive. One example are floodplain levees designed to limit losses from relatively frequent floods below a certain magnitude: when such levees are overtopped, actual losses may be catastrophic (White et al. 2001).

Extreme geophysical events are distributed unevenly across the globe, with a tendency towards more frequent exposure in areas affected by plate tectonics and low-lying coastal areas. A world map has been compiled providing a rough impression of such exposure (Berz et al. 2001; map available from ww.munichre.com, 'Weltkarte der Naturgefahren'). Global trends in man-made environmental hazards are less clear. In a densely populated country like the Netherlands, however, with important economic activity in the fields of chemical industry and transport, the overall hazard in terms of (expected) loss of life may actually be on the increase (RIVM 2000).

6.6 Allocation of Physical Space

6.6.1 General Description

Space is a many-sided concept. In part it is used in the context of physical access to resources or places where a relatively good life can be led. The last World War (1939–1945) was started by Germany to expand the 'Lebensraum' ('living space') of the Germans. Used in this sense, 'space' has been and still is instrumental in leading not only to war but also to migration. The latter includes the emigration out of Africa of the remote ancestors of the current inhabitants of the developed world, the mass migrations that led to the collapse of the Western Roman Empire, and current environmental refugees, who have left their traditional homes faced with a deterioration of natural resources such as arable land.

Among sedentary, peaceful people, too, access to physical resources may be a matter of concern and may be deemed an environmental problem. This is illustrated by the case of windmills. For centuries windmill owners have sought to have tall trees removed and agitated against other mills potentially affecting the amount of power generated by theirs. In response to this kind of 'incompatibility' between neighbouring objects or activities, over the ages governments have often instituted some form of 'physical planning', especially in densely populated areas. In countries like the Netherlands physical planning provisions relating to windmills have in fact been in existence for centuries. So, too, have city ordinances setting restrictions on the location of foul-smelling production facilities, for example. Similar ordinances often ensured that production and storage of explosives were located outside city walls, as were hospitals for treating contagious diseases. The general principle will be clear: given that specific uses of physical space may conflict with one another, there may be a need for policy measures designed to limit or eliminate such conflicts.

An associated problem is efficient use of limited space, an issue that already began to emerge in the very first cities we know of. In a world that was far from peaceful, cities needed walls or similar defences. This was true of ancient cities like Jericho and was still the case for many 18th century cities in Europe. Within the spatial constraints of city walls, efficient use of space was often of the essence. In classical Rome the rich could afford to live fairly inefficiently in this regard, while the poor were housed in four or five story tenement buildings. These days skyscrapers, underground metro systems and offices spanning motorways are evidence that efficient use of limited space is still an important issue.

Relatively new historically are problems concerning the allocation of space for 'green' purposes like nature conservation or, more recently, 'nature creation' i.e. converting of former farmland or derelict urban areas into nature. An early example of such problems befell the Dutch 'Father of the Fatherland' William of Orange, who intended to cut down the city forest of The Hague to finance the war against the Spanish, an effort in which he was successfully frustrated by the city populace. More generally, nature conservation became an important space allocation issue in industrialising countries during the 19th and 20th centuries, with countries like the United Kingdom, the USA and Germany leading the way. Arguments in favour of reserving certain areas for nature have related largely to the 'functions' nature is said to fulfil: recreational functions, being a healthy antidote to city life, educational functions, purification of air, and so on.

Also relatively new is the claim that certain spaces should be preserved as a vehicle for memories of the past. Largely a product of the 20th century, in practice such spaces are generally products of human culture such as buildings, infrastructure works, gardens, parks, agricultural lands and landscapes.

6.6.2 Recent Trends

As the world's population grows, and with it economic output and consumer spending power, problems of space allocation have become progressively more acute. The tensions between agriculture and nature have already been discussed in Section 6.3. Suburbanisation in rich countries and the spread of squatter settlements in developing countries have often proceeded largely uncontrolled and have exacerbated tensions further. Changes in the natural environment have not helped either. Thus, abandonment of land spoiled by erosion, salinisation or chemical pollution has increased pressure on land elsewhere. Similarly, climate change and deforestation are intensifying claims on land for the purposes of water management. All in all, then, allocation of available physical space is becoming increasingly problematical (see for example UNEP 2002).

In the European Union the main loser in the competition for space has been agriculture. Between 1980 and 2000 farmland acreage in the European Union decreased overall by 5%, mainly to accommodate urban expansion (EEA 1999; EEA 2006). Worldwide, though, agricultural expansion is a major driving force, with nature the main loser (see Fig. 6.6).

Indeed, the greatest challenge facing society today is probably the allocation of space to nature and key historic landscapes. In developing countries population growth is the main driving force in the rollback of nature. In industrial countries an important problem appears to be that the functions of nature are not (or only poorly) translated into monetary terms ('monetarized'; see Sala 2001), in contrast to such functions as agriculture, housing, transport and industry. In competing for space this makes 'green' purposes soft. This lack of proper monetarization of environmental goods is an important factor determining the scale of many other environmental problems (see Chapter 11).

References

AFEAS. (2002). *Production, sales and atmospheric release of fluorocarbons through 2000*. Washington DC: Alternative Fluorocarbons Environmental Acceptability Study (AFEAS) Program Office.

Akesson, A., Lund, T., Vather, M., Bjellerup, P., Lidfelt, J., Nerbrand, C., et al. (2005). Tubular and glomular kidney effects in Swedish women with low environmental cadmium exposure. *Environmental Health Perspectives, 113*, 1627–1631.

Anawar, H. M., Akai, J., Mostofa, K. M. G., Dafillah, S., & Tareq, S. M. (2002) Arsenic poisoning in groundwater. Health risk and geochemical sources. *Environment International, 27*, 579–604.

Aneja, V. P., Agarwal, A., Roelle, P. A., Phillips, S. B., Tong, Q., Watkins, N., et al. (2001). Measurement and analysis of criteria pollutants in New Delhi, India. *Environment International, 27*, 35–42.

Archer, D. (2005). Fate of fossil fuel CO_2 in geologic time. *Journal of Geophysical Reasearch, 110*, Co9 S05.

Arrigo, K. R. (2005). Marine microorganisms and global nutrient cycles. *Nature, 437*, 349–355.

Bermejo, M., Rodriguez-Teijero, J. D., Illera, G., Braaoso, A., Vila, C., & Walsh, P. (2006). Ebola outbreak killed 5000 gorillas. *Science, 314*, 1564.

Berz, G., Kron, W., Loster, T., Rauch, J., Schimetschek, J., Schmieder, J., et al. (2001). World map of natural hazard – A global view of the distribution and intensity of significant exposures. *Natural Hazards, 23*, 443–465.

Bouwman, A. F., & Van Vuuren, D. (1999). Global assessment of acidification and eutrophication of ecosystems. United Nations Environment Programme, *Technical report UNEP/DEIA&EW/TR.99–6*/National Institute of Public Health and the Environment *Technical report 4002001012*, Nairobi/Bilthoven.

British Petroleum. (2003). *BP statistical review of world energy*. London: British Petroleum.

Calabrese, E. J., & Baldwin, L. A. (2001). Scientific foundations of hormesis. *Critical Reviews in Toxicology, 31*, 349–681.

Cardinale, B. J., Srivastava, D. S., Duffy, J. E., Wright, J. P., Downing, A. L., Sankaran, M., et al. (2006). Effects of biodiversity on the functioning of trophic groups and ecosystems. *Nature, 443*, 989–992.

Darmstadter, J. (1971). *Energy in the world economy*. Baltimore: John Hopkins.

Disse, M., & Engel, H. (2001). Flood events in the Rhine Basin: Genesis, influences and mitigation. *Natural Hazards, 23*, 271–290.

EEA. (1999). *Environment in the European Union at the turn of the century*. Luxembourg: Office for Official Publications of the European Communities.

EEA. (2001). *Late lessons from early warnings*. Copenhagen: European Environment Agency.

EEA. (2006). *Urban sprawl in Europe – The ignored challenge*. Copenhagen: European Environment Agency.

Emanuel, K. (2005). Increasing destructiveness of tropical cyclones over the past 30 years. *Nature, 436*, 686–688.

Etemad, B., Bairoch, P., Luciani, J., & Toutain, J.-C. (1991). *World energy production 1800–1985*. Geneve: Libraire Droz.

Falkenmark, M. (1998). Dilemma when entering the 21st century – Rapid change but lack of sense of urgency. *Water Policy, 1*, 421–436.

FAO (1997). *Review of the state of the world's fisheries resources*. Rome: World Food and Agriculture Organization.

FAO. (2002). *The state of the world's fisheries and aquaculture*. Rome: FAO.

Fenger, J. (1999). Urban air quality. *Atmospheric Environment, 31*, 4877–4900.

Fritts, T. H., & Rodda, G. H. (1998). The role of introduced species in the degradation of land ecosystems: A case history of Guam. *Annual Review of Ecology and Systematics, 29*, 113–140.

Galloway, J. N., Schlesinger, W. H., Levy, H. (II), Michaels, A., & Schnoor, J. L. (1995). Nitrogen fixation: Anthropogenic enhancement–Environmental response. *Global Biogeochemical Cycles, 9*, 235–252.

Gordon, R. B., Koopmans, J. J., Nordhaus, W. D., & Skinner, B. J. (1987). *Toward a new iron age?* Cambridge, MA: Harvard University Press.

Graedel, T. E. (2002). Material substitution, a resource supply perspective. *Resources, Recycling and Conservation, 34*, 107–115.

Graedel, T. E., & Crutzen, P. J. (1993). *Atmospheric change*. New York: W. H. Freeman.

Groombridge, B., & Jenkins, M. D. (2000). *Global biodiversity: Earth's living resources in the 21st century*. Cambridge: World Conservation Monitoring Centre.

Groombridge, B., & Jenkins, M. D. (2002). *World atlas of biodiversity*. Berkeley: UNEP World Conservation Monitoring Centre/University of California Press.

Haberl, H., Erb, K. H., Krausmann, F., Gaube, V., Bondeau, A., Plutzar, C., et al. (2007). Quantifying and mapping the human appropriation of net primary production in earth's terrestrial ecosystems. *Proceedings of the National Academy of Sciences of the USA, 104*, 12942–12947.

Han, F. X., Banin, A., & Su, Y. (2002). Industrial age anthropogenic inputs of heavy metals in the pedosphere. *Naturwissenschaften, 89*, 497–504.

Han, X., & Naeher, L. P. (2006). A review of traffic-related air pollution exposure assessment studies in the developing world. *Environment International, 32*, 106–120.

Hittelman, A. M., Lockridge P. A., Whiteside, L. S., & Lander, J. F. (2001). Interpretive pitfalls in historical hazards data. *Natural Hazards, 23*, 313–338.

Holland, E. A., Dentener, F. J., Braswell, B. H., & Sulzman, J. M. (1999). Contemporary and pre-industrial global reactive nitrogen budgets. *Biogeochemistry, 46*, 7–43.

Hubbert, M. K. (1956). *Nuclear energy and fossil fuels, drilling and production practice*. Washington, DC: American Petroleum Institute.

IEA. (2002). CO_2 *emissions from fossil fuel combustion, 1971–2000* (2002 ed.). Paris: International Energy Agency.

IEA. (1971–2001). Energy statistics of OECD and non-OECD countries, various editions, Paris.

Imhoff, M. L., Bounoua, L., Ricketts, T., Loucks, C., Harriss, R., & Lawrence, W. T. (2004). Global patterns in human consumption of net primary production. *Nature, 429*, 870–873.

IPCC. (2001). *Climate change 2001. Synthesis report*. Cambridge: Cambridge University Press.

IPCC. (2007). *Climate change 2007*. Cambridge: Cambridge University Press.

Jackson, J. B. C., Kirby, M. X., Berger, W. H., Bjorndal, K. A., Botsford, L. W., Bourque, B. J., et al. (2001). Historical overfishing and the recent collapse of coastal ecosystems. *Science, 293*, 629–632.

Krug, E. C., & Winstanley, D. (2002). The need for comprehensive and consistent treatment of the nitrogen cycle in nitrogen cycling and mass balance studies: 1. Terrestrial nitrogen cycle. *The Science of the Total Environment, 293*, 1–29.

Laidlaw, M. A. S., Mielke, H. W., Filippelli, G. M., & Johnson, D. L. (2006). Blood lead in children. *Environmental Health Perspectives, 114*, A19.

Lamphear, R. P., Dietrich, K., Auinger, P., & Cox, C. (2000). Cognitive defects associated with blood lead concentrations <10 microgram/dl in United States children and adolescents. *Public Health Reports, 115*, 521–529.

Lastovicka, J., Akmaev, R. A., Beig, G., Bremer, J., & Emmert, J. T. (2006). Global change in the upper atmosphere. *Science, 314*, 1253–1254.

Lomolino, M. V., & Perrault, D. R. (2001). Island biogeography and landscape ecology of mammals inhabiting fragmented temperate rain forests. *Global Ecology and Biogeography, 10*, 113–132.

Lovelock, J. E. (1987). *Gaia: A new look at life on earth*. Oxford: Oxford University Press.

MNP, UNEP, WCMC and UNEP/GRID Arendal (2006). *Crossroads of Planet Earth's Life. Exploring means to meet the 2010 biodiversity target*. Study performed for the Global Biodiversity Outlook2. Netherlands Environmental Assessment Agency, UNEP World Conservation Monitoring Centre; UNEP/GRID-Arendal. MNP report 555050001/2006.

Monson, R. K., & Holland, E. A. (2001). Biospheric trace gas fluxes and their control over tropospheric chemistry. *Annual Review of Ecology and Systematics, 32*, 547–576.

Muradian, R. (2001). Ecological thresholds: A survey. *Ecological Economics, 38*, 7–24.

Murray, C. J., & Lopez, A. D. (1997). Global mortality, disability and the contribution of risk factors: Global burden of disease study. *The Lancet, 349*, 1436–1442.

OECD, Working Party on Environmental Performance. (2003). *Environmental performance review of the Netherlands*. Paris: OECD.

Olivier, J. G. J., & Bakker, J. (2000). Historical global emission trends of the Kyoto gases HFCs, PFCs and SF_6. *Proceedings of Conference on SF_6 and the Environment: Emission Reduction Strategies*, November 2–3, 1999, San Diego. EPA, Washington DC.

Olivier, J. G. J., & Berdowski, J. J. M. (2001). Global emissions sources and sinks. In J. Berdowski, R. Guicherit, & B. J. Heij (Eds.), *The climate system* (pp. 33–78). Lisse, The Netherlands: A. A. Balkema/Swets & Zeitlinger.

Olivier, J. G. J., Berdowski, J. J. M., Peters, J. A. H. W., Bakker, J., Visschedijk, A. J. H., & Bloos, J.-P. J. (2001). Applications of EDGAR. Including a description of EDGAR 3.0: Reference database with trend data for 1970–1995. RIVM report 773301 001/ NOP report 410200 051, RIVM, Bilthoven.

Pauly, D., & Christensen, V. (1995). Primary production required to sustain global fisheries. *Nature, 374*, 255–257.

Pauly, D. (2007). The sea around us project: documenting and communicating global fisheries impacts on marine ecosystems. *AMBIO, 34*, 290–295.

Pesaran, M. H., & Samier, H. (1995). Forecasting ultimate resource recovery. *International Journal of Forecasting, 11*, 543–555.

Pimentel, D. (2001). The limitations of biomass energy. In Meyers, R. [Ed.]. *Encyclopedia on physical science and technology* (pp. 159–171). San Diego, CA: Academic Press.

Platt, J., Smith, D., Smith B., & Williams, L. (2001). Environmental geochemistry at the global scale. *Applied Geochemistry, 16*, 1291–1308.

Prüss, A., & Havelaar, A. (2001). The global burden of disease; study and applications in water, sanitation and hygiene. In L. Fewtrell & J. Bartram (Eds.), World Health Organisation (WHO), *Water quality: Guidelines, standards and health*. London: IWA.

Raga, G. B., Baumgardner, D., Castro, T., Martinez-Arroyo, A., & Navarro-Gonzalez, R. (2001). Mexico city air quality: A qualitative review of gas and aerosol measurements (1960–2000). *Atmospheric Environment, 33*, 4041–4058.

Rahmstorf, S. (2003). The current climate. *Nature, 421*, 699–703.

Raskin, P., Gleick, P., Kirshen, P., Pontius, G., & Strzepek, K. (1997). *Water futures: Assessment of long-range patterns and problems*. Stockholm: Stockholm Environment Institute.

Reynolds, D. R. (1999). The mineral economy: How prices and costs can falsely signal decreasing scarcity. *Ecological Economics, 32*, 153–166.

RIVM. (2000). *Nationale milieuverkenning 2000–2030 (national environmental outlook 2000–2030)*. Alphen aan den Rijn: National Institute for Environment and Public Health, Bilthoven and Samson.

Rodrigues, A. S. L. (2006). Are global conservation efforts successful? *Science, 313*, 1051–1052.

Ruiz, G. M., Foffonoff, P. W., Carlton, J. T., Wonham, M. J., & Hines, A. H. (2000). Invasion of coastal marine communities in North America. *Annual Review of Ecology and Systematics, 31*, 481–531.

Sala, O. E. (2001). Price put on biodiversity. *Nature, 412*, 34–36.

Samet, J. M., Dominici, F., Curriero, F., Coursac, I., & Zeger, S. L. (2000). Fine particulate air pollution and mortality in 20 US cities 1987–1994. *The New England Journal of Medicine, 343*, 1742–1749.

Silveira, L. da & Sternberg, L. (2001). Savanna-forest hysteresis in the tropics. *Global Ecology and Biogeography, 10*, 369–378.

Simberloff, D., & von Holle, B. (1999). Positive interactions of non-indigenous species: Invasional meltdown? *Biological Invasions, 1*, 21–32.

Smil, V. (1999). Nitrogen in crop production: An account of global flows. *Global Biogeochemical Cycles, 13*, 647–662.

Smith, K. R. (2000). National burden of disease in India from indoor air pollution. *Proceedings of the National Academy of Sciences USA, 97*, 13286–13293.

Smythe, K. D. (2000). Production and distribution of SF_6 by end-use application. *Proceedings of the Conference on SF_6 and the Environment: Emission Reduction Strategies*, November 2–3, San Diego. EPA, Washington DC.

Svensson, B. H., & Söderlund, R. (Eds.). (1976). *Nitrogen, phosphorus and sulphur: Global cycles*. Stockholm: Ecological Bulletins 32.

Thomas, C. D., Cameron, A., Green, R. E., Bakkenes, M., Beaumont, L. J., Collingham, Y. C., et al. (2004). Extinction risk from climate change. *Nature, 427*, 146–148.

UK Met Office, Hadley Center (2007). *Annual global near-surface temperatures*, available at: www.metoffice.gov.uk/research.

UNEP. (2002). *Global environment outlook 3*. London: United Nations Environment Programme (UNEP), Nairobi, Kenya and Earthscan Publications.

UNESCO. (2003). *Water for people, water for life*. United Nations Educational, Scientific and Cultural Organization New York: (UNESCO)/Berghahn Books.

UNFPA (United Nations Population Fund) (2007). *State of the world population 2007*. available at: www.unfpa.org

Van Vuuren, P. P., Strengers, B. J., & de Vries, H. J. M. (1999). Long-term perspectives on world metal use – A systemdynamics model. *Resources Policy, 25*, 239–255.

Van Wilgen, B. W., Richardson, D. M., Le Maitre, D. C., Marais, C., & Magadlela, D. (2001). The economic consequences of alien plant invasions. *Environment, Development and Sustainability, 3*, 145–168.

Velders, G. J. M., Andersen, S. O., Daniel, J. S., Fahey, D. W., & McFarland, M. (2007). The importance of the Montreal Protocol in protecting climate. *Proceedings of the National Academy of Sciences, 104*, 4814–4819.

Vellinga, P. M., & Swart, R. (1991). The greenhouse marathon: A proposal for a global strategy. *Climatic Change, 18*, vii–xii.

Vitousek, P. M., Ehrlich, P. B., Ehrlich, A. H., & Matson, P. A. (1986). Human appropriation of products of photosynthesis. *BioScience, 38*, 368–373.

Vitousek, P. M., Aber, J. D., Howarth, R. W., Likens, G. E., Matson, P. A., Schindler, D. W., et al. (1997). Human alteration of the global nitrogen cycle: Sources and consequences. *Ecological Applications, 7*, 737–750.

Walsh, P. D., Abernethy, K. A., & Nermejo, M. (2003). Catastrophic ape decline in western equatorial Africa. *Nature, 443*, 611–614.

Wartenberg, D. (2001). Residential EMF exposure and childhood leukemia: Meta-analysis and population attributable risk. *Bioelectromagnetics Supplement, 5S*, 86–104.

White, G. F., Kates R. W., & Burton, I. (2001). Knowing better and losing even more: The use of knowledge in hazards management. *Global Environmental Change Part B: Environmental Hazards, 13*, 81–92.

Witte, F., Msuku, B. S., Wanink, J. H., Seehausen, O., Katunzi, E. B. F., Goudswaard, P. C., et al. (2000). Recovery of cichlid species in Lake Victoria: An examination of factors leading to differential extinction. *Reviews in Fish Biology and Fisheries, 10*, 233–241.

Wolff, W. J. (2000). The S.E. North Sea, loss of vertebrate fauna during the past 2000 years. *Biological Conservation, 95*, 209–217.

World Water Council. (2000). *World water vision*. London: Earthscan.

Worm, B., Barbier, E. B., Beaumont, N., Duffy, J. E., Folke, C., Halpern, B. S., et al. (2006). Impacts of biodiversity loss on ocean ecosystem services. *Science, 314*, 787–790.

Zhang, J., Song, H., Tong, S., Li, L., Liu, B., & Wang, L. (2000). Ambient sulphate concentration and chronic disease mortality in Beijing. *The Science of the Total Environment, 262*, 63–71.

Zock, J. P., Jarvis, D., & Luczynska, C. (2002). Housing characteristics, reported mold exposure and asthma in the European Union respiratory health survey. *Journal of Allergy and Clinical Immunology, 110*, 285–292.

Further Readings

General Overviews

GEO reports (Global Environmental Outlook UNEP; background reports; and www.unep.org for underlying data).

OECD Environment Outlooks. Paris: Organisation for Economic Co-operation and Development (OECD).

Smil, V. (2003). *The earth's biosphere: Evolution, dynamics and change*. Cambridge: MIT Press.

The five-yearly comprehensive reports of the European Environment Agency (EEA), Copenhagen.

Vital Signs (World Watch Institute).

World Resources reports. Washington, DC: World Resources Institute (WRI).

On Specific Themes

AMBIO (2002). Special issue. Optimizing nitrogen management in food and energy production and environmental change, *AMBIO*, XXXI, 2, March 2002.

Bruinsma, J. (Ed.). (2003). *World agriculture: Towards 2015/2030. A FAO perspective*. Rome/London: World Food and Agriculture Organization (FAO)/Earthscan.

FAO Global Forest Resources Assessments.

FAO The State of the World's Fisheries and Aquaculture.

Intergovernmental Panel on Climate Change, Assessment Reports

Global Biodiversity and the World Atlas of Biodiversity (Groombridge and Jenkins 2000 and 2002).

Global Biodiversity Outlook and underlying reports

Millennium Ecosystem Assessment (www.millenniumassessment.org)

Water for People, Water for Life (UNESCO 2003).

World Water Vision (World Water Council 2000).

Molden, D. (Ed.). (2007). *Water for food, Water for Life. A comprehensive assessment of water management in agriculture*. London: Earthscan. Summary downloadable from www.iwmi.cgiar.org

Part II
Principles and Methods

Chapter 7
General Principles

Craig Dilworth

Contents

7.1	Introduction	75
7.2	The Principle of Sustainable Development	75
7.3	The Principle of the Conservation of Energy	76
7.4	The Principle $E = MC^2$	76
7.5	The Principle of the Conservation of Matter	76
7.6	The Entropy Principle	77
7.7	The Principle of Evolution	78
7.8	The Notion of a System	78
7.9	The Principle of Ecology	80
7.10	The Principle of Population	80
7.11	The Vicious Circle Principle	82
References		83

7.1 Introduction

When one speaks of the *principles* of a particular science or scientific discipline, one means that which is *presupposed* in all activities of the discipline. Such presuppositions are essentially of one of two kinds, namely principles of action, and ontological principles. *Principles of action* are the rules saying how the discipline is to be pursued, while *ontological principles* are presuppositions made in the discipline about the fundamental nature of reality. It is a discipline's principles that determine just what the discipline is.

C. Dilworth (✉)
Uppsala University, Sweden
e-mail: Craig.Dilworth@comhem.se

Principles of action can themselves be divided into two groups, one concerned with the *goal* of the action, and the other with the *means* for reaching the goal. In the pure sciences the goal is epistemological, that is, it is to acquire knowledge or understanding for its own sake. In the *applied* sciences, such as environmental science, knowledge and understanding are not goals in and of themselves, but are rather to be (part of) the *means* for reaching some other goal.

For environmental science, that other goal may be said to be the attaining of an ecologically stable situation for humankind, as discussed in the next section.

7.2 The Principle of Sustainable Development

The principle of sustainable development is a principle of action concerning the whole of humanity, and is officially accepted by virtually all decision makers in business and politics. This principle states that:

We humans must move towards the adoption of a lifestyle that can continue indefinitely.

In other words, the human species must stop undermining the very preconditions of its own existence. The adoption of this principle includes the tacit understanding that our present lifestyle cannot continue indefinitely – is not sustainable – as well as that it can be changed to a sustainable lifestyle through the exercise of free will based on reason.

While everyone agrees to the *goal* of sustainability, there is much disagreement regarding the *means* for achieving the goal. (Some people might question whether environmental science was part of the means.) As regards this question there are essentially two main camps, which may be called the growth and the no-growth camps.

The *growth* camp sees sustainability as being reached through economic growth and free trade, which are to generate the wealth necessary to pay for developing and using the technology needed to bring us into harmony with our biological and physical environment. The *no-growth* camp maintains that economic growth is itself a major cause of our environmental problems, and that such growth is leading us away from, rather than towards, a sustainable society. Virtually all economic actors, political decision makers and neoclassical economists belong to the growth camp, while virtually all biologists and most other natural scientists belong to the no-growth camp.

It has been suggested by persons in the no-growth group that the difference between the two views is fundamental, each of them constituting a *paradigm*, in Thomas Kuhn's sense of the term. According to them, the transformation necessary to get humanity on the path to sustainability thus requires decision makers of the growth camp going through a *paradigm-shift*.

Some of the classic works presented by the no-growth group include Paul Ehrlich's *The Population Bomb* (Ehrlich 1968), Dennis Meadows and others' *Limits to Growth* (Meadows et al. 1972), and Ernst Friedrich Schumacher's *Small is Beautiful* (Schumacher 1973). The writings of the growth party are more diffuse and, except when occasionally responding to direct criticism, generally do not recognise the growth/no-growth issue at all. Important efforts in this vein are the UN report *Our Common Future* (WCED 1987) and the Rio Declaration on Environment and Development (UNCED 1992).

Growth *economists* doing environmental research are paradigmatically represented by *environmental* economists, such as David Pearce and Robert Costanza, while ecological economists, including Nicholas Georgescu-Roegen, Ernst Friedrich Schumacher and Herman Daly, represent no-growth advocates. As far as research is concerned, the former advocate such techniques as cost-benefit analysis and contingent valuation, with the intention of being able to place a price on the environment. Those of the no-growth camp, on the other hand, would presumably advocate as a first step the creation of a coherent steady-state alternative to neoclassical economics. The growth/no-growth issue is a fundamental one in environmental research, and it is important for an environmental scientist to know where he or she stands on it, and why.

Where the principle of sustainable development is the most important *principle of action* of ecological science, determining its goal, the principles to follow are *ontological*, stating something about the nature of reality as it is assumed to be by environmental science.

As an applied science, environmental science presupposes other, intellectually more fundamental, disciplines. These include physics, chemistry, biology, ecology, and human ecology, where chemistry presupposes physics; biology presupposes chemistry, and so on.

7.3 The Principle of the Conservation of Energy

The most important principle of *physics* is the principle of the conservation of energy. R.J. Mayer first enunciated this principle in 1842. It is also known as the First Law of thermodynamics. It says that:

Quantity of energy is constant.

In other words, energy can be neither created nor destroyed, but only *transformed*.

7.4 The Principle E = MC²

Another physical principle of consequence to environmental science is:

$E = MC^2$, or 'energy equals mass times the speed of light squared.'

The aspect of this principle – propounded by Albert Einstein in 1905 – that is important for environmental science is that *matter is a form of energy*.

7.5 The Principle of the Conservation of Matter

This is the basic principle underlying the science of *chemistry*, and was first expressed by Antoine Laurent Lavosier in 1789. It says that:

Quantity of matter is constant.

In other words, *energy cannot be transformed from a material form to a non-material form or vice versa.* Matter can change from being in a state of high potential energy to being in one of low potential energy, but the total amount of matter in both states is the same. Though this principle does not apply to the production of atomic energy, it is considered to apply to all other forms of energy transformation, thus generally making matter a form of energy which is itself like energy as a whole in that it can never go into or out of existence.

Where the principle of the conservation of energy is the basic principle of physics, the principle of the conservation of matter is the (physically less-universal) basis of chemistry.

The fundamental way in which energy is transformed is captured by:

7.6 The Entropy Principle

This principle, first put forward by Sadi Carnot in 1824, is also known as the Second Law of thermodynamics. It states that:

Energy tends to dissipate.

Over *time*, energy tends to spread in *space*. This principle may also be expressed in terms of systems, viz.: *systems tend towards disorder*; or, *the amount of entropy in a system tends to increase*. It can also be expressed as: *the quality of the energy in a system tends to decline*. We can also say that *the degree of 'organisation' of matter in a system will tend to decrease*; or, *the potential energy of a body will tend to diminish*.

As far as environmental science is concerned, this means that whatever concentrations of energy or chemical elements as presently exist on earth will tend to dissipate over time. This natural process is being given a gigantic push by humankind, through its mining and use of fossil fuels and minerals. The entropy principle implies that there will be a day when humans cannot use fossil fuels, uranium or metals, the ultimate limit for fuels being where the amount of usable energy required to extract them exceeds the amount of usable energy they provide.

Like fossil fuels, minerals constitute a finite resource. Unlike sources of usable energy, however, most minerals can be recycled. But the Second Law makes it clear that their level of concentration – their degree of 'organisation' as physical systems – will diminish with each recycling. Eventually there will be nothing big enough left to recycle. The *amount* of matter will not have changed, but its organisational *quality* will have deteriorated to the point of non-existence. The sustainable society will have to do without these resources.

Resource use has three stages: acquiring the resource, using it, and then getting rid of the remains. The place from which the resource is acquired is its *source*; its *use* involves taking usable energy from it and thereby increasing its entropy; and its place of disposal is its *sink*. As regards the growth and no-growth views of sustainability, the growth view tends to concentrate on problems as they manifest themselves at the sink, while the no-growth view looks at them with regard to the source. The principle of the conservation of matter tells us that the total quantity of material that goes into a sink is the same as that which comes from the source, and the principle of the conservation of energy tells us that the total amount of energy before and after using the material will also be the same. But the entropy principle tells us that the material's organisational quality will be lower: there will be less *usable* energy (concentrated material) after the resource has been used. The entropy increase incumbent upon using a resource means that eventually *everything* that comes from a source will end up in a sink.

The whole of human resource use can itself be seen as constituting a system, which, in a sustainable society, must be in *equilibrium* with its surroundings. For this equilibrium to come about, the input of usable solar energy into the system must be of the same quantity as the usable energy lost through resource use. This is a prerequisite for a sustainable society, and the fact that it is not presently being met is one of the main reasons for the enunciation of the principle of sustainable development.

The entropy principle also applies to the sun, which can be viewed as a system that is constantly losing energy. But the loss of energy in the case of the sun is happening so relatively slowly that it is not taken into account with regard to the issue of attaining a sustainable society. We may think of the earth as continuing to receive a constant influx of solar energy in the future, counteracting entropy and making possible the existence of human and other forms of life.

In the past a tiny portion of this influx of energy was conserved in the accumulation of fossil fuels beneath the earth's surface. The viability of the entropy principle means that a sustainable society can exist only on what it can reap from the constant stream of solar energy on a *year-to-year* basis. There can be no dipping into stores of energy or material amassed over long periods in the past. More generally, in a sustainable society there can be no mining.

Energy is necessary to life on earth. The nature of this life, as studied in *biology* and presupposed by environmental science, is determined by the following principles:

7.7 The Principle of Evolution

This is the fundamental principle of modern *biology*, first presented by Charles Darwin in 1859. In its modern form it states that:

> *Life forms on earth have evolved from a common source, each surviving as a species as a result of its being genetically adapted to its biological and physical environment.*

One implication of the principle of evolution is that humans are animals, and constitute a part of nature just as other animals do. Accordingly, we do not stand above other life forms, but amongst them. And, just as is the case with other life forms, if or when the human species proves unable to adapt to its environment, it will become extinct.

Relative to other life forms, humans are an extremely recent arrival. Our line evolved into modern humans only about 150,000 years ago. This may be compared with other species, an extreme example being that of certain shellfish, which evolved into its present form some 500,000,000 years ago. This species of animal has existed on earth more than 3,000 times as long as we have.

The human species has yet to come into a stable relation with its environment, a prerequisite for its continued existence. It is just this stable relation that the principle of sustainable development is to lead us to.

Adapting to the environment means fitting into a system.

7.8 The Notion of a System

A *system* in its simplest form is any group of entities among which there is an ongoing cause-and-effect relationship. The existence of the form of the relationship over time is what makes the system a system and not simply a one-time event. Thus a raindrop falling on a stone is not a system, but a constant dripping of water on one part of the stone is a system.

What more is required of a state of affairs in order for it to be considered a system is that its constituents *interact*. Given the physical notion of cause, however, any *continuing* causal relation constitutes a system. This is thanks to *Newton's third law* – another principle, already expressed by Aristotle – which states that physical cause and effect are equal and opposite, or, to every physical *action* corresponds an equal and opposite physical *reaction*.

This interaction may be more or less noticeable in different cases. A single stone constitutes a system, since its constituents interact, holding it together. (In fact, from one point of view, any situation that does not evince total entropy constitutes a system.) But their interaction is not obvious, in the sense that it does not involve change. In some systems the cause may be obvious while the effect is hidden, or vice versa. In the case of water dripping on a stone, the collision of the water drops with the stone causes the water molecules to disperse rapidly, while the stone molecules disperse much more slowly, eventually giving rise to a hollow in the stone. The effect of the stone on the water drops is more noticeable than the effect of the water drops on the stone, though they are physically equal.

The *solar system* constitutes a relatively simple system, where the gravitational force (cause) exerted by the sun on the planets is exactly equal and opposite to the force exerted by the planets on the sun. The effect of the sun on the planets appears to be larger however. One thinks of their elliptical orbits as resulting from their attraction to the sun, when they are just as much a result of the sun's attraction to them.

The cause-and-effect relation or relations in a system give the system its *structure*, and determine how it is *organised*. You can say that the *structure* of a system can change while it remains the same system, but with a change of *organisation* there is also a change of system. The distinction between structural change and organisational change is not a black-and-white affair, though it is vital to understanding the nature of systems. So long as the processes in a system are not leading to organisational change, the system is in a state of *equilibrium*. When there is no structural change the equilibrium is *static*; when there is structural change, it is *dynamic*. A system is *out of equilibrium* when it has earlier experienced static equilibrium and is in the process of failing to exhibit dynamic equilibrium, at a time before there is a change in its organisation (after which it is no longer the same system).

These concepts can be clarified with our example of the solar system. The solar system may be seen as an instance of a system in *static equilibrium*: there is an interaction amongst the elements in the system, while their relation to one another remains unchanged. Imagine now a piece of matter the size of the earth entering the solar system and going into orbit around Jupiter. The system is no longer static, but dynamic. The orbit of Jupiter as well as the orbits of the other planets are affected by this new source of gravity, and become very irregular. When the orbits eventually stabilise, the system is once again static, and has a different structure: the orbits of all the planets have undergone a change. But we need not say that the *organisation* of the system has changed, so long as, say, the relative positions of the planets with respect to the sun has not changed. If it *has* changed – depending on how we conceive of the system's being organised– we might say that the old system has ceased to exist and a new system has come into being. In other words, the entrance of the new mass has had such an effect on the original system that it was unable to maintain *dynamic equilibrium*. In the event that the planets maintain their relative positions after settling into new orbits, we could say that dynamic equilibrium has been maintained: the system's organisation has remained the same despite structural change in the system.

If the relative positions of the planets should change, we could say that there has been a change of system. After the arrival of the piece of matter from outer space, the solar system lost not only its static equilibrium but also its dynamic equilibrium; this it did at one and the same time, namely, when the orbits of the planets started to change. The organisation of the system – and consequently the system itself – can be said to have changed when the planets adopted their new orbits. And the original system can be said to have been *out of equilibrium* between these two times.

Note that what one takes to be static as compared to dynamic equilibrium is to some extent arbitrary. For example, you *could* say that the solar system would still be essentially the *same* system even if the relative positions of the planets were to change, or even if all the planets were to spiral into the sun. In the case of living beings however, the notion is much less arbitrary. In the case of *individuals* seen as systems, a loss of dynamic equilibrium results in death; and in the case of *species*, it results in their extinction. In the case of an ecological system or *ecosystem*, however, the distinction is not so clear. If a tropical forest becomes a semi-arid desert, it is clear that the dynamic equilibrium of the original system has been lost – and so there has been system change – even if some of the original species remain. Though it is difficult to specify at what point an ecosystem may be said to lose dynamic equilibrium, there do exist criteria for such an event, the most important of which pertain to species loss. (In this regard it may be noted that the world's sixth period of massive extinctions started about 15,000 years ago. The last of the previous five occurred about 65 million years ago and wiped out the dinosaurs. Its cause was an asteroid hitting the earth; the cause of the present period of extinctions is the activity of humans.)

Three points are passed in the demise of any system. The first is that at which the system's structural change becomes organisational change, after which the system is out of equilibrium. The second is the 'point of no return' after which equilibrium cannot be regained. In purely physical systems these points coincide; in biological systems they tend not to. And the third point is that after which the system no longer exists.

In our solar-system example, as soon as the body from outer space began to interact with the solar system, it became a part of it, and the structure of the system was altered. If the body were simply to have passed through the system, at some point it would no longer have been a part of it. Though such an event is bound to cause permanent structural change in the system, it may or may not lead to loss of equilibrium (here, a 'wobbling' of the system). If it does, the system will either experience organisational change or it will not. If it does, there will have been a change of system. But even before the new matter enters the system, both its future behaviour and that of the system are already determined, thanks to the regular operation of gravity. So in this case the point at which equilibrium is lost, and the point after which system change cannot be avoided, are the same.

In the case of *ecosystems*, the demise of a system can be avoided despite loss of equilibrium thanks to the *regenerative* ability of such systems. This regenerative ability is the effect of the fact that ecosystems, unlike physical systems, tend to *decrease* in entropy, thanks to the energy they take in from the sun. Thus not only is the point after which the system is doomed later than that at which it loses equilibrium, but it is at least as hard to determine, though species extinction should again function as the prime indicator.

If a new system has been created, it may be one having the *same* level of organisation as the old: the planets continue to orbit the sun but follow radically different paths; a new biotope emerges which has the same biotic mass and variety of species as the old, though many of the species are different. Or it may be of a *new* level of organisation: the planets might all fall into the sun; a rainforest may become a desert; a desert may become a rainforest.

The entropy principle says that systems-change generally will be towards *lower* levels of organisation. But in our solar-system example, energy was *added* to the system in the form of new matter in motion, which meant that the system's level of organisation could increase. Of course as regards the biosphere it should be remembered that the sun is constantly contributing usable energy, so, for example, a shift from desert conditions to rainforest conditions is quite possible.

Other important systems notions that might be mentioned here are those of open, closed, and isolated systems. As depicted in our example, the solar system is an *open* system, taking in matter and therefore energy from outside, as well as giving off solar energy to the outside. A *closed* system is one which *matter* neither enters nor leaves; and an *isolated* system is one which *energy* neither enters nor leaves.

The biosphere, for all intents and purposes, is a *closed* system (as is the solar system, normally considered). No appreciable quantities of matter enter or leave the system, while there is a constant influx of solar energy (all of which is sooner or later reflected back into space). Particular species, on the other hand, are *open* systems, each of which interacts with the other systems on the surface of the earth.

7.9 The Principle of Ecology

If we were to suggest a principle for the discipline of *ecology*, it could well be that:

> *Various groupings of living beings constitute systems, each of which may or may not be in equilibrium with the other systems constituting its environment.*

The systems with which the human species interacts are physical and (physico-) biological. DNA strands, cells, and the organs of living beings are each biological systems. Individual living beings such as plants and animals are also biological systems. And populations of living beings constitute systems, which are subsystems of even larger systems, namely *ecosystems*. Eventually we get to the largest ecosystem, the *biosphere*, which has all other biological systems as subsystems.

The existence of each species, like the existence of the biosphere as a whole, is dependent on solar energy. In terms of systems, the maintenance of the population of each species as a system requires an influx of solar energy in order (temporarily) to counteract the effects of the entropy principle. In the case of plants this input is at least partly direct, via photosynthesis. For humans and other animals it consists in taking usable energy from other material (biological and physical) systems, thereby increasing their entropy – in accordance with the entropy principle.

As compared with physical systems, systems involving living beings are exceedingly complex; and systems involving humans who have free will are even more so, their behaviour being impossible to predict in detail. Nevertheless, limits can be set on what behaviour is possible, thanks to the principles being presented here. For example, the entropy principle makes it impossible for a sustainable society to continue using a finite resource indefinitely.

Another difference between populations of living systems and purely physical systems, as taken up in the previous section, is that each *living* system tends to increase, not decrease, in its degree of organisation. Living systems could be physically defined as open systems in which entropy tends to decrease. Against the background of the entropy law, this means that living systems must exist in larger physical systems in which the *total* entropy is increasing.

This energy-accumulating aspect of life lies behind the principle of population.

7.10 The Principle of Population

The principle of population, first enunciated by Thomas Malthus in 1798, may be seen as being a corollary to the biological principle of evolution. It says that:

> *There is a tendency for the human population (and that of any species) to be as large as its environment will allow.*

We shall call this the first formulation of the principle. The second is:

If there were no checks to population size, that of the human (or any) population would tend to increase indefinitely.

Note that this principle, in either of its formulations, is to apply at all times, and not only at some future time. The misunderstanding of the principle in this regard has recurred ever since the time of Malthus.

Both formulations of the principle imply that, so long as the checks on population growth consist in scarcity of resources (paradigmatically, *food*), when those resources increase or decrease, the size of the population will follow suit, and generally be such as to utilise *all* the available resources. Note that, in being a corollary to a biological principle, the principle of population is itself a biological principle, and thus applies to virtually all species, and not just humans – though Malthus and his commentators apply it only to humans. In fact, in accordance with the principle of evolution, the principle of population should apply to the biosphere as a whole, such that it always tends to expand.

A population constitutes a system interacting with other systems, some of which are other living systems and some of which are physical. The usable energy in all of the purely physical systems is decreasing. In living systems usable energy *tends* to increase, in a way we still do not understand, but which is physically associated with the replicability of the DNA molecule. Thus another way of expressing the first formulation of the population principle could be: *the population of each species tends to acquire as much solar energy as it can, thereby increasing its usable energy and decreasing its entropy* – all this, of course, within a broader context (system) in which the total entropy is increasing. For species populations, this has the effect of promoting an increase in their numbers. The limit of this expansion is directly affected by the expansion of the populations of other species: the expansion of 'prey' species will support the expansion of the species in question, and the expansion of 'predator' species will lead to its diminution. There is a point at which the size of each of the populations of the biological systems tends to stabilise. According to the first formulation of the principle of population, every population, including the human, will tend to that size, the *cause of expansion* stemming from within the population, and the *cause of contraction* from without.

While the first formulation of the principle of population is a natural corollary to the principle of evolution, anthropological and archaeological research has shown that it need not hold for all populations. For example, certain human societies, in particular those of hunter-gatherers, have maintained themselves *below* the carrying capacity of their environment through the use of *cultural* checks to population growth (e.g. infanticide, taboos on intercourse). In these societies individuals do not suffer greatly for lack of food (nor do they in most nonhuman populations, which lack cultural checks).

Given this, however, Malthus' principle is still correct, as expressed in the second formulation. Even the human population has a constant tendency to increase, and is constantly constrained from doing by the operation of checks. It is just that (some of) the checks can originate from *within* the population. In other words, some of the energy accumulated in the system can be employed to stop the system from expanding as much as it otherwise would.

Malthus himself saw the problem with the first formulation of the principle, and in the second and later editions of his treatise explicitly introduced cultural checks in the form of late marriages, and went on to advocate their use so as to reduce the size of the population and thereby the suffering of the poor. Such checks to population as reduced number of *births*, Malthus termed *preventive*; checks to population in the form of *death*, he termed *positive*. Where preventive checks to population are normally *internal* and *cultural* (e.g. late marriage, long lactation, use of contraceptives, abortion), positive checks are normally *external* and *natural* (disease, starvation); however there exist internal, cultural positive checks as well (infanticide, murder, execution, war).

It may be noted in passing, however, that it may well be the case that nonhuman species also regularly employ intra-systemic population checks (e.g. infanticide). But such checks are paradigmatically based on *instinct*, where many human checks have been *cultural*. We also take this opportunity to note that some people might prefer to consider abortion a positive check; and that war, while internal to the human population as a whole, is external to each of the particular populations participating in it.

The causes of both population expansion and population contraction are necessary for the population to constitute a system in equilibrium. (The general effect of this tendency of populations towards stability is the tendency of the biosphere as a whole towards equilibrium.) We can see this as an application of Newton's third law in the biological realm – the forces tending to increase the size of a biological system are equal and opposite to those tending to restrict it. If the internal forces increase, the system will tend to grow until it meets equally strong forces from without. If the forces from without become greater than those from within, the system begins to contract, in the worst of cases disappearing entirely. If this happens for all populations of a particular species, the species becomes extinct. Causes restricting the population size to the point of the extinction of the species are a manifestation of the species being out of equilibrium with its environment. But this disequilibrium can have been brought about by a period during which the checks to population growth were particularly *lax*. Overly lax checks lead to overpopulation and a consequent disarming of the environment, which could result in the eradication of the species' resources.

The populations of nonhuman species have generally followed the first, stronger, formulation of the principle of population. Though some species have employed internal checks, from the point of view of the principle of evolution, such checks can be seen as promoting the overall size of the population in the long run. In any case, nonhuman populations tend to increase and decrease in cycles. The greater the numbers of a species, the greater the food consumption, ultimately leading to the population's experiencing scarcity. This in turn necessitates reduced consumption, which leads to a reduction in population size – paradigmatically through an increase in infant mortality – allowing the resource to recover. Note that this process presupposes *renewable* resources. Once non-renewable resources are introduced, the chance of the system experiencing disequilibria through overpopulation increases.

While cultural population checks have been sufficient to keep the populations of many hunter-gatherer societies from pressing against their environmental limits for thousands of years at a time, in a broader perspective the human population as a whole, like the populations of other species, has in fact tended to press against the limits of its environment. Thus Malthus' principle in its stronger formulation has been the rule *in the long term* even as regards humans and earlier hominids. Just as in the case of other species, there has been a constant tendency of the hominid strain to increase the number of individuals that can occupy the earth. In the case of humans we see a particularly poignant indication of this in the fact that the human population as a whole has not only been constantly growing, but has been doing so at a constantly accelerating rate.

Note, however, the importance of the hunter-gatherer example. Humans, unlike all other species, have the ability *consciously* to set their *own* checks on their population size. One way of seeing the principle of sustainable development is as a cry to humankind to employ its own checks to keep its population size within what the earth can sustain in the long term.

7.11 The Vicious Circle Principle

All of the ontological principles presented above apply just as well to other species as to humans. But once the notion of cultural checks has been introduced, we begin moving into an area specially reserved for humans. Is there a particular ontological principle applying only to humans – a principle of *human* ecology? The vicious circle principle is a candidate:

> *Human development consists in an accelerating movement from situations of resource scarcity, to technological innovation, to resource increase, to increased consumption, to population growth, and back to resource scarcity.*

Note that this principle, like the principle of population, is one that may be used to explain human development, in the main, to date. But both principles leave the door open to a contrary form of development, as is presupposed by the principle of sustainable development. For example, thanks to free will, reason, and certain values, humanity *could* learn to set its own limits on its population size, and consciously decrease its use of resources.

Where the population sizes of other species have been forcibly reduced when the population has experienced a shortage of resources, the human species is unique in sometimes reacting to resource shortage, or threat of resource shortage, with innovative technology. But where other species have settled down to roughly stable population sizes within certain limits (until losing equilibrium and becoming extinct), we humans have as yet not come to the point where our population has stopped growing for the first time. We have not only been pushing against the limits of our environment, we have constantly been *stretching* those limits. In other words, the human species, as a system, has yet to attain the static equilibrium necessary for its dynamic equilibrium even to be tested. In terms of our astronomical analogy, where other species can be seen as each being in certain respects analogous to the solar system, the human species is more closely analogous to a super-nova. Will the human super-nova settle down into a planetary system, or become a black hole?

One could say that the successive populations of the human species actually do *not* constitute a system, as they have as yet to come into equilibrium with the other systems constituting their environment. In terms of our water and stone analogy, the physical manifestation of the human species is but a *drop*, and has yet to become *dripping*. We are only a candidate for becoming a lasting part of the biosphere, and have still to find our biological niche. We are the result of a genetic mutation that has yet to show itself to be resilient.

Unlike in the case of other species, for humans and earlier hominids the technological development of such things as hand-axes, fire, clothing, the bowl, boats, spears, bows and arrows, the hoe, the plough, irrigation systems, windmills, sailing vessels, various engines, and atomic energy, have led to an *exponential increase* in total human resource consumption as well as population size right from when we first came into existence as a species.

As regards resource use it may be noted that our consumption of energy has not only been absolutely accelerating throughout our existence, but it has even been accelerating per capita. And to this we can add that, particularly since the Industrial Revolution, the resources fuelling human population growth have become predominately *non-renewable* (coal, oil, natural gas, uranium), their use presently constituting more than 90% of total human energy use. Reliance on non-renewable resources means that the population is greater than can be maintained in the long run.

The vicious circle principle tells us that this state of affairs is the result of a process in which technology has made it possible for us to use non-renewable resources to promote the growth of our population. Since the resources being used are finite, this process will stop sooner or later; but the later it stops, the greater will be the likelihood of our species becoming extinct. In this light, what the principle of sustainable development is telling us is to get off the vicious circle as fast as we can.

References

Ehrlich, P. R. (1968). *The population bomb*. New York: Ballantine Books.
Meadows, D. H., Meadows, D. L., Randers, J., & Behrens III, W. W. (1972). *The limits to growth*. New York: Universe Books.
Schumacher, E. F. (1973). *Small is beautiful. A study of economics as if people mattered*. London/New York: Blond & Briggs/Harper & Row.
UNCED. (1992). *Rio Declaration on Environment and Development*. Rio: United Nations Conference on Environment and Development.
WCED. (1987). *Our common future*. Oxford: Oxford University Press.

Further Reading

Nelissen, N., van der Straaten, J., & Klinkers, L. (1997). *Classics of environmental sciences*. Utrecht: International Books.

Chapter 8
Specific Principles

Lucas Reijnders

Contents

8.1 Introduction 85
8.2 The Case-by-Case Approach 86
8.3 Beyond the Case-by-Case Approach 86
8.4 Guiding Principles Originating
     in Economic Thinking 88
8.5 Guiding Principles Originating
     in Law and Its Philosophy 90
8.6 Guiding Principles Originating
     in the Life Sciences 91
8.7 Guiding Principles Originating
     in the Social Sciences 92
8.8 Importance of Guiding Principles
     for the Future 94
8.9 Changes in Worldview 94
References 95

8.1 Introduction

Among the environmental problems identified as such today are some that already have a very long history. Pleistocene hunter-gatherers in Australia changed Australian ecosystems greatly, may have contributed to the extinction of some plant and animal species and caused significant pollution especially by their use of fire (Goudsblom 1992; Reijnders 2006). The problems of soil erosion and salinisation have plagued agricultural communities for millennia, as pointed out by Ponting in Chapter 5. Furthermore, as again noted by Ponting,

L. Reijuders (✉)
Environmental science (s) University of Amsterdam and the Open University of the Netherlands, The Netherlands
e-mail: l.reijnders@uva.nl

agriculture has from its very inception 'rolled back' living nature, with a substantial impact on natural ecosystems and biodiversity (also Roberts 1989). From their earliest beginnings mining and metalworking have been associated with toxic impacts (Lucretius 1951; Hughes 1980). Similarly, indoor use of open indoor fires must have had a negative impact on lung function from the very outset. And in cities there is a long tradition of nuisances associated with productive activities, transport and wastes (Hoesel 1990). This does not mean that what we now consider to be environmental problems were viewed as such in the past or indeed even recognized at all. It does seem likely, however, that there has been substantial continuity in how certain environmental problems have been perceived and managed over the ages. A case in point is the nuisance caused by excessive noise and bad smells, a long-standing source of irritation for city-dwellers. References to such forms of nuisance are to be found in contemporary descriptions of Roman and Judaic cities of the classical era and the same holds for the cities of medieval Europe. There is a long tradition of pragmatic efforts to limit such nuisance, again dating back to classical Roman and Judaic times and continuing in Europe throughout the Middle Ages. Abatement measures included imposition of limits on the offending activities (e.g. limited access for carriages), physical planning (e.g. separating odorous production facilities from dwellings) and technological change (e.g. using horse-drawn sleighs instead of wheeled carriages). Legal means were usually invoked to implement such measures, in these cases Roman law, Judaic Halacha and medieval city ordinances.

Interestingly, similar basic approaches to environmental problems are still prevalent in much of environmental policy today. At the same time, though, major discontinuities are also apparent. From a historical perspective, some of the views currently held in the west about the relationship between man and living nature may seem relatively eccentric. Animistic religions often attribute supernatural powers to elements of their natural surroundings. A theocentric religion like Judaism sees nature as being in God's hand. The God of Judaism treats nature as he sees fit (Gerstenfeld 1999). Within such systems of belief there are limits as to how living nature may decently be treated. In all cases, though, these limits are of supernatural origin. Contemporary arguments in favour of nature conservation, for instance that natural species have intrinsic value or legal status in and of themselves, are very different and quite alien to theocentric and animistic systems of belief. Such notions are likewise at variance with the important tradition in classical Mediterranean and Near Eastern beliefs that nature was created for mankind (Cohn 1999). Furthermore, deterioration of natural resources has traditionally often been viewed as supernatural revenge for transgression of divine precepts or as the work of evil supernatural powers, again in contrast to current opinion that such deterioration is in many cases due to human activity (Cohn 1999). In addition, traditional concepts of pollution often stress the element of ritual impurity rather than negative impacts in the physical sense. Thus, in the Judaic tradition adding salt to fresh water is seen as a means of purifying the water (Gerstenfeld 1999), while today it is generally perceived as pollution.

It should be noted, though, that discontinuities in thinking about environmental affairs may in fact be less pronounced than they appear at first glance. It has been argued, for instance, that although our contemporary concept of pollution originates in an awareness of adverse physical impacts, it still has important overtones of impurity (Douglas and Wildavsky 1982). And though the supernatural has now largely disappeared in western countries as the 'sacrosanct' origin of limits, today 'the market' and 'competitiveness' have acquired an aura of sanctity which actually legitimises damage to nature. In addition, we seem to have blind spots and weaknesses of our own. The prevailing idea that "healthy, sustained economic growth is a prerequisite for environmental protection" (Dryzek 1997) will probably prove to be incorrect. In our attitudes towards living nature today there is usually considerable interest in its cultural value and its value as a potential natural resource, but little affinity with its significance for biogeochemical cycles. In conclusion, then, we are no exception to the rule that every culture has its own biases in perceiving the environment, environmental problems and their solution (Douglas and Wildavsky 1982).

8.2 The Case-by-Case Approach

As we have seen, in the case of nuisances like noise and bad odour there has been substantial historical continuity in the perception and handling of problems. Here, the approach taken is clearly problem-based. It may be termed a pragmatic approach, with individual environmental problems being specifically identified and efforts to control them implemented on a case-by-case basis. It may in fact be argued that this pragmatic approach has broader significance than it did in the past, as it now tends to be adopted for more kinds of environmental problems than just nuisances. In many countries it is currently the adopted strategy for tackling pollution problems that are not readily visible, but require scientific analysis for their identification. Pragmatic, case-by-case approaches have also emerged in the realms of natural resource protection and nature conservation.

Over the ages, though, there have also been substantial shifts of emphasis as well as innovations in the instruments employed to manage environmental problems. Thus, technological measures are now probably more important than they were in the past, and a variety of economic instruments as well as 'voluntary' agreements negotiated between government and industry have taken their place alongside traditional custom and the law.

8.3 Beyond the Case-by-Case Approach

The case-by-case approach to environmental problems is not without its critics. It may be argued that it leads to myopic interventions, dissimilar treatment of similar problems and suboptimal or even unproductive solutions (Dryzek 1987; Ayres 1994). Such criticisms have fostered the emergence of environmental problem-solving strategies that go beyond a pragmatic case-by-case or problem-by-problem approach. These strategies essentially reflect different heuristics as to environmental problems and/or their underlying causes and generally have major implications for the ways those problems are solved. To some extent at least, they are the modern equivalents of traditional belief systems relating to the proper relationship between man and his environment.

Some of these approaches have a very wide sweep that may lead to a highly relativistic view of environmental problems. Within cultural theory, for instance, the environmental problem may be considered a social construction descended directly from earlier constructions defining (im)purity in socio-religious settings (Douglas and Wildavsky 1982). Similarly, taking one's starting point in current cosmology and evolutionary theory, one may take the 'such is life' point of view that *Homo sapiens* is a species with a finite lifetime on a planet where life is ultimately doomed anyway as a result of solar expansion.

There are, however, other heuristics in contemporary Western thought that are rooted in more weighty perceptions of what are to be considered environmental problems. In several of these heuristics the focus is on broader, systems approaches rather than on individual environmental problems. In such heuristics proper relations – i.e. relations as they 'should be' – take centre stage. Equilibrium relations may be favoured in such system orientations. Such is the case, for instance, in heuristics based on 'ecosystem health' or a 'steady state economy', the latter signifying a dynamic equilibrium between the economy and its environment. The scope of these systems approaches often extend beyond the environment, as is the case with the steady state economy and its intellectual descendant, sustainability. These are not only considered to be a solution to environmental problems but are also argued to contribute to social development, justice and/or moral improvement (Mill 1848; WCED 1987). Similarly, ecotaxation is often advocated not only for its environmental benefits but also with reference to reduced labour costs, which may in turn be advantageous for job creation, creating a so-called 'double dividend'.

And if the systems approach at first sight does not go beyond the environmental realm, as in the case of 'industrial ecology' or 'industrial metabolism', the proper way to do things in industry is more central than the solving of specific environmental problems. On the other hand, there are also strategies going beyond the pragmatic, problem-by-problem approach that are still predominantly problem-oriented. The Polluter Pays Principle, environmental liability and nature compensation fall in this category.

There are several ways of grouping heuristics that go beyond case-by-case tactics. As a first approximation, we shall here associate them with academic disciplines.

Some heuristics essentially take their cue from economics. In this category, for instance, come concepts like a steady state economy, increased resource productivity, dematerialisation, the Polluter Pays Principle, and environmental utilisation space. Other ways of looking at environmental issues take their heuristics from the law, including its philosophy. This then leads to the use in environmental discourse of such concepts as intergenerational justice, property rights, environmental security and environmental liability. Still other approaches take their paradigm from life sciences like biology and medicine, employing such concepts as ecosystem health, carrying capacity, industrial ecology and industrial metabolism. 'Nature compensation' to offset man-made damage to nature may also be included in this category. Yet other heuristics have their origin in concepts developed in the social sciences. Within this category come the social constructivist and anthropological approaches, according to which environmental problems are collectively constructed. In the social constructivist view these problems should be resolved by discursive means such as negotiation and consensus-building. In the realm of solutions offered, there may also be an emphasis on such matters as cultural and organisational change. The final group of heuristics going beyond the pragmatic, case-by-case approach originate in decision theory and game theory and stress issues like social dilemmas, acceptable risk and the 'precautionary principle'.

All these heuristics are within the realm of current western thought. At the same time they have not gone uncontested. It has been argued, for example, that they show a lack of reflexivity (Smith 1995). And conservation efforts by the Chipko 'tree huggers' movement in India, for example, who attribute spiritual value to the forest and land (Shiva 1989), are clearly beyond their line of vision. Thus, a final category of guiding principles is acknowledged that stresses the overriding importance of changes in worldview.

It should be noted that the subdivision into approaches taking their cue from various fields of thought is less strict than might appear from the ordering outlined above. For instance, sustainability has its roots in both economic thinking ('steady state economy') and in the philosophy of law ('intergenerational justice'). Industrial metabolism (with strong life-science connotations) stresses increased resource productivity (with strong economic overtones). Eco-efficiency is a concept that seems to hover somewhere between economy and ecology. As the term eco-efficiency is usually employed in the context of improving the environmental performance of companies and economies, here the concept is somewhat arbitrarily categorised as 'economic'. A final example is the notion of (environmental) liability, which has its roots in the law but is also seen an 'economic' instrument in environmental policy.

The following few sections review some of the principal guiding principles originating in different heuristics. It should be noted that inclusion of a guiding principle in this overview does not mean that its use is uncontested. In fact the opposite seems to be true. Polluters regularly oppose the Polluter Pays Principle, which is deemed to constitute an undue burden in view of limited financial resources or competition. The notion of intergenerational justice has been opposed using an argument first voiced by one of the Marx brothers: what has the future done for us? And it has been called irrelevant on the grounds that human inventiveness is boundless and that thus what we currently do is not at the expense of future generations anyway (Simon 1981). The precautionary principle, invoked in decision-making when there is uncertainty about environmental consequences, has been called lunacy (Wildavsky 1995). The normative slant of concepts like industrial metabolism and industrial ecology may be opposed by those who feel that the existence of a given biological reality does not automatically sanction its use as an 'ought' in industrial policy. And so on. Neither does inclusion in the overview mean that a guiding principle is necessarily consistent with any or all of the others. Thus, those arguing in favour of the Polluter Pays Principle may be opposed by others who favour legal prohibition of the pollution in question, in order to uphold the freedom of the general population from negative interference by polluters. And those arguing from the point of view of 'acceptable risk', may find others in their way who hold that this violates the principle of intergenerational justice and are against discounting for future risks. In addition, opposing views may emerge from within one and the same field of thought. Within the framework of the law, for example, advocates of the principle of 'common heritage' are opposed by those preferring 'private property' as a construct for safeguarding the environment (Vogler 1995). Similarly, proponents of a steady state economy may be at odds with economists who focus exclusively on external costs, on the grounds that incorporating these costs in prices may still not prevent damage to natural capital that is incompatible with a steady-state economy.

8.4 Guiding Principles Originating in Economic Thinking

Certain environmental problems have already been attracting the interest of economists for a long time. The potential depletion of natural resources was a matter of concern to such 19th century economists as Mill (1848) and Jevons (1865), while inclusion of environmental costs in external costs was a major topic in the work of Pigou in the 1920s. Early in the modern environmental debate capitalism was argued to be the root cause of environmental problems. For a section of the environmental movement, today part of the wider movement for 'globalisation-from-below', this is still the case. And in recent years there has been a rapid proliferation of environmental and ecological economics. So it will come as no surprise that the guiding principles with their origins in economics vary widely in their age.

The oldest guiding principle from economics is that of the 'stationary economy'. Many 19th-century economists saw such an economy as the inevitable final stage of economic development in which economic growth comes to an end. It was held to reflect an equilibrium between an economy and its natural resource base. Opinions among economists as to its attractiveness varied widely. Ricardo, for instance, deplored a stationary economy. John Stuart Mill was of the opposite opinion, however. In his view, man would be wise to consciously pursue a stationary economy, long before the economy takes this course by necessity (Mill 1848). To a large extent this reflected his views on intergenerational justice. An economy that became stationary sooner rather than later would allow a relatively large stock of natural capital (including living nature) to remain available indefinitely. It would, according to J.S. Mill, also channel man's efforts into moral progress rather than into what he considered to be the unpleasant social practices underpinning economic growth. This view re-emerged forcefully in the 1970s, be it in a somewhat different terminology. In 1973 Herman Daly published 'Toward a steady state economy', in which 'steady state' was essentially used as a synonym for 'stationary'. Steady state referred to the relationship between economy and the environment. Knowledge, technology and the distribution of income are not constant in a steady state economy but allowed to develop, just as the planet Earth develops without growing (Daly 1995). And in 1980 the International Union for the Conservation of Nature (IUCN) began to promote the notion of sustainability, with essentially the same meaning as Mill's stationary economy (IUCN 1980) (see Box 8.1).

Since then, and especially after 1987 when the so-called Brundtland Commission published its report 'Our Common Future' (WCED 1987) 'sustainability' has become a much less precise concept. 'Our Common Future' coined 'sustainable development' as being central to mankind's future development. In defining sustainable development reference was made to intergenerational justice: "(sustainable development) meets the needs of the present without compromising the ability of future generations to meet their own needs". The Brundtland Report also stressed the need to close the gap between the rich and the poor. The political debate emerging in the wake of its publication essentially made sustainability into a political buzzword. Today, there are scores of definitions of sustainable development (Holmberg and Sandbrook 1992; Dryzek 1997). They range from being synonymous with Mill's 'stationary economy' to being environmentally essentially meaningless.

Other concepts closely allied to sustainability and the stationary or steady state economy include environmental utilisation space, ecospace and ecocapacity, denoting the actual use of the environment that is permissible when the economy has attained such a state (Weterings and Opschoor 1992; Opschoor and Weterings 1995).

Box 8.1 Aspects of a *steady state economy*, an economy in equilibrium with its environment

- Natural capital should not be diminished, thus:
  - The dissipation of natural resources should not exceed their rate of formation.
  - The functions of living nature having utility value to *Homo sapiens* should be preserved at current levels.
- Levels of persistent environmental pollutants should not increase.
- Levels of mutagenic pollutants should be such that the mutation load on the human germ line does not increase.

Another line of economic thinking that has led to guiding principles revolves around the internalisation of external costs. It was heavily influenced by publications of the economist Pigou in the 1920s, while Marshall and Scitovsky have also contributed substantially to this tradition (Hueting 1980). 'External costs', for instance those of production, are costs that are neither borne by the producer, nor by the consumer of his products, but are passed on to third parties. These 'third parties', while obviously including contemporary generations, are, increasingly, held to include future human generations as well. In a first approximation, however, the concept of third parties cannot be extended to include either present or future organisms belonging to other species, as these do not participate in markets (Hueting 1980). At the same time, though, other species or biogeochemical functions provided by ecosystems may emerge in discussions about external costs, as human beings may put a monetary value on their existence because of their usefulness or value to human society. Especially biogeochemical functions have attracted much research. In this research they have been described in terms of 'ecosystem services', 'ecosystem goods' and 'natural capital'. There is lively debate about their monetary valuation, reflected in two issues of *Ecological Economics*: volume 25, issue 1 (1998) and volume 41, issue 3 (2002).

Thinking in terms of external costs was forcefully applied to environmental matters in E.J. Mishan's 'Economic Costs of Economic Growth' (1971). Many examples of external costs in the environmental field can be given, but one may suffice to illustrate. Ozone depletion due to halocarbon emissions increases the risk of skin cancer. However, the costs of skin cancer are borne by producers and consumers of halocarbons to a very limited extent only. They are in fact passed on largely to others, both contemporaries and future human beings.

To counteract these externalities, early on in the modern environmental debate the Polluter Pays Principle emerged. This was viewed as a means of internalising external costs in the costs of production and consumption. Other ways to internalise external costs include environmental levies, ecotaxation, compensation for environmental damages and environmental liability. The last of these proceeds from the principle that those causing environmental deterioration are also liable, both legally and financially. Environmental levies, raised on economic activities held to cause unacceptable environmental damage, are used specifically to finance abatement and elimination. Ecotaxation is the name given to taxes used to finance general government expenditure. Ecotaxes may be levied on such activities as fossil fuel burning or pesticide use.

The last group of guiding principles from the economic field originated from a closer analysis of economic productivity. It was found that both labour productivity and capital productivity had increased dramatically in the course of the 20th century. Analogously, it was felt that there was major scope for increasing the productivity of the physical inputs to the economy, i.e. resources like mineral ores and hydrocarbons. Increased natural resource productivity and 'dematerialisation', i.e. a reduction of the material flows associated with a given monetary output, are two normative concepts to have emerged from this line of thinking (Herman et al. 1989; Schmidt-Bleek 1993). Related to concepts like resource productivity and dematerialisation is the notion of 'eco-efficiency', developed by the United Nations Environmental Programme (UNEP) and the World Business Council for Sustainable Development (1996). On occasion the terms are used synonymously. However, the notion of eco-efficiency may also primarily reflect environmental impact in a wider sense, including such aspects as pollution and damage to living nature (Reijnders 1998). Improving eco-efficiency has been termed ecological modernisation (Huber 1982).

8.5 Guiding Principles Originating in Law and Its Philosophy

The second field of thinking which contributes to guiding principles is that of the law and its philosophy. This is not surprising, as much of the wave of environmental interest that emerged in the 20th century was reflected in government interventions that made much use of the law. Important to the law are issues of justice and freedom. Because most, if not all, of the negative environmental consequences of human activity impinge upon human beings and other organisms deriving no direct benefit from that activity, the importance of considerations of justice is evident. With respect to freedom, an important distinction is made between positive and negative freedom, the former being freedom to act, the latter freedom from (unwanted) interference by others. Here the law obviously has to do a balancing act, as positive freedom may often be restricted by the negative freedom of third parties.

Several guiding principles have emerged from considerations of justice and freedom. One important principle is that of justice between the generations, i.e. intergenerational justice, which underpins the concepts of sustainability and a steady state economy. Contemporary third parties have also emerged with rights vis-à-vis those who undertake or plan to undertake activities having a potentially negative environmental impact. Such rights may include the right to be formally consulted about plans for such activities and injunction rights at the courts. Another guiding concept that is emerging is that of legal standing for organisms other than *Homo sapiens*. This is based essentially on the opinion that there should be respect for other organisms and that it is right to preserve the integrity of the biotic community or members thereof (Leopold 1949). There has been a vexed debate as to whether such legal standing implies rights for such organisms or rather entitlement to protection. There is no doubt, however, that the affording of legal status to certain species of wild flora and fauna has had an important impact on environmental policy, in particular through implementation of Endangered Species Acts, serving as a kind of legal Noah's Ark. Increasingly, legal status for non-human organisms is being defended on the grounds that nature and biodiversity have an intrinsic value, irrespective of their utility value to *Homo sapiens* (see Box 8.2). This highlights an interesting difference from guiding principles in the economic field, where value is attributed to non-human organisms only to the extent that human beings assign a monetary value to them.

A further concept to have emerged from the legal field is that of environmental liability, the liability to bear the costs of environmentally damaging activities. There are several varieties of environmental liability.

Box 8.2 Possible implications of assigning an *intrinsic value to other species* (here: animals)

- The killing or genetic modification of animals is permitted only when overriding human interests are at stake.
- When caught, sentient animals should be treated in such a way that their well-being is not adversely affected.
- Human use of animals should be subject to safeguards ensuring the survival of wild populations.

In the minimum case, liability refers to the environmental costs as they might reasonably have been foreseen at the time of undertaking the activity in question. In the industrialised countries of the west, however, it is now more common for liability to relate to the actual environmental costs emerging in practice. The concept of liability has also been extended to cover post-consumer wastes, for example. In this case a producer is also held liable for the costs associated with his product once the consumer decides to discard it.

Another group of guiding principles from the field of law revolves round property rights, an ambiguous concept, because it may refer to a right of exclusion, to a right of access, or to both (Gray 1995). There is a long-standing debate on the appropriateness of various forms of property rights for preventing or restricting environmental damage, one side arguing for private property and the right of exclusion, the other for a notion of 'common heritage' and the right of access.

Historically, many resources have been considered nobody's property (*res nullius*) or common property (Vogler 1995). Such resources include common pastures, marine resources (e.g. fish) outside territorial waters and the genetic pool of crop plants. At the local level, custom-based regimes have generally developed to safeguard small-scale common property resources such as local fisheries, irrigation water, woodland and pastures (Vogler 1995). In dealing with common property and *res nullius* resources, use of the law is becoming increasingly common. For resources transcending national borders, international law is being invoked more and more. National or sub-national regulations may apply to more localised common property resources. In general, the law may be used to limit access or establish quota systems.

On the other hand, following Hardin's argument that use of the commons 'automatically' leads to overexploitation (Hardin 1968), private property has been argued to be more appropriate for purposes of conservation. Opponents thereof have stressed the occurrence of many 'tragedies of enclosure' in which exclusionary private property has led to over-exploitation (Bromley 1991) and have argued that many commons have proved sustainable (Kirby et al. 1995). This controversy over property rights has had an impact on the political debate about nature conservation in a number of countries, including the United States (Anderson and Leal 1991; Dryzek 1997). It has also led to the factual demise of the 1983 Declaration on Genetic Resources of the UN Food and Agricultural Organisation, FAO, which at the time took for granted its grounding in the notion of common heritage. The importance of property in current environmental thinking can also be gleaned from the emergence of tradable permits or emission rights. While 'permits' originally defined the physical limits that companies were not to exceed in view of permissible damage to third parties, today, especially in the industrialised countries of the west, such permits are perceived increasingly as tradable property or emission rights. A system based on such tradable emission rights is argued to be more efficient in limiting pollution than more traditional ways of using the law (Dryzek 1997; Cerin and Karlson 2002).

A final guiding principle from the realm of the law, and especially international law, is that of environmental security. In the context of water resources security is a longstanding concern, and increasingly an international concern (Wolf 1998; Haftendorn 2000). The scope of environmental security concerns has also widened, for instance due to climate change. The importance of environmental security was emphasised by a 1992 statement by the United Nations Security Council declaring, *inter alia*, that non-military sources of instability in the realm of ecology have become threats to peace and security. Environmental security is being modelled along the lines of existing international security arrangements, in particular the treaties and institutions established since the Second World War (Sands 1995; Vavrousek 1995). Such arrangements to tackle specific issues may be termed (cooperative) regimes and often contain a set of principles, norms, rules and decision-making procedures (Vogler 1995).

8.6 Guiding Principles Originating in the Life Sciences

A third field of thinking that has contributed substantially to the guiding principles and concepts used in environmental science, is that of the life sciences, including medicine, biology and physiology. In this case concepts such as 'ecology' or 'health' are taken out of their life science context and used normatively to refer to the functioning of the economy, society or its institutions.

Thus, 'industrial ecology' (Frosch and Gallopoulos 1989; Ayres and Ayres 1996) takes its cue from ecology, where individual organisms and their waste form the material basis on which other organisms live. It does so with a normative slant. It 'prescribes' that current non-product outputs of production and consumption should be used as inputs to other economic activities. The key aim is not to minimise the waste associated with an individual production unit or individual product, but to minimise the sum total of waste in the industrial equivalent of the ecological food web (Allenby and Richards 1994) (see Box 8.3).

A related guiding principle is that of 'industrial metabolism' (Ayres 1994). As currently used, it is to an extent equivalent to industrial ecology in terms of its interest in the interrelations between industrial activities. Thus, non-product outputs of industries as well as the consumer cycle should serve as industrial inputs. At the same time, though, the concept of industrial metabolism articulated by Ayres is also applicable to the operations of individual manufacturers or firms. In this context it is eco-efficient internal operation that is stressed, the aim being to increase materials productivity and reduce non-product output (Ayres 1994).

'Ecosystem health' transfers the medical concept of health to ecology. It requires abiotic conditions conducive to the permanence of specific ecosystems.

Box 8.3 Aspects of *industrial ecology*

- The wastes of one company should be used as resource inputs by another.
- Resources should be used analogously to the cascades occurring in nature, where water falls from levels of higher to lower potential energy, in such a way as to maximise the product of quality and time in use, over the economic life of the resource.
- The following types of industry should be present:
  - Repair facilities to ensure that products are maintained at a high quality
  - Decomposer industries to dismantle obsolete and discarded products
  - Remanufacturing plants to bring discarded products or parts thereof back to specification
  - Scavenger industries to 'gobble up' wastes left by other activities

'Carrying capacity', a concept used in biology, has been transferred to the wider environmental discourse. An example is to be found in Hardin (1993), who holds that the environmentalist's eleventh commandment is: 'Thou shalt not transgress the carrying capacity'. Similarly, Arrow et al. (1995) hold that sooner or later economic growth must inevitably face limits associated with the environment's carrying capacity. In biology this capacity is expressed in terms of the population of a given species an ecosystem can sustain. In more general environmental discourse it has become akin to the previously mentioned concepts of ecospace and environmental utilization space, denoting a use of the environment that is compatible with a steady state economy. It then usually refers to the number of people that can be sustained indefinitely on a particular area of land at specified levels of production, consumption and technology. A related concept is that of 'critical load'. Critical load is a quantitative value regarding exposure to pollutants below which significant harmful effects on specified elements of the environment do not occur according to current knowledge. This concept has found its way to policy, especially in the context of exposure to nitrogen and sulphur compounds (Porter et al. 2005).

One last, normative guiding principle to have emerged from the field of biological thought is 'nature compensation', which proceeds from the desirability of keeping nature's 'value' constant. Thus, when nature or a natural resource somewhere is destroyed or reduced in 'value' to permit expansion of human activities, this should be compensated for by creating equivalent 'new' nature or natural resources (NOAA 1996). It should be noted that the term 'value' as used here is a contested concept, as its actual meaning in practice is to a considerable extent subjective. As a result its substantive content may suffer. Compensation for lost biogeochemical functions, for example, is a rare occurrence in nature compensation.

8.7 Guiding Principles Originating in the Social Sciences

A radically different approach may be traced back to the social sciences. In both anthropology and sociology the focus is not on environmental problems in the physical sense but on the cultural or social constructs thereof that have emerged in particular cultures and societies. The anthropological approach stresses the importance of culture and cultural diversity in the perception of environmental problems (Douglas and Wildavsky 1982; Milton 1996). It views environmental risks essentially as collective constructs. Sociology has given rise to a similar, social-constructivist view of environmental problems, in which these problems are regarded as social constructs that are subject to changing meaning and social (re)negotiation (Van Dongen et al. 1996) (see Box 8.4). To be meaningful socially, according to social constructivists, the existence and seriousness of such problems should be duly discussed, hopefully leading to a consensus. Solutions to environmental problems should essentially be measured on the basis of stakeholder satisfaction.

Cultural theory approaches to environmental problems may stress the incommensurability of the different cultural views (Schwarz and Thompson 1990). Focussing on the social aspects of environmental problems also leads to suggestions as to their possible solutions. In this context both cultural change, essentially extending the civilizing process to environmentally relevant behaviour (Goudsblom 1992), and organisational change to bring organisational performance more into line with perceived environmental requirements (Tellegen and Wolsink 1998) have been proposed. Finally, decision theory has emerged as a source of guidance in the environmental field.

Box 8.4 A social constructivist view of a *sustainable company*

By stressing that the content of sustainability is subject to negotiation, social constructivists allow for a variety of views of what a sustainable company may be. A possible example of one such view is the following:

- Companies should maintain transparency and dialogue vis-à-vis workers and outside stakeholders, including environmental non-governmental organisations.
- In determining sustainability, a 'triple bottom line' approach should be followed (giving due consideration to 'people, profit and planet').
- Sustainability in the social and environmental sense should be determined on the basis of stakeholder satisfaction.

Especially important have been the characterisation of environmental problems as social dilemmas and the concept of 'acceptable risk'. The social dilemma aspect of environmental problems emerges from the game theory component of decision theory and has become important in strategic thinking about solving environmental problems (Tellegen and Wolsink 1998). Many environmental problems can be viewed as social dilemmas. This point was brought home forcefully in one of the classics in environmental thinking: 'Tragedy of the Commons', in which G. Hardin argued that people acting on their own and in their own interest will exploit common resources to such an extent that everyone will eventually lose out. The only two solutions to this dilemma that he saw were private property or forceful state intervention. Since then many other game theoretical dilemmas have been proposed that reflect specific environmental problems (Tellegen and Wolsink 1998). This in turn has led to solution-oriented strategies for environmental problems based on game theory (Ostrom 1990; Ostrom et al. 1994; Vogler 1995). Many of these strategies stress cooperative solutions.

There have also been efforts to base large segments of environmental policy on the concept of 'acceptable risk' (USEPA 1992; Wynne 1987; Lofstedt and Frewer 1998). Although at first glance a simple concept, acceptable risk is probably a good match for sustainability in terms of the number ways in which it has been operationalised (Lofstedt and Frewer 1998; Tellegen and Wolsink 1998). While some of these operationalisations focus on substance, others emphasise the procedure to be followed to establish 'acceptable risk'. With respect to the former, some essentially follow the actuarial definition of risk (chance times effect, used by insurance companies), whereas others extend operationalisation to include a variety of other factors, varying from the extent of voluntariness to the perceived catastrophality of effects, and from perceived controllability to the dread associated with consequences. There are ways to define acceptable risk in which future generations are excluded and included, and for the latter again a wide variety of operationalisations exists. There is also wide variation with regard to the inclusion of organisms belonging to other species. And finally, estimates of risks are often highly uncertain (Wynne 1987; Lofstedt and Frewer 1998).

As to the element 'acceptability', there is a wide variety of opinion as to the factors involved and the procedure to be used for determining its actual meaning.

Some see 'acceptable' as the outcome of risk-benefit considerations, others as the outcome of social negotiations, while others again stress the rights of those at risk as being the overriding determinant of what is acceptable. There is also controversy on the issue of whether the informed consent of those at risk should be sought and on the implications of uncertainty in all its different manifestations, ranging from ignorance to parameter uncertainty.

One of the factors to emerge in this context has become a guiding principle in its own right: the Precautionary Principle. This principle holds, minimally, that in decision-making under conditions of uncertainty, abatement and similar measures should not necessarily wait until there is certainty about a given risk if there are indications of grave and, especially, irreversible effects. At the Earth Summit held in Rio de Janeiro in 1992 this principle was formulated as follows: "Where there are threats of serious or irreversible damage, lack of full scientific certainty should not be used as a reason for postponing cost effective measures to prevent environmental degradation". There are also stronger versions of the principle, in favour of abstaining from activities that may carry serious risk in the absence of sufficient scientific certainty.

8.8 Importance of Guiding Principles for the Future

The guiding principles described vary enormously in scope. Some are very general and cover many of the recognised environmental impacts of human behaviour. Examples here are the twin guiding principles of a steady state economy (Section 8.4) and intergenerational justice (Section 8.5). Others are more limited in scope, as with the concepts of liability, in the sense of an obligation to bear the environmental costs that might reasonably have been foreseen (Section 8.5), or organisational change, with the aim of aligning organisational performance more with environmental requirements (Section 8.7). Limitation of scope does not detract from usefulness in specific circumstances.

Mankind's future well-being, however, seems to hinge largely on the question of whether we will be able to put into practice the twin guiding principles of a steady state economy (Section 8.4) and intergenerational justice (Section 8.5). If such principles are implemented, other species will most likely also fare better than under 'business as usual'. The ultimate future survival of these species will depend mainly, though, on the extent to which their legal status (see Section 8.5) is improved and realised in practice.

8.9 Changes in Worldview

There is a substantial array of authors that find the above-mentioned guiding principles shallow if not misguided. Early on in the modern environmental debate the Christian system of belief was argued to be the root cause of environmental problems (White 1967). V. Shiva, associated with the Indian Chipko or 'tree huggers' movement, sees science as one of the root causes of environmental destruction (Shiva 1989). A variety of others have also viewed 'the mastering of nature' attitude, prevalent amongst protagonists of natural science since Francis Bacon, as a major cause of current environmental problems. It has also been argued that a nature-based spirituality is the proper response to the environmental predicament, as in the perspective of ecocentric 'deep ecology' (Naess 1973; Dryzek 1997). And several eco-feminists have argued the overriding importance of a non-instrumental, non-dominating, intuitive relationship with nature (Devall and Sessions 1985; Spretnak 1986).

The relationship between major belief systems and environmental problems is by no means an easy one, especially as the former can on closer inspection be interpreted in more than one way. The Jewish tradition, which subscribes to the view that nature was created for man (Cohn 1999) has restrictions on how *Homo sapiens* may actually treat nature (Gerstenfeld 1999; Boersema 2001). The same holds for Islam (Khalid and O'Brien 1997). The Christian catholic tradition, which in the main likewise holds that nature was created for man, has the Church father Dionysius, who held that plants and animals were less alienated from God than mankind, and saints like St. Ambrosius, who praised the animals because they had not sinned, and St. Francis, who conversed with the birds. There may, moreover, be only a tenuous relationship between belief and everyday practice. Polynesians considered living organisms as one big family, but drove many bird species to extinction.(Steadman 1989). Over 2,000 years ago the idea that nature cooperated with good men became the dominant Chinese tradition, but nevertheless, as pointed out by Ponting (Chapter 5), by the 20th century little more than 3% of China's original forests still survived.

Similarly, the relation between science, technology and environmental problems is a complex one. These days science is a broad church, in which the 'mastery of nature' attitude may well be contested. And technology has been invoked as a major solution to our environmental woes (Reijnders 1998). Against this background a more precise focusing, exemplified by the guiding principles discussed in Sections 8.4–8.7, is no luxury.

References

Allenby, B. R., & Richards, D. J. (1994). *The greening of industrial ecosystems*. Washington, DC: National Academy Press.
Anderson, T. L., & Leal, D. R. (1991). *Free market environmentalism*. Boulder, CO: Westview Press.
Arrow, K., Constanza, R., Dagupta, P., Folke, C., Holling, C. S., & Pimentel, D. (1995). Economic growth, carrying capacity and the environment. *Science, 268*, 520–521.
Ayres, R. U. (1994). Industrial metabolism: Theory and policy. In B. R. Allenby & D. J. Richards (Eds.), *The greening of industrial ecosystems*. Washington, DC: National Academy Press.
Ayres, R. U., & Ayres, L. W. (1996). *Industrial ecology. Towards closing the materials cycle*. Cheltenham: Edward Elgar.
Boersema, J. J. (2001). *The torah and the stoics on humankind and nature*. Leiden: Brill.
Bromley, D. W. (1991). *Environment and economy: Property rights and public policy*. Oxford: Blackwell.
Cerin, P., & Karlson, L. (2002). Business incentives for sustainability: A property rights approach. *Ecological Economics, 40*, 13–22.
Cohn, N. (1999). *Cosmos, chaos and the world to come*. New Haven, CT: Yale University Press.
Daly, H. (1973). *Toward a steady state economy*. San Francisco: W. H. Freeman.
Daly, H. (1995). The steady-state economy: Alternatives to growthmania. In J. Kirby, P. O'Keefe, & L. Timberlake (Eds.), *Sustainable development* (pp. 331–342). London: Earthscan.
Devall, B., & Sessions, G. (1985). *Deep ecology: Living as if nature mattered*. Salt Lake City, UT: Peregrine Smith.
Douglas, M., & Wildavsky, A. (1982). *Risk and culture*. Berkeley, CA: University of California Press.
Dryzek, J. S. (1987). *Rational ecology: Environment and political economy*. Oxford: Basil Blackwell.
Dryzek, J. S. (1997). *The politics of the earth*. Oxford: Oxford University Press.
Frosch, R. A., & Gallopoulos, N. E. (1989). Strategies for manufacturing. *Scientific American, 264*(3), 94–102.
Gerstenfeld, M. (1999). *Judaism, environmentalism and the environment*. Jerusalem: Rubin Mass.
Goudsblom, J. (1992). *Fire and civilisation*. London: Allan Lane.
Gray, K. (1995). The ambivalence of property. In J. Kirby, P. O'Keefe, & L. Timberlake (Eds.), *Sustainable development* (pp. 223–226). London: Earthscan.
Haftendorn, H. (2000). Water and international conflict. *Third World Quarterly, 21*, 31–68.
Hardin, G. (1968). The tragedy of the commons. *Science, 162*, 1243–1248.
Hardin, G. (1993). *Living within limits; ecology, economics and population taboos*. Oxford: Oxford University Press.
Herman, R., Ardekani, S. A., & Ausubel, J. A. (1989). Dematerialisation. In Ausubel, D. H., & H. E., Sladovich [eds.] *Technology and environment* (pp. 50–69). Washington, DC: National Academy Press.
Hoesel, G. (1990). *Unser Abfall aller Zeiten*. München: Kommunalschriften Verlag.
Holmberg, J., & Sandbrook, R. (1992) Sustainable development: What is to be done? In J. Holmberg (Ed.), *Policies for a small planet* (pp. 19–38). London: Earthscan.
Huber, J. (1982). *Die verlorene unschuld der ökologie, neue technologien und superindustrielle entwicklung*. Frankfurt am Main: Fischer Verlag.
Hueting, R. (1980). *New scarcity and economic growth*. Amsterdam: North Holland.
Hughes, J. D. (1980). Early Greek and Roman environmentalists. In L. J. Bilksi (Ed.), *Historical ecology* (pp. 45–60). London: Kennicat Press.
IUCN. (1980). *World conservation strategy: Living resource conservation for sustainable development*. Gland: IUCN.
Jevons, W. S. (1865). *The coal question*. London: MacMillan.
Khalid, F., & O'Brien, J. (1997). *Islam and ecology*. London: Cassell.
Kirby, J., O'Keefe, P., & Timberlake, L. (1995). *Sustainable development*. London: Earthscan.
Leopold, A. (1949). *A sand county almanac*. Oxford: Oxford University Press.
Lofstedt, R., & Frewer, L. (1998). *Risk and modern society*. London: Earthscan.
Lucretius. (1951). *On the nature of the universe* (R. E. Latham, Trans.). Harmondsworth: Penguin.
Mill, J. S. (1848). *Principles of political economy*. Reprinted by Harmondsworth: Penguin.
Milton, K. (1996). *Environmentalism and cultural theory*. London: Routledge.
Mishan, E. J. (1971). *Economic costs of economic growth*. Harmondsworth: Penguin.
Naess, A. (1973). The shallow and the deep, long range ecology movement. *Inquiry, 16*, 95–100.
NOAA (National Oceanic and Atmospheric Administration). (1996). *Damage assessment and restoration program, habitat equivalency analysis*. Washington, DC: US Government Printing Office.
Opschoor, J. B., & Weterings, R. A. P. M. (1995). Environmental utilisation space: An introduction. *Milieu, 9*, 198–206.
Ostrom, E. (1990). *Governing the commons: The evolution of institutions for collective action*. Cambridge: Cambridge University Press.
Ostrom, E., Gardner, R., & Walker, J. (1994). *Rules games and common pool reserves*. Ann Arbor, MI: Michigan University Press.
Pigou, A. C. (1920). *The economics of welfare*. London: MacMillan.

Porter, E., Blett, T., Potter, D. U., & Huber, C. (2005). Protecting resources on federal lands: Implications of critical loads for atmospheric deposition of nitrogen and sulfur. *BioScience, 55*, 603–612.

Reijnders, L. (1998). The factor X debate. *Journal of Industrial Ecology, 2*, 13–22.

Reijnders, L. (2006). Is increased energy utilization linked to greater cultural complexity? Energy utilization by Australian Aboriginals and traditional swidden agriculturalists. *Environmental Sciences, 3*, 207–220.

Roberts, N. (1989). *The holocene. An environmental history*. Oxford: Blackwell.

Sands, P. (1995). Enforcing environmental security. In J. Kirby, P. O'Keefe, & L. Timberlake (Eds.), *Sustainable development* (pp. 259–266). London: Earthscan.

Schmidt-Bleek, F. (1993). MIPS re-visited. *Fresenius Environmental Bulletin, 2*, 407–412.

Schwarz, M., & Thompson, M. (1990). *Divided we stand. Redefining politics, technology and social choice*. London: Harvester Wheatsheaf.

Shiva, V. (1989). *Staying alive: Women ecology and development*. London: Zed Books.

Simon, J. L. (1981). *The ultimate resource*. Oxford: Martin Robertson.

Smith, M. A. (1995). Green thought in a green shade. A critique of the rationalisation of environmental values. In Y. Gerrier (Ed.), *Values and the environment*. Chichester: Wiley.

Spretnak, C. (1986). *The spiritual dimension of green politics*. Santa Fe, NM: Bear.

Steadman, D. W. (1989). Extinction of birds in Eastern Polynesia. A review of the record and comparison with other Pacific island groups. *Journal of Archaeological Science, 16*, 177–205.

Tellegen, E., & Wolsink, M. (1998). *Society and its environment*. Amsterdam: Gordon & Breach.

Van Dongen, H. J., de Laat, W. A. M., & Mas, A. J. J. A. (1996). *Een kwestie van verschil. Conflicthantering en onderhandeling in een configuratieve integratietheorie*. Delft: Eburon.

Vavrousek, J. (1995). Institutions for environmental security. In J. Kirby, P. O'Keefe, & L. Timberlake (Eds.), *Sustainable development* (pp. 267–273). London: Earthscan.

Vogler, J. (1995). *The global commons*. Chichester: Wiley.

UNEP and World Business Council for Sustainable Development. (1996). *Eco-efficiency and cleaner production*. Paris: UNEP.

USEPA. (1992). *Environmental equity: Reducing the risk for all communities*. Baltimore: USEPA.

WCED. (1987). *Our common future*. Oxford: Oxford University Press.

Weterings, R. A. P. M., & Opschoor, J. B. (1992). *Eco-capacity as a challenge to technological development*. Rijswijk: RMNO.

White, L. (1967). The historical roots of our ecologic crisis. *Science, 155*, 1203–1207.

Wildavsky, A. (1995). *But is it true? A citizen's guide to environmental health and safety issues*. Cambridge, MA: Harvard University Press.

Wolf, A. T. (1998). Conflict and cooperation along international waterways. *Water Policy, 1*, 251–265.

Wynne, B. (1987). *Risk management and hazardous waste*. Berlin: Springer.

Chapter 9
Social Science and Environmental Behaviour

Linda Steg and Charles Vlek

Contents

9.1 Introduction and Preview 97
9.2 Why the Need for a Behavioural Science Perspective? 98
9.2.1 Actor Versus Impact Perspective 98
9.3 Behavioural Science Methods and Principles for Environmental Research 99
9.3.1 Basic Types of Measurement Scale 99
9.3.2 Observational, Self-Report and Performance Measures 100
9.3.3 Study Designs and Data Analysis 102
9.3.4 How to Design and Conduct an Empirical Study 104
9.4 Studying Environmental Impacts on Human Well-Being 106
9.4.1 Environmental Stress and Protection Motivation 106
9.4.2 Environmental Risk Perception 109
9.4.3 Appreciation of Nature 110
9.4.4 Evaluation of Common Goods 111
9.5 Studying Behaviours Affecting Environmental Qualities 113
9.5.1 Categorising Environmental Behaviour 113
9.5.2 Measuring Intent- and Impact-Oriented Behaviour 114
9.6 The Commons Dilemma as a General Model of Environmental Exploitation 115
9.6.1 Definition of the Commons Dilemma 116
9.6.2 Twelve Key Issues for Researchers and Policy-Makers 116
9.6.3 Macro, Meso and Micro Perspectives 117
9.6.4 Social and Psychological Factors Governing 'Cooperation' and 'Defection' 118
9.7 Psychological Diagnosis: Behaviour Limitation and Determination 119
9.7.1 Cognitive and Affective 'Short-Sightedness' 119
9.7.2 Needs, Opportunities and Abilities 120
9.7.3 Four Different Behavioural Processes 120

9.8 Changing Environmentally Significant Behaviour 122
9.8.1 Strategies for Behavioural Change 122
9.8.2 A Three-Dimensional Policy Model 124
9.8.3 Territorialisation and Tradable Exploitation Rights 125
9.8.4 Principles and Practical Steps for Intervention Planning 126
9.9 Environmental Values, Beliefs and Quality of Life 127
9.9.1 Value Orientations and General Environmental Beliefs 128
9.9.2 Human Needs and Quality of Life 131
9.10 Societal Analysis and Management of Environmental Problems 132
9.10.1 Driving Forces of Environmental Decline 133
9.10.2 'Sustainable Development' 137
9.11 What Behavioural Scientists Can Do to Enhance Environmental Qualities 138
9.11.1 Positivistic and Constructive Behavioural Science 139
9.12 Conclusion 139
References 140

9.1 Introduction and Preview

This chapter seeks to provide a behavioural science perspective on environmental problem analysis and management. Why is there a need for such a perspective and what does it involve? How can a behavioural science perspective be aligned with a physical science perspective? Our emphasis in this chapter will be on the scientific methods available for studying environmental behaviour. A method is a means of doing something or getting somewhere. This means that before any particular method is adopted, the researcher must first be clear about the *rationale* of

L. Steg (✉)
University of Groningen,
The Netherlands
email: E.M.Steg@rug.nl

the study, answering such questions as 'what?', 'why?', 'to what end?', 'with or for whom?', 'with what intended results?', 'to be used for what purpose?'.

Environmental problems may relate to *sources* (of energy, water, raw materials), *sinks* (for absorbing emissions and waste) or *ecosystems* (for supporting diverse forms of life). Interaction between the human species and the natural environment is a two-way process, with environmental quality affecting human functioning and well-being, and human behavioural patterns impacting on environmental quality. To clarify the nature of this interaction, we shall discuss various behavioural science conceptions of environmental problems.

The chapter is structured as follows. Section 9.2 briefly explains the need for a behavioural science perspective. The longer Section 9.3 provides a general overview of behavioural science methodology, applicable to any topic of study. Sections 9.4 and 9.5 follow this on the study of the two-way relationship between human behaviour and the environment. Section 9.6 introduces the 'commons dilemma' as a basic paradigm for thinking about the tensions between individual interests (such as freedom of consumption) and collective goods (including, specifically, environmental quality). From within this framework a four-component management model is presented, embracing diagnosis, decision-making, intervention and evaluation.

Section 9.7 offers a psychological diagnosis of behavioural processes and determinants. Section 9.8 considers some of the strategies and policy instruments available for effectuating behavioural change. Section 9.9 discusses pro-environmental values and general beliefs, along with the evaluation of changes in human quality of life resulting from environmental conditions or policies. Section 9.10 examines the driving forces operative in society as well as the concept of sustainable development. Section 9.11 presents some of the different ways in which behavioural scientists may help to enhance environmental qualities. Finally, Section 9.12 presents some general conclusions.

9.2 Why the Need for a Behavioural Science Perspective?

Environmental problems are deeply intertwined with human society, the economy and human behavioural patterns. Therefore, in tackling these problems it is not sufficient for policy-makers to focus solely on physical or technical measures like exhaust filters, spatial zoning around high-impact industries, or noise insulation of buildings along inner-city roads. Growth in consumption levels, for example resulting from population and economic growth, may mean that the aggregate environmental burden is *not* reduced. There is thus a real need to complement physical and technological solutions by policies ensuing from socio-behavioural analyses of cultural premises, social values, human beliefs, widespread cultural habits and the preferred choices of people and organisations. These choices relate to what they do and do not do in their private, domestic lives, their work and education, their transport, holidays and other leisure activities. Physical and technical strategies must be accompanied by behavioural and social strategies. For one thing, innovative technology always needs to be accompanied by appropriate human understanding, acceptance and mode of utilisation, all areas in which behavioural science research can offer useful insights and support.

9.2.1 Actor Versus Impact Perspective

One difficulty in communications between behavioural and physical scientists lies in the different perspectives they have on environmental problems. While a physical scientist would tend to address environmental effects and physicochemical processes 'from source to sink', a social scientist would be inclined to focus on human needs, goals, activities and determinants of behaviour with environmental impact. This fundamental difference in perspective can be seen as a difference between an impact orientation and an actor orientation towards environmental problems, as shown alongside one another in Fig. 9.1, which highlights two particular interlinkages. When addressing energy issues (reduced use of fossil fuels, for example), policy-makers may be led back from the impact side to a broad spectrum of actors and energy-relevant behaviours. Conversely, policy-makers starting from the actor side (households, say) may be prompted to consider a broad range of environmental effects.

If one were to define environmental sustainability in terms of fossil fuel consumption (on the right of the figure), one would have to simultaneously address – retroactively – a range of very different energy con-

Fig. 9.1 Actors versus impacts perspectives on environmental research

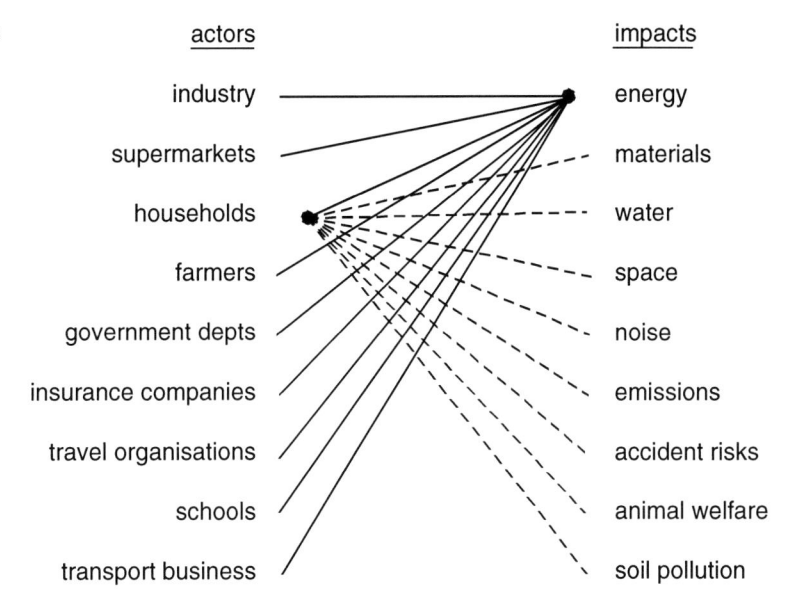

```
actors                            impacts
industry                          energy
supermarkets                      materials
households                        water
farmers                           space
government depts                  noise
insurance companies               emissions
travel organisations              accident risks
schools                           animal welfare
transport business                soil pollution
```

sumers. If, on the other hand, 'sustainability' were defined in terms of behavioural patterns (of households, supermarkets, farmers and so on; on the left), then one would have to simultaneously address – proactively – a range of very different environmental impacts. Either of these perspectives can provide a useful starting point for policy development. Ideally, though, a combination – i.e., consistently addressing a variety of actors and considering their spectrum of environmental impacts – seems to offer the greatest potential. Thus, one should focus on environmental impacts *and* their behavioural causes.

9.3 Behavioural Science Methods and Principles for Environmental Research

This section briefly reviews some of the main approaches, methods and principles of behavioural science research on environmental problems. Additional methodological information will be provided later on, as particular topics are discussed. There are three basic types of behavioural research, employing descriptive, correlational and experimental methods. Today, computer simulation studies, using algorithms derived from one or more theoretical models, provide a useful additional research tool. By way of general introduction let us first discuss the four types of scale available for measuring phenomena of interest, taking as practical examples motorist speeding, office elevator use and recreational use of parks. To conclude this methodological exposé, in Section 9.3.4 an approximate recipe is provided for designing and conducting an empirical study.

9.3.1 Basic Types of Measurement Scale

Scientific measures can be defined as variables on a nominal, ordinal, interval or ratio scale. *Nominal* measures are suitable for description and qualitative categorisation, but are not very amenable to mathematical computation. An example is the subcultural variety of transport users or park visitors: young females, handicapped people or male business executives, for example (which may be overlapping groups).

Ordinal measures are amenable to numerical analysis, especially in multidimensional scaling aimed at revealing underlying structures of meaning, but their ranking properties do not allow for summation or averaging since the distance between ranks 1 and 2 may be entirely different from that between ranks 4 and 5, for instance. A case in point is the ranking of transport modes according to apparent user popularity. In the Netherlands, the ranking car-train-bus generally means the car is preferred to the train, with far fewer people opting for the bus.

Interval scales have the property of scale differences being preserved under any linear transformation of scale. For example, the energy required to lift an elevator three floors is similar between floors 1 and 4 and between 3 and 6 (for a given number of passengers). At the same time, one could not maintain that twice as much energy is needed for reaching the sixth floor as for reaching the third. Subjective judgements about policy preferences or personal satisfaction in life, for example, are well amenable to rating on an interval scale.

Measures on a *ratio* scale allow for comparison of both intervals and ratios (and, of course, rankings). An example is vehicle speed measurement, which allows for statements like 'this vehicle is twice as fast as that one', but also 'the speed difference between 100 and 80 km/h is the same as that between 70 and 50 km/h'.

9.3.2 Observational, Self-Report and Performance Measures

The behavioural sciences are concerned with three categories of research data, specified in terms of observational, self-report and performance data.

9.3.2.1 Observational Measures

A theoretically straightforward but often problematical approach in the behavioural sciences involves the researcher directly and systematically observing how subjects act in a particular setting, carefully noting their behaviour, as (s)he perceives it. For example, one may wish to observe the road and traffic conditions that induce car drivers to speed on city streets. Or one may be interested in everyday use of office elevators by different types of people at different hours of the working day. The main advantage of observation compared with other methods is that it allows the researcher to study behaviour as it occurs naturally, in realistic settings. This makes it relatively easy to collect data unobtrusively, as people need not be aware their behaviour is being studied and are therefore less likely to exhibit a biased response (behaving strategically or according to assumed expectations) than in strict experimental settings.

However, observational methods put a relatively high burden on the definition, processing and analysis of the data collected. Some form of prior study rationale based on non-trivial research questions must therefore be set out beforehand to help the researcher decide which behavioural and situational aspects to observe and how to record them. After responses have been collected, they often require further coding to ensure they can be interpreted unambiguously and are broadly consistent across different subjects. Also important is consistency of judgement among response coders or judges, or what is known as the inter-rater reliability of a test method. For example, an observation of 'sudden acceleration' with respect to cars on city streets must have roughly the same meaning at each instance of observation and for all response coders, as well as for all drivers observed. The more precisely observational data are defined and the more consistent their meaning across the subjects studied, the greater the likelihood of the researcher conducting meaningful data analyses and ultimately publishing useful conclusions. As observational methods focus on observable behaviour, they only reveal how people act, but do not clarify why they do so. Thus, underlying motivations are not known.

9.3.2.2 Self-Report Measures

It is often convenient to assess human behaviour by asking subjects to report on how they usually act or have acted in the past, why they do or did so, how they feel about some topic, object or person, what satisfaction they derived, and so on. The kind of concepts and measurement variables used in self-report studies are shown in Table 9.1, with some example items.

While often useful, self-report measures have their disadvantages. In the first place, subjects do not always report correctly or truthfully about their own meter readings, behaviours, attitudes, preferences or satisfaction. If the issue being studied is controversial, people may be tempted to give answers that are socially desirable, i.e., what they think other people would approve of. For example, few people would be prepared to admit that driving a particular type of car or driving fast are status symbols. In self-report studies, subjects may also be tempted to give so-called strategic responses. For example, people strongly opposed to an increase in car fuel prices might dishonestly state it would drastically reduce their quality of life, in the hope that such a response might prevent implementation of the policy.

A second problem of self-report measures is that subjects may not interpret research questions or response options in the same way. For example, reported frequencies of park visits may be of only limited meaning in cases where it can be safely assumed that visits are of varying duration, are made for very different reasons and can encompass a diversity of activities. In such cases no valid conclusions can be drawn, for it remains unclear what general pattern these frequency data reflect. Data interpretation problems of this kind should be minimised by ensuring that research questions are carefully formulated beforehand.

An important advantage of self-report measures is that sampling data are relatively easy to collect. For certain issues, self-reporting is in fact the only option, as when assessing the perceived attractiveness of cars, parks or elevators, for example. For evaluating the kind of complex policy scenarios involved in sustainable development, in particular, the self-reported judgements and preferences of those involved are crucial. Standardised methods have therefore been developed to maximise the validity and reliability of what is being measured and to minimise response bias.

Interviews and questionnaire studies are the commonest ways of collecting self-report data. *Individual interviews* can be conducted at people's homes or by telephone. A major advantage of interviews is that they give the researcher greater control over how questions and response options are interpreted. On the other hand, an interview situation may threaten people's desire for anonymity. There is a good chance that the interviewer will influence the elicited responses, for respondents may again be tempted to give socially desirable answers, especially if the interview is 'one-on-one'. In practice, then, interviews are less frequently used, being rather costly and time-consuming compared with mass questionnaire studies. In this respect, telephone interviews occupy an intermediate position.

Questionnaires are easy to administer, relatively inexpensive to produce and distribute, and can be presented simultaneously to large numbers of subjects. In addition, they can accommodate people's desire for anonymity, by not requiring respondents to provide their name or address. However, designing a good, self-supporting questionnaire requires a great deal of experience and prior testing, to ensure that questions and response options can be clearly understood and unambiguously processed. Individual questionnaires can be distributed by traditional mail or in small group sessions, where administration can be better standardised than in respondents' own homes, without the investigator present. An interesting new medium for questionnaire studies is the Internet, which allows for efficient electronic handling of both questions and responses.

An alternative to both interviews and questionnaires are *focus group meetings*, where subjects discuss the topic of interest in a small group. One advantage of focus group discussions is the social interaction they involve, spurring participants to elaborate on issues of interest at greater length and allowing the researcher to elicit individualised responses or have subjects perform pre-defined assessment tasks. Minutes prepared from the group discussion provide an essential source of data and can also aid interpretation of the information collected from individual subjects. A disadvantage of focus group discussions is that individuals may in some way be prejudiced by the group's deliberations prior to giving their own personal response. Also, dominant individuals can significantly influence the opinions of other group members. There are, however, useful procedures for reducing negative group phenomena. One example is the 'round robin' technique, whereby each group member is given a fixed amount of time to express his or her opinion. The composition of focus groups therefore requires a great deal of care and should match the purposes of the study. This also means that the procedures to be followed should be well designed and laid down precisely in advance. Focus groups are especially useful for studying 'new'

Table 9.1 Overview of self-report concepts and measures of environmental behaviour, with examples

Qualitative descriptions:	'How would you describe…?'
Reported meter readings:	'How many car-km, m^3 gas, kWh electricity has your household used?'
Behavioural frequencies:	'How often do you…?'
Ratings:	'How attractive do you find…?'
Attitudes:	'How much do you agree with..? how do you feel about…?'
Rankings:	'Which option is more commendable…?'
Choices:	'What or whom would you prefer…?'
Satisfaction:	'How satisfied are you with…?'

phenomena, to examine which aspects should be considered to better understand such issues.

9.3.2.3 Performance Measures

In behavioural science research, subjects can not only be observed 'in natural surroundings', or report on their own behaviours, attitudes and judgments. They can also produce a priori defined performance measures in a deliberately designed task or measurement situation. For example, speeding on city streets can be studied in an experimental laboratory car equipped to record the subject's head and eye movements, heart rate and skin conductance and the distance observed with respect to the preceding car. Similarly, elevator users can be closely monitored for personal fitness, walking speed and cardiovascular health. Thus, 'judgement-free' measurements may relate to psycho-physiological and behavioural data.

At first sight, performance measures appear much the same as the observational measures discussed above. However, the data collected is of a much higher quality, as the measured quantities are strictly defined beforehand, often on the basis of some theoretical model of the behavioural process under study. In the case of speeding on city streets, for example, there is an inevitable trade-off between vehicle speed and accuracy of driving. There is also heightened arousal and sensation as speed increases, which may turn into anxiety and stress above a certain speed. To adequately assess these processes one would have to collect several types of data in a precisely defined manner. Another area where performance measures are highly relevant is in studies of environmental stress (see Section 9.4.1). While the researcher may logically presume that urban noise, say, constitutes a nuisance and ask respondents to state their views on the matter, it is quite another thing to demonstrate scientifically that specific noise levels actually cause deterioration of school or office performance or loss of sleep at night. Again, it is essential that the research be meticulously designed and planned and appropriate data collected. The conclusion here is that human performance measures under controlled (often experimental) settings are often valuable because of their greater precision and significance. At the same time, though, they are more difficult to define and acquire, because of the stricter circumstances under which they must be recorded. Because of their specificity, moreover, they may be more laborious to work with than self-report measures. For example, a subject's satisfaction with his or her life or the perceived acceptability of alternative policy scenarios can be more accurately determined using self-report measures than with one or several performance measures, however expressive. Which specific type of measure is most appropriate generally depends on the research question being addressed.

9.3.3 Study Designs and Data Analysis

Collection of raw data may be attuned to essentially three kinds of data analysis. First there is descriptive analysis, which yields qualitative conclusions about what has been observed, what subjects have reported about themselves or the behaviour they have displayed and how often. In the second place, conclusions from systematically designed experiments are frequently obtained using analysis of variance. Third, there is correlational analyses, whereby conclusions about the relationships among measured variables are obtained using correlation plots and matrices, multiple regression analysis, factor analysis and other forms of relational analysis.

9.3.3.1 Descriptive Analysis

In some cases a researcher's sole interest is to report on the responses occurring in a particular setting, according to what he or she takes to be an accurate description of reality. For example, one may wish to describe the street-crossing behaviour of pedestrians, in an attempt to capture the conditions and strategies characteristic of different kinds of people. Generally, descriptive analysis implies that one does not need or want to infer causality or association. Descriptive methods are particularly suitable for studying 'new' phenomena, as when the behavioural patterns occurring in a natural setting must first be characterised before some kind of theoretical model can be developed and more sophisticated methods applied to identify underlying mechanisms and causes.

Descriptive analysis is naturally coupled to the observational data discussed in Section 9.3.2. The main requirement of descriptive methods is that the raw data be valid, i.e., adequately reflect what they are supposed to measure, as well as reliable, i.e., recur when data collection is replicated. Further data analysis based on systematic observations in realistic settings depends on the type of data eventually available after the coding of raw materials. This is the simplest form of data analysis, permitting qualitative description. Also, group differences in the variables of interest may be explored. For example, one may examine differences between older and younger pedestrians in total street-crossing duration. Alternatively, the researcher can report how many people of different genders and age groups have crossed the road.

9.3.3.2 Experimental Analysis

The ideal way to test a scientific theory is to conduct a series of controlled experiments in which a limited number of independent variables are systematically varied to measure the effect on one or more dependent variables. For example, one may wish to examine the extent to which an information campaign influences the environmental awareness and littering behaviour of park visitors. At least two levels of an independent variable are usually distinguished, here perhaps: 'no information', 'encouraging information' and 'discouraging information'. Similarly, the dependent variable(s) should be clearly defined and relevant as something at least potentially affected by the information provided.

Because they require tightly controlled conditions, experimental methods are typically employed in a laboratory setting. If carefully designed, though, field experiments in real-life settings are well feasible, provided the researcher ensures that it is only the independent variable(s) that differ(s) between experiments, with all other possibly relevant factors remaining constant (or comparable). It is essential to avoid such 'confounding' of study variables if unambiguous research conclusions are to be later drawn. This also implies that subjects should be randomly assigned to the different experimental conditions, to avoid any specific association between experimental condition and group of subjects. If other (situational or individual) factors were also to vary across conditions, it would no longer be possible to unravel the causes of any observed differences in the dependent variables. Because of their special character, experimental methods are very useful for identifying the specific causes of an observed phenomenon.

Experimental analysis has three important drawbacks. First, the strict experimental controls may create artificial settings that are not representative of real life. Second, because of the high precision and specificity required, a single series of experiments can never encompass more than a handful of independent variables and a similar number of dependent variables, perhaps a few more. This limits the scope for investigating complex phenomena in multivariate settings. The third drawback is that experimentation is often only suitable for investigating short-term effects, precluding the study of long-term, dynamic phenomena in which subjects are continually adapting to changing circumstances. One response to these problems encountered in traditional experimentation lies in the development of computer simulation, using programs based on behavioural theory, as discussed below.

9.3.3.3 Correlational Analysis

In this form of analysis the aim is to identify any correlations between the variables one has been able to measure on at least an ordinal, but preferably an interval or ratio scale. For example, one may want to know whether elevator use in office buildings is related to people's awareness of environmental problems. These two variables obviously cannot be studied in isolation, for a range of other variables, such as the subject's physical condition or age, may affect both. In most cases, correlational analysis does not permit causal inferences to be made, because there can be no certainty which variable is the antecedent and which the consequence. Suppose a significant negative correlation is observed, then do people often use elevators because they are environmentally relatively unaware, or are they environmentally unaware because they often use elevators? Given these limitations, correlational analysis usually cannot provide critical tests of behavioural theories. However, in some cases causality is crystal clear. For example, crime rates appear to be related to ambient temperature. Here, causality is obvious, as it is not likely that crime rate would affect ambient temperature.

Despite this drawback, correlational analysis methods provide scientists studying environmental behaviour with an important tool. In many cases it is impossible or unethical to manipulate the environmental conditions whose influence one wishes to study, which would be necessary for examining causal relationships. One could nonetheless study their mutual influences. For example, it is hardly feasible to assess the effectiveness of a disincentive for car use involving a fuel price rise of 50% for one group and 100% for another simply by implementing it and recording transportation behaviour over an extended period. In such cases, behaviour must be studied in real-life situations, if authentic and valid raw data are to be obtained. For example, one may compare the level of car use between countries in which fuel costs vary significantly. However, such data reflect only limited policy conditions and generalising the results to other situations may be problematical, as the behaviour may also have been influenced by other factors.

9.3.3.4 Computer Support

In environmental research computers can be a great aid for tackling a variety of problems, and not only in raw data analysis. Data collection involves three basic tasks: (1) preparation of presentable materials, (2) selection of the statistical population to be studied, and (3) recording the subjects' responses. With each of these tasks the computer can help.

First, computer programs can be used to design and present materials to elicit responses from subjects of interest. For example, preferences for urban park design might be examined by asking subjects to evaluate a set of pictures, colour slides or videos in which important features like trees, water or buildings are varied systematically. Computer programming of stimulus composition can enable one to generate a wide range of possible variations that can be readily presented to subjects. Equipped with the findings, researchers can then identify the factors having greatest influence on preferences.

The second area in which computers can fruitfully be used is for recording responses, particularly in multivariate laboratory studies where a range of specific behavioural data are being collected and filed for later analysis. In environmental field studies, too, computers may be a very convenient means of administering a questionnaire or personal interview.

A third use of computers is aimed at simulating human behaviour, as a surrogate for 'real' respondents. This research strategy combines theoretical

Table 9.2 Overview of behavioural science methods and principles, with example items

Types of measurement scale	*Nominal*: types of park visitor
	Ordinal: preferential ordering of cycle, car and rail transport
	Interval: elevator energy use by floor; environmental attitude ratings
	Ratio: weight of cars in kilograms, distance in kilometres
Behavioural measures	*Observational*: observers' descriptions, counts, ratings
	Self-report: via questionnaires, interviews
	Performance: model-behaviour recordings, physical stress measures
Designs and data analysis	*Descriptive analysis*: coding, categorisation, counting
	Experimental analysis: test of significant effects, causal relationships
	Correlational analysis: patterns of interrelationship
	Computer simulation of behavioural processes

modelling, (simulated) behavioural processing and data analysis. In scientifically studying consumer environmental behaviour, for example, there may be a need to go beyond controlled experiments with human subjects, because of the multitude and complexity of factors and time scales involved. Proceeding from a theoretical set of behavioural processes (see Section 9.7.3 below) a set of programming algorithms can be designed to simulate consumers under different configurations of behavioural determinants (examples are provided by Jager and Mosler 2007). This work has demonstrated that behavioural theory can be used to develop computer software that can rapidly (virtually on command) yield highly relevant conclusions about the fate of a common resource when artificial consumers use that resource under a range of different conditions.

To summarise the main approaches, methods and principles used in behavioural science research, Table 9.2 provides an overview of what we have discussed so far.

9.3.4 How to Design and Conduct an Empirical Study

To conclude this section on general methodology, let us briefly examine how a specific environmental problem might be empirically studied using the methods of behavioural science. Let us assume we are interested in why some visitors to wildlife parks show little sensitivity to the animal life to be found there, which visitors show least sensitivity in this regard, and how the situation might be improved. Our working hypothesis might be that 'insensitive' park visitors typically have a long-standing urban background and little knowledge or experience of the conditions conducive to wildlife. An empirical study might then be designed as follows:

1. Formulation of study rationale. What is the precise problem being researched? How can the problem be theoretically modelled? What issues are driving the research? Can any prior hypotheses be formulated based on prevalent theories and tested?
2. *Design of measuring instrument(s)*. This involves specifying the raw data you need to collect and the methods and procedures to obtain them. The chosen means of data collection should be well-organised, of clear content and properly tailored to the data sources being targeted, so that the 'ins and outs' of actual data collection are familiar from the very outset. Pre-testing of study materials and measuring instruments is essential and can be done with the help of a limited number of trial subjects.
3. *Design of subject categories and groupings*. Knowing what data to collect still gives no indication as to where that data should be collected. The significance of the raw data and the eventual study conclusions are determined just as much by the means of measurement adopted by the researcher as by the sample of subjects chosen for study. So, be explicit about how many and what kinds of respondents should be involved and anticipate what their responses may mean to you, in terms of whether the results can be generalised, for example. Anticipate the formation of subgroups of subjects, so that due consideration can be given to possibly relevant differences in response patterns. Be clear about the subject variable(s) on which you intend to draw general conclusions.
4. *Practical preparations for data collection*. Study subjects, once identified, need to be conscientiously approached with an announcement of the study and an invitation to participate. Depending on the nature of the research and the number of respondents envisaged, careful thought may have to be given to how respondents are contacted, appointments made and practical sessions scheduled and monitored. If interview assistants are involved, their role should be clearly specified and they should be given appropriate prior training and careful instructions on how to behave towards interview subjects.
5. *Actual data collection and validation*. This often entails the researcher going out 'into the field' to obtain raw data, presenting human subjects with study materials and tasks. If such research is thoroughly prepared and well planned, actual implementation may take (far) less time than the preliminary work of study design. Obviously, subjects should be treated with proper respect and be given an opportunity to validate the data they provide before it is filed for further processing and analysis.
6. *Data analyses*. This may be another rather time-consuming phase of the research. As methodological expertise is gained, however, and with a certain anticipation of data analysis during the planning phase, one can efficiently carry out a varied sequence of data analyses to answer the research questions and test the hypotheses. Further data analyses can be performed later on, to answer supplementary questions about the meaning of study outcomes.
7. *Draft reporting, report revision and finalisation*. This stage speaks for itself. However, researchers should not underestimate the time required to wrest from the data analyses a comprehensive picture of what the study has yielded. At first sight, the analyses may give a chaotic idea of what there is to be learned. Combining the results of data analysis with the original problem analysis and some (possibly additional) theorising may eventually permit a consistent as well as convincing set of conclusions and interpretations to be formulated. This will not succeed immediately, however. Re-reading and revising one's text as well as additional writing may be necessary to achieve the desired, satisfactory end result.

It is always important to establish beforehand what exactly one can and needs to measure, and avoid an overly 'rich' database that later allows for all kinds of data analyses that may or may not provide convincing answers to the original research questions. If the researcher is to keep on track and achieve his or her stated goals it is therefore essential that the procedures adopted for data collection and data analysis are strictly in line with the research questions originally formulated.

Having completed this basic review of research methodology, let us now turn to the substance of behavioural science perspectives, studies and findings with respect to environmental problems. Below, we first discuss the study of environmental impacts on human well-being (Section 9.4) and the study of environmentally harmful behaviours (Section 9.5). We then provide a more general view of human behavioural interactions with the environment, taking as our point of departure the 'commons dilemma' paradigm discussed in Section 9.6.

9.4 Studying Environmental Impacts on Human Well-Being

Environmental conditions can significantly affect the well-being and functioning of human individuals and societies. Environmental deterioration and its consequences may therefore provide solid reasons for policy intervention. The fundamental environmental qualities encompassed by climate, soil fertility, water resources, mountains and wilderness areas may play an important role in economic development, security and wealth. More specific qualities may also significantly influence human well-being and behaviour, either positively or negatively. The mediating variables here include stimulation, arousal, sensation, stress and anxiety on the one hand, and comfort, relaxation and restoration on the other. Here again, the measured variables are meaningful only to the extent that they derive from clear definitions and conceptual models of the phenomena under study. As an illustration, let us consider practical measurement of three concepts: environmental stress, perceived risk and appreciation of nature, which are of central importance for assessing the human side of variations in environmental quality. In addition, contingent valuation is explained as a method for valuing common (environmental) goods.

9.4.1 Environmental Stress and Protection Motivation

Environmental degradation manifests itself by way of a variety of so-called 'environmental stressors'. Key examples are air pollution, malodour, noise, crowding, risk of accident and loss of nature and biodiversity. These may be either toxic or harmful (yielding health effects), threatening (reason for worry) or annoying (as an ambient condition), and they may occur at varying scale levels, from local to global. Health psychologists find it useful to distinguish between life events (disasters), risks (of accidents) and daily hassles (e.g., a noisy work environment).

Stress arises when human capabilities are overtaxed. The human stress and annoyance arising from environmental conditions may be conceived as resulting from a discrepancy between environmental demands on human functioning (e.g., being distracted by malodorous air or intrusive noise) and the subject's ability to cope with such a demanding environment (e.g., having to live with the stressor). Human stress responses may manifest themselves physiologically (e.g., in heart rate variability or skin conductance), emotionally (e.g., in worries, anger, frustration) and/or behaviourally (e.g., in resignation, aggression, or functional counteraction). Coping depends on the size (and sign) of the discrepancy between environmental demands and personal capacities.

As an example, Fig. 9.2 shows a so-called transactional model of stress and coping with environmental *noise*. This model indicates that the balance between 'primary appraisal' (total annoyance) and 'secondary appraisal' (perceived control) underlies the kind of (negative or positive) activation experienced. This in turn influences subsequent coping behaviour, which may be problem-oriented, with the subject seeking to effectively reduce the environmental stressor, avoidance-oriented (possibly involving indifference, denial or helplessness), or emotion-oriented, with 'comforting cognitions' involving some kind of self-assurance that the stressor is not that bad after all. Through his or her actual coping responses the subject can change the stress situation in three basic ways: by reducing the external stressor (here: noise), by altering their primary appraisal (e.g., by reducing their vulnerability) and/or

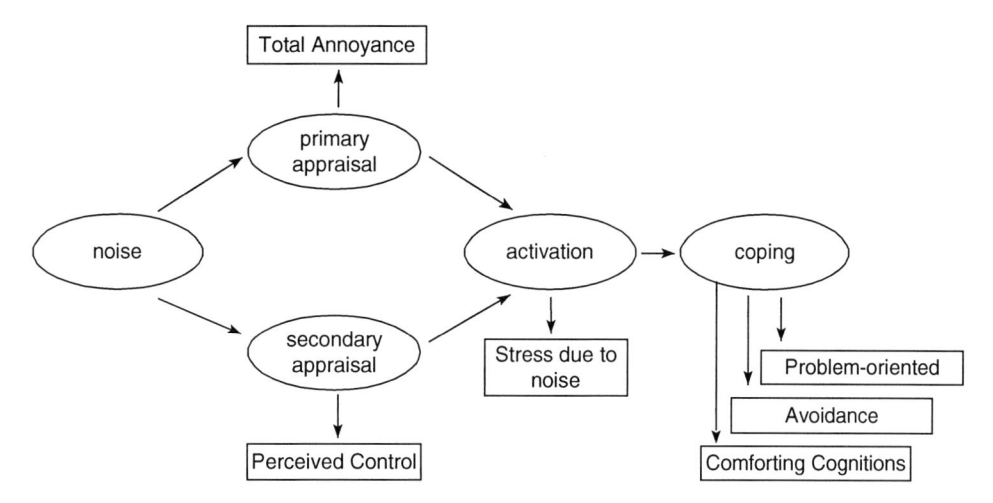

Fig. 9.2 Theoretical model of stress and coping due to environmental noise

by altering their secondary appraisal (e.g., by empowering themselves). The three relevant feedback loops are not shown in Fig. 9.2.

To demonstrate the practical applicability of this model one would need to:

- Measure the physical environmental conditions (e.g., the nature and level of noise)
- Assess the subject's primary appraisal of the stress factor: 'how burdening is this for me and what adaptive behaviour and/or acceptance does it demand?'
- Assess his/her secondary appraisal: 'how well and by means of what capacities am I able to meet these demands?'
- Assess the subject's coping response, i.e., any emotional responses (denial, anger, frustration) or problem-oriented actions (to alleviate, evade or avoid the stressor)
- Determine the degree to which such responses lead to adaptation or revision of primary and/or secondary appraisal and observe the further coping response pattern as it develops

Thus, to specifically validate the kind of stress-and-coping model shown in Fig. 9.2 in the concrete context of, e.g., air pollution, malodour, noise, crowding or accident risk, the researcher must carefully specify the relevant details of the environmental setting, the subject's primary and secondary appraisal, and his or her emotional and/or problem-oriented coping behaviour. In this way the model may be practically useful for explaining, predicting and possibly managing environmentally stressful situations.

Protection-motivation theory (PMT) is strongly related to the double-sided stress model just discussed. Following this individual-oriented framework any risk situation is approached by performing an external threat appraisal on the one hand (cf. the 'primary appraisal' above), and an appraisal of the subject's own coping potential on the other ('secondary appraisal' above). It is the balance of threat appraisal and coping appraisal that determines the direction and the extent of the subject's protection motivation: 'Actively trying to reduce the threat (or avoid it) or passively trying to live with it?'. Using Gardner and Stern's (2002, Chapter 9) adapted version of PMT, we have further adapted their model and also made it more explicit with respect to the temptation to accept the threat (because of benefits involved) and the motivation (not just the capacity) to cope with it, respectively. This further adapted version of PMT is shown in Fig. 9.3 which implies that the balance of threat appraisal and coping appraisal may lead to problem-focused and/or emotion-focused coping strategies.

As Fig. 9.3 shows, both threat appraisal and coping appraisal rest on several specific variables which for many people may be accompanied by specific feelings or emotions. Under threat appraisal, the severity of the threat and the subject's vulnerability are evaluated, along with the benefits associated to the threat; this is done against the background of the subject's current needs, values and experiences, and it yields feelings of anxiety and (some) temptation. Under coping appraisal, the subject judges general response efficacy ('could anything be done?'), his or her own self-efficacy ('what could I do myself'?) and the costs and barriers of appropriate action; this is done against the background of one's knowledge, skills and social support, and it yields feelings of perceived control and a certain motivation to act. A more detailed exposition of perceived risk (threat as well as coping) variables is given in the next section.

The essential message of this adapted PMT is that environmental risk management always has two sides. On the one hand, you may work on reducing the threat and the demands of the external environment which, however, may tempt you via associated benefits. On the other hand, you may work on reducing the vulnerability and enhancing the self-efficacy of the (potential) victims as well as overall response efficacy, which, however, may be costly and difficult.

Currently, there is only limited research about subjects' responses to large-scale and long-term (i.e.,

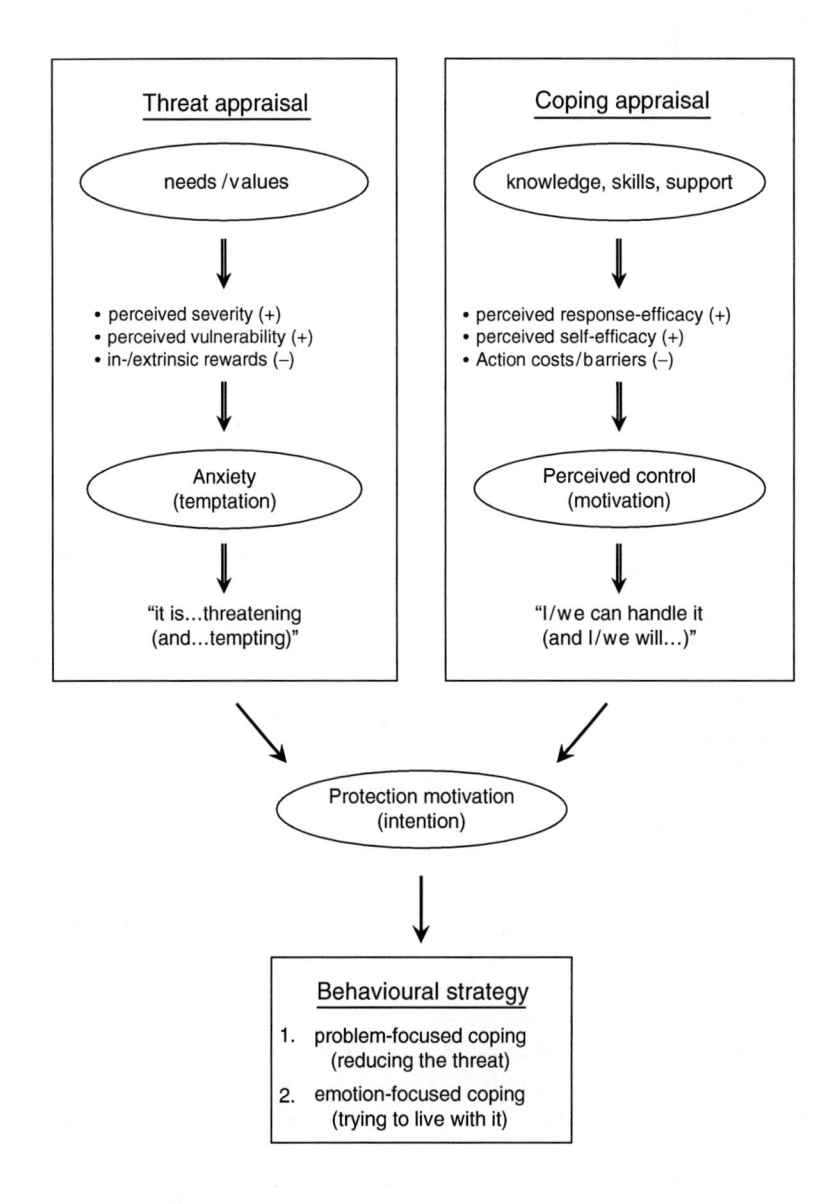

Fig. 9.3 Schematic representation of protection motivation theory

9.4.2 Environmental Risk Perception

Several decades ago, environmental risk (initially from industrial activities only) was defined as the probability of a major accident. Safety was considered to be equivalent to a sufficiently low probability of accident. The kind of accident and the processes leading up to and following accidents were of little relevance for permit decisions about major industrial activities such as nuclear power generation or road transport of hazardous chemicals. However, this limited, engineering view of risk analysis and acceptability changed when behavioural scientists entered the debate. Since then, behavioural science research on technological and environmental risk has taught us:

- That 'risk' is a multidimensional concept involving much more than just accident probability.
- That risk acceptance is highly dependent on the expected benefits and perceived 'controllability' of events, rather than on a maximum tolerable probability or maximum credible accident.
- That safety management is often a multi-stage and multi-actor affair, with several parties sharing responsibilities in both early and late stages of overall 'risk generation'.
- That external safety measures may be undermined by subjects' risk *homeostasis*, i.e., by subjects exploiting increased safety to intensify their activities and obtain greater benefits, thereby returning to the original risk level and accident frequency.
- That environmental risk management requires multi-party, participative procedures, because 'risk' cannot be sufficiently objectified and risk management requires contributions from multiple parties.

One notorious problem in the risk debate is the existence of many different conceptions of risk: in statistics, economics, psychology and cultural anthropology, for example. These differences revolve around two main concepts: 'probability' and 'risk'. The notion of probability may be approached in terms of frequency ('how often has this happened in the past?'), personal expectation ('how often do you believe this will happen in the future?') or even as a social construction ('what is our consensus about what might happen and how likely that is?'), according to which scientific discipline is being pursued. There is parallel divergence with respect to what is meant by 'risk', ranging from probability *per se* to a consideration of comprehensive decision problems where risk is a major factor but not the only one and perhaps not even a dominant one. For example, expected benefits, continuation of current practice and/or avoidance of social conflict may be more important.

The multidimensional character of the risk concept is represented in Table 9.3, which lists 11 dimensions of perceived riskiness that have emerged from various empirical studies.

The message of Table 9.3 is twofold. First, scoring a risky activity or situation along these dimensions may help understand (better) why a particular activity or situation is considered risky by some people but not by others; note that expected beneficiality (no. 9) mitigates perceived riskiness overall. Second, the respective dimensions suggest ways and means of rendering a risky activity or situation safer or *apparently* safer. Thus, the actual and/or apparent risk posed by a potential accident may be reduced by limiting its lethality, confining the possible physical and social damage, reducing its likelihood, enhancing its (perceived) controllability, expanding freedom of choice and/or increasing anticipated benefits.

Perceived risk may thus be measured by rating an activity or situation with reference to a specific set of concrete variables derived from the dimensions listed in Table 9.3. As the term indicates, 'perceived risk' is often in the eye of the beholder, who is the only one

Table 9.3 Basic dimensions of perceived riskiness

1	Potential degree of harm or fatality
2	Physical extent of damage (area affected)
3	Social extent of damage (number of people involved)
4	Time distribution of damage (immediate and/or delayed effects)
5	Probability of undesired consequences
6	Controllability (by self or trusted expert) of consequences
7	Experience with, familiarity, conceivability of consequences
8	Voluntariness of exposure (freedom of choice)
9	Clarity and importance of expected benefits
10	Social distribution of risks and benefits
11	Intentionality of harm

able to appraise the extent to which a given activity or situation is, to him or herself, familiar, controllable, dreadful, voluntary and suchlike.

Given the long list of Table 9.3, one prior note of warning is that multidimensional risk comparisons of such diverse affairs as LPG storage, electrical power generation and car driving, say, are very difficult to make. It is in fact advisable to restrict risk comparisons to activities or situations that can be meaningfully posited as alternative options within one and the same decision problem, as when assessing the perceived riskiness of electrical power generation using oil, coal, wind or nuclear energy.

9.4.3 Appreciation of Nature

Nature can be demanding and threatening to us as well as supportive and benign. Rewarding experiences in nature are essential to human well-being and health. Below, we discuss landscape evaluation and restorative nature experiences.

To measure 'human appreciation of nature' first of all requires definitions and some basic taxonomy of natural situations. Natural landscapes may be characterised as more or less extensive, mountainous, 'wet', vegetationally varied, rich in animal life, cultivated through human effort, and so on. The subjective counterparts of such variables depend on the evaluative judgements and actual behavioural patterns of different groups of subjects. The question is how relevant the 'objective' variables in fact are when it comes to people's own appreciation of landscapes.

The next step, then, is to study the perceptions and evaluations of relevant subject groups. Such appraisal is affected by variables relating to people's needs for understanding and explanation, and often also by a need to use a given natural landscape for certain ends. In empirical research subjects have been presented with colour photographs or slides which they were asked to evaluate in a general sense (e.g., by ranking or rating them according to 'beauty') and subsequently with respect to a number of specific landscape characteristics. Such research has shown that landscape complexity, mystery, legibility and coherence are important evaluative variables that serve as predictors of overall landscape attractiveness. It has also been found that most subjects consistently sort a wide variety of landscapes into artificial (or built) or natural environments, while indicating a general preference for the latter. When it comes to assessing or predicting the perceived 'usefulness' of a landscape, relevant action variables can be defined in terms of a landscape's suitability for such activities as farming, fishing, hunting, hiking and other recreational sports like camping, cycling and mountain-climbing. In this functional context of 'nature appreciation', exclusively economic activities such as mining and logging are left out of consideration. One may also record emotional responses to various landscapes, to assess the pleasure or arousal engendered by a given scene, for example.

As in research on environmental stress and risk (see above), here too the investigator is confronted with a difference in perspective between objective and subjective assessment. In seeking to explain such variables as 'scenic beauty', 'usefulness' (from the subject's perspective) and 'worthiness of preservation' (from a long-term policy perspective), then, one may either use objective (i.e., landscape) characteristics or subjective (i.e., perceivers' or users') appraisals, which may in practice be only loosely related. The problem of 'objective' versus 'subjective' variables for landscape evaluation helps us understand systematic differences in landscape perception and evaluation by different subject or user groups. Clarifying the structure of experts' as well as laypersons' judgements may help understand differences of opinion and may provide suggestions for resolving conflicts about landscape development.

A fundamental hypothesis regarding environmental effects on human behaviour posits a basic human 'biophilia', or 'friendship with nature', without which we would be unhappy, grow depressed or otherwise function in suboptimal fashion (Ulrich 1993). Formulated differently, human beings evolved as, and are thus, part of nature: they cannot live without it, and it would be unwise for human societies to seek to live in artificial, built environments in complete isolation from (other) nature. There is much empirical evidence for the conclusion that intimate contact with natural settings enhances human health. However, contact with nature should be safe and rewarding enough to prevent 'biophobia', borne of experiences that frighten and make people withdraw from nature or seek refuge in artificial settings. Throughout history, people have been confronted with the friendliness of nature at some times, and its harshness or hostility at others. To maintain, restore and further develop biodiversity around human settlements, it is important to know which kinds of nature (landscapes, vegetation, animal life) evoke biophilia, and which can induce biophobia.

Valid explanations and/or predictions of 'typical' human responses to various kinds of landscapes can be sought through rigorously designed experiments (see Section 9.3.3). Systematically configured landscapes are presented – in reality or virtually, in photographs or videos – to invited subjects from one or more relevant groups, for example, farmers, local villagers and urban tourists from further away. The respondents' perceptions of landscape characteristics and their evaluative judgements regarding scenic beauty and so on can then be studied, as well as their actual or imagined behaviours and the possibly emotional response a given landscape might elicit. Alternatively, these factors can be studied in real-life situations, as when schoolchildren spend summer camp in a wilderness area or 'nine to five' office workers go for an evening hike along a nearby river. Such experimental fieldwork is far from easy, however, because the independent variables (landscape characteristics) must be fairly tightly controlled, while the dependent judgemental and behavioural variables must be carefully specified and measured on the spot.

9.4.4 Evaluation of Common Goods

In Section 9.6 below, the common-resource dilemma is discussed as a paradigm for understanding conflict between individual and collective interests. In assigning a value to (environmental) goods, it is obviously essential to try and transcend a purely individual perspective, driven by personal interest or simple indifference. While purely individual goods can be readily valued on the basis of their economic, emotional and social significance to the subject, the meaning and value assigned to collective goods like a park, urban air quality or public streets are far more remote because these goods are shared, extend beyond one's 'own backyard' and often have a significance far exceeding an individual lifetime; see also Ekins (Chapter 11, this volume) on resource and environmental accounting, and on cost-benefit analysis. A crucial issue when it comes to managing any common resource, then, is how it is to be valued from a *social* (i.e., multiple interests) perspective that properly accounts for the different elements of its 'value'. The latter can be broken down into *present* (market) value, *future* (option) value, and long-term *existence* value (independent of human economic and/or social use).

There are several approaches towards assigning a value to common goods. One is a direct economic approach, which in fact applies only to goods that can in principle be traded. For example, the overall value of a zoological garden or museum can be defined as the annual number of visitors multiplied by the average price of an admission ticket. However, this ignores the fact that long-term preservation of the collective good might well require additional public funding for renovation and possible expansion.

An alternative, indirect economic method, useful for valuing freely accessible goods like nature reserves, mountains and beaches, is to multiply the annual number of visitors by their average expenditure on getting to the area and enjoying it, i.e., the cost of travel and accommodation and other expenses specifically attributable to the visitor's 'consumption' of the relevant good.

To assess the value of non-tradable and non-'costable' common goods such as a coastal island chain, a mountain range or an historical rural landscape, the method of *contingent valuation* is available. In this approach, the common good in question must first be strictly delineated and described. Then one or more samples of respondents are invited to state, as a citizen, the highest amount of money they would be willing to pay to obtain, preserve or restore the common good. The results are then generalised across the relevant population of citizens, and the sum total of amounts offered is taken as the 'contingent value' of the good. The basic (economic) idea, of course, is that a common (environmental) good cannot be deemed valuable if the relevant population of citizens proves unwilling to pay much, if anything, for its maintenance or acquisition. Contingent valuation is not as easy as it may seem. Table 9.4 lists the main steps followed in using the method.

The contingent valuation approach has been criticised for resting on the assumption that respondents can grasp the full meaning of a common (environmental) good and provide answers reflecting what they would actually do, financially speaking, if the relevant environmental policy were indeed to be implemented.

Table 9.4 Methodological steps in contingent valuation of common environmental goods

1. Delineate and circumscribe the relevant environmental good (location, size, conditions, temporal extent)
2. Identify and clarify assumptions about intended expenditures, expected cooperation and time perspective
3. Indicate and define the relevant population of citizens (users, visitors, people concerned)
4. Select samples of citizens and invite respondents for questionnaire, interview or focus group activity
5. Ensure that respondents can grasp the extent, meaning and importance of the good (various aspects)
6. Define and explain the method and procedure; collect data on respondents' background
7. Explain how citizen responses would translate into policy effectiveness (e.g. 'higher value = greater protection')
8. Instruct respondents to consider and evaluate the environmental good from a social, not a personal perspective
9. Create and supervise conditions for serious responding to the questions asked
10. Conduct data analysis taking account of the purposes and assumptions of the method, and of weaknesses in the raw data
11. Make a compact, comprehensible summary of results and present these along with clear conclusions

General reservations have thus been expressed about the validity and reliability of respondents' judgements vis-à-vis common goods.

From a specifically psychological perspective, too, there has been criticism of the very concept of 'willingness to pay' (or WTP) as representing no more than a statement of *attitude*, driven by people's desire for moral satisfaction. As such, the attitude expressed would be a measure of people's awareness of environmental problems and their willingness to donate to the common good rather than purchase their part to help enhance or preserve it. The results yielded by contingent valuation are sensitive to the method employed. For example, asking people to pay (WTP) for preserving a particular good yields a lower figure than asking them how much they would want to be compensated for loss of the same good (WTA: 'willingness to accept'). This can be explained by people's greater sensitivity to losses (WTA) than to comparable gains (WTP).

Having devoted extensive empirical research to the issue, psychologists Ritov and Kahneman (1997) came to the following five conclusions:

1. Different measures of attitude tend to yield very similar rankings of (environmental) issues.
2. Contrary to a (consequentialist) model of economic value, measures of attitude, including willingness to pay, towards public issues are determined mainly by the severity of the problems, not by the nature of proposed interventions.
3. Extensional aspects, regarding size or 'numerosity' of the good, have little effect on evaluation.
4. Ecological problems are seen as less important than human-health problems.
5. The true importance of ecological problems tends to be overestimated if considered without reference to problems of human welfare.

The last of these conclusions echoes the three-dimensional concept of sustainable development as encompassing economic wealth and social well-being, alongside environmental quality, discussed in Section 9.10.2.

9.4.4.1 Multi-attribute E.valuation

Contingent valuation need not be as straightforward as simply stating willingness to pay for a common good. From scientific decision analysis we can adopt the well-established method of *multi-attribute utility assessment*, also often used in everyday life for ranking a limited number of dwellings, transport vehicles or job candidates in order of preference. The basic idea here is that the assessor identifies relevant value attributes, rates each alternative option with respect to each attribute, then weights the attributes according to their relative importance, and finally compares the weighted, summed rating of each alternative. In this way all the alternatives under consideration can be ranked in order of preference. Table 9.5 lists the main steps of a multi-attribute evaluation procedure, which may be useful in managing problems relating to common environmental goods.

Multi-attribute evaluation is not free of some of the problems of principle and method discussed above. Utility judgements may co-vary with attribute weights, the weights themselves may be prejudiced by the preferred outcome, and ranking may in practice be determined by just two or three overriding attributes, depending on the subject's value orientation or basic beliefs (see Section 9.9.1); compensatory weighting of attributes may only occur when subjects are highly involved. Nevertheless, the overall approach helps respondents elaborate on the common

Table 9.5 Main steps in multi-attribute utility evaluation of common goods

1. Assess the policy (decision) problem and the wider context of evaluation
2. Establish whose (or: which groups') multi-attribute utility is to be determined
3. Delineate the set of alternatives available
4. Inventory the relevant value attributes; select an essential, non-redundant set for practical use
5. Rank value attributes according to importance and assign (normalised) ratio-scale weightings
6. Score or rate each alternative with respect to each value attribute
7. Aggregate the weighted scores/ratings for each alternative and rank them
8. Perform sensitivity analysis by varying ratings, weights and/or attributes, and recompute overall ranking

Note: In sensitivity analysis (no. 8) one varies elements of the set of attributes, their weighting and/or the rating of alternatives in order to assess corresponding changes in the computed final ranking.

good presented to them, appreciate that its value or utility depends on various elements and/or attributes, and ensure their overall judgement is deliberately constructed on the basis of attributes, weights and ratings. Moreover, subjects may learn to acknowledge that attributes, weights and ratings may each be somewhat arbitrary or unstable, so that sensitivity analysis (the decision analyst's mightiest weapon) can help improve the subject's insight into what serious evaluation actually involves. Differences in preference among stakeholder groups can be better understood, furthermore, for the considerations underlying preferential ordering are rendered explicit.

Multi-attribute utility evaluation brings us into the sphere of complex decision making. Cost-benefit analysis (CBA) of alternative options or policies is a well-known economic approach here. CBA entails a description and monetary evaluation of all expected costs and benefits of any major possible course of action, and a cost-benefit comparison of available alternatives in order to identify the option with the highest benefit-cost ratio. CBA and other economic evaluation methods are further discussed by Ekins (Chapter 11, this volume). For environmental decision making, however, CBA has several weaknesses. It is often difficult, for a start, to estimate the likely environmental effects and side-effects of policy measures. The social and economic benefits of such environmental effects and side-effects may be hard to assess in themselves, and their monetary evaluation may be rather artificial

or plainly impossible. Trade-offs between short-term costs and long-term benefits may be controversial (should long-term, collective consequences be discounted, and by what rate?). And the social distribution of policy costs and benefits may be hard to describe, let alone to evaluate.

9.5 Studying Behaviours Affecting Environmental Qualities

Let us define 'environmental behaviour' as any form of human activity having either a positive or a negative effect on environmental sources, sinks and/or ecosystems.

9.5.1 Categorising Environmental Behaviour

Behavioural and social science research to date has sought to clarify individual environmental behaviour as well as the role of institutions and culture in shaping that behaviour. Here we focus on individual behaviour and how it is embedded in the social fabric. Sustainable patterns of behaviour are often characterised with reference to the 'five Rs': *reduce, repair, reuse, recycle* and *redesign*. For example, the environmental impact of consumerist lifestyles can be decreased by reducing energy use, repairing appliances, reusing products (perhaps for other purposes), recycling waste materials and redesigning production processes. There is obviously a huge diversity of forms of 'environmental behaviour', with very different impacts. Refusing plastic bags in stores or buying recycled paper have far less environmental impact than cutting back on car or air travel. Given this diversity, various attempts at classification have been undertaken. Our exposition below is based on three separate dimensions.

First, we can distinguish between *intent*-oriented definitions of behaviour, which focus on what people do to benefit the environment (e.g., buying wholefood or recycling), and *impact*-oriented definitions, which focus on specific environmental impacts of behaviour, cutting energy and/or materials consumption or waste production, for example. While orientation towards action provides no guarantees of environmental sustainability (for lack of sufficient knowledge about actual effects), impact orientation leaves the relevant behaviour unspecified (possibly for lack of sufficient knowledge about actual sources).

A second distinction is between curtailment behaviours and efficiency (decision) behaviours. Curtailment generally implies limiting the use of existing appliances, by turning down heater thermostats, for example, or reducing car use. Efficiency behaviours include such actions as switching to a more fuel-efficient car or energy-efficient domestic central heating system. Curtailment behaviours typically imply repetitive action and a corresponding alertness or newly established habit. Such behaviours are often associated with increased effort and/or reduced comfort. Efficiency behaviours generally call for initial investments but may in the long term save costs. Efficiency behaviours are generally quite effective in reducing environmental impact, and such everyday efficiency gains require little further attention.

The third distinction is between forms of behaviour having a direct or proximal environmental impact and those that influence environmental qualities indirectly, by shaping the context in which choices with a direct environmental impact are made. Examples of the former are household waste disposal and car-driving, while the latter include acceptance of higher energy taxes, voting for a green party candidate, petitioning on environmental issues or donating to environmental protection organisations.

The message of this threefold taxonomy is that it is important to properly account for the varying dimensions of environmental behaviour, each of which is affected differently by relevant psychological, sociodemographic and situational factors. Focusing on intentions is different from thinking about impacts. For example, recycling (intention) is more strongly related to environmental attitudes, while energy use (impact) is more strongly related to household income. Repetitive curtailment behaviour differs from one-time efficiency (decision) behaviour. For example, the latter is more acceptable to high income groups than to low income groups. Likewise, changing behaviours with a direct environmental impact is not the same as changing behaviours of only indirect relevance. For example, environmental attitudes may correlate with environmental organisation membership but not with car use or air travel.

9.5.2 Measuring Intent- and Impact-Oriented Behaviour

Besides studying specific issues such as recycling or private vehicle usage, researchers have also sought to identify coherent patterns of environmental behaviour. A comprehensive measure of environmental behaviour can reveal whether people behave more (or less) pro-environmentally across a range of different behaviours, and consequently, what environmentally sound behavioural patterns entail. Moreover, a comprehensive yardstick would enable researchers to examine any common antecedents of environmentally (un)sound behavioural patterns instead of studying each behaviour separately. Such studies may reveal interventions that might successfully move people to behave pro-environmentally, by addressing key antecedents of a more encompassing behavioural pattern. Aggregate measures of environmental behaviour can be defined from either an intent-oriented or impact-oriented perspective. The feasibility of each of these perspectives depends on the particular research questions being addressed.

9.5.2.1 Intent-Oriented Measures of Behaviour Patterns

These are especially useful for examining what intentional actions people undertake to protect or improve environmental quality, why some people act more pro-environmentally than others, and what can be done to persuade people to behave more pro-environmentally. Intent-oriented measures can be included in questionnaires or interview studies in which respondents indicate how often they perform specific environmental behaviours, such as buying wholefood or recycling waste materials. Such questionnaires often also include questions on possible antecedents of such behaviours. Aggregate intent-oriented measures are typically based on observed correlations between behaviour frequencies or intensities, from an individual consumer's point of view. For example, if someone often returns plastic bags to store counters, turns off lights in unused rooms and closes all the curtains in the evening, their pattern of behaviour can be rated 'pro-environmental'. Aggregate measures can be constructed by means of factor analysis identifying behavioural clusters which reveal the extent to which respondents often performing behaviour A are also likely to perform behaviour B but not behaviour C, and so on. Strongly correlated behaviours belong to the same environmental-behaviour cluster (e.g., recycling), while low-correlating behaviours may belong to a different cluster (e.g., buying whole food).

Empirical research has shown that most people are fairly inconsistent in their environmental behaviour. People may be pro-environmental in some areas of their life (recycling of household waste, for example), but certainly not in others (such as transportation). The explanation for this may lie in the fact that factors determining behaviour extend far beyond environmental attitudes and vary substantially from domain to domain of human activity. For example, people may also consider how much effort is involved, whether others would approve of the behaviour, or how costly the behaviour is. Also, feasible pro-environmental behavioural options should be available.

The general problem with intent-oriented measures of environmental behaviour is that they can reveal much about people's intentions, motives and attitudes towards the environment, but little about the actual environmental impacts of their behaviour. For example, many studies conclude that someone who performs four out of five low-impact behaviours is more environment-friendly than a person who performs only two of them. However, such a conclusion need not be valid if the environmental impacts of the five behaviours differ to any great degree.

9.5.2.2 Impact-Oriented Measures of Behaviour Patterns

These measures enable researchers to identify behaviours associated with a significant environmental impact and examine ways of reducing that impact. For example, a researcher may wish to measure the total amount of electricity or water that people use or their total car mileage from meter records or readings. From an environmental point of view, such measurements are highly relevant because they provide a ready indication of the actual environmental impact of domestic behavioural patterns. However, this kind of measure does not reveal which specific behaviours contribute most, under what specific circumstances, to energy-extensive lifestyles, nor indicate the key behavioural differences between households with high and low total energy or water consumption. From the point of view of public education this may be problematical, for people do not know what or whose changes in behaviour, under what circumstances, hold out greatest promise. Moreover, in the case of experiments, people cannot be provided with specific feedback on the results of changes in their behaviour.

9.5.2.3 Intent and Impact Combined

To overcome the problems of both intent- and impact-oriented measures of environmental behaviour, more sophisticated 'intent-impact' measures are being developed that consider specific behaviours, their determinants and their environmental effects in combination. These measures are based on an aggregation of environmental impacts across a set of behaviours. One example is computation of total household or individual energy use. For this purpose, an interview or questionnaire is administered in which respondents first indicate which household appliances, goods and/or vehicles they own and how often they use them. Next, the researcher estimates the direct energy requirements (electricity, natural gas and car fuels) of the various goods and activities, as well as the indirect energy requirements for production, distribution and disposal of the goods in question. Finally, the sum of all specific energy requirements is computed over the total set of goods and services owned or used.

Depending on the set of goods and appliances and uses covered, the resulting measure of energy intensity may be a reasonably valid proxy for total household energy consumption, good enough at least for investigating differences between households and relationships with relevant antecedent conditions. In this approach, the energy intensity of different household activities (e.g., lighting, heating, washing or cleaning) can be assessed and possibly explained in order to improve understanding of how environmental impacts and pro-environmental behaviour come about.

9.6 The Commons Dilemma as a General Model of Environmental Exploitation

The 'commons dilemma' paradigm is a systematic and comprehensive way of thinking about the generation and management of environmental risks via the external effects of many individuals' activities and behaviours. The paradigm yields useful concepts, models and methods for environmental problem diagnosis, policy decision-making and actual intervention and evaluation in terms of individual behavioural change to preserve and enhance common goods.

9.6.1 Definition of the Commons Dilemma

A commons dilemma[1] is a situation of conflict between an aggregate collective interest and numerous individual interests. In pursuing their own personal interests, many individuals tend to shift the (possibly limited) negative impact of their behaviour onto their common environment. The cumulative effect of these numerous small impacts may be disastrous, however, leading to a serious deterioration of collective environmental qualities. There are four essential elements of any commons dilemma:

1. Each individual receives a higher pay-off for a socially 'defective' (i.e., environmentally harmful) choice than for a 'cooperative' choice, no matter what other individuals in a society do.
2. Compared to a cooperative choice, a defective choice always detracts from others' benefits.
3. All individuals are better off if they all cooperate than if they all defect.
4. The outcomes of individual behavioural choices are also governed by the choices made by others.

An example here is the urban air pollution resulting from accumulation of exhaust fumes from numerous individual cars. In the short term each individual may be better off driving a car than taking the bus, because the former option is more comfortable and flexible. However, if everyone drives his car, environmental quality will deteriorate in the longer term. All individuals would be better off if everyone reduced their car use 'now'. However, the actual outcomes of many individual behavioural choices are (also) governed by the choices made by others. For any particular individual, reducing car use is worthwhile to the extent that others do the same. Thus, a lack of trust in others' cooperation may lock individuals into favouring self-interested choices and using their cars. This kind of socially driven process of environmental decline often proceeds gradually, in contrast to the immediacy of the benefits to individual polluters or resource users. It is the combined occurrence of the negative external effects of many individual actions, either immediate or delayed, which constitutes a collective cost or risk.

9.6.2 Twelve Key Issues for Researchers and Policy-Makers

Given that a commons dilemma by definition exceeds the physical, cognitive and motivational scope of individual actors at the micro-level of society, the basic question is how the collective environmental cost or risk can be validly assessed, effectively communicated and acceptably managed so as to meet due quality criteria. Environmental risk management is a matter of decision-making about the acceptability of collective risk and of practical interventions to control that risk by changing the behaviour of the contributing individuals. Risk management is most effective when it dovetails with policy-makers' diagnosis of the processes creating or exacerbating the risk in question. Monitoring and evaluation are also essential, so that the impact and knock-on effects of policies can be fed back to relevant actors and policies can be revised when necessary. Thus, understanding commons dilemmas and managing the collective (environmental and social) risks generated by individual activities revolves around problem diagnosis, policy decision-making, practical intervention and evaluation of policy effectiveness. Together, these can be elaborated into 12 focal issues, listed in Table 9.6.

The four divisions of Table 9.6 embrace the key problems involved in understanding and managing common resource dilemmas: appreciating the collective risk, weighing the risk against (total) individual benefits and deciding on a course of action, promoting behavioural change and evaluating the actual consequences of policy interventions.

---

[1] The term 'commons dilemma' goes back to Hardin's (1968) inspiring analysis of the disastrous consequences of common pasture land ('the commons') being overgrazed by ever more cattle owned by individual farmers trying to improve their individual income.

Table 9.6 Twelve key issues for research and policy-making on commons dilemmas

I. Problem diagnosis	Analysis and assessment of collective risk, annoyance and stress
	Analysis and understanding of socio-behavioural factors and processes underlying risk generation
	Assessing problem awareness, risk appraisal and actors' individual values and benefits
II. Policy decision-making	Weighing of collective risk against total individual benefits ('need for change?')
	If 'risk unacceptable': setting objectives for reducing environmental and/or social risks
	Translation of risk reduction objectives into individual behavioural goals
III. Practical intervention	Focusing on individual target groups and considering essential conditions for policy effectiveness
	Specifying feasible behavioural alternatives and selecting effective policy instruments
	Target group-oriented application of strategic programme of behavioural change
IV. Evaluation of effectiveness	Designing a monitoring and evaluation programme to determine policy effectiveness
	Systematic, comprehensive evaluation of observable effects and side-effects
	Intermittent and *post hoc* policy feedback, and possible revision of policies

9.6.3 Macro, Meso and Micro Perspectives

Real-life commons dilemmas often involve several different types of actors, each prone to different perceptions, motivations and responsibilities with regard to the external effects that together accumulate into collective risks. Let us consider three different actor perspectives, at the macro, meso and micro level of society.

9.6.3.1 The Macro Level or Government Perspective

For a political authority at the macro level of society, a commons dilemma – if recognised as such – presents itself as a permanent contrast between a (changing) collective risk, such as air pollution, and a large collection of (changing) individual benefits, from car driving, for example. Minimisation of risk and maximisation of benefits are incompatible social goals between which a trade-off must be made. Or, more concretely, one cannot have clean air *and* unlimited motorised traffic, produce butter *and* guns, or maximise individual freedom *and* social equality. In many domains a balance must be struck. However, the political authority may be tempted to do little or nothing about the collective risk, being elected to office for only a limited term, possibly by voters from the very population of individuals whose behavioural freedom is to be checked. In addition, the political authority may itself have an interest in continuing the collectively risky activity, due to considerations of tax revenue, employment or ownership of public infrastructure, for example. Motorised transport may serve as an example.

9.6.3.2 The Perspective of Meso-level Organisations

Representing the meso level of a commons dilemma are groups and organisations, such as consumer groups and businesses, and their governing bodies. These may well play a double role. On the one hand, they may defend and promote the individual interests of their micro-level constituency by protecting (often acquired) individual rights to shift externalities onto the collective, or even by demanding yet more scope in this respect. On the other hand, organisational leaders may support macro-level authorities in the agenda-setting of environmental problems and in applying strategies to reduce collective risks through individual behavioural change.

9.6.3.3 The Micro Level or Citizen Perspective

The individual actor's perspective at the micro level of society may be focused on his or her own benefits 'here and now'. There are several reasons why individuals may not recognise a commons dilemma as such:

- Because they are unaware of any collective damage
- Because they do not appreciate their own responsibility for collective problems
- Because they do not feel the long-term collective risk is serious enough in relation to the numerous short-term benefits and/or

- Because they know there is a collective risk but feel little can be done about it (because of a lack of feasible alternatives and/or a lack of trust in others' cooperation)

The combination of personal benefit, 'short-sightedness' and perceived lack of control over the situation, in particular, may readily lead to a denial or belittling of the risk and thus to a refusal to participate in risk control efforts. Factors facilitating cooperation are discussed in more detail below. Section 9.7 provides a more elaborate framework for diagnosing and modifying environmentally harmful behaviours.

9.6.4 Social and Psychological Factors Governing 'Cooperation' and 'Defection'

A multitude of experiments and several field surveys have been conducted to lay bare the factors governing people's tendency to either cooperate or 'defect' (i.e., behave selfishly) in a commons dilemma situation.[2] For the sake of convenience a distinction is made between structural (environmental and/or social) factors and individual psychological factors.

9.6.4.1 Structural Factors Affecting Cooperation

People involved in a conflict between personal and collective interests tend to cooperate for the common good (or to prevent the common 'bad') more easily within relatively small groups or communities. The smaller the collective, the more strongly one identifies with the collective interest, the easier it is to perceive what 'the others' do, the more an own contribution seems worthwhile, the more feasible it is to cooperate and/or the more expensive it is (socially) to defect. This seems to amount to four major drivers of cooperative behaviour: (1) a sense of community along with social control, (2) a sense of control over the outcomes of behaviour choices, (3) a balance between individual and collective costs and benefits, and (4) general feasibility of cooperation. The latter depends on the availability of 'cooperative' behavioural alternatives and the expense or effort required to adopt those alternatives. 'Expense or effort' may here be taken broadly to mean not only monetary costs or physical labour, but also lack of social support or even disapproval by significant others. An ultimate pressure on cooperation is legal obligation, provided this is effectively monitored and enforced.

9.6.4.2 Individual Psychological Factors

What we are concerned with here are personal characteristics that may vary among individuals involved in any commons dilemma. Problem awareness and basic insight into the process of collective resource depletion may induce a cooperative attitude. A feeling of co-responsibility for the problem is also helpful, as well as a moral inclination to sacrifice individual freedom (of unlimited consumption) for the preservation of common goods. Obviously, the greater the perceived relative attractiveness of a (feasible) cooperative behavioural alternative, the greater the likelihood of cooperation.

Another mechanism in commons dilemmas is people's tendency to compare their own behaviour with that of others, which they may mimic or conform to so as not to stand out as a 'sucker' (see also Section 9.7.3 on social comparison). Thus, personal trust in other people's cooperation is an important pillar of collective behaviour, as it makes individuals feel they are not acting alone and can together respond effectively to the situation at hand. Conversely, deliberate and conspicuous 'free riders' who take advantage of others' leniency may function as 'rotten apples' inducing (many) other individuals to adopt a non-cooperative behavioural pattern, too. People may also be uncertain about the size and regenerative capacity of the common resource, as with water supplies or air quality, for example. This kind of environmental uncertainty will also reduce people's inclination to exhibit cooperative forms of behaviour.

A final important aspect of personality is *social value orientation*. Research in this area points to the existence of four basic personality types, geared respectively to altruism, cooperation, individualism

---

[2] Much experimental laboratory work has been based on small-scale *social* dilemma games, as when two friends want to run a race without completely exhausting each other (see, e.g., Dawes 1980; Liebrand and Messick 1996). For environmental science, however, large-scale *commons* dilemmas are more relevant (e.g., Edney 1980; Ostrom et al. 2002). Although some material in this section comes from small-scale social dilemmas research, the results may generally hold for commons dilemmas, too.

and competition. Defined operationally, *altruistic* people tend to maximise others' gains; *cooperative* people are inclined to maximise behavioural outcomes for both themselves and relevant others, *individualists* simply maximise their own outcomes (whether or not this suits others' interests), while *competitive* people tend to maximise the difference between their own and other people's outcomes: in other words, they are keen to have the most, biggest or best of all. Altruistic and cooperative people are thus more likely to protect the collective interest, while individualists and competitive people are more likely to behave selfishly. Table 9.7 reviews the structural and individual psychological factors governing cooperation in commons dilemmas.

9.7 Psychological Diagnosis: Behaviour Limitation and Determination

Following on from the previous section, what we now need is a psychological diagnosis of environmentally harmful behavioural patterns in order to clarify underlying factors and mechanisms, identify promising starting points for behavioural change and assess potential barriers to cooperation. This section therefore first considers some of the basic factors determining and limiting human sensitivity to environmental risks, at both an individual and a social level. Second, it presents a general conceptual model of consumer environmental behaviour. And third, it sets out a fourfold taxonomy of behavioural processes and briefly mentions the corresponding theories of human behaviour.

9.7.1 Cognitive and Affective 'Short-Sightedness'

From the previous summary of social and individual factors governing cooperation in commons dilemmas it may justifiably be inferred that 'small is beautiful' when it comes to the management of common resources. For example, cooperation is easier when the group of people using the resource is small and there is strong identification with the group, when group members feel co-responsible and have faith in others' cooperation, and when their social value orientation is non-competitive; see Table 9.7. The apparent desirability of

Table 9.7 Factors facilitating cooperation in commons dilemmas

Structural factors	– Smaller group
	– Stronger group identity
	– Higher visibility of others' behaviour
	– Availability of cooperative alternatives
	– Lower cost of cooperative choice
	– Higher cost of defective choice
	– Legal obligation to cooperate
Individual-psychological factors	– Environmental problem awareness, insight into causal processes
	– Feeling of responsibility, moral obligation
	– Perceived (relative) appeal of cooperation
	– Self-efficacy in making cooperative contributions
	– Trust in others' cooperation (low social uncertainty)
	– Ability to resist bad social examples ('rotten apples')
	– Perceived environmental effectiveness of cooperation (outcome efficacy)
	– Altruistic and cooperative social value orientation

social 'small-scaledness' is related to 'short-sightedness'. When it comes to conceiving of plans and developments for the longer term, we may *think* we are responsible decision-makers. In fact, though, most people are inevitably caught in an 'us, here and now' complex of perceptions and evaluations, which may severely hamper long-term problem awareness, strategic planning and policy decision-making. Human short-sightedness can be explained in terms of four different survival dilemmas which lead people to discount behavioural consequences that are risky, spatially remote, temporally delayed and befalling to others, in view of the immediate benefits to themselves (Vlek and Keren 1992).

The *benefit-risk dilemma* refers to the human tendency to discount risks to the extent that the associated benefits are necessary or attractive. For example, the risks of air pollution tend to be underestimated because of the attractiveness of motorised transport.

The *spatial survival dilemma* refers to the tendency to discount events occurring farther away compared to events closer by. For example, a remote scenic coast may be valued less than one close to home (what is worse: a small oil spill close by or a bigger one farther away?).

In the *temporal survival dilemma* delayed consequences tend to be discounted over short-term consequences, as when a minor environmental improvement 'now' has to be weighed against a greater improvement later. However, research revealed that temporal discounting is less likely in the environmental domain, i.e., environmental problems are not valued less when they occur in the long term rather than on the short term.

Finally, the *social survival dilemma* refers to the fact that egocentrism may lead one to discount things happening to people whom the subject regards as more or less 'other'. Social dilemmas may require weighing up a small gain for oneself against a greater gain for a relative stranger, for example. Table 9.8 lists the four survival dilemmas and the respective 'temptations' to discount.

Altogether, human limitations and here-and-now biases in judgement and decision-making are important factors in the generation as well as management of environmental risks. Studies of human responses to environmental stress revealed that people are inclined to deny and remain passive about risks they believe to be uncontrollable (cf. Section 9.4.1 on environmental stress). To compensate for these shortcomings and tendencies, it seems useful to organise and communicate both physical and behavioural science analyses of environmental risks concisely and clearly, presenting the problems as something citizens or other parties can *do* something about, and formulating the results and conclusions of studies in an incisive manner to induce an adequate response or responses by the parties in question; see further Section 9.8.

9.7.2 Needs, Opportunities and Abilities

A useful structuring of internal and external behavioural determinants is provided by the 'NOA model' depicted in Fig. 9.4. Here, consumer environmental behaviour can be regarded as being governed by the *Needs* (N), *Opportunities* (O) and *Abilities* (A) at hand for undertaking a particular resource-consuming activity. Needs and Opportunities interact to shape people's *Motivation to Perform* (MP) an activity: if there is a need as well as an opportunity, you *want* to consume. Opportunities and

Table 9.8 Four survival dilemmas and discounting temptations

Benefit-risk dilemma:	Discounting of risks (when benefits are attractive)
Spatial dilemma:	Discounting of more spatially remote consequences
Temporal dilemma:	Discounting of delayed, future consequences
Social dilemma:	Discounting of consequences for relative strangers

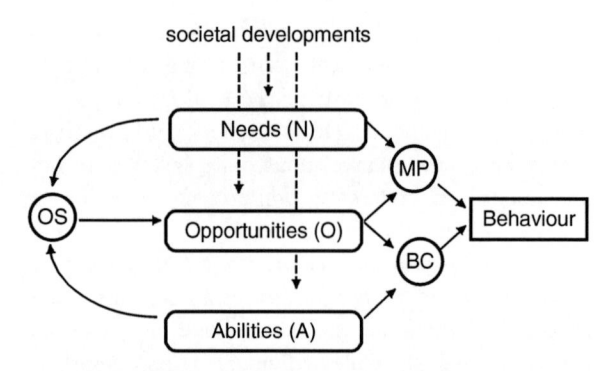

Fig. 9.4 A model of needs, opportunities and abilities underlying environmental behaviour

Abilities together determine *Behavioural Control* (BC): if there is an opportunity which you feel able to use, you *could* consume. And Needs and Abilities together underlie a subject's *Opportunity Search* (OS): if there is a need and you are capable of fulfilling it, you *seek* an opportunity to consume.

Hence, changing consumer behaviour would involve changing people's needs, their (physical, technical or social) opportunities and/or their ability or capacity (physically, mentally or financially) to engage in the relevant behaviour. This means that environmental policy-making may be oriented towards the 'inside' worlds of needs and abilities, the 'outside' world of opportunities, and the 'mixed' worlds of motivation arousal (e.g., through marketing), enhancement of behavioural control (e.g., through education) and opportunity seeking (e.g., through finding feasible ways to fulfil needs).

9.7.3 Four Different Behavioural Processes

To design an effective programme of behavioural change an understanding of how human behaviour comes about is needed. In psychology there are two 'grand' theoretical frameworks that are relevant here. One is *cognitivism*, which maintains that human behaviour stems from more or less reasoned cognitive processes involving learning, feeling and thinking. These processes are sensitive to information, persuasion and social examples. Cognitivism implies that human behaviour can be changed by providing information aimed at *guiding* choices to be (otherwise) made in freedom (see also Section 9.8.1).

The other grand theoretical framework is *behaviourism*, which assumes that human behaviour is governed largely by physical, economic and social *environmental* conditions or manipulations. An ingenious demonstration of this is when animals are taught tricks by applying a consistent scheme of reinforcement in the form of rewards and punishments. The latter is usually called the incentive structure of the behaviour. Behaviourism implies that human behaviour can be changed by environmental manipulations *forcing* choices that can not or only scarcely be evaded. The cognitivism-behaviourism dichotomy runs parallel to the dichotomy between reasoned and automatic behaviour, or between deliberation and habit.

A second major dichotomy is between private, or individual, and public, or social, behaviour. Public behaviour is strongly dependent on the influence of other persons, tending to make it a rather different kind of process from private behaviour. Combining these two dimensions implies that at least four fundamentally different behavioural processes can be distinguished: deliberation, repetition (habit), social comparison and imitation, as shown in Table 9.9. For each of these processes theories are available that specify behavioural determinants, how these operate and the external manipulations through which they might be changed.

To explicate, according to the theory of planned behaviour 'deliberation' may involve a process of individual reasoning about possible behavioural alternatives and their feasibility (yielding feelings of control), the pros and cons of the alternatives (yielding certain attitudes), and social approval or disapproval by significant others (yielding social norms), and a final intention to act. Here, the main motive for deliberative behaviour is maximisation of the expected value of (the consequences of) that behaviour, taking into account non-monetary consequences such as time or effort spent or social approval as well. Some forms of deliberation may lead to environmental behaviour based on the activation of a personal norm, as proposed by the norm activation model. This model proposes that personal norms are activated when an environmental good is perceived as being under threat, the intended behaviour is perceived as potentially effective to reduce this threat, and one feels responsibility for the action to be taken. Norm activation has been extensively studied in relation to altruistic (helping) behaviours. More recently, it has also been applied to explain environmental behaviour.

'Repetition', or habit, is an individual process of automatic behaviour, established over time through recurrent positive reinforcements and the absence of major disincentives or punishments (and/or the presence of disincentives for alternatives). Repetitive, habitual behaviour is thus driven primarily by positive rewards, but note that part of the reward lies in the absence of deliberate effort. Not having to think about what to do in everyday situations like travelling to and from work saves cognitive energy for other, more demanding behavioural decisions.

'Social comparison' is a process of attuning one's behaviour to what one observes in or assumes about significant others, i.e., family, friends, colleagues, teachers or other authorities whose opinions, abilities and accomplishments one holds in esteem. The main motive for such behaviour is to be similar to important others, while at the same time distinguishing oneself in some positive way from comparable others and to avoid feelings of relative deprivation. Owning a 'respectable' car is an example. In many areas of society there is an upward drive towards 'better' opinions, greater abilities, nicer possessions and improved performance. Hirsch (1976) has proposed that, as average income rises, an increasing portion of expenditure is allocated to social status seeking. To the extent this is expressed in environment-burdening activities (e.g., living spaciously, driving a fuel-intensive car, making frequent airplane trips), this contributes to increasing environmental degradation. Social comparison and the competition it involves tend to be strongest when it comes to resources or goods that are (becoming) scarce, so that fewer people can afford them.

Table 9.9 Four different behavioural processes along two dimensions, with relevant theories

	Private (individual)	Public (social)
Reasoned ('cognitivism')	DELIBERATION	SOCIAL COMPARISON
	Theory of planned behaviour	Social comparison theory
	Norm-activation theory	Relative deprivation theory
Automatic ('behaviourism')	REPETITION (HABIT)	IMITATION
	Pavlovian conditioning theory	Social learning theory
	Instrumental learning theory	Theory of normative conduct

'Imitation' is a behavioural process of automatically copying other people's (visible) behaviour or their behavioural norms or expectations. Like repetitive behaviour, imitation also goes without explicit reasoning. Simply copying what others do or did in similar situations is not only cognitively easy ('if others do so, it must be a right way to act'); it also shields the subject from possibly critical responses or admonitions by others.

Taken together, the four behavioural processes shown in Table 9.9 indicate that environmental behaviour may be determined by 'reasoned' as well as 'automatic' factors, and by 'private' as well as 'public' factors. Actually, in the course of time a given type of behaviour may swing back and forth between automatic repetition and deliberate decision, or between simple imitation and thoughtful social comparison.

9.7.3.1 Behaviour May Influence Attitudes and Opinions

As the lower row of Table 9.9 already indicates, people do not always think before they act. Beliefs and attitudes may also be formed or adjusted *after* a given behaviour has been performed. Self-perception theory holds that people derive their attitudes, *post hoc,* from their behaviour. People often do not know what their attitudes are, and may have no clear, firm attitude towards certain types of behaviour. This may be because the behaviour is unimportant to them (there is low involvement) or because they have never explicitly thought about it. In many cases, when asked about their attitudes people will just 'think up' an attitude which is in line with their prior behaviour. Another explanation for the influence of behaviour on attitudes lies in the notion of cognitive dissonance reduction. People feel most comfortable when they have consistent cognitions (knowledge, beliefs, expectations). When behaviour and cognitions are not in line, or when different attitudes are in conflict, people will try to reduce this cognitive dissonance by either changing their behaviour or changing their attitude. People are generally more inclined to change their attitudes rather than their behaviour, the former being easier.

9.8 Changing Environmentally Significant Behaviour

Reducing environmental harm and risk hinges upon changes in the behaviour of actors at the micro, meso and macro levels of society (see Sections 9.6.3 and 9.10). For behavioural change to occur there must be clear goal-setting, feasible behavioural alternatives and motivating factors that steer behaviour in the desired ('sustainable') direction. Below, we first examine four basic strategies for behavioural change, subsequently developing a three-dimensional policy model. Finally, a number of general principles and guidelines are given for planning and implementing effective policy campaigns for behavioural change.

9.8.1 Strategies for Behavioural Change

As discussed in Section 9.6, common-resource dilemmas reflect persistent conflicts between many individual (producer and consumer) interests on the one hand, and a limited number of (large-scale) collective interests on the other. As dilemmas, they can be 'resolved' only by achieving a safer, sustainable *balance* between individual and collective benefits and risks, through making choices in the collective interest more attractive and likely. Various strategies for behaviour change to reduce environmental impact have been proposed. Most of these strategies can be categorised under two general headings. First, structural strategies are aimed at changing the context in which decisions are made and behaviours take place, thereby altering the basic nature of a commons dilemma. The positive and/or negative consequences of behavioural options may be changed, specific behavioural options may be provided or inhibited, or the entire choice setting may be modified. The effectiveness of structural strategies is based on behaviouristic principles, on the assumption that human behaviour is strongly dependent on environmental conditions (see Section 9.7.3).

Second, informational strategies are aimed at changing individual perceptions, motivations, and preferences, inducing actors to behave in a cooperative (i.e., collectively rational) manner, while leaving the basic nature and the pay-off structure of the commons dilemma unchanged (although the *perceived* pay-off structure may change). Informational strategies are based on cognitivistic principles, on the assumption that behaviour may be changed via reasoning processes (see Section 9.7.3). Below, we discuss three structural and one informational strategy for behaviour change. The four strategies and some typical examples are given in Table 9.10.

Modifying behavioural choice-setttings implies offering individual actors new or alternative behavioural settings, or inhibiting or introducing specific behavioural options. This strategy may involve changes in physical, technical, and/or organisational systems. First, examples of physical changes are urban design, the construction of new infrastructure for public transport, the provision of recycling bins, the design of park-and-ride facilities, or closing town centres for motorised traffic. Second, technological innovations may significantly reduce the environmental impact per unit of product or service. Examples are energy-efficient heating systems, water-saving shower heads, and cleaner and lighter cars. Third, organisational changes may involve, among others, goods delivery systems by stores, company transport services, or carpool systems. The underlying assumption here is that individual behaviour is shaped by the physical, technical, and organisational systems in which it takes place.

Regulation-and-enforcement generally serves to direct individual actors' choices. This strategy may be based on legal measures such as laws, regulations and standards issued by the political leadership of the collective. Violations of these norms – if detected – usually meet with some kind of punishment, fine or disapproval. Also, an agreed system of self-regulation, e.g., via accepted quotas or resource privatisation, may be introduced. Covenants are an example of a voluntary agreement between government and actors (often industry) concerning the reduction of specific environmental impacts, such as CO_2 emissions or hazardous waste. The underlying assumption is that laws, regulations, standards and voluntary agreements are eventually internalised under threat of punishment and/or social disapproval. Such strategies imply a need for supervision, monitoring and enforcement.

Financial-economic incentives typically alter the pay-off structure of any commons dilemma and, in doing so, may change its very nature. Financial-economic strategies are designed to reward people financially for choosing certain behaviours with little or no external effect on the collective, or to punish them financially for behaving in a collectively burdensome fashion. The relevant policy instruments are subsidies, discounts, levies, taxes, fines, tolls and the like. The basic underlying assumption is that the behaviour of individual actors is subject to the workings of the 'price mechanism' and that the demand-price elasticities involved are reasonably high.

Information, education and communication may involve providing information, education, arguments, prompts and admonitions aimed at and strengthening actors' inclination and ability to adopt other, less harmful behaviours. This strategy can serve three general goals: increase actors' knowledge, persuade actors to act pro-environmentally, or provide role models. First, actors' knowledge may be increased to heighten their awareness of environmental problems, their awareness of environmental impacts of their behaviour, and/or to increase their knowledge of behavioural alternatives, for example. Second, persuasion may be aimed at, for example, influencing actors' attitudes, strengthening their altruistic and ecological values, and/or strengthening their concern about future consequences. Third, social support and role models may be provided to strengthen social norms and get social comparison processes going, to inform actors about the perceptions and efficacy of others, and/or to inform actors about the pro-environmental intentions and behaviour of others, thereby increasing their trust in others' cooperation. This strategy is based on the assumption that the behaviour in question is 'reasoned' (cf. Table 9.9) and that influencing the cognitive processes underlying behavioural choices will also affect the choices themselves.

Table 9.10 General strategies for behavioural change in managing common resource dilemmas

1 **Provision of physical alternatives, (re)arrangements**
   [adding/deleting/changing behavioural options]
2 **Regulation-and-enforcement**
   [enacting laws, rules, setting/enforcing standards, norms]
3 **Financial-economic incentives**
   [rewards/fines, taxes, subsidies, discounts, levies, tolls]
4 **Information, education, communication**
   [about risk generation, types and levels of risk, others' perceptions and intentions, risk reduction strategies]
5 **(By default) 'Wait and See'**
   ['do nothing, the quay will turn the ship']

The four general strategies just discussed fall (primarily) under the interests and competence of various disciplines, such as architects and civil engineers (strategy 1a), technologists and product designers (1b), management scientists (1c), lawyers (strategy 2), economists (3), and psychologists/sociologists (4a–c). From a behavioural and social science perspective, however, the key question is: How does any strategy actually address major behaviour determinants and how effective is it, or could it be in changing environmentally harmful activities? To further clarify the basic assumptions underlying each strategy, the practical workings and the actual effects and side-effects of the various strategies just discussed, stronger cooperation of behavioural scientists with their colleagues from other disciplines seems desirable.

Which strategy will be most effective in changing behaviour and reducing environmental impact depends on which are the key determinants of the behaviour at stake. To put it differently: specific strategies for behaviour change will be more effective if they target important factors inhibiting or promoting proenvironmental behaviour. Informational strategies are generally easier to design and apply than structural strategies. Informational strategies are especially effective in increasing proenvironmental behaviour when the instructions are behaviour-specific and given in close physical and temporal proximity, when people have opportunities to engage in the target behaviour, when behaviour is not habitual, and when engagement in the behaviour is relatively convenient and not very costly in terms of money, time and effort. When changes in target behaviours are relatively inconvenient and unattractive, structural strategies are generally more effective. However, structural strategies are not always feasible, nor easy to implement.

One solution-by-default to a collective environmental risk problem is nicely expressed in a Dutch saying that translates as 'the quay will turn the ship', that is, strategy 5 in Table 9.10: 'Wait and see', will automatically elicit adaptive behavioural responses, but these can be anything 'too little, too late'. Making such unsustainability scenarios palpable – for example, via scenario studies – may allow individual actors to become (better) able to 'live under the shadow of the (common) future' and undertake due preventive action. Business-as-usual scenarios may thus initiate 'self-destroying prophecies', as society tries to steer away from them.

9.8.2 A Three-Dimensional Policy Model

In practice just about every policy, defined as 'a concrete measure to achieve one or more policy goals', embodies several of the above strategies simultaneously. For example, if the government decides to reduce the traffic speed limit, it must change the law (regulation and enforcement), communicate this measure to the public (provision of information) and introduce fines for people found in violation (financial-economic incentive).

As a single strategy (e.g., pricing) or as a combination of strategies (e.g., new technology, information and pricing), any given policy can be characterised along three dimensions. In the first place a distinction can be made between directive and constitutive policies. Directive policies seek to influence behaviour directly, through specific instructions, prohibitions or price incentives, for example. Constitutive policies influence behaviour indirectly, by creating the preconditions for specific behavioural changes, through education, for example, or by providing new infrastructure or equipment.

The second dimension differentiates between collective and individualised policies. Collective policies seek to simultaneously influence many different individuals in different situations. Examples include general legislation, price regulation and mass-media campaigns. Individualised policies, in contrast, seek to change the behaviour of specific groups of individuals in particular situations, by means of licences, levies or advice, for example. Personal computers and the Internet offer new possibilities for providing individualised, 'tailor-made' information and advice about environmental behavioural patterns.

The third dimension of this policy model distinguishes between 'push' and 'pull' policies, restricting and extending freedom of choice, respectively. 'Push' measures are restrictive, aiming to make environmentally harmful behaviour less attractive, through prohibitions, price rises or 'discouraging' information campaigns, for example. 'Pull' measures extend freedom of choice by making environmentally sound behaviour more attractive. Examples are subsidies or 'encouraging' information campaigns, for example. In general, a combination of punishing undesirable behaviour ('push') and rewarding desirable behaviour ('pull') is likely to be most effective.

The effectiveness of push and pull measures depends on four factors. First, rewards and punishments are more powerful if they are guaranteed and experienced directly and if there is a self-evident link between the reward or penalty and the relevant behaviour. Second, incentives and disincentives should be just sufficient to make people reconsider their behaviour, but not so strong as to provide complete justification for the effort and thereby suppress feelings of personal responsibility and reduce intrinsic motivation vis-à-vis environmental protection. Third, the incentive or disincentive should be relevant, that is, aimed at changing consequences of behaviour that are significant to the target group. For example, monetary rewards will only be effective if people do value the price of the behaviour in question. Fourth, an important issue is who is to be responsible for rewards and/or penalties. Incentives and disincentives will generally be more effective when they are implemented by agencies with which subjects can identify.

9.8.2.1 Tuning Policies to Target Groups

Policies will generally be more effective and acceptable if they address selected target groups, whose specific needs, abilities and opportunities will then be easier to address (cf. the NOA model in Fig. 9.4 above). Across-the-board measures will often provoke unnecessary resistance because they cause inequality, with certain groups being restricted more or offered better opportunities than others. Target groups can be distinguished on the basis of demographic characteristics, observed behaviour (e.g., people who do and do not act pro-environmentally), opportunities (e.g., people who own or rent a house), abilities (e.g., poor and rich households) and needs and attitudes (e.g., the environmentally concerned and unconcerned). In general, it is easy to define target groups on the basis of demographic characteristics and observed behaviour, because both are readily observable. In many cases, however, it will not immediately be clear how such target groups should be addressed, because there may be no clear prior idea about the factors influencing their behaviour. Conversely, it will be easier to design an effective policy for groups whose needs, opportunities and abilities have been clarified beforehand.

9.8.3 Territorialisation and Tradable Exploitation Rights

Given the social and behavioural limitations of human resource users discussed in Section 9.7.1, it is not surprising that one strategy that experts have recommended for environmental management is 'territorialisation': making local communities responsible for the availability and quality of the agricultural, natural and industrial resources on which their livelihoods depend. In many cases, of course, such a strategy may not be feasible, as in the case of air quality or river management. It may also be inefficient because it foregoes the benefits accruing from economies of scale, in urbanisation, international trade, transportation, tourism and socio-cultural interaction, for example. However, as the scale of collective environmental problems increases, such as, e.g., in anthropogenic climate change or in decreasing biodiversity, territorialisation may somehow be forced upon policy makers who would otherwise be unable to come to grips with the sources and processes underlying the problem. This might, for example, involve the envisaged collection of all major CO_2 emitters in the 25 countries of the European Union, to be gradually subjected to the international system of CO_2 emissions reduction the European Commission is currently deploying.

One particular combination of government regulation (strategy 2 in Table 9.10), payment according to resource use (no. 3), information provision (no. 4) and privatisation (under no. 1, organisational change) may prove to be an effective as well as cost-efficient approach towards managing common goods, if applied to a suitably defined territory. This is the management system known as 'tradable emission permits' or, more generally, *tradable exploitation rights* (see Tietenberg 2003). The idea here is that, for the collective as a whole, a central authority determines a maximum sustainable level of exploitation of some social or environmental resource (to be consumed, enjoyed or polluted) and then distributes among its constituents a collection of 'tickets' each representing a fixed, limited exploitation right. Tickets can be distributed either by auctioning them out or by following a predetermined distribution; the latter may be perfectly uniform across all users, or it may be proportional to users' perceived (or self-reported) needs, or it may follow their historic distribution of environmental exploitation.

As tickets are tradable, they can be purchased by those seeking above-average resource use, from those whose demand is below-average. Ticket prices are established under free-market laws of supply and demand, but with a ceiling set on total resource use by any party, corresponding to its ownership of a certain number of tickets. Theoretically at least, through the exchange of tickets in a perfectly competitive market, 'permitted' environmental exploitation will eventually be distributed such that its overall benefits are highest; for example, emission reductions will eventually be highest where they are cheapest. If necessary, the central authority can reduce the amount of exploitation represented by a single ticket. Alternatively, it can sanction increased exploitation if resource regeneration is faster than originally estimated. This system of risk management thus embodies a balanced combination of free-market operation and government guidance, including constraints on overall resource use.

Tradable exploitation rights as a policy instrument may be problematic for one of the following reasons.

- Putting an upper limit on aggregate resource exploitation reflects a decision about what would be sustainable in the long run. Too low an exploitation ceiling may imply too much constraining of current activities, either because of user-shortage of tickets or because of increased costs of exploitation. Too high a ceiling, however, may lead to massive ticket dumping and subsequent collapse of their market price.
- The initial distribution of exploitation rights may be unnecessarily egalitarian (e.g. when rights are allocated to those who don't need them), it may be unfair, or it may confirm existing differences (among 'grandfathers') in market power.
- Full market operation, so that tickets can be freely exchanged, and equilibrium prices can be established, may be hard to secure because of insufficient information about demand and/or supply, and/or about actual ticket prices.
- If exploitation rights are owned by and may be exchanged between business companies, the individual consumers of their goods and services may be (kept) unaware of the producers' actual costs of resource use.
- The central authority may be unable and/or insufficiently motivated to supervise the tradable-permit market and control for the actual amount of environmental exploitation in view of acquired tickets; supervision and control need a sanctioning instrument whereby acquired rights can be effectively enforced.

Given the common-resource dilemma model of collective environmental problems (Section 9.6), tradable exploitation rights is, theoretically, an almost ideal system of risk management, because it seeks to achieve the necessary *balance* between individual and collective interests, between 'market' and 'government', and (thus) between individual freedom and social equality. In practical terms, though, it means making a number of serious political choices with regard to at least the total ceiling on exploitation, the initial distribution of rights, and government supervision and sanctioning.

As an environmental management system, 'tradable exploitation rights' lends itself well to experimental evaluation and hypothesis testing. Are equilibrium ticket prices reached more rapidly when the initial collection of tickets is auctioned out or when it is uniformly distributed among users? What happens to the unit price of the resource when aggregate exploitation is (further) limited? How is trading affected when users have only partial information about ticket supply and/or demand? Does users' acceptance and cooperation depend on the way revenues are (re-)distributed? To what extent does actual environmental exploitation remain within aggregate limits under different government supervision and control regimes? Under which circumstances would users start to evade the predefined 'territory' and move into ticket-free exploitation elsewhere? Overall policy questions are: how feasible is this instrument; how much environmental protection does it actually deliver; and what are its economic effects for participants? Behavioural scientists can help clarify several of these points and may thereby contribute to improving the quality of the instrument.

9.8.4 *Principles and Practical Steps for Intervention Planning*

In designing and implementing effective behavioural intervention campaigns, it is useful to work according to the four so-called 'DO IT' principles, viz.:

1. *Define* the target behaviour. The two main criteria here are that the desirable behaviour should contribute significantly to environmental quality, and that it is acceptable to the target group.
2. *Observe* and describe current behaviour(s). Factors encouraging or upholding environmentally harmful behaviour and promoting pro-environmental behaviour should be analysed to identify policy interventions holding out greatest promise. It is also important to record current, baseline behaviour, so that the effects of actual intervention can later be duly assessed.
3. *Intervene*: design and implement practical interventions aimed at modifying the behaviour-determining factors identified under principle 2.
4. *Test* the success of the intervention(s). Field experiments can be conducted to gather information on the effects and side-effects of actual intervention(s). This information is needed to improve, refine or replace a chosen intervention strategy, and to provide feedback to those targeted as to make clear whether their efforts have been effective.

These four principles can be aligned reasonably well with the four divisions of the commons dilemma paradigm illustrated in Table 9.6. Thus, from a more elaborate perspective, intervention planning can be designed using the fourfold model set out in Section 9.6. Table 9.11 lists 15 steps for systematic intervention planning, following the four divisions of Table 9.6. Note that Table 9.11 is focused on selected target groups, while Table 9.6 is a more general representation from the policy maker's perspective.

A systematic approach to intervention-planning according to the scheme of Table 9.11 can best be elaborated by a multidisciplinary team of researchers and policy-makers made up of physical as well as social scientists. Interventions should be based on a careful diagnosis of the current behavioural patterns of specific target groups and the factors underlying them. The decision concerning the need for intervention is a political decision and must be politically justifiable. It is important that clear and consistent goals are set and that feasible behavioural alternatives are (made) available. In addition, careful monitoring and communication of the impacts of interventions are necessary to assess the acceptability of effects and possibly refine interventions, in a process of continual improvement. Thus, intervention planning to promote sustainable development requires expert input from several disciplines, and from scientists as well as practitioners.

9.9 Environmental Values, Beliefs and Quality of Life

As we have seen, human environmental behaviour, i.e., behaviour impinging on environmental sources, sinks and ecosystems, is primarily directed by human needs and values and by opportunities and abilities with respect to resource use; see the NOA model in Fig. 9.4 above. In the words of the World Commission on Environment and Development (WCED 1987: 43), sustainable development is:

development that meets the needs of the present without compromising the ability of future generations to meet

Table 9.11 Fifteen steps for practical intervention planning, derived from the four-component model of Table 9.6

I. Problem diagnosis
 1 Physical-technical analysis of 'unsustainable' activities ['what is the problem?']
 2 General information and communication about environmental risks ['knowledge, problem awareness?']
 3 Rough categorisation of target groups from relevant population ['who?']
 4 Specific behavioural analysis per target group ['behavioural determinants, mechanisms?']

II. Policy decision-making
 5 Specification of 'sustainable' behavioural options ['what is possible ?']
 6 Assessment of feasibility and acceptability of sustainable behaviours ['feasible, acceptable?']
 7 Formulation of clear and consistent policy goals ['to what end?']
 8 Translation of policy goals into behavioural goals for target groups ['what exactly, with whom?']

III. Practical intervention
 9 Selection and design of practical measures and strategies ['what, how?']
 10 Assessment of policy effects and side-effects ['expected effects?']
 11 Selection and approach of target groups ['whom, via which media, when?']
 12 Time scheduling and division of tasks for interventions ['who does what, when and how?']

IV. Evaluation of effectiveness
 13 Application of interventions and evaluation of environmental and social effects ['is it going well?']
 14 Description and communication of environmental, economic and social effects ['actual effects?']
 15 Feedback to target groups regarding intervention effects ['knowledge, awareness of policy effects?']

their own needs. It contains within it two key concepts: (-) the concept of needs, in particular the essential needs of the world's poor (..); and (-) the idea of limitations imposed by the state of technology and social organization on the environment's ability to meet present and future needs.

It is thus important to examine human (basic) needs and values, beliefs and worldviews, as well as different aspects of human quality of life. This can deepen our understanding of different forms of environmental behaviour and help us assess the likely impact of key environmental policies on quality of life. In this section we consider these issues from the perspective of the individual; in Section 9.10 we shall be considering sustainable development from the perspective of society as a whole.

9.9.1 Value Orientations and General Environmental Beliefs

People think and act differently with respect to the natural world and the environment in which they live. As people grow up and develop, their values and beliefs may take on enduring significance for their everyday behaviour. Below, we discuss values and general beliefs that are relevant in human-environment interactions: value orientations, the New Environmental Paradigm, and myths of nature, respectively.

9.9.1.1 Value Orientations

As discussed in Section 9.6.4, research on individual behaviour in commons (or social) dilemmas points to four basic 'social' orientations towards common resources: altruistic and cooperative (or pro-social), and individualistic and competitive (or pro-self). From a broader perspective, the values people cherish have been extensively studied internationally. *Values* can be defined as 'enduring beliefs that a specific mode of conduct or end state is personally or socially preferable to an opposite or converse mode of conduct or end state'. Research by Schwartz (e.g., 1992) and others revealed that a variety of human values can be plotted in a two-dimensional space. One dimension reflects Conservation (tradition, conformity, security) versus Openness to Change (self-direction, stimulation, hedonism). The other dimension represents Self-Transcendence (universalism, benevolence) versus Self-Enhancement (power, achievement). Later work by other researchers has confirmed that environmental beliefs and behaviour are especially related to Self-Transcendence. Environmental psychologists have argued that the self-transcendence cluster may be further divided into an altruistic and a biospheric value orientation. Those who strongly endorse altruistic values especially consider consequences for other people, while a biospheric value orientation reflects concern for nature and the biosphere. Thus, an egocentric, altruistic and biospheric value orientation may be distinguished. Research has revealed that altruistic and biospheric values are most strongly related to environmental beliefs and behaviour. However, both value orientations may affect environmental behaviour differently, especially when people have to choose between socially sound (e.g., not involving child labour) and environmentally sound (resource-extensive) products.

9.9.1.2 General Environmental Beliefs: The New Environmental Paradigm

The New Environmental Paradigm, or NEP, scale provides a general measure of a person's environmental concern and stands for his or her general beliefs about the nature of the Earth and humanity's relationship with it. The measure was developed in the late 1970s to examine the extent to which environmentalism was developing as a new paradigm in society, where human survival is understood to depend on the quality of the global environment. The NEP measure has been used extensively in questionnaire studies to examine relationships between NEP scores and actual environmental behaviour. The NEP appears to be related to environmental behaviour, although relationships are generally not strong. Table 9.12 lists the items included in the NEP scale in its recent revision.

9.9.1.3 Myths of Nature

General environmental beliefs may also be associated with the definitions and images of nature people hold. One conception here is a model of 'Myths of Nature' seeking to reflect different archetypal views on the vulnerability of nature. Four Myths are distinguished: Nature Capricious, Nature Tolerant, Nature Benign and Nature Ephemeral. These are conceived

9 Social Science and Environmental Behaviour

Table 9.12 The New Environmental Paradigm scale

1	We are approaching the limit of the number of people the earth can support.
2	Humans have the right to modify the natural environment to suit their needs.
3	When humans interfere with nature it often produces disastrous consequences.
4	Human ingenuity will ensure that we do *not* make the earth uninhabitable.
5	Humans are severely abusing the environment.
6	The earth has plenty of natural resources; we just need to learn how to develop them.
7	Plants and animals have as much right to exist as humans
8	The balance of nature is strong enough to cope with the impacts of modern industrial nations.
9	Despite our special abilities, humans are still subject to the laws of nature.
10	The so-called 'ecological crisis' facing humankind has been greatly exaggerated.
11	The earth is like a spaceship with very limited room and resources.
12	Humans were meant to rule over the rest of nature.
13	The balance of nature is very delicate and easily upset.
14	Humans will eventually learn enough about how nature works to be able to control it.
15	If things continue on their present course, we will soon experience a major ecological catastrophe.

Source: Dunlap et al. (2000)

Note: Ratings (disagree-agree on, say, a 5-point scale) on the seven even-numbered items must be reversed before summing the scores.

of as being embedded in four different lifestyles, respectively: fatalistic, hierarchical, individualistic and egalitarian. Each Myth of Nature can be visualised as a ball in a basin-shaped landscape, as presented in Fig. 9.5. The landscape symbolises the vulnerability of nature, the ball environmentally risky behaviour. Within each landscape, the ball is in equilibrium. The four Myths of Nature are described in more detail in Box 9.1.

These four Myths of Nature reflect individual perceptions of risk as well as preferred strategies for risk reduction. This paradigm has often been used to philosophise about long-term policy scenarios. It is also used as a tool to systematically analyse decision-making and negotiation processes on environmental issues, particularly as it can help clarify why certain parties have difficulty coming to an agreement. Adherence to the four Myths of Nature has been assessed in a number of questionnaire studies, too, although empirical studies on this topic are scarce.

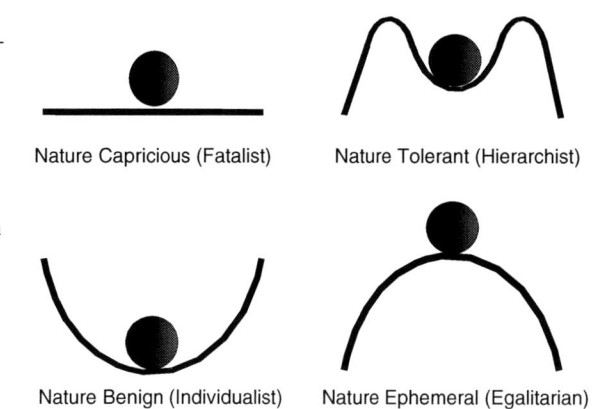

Fig. 9.5 Pictorial representation of four Myths of Nature
Source: Steg and Sievers (2000, after Douglas' & Thompson's cultural theory)

9.9.1.4 Value-Belief-Norm Theory of Environmentalism

Using several of the elements discussed above, Stern (2000) and colleagues have developed a Value-Belief-Norm, or VBN, theory linking general values, general environmental beliefs, behaviour specific beliefs, personal norms, and behaviour – from general to specific. As in norm-activation theory (Section 9.7.3, Table 9.9) it is assumed that people will only act pro-environmentally if they feel morally obliged to do so. Personal norms about pro-environmental behaviour are activated by beliefs that environmental conditions threaten things that are valued by the individual, as well as by beliefs that the individual can do something to reduce this threat. Such specific beliefs depend on general beliefs about the nature of the Earth and humanity's relationship with it (as measured on the NEP scale, for example; see Table 9.12) and on relatively stable value orientations (such as egoistic, altruistic, and biospheric values; see above). Figure 9.6 gives a schematic representation of VBN theory, in which each variable is assumed to directly affect the next variable in the causal chain, and possibly others further down the chain as well.

VBN theory has been tested in several questionnaire studies, and has proven useful in explaining a range of environmental behaviours, particularly low-cost behaviours (Section 9.5) such as waste recycling, and political action in support of the environment. Several studies suggest that the model is less successful in explaining high-cost behaviours such as reducing

Box 9.1 Characteristic descriptions of the four Myths of Nature

Nature Capricious represents nature as an unmanageable and inefficacious system: one never knows which way the ball will roll or what the consequences will be. Nature as well as the availability of resources are seen as unpredictable; neither needs nor resources are held to be controllable. Risk perception is based on the belief that 'what you don't know can't harm you'. Since all events are taken to happen by chance, learning and management are impossible. The best one can therefore do is just cope with erratic events. Nature Capricious justifies fatalistic social relations. Fatalists view life as a lottery. Their preferred risk management strategy is therefore 'Why bother?' This attitude rationalises isolation and resignation to stringent control on personal behaviour. Fatalists do not differ systematically or consistently from those adhering to some other Myth of Nature, because they are not consistent in their thought and action; the fatalist simply copes with prevailing circumstances.

Nature Tolerant conveys a robust system, but only up to a certain point. The ball will find its way to the bottom of the basin, but only as long as we respect the limits set by experts and authorities. Nature is seen as being in unstable equilibrium, with resources scarce. Nature Tolerant is associated with an intermediate level of environmental concern. Risks are accepted as long as these are sanctioned by the experts; life is a matter of dealing with 'acceptable' risks. Environmental problems can be controlled by government regulations. Nature Tolerant legitimises hierarchical social relations. The preferred risk management strategy is sustainable growth, that is, a policy taking advantage of the perceived resilience of nature. This kind of policy seeks to control common resources, for hierarchists believe that human needs cannot be controlled. Nature Tolerant justifies having rules and regulations 'flow down' (because only experts know the limits of growth), while compliance to regulations is supposed to 'flow up'.

Nature Benign conveys a robust and resilient system. The ball will always find its way back to the bottom of the basin. Nature is seen as a stable and global equilibrium, and resources are expected to remain abundant. Resources as well as needs are assumed to be controllable. Environmental risks are seen as being associated with opportunities; new technological solutions will emerge to control them. Nature Benign goes together with a low degree of environmental awareness. Adherents to this view are held to be environmental risk-takers. Nature Benign serves to legitimise individualistic social relations. Individualists strive for freedom to bid and bargain in self-regulated networks with few restrictions or obligations. They are opposed to collective resource management. Their preferred risk management strategy is therefore unhampered operation of the market (as opposed to government regulation); they strongly believe in market forces and in equal opportunity for all.

Nature Ephemeral represents a precarious and delicate balance; the least jolt may lead to disastrous consequences. Nature is in limited equilibrium and natural resources are assumed to be in irreversible decline. Unlike human needs, resources cannot be controlled. Nature Ephemeral is associated with risk aversion, to the extent that risks are perceived as being hidden, irreversible and inequitable. Adherents of the Nature Ephemeral view are very concerned with environmental problems and consider they can and should moderate their needs. Nature Ephemeral legitimises egalitarian social relations. Egalitarians use their claim that nature is fragile to justify equal sharing of what they see as a single, finite Earth. They prefer risk management strategies fostering equality of outcomes, for present as well as future generations. Because resources are taken to be in decline (and uncontrollable), the only solution is to control human needs. They therefore argue for radical changes in human behaviour and in social systems.

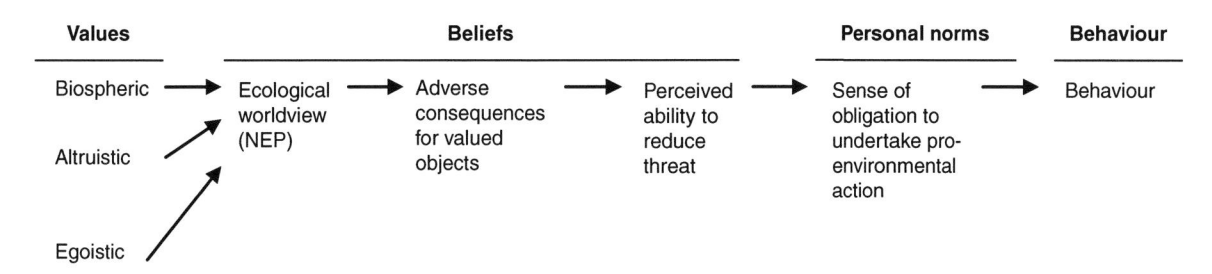

Fig. 9.6 Value-Belief-Norm theory of environmentalism Source: Steg et al. (2005).

car use or household energy consumption. In these domains, other factors like self-efficacy ('can I do it?') and non-environmental beliefs (such as comfort, effort, costs, and status) may play a more important role.

9.9.2 Human Needs and Quality of Life

The *social* dimension of sustainable development refers to social well-being or human Quality of Life (QoL). This concept can be measured objectively in terms of QoL indicators relating to society as a whole, such as health, education, women's rights, employment, leisure time, availability of goods and services, income and personal safety. Quality of life can also be assessed at an individual level, by measuring people's own evaluation of their overall satisfaction with life as well as their evaluation of specific QoL aspects of perceived living conditions. Subjective QoL measures are an indispensable complement to objective, social QoL indicators. For one thing, they may be helpful in examining the extent to which higher-level economic, social and environmental conditions influence the personal well-being of different citizen groups, as they see it themselves. Quality of life is a multidimensional construct which depends on the extent to which important (basic) needs and values are fulfilled. In this respect, economic and environmental variables are also important.

Basic needs, presumed universal, differ from 'needs' as used in everyday parlance and from needs for specific forms of consumption (e.g., 'I need a car'). The latter are in fact 'wants' and refer to preferences for certain ways of satisfying needs, which may be culturally and temporally determined. With respect to basic needs, a distinction is generally made between:

- Survival needs (e.g., subsistence, safety, physiological needs)
- Social needs (e.g., sense of belonging, affection, participation) and
- Needs for personal growth (e.g., recognition, self-actualisation, understanding, identity, freedom)

One well-known taxonomy of human needs was proposed by Maslow (1954, who distinguished seven levels of need, viz. for: (1) physiological comfort, (2) safety, (3) belongingness, (4) social approval and self-esteem, (5) knowledge and understanding, (6) aesthetic beauty and (7) self-actualisation. Levels 1 and 2 are survival needs, 3 and 4 are social needs and 5–7 are personal-growth needs.

Quality of life can also be considered, hierarchically, in terms of (a) universal needs, (b) instrumental goals, (c) activities/endowments and (d) available resources (Ormel et al. 1997). Social Production Function (SPF) theory specifies how well-being is 'produced' by individuals trying to optimise their achievement of universal goals within the set of resources and constraints they face. According to SPF theory, quality of life refers to psychological or emotional well-being, which is dependent on the extent to which universal needs are met. Two universal goals are distinguished: *physical* and *social well-being*. These needs are met through activities satisfying derived instrumental goals that are intrinsically rewarding. *Physical well-being* results from activities producing stimulation, internal comfort and external comfort. *Social well-being* results from activities producing status, behavioural confirmation and affection. Thus, overt activities may realise instrumental goals in fulfilling universal needs. Individuals differ in their preferences for instrumental goals and may therefore fulfil the universal needs of physical and social well-being in different ways. Figure 9.7 (adapted from Ormel et al. 1997) shows the framework of SPF theory.

Overall utility
psychological well-being
↑
Universal needs
Physical well-being, social well-being
↑
Instrumental goals
Stimulation, internal/external comfort
status, behavioural confirmation, affection
↑
Activities and endowments
e.g., householding, work, sports, dwelling, vehicles, capital
↑
Resources
active resources: e.g., time, effort, skills;
latent resources: e.g., inactive kinship ties, skills

Fig. 9.7 Conceptual structure of Social Production Functions theory

Table 9.13 Seven quality-of-life (QoL) factors summarising 21 out of 22 specific QoL variables (in brackets)

Quality-of-life factor (and specific aspects bundled)	Mean importance factor	% explained variance
Health, family and safety (health, partner/family, safety)	4.69	7.0
Self direction (freedom, privacy, leisure time)	4.35	9.8
Achievement (education, work)	4.22	6.4
Environmental quality (environmental quality, biodiversity, aesthetic beauty)	3.93	9.9
Maturity (identity/self-respect, security, spirituality/religion)	3.77	7.6
Openness to change (social relations, change/variation, excitement)	3.57	8.5
Self enhancement (money/income, comfort, status, material beauty)	3.20	11.2

Source: From Poortinga et al. (2004)

Note. From top to bottom, the factors bundle QoL variables of decreasing importance. Ratings could range from 1 to 5 (= 'most important'). The singular aspect 'social justice' (mean rating 4.65) fell outside the above factor structure.

This theory and other literature on human needs and values allow us to draw up a comprehensive list of essential QoL variables. Table 9.13 specifies 22 such variables and illustrates how 21 of them can be statistically bundled into key 'QoL factors', based on their relative importance as reported in a survey of 455 Dutch respondents.

As Table 9.13 shows, 'family, health and safety', 'self direction' and 'achievement' were judged more important than 'maturity', 'openness-to-change' and 'self enhancement'. The total list of 22 specific QoL variables can be used in interview or questionnaire studies to: (a) assess respondents' current satisfaction in life, (b) their profile of (most) desired improvements in QoL, and/or (c) the degree to which they anticipate changes in QoL factors with changing economic, social and/or environmental conditions or as a consequence of environmental policies, including their own consumption pattern.

Assessment of quality of life is an important part of designing and evaluating campaigns of behavioural change in the context of sustainable development. This may answer such questions as: in what respects does growing material wealth increase human well-being; would people in fact experience greater or lesser life satisfaction if they adopted an environmentally friendly lifestyle, and why? This may facilitate attempts to find a proper balance between economic, social and environmental interests and to assess, for a particular location, region or country, what kinds of development would be sustainable. In any case, sustainable development should not imply that people's quality of life is grossly reduced.

9.10 Societal Analysis and Management of Environmental Problems

In Section 9.6 the commons dilemma was discussed as a model capturing the tension between individual interests and collective goods. Large-scale environmental problems were seen as resulting from the accumulation of the negative external effects of numerous individual activities. Also discussed were the roles played by human actors at the macro, meso and micro level of society in the continued creation as well as management of environmental problems.

Next, Sections 9.7 to 9.9 were devoted to a psychological analysis, mostly at the micro, i.e., individual level of society. These sections covered the processes and determinants of environmental behaviour, different strategies of behavioural change, and individual values and beliefs related to environmental qualities and individual quality of life.

In the present section we discuss the key factors at work at the macro and meso level of large-scale commons dilemmas, which we shall refer to here as the 'societal' perspective. This level of analysis is a necessary complement if we are to understand the power of the broader system or context in which individual environmental behaviour is conditioned and which helps determine the extent to which that behaviour can or cannot be changed (cf. strategy 5 in Table 9.10). Behavioural science studies into the societal driving forces of environmental exploitation and (un)sustainable development are far wider in scope than research on individual environmental behaviour, and often involve a variety of scientific disciplines such as sociology, economics, psychology and (physical) environmental science itself. Here, as in regard to individual environmental behaviour, any driving factor may best be considered from the four perspectives of problem diagnosis, policy decision-making, practical intervention and evaluation of policy effectiveness (cf. Table 9.6).

9.10.1 Driving Forces of Environmental Decline

In a classic analysis focusing on rapid population growth, Ehrlich and Holdren (1971) proposed the well-known 'IPAT formula': $I = P \times A \times T$, which states that, for any geographic unit, total environmental impact (I) is a product of population size (P), average per capita level of affluence (A) and the

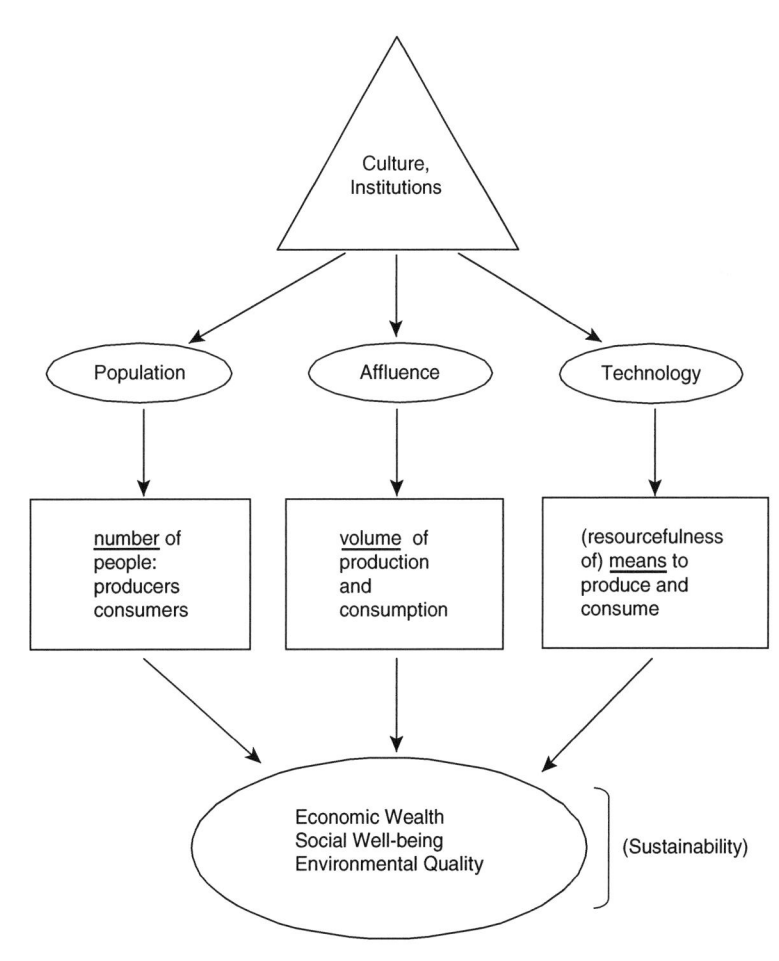

Fig. 9.8 Societal driving forces of unsustainable development
Source: Vlek (2002)

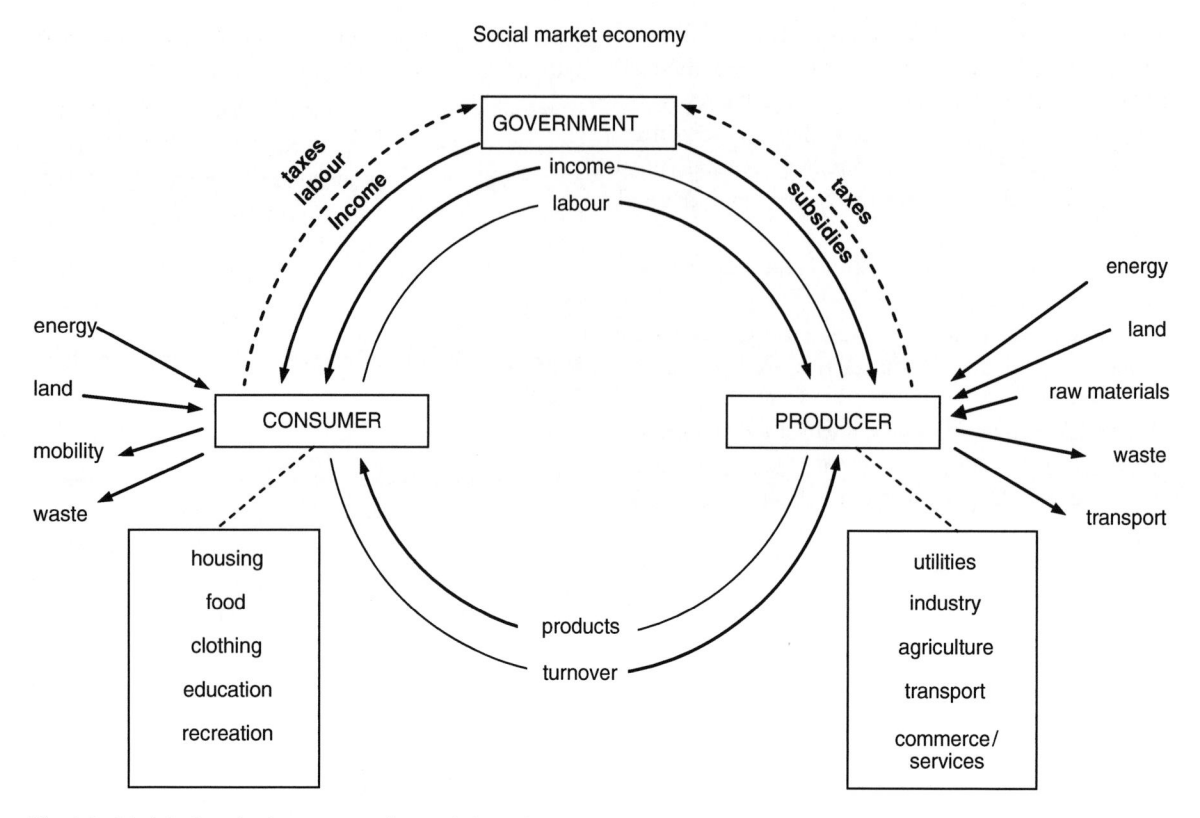

Fig. 9.9 Model of production-consumption cycle in social market economy Source: Vlek (2000)

environmental resource intensity of the technology (T) used to produce one unit of affluence (see also Chapter 1). Since developments in P, A and T are themselves subject to prevalent cultural beliefs and values as well as the institutions based on them, the IPAT formula gains in significance if it is extended by the two more general driving forces of Culture and Institutions. As Fig. 9.8 indicates, developments in P, A and T are influenced by C and I, and the entire complex of forces is related to economic wealth, social well-being and environmental quality. The latter, it should be noted, are the three acknowledged dimensions of the concept of sustainable development (see Section 9.10.2).

The significance of Fig. 9.8 for environmental policy-making is that policies aimed at sustainable development may be elaborated and applied on five fronts. From relatively easy to hard to change, these fronts are: technology, affluence (consumption), population, institutions and culture. Let us consider each of these driving forces in more detail.

9.10.1.1 Technology

Since about 1850 technological developments have led to widespread mechanisation and motorisation of human activities. All this involves ever-growing worldwide consumption of raw materials and fossil fuels, and it is causing a multitude of environmental problems among which pollution, noise, land use and waste.

On the basis of limited behavioural research (see Midden et al. 2007), reasonable advice can be given on developing sustainable producer and consumer behaviour through smart design, production, marketing, use and ultimate disposal of products and machinery. There is in fact huge technical potential for improved materials efficiency, energy saving, waste reduction and lower ambient noise levels. However, technical changes and innovations do not come about of their own accord. New or modified installations, products and services must first be widely known to exist and be understood, accepted and properly used before their environmental benefits can be reaped. To elaborate the communication, education, training and

Table 9.14 Aspects of sustainable technology development that may benefit from behavioural science support

1. Designing installations, products and services (briefly: technologies) that have no unsustainable environmental impacts when used by various client groups
2. Introducing technologies in such a way as to engender wide familiarisation, comprehension and acceptance in regular use
3. Supporting producer and consumer decisions about technologies, thereby ensuring due weight is given to environmental considerations
4. Supervising and promoting the environmentally benign use of technologies and providing useful feedback about their actual environmental impacts
5. Finding socially acceptable ways to dispose of, trade in, resell, repair, re-use and/or recycle technologies at the end of their (initial) lifetime

support activities this necessarily requires professional expertise and advice. Development and large-scale implementation of technologies that are less resource-intensive and more environmentally benign therefore demand input from the behavioural sciences, particularly with respect to the issues cited in Table 9.14.

This kind of behavioural science support could best be offered in close collaboration with physical scientists and engineers, so that attention can be paid to the specific ways in which a given technology is perceived by different groups of people and the forms of use that might be associated with relevant products and services.

9.10.1.2 Affluence (or Consumption)

Environmental resource use, harmful emissions and damage to ecosystems are generated in numerous so-called production-consumption cycles. In this context producers seek to meet the demands of consumers as well as their own needs for profit and continuity. Consumers, in turn, need paid work to develop purchasing power and pay taxes. It is the government's task to oversee economic processes and take responsibility for collective goods and services. Figure 9.9 illustrates this intricate interdependence schematically for a social market economy in which a regulating government co-determines 'what goes in and out'.

Production-consumption cycles may be modelled for a wide range of products and services: fruit and vegetables, electrical power, financial services, household furniture, transportation, leisure events and holiday tours, and many other domains. Each of these can be described in terms of:

- The major actors involved, each with their own beliefs, needs and values
- The products, technologies and/or apparatus involved
- Existing behavioural preferences and habits
- The environmental impacts caused and/or accepted and
- The satisfaction or quality of life deriving from the particular domain of production and consumption

The message of Fig. 9.9 is that environmental impacts are generated in an entire *system* involving producers, consumers, government agencies and also media actors. The mutual dependencies and influences among these various parties are such that individual households and businesses have only limited freedom to change their behaviour. To identify effective means of reducing environmental impacts requires comprehensive analysis of the relevant production-consumption cycle, clarifying the technical as well as social factors underlying current modes of production and consumption. Life-cycle analysis or LCA of a product or service is a suitable way of charting the individual contributions of relevant actors (and their savings potential) with respect to the exploitation of environmental sources, sinks and ecosystems.

Moves towards environmental sustainability can be accomplished by a variety of strategies for behavioural change (see Section 9.8), with all major actors having their own particular opportunities and responsibilities. For example, 'eco-efficiency' is typically a producer-oriented strategy aimed at introducing products and services that are less energy- and materials-intensive. However, the reduction in resource intensity achieved with eco-efficiency is only per unit of product or service. Due to increasing consumption levels, no absolute improvement in environmental impact may actually occur. Hence, eco-efficiency would have to be supplemented by consumer *sufficiency* (instead of insatiability, as neoclassical economic theory implies). This may yield a balanced form of environmental management contributed to by producers and consumers alike, which may be stimulated and supported by governments (cf. Fig. 9.9).

9.10.1.3 Population

According to recent reports by the Worldwatch Institute the world's population currently exceeds 6.5 billion and is expected to rise by some 76 million a year.

Around 1800 there were 1 billion people worldwide, around 1900 1.6 billion, and around 1950 some 2.5 billion. The rapid growth in recent decades is due mainly to improved health and sanitation. However, lagging economic security, low status of women, and poor birth-control services in the poorer countries are causing continuing high population growth, while populations in the wealthier industrialised countries are stabilising. Population pressures in various countries contribute to environmental problems associated with fresh water resources, agricultural land, (de)forestation and biodiversity. Experts believe that a reduction in population growth will follow in the wake of improvements in people's standard of living and increased economic security.

What can behavioural scientists do to analyse and support reproductive autonomy, especially that of young women in the poorer countries? Bandura (2002) demonstrates how social cognitive theory can be helpful in designing 'serial dramatisations' for mass media communications in Asia, Africa and Latin America. This kind of audiovisual broadcast can help people improve their understanding of, and control over their sex life, family planning and child rearing under better living conditions. Key ingredients are information provision, role models, goal-setting, self-efficacy, expected outcomes and support from the social environment.

> *one of the central themes (...) is aimed at raising the status of women so they have equitable access to educational and social opportunities, have a voice in family decisions and child bearing, and serve as active partners in their familial and social lives. This involves raising men's understanding of the legitimacy of women making decisions regarding their reproductive health and family life.* (Bandura 2002: 222)

From a more general point of view, behavioural scientists can analyse possible changes to traditional ('survival') inclinations to mate, reproduce and raise children. How worthwhile is it to live in partnership without having children? What kind of motivation and support are needed to secure a rewarding old age? How can status of women be enhanced? Can we further clarify the processes by which teenagers develop an image of parenthood, young adults decide to have children and older people evaluate their own ongoing parenthood, so that we can distinguish between basic human needs and sociocultural fashions and suppositions? How great a role in behaviour determination is played by expected utility, perceived risk, family habit, social comparison and conformity to social norms (cf. Table 9.9)? There is a rich domain for intercultural and intergenerational comparison here.

9.10.1.4 Institutions

One of the reasons why many environmental problems continue to persist is that institutions are generally focused on the achievement of one or several key goals. For example, government ministries of agriculture and fisheries are primarily concerned with promoting the cattle, poultry and fishing sectors of the economy. Likewise, the work of national tourist boards is geared to increasing the number of foreign tourists and the pattern of (paid) activities during their holiday period. In many sectors of society the desire for economic growth often means that less attention is being given to immediate as well as wider-spread and longer-term environmental impacts.

Given the recent retreat of liberal-democratic governments in free-market economies, the question arises whether new institutional arrangements are needed to effectively deal with common environmental problems. For example, reducing local environmental noise problems around airports, road junctions or railway yards could be accomplished collaboratively, by involving users, local government representatives and neighbourhood organisations, all of whom have an interest in improving their common environment. As another example, the scope of national tourist board activities might be widened by the participation of environmental groups, clean-technology experts and recreational planners ready to suggest sustainable tourism alternatives.

Behavioural science research on the environmental impacts (direct as well as indirect) of social and economic institutions may clarify the perceptions, attitudes, preferences and strategies of those institutions with respect to various environmental qualities. Such research may assist in developing ways and means of supporting internalisation of environmental sustainability, so that weighing common environmental interests becomes self-evident in the institutions' long-term plans and everyday practices.

9.10.1.5 Culture

The fifth and most difficult-to-influence driving factor of environmental decline steers the focus of research towards cultural premises, enduring beliefs and fundamental values, but also to ways of growing up, socialisation and familiarisation with the natural world and the importance of common environmental qualities.

'Culture' manifests itself in many different ways. It often determines gut feelings or emotional responses towards persons, objects or events. Culture involves self-evident knowledge, basic beliefs and fundamental values, which serve as the basis of many different kinds of behaviour. Hence, if the 'culture' regarding the natural environment were changed, many forms of explicit regulation (via laws, pricing policies or information provision) would become superfluous.

Given that the concept of culture embraces a great deal, behavioural science research in this area is rather broad in scope. It covers the micro, meso, and macro levels of society (e.g., in assessing basic beliefs and mutual influences), it looks into the origins of cultural expressions (e.g., in child-rearing and education) and it attempts to link specific behaviours to more fundamental beliefs and values. The Value Orientations, New Environmental Paradigm and Myths of Nature discussed above are very relevant here, as is Value-Belief-Norm theory; see Section 9.9.

However, what seems missing in research to date is a dynamic-temporal perspective: how and on what grounds do people gradually develop a particular set of values, beliefs, preferences and behavioural patterns vis-à-vis the natural environment? What 'alienation from nature' has occurred through educational practices and the fact that more and more people are born and bred in large urban areas, surrounded by motorised traffic? How could society organise everyday life in such a way that we become, or remain, properly aware of the relevance and importance of natural environmental qualities for the overall quality of human life? Understanding the meaning and day-to-day functioning of 'culture' is an important element of coming to grips with the sustainable development of population, consumption and technology as well as the institutions supporting them.

9.10.2 'Sustainable Development'

Sustainable development is a process aimed at achieving a sustainable balance between economic, social, and environmental qualities. For example, high economic growth may not be sustainable if is accompanied by severe environmental degradation. Similarly, environmentally benign consumption patterns may not be sustainable if they significantly reduce people's well-being. For each dimension, a minimum acceptable quality level can be defined, implying a need for trade-offs between dimensions above these levels. Minimum acceptable levels as well as trade-offs are, of course, a matter of political decision-making. For example, in industrial areas more emissions and noise may be tolerated for economic reasons, as long as living and working conditions remain at an acceptable level. Similarly, in natural areas economic activities causing significant environmental impacts might not be allowed unless local human livelihoods begin to fall below a reasonable subsistence level.

The concept of sustainable development has been much debated, used and misused to justify environmental policies as well as the acceptance of certain environmental problems. Important systematisation came about in the mid-1990s, as expressed in Table 9.15 for example.

Sachs (1997) has pointed out that the ecologist's concept of environmental sustainability has been combined with the 'Third World' issue of economic development. Thus the meaning of sustainable development goes back and forth between reducing environmental impact (a 'northern' issue) and limiting population growth (a 'southern' issue). Sachs distinguishes three different perspectives on sustainable development. According to the *contest* perspective, economic growth is inevitable and may (or will) go on indefinitely, but not equally in all parts of the world. The *astronaut's* perspective takes the earth as an integral system with a limited capacity to sustain its inhabitants. If the entire system is carefully monitored and assessed, the maximum or optimum combination of population, affluence and technology use can be inferred. The *home* perspective, finally, is about neither economic excellence nor biospheric stability, but is concerned with the sustainability of local livelihoods across the world, rooted in outside respect for local autonomy (in particular by 'developers') and local self-sufficiency, involving no significant resource extractions or waste disposal elsewhere. There is a certain degree of similarity between

Table 9.15 Three dimensions of sustainability

Economic sustainability	Social sustainability	Environmental sustainability
Efficient production/ consumption	Human well-being	Natural resources
Income growth	Consultation/ empowerment	Limited pollution
Stable development	Culture/heritage	Biodiversity/ resilience

Source: After Munasinghe (1993)

Sachs' perspectives and the Myths of Nature discussed in Section 9.9.1, with Sach's *contest* resembling *Nature Benign*, his *astronaut* sharing features with *Nature Tolerant*, and his *home* perspective coming close to *Nature Ephemeral*.

Methodologically, the three dimensions of sustainable development: economic, social and environmental (Table 9.15), can each be elaborated in terms of a limited number of issues and corresponding (more or less aggregate) measurement variables which can be used to assess the sustainability profile of a household, business, region or country. For example, a business company's *economic* performance can be charted in terms of employment, turnover, profits, share value and unpaid external costs. The company's *social* performance can be assessed by examining the well-being (or quality of life; cf. Section 9.9.2) of its employees, customers, suppliers, neighbouring residents and competitors. Its *environmental* performance, finally, can be measured in terms of energy and raw materials consumption, emissions and waste production, land use and transport demands.

Evaluating the sustainability of a particular entity can then be based on a two-step judgment. First, the (multivariate) assessment of economic, social and environmental performance is tested against a set of minimum quality standards with regard to, say, income or profits and amount of work or turnover (for 'economy'), quality of housing and community safety (for 'society'), and energy consumed and waste disposed of (for 'environment'). Second, insofar as these minimum quality levels are exceeded, trade-offs may be made between economic, social and environmental performances. This implies that the 'sustainable-quality profile' of different localities, regions or countries may differ, provided each satisfies the minimum economic, social and environmental quality standards. For example, one region may be stronger in an economic sense, while another may score higher on criteria of environmental quality.

9.11 What Behavioural Scientists Can Do to Enhance Environmental Qualities

Understanding and managing environmental problems is basically a four-step process, consisting of problem analysis, policy decision-making, practical intervention and evaluation of policy effectiveness. By its very nature, this process means that scientific expertise must be brought together from various sides. For their part, behavioural scientists have much to contribute in supporting the management of environmental problems at different scale levels. The focus of behavioural research and advice in this respect may be population, consumption and technology, as discussed in Section 9.10.1. Within these broad areas, behavioural scientists can invest their efforts in four different but interrelated kinds of activity, as follows.

First, behavioural scientists should 'look over the fence', moving outside their own basic body of theory and methodology and familiarising themselves with specific domains of environmental problems. Examples include family planning, agriculture, motorised transport, recreation and tourism, waste management and military defence activities. Without such familiarisation, behavioural scientists will be unable to explain their own meaningful contribution. Nor will they be listened to or taken seriously when they draw policy-oriented conclusions. Behavioural scientists, too, should therefore know their population statistics, energy sources, greenhouse gases, fertilisers and pesticides, clean-up technologies, waste handling facilities, energy-efficiency strategies and innovative transport options. This involves physical-environmental and technological 'homework'. It also involves transdisciplinary networking and smart use of knowledge that other experts can readily provide.

Second, behavioural scientists have the capacity to conceptualise and analyse environmental problems as socio-behavioural problems. For example, urban air pollution may be conceived as a problem due to many individual emitters of harmful exhaust gases whose motivation rests largely on the immediate needs and benefits involved, as well as on perceived social norms. At the same time, less weight may be assigned to their personal contribution to a collective environmental problem, which may be seen as 'uncontrollable anyway' (see especially Section 9.6.4). The researcher's main job here is to demonstrate that air pollution problems are man-made problems, the driving forces of which can be unravelled, and that such problems can be alleviated by means of programmes of behavioural change that are attuned to the various specific causes and determinants of air-polluting behaviours. This kind of message should also be taken 'across the fence', with a view to effectively reaching researchers and policy-makers from other disciplinary backgrounds.

Third, behavioural scientists can conduct empirical studies to test and revise conceptual models and hypotheses about specific environmental problems. For example, research may be able to clarify the reasons behind rising household energy consumption, personal opinions about and (lack of) potential for energy-saving behaviour, and people's expectations about future quality of life if fossil energy consumption were to be seriously discouraged. Behavioural studies may also be geared towards explaining the intensification of motorised transport, users' perceptions of the associated environmental problems, and the feasibility of sustainable-transport policies involving reduced use of private motor vehicles. There is already a certain tradition in this kind of work, but there is room for improvement in the scope of the research as well as its conceptual basis. In particular, there is a need for further interlinking of the psychological, economic and environmental-science approaches.

Fourth, behavioural scientists can assist policy-makers and other professionals in designing, applying and evaluating strategies for the development of sustainable behaviour patterns, lifestyles and business cultures. Policywise, this is the key issue: given our ever-growing environmental problems and the serious risks ahead, what might reasonable sustainability goals entail, what behavioural options might be feasible (and for whom?) and how might the required changes in current behavioural practices actually be brought about? Behavioural theory, and methodology can be of great help in clarifying these questions and in providing meaningful answers, once the goals are set by policy makers. Policy-design research is needed in which technical as well as behavioural change scenarios are developed and tested for their feasibility, acceptability and eventual effectiveness.

9.11.1 Positivistic and Constructive Behavioural Science

The last of these categories of behavioural-science work brings us to a distinction that is of fundamental importance for environmental-behavioural research. This is the difference between a pure, 'value-free', positivistic approach to sustainability problems and an applied, 'service-driven', constructive approach. A *positivistic* behavioural science of sustainable development means that theory and research should be given a broader-than-usual scope anyway, in order to describe, explain and predict environmental quality as a function of human behaviours and as a prerequisite for human quality of life. Alternatively, a *constructive* behavioural science of sustainable development implies lifestyle visions, societal scenarios and behaviour-change strategies aimed at enhancing environmental quality as an essential condition for human well-being. Positivistic and constructive environmental-behavioural science should supplement and support one another rather than stand in opposition. Given the scale and gravity of environmental problems our world is facing, classical, neopositivistic research may not be enough. Policy scenarios need to be designed, evaluated and tested for their practical significance not only in the short term, but in the long term as well.

9.12 Conclusion

In the present chapter we have tried to order and explicate the principal elements of a behavioural science perspective on environmental problems and the management of sustainable development. As psychologists we have inevitably paid greatest attention to the characteristics, behavioural processes and potential for change of individual actors (possibly as members of households, business companies, or government agencies). We acknowledged that individual attitudes and behaviours are often embedded in larger social and organisational structures, which may be steering their own course under the influence of wider driving forces such as population, affluence and technology. These social and organisational developments may be the subject of strategic decisions and actions by governments and/or businesses. Thus, behavioural and social research may well cover human perceptions, attitudes, decisions and actions not just at the micro level of society, but also at the meso and macro levels.

A behavioural science perspective on environmental problems can yield useful knowledge and understanding of the human sources, behavioural processes and specific determinants of environmental change. Moreover, it may permit one to predict the likely effects of environmental policies, directly or indirectly, on people's future quality of life: on their health, autonomy, personal development and material wealth, for example. There are many questions open for further research and theorising, for example about environmental risk perception, the appreciation of common goods, the effectiveness of strategies for behavioural change, and the mobilisation of social motivation for achieving long-term sustainability. Thus – as we hope to have shown – behavioural scientists are in a fruitful position to fill important gaps in the environmental sciences as these have developed over the past 30 years.

References

Note: Throughout this handbook detailed use of references has been avoided. For the present chapter the selection below covers the actual text references as well as some major review articles and books on the various topics treated in the text. The items below have not been otherwise classified; we trust the titles of articles and books speak for themselves.

Bandura, A. (2002). Environmental sustainability by sociocognitive deceleration of population growth. In P. Schmuck & W. Schultz (Eds.), *Psychology of sustainable development* (pp. 209–238). Norwell, MA: Kluwer.

Dawes, R. M. (1980). Social dilemmas. *Annual Review of Psychology, 31*, 169–193.

Dunlap, R. E., Van Liere, K. D., Mertig, A. G., & Jones, R. E. (2000). Measuring endorsement of the new ecological paradigm: A revised NEP scale. *Journal of Social Issues, 56*(3), 425–442.

Edney, J. J. (1980). The commons problem; alternative perspectives. *American Psychologist, 35*, 131–150.

Ehrlich, P. R., & Holdren, J. P. (1971). Impact of population growth. *Science, 171*, 1212–1217.

Gardner, G. T., & Stern, P. C. (2002). *Environmental problems and human behavior* (2nd ed.). Boston: Pearson Custom.

Hardin, G. (1968). The tragedy of the commons. *Science, 162*, 1243–1248.

Hirsch, F. (1976). *Social limits to growth*. Cambridge, MA: Harvard University Press.

Jager, W., & Mosler, H. J. (2007). Simulating human behavior for understanding and managing environmental resource use. *Journal of Social Issues, 63*(1), 97–116.

Liebrand, W. B. G., & Messick, D. M. (Eds.). (1996). *Frontiers in social dilemmas research*. Berlin: Springer.

Maslow, A. H. (1954). *Motivation and personality*. New York: Harper & Row.

Midden, C., Kaiser, F., & McCalley, T. (2007). Technology's four roles in understanding individuals' conservation of natural resources. *Journal of Social Issues, 63*(1), 155–174.

Munasinghe, M. (1993). *Environmental economics and sustainable development*. Washington, DC: World Bank.

Ormel, J., Lindenberg, S., Steverink, N., & Vonkorff, M. (1997). Quality of life and social production functions: A framework for understanding health effects. *Social Science and Medicine, 45*(7): 1051–1063.

Ostrom, E., Dietz, T., Dolšak, N., Stern, P. C., Stonich, S., & Weber, E. U. (Eds.). (2002). *The drama of the commons*. Washington, DC: National Academy Press.

Poortinga, W., Steg, L., & Vlek, C. (2004). Values, environmental concern, and environmental behavior. *Environment and Behavior, 36*, 70–93.

Ritov, I., & Kahneman, D. (1997). How people value the environment: Attitudes versus economic values. In M. H. Bazerman, D. M. Messick, A. E. Tenbrunsel, & K. A. Wade-Benzioni (Eds.), *Environment, ethics and behavior. The psychology of environmental valuation and degradation* (pp. 33–51). San Francisco: The New Lexington Press.

Sachs, W. (1997). Sustainable development. In M. Redclift & G. Woodgate (Eds.), *The international handbook of environmental sociology* (pp. 71–82). Cheltenham: Edward Elgar.

Schwartz, S. H. (1992). Universals in the content and structure of values: Theoretical advances and empirical tests in 20 countries. In M. P. Zanna (Ed.), *Advances in experimental psychology* (Vol. 5, pp. 1–65). New York: Academic Press.

Steg, L., Dreijerink, L., & Abrahamse, W. (2005). Factors influencing the acceptability of energy policies: Testing VBN theory. *Journal of Environmental Psychology, 25*(4), 415–425.

Stern, P. C. (2000). Toward a coherent theory of environmentally significant behaviour. *Journal of Social Issues, 56*(3), 407–424.

Tietenberg, T. (2003). The tradable-permits approach to protecting the commons: Lessons for climate change. *Oxford Review of Economic Policy, 19*(3), 400–419.

Ulrich, R. S. (1993). Biophilia, biophobia and natural landscapes. In S. R. Kellert & E. O. Wilson (Eds.), *The biophilia hypothesis* (pp. 73–137). Washington, DC: Island Press.

Vlek, C. (2002). Environmental psychology/perception. In R. E. Munn (Ed.), *Encyclopedia of global environmental change* (Vol. 5, pp. 257–269). New York: Wiley.

Vlek, C. (2000). Essential psychology for environmental policy making. *International Journal of Psychology, 35*(2), 153–167.

Vlek, C., & Keren, G. B. (1992). Behavioral decision theory and environmental risk management: Assessment and resolution of four 'survival' dilemmas. *Acta Psychologica, 80*, 249–278.

WCED (World Commission on Environment and Development). (1987). *Our common future*. New York: Oxford University Press.

Further Readings

Abrahamse, W., Steg, L., Vlek, Ch., & Rothengatter, J. A. (2005). A review of intervention studies aimed at household energy conservation. *Journal of Environmental Psychology, 25*, 273–291.

American Psychologist. (May 2000). Special section on psychology and sustainable development (contributions by S. Oskamp, G. Howard, D. DuNann Winter, P. Stern, & D. McKenzie-Mohr).

Bartels, G., & Nelissen, W. (Eds.). (2002). *Marketing for sustainability: Towards transactional policy making*. Amsterdam/Burke, VA/Leipzig/Oxford/Tokyo: IOS Press.

Bazerman, M. H., Messick, D. M., Tenbrunsel, A. E., & Wade-Benzioni, K. A. (Eds.). (1997). *Environment, ethics and behavior. The psychology of environmental valuation and degradation*. San Francisco: The New Lexington Press.

Bechtel, R. B., & Churchman, A. (Eds.). (2002). *Handbook of environmental psychology*. New York: Wiley.

Bell, P. A., Greene, T. C., Fisher, J. D., & Baum, A. (2000). *Environmental psychology* (5th ed.). Fort Worth, TX/New York: Harcourt Brace College.

Bonnes, M., & Secchiaroli, G. (1995). *Environmental psychology: A psycho-social introduction*. London: Sage.

Cobb, G. W. (1998). *Introduction to design and analysis of experiments*. New York/Berlin/Heidelberg: Springer.

Cummings, R. G., Brookshire, D. S., & Schulze, W. D. (1986). *Valuing environmental goods: An assessment of the contingent valuation method*. Totowa: Rowman & Allanheld.

Diekmann, A., & Franzen, A. (1995). *Kooperatives umwelthandeln*. Chur/Zürich, CH: Verlag Rüegger.

Evans, G. W., & Cohen, S. (1987). Environmental stress. In D. Stokols & I. Altman (Eds.), *Handbook of environmental psychology* (Vol. 1, pp. 571–610). New York: Wiley.

Gatersleben, B., Steg, L., & Vlek, C. (2002). Measurement and determinants of environmentally significant behavior. *Environment and Behavior, 34*, 335–362.

Geller, E. S. (2002). The challenge of increasing proenvironmental behavior. In R. B. Bechtel & A. Churchman (Eds.), *Handbook of environmental psychology* (pp. 525–540). Chichester/New York: Wiley.

Geller, E. S., Winett, R. A., & Everett, P. B. (1982). *Preserving the environment: New strategies for behavior change*. Elmsford: Pergamon.

Gifford, R. (2002). *Environmental psychology: Principles and practice* (3rd ed.). Colville, WA: Optimal Books.

Hoevenagel, R. (1994). *The contingent valuation method: Scope and validity*. Amsterdam: Free University Press (doctoral dissertation, Department of Economics).

Journal of Social Issues. (1995, 2000, 2007). Special issues on 'Promoting environmentalism' (51, 4), 'Psychology and the promotion of a sustainable future' (56, 3), and 'Human behavior and environmental sustainability' (63, 1), respectively.

Kaplan, S., & Kaplan, R. (1989). *The experience of nature: A psychological perspective*. New York/London: Cambridge University Press.

Kerlinger, F. L., & Lee, H. B. (2000). *Foundations of behavioral research*. London: Wadsworth Thomson Learning.

Koelega, H. S. (Ed.). (1987). *Environmental annoyance: Characterization, measurement and control*. Amsterdam: Elsevier.

Krimsky, S., & Golding, D. (Eds.). (1992). *Social theories of risk*. Westport, CT/London: Praeger.

Max-Neef, M. (1992). Development and human needs. In P. Ekins & M. Max-Neef (Eds.), *Real-life economics: Understanding wealth creation* (pp. 91–108). London/New York: Routledge.

Noorman, K. J., & Schoot Uiterkamp, T. (Eds.). (1998). *Green households? Domestic consumers, environment and sustainability*. London: Earthscan.

Ostrom, E. (1990). *Governing the commons: The evolution of institutions for collective action*. Cambridge/New York: Cambridge University Press.

Robinson, J. (2004). Squaring the circle? Some thoughts on the idea of sustainable development. *Ecological Economics, 48*, 369–384.

Spaargaren, G. (2003). Sustainable consumption: A theoretical and environmental policy perspective. *Society and Natural Resources, 16*, 687–701.

Steg, L., & Gifford, R. (2005). Sustainable transport and quality of life. *Journal of Transport Geography, 13*, 59–69.

Steg, L., & Sievers, I. (2000). Cultural theory and individual perceptions of environmental risks. *Environment and Behavior, 32*, 250–269.

Stern, P. C., Dietz, T., Ruttan, V. W., Socolow, R. H., & Sweeney, J. L. (1997). *Environmentally significant consumption; research directions*. Washington, DC: National Academy Press.

Schmuck, P., & Schultz, W. (Eds.). (2002). *Psychology of sustainable development*. Norwell, MA: Kluwer.

Schwartz, S. H. (1977). Normative influences on altruism. In L. Berkowitz (Ed.), *Advances in experimental social psychology* (Vol. 10, pp. 221–279). New York: Academic Press.

Stokols, D., & Altmann, I. (Eds.). (1987). *Handbook of environmental psychology*. New York: Wiley.

Svendsen, G. T., & Vesterdal, M. (2003). How to design greenhouse gas trading in the EU? *Energy Policy, 31*, 1531–1539.

Van den Berg, A. E., Hartig, T., & Staats, H. (2007). Preference for nature in urbanized societies: Stress, restoration, and the pursuit of sustainability. *Journal of Social Issues, 63*(1), 79–96.

Van Vugt, M. (Ed.). (2000). *Cooperation in modern society: Promoting the welfare of communities, states and organizations*. London: Routledge.

Vlek, C. (2004). Environmental versus individual risk taking: Perception, decision, behavior. In C. Spielberger (Ed.), *Encyclopedia of applied psychology* (Vol. 1, pp. 825–840). Amsterdam/Boston: Elsevier.

Von Weizsäcker, E. U., Lovins, A. B., & Lovins, L. H. (1997). *Doubling wealth, halving resource use; the new report to the Club of Rome*. London: Earthscan.

Winter, D. D., & Koger, S. M. (2004). *The psychology of environmental problems* (2nd ed.). Mawhah, NJ: L. Erlbaum.

WWI (Worldwatch Institute). (2006). *Vital signs 2006–2007*. Washington, DC: WWI. www.worldwatch.org (accessed August 2007).

Yoon, K. P., & Hwang, C.-L., (1995). *Multiple attribute decision making, an introduction*. Thousand Oaks/London/New Delhi: Sage (University Paper no. 104).

Chapter 10
The Natural Environment

Piet H. Nienhuis with contributions of Egbert Boeker (Sections 10.2 and 10.3)

Contents

10.1	**Introduction**	143
10.2	**Principles of Physics**	144
10.2.1	Mass Balance	144
10.2.2	Energy Balance	145
10.2.3	Mass-Transport Processes	148
10.3	**Principles of Chemistry**	149
10.3.1	Chemical Kinetics and Thermodynamics	149
10.3.2	Chemical Equilibrium	151
10.3.3	Oxidation and Reduction	151
10.3.4	Photochemical and Biochemical Transformations	152
10.4	**Principles of Biology**	154
10.4.1	Genetic Variability	154
10.4.2	Population Dynamics	154
10.4.3	Ecosystem Structure and Function	155
10.4.4	Energy Flow	156
10.4.5	Materials Cycling	157
10.5	**Spatial and Temporal Dimensions of the Environment**	159
10.5.1	The Question of Spatial and Temporal Scales	159
10.5.2	Environmental Distribution of Chemicals	161
10.5.3	Distribution of Biodiversity Units – Mapping and Monitoring	162
10.5.4	Geographical Information Systems: GIS	163
10.5.5	Remote Sensing	165
10.6	**Integrated Indicators for Environmental Quality Assessment**	167
10.6.1	Index of Biotic Integrity – IBI	167
10.6.2	The TRIAD Method	167
10.6.3	Ecosystem Health	168
10.6.4	The AMOEBA Method	172
10.7	**Summarising Conclusions**	174
References		175

Piet H. Nienhuis (✉)
Department of Environmental Science, Institute for Wetland and Water Research, Radboud University Nijmegen,
P.O. Box 9010, 6500 GL Nijmegen, The Netherlands
e-mail: p.nienhuis@science.ru.nl

10.1 Introduction

This chapter discusses a range of concepts and methods for analysing 'the natural environment', here considered as the physical, chemical and biological (i.e. living and non-living) environment and as the 'resource base' of human society, to which it thus bears a reciprocal relationship (Boersema et al. 1991: 22). This definition does not include the social environment, which is treated in Chapter 9. Although the natural and engineering sciences potentially provide a plethora of methods for studying the environment, we shall here restrict ourselves to those used specifically for analysing and resolving environmental problems, which we shall here take to mean an actual or potential deterioration of the quality of the environment, or a disturbance of the relationship between environment and human society (Boersema et al. 1991). The term 'environmental quality', in turn, is taken to comprise the structural and functional properties of the environment in the context of human appreciation, either positive or negative (Boersema et al. 1991).

In this chapter the physical environment is considered as the object of study of the environmental scientist. By implication, then, the focus will be on the application of knowledge from the natural sciences for solving environmental problems. Every environmental scientist should thus have some basic working knowledge of the 'laws' of nature. For the basics of physics and chemistry this chapter relies heavily on the writings of Mihelcic (1999), while for biology – here mainly ecology – various other sources have been consulted. The source of knowledge from the natural sciences to be applied in environmental science originates from practical observation and experimentation in a wide range of component disciplines. Field and laboratory studies are fundamental components of undergraduate training in most aspects of environmental science. The skills required range from accurate observation and recording of field data to operation of sophisticated field and laboratory equipment. Students are expected to design practical investigations, keep records of their work, analyse data and present and communicate their results in writing as well as orally. Jones et al. (2000) have published a useful book on the practical skills required by the environmental scientist. The book's main emphasis is on fieldwork and taking field measurements and samples, but it also includes sections on basic laboratory procedures and transferable skills such as computing and study skills.

The methods from the natural sciences to be applied in environmental science range between purely technological applications to normative societal applications. Metaphorically, technology has no 'conscience'. When engineers perform their observations and experiments they are 'merely' counting, measuring and weighing. The limits to technology and the standards defining those limits must be set by society. A recent example of technologically driven engineering is recombinant DNA technology, i.e. genetic engineering of plants and animals, including those used as human food. This kind of engineering can potentially lead to serious deterioration of environmental quality and hence to environmental problems. Although these problems are within the domain of the environmental scientist, they will not be discussed in this introductory chapter, however, because the methods of genetic engineering are not specifically designed for solving environmental problems. Similar considerations hold for the specific methodologies applied in such fields as biochemistry, endocrinology and physiology. The criterion used in selecting the methods explained in this chapter, in other words, is their usefulness for analysing or resolving environmental problems. Priority has thereby been given to methods that are particularly useful in multi- and interdisciplinary research and consequently of greatest relevance to society as a whole.

This chapter thus has two aims: to set out the basic principles of physics, chemistry and biology, to the extent that these are relevant for analysing and solving environmental problems, and to review a number of key multi- and interdisciplinary concepts and methods used to this same end.

10.2 Principles of Physics

10.2.1 Mass Balance

Physical processes are of key importance in the movement of pollutants through the environment, as well as in the processes used to treat and control pollutant emissions. An understanding of mass and energy balances is therefore an essential prerequisite for solving a wide range of problems in environmental engineering and environmental science.

> **Box 10.1** Mass balance of the North American Great Lakes
>
> The North American Great Lakes contain approximately 18% of the world's supply of freshwater, making them the largest freshwater system on earth. The first humans arrived in the area approximately 10,000 years ago. Around 6,000 years ago copper mining began along the south shore of Lake Superior and hunting and fishing communities were established throughout the area. In the 16th century the regional human population is estimated to have been between about 60,000 and 117,000, small enough to leave the natural environment relatively undisturbed. Today, the combined Canadian and U.S. population of the region exceeds 33 million. The increased intensity of settlement and exploitation over the past two hundred years has disturbed the ecosystem here in numerous ways. The outflow from the Great Lakes is less than 1% per year and any pollutants entering the system via the air, direct aqueous discharges or non-point pollution sources such as motorised traffic, can consequently remain in the lake system for a long time. The 'retention time' of Lake Michigan is 136 years, that of Lake Ontario, connected to the St. Lawrence River, 8 years. As this example shows, methods from the realm of physics can be used to generate insight into the mass balance of lakes and hence the magnitude of the environmental problems requiring a solution.
> Source: Honrath and Mihelcic (1999)

The *law of conservation of mass* states that mass can be neither created nor destroyed. This notion of *mass balance* means that if there is an increase in the amount of a chemical present in a given environment such as a lake, this cannot be due to some sort of 'magical' process of formation. The chemical must either have been transported into the lake from elsewhere or produced by a chemical or biological reaction from other compounds already present in the lake. Similarly, if these reactions were responsible for the mass increase of this particular chemical, they must also have caused a corresponding decrease in the mass of one or more other compounds. Thus, conservation of mass provides a basis for compiling a mass budget of the chemicals of interest. In the case of the lake, this budget keeps track of the amount of chemical entering the lake, the amount leaving and the amount created or destroyed by chemical or biological reaction. This budget can be balanced over a given period of time, just like a bank balance. To have any meaning, a mass balance must obviously be specific to a well-defined region of space, with boundaries across which the mass flux in and out of the system can be determined (Honrath and Mihelcic 1999; see Box 10.1 and Chapter 12, Section 12.4 on Material Flow Analysis).

10.2.2 Energy Balance

The movement of *energy* (i.e. the ability to do work) and changes in its form can be tracked by means of an *energy balance*, analogous to a mass balance. The first law of thermodynamics states that in an isolated closed system, energy will be conserved and can be neither created nor destroyed. Just as the law of conservation of mass permits mass balances to be drawn up, conservation of energy provides a basis for energy balances.

A complication in applying an energy balance is that energy may appear in several forms. In the environment the most important ones are potential energy, kinetic energy, sensible heat, latent heat and radiation. It is total energy that is conserved, not necessarily the various forms by themselves. We give an example of each form and terminate this discussion with the radiation balance of the Earth and the atmosphere with incoming solar radiation, which leads to the discussion of the greenhouse effect.

Potential energy is related to an external force. A rock on a mountain has a potential energy with respect to the valley below, due to the gravitation of the Earth. When the rock moves down, it gains speed; the rock gets kinetic energy until down in the valley all potential energy has been converted into kinetic energy. In this case the sum of potential energy and kinetic energy is conserved.

When one puts a kettle of water on a fire, the energy of the fire is converted into sensible heat and latent heat. The first expresses that the temperature of the water goes up, which one may feel with the hands. The second, latent heat, expresses that part of the water will evaporate. One may observe this on a cold glass pane, where the water condenses. The invisible water vapour contains latent heat, which is liberated when condensing, increasing the temperature of the glass somewhat. See also Box 10.2.

Water evaporates, even when it does not boil, as one experiences when drying wet clothes in the air. When the sun is shining, it is irradiating the surface of the oceans. Part of the radiation energy is used to heat the oceans and therefore is converted into sensible heat. Part of the radiation will evaporate the water and is converted into latent heat of the water vapour. This vapour disperses in the air, and may join the air in an upward movement, where high up it cools, condensates and forms clouds.

The concepts introduced here are used to study the radiation balance for Earth and atmosphere. A simpli-

---

Box 10.2 Sensible heat and latent heat

In order to get a an idea of the relative importance of sensible heat and latent heat, consider 1 m^3 of dry air with a temperature of 10°C just above the ocean surface. Let us assume, for the sake of the argument, that it contains no water vapour. After a few hours of sunshine the air has got a temperature of 20°C, say, and has got a humidity of 80%, for example, which means that it has 80% of the maximum amount of water vapour that air may contain without condensation. Consulting tables (Kaye and Laby 1995) one finds that the warm air has a mass of 1.23 kg and contains 13.9 g of water vapour, with a latent heat of 34.1 kJ.[1] The sensible heat of the air has increased by 12.4 kJ. Therefore, in this example, the solar irradiation was converted more into latent heat of the water vapour than in sensible heat of the air. Besides it is, of course, heating the ocean waters as well.

---
[1] kJ is pronounced as kilo Joule and is a unit of physical energy. It is roughly the kinetic energy which 1 kg of rock gains when it falls down 100 m.

fied summary of the experimental data is given in Fig. 10.1. We use this figure to illustrate the way climate models work and therefore will introduce a few relevant concepts while discussing the balance.

The radiation balance is deduced from experimental data and is written down as an average over the globe and over the year. The sunlight entering the atmosphere is put as 100% and comprises visible light as well as infrared and ultraviolet radiation. A fraction $a_a = 0.30$ is scattered back to outer space. Clouds, gases and particles in the air cause this scattering and reflection. Another fraction is absorbed by the atmosphere, leaving a transmission t_a reaching the Earth surface, where most of the sunlight (0.47) is absorbed but a fraction $a_s = 0.11$ is scattered back, much of it by snow and ice, but part of it also by forests and crops. The absorption by the surface therefore amounts to $(1 - a_s)t_a$, which gives a transmission $t_a = 0.53$.

Part of the absorbed energy is used for evaporation, resulting in latent heat within the atmosphere and another part heats the lower part of the atmosphere, resulting in sensible heat and rising air, a phenomenon called convection. In a simple model one may assume that both are proportional to the temperature difference $(T_s - T_a)$ between the surface of the Earth and the atmosphere, resulting in an upwards-going energy of $c(T_s - T_a)$. This process heats the atmosphere, which already is heated by the direct absorption of sunlight, discussed above.

In physics any body with a temperature T_s or T_a above the absolute zero of $T = 0$ (or $-273°C$) is radiating energy, so-called thermal radiation. For temperatures like we have on Earth this radiation is invisible, in the far infrared. One may feel it however by putting the hand below a hot car exhaust. There one cannot feel the rising hot air, so one feels the radiation. Numerically the radiation can be approximated by the formulas σT_s^4 or σT_a^4, where σ is the Stefan-Boltzmann constant. In Fig. 10.1 the Earth surface is emitting a fraction 1.14 of the incoming solar energy upwards as thermal radiation. This is more than the total of the incoming sunlight, which is only possible by the presence of the atmosphere. The data at the right hand side of Fig. 10.1 suggest that the atmosphere emits a fraction 0.63 of the incoming sunlight up and down, while a fraction 0.35 originating from the surface radiation is scattered down. This results in a fraction $a'_a = 0.31$ of the Earth radiation (which equals 1.14) coming back down. Most of the thermal radiation of the Earth is absorbed by the atmosphere but 0.07 out of 1.14 is transmitted, resulting in a transmission coefficient of $t'_a = 0.06$.

It follows that the existence of the atmosphere makes it possible for the Earth surface to have a much higher temperature than without an atmosphere. Without atmosphere the thermal emission of the earth could not be higher than 100% or 1.00, and the result would be a much lower temperature. This heating effect of the atmosphere is called the *greenhouse effect*. In order to calculate the consequence of human induced changes on the greenhouse effect, one needs physical models. The simplest possible model is given in Fig. 10.2, where one will recognize the parameters introduced above.

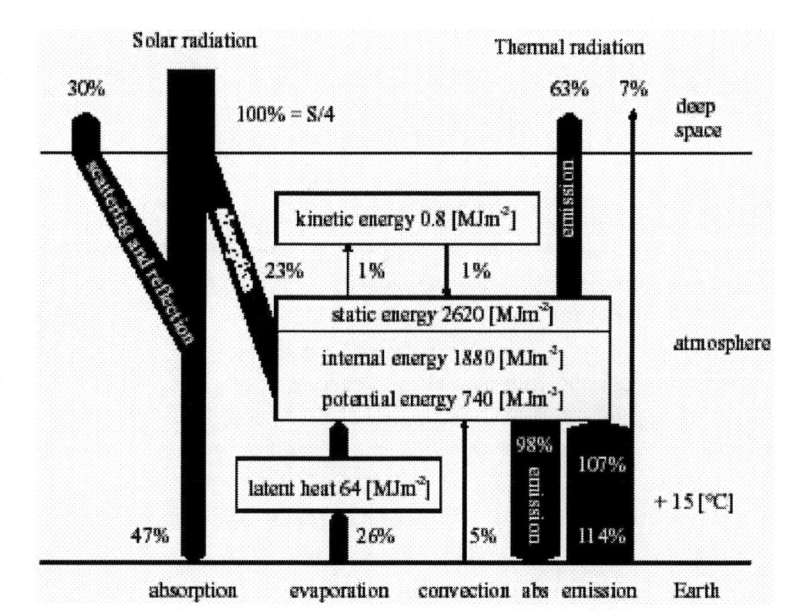

Fig. 10.1 Radiation balance for the Earth and the atmosphere. The solar radiation on the left is put as 100% and its destiny displayed. On the right, long wave radiation by Earth and atmosphere is presented. In the blocks one finds the energy contents of the atmosphere

Source: Reproduced by permission of the Department of Geography, ETH Zürich from Hutter et al. (1990)

The main effect of human interference up to the present is the combustion of fossil fuels, with their emissions of CO_2. This makes the atmosphere even less transparent and reduces the parameter t'_a. Combustion also brings more particles in the air, which will result in a smaller transmission t_a of solar radiation and perhaps more back radiation a'_a. The effect of these changes may be studied by putting the model in a computer.

The model of Fig. 10.2 is much too simplified. For example, it does not take into account the fact that the ocean will heat only slowly, delaying the effects of climate change. A first approximation of the effect of the oceans may be found on the website, mentioned above. More realistic models will imitate the geography of the Earth, divide the atmosphere in ten or more layers as well as the oceans, and take all kinds of horizontal movements of the air, such as the winds, into account. Sophisticated climate models are being improved all the time and the Intergovernmental Panel on Climate Change (Houghton et al. 1995) regularly discusses their results. Website: www.ipcc.ch. A summary is given in Table 10.1.

In the second column of Table 10.1 one will find the gases, which are responsible for the greenhouse effect. In the fourth column their temperature effect is displayed. They add up to 33°C, which is the reason why the Earth is habitable. The *greenhouse gases*, indicated with an asterisk in the first column, are mentioned in the Kyoto Protocol of 1997. The point is that all of them are increasing by the emissions of our industrial society. That increase is indicated in the sixth column and will result in a human induced global warming, the extra on top of natural variations. In order to obtain policy agreements one needs a single figure, which is taken as the equivalent addition of CO_2. Most of the greenhouse gases have a stronger greenhouse effect than CO_2; the multiplication factor with respect to CO_2 depends on the time scale and is given in the last two columns of Table 10.1, for 20 years and for 100 years. For methane (CH_4) for example the multiplication factor, called Global Warming Potential, GWP, is 56 on a time scale of 20 years and 21 on a scale of 100 years. The Kyoto Protocol does not give a time scale but leaves that to future negotiations.

In order to predict a global warming in 2100, say, one needs two sets of models. First an economic model, which gives the economic growth and the resulting concentration of all greenhouse gases, including reductions instigated by governments. Second one has to calculate the effective CO_2 concentrations and compute their effects in 2100. In all these calculations there are uncertainties and best guesses. The IPCC forecasts for 2100 therefore vary between a temperature increase of 1°C and 6°C, while most of the models indicate a value around 3°C.

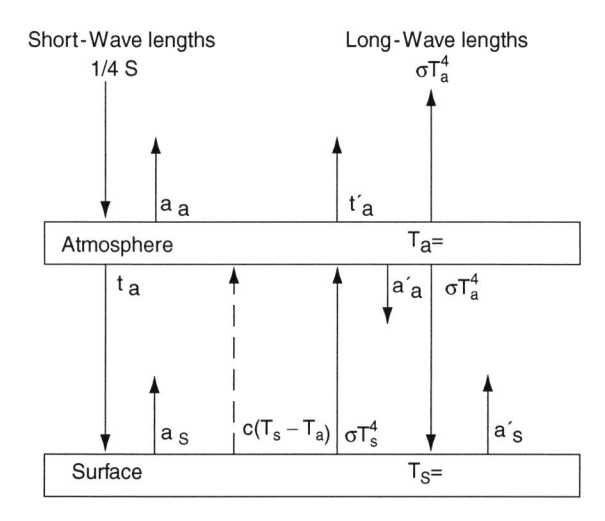

Fig. 10.2 A zero-dimensional greenhouse model, based on an energy balance and easy to computerize
Source: With permission of Wiley reproduced from Boeker and Van Grondelle (1999)

Table 10.1 Greenhouse gases and their effects

K	Gas	Conc. [ppmv]	Warming [°C]	Lifetime [year]	Increase [% year⁻¹]	GWP (20 years)	GWP (100 years)
	H₂O vapour	5 × 10³	20.6				
*	CO₂	358	7.2	50–200	0.45	1	1
	O₃ tropo-sphere	0.03	2.4				
*	N₂O	0.3	1.4	120	0.25	280	310
*	CH₄	1.7	0.8	12 ± 3	0.47	56	21
*	HFC's			1–200	5?	400–9,000	100–12,000
*	PFC's		0.6	−50,000	2?	4,000–6,000	6,000–24,000
*	SF₆			3,200	0.6	16,300	23,900

Source: From IPCC and other sources; see Boeker and Van Grondelle 2001. Greenhouse gases (from Kyoto Protocol 1997) are indicated by an asterisk. 'GWP' = Global Warming Potential

10.2.3 Mass-Transport Processes

Mass-transport processes are also physically driven. *Convection* occurs when a gas or fluid gains heat and hence expands and moves from one place to another. The processes of *advection* and *dispersion* are responsible for the transport of pollutants through the environment. *Advection* refers to transport in the mean direction of flow of the air, water or another fluid. If the wind is blowing from the west for example, advection will carry any pollutant present in the atmosphere to the east. Similarly, if a bag of dye is emptied in the middle of a river, advection will carry the resulting spot of dye downstream. Environmental pollutants can be transported advectively over major distances by the movement of wind and water currents.

In contrast, *dispersion* refers to the transport of compounds as a result of random motion. Dispersion works to eliminate sharp discontinuities in pollutant concentrations and leads to smoother, randomly distributed concentration profiles. Advective and dispersive processes can usually be modelled independently. In the case of the dye in the river, while advection moves the centre of the mass downstream, dispersion acts to spread out the concentrated spot of dye over a larger area. This type of dispersion is called *turbulent dispersion*, or eddy dispersion.

The simplest model for dispersion without advection refers to a drop of ink of 1 kg, which is put in the middle of a narrow tube, which is indicated as an x-axis. The ink will disperse to the left and the right and may be described by a Gauss shape. Mathematically this is expressed by:

$$f(x) = \frac{1}{\sigma\sqrt{2\pi}} e^{-x^2/2\sigma^2} \qquad (10.1)$$

The factor in front of the exponential function guarantees that the total surface area of the function equals one. This makes it possible to interpret the value $f(x)\,dx$ as the amount of the ink between x and $x + dx$. The function (10.1) is displayed in Fig. 10.3 for several values of the parameter σ. One will notice that for increasing values of σ the function becomes lower and wider. This is precisely what happens in a dispersing drop of ink. Therefore σ should be connected with the time since the drop started to disperse. It is expressed as:

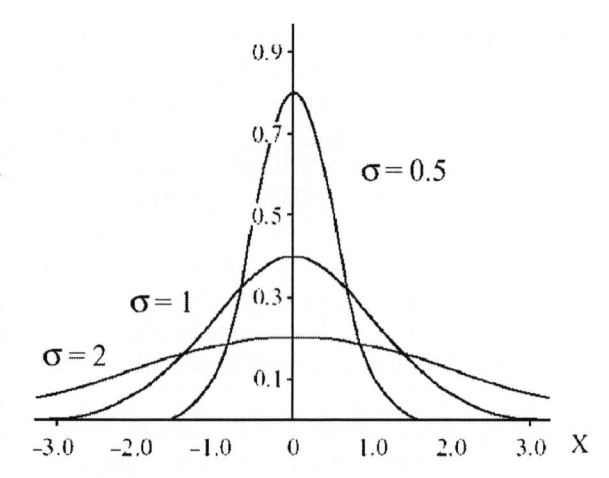

Fig. 10.3 The Gauss function of Equation (10.1) for three different width parameters σ

$$\sigma = \sqrt{2Kt} \qquad (10.2)$$

where K is a dispersion coefficient. In three dimensions one may generalize Equation (10.1) by assuming that an amount Q kg of pollutant is emitted at time $t = 0$ and disperses into the three directions x, y and z. One finds:

$$f(x,y,z) = \left(\frac{1}{\sigma\sqrt{2\pi}}\right)^3 e^{-(x^2+y^2+z^2)/2\sigma^2} \qquad (10.3)$$

Again, Equation (10.2) holds, while in a turbulent medium K is called the coefficient of turbulent dispersion. Its value is often determined semi-empirically.

Up to this point, advection was not considered. Advection occurs when there is one dominant direction of transport, for example the direction of a prevalent wind in the atmosphere or the direction of a stream in a river. This dominant direction of this flow of air or water usually is taken as the x-direction. In that direction, advection, which may be seen as the 'gluing' of the pollutants to the flow particles, is the main means of mass transport. A complication is that physically the turbulence in the three dimensions, the horizontal x, the other horizontal y perpendicular to the flow and the altitude z are not equivalent. For it is clear that both in air and in rivers horizontal dispersion y has fewer constraints than vertical dispersion z. This leads to different dispersion coefficients K in the three directions.

For the accidental discharge of a poisonous pollutant in a river of Q kg, the discussion may be simplified.

10 The Natural Environment

For one may assume that the pollutant is quickly dispersed over the y- and z-directions, leaving the concentration as a function of x, the distance along the river. Advection will be the major means of mass transport, while dispersion in the x-direction levels out the concentration somewhat. If the velocity of the main stream equals u, combining advection and dispersion leads to Equation (10.4).

In that case Equation (10.1) with the required substitution $\sigma = \sqrt{2Kt}$ gives

$$C(x,t) = \frac{Q}{\sqrt{4\pi Kt}} e^{-(x-ut)^2/4Kt} \quad (10.4)$$

Advection is taken into account by replacing x by x - ut in Equation (10.1). In fact, the peak of Equation (10.1) occurs at $x = ut$, where the exponent is zero and the exponential reaches its highest value of one. The equation $x = ut$ indeed describes propagation with a velocity u.

At smaller scale levels, however, dispersion due to the random motion of molecules, or *molecular diffusion*, is often more important than turbulent dispersion. Transport of polychlorinated biphenyls (PCBs) from the atmosphere into the water mass of a lake is of concern, for example, because of ecological and health risks to aquatic life and human beings and wildlife consuming fish from the lake. PCB transport is limited by molecular diffusion across the thin air-water interface at the lake's surface.

Transport processes may also involve the movement of particles in fluids and, in a reversal of that process, the movement of fluids through the soil. The speeds with which particles settle out and rates of groundwater flow are both governed by the interplay of forces acting on the particles or body of groundwater. All over the world groundwater is used extensively as a water source for agricultural, industrial and domestic (i.e. drinking water) purposes. In comparison with surface water flows, groundwater flow is very slow; once an aquifer becomes contaminated, the pollution will move only gradually. Because the volumes concerned are often large, though, once an aquifer is contaminated it may be very difficult to clean up.

As a realistic example consider Fig. 10.4, which shows a Chromium plume on Long Island, USA, measured 13 years after the disposal of Chromium at the site indicated in the figure. One will notice that the plume follows the main flow and disperses over 1,300 m in 13 years. This value of 100 m per year is

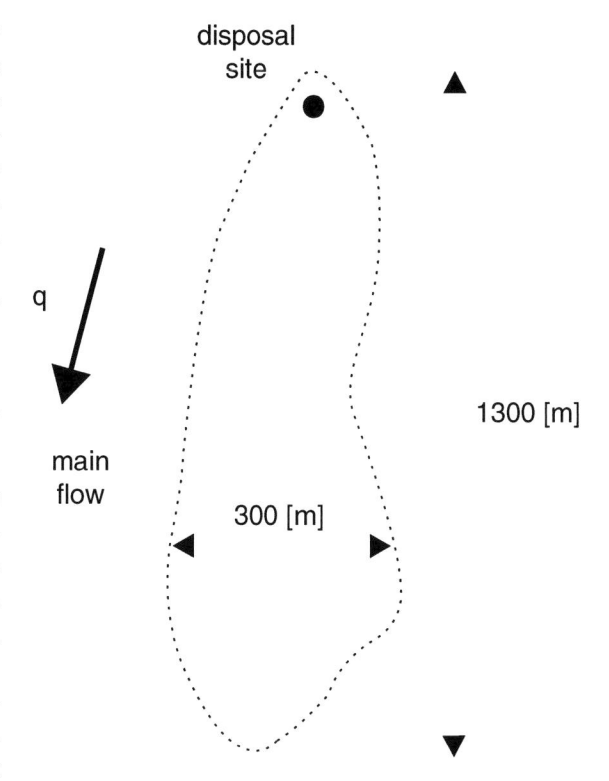

Fig. 10.4 Chromium plume in Long Island, USA, in an aquifer of sand and gravel after 13 years
Source: With permission of Wiley taken from Devinny et al. (1990)

typical for groundwater movement. In practice velocities may go up to a few hundreds of meters per year. These velocities do not seem large, but for instance in case of radioactive waste, which is dangerous for 10,000s of years, advection with the groundwater must be avoided.

Another interesting feature of Fig. 10.4 is the dispersion perpendicular to the main flow. It still amounts to 300 m in 13 years, which is not negligible.

10.3 Principles of Chemistry

10.3.1 *Chemical Kinetics and Thermodynamics*

Chemistry is an important subject for environmental scientists because the ultimate fate of many pollutants discharged to water, air and soil is controlled by the way chemicals react and by their *'chemical speciation'*, the process of formation of the numerous chemical forms in which an element or a combination of elements can exist. To identify the fate of a chemical pollutant in the environment or assess the effectiveness of a treatment process either a *kinetic approach* or an *equilibrium approach* can be adopted, focusing respectively on chemical reaction rates and the ultimate, steady-state result of such reactions.

It is often useful to express a chemical reaction in terms of the time required for it to proceed to completion; the half-life of a chemical is thus defined as the time required for the concentration initially present to decrease by one half. The rates of most reactions (except radioactive decay) are dependent on temperature (rate constants). In biological systems (the communities of microbes used in wastewater treatment plants, for example) one often observes a doubling of biological reaction rate with every 10°C increase in temperature (Q10) up to a certain point, after which it begins to decline.

The equilibrium approach is based on the fundamental principles of *thermodynamics*. Thermodynamics provides a well-defined set of laws that determine, first, whether or not a reaction can occur and, second, in which direction it will proceed: forward or reverse. Equilibrium, as defined by thermodynamics, is a condition that can be predicted. Thermodynamics is concerned with the conversion of energy from one form to another. Physical chemists have defined a quantity, the *Gibbs free energy* (*G*), as that energy available within a given system for doing work. The formal definition of Gibbs free energy is:

$$G = H - T \times S \quad (10.5)$$

Gibbs free energy is related to the system's enthalpy (H), entropy (S) and temperature (T). The term *enthalpy* designates the energy of the inter- and intra-molecular bonds that bind the system's atoms and molecules together, while *entropy* refers to the 'disorder' of the system. The Equation (10.5) indicates that the more disorder there is, i.e. the larger the value of S, the less energy there will be available for doing work. Conversely, the more disorganised a system, the more work will be required to impose order on it. The implications of this law of thermodynamics are far-reaching: any chemical reaction resulting in conversion of a liquid or solid to gaseous form will lead to a large increase in entropy. An increase in entropy is therefore equivalent to a decrease in the free energy of the system, or a decrease in the amount of work accomplished.

In closed systems the entropy remains at best constant, but usually increases. The consequence is that complete conversion of heat into mechanical work is not possible. Power stations where heat is used to rotate turbines in order to produce electricity therefore always leave waste heat and their efficiency by necessity is smaller than 100%. For the best coal power stations the efficiency may amount to 35% or 40% and for gas power stations an efficiency of 60% may be reached. In all cases the waste heat comes into the environment.

For many processes the changes in enthalpy *H* or Gibbs free energy *G* may be found from tabulated data (Kaye and Laby 1995; Lide 1996). Table 10.2 shows the enthalpy and Gibbs free energy of a few compounds. We will use these data to discuss a few numerical examples of reactions that are of environmental relevance.

Natural gas consists for a major part of CH_4 for which the process of combustion may be described by:

$$CH_4 + 2O_2 \rightarrow CO_2 + 2H_2O \quad (10.6)$$

From Table 10.2 one may deduce that one has to add 74.6 kJ/mol for the methane to separate into its building blocks. Next these blocks rearrange to CO_2 and $2H_2O$ and produce 393.5 kJ/mol and 2 times 241.8 kJ/mol if the water ends up in the gas phase. Combustion of 1 mol, or 16 g, of methane in this way produces 802.5 kJ. If, however, one lets the water vapour condense in the furnace one also gains the condensation heat and ends up with 890.5 kJ. This extra 11% is used in advanced central heating systems with a high efficiency. For the same amount of the greenhouse gas CO_2 one gains more useful heat, or for a certain amount of heat one produces a smaller amount of greenhouse gases.

Table 10.2 Enthalpy *H* and Gibbs free energy *G* for some chemical compounds, relative to C, O_2 and H_2. The data hold for a temperature of 25°C and a pressure of 1 atm

	Enthalpy *H*	Gibbs free energy *G*
CH_4	−74.6 kJ/mol	
CO_2	−393.5 kJ/mol	
H_2O gas	−241.8 kJ/mol	−228.6 kJ/mol
H_2O liquid	−285.8 kJ/mol	−237.1 kJ/mol

Source: Lide 1996

One may try to avoid the CO_2 by using a catalytic chemical process, which gives:

$$CH_4 \rightarrow C + 2H_2 \quad (10.7)$$

The hydrogen then may be combusted, while the C is removed. From Table 10.2 one will find that now only half the heat in kJ/mol is produced. One is therefore playing with the idea to combust the C separately in a controlled way and isolate the produced CO_2 in order to put it away (sequestration).

As mentioned before, one looses a considerable amount of energy if electricity is produced by the detour of combustion. It is possible to convert the energy of hydrogen directly into electricity by a fuel cell following the reaction:

$$2H_2 + O_2 \rightarrow 2H_2O \quad (10.8)$$

The Gibbs free energy of water could in principle be completely converted into electric energy, which according to Table 10.2 amounts to 237 kJ/mol for liquid water. The maximum amount of heat, which might be liberated is given by the enthalpy, which equals 285 kJ/mol. The efficiency of the process therefore would be 0.83. Problems with the use of the fuel cell in cars and buses are the expensive catalysts needed and the expensive way to store the H_2 in the vehicle until needed.

An example will serve to illustrate the importance of thermodynamics and chemical rules for environmental scientists. Acid deposition can originate from a variety of source, including the burning of fossil fuels, volcanic eruptions and the spreading of manure. In the case of fuels, the sulphur contained is oxidised during combustion to form sulphur dioxide, SO_2. In the atmosphere this SO_2 dissolves in water droplets to form sulphurous acid, H_2SO_3, which may in turn be oxidised by other substances in cloud water such as peroxide, iron or copper to form sulphuric acid, H_2SO_4. The question of interest to the environmental scientist is then whether (1) the combustion of FeS_2 (pyrites) in the coal to yield SO_2 and (2) the oxidation of SO_2 by Fe^{3+} in the clouds to yield H_2SO_4 can proceed spontaneously under standard conditions? After appropriate calculations using the tabulated free energy values of the transformations of the chemical species in question under these conditions, the answer to this question is found to be affirmative. Both reactions result in a decrease of free energy and both will therefore proceed spontaneously under standard conditions (Mihelcic et al. 1999).

Due knowledge of the laws of thermodynamics and prudent application thereof in environmental science should encourage policy-makers to move away from effect-based to source-based measures, in other words measures addressing essential causes.

10.3.2 Chemical Equilibrium

In environmental systems there are a number of important *equilibrium processes* at play, in particular those between solid compounds and air, and between water and air. *Sublimation* is the term used for transformation of a compound from the solid state to the gaseous state, without the intermediate formation of a liquid, frequently followed by the reverse reaction. *Evaporation* is the transformation of a compound from a liquid state to a gaseous state. The reverse reaction, the formation of a liquid from its vapour is called *condensation*.

One feature of environmental problems is that they are seldom confined to just one environmental medium. For example, although mercury is emitted mainly as an air pollutant, its most damaging effects occur after it has passed through the atmosphere and been deposited in lakes, where it undergoes a biological transformation process called methylation. A second example are persistent organic pollutants ('POPs') such as DDT and PCBs which, though banned in many developed countries, are still manufactured for export and remain in widespread and largely unregulated use in developing countries. In the warmth of the tropics and subtropics these chemicals volatilise more readily, subsequently condensing out and being deposited at high latitudes, where it is cooler (Wania and MacKay 1996). This process of toxic chemicals first being exported and then returned through atmospheric transport has been termed "the circle of poison" (Mihelcic et al. 1999).

10.3.3 Oxidation and Reduction

Two of the major pathways by which both inorganic and organic chemicals undergo transformation in the environment are *oxidation* and *reduction*, particularly important in the context of photochemical and biologically mediated transformation processes, discussed below. The easiest way to recognise whether a chemical compound has been oxidised or reduced during a chemical reaction is to establish whether there has been a net change in the oxidation state(s) of the atom(s) involved. For example, when one of the chlorine atoms in an organic molecule is replaced by a hydrogen atom, as in the transformation of the insecticide DDT to DDD, the oxidation state of the carbon atom at which the reaction occurs changes from +III to +I, with the oxidation states of all other atoms remaining the same. Hence, conversion of DDT to DDD requires a total of two electrons to be transferred to the DDT from a so-called 'electron donor'. This type of reaction is termed reductive dechlorination. Note that the chemical compound donating the electrons is oxidised in the process. Thus, in any electron transfer reaction one of the reactants is oxidised, while the other is reduced. These are consequently termed *redox reactions:* transformation reactions in which there is a net transfer of electrons from (in oxidation) or to (in reduction) the chemical compound of interest (Schwarzenbach et al. 1993).

10.3.4 Photochemical and Biochemical Transformations

When a chemical compound (an organic pollutant in the aquatic environment, for example) absorbs light it may undergo *photochemical transformation*, either directly or indirectly. The direct reaction occurs when a given chemical (pollutant) absorbs light and is thereby transformed. Well-known examples are the photochemical reactions leading to urban smog (i.e. high concentrations of ozone in the troposphere), which is therefore commonly referred to as photochemical smog, and to depletion of the stratospheric ozone layer (i.e. the 'ozone hole'). Specific photochemical reactions may occur in natural waters where light-induced processes may lead to transformation of an (organic) pollutant. Such processes are initiated by the absorption of light by other chemicals present in the system. This process is referred to as *photolysis*.

Chemical compounds may also be removed from the environment by *biochemical transformation*, i.e. by reactions mediated by natural organisms. As with chemical and photochemical reactions, these biochemical processes change the structure of the chemical of interest, thereby removing the particular compound from a given environmental system. Many such reactions, although thermodynamically feasible, occur only slowly because of kinetic constraints. Biochemical transformations are initiated by organisms secreting special proteins called *enzymes*, which act as catalysts. In quantitative terms, micro-organisms are the most important life-forms involved in the environmental degradation of chemical compounds. They come in many forms, including bacteria, fungi, protozoans and micro-algae, and are present virtually everywhere in the natural and man-made environment (Schwarzenbach et al. 1993).

Biochemical transformation processes are widely occurring in nature. One biologically mediated transformation process closely allied with major environmental problems is *biodegradation*, the chemical and biological breakdown of inorganic and organic complexes. Biodegradation is particularly relevant to marine oil spills that threaten coastal wildlife. The chemical composition of crude oil, or petroleum, is fairly complex and variable. Typically, 50% to 90% is constituted by hydrocarbons, i.e. compounds made up of only carbon and hydrogen. In addition, though, crude oil contains compounds of sulphur, nitrogen and oxygen. Oil spills in a natural body of water may be subject to a number of chemical transformation and degradation processes. In principle, most petroleum constituents can be degraded by a variety of micro-organisms, such as hydrocarbon-oxidising bacteria and fungi. Whether or not this occurs in practice in the marine environment depends on a range of biological, chemical and physical factors, such as the number and type of micro-organisms present, oxygen and nutrient availability and temperature. Petroleum is a natural and long-standing part of our environment and it should come as no surprise that hydrocarbon-oxidising micro-organisms have evolved. The problem with coastal oil spills is that ecological communities living and feeding in the intertidal zone are particularly vulnerable to physical coating with oil and subsequent adverse effects. Much of the public concern about oil spills focuses on the plight of seabirds. If the plumage of these birds becomes contaminated with hydrocarbons, the trapped air providing thermal insulation and buoyancy is lost, leading eventually to the birds' death (Connell 2005). Whether or not cleaning agents, commonly referred to as detergents, should be used to dislodge the oil from the coast and its flora and fauna and suspend the oil particles in the water column is an important question that demands a timely decision by policy-makers, for aquatic organisms are sensitive to detergents, and the 'cure' could be worse than the disease.

These examples demonstrate the basic importance of the methods and concepts of environmental chemistry in environmental science. There are many more applications of chemistry in our discipline, however (see for example Box 10.3). For further reading the student is referred to standard textbooks of environmental chemistry (e.g. Baird and Cann 2004; Connell 2005).

Box 10.3 Environmental engineering and environmental science

Environmental engineering is the field in which the fundamentals of mathematics, physics, chemistry and biology are applied to protect human health and the environment. Many practical, everyday environmental problems can be analysed and often resolved by the environmental engineer. Let us consider one example in which the environmental engineering toolkit is indispensable.

Municipal wastewater treatment
The questions to be asked in this case can be:

- Does the water need to be treated in a wastewater plant? If 'yes', proceed to next question.
- Given the need for continuous treatment of a certain volume of wastewater, with specific pollutant characteristics, what is the best design for a treatment plant? Is chemical or biological treatment preferable? How is the plant to be built and operated?
- What capacity should the plant have, and can chemical or biological reaction rates be increased to reduce plant size and cost?
- If there are several types of plant to choose from, which works best for a given treatment objective?
- How is the design and operation of a particular treatment process influenced by daily and seasonal variations in the flow of wastewater supplied by households and industry?
- Are there special considerations pertaining to the river or lake receiving the treatment plant effluent that are of influence on plant design and operation? (In nutrient-sensitive waters more stringent phosphorus discharge standards may be in force, for example.)
- How much of a particular waste can be discharged to the receiving water body; how will environmental mixing and transport influence the fate of the waste; and will the discharge adversely affect the chemistry or biology of the water body?
- How can the treatment process be optimised for more efficient handling of the wastewater?
- Should the wastewater be biologically treated in the presence or absence of oxygen, i.e. aerobically or anaerobically, and how do the methods differ?
- If aerobic treatment is to be preferred, what are the most efficient methods of supplying the required oxygen?
- How will seasonal temperature changes affect the treatment process, and how will they affect naturally occurring recovery processes in the receiving body of water?
- And so on.

In the industrialised countries municipal wastewater treatment has successfully solved what until the 1960s of the 20th century was a major environmental problem. In many developing countries it still is a pressing problem: nearly two billion people on the planet still lack an adequate sewerage system, are drinking contaminated drinking water and are consequently exposed to infectious disease. Most of the questions posed above are basic questions for engineers, whether they are environmentally involved in their job or tackling the problems from a strictly technical perspective. The combination, however, of questions, the integration of knowledge required to answer them and the combination of purely technical skills with the normative choices made by society, in this case with regard to its wastewater treatment, make environmental engineering an important supplier of basic tools in environmental science.

Source: Mihelcic (1999)

10.4 Principles of Biology

10.4.1 Genetic Variability

In nature plant and animal individuals are grouped in populations. A *population* is a group of individuals of one species in an area, though the size and nature of that area is defined, often arbitrarily, for the purpose of the study being undertaken. Important processes influencing the size of a population of a given plant or animal species are demographic processes, capable of changing the size of a population, viz. birth, death or migration. Temporal or spatial variability in these processes is called *demographic variability*. Each individual in a population has a specific chance of reproducing and surviving; the smaller the number of individuals in a population, the greater the chance that none of them will reproduce or that all of them will die within a relatively short period. Demographic variability is of paramount importance for the survival of small populations: less than 50 individuals of an orchid species in a fragmented wetland or of an amphibian species in an isolated water pool, for example.

Environmental variation is mainly of influence on the occurrence of a particular species or complex of populations in a specific geographic area and affects all the individuals of a population concurrently. Besides natural factors like seasonal changes, predator presence and disease, environmental variation may today also be triggered by anthropogenic factors like toxic emissions, coastline oil spills or motorway noise.

Any decline in population size and increase in geographic isolation (*habitat fragmentation*) may lead to a decrease in *genetic variability*, a process known as **genetic erosion**. Small populations stand a higher chance of being affected by two processes causing a decrease in genetic variation, viz. genetic drift and inbreeding. *Genetic drift* is the chance that a genetically fixed characteristic will disappear from the population, leading to a more uniform and hence more vulnerable population. *Inbreeding* results from the reproduction of closely related individuals of the same species. Genetic erosion influences a population's chance of survival and is therefore of great importance for nature management. Specific land use measures may lead to the decline or even extinction of local populations of protected species as well as geographic isolation of these and other populations owing to habitat fragmentation. Consequently, nature management is also gene management.

With the ongoing fragmentation occurring in today's human-moulded, cultivated landscapes, many plants and animals that had a fairly continuous regional distribution in the past are becoming segregated in isolated patches that together function as a metapopulation. A *metapopulation* is defined as a composite of local populations, whose continued survival is due to an ever-shifting balance between the continuous extinction of local populations and subsequent recolonisation of open niches. A species restricted in its distribution by fragmentation of its habitat will not always form a metapopulation, however, for its dispersion capacity may be insufficient.

10.4.2 Population Dynamics

Plants and animals never live in isolation but are in continuous interaction with members of their own species and other species and with their environment. *Competition* is interaction between individuals and, more precisely, competing for a (limiting) resource, whereby one individual uses a resource that could also have been used by another individual. Resources are in this context food for animals, living space for plants and animals, light for plants, etc.

Intraspecific competition occurs between individuals within a population of a plant or animal species. Its intensity depends on population density, which means the reproduction and mortality rates of the population are determining factors. Individuals of a given species make roughly the same claims on the same resource(s) and collective pressure may easily lead to exhaustion thereof, whether it be a food source or physical habitat. The size of the population is determined ultimately by the *carrying capacity* of the area in question, governed in turn by the availability of renewable (e.g. food) and non-renewable (e.g. space) resources, which affect population size (biomass) through reproduction, growth and survival.

Interspecific competition occurs between individuals of separate species, with strong competitors driving weaker competitors away from resources. This type of competition may be severely aggravated by human disturbance of the environment (pollution, habitat loss, etc.), which may favour aggressive, colonising species, driving off vulnerable species to undisturbed habitats (Bakker et al. 1995).

10.4.3 Ecosystem Structure and Function

From its original conception in 1935 to the present day, Tansley's concept of the ecosystem has remained a key element of the ecological sciences. An *ecosystem* is a plant and animal community in its physical environment together with all the relationships between them, i.e. considered as a system. In principle there are no limits to the size of an ecosystem, which may range from a pot-plant to a forest or lake or the entire biosphere. The ecosystem concept, the idea that flora and fauna interact with their environment to form an ecological complex, has long been central to the public perception of ecology and to a growing awareness of ongoing environmental degradation (Golley 1993).

The biotic components of an ecosystem can be conceptually organised in terms of *species*, *populations* and *communities*. A biotic *community* is any assemblage of populations living in a given area or physical habitat. Communities can be conveniently named and classified according to (1) their major structural features (e.g. dominant or 'indicator' species), (2) the physical habitat of the community, or (3) functional attributes such as the type of community processes (e.g. production and respiration, see Energy flow). No precise rules have been drawn up for naming communities, as has been done for naming and classifying organisms into species. Functional attributes provide a good basis of comparison for communities of widely differing habitats (e.g. terrestrial, marine and freshwater).

There is much debate about the relationship between ecosystem stability and complexity. Ecosystems are not static entities, but dynamic systems that are constantly altering and shifting. The *stability* of an ecosystem refers to its capacity to remain 'constant', i.e. basically unchanged over time, returning to an equilibrium state following any temporary disturbance. *Resilience* refers to the velocity with which an ecosystem returns to its equilibrium state. The feeding relations between animals and plants are fundamental to ecosystem organisation and functioning, facilitating the throughput of energy and the recycling of nutrients via decomposition processes. Before 1970 the generally accepted view among ecologists was that ecosystem stability was positively correlated with ecosystem complexity. Today, though, we know that optimal complexity is not expressed under constant, unchanging environmental conditions but under slightly disturbed conditions, which constantly create new niches for colonising species. Once disturbance increases beyond a certain level, however, ecosystem complexity quickly drops. These concepts are important in environmental science, and specifically in the realms of nature management and nature development.

Huston (1979), has developed a dynamic equilibrium model, shown modified in Fig. 10.5, that integrates the intermediate disturbance hypothesis (Connell 1978) and the intermediate productivity hypothesis (Grime 1973). This model predicts that the potential number of competing species in a habitat varies according to the relationship between disturbance and productivity (resources). At high levels of disturbance, biodiversity is maximised in habitats with abundant resources (translated into a high rate of population growth), whereas at low levels of disturbance, maximum diversity is attained in habitats with sparse resources. Stated another way, the level of 'intermediate' disturbance necessary to sustain maximum biodiversity changes as a function of resource level (i.e. population growth), with progressively greater perturbations needed to prevent com-

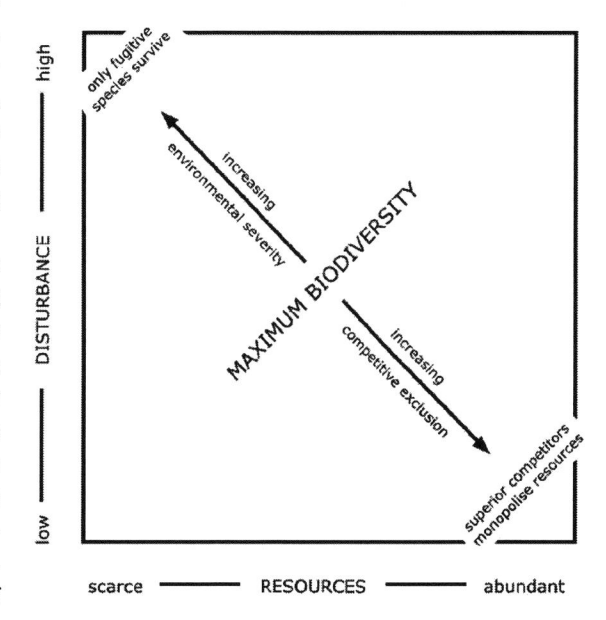

Fig. 10.5 The hypothesized relationships between disturbance, resources and biodiversity
Source: In Ward et al. (1999), modified after Huston (1979)

petitive exclusion from being realised as resource increase (Ward et al. 1999).

The complexity of an ecosystem can be measured in several ways, by the number of species present, for example, or the number of interactions between the species. *Species richness* refers to the number of species present in a particular community. However, this measure disregards the relative abundance of individual species, assigning the same weight to rare and common species alike. *Species diversity* takes into account both species richness and the relative abundance of constituent species. The species composition of a particular community can be described by identifying all its constituent species. The more recent term *biodiversity* embraces more structural components than mere species diversity, as it refers to the overall qualitative and quantitative variability of the biosphere, or some part thereof, covering genes, species, communities, ecosystems and landscapes.

The complexity of food webs within and between ecosystems can be expressed in terms of *connectance*, defined as the actual number of interspecies interactions in the food web (i.e. one species eating the other) divided by the potential number of such interactions. A higher connectance value thus signals a more complex food web.

10.4.4 Energy Flow

Energy is defined as the ability to do work. As explained earlier, the energetic behaviour of substances and compounds is described by the first and second laws of thermodynamics. The first law states that energy cannot be created or destroyed but can only be converted from one form to another. Applied to an ecosystem, this law suggests that no organism can create its own, autonomous energy supply. Plants, for example, rely on the sun for energy, while grazing animals are dependent on plants and thus indirectly on the sun. Organisms use the food energy they produce or assimilate to meet their metabolic requirements for performing the work of cell maintenance, growth, reproduction and so on. Ecosystems must therefore import energy, and the needs of individual organisms is met by transforming that energy into an appropriate form. The second law of thermodynamics states that whenever energy is transformed some is lost as heat and thus becomes unavailable for doing work. In food webs only a very small portion of the sun's radiant energy is 'fixed' by *autotrophs*, or *primary producers*: mainly green plants. The other members of the food web, *secondary producers*, also referred to as *heterotrophs* or *consumers*, animals of all types and sizes, depend directly or indirectly on autotroph activity. To understand the pathways by which energy flows and nutrients are cycled within and among ecosystems, ecologists have assigned groups of plants and animals to various feeding levels, or *trophic levels*. Primary producers occupy the first trophic level, consumers the second, third and higher levels.

Animals consume food, i.e. organic and inorganic matter, as a source of energy, the fate of which can be represented in terms of an *energy budget* (see Equation (10.9), below). Most of the food ingested (*consumption = C*) by animals is assimilated (*assimilation = A*) (Equation (10.10)). Energy assimilated but not respired is devoted to *production (P)*, that is, invested in the organism's growth or reproduction. The major losses of energy ingested with food are due to expenditures required to sustain metabolic and other vital processes (*respiration = R*). In chemical terms respiration is the oxidation of organic matter, resulting in the release of energy (heat) and CO_2. A substantial amount of organic matter may be lost by consumers during feeding. Such losses during 'messy' feeding are not available for assimilation, and the same holds for losses of faeces and urine and other excretions of metabolised compounds. These losses are together referred to as *egestion (E)* (Valiela 1995).

$$C = P + R + E \quad (10.9)$$

$$A = P + R \quad (10.10)$$

The *production efficiency* of an animal population is the proportion of assimilated energy devoted to production ([P/A]100); it varies from 2% for birds and mammals to 20% for fish and invertebrates. The *ecological efficiency* of an animal population ([P/C]100) has even greater relevance for environmental science, because it provides an idea of the proportion of consumed energy available to the next link in the food chain, whether it be another predator or a fleet of fishing vessels. Ecological efficiencies in natural water columns should be about 10–20% but are generally less than 10%. This is a fundamental property of food

> **Box 10.4 Energy flow and the American diet**
>
> The principles of thermodynamics and energy transfer tell us that consumption patterns at the top of the food pyramid are tremendously inefficient. For example, Americans consume 90 kg of meat a year on average. If they reduced their meat intake by just 10%, over 60 million people could be fed on the fodder grain saved. As the human population continues to rise worldwide, so too will the demand for land and the food it yields, as well as the environmental problems associated with conversion of natural ecosystems to agricultural systems. The United States, Western Europe and the rest of the developed world will have to choose between helping less developed countries and maintaining good relations with them, or isolating themselves in a life of relative splendour. The first mentioned alternative implies investment in changes in lifestyle, consuming food from lower down the food chain and reducing energy demand with all its attendant environmental impacts.
> Source: Auer et al. (1999)

chains and is of interest to fisheries. Because of the inherent inefficiencies of energy transfer processes, a relatively small fraction of the energy originally fixed by primary producers (plants) remains available for transfer to higher trophic levels. This explains why it takes a large number of primary producers to support a single organism at the top of the food chain, such as a carnivore or omnivore (Box 10.4).

10.4.5 Materials Cycling

Given the right basic physical setting (in terms of light and temperature, for example), organisms require only two things from their environment: energy to power them and chemicals to give them substance, i.e. biomass. Energy does not move in a cycle but flows through ecosystems, being dissipated as heat and lost as useful energy forever. Chemical elements, in contrast, are used over and over again in ecosystems during processes of growth and decomposition. Chemicals are recycled both within and between ecosystems and the rate of cycling is an important factor determining ecosystem processes. Some chemical elements – carbon, hydrogen, nitrogen, oxygen and phosphorus – are required in relatively large amounts for the synthesis of organic matter and are termed *macronutrients*. other elements like calcium, zinc and magnesium, in contrast, are needed in only very small quantities. The availability of these *micronutrients*, or *trace elements*, may nonetheless act to regulate the productivity of an entire ecosystem. The principle of nutrient limitation has been formulated as '*Liebig's Law of the Minimum*', which states that the production (see Energy flow) of any organism will be determined by the abundance of the substance that in relation to the needs of the organism is least abundant in the environment. Ecosystem production is thus governed by the availability of one or other chemical element, acting as a limiting factor in several respects, viz. in terms of: (1) the chemical form(s), or species, in which the element occurs locally, (2) the physical location of the materials in question, and (3) the rate at which they are recycled.

Living organisms are important participants in material cycles, mediating the transfer of chemical elements from one chemical species to another (e.g. from CO_2 to $C_6H_{12}O_6$), from one location to another (e.g. from N_2 in the atmosphere to amino acids in the cells of an organism) and from one reservoir to another (e.g. from assimilation in a living primary producer to decomposition in dead organic matter). The cycles of chemical elements through living organisms and through the abiotic environment are called *biogeochemical cycles*. Although every chemical element has its own specific role in global material cycles, there are a small number of biogeochemical cycles which, in quantitative terms, are of paramount importance for a better understanding of the functioning of the environment, from the local to the global scale: the cycles of carbon, oxygen, nitrogen, phosphorus and sulphur, and those of sodium, potassium, calcium and magnesium. The *hydrological cycle*, or water cycle, is also of major importance because of its crucial role in transporting chemical elements through the biosphere. Environmental problems, the key focus of environmental science, are often associated with a disruption of one or more biogeochemical cycles.

Part of one of these cycles, the *nitrogen cycle* in a marine environment, is shown in Fig. 10.6 (Valiela 1995). Nitrogen is transformed and transported in

> **Box 10.5 Eutrophication in a nutshell**
>
> Eutrophication is a serious environmental problem that needs to be addressed by means of a multidisciplinary approach. It started with the nuisance caused by blue-green algae proliferation in lakes due to the influx of growing volumes of domestic wastewater. The 'solution' adopted in the 1930s was to add a selective poison to the lake to kill the algae. When it was subsequently realised that the phosphorus load from detergents was the primary cause, opinions became divided in the 1950s and 1960s on what action should be taken: wastewater treatment or a ban on phosphorus in detergents. When it was later realised that nitrogen was limiting algal growth in estuarine and coastal waters, there was disagreement among experts as to whether P or N was the limiting factor. In 1987 the Danish government took a far-reaching decision. With experts remaining divided on which was the limiting nutrient, and given a political situation demanding effective action, it was decided to limit pressure on the aquatic environment by reducing Danish loads of both P and N, by 80% and 50% respectively.
>
> This is an example of the precautionary principle being applied: here, scientific uncertainty calls for more preventive measures than can be proven to be effective. In Denmark the P-load has been successfully reduced, as has the amount of N in domestic and agricultural wastewater. The reduction of nutrient inputs from diffuse sources has been rather less successful, though. The debate as to whether P or N can most effectively be made the limiting factor for algal growth will be with us for some time. The debate on whether it is right to apply the precautionary principle here is also still open, because addressing the causes of the nutrient influx from agricultural fertilizers impinges on an activity as fundamental as human food production. Given the complexities of the eutrophication issue, then, it is important to adopt a multidisciplinary approach that addresses not only the scientific dimension but the social and political aspects as well. One such approach that has been adopted in many western countries is known as 'integrated modelling' or 'integrated assessment' (see Chapters 17 and 18).
>
> Source: Harremoës (1997)

the marine environments in a complex pattern. *Nitrogen fixation*, the conversion of N from atmospheric nitrogen gas into organic nitrogen by a variety of micro-organisms, is a source of new nitrogen for such ecosystems. Another source is run-off from land via rivers, which are often 'enriched' with effluents from sewage treatment plants (see Boxes 10.3 and 10.5). Plants and animals grow by feeding on sources of inorganic and organic nitrogen, in addition to other elements. As a result of egestion during lifetime and final decay after death a range of nitrogen-containing derivates (particulate organic nitrogen [PON], dissolved organic nitrogen [DON]) become available in the environment. *Decomposition* of organic nitrogen compounds is accomplished by complex assemblages of micro-organisms, employing a range of biological, enzymatically mediated redox processes. Figure 10.6 shows the conversion of organic derivates to ammonium (*ammonification*), the conversion of ammonium (NH_4) to nitrate (NO_3) (*nitrification*) and finally nitrate reduction (*denitrification*), by means of which NO_3 is converted into gaseous N_2. In short, this step closes the nitrogen cycle in the marine environment.

A number of environmental problems arising in marine and coastal ecosystems can be neither understood nor cured without a knowledge of the nitrogen cycle, as is illustrated by two examples. The best studied of these is *eutrophication*, the enrichment of surface waters with plant nutrients, in particular nitrogen and phosphorus (see Box 10.5). In the case of marine and coastal waters, excessive nutrient influx from rivers and land run-off can readily trigger a series of unwanted environmental changes: primary production increases, the waters become turbid, macrophytes (i.e. macroscopic plant species) disappear, benthic (i.e. bottom-living) fauna die off and finally the ecosystem as a whole may collapse. Denitrification may provide an important means of eliminating excessive discharges of these anthropogenic nitrogen compounds. The process of dinitrification occurs pri-

Fig. 10.6 Biogeochemical cycle, the simplified scheme of major transformations and transport of nitrogen in marine environments (Valiela 1995). DIN = Dissolved Inorganic Nitrogen; DON = Dissolved Organic Nitrogen; further explanation see text

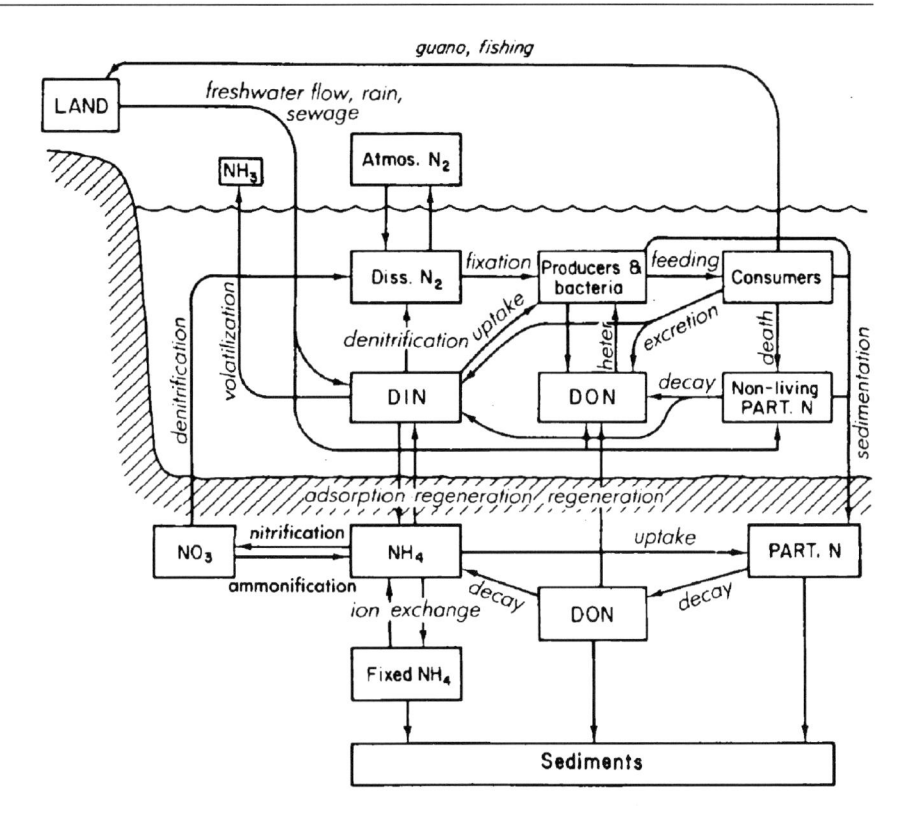

marily where there is a juxtaposition of oxic and anoxic conditions, because while denitrifying bacteria are anaerobic, nitrification requires oxygen. Higher denitrification rates are therefore found in layers of water and sediment intermediate between oxic and anoxic conditions (Valiela 1995). Knowledge of this process has led to improvements in sewage treatment design and operation.

The second example is less well understood. Treating polluted river water before it is discharged into an estuary may help improve water quality and hence boost oxygen levels in the water column. Superficially, then, the policy target ('a cleaner river') is met. With a knowledge of mass-transport processes, however, it can be predicted that these improvements in water quality will trigger the leaching of P- and N-compounds from old deposits of organically rich sediment. Even worse are the unwanted side-effects of bottom sediments becoming oxygenated, which may trigger mobilisation of heavy metals.

Concise descriptions of the main biogeochemical cycles can be found in Auer et al. (1999) and Jackson and Jackson (2000).

10.5 Spatial and Temporal Dimensions of the Environment

10.5.1 The Question of Spatial and Temporal Scales

There are many physical, chemical and biological indicators with which small pieces (e.g. a biochemical reaction, a response of an individual to pollution) of an ecosystem can be measured directly, often with great precision. When it comes to measuring overall environmental quality, however, these kind of indicators are of little use, as they each provide only a partial picture of reality and a relatively arbitrary one at that. Ideally, environmental quality indicators should be comprehensive in scope, as relevant as possible for policy-makers and easy to manage. Precisely because of their comprehensiveness, though, such indicators are unfortunately not the easiest to measure (see also Fig. 10.12 in Section 10.6.3). This outlines the basic dilemma of 'scaling': how to balance the availability of sound and reliable data on the one hand with understanding of an ecosystem, and hence its manageability, on the other.

Holling (1978) provides a useful classification putting the availability of data and the understanding of the ecosystem into a modelling context. In Fig. 10.7, the horizontal axis represents the degree of scientific understanding of the system and hence increasing relevance for environmental management purposes, the vertical axis the amount of system data available. Domain 3 is where we want to be: good or possibly even thorough understanding and proper management, supported by a wealth of data. When the object of study is the functioning of an entire ecosystem, as with coastal and marine sciences, for example (Gordon et al. 1996), we are rarely in domain 3. This is frequently the domain of engineering and physics, however, where processes are generally limited in number and complexity and often well understood, and data relatively easy to obtain. In these disciplines highly accurate numerical models have been developed that are applied routinely on an everyday basis. For instance, tides around the world can be predicted with a high degree of speed and accuracy. Models of the global water cycle are also approaching this domain.

Domain 1 comprises those systems on which scientists have a wealth of data but about which they have little understanding. This may sometimes be the case in marine science, where plenty of monitoring data are available as well as data from a variety of scientific sources. These are usually to be found scattered across government archives and departmental files, or published in scientific papers – if the researchers are lucky – and will have been collected for many different purposes. For instance, fishery scientists often collect data on hydrography, chemistry, biology and economics in addition to data on fish. Hydrologists can provide data on freshwater and even nutrient inputs to the coastal zone. Harbour authorities may be able to provide additional data on water depth (bathymetric data), sea level variations and sediment movements. In this kind of situation, compiling and structuring the data for management purposes may be sufficient to yield a useful level of understanding.

A natural first step in any study of an ecosystem is to organise the available data into a common framework – in a database, for example – with the aim of describing the system's basic spatial and temporal characteristics. Statistical techniques can then be applied to identify significant relationships and develop hypotheses. Today GIS (Geographic Information System) methods are being used with growing frequency for graphically describing observed spatial patterns. These methods are considered in a separate section below.

Domain 4 is the usual domain of biogeochemical research (see Section 10.4.5: 'Materials Cycling') in coastal environments, the domain of knowledge concerned with the environmental pathways of chemical elements and materials cycling, so important for a proper understanding of many environmental problems involving hazardous chemicals. In this domain there is a paucity of data and understanding. Sometimes there is a good conceptual idea of how a system operates, based on studies on similar systems elsewhere, but little data to support these hypotheses, as in domain 2. It is in these two domains, 2 and 4, where mathematical modelling is most useful.

Figure 10.8 shows some of the phenomena encountered in most coastal and marine ecosystems, along with their spatial and temporal scales. Molecular processes as well as individual movements of biological organisms are generally underpinned by ample data (domain 1 in Fig. 10.7) and are sometimes well understood (domain 3). Large-scale storms, gyre circulations (more or less circular circulation of water masses, like in the Sargasso Sea) and seasonal biomass cycles, on the other hand, are difficult to predict and lack an adequate database and management perspective (domain 4).

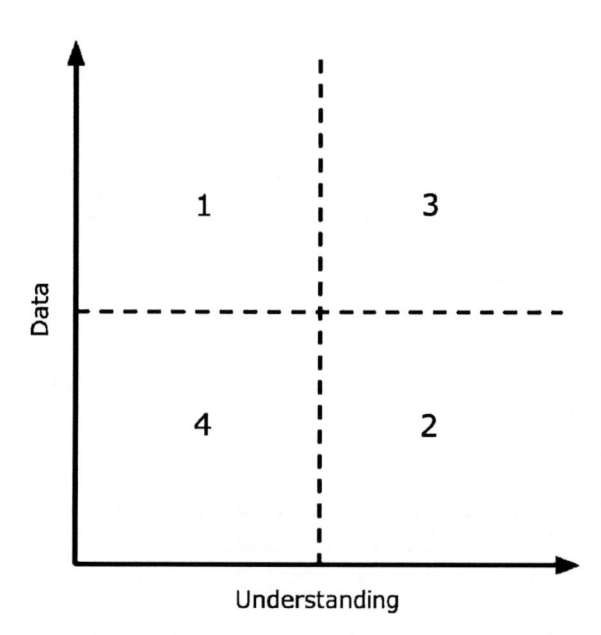

Fig. 10.7 Data and understanding requirements for modelling (Holling 1978). For explanation, see text

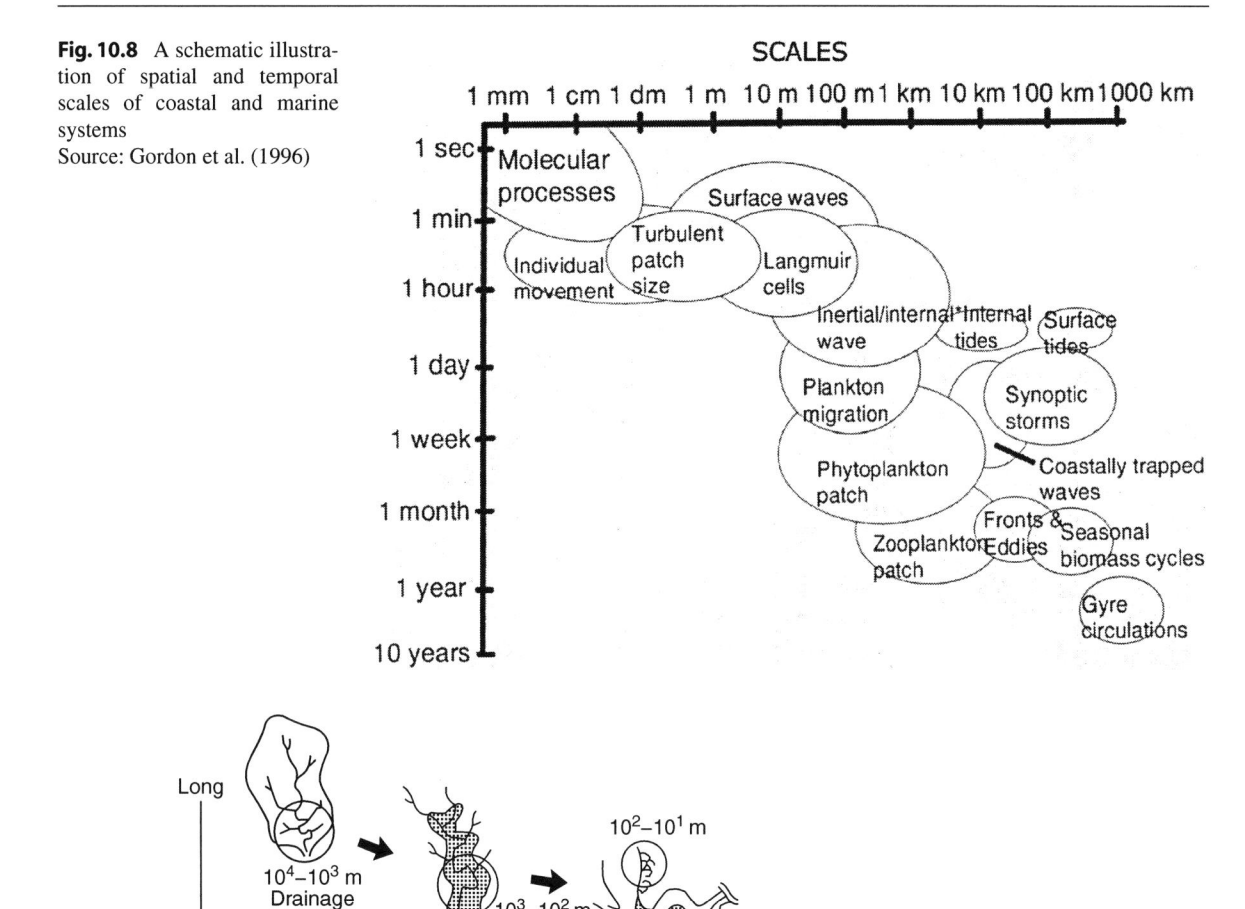

Fig. 10.8 A schematic illustration of spatial and temporal scales of coastal and marine systems
Source: Gordon et al. (1996)

Fig. 10.9 Hierarchical organization of a river system in relation to sensitivity to disturbance and recovery time
Source: Frissell et al. (1986)

In Fig. 10.9 (from Frissell et al. 1986) the spatial and temporal scales to be distinguished in river basins are mapped on two axes: sensitivity to disturbance and subsequent recovery time. The larger and more complex a component system is, the longer it takes to recover, but the lower the sensitivity to disturbance of the river system as a whole. A long recovery time in combination with low sensitivity requires ever greater integration of scientific knowledge, in turn providing greater management leverage. It is at the level of the floodplain and drainage basin that environmental rehabilitation measures are most effective.

10.5.2 Environmental Distribution of Chemicals

Today, man-made chemicals are to be found throughout the natural environment and predicting the full range of adverse effects on human health and natural ecosystems is now an important element of environmental management. A vital step in the procedures used for this purpose is to quantify actual and potential exposure of biota to contaminants. This requires methods to understand and estimate the patterns of distribution of the contaminant in the environment. Before a company is given a license to discharge a particular chemical, for example, the issuing authorities should have a precise knowledge of the compound's environmental fate: where will it be deposited, and how much of it will either physically or biologically become mobile in the environment and potentially harmful? The first step towards understanding the environmental distribution of chemicals is to simplify the complexity of the real-world environment. To this end the environment can be considered to consist of discrete phases, or compartments, that are physically distinct and relatively homogeneous and within which a given chemical exhibits uniform behaviour. The environment is commonly taken to consist of six compartments: air (the atmosphere), water, soil, (underwater) sediment, suspended sediment (in water) and biota. A compound's distribution between these phases can be seen as the result of a set of *two-phase partition processes*. Some of the fundamental aspects of partitioning and the properties influencing these processes have been described above, in the Sections 10.2: 'Principles of Physics', and 10.3: 'Principles of Chemistry'.

Unfortunately, our capacity to assess the impacts of chemical discharges on the various compartments, and thus on whole ecosystems, is not particularly impressive. Greatest success has to date been achieved using approaches based on *ecotoxicology*, the science concerned with the environmental fate and toxic effects of chemicals in natural and disturbed ecosystems. Today an huge number and variety of toxic chemicals are being emitted to the environment and the organisms in any given ecosystem may show differing susceptibilities to each. *Bioavailability* is the ability of a substance to affect organisms; it is a measure of the physicochemical access that a toxicant has to the biological processes of an organism. The less the bioavailability of a toxicant, the less its toxic effects on an organism. For example, while herbicides are selectively toxic to plants and insecticides to insects, within these broad groups of organisms individual species exhibit a very wide range of susceptibilities. The upshot is generally a reduction of species diversity in the ecosystem and a consequent change in community structure. Loss of biodiversity is often a slow and complicated process detectable only by experts. Intensive monitoring of ecological changes at selected field sites can be used to establish trends, whether negative or positive. *Biomonitoring* refers particularly to the study of changes in the levels of toxics in selected organisms over time, or changes in (groups of) individuals from plant and animal populations exposed to harmful physical or chemical conditions. Predicting the specific effects of a toxic substance on a particular ecosystem is an area of ecotoxicology that is in urgent need of further development (Van de Meent and Huijbregts 2005).

Nonetheless, the adverse impacts of a broad range of chemicals have now been identified, in both field and laboratory experiments. Many of the toxic compounds in use today pose a direct risk to human health and ecosystem functioning. *Risk* can be defined as the probability that a potential *hazard* will indeed result from exposure to a chemical or other agent. The concept of risk itself poses many difficulties in interpretation and understanding, as explained in Section 12.3. *Risk assessment* consists of identifying the hazard (e.g. discharge of persistent pesticides into an aquatic ecosystem) and then quantifying the risks posed to the ecosystem by that hazard (e.g. the risk of massive fish mortality in that particular water body). The reader is referred to Chapter 9, where there is extensive discussion of risk assessment and risk management as important concepts and tools in environmental science.

10.5.3 Distribution of Biodiversity Units – Mapping and Monitoring

The idea of preservation and restoration of biodiversity, the overall variety and variability of the biosphere, has gained momentum since the UN Conference on Environment and Development held in Rio de Janeiro in 1992. One of the prerequisites for effective management of biodiversity is adequate information on the physical, chemical and biological status of the environment. Two basic approaches to gathering environmental information will be distinguished here: mapping and monitoring.

Mapping establishes the value of variables in spatial terms and is sometimes called pattern-oriented research. This kind of research focuses on the cause and effect relationships holding at a particular time in physical space. For example, what is the observed variation in breeding bird numbers between areas with virtually the same landscape structure but under different pressures from tourism?

Monitoring establishes the values of parameters as they change in the course of time and is sometimes called process-oriented research. Monitoring can also be considered as repeated mapping or inventorying of variables over time, at either one or several locations. This kind of research focuses on relationships of cause and effect as they unfold over time. For example, what changes can be observed in breeding bird numbers in the years following closure of a nature reserve to tourists?

There are numerous types of *maps* available on widely differing scales: geological maps, hydrogeological maps (groundwater potential maps), soil survey maps, vegetation maps, topographic maps, land use capability maps (agricultural potential maps), environmental geology maps (environmental potential maps), bathymetric maps, to name the most common. Geological and soil survey maps provide information on altitude (contours) and on surface and sub-surface distribution of rocks and soil types, hydrological maps on the distribution of aquifers, water table depth, distribution of boreholes, borehole groundwater yield and quality, and other physical and chemical parameters. Topographic maps are produced at a number of standard scales, the most useful of which in practical environmental science are 1:50,000, 1:25,000 and 1:10,000. A 1:50,000 map is the general-purpose scale for route navigation when motoring, locating roads, railways, cities, towns, villages, farms, rivers and so on. Topographic contours are drawn at 10 m intervals, enabling mountains, hills, valleys and other physical features of the landscape to be readily identified. Maps at the 1:25,000 and 1:10,000 scale provide greater detail of geomorphological, botanical, archaeological, historical and urban features. The location of footpaths, wells and springs is shown and these maps are useful for navigating on hikes and for environmental surveys in which only a moderate level of positional accuracy is required. The 1:10,000 maps provide additional scope for field mapping of geological and geomorphological features. In towns and cities the names of the major streets are given and individual houses or blocks identified. Such maps are therefore ideal for planning and property management purposes (Jones et al. 2000). For the monitoring of vegetation development in ecotopes, or permanent plots, even smaller scales are used (e.g. 1:1,000 to 1:10). An *ecotope* is a spatial unit homogeneous with respect to vegetation structure, succession stage and abiotic factors governing plant growth. A permanent plot or *permanent quadrate* (*PQ*) is a small, marked spatial unit chosen for regular recording of the changes in vegetation (or fauna) composition and abundance.

For analysing the biotic community of a given geographical region or landscape, two contrasting approaches are available, based on *zonation*, in which discrete communities are recognised and classified, and *gradient* analysis, in which populations are arranged along a unidimensional or multidimensional environmental gradient. *Classification* refers to the identification of discrete communities, while *ordination* refers to the indiscrete ordering of species and communities along gradients, or continuums.

Reports on the current state of the natural environment are published regularly by a number of global, national and international monitoring networks, including the World Conservation Monitoring Centre (WCMC; www.unep-wcmc.org) and the World Conservation Union (IUCN; http://iucn.org).

10.5.4 Geographical Information Systems: GIS

Over the last two decades there has been a growing recognition of the need to develop new concepts in natural resource management. Many local and global resource problems have obvious spatial and temporal dimensions and consequently there is a rapidly growing interest in the use of methods for integrating data, and knowledge in models that share a high level of precision and a high level of management relevance (see also Fig. 10.12 in Section 10.6.3). A *Geographical Information System* (*GIS*) is the total sum of hardware, software and procedures supporting the collection, manufacturing, transformation and presentation of geographical data. GIS is used in disciplines concerned with the spatial and temporal aspects of the earth's surface, regardless of the scale of observation. As the environmental sciences are by definition concerned with the physical environment, i.e. some part of the earth's surface, GIS is an important tool for the environmental scientist.

The need for integrating domain-specific knowledge with other (spatially related) data and models has fuelled development of (GIS-related) digital information technologies and computer-assisted approaches, such as Artificial Intelligence (AI), Expert Systems (ES) and Decision Support Systems (DSS). GIS is thus being increasingly integrated with various other fields of science using computer vision as an interface to the user. This development has led the way to a new discipline of *geo-informatics*. The development of geo-informatics considers didactic aspects, and educational programs using GIS are well suited for project-oriented approaches, in which a problem-based approach to learning is adopted. By implementing geo-informatics in a user-friendly environment, GIS students, also from interdisciplinary fields, can access a powerful tool for spatial problems in policy, management and operational tasks (DeMers 2002; Gumbricht 1996).

GIS first emerged in the nineteen-sixties, in the form of simple overlays of geographical maps of different land features, such as soil type, hydrology and elevation. Aerial photography and satellite imagery, together with the increasing precision with which field samples were being taken, opened up major new opportunities for GIS applications, at the same time making substantial demands on data processing capacity. Today, the management of pure data as well as complicated graphics and maps has been facilitated enormously by the vastly greater capacity of modern computers.

The Geographic Information Systems in use today can be seen as comprising three 'layers': (1) a database layer (data collection and construction), (2) an application layer (data transformation and application), and (3) a presentation layer (data presentation). In terms of procedure, then, GIS comprises the following steps:

- Data collection: aerial photographs and satellite imagery, field data (physical, chemical, biological), data on human use of the environment (various).
- Data manufacturing: data storage and processing in the computer, digitalisation of data input, input of geographical coordinates, rectification of satellite imagery to GPS (global/geographical positioning system).
- Data transformation: transformation does not change the meaning of the data, merely the 'form' in which they are described. First, a local example:

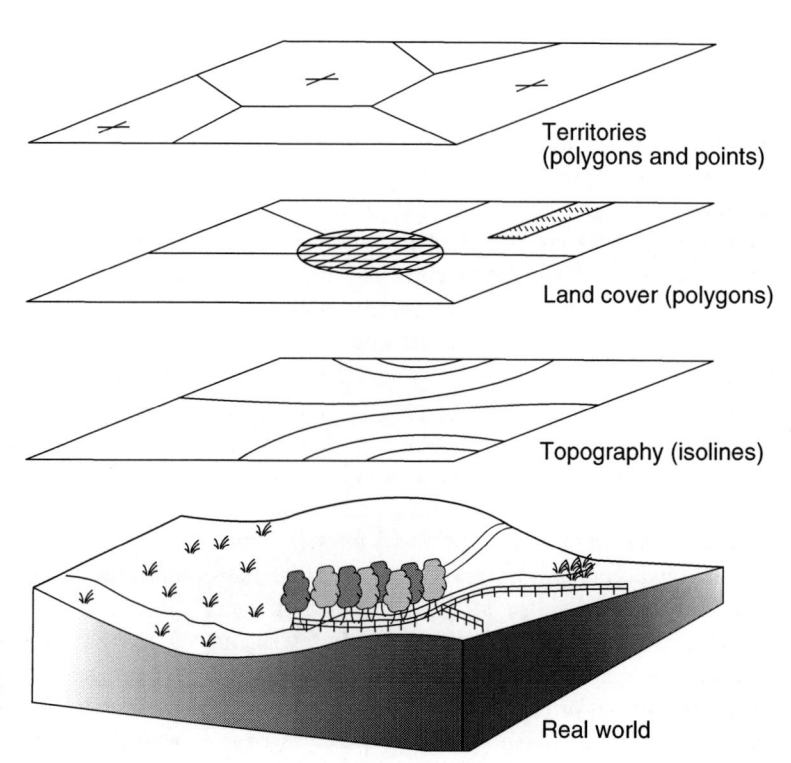

Fig. 10.10 GIS concept of using layers to represent the real world
Source: Wadsworth and Treweek (1999)

drilled soil samples for use in a vegetation survey of 1 m² have a surface area of only a few cm². To be properly interpreted and projected onto a 1:10,000 map, the data must be transformed, based on parameters like density of the sampling points and heterogeneity of the geographical unit. Second, a European example: national soil maps constructed using national soil units need to be transformed into international units if they are to be used for Europe-wide applications.
- Data presentation: in maps, tables, statistical reports, computer databases, computer applications, etc.

The most common type of data model used in GIS software is the 'field' model, which is essentially a mapping onto physical, geographic space. The basic idea of a field model is illustrated by the simple example shown in Fig. 10.10, in which the real world is represented by three layers, each representing one particular 'theme' (*'thematic layers'*). The first layer represents topography or elevation, in this case portrayed as isolines or contours, the second different types of vegetation and the third the territorial boundaries of selected species. Layers can be directly compared because they share a common framework located in the real world (Wadsworth and Treweek 1999).

10.5.5 Remote Sensing

Remote sensing of the earth by unmanned satellites began in April 1960 with the launch of TIROS-1, a forerunner of today's low-orbiting meteorological satellites. Landsat-1 was sent into space in July 1972, the first member of the Landsat series that has provided the most widely used remote sensing data outside the meteorological community. Landsat-1 was provided with a scanner for visible and near-infrared light, the Multispectral Scanner (MSS), which has pixels about 80 m across. Landsat-5, launched in 1984, was equipped with a new scanner system, the Thematic Mapper (TM), which extended further into the reflected infrared spectrum than the old MSS and has a smaller pixel size of 30 m. Throughout the 1970s and early 1980s Landsat data were distributed cheaply and on an equal basis to scientists and other users throughout the world. In 1985, however, Landsat operations were sold off to private business and prices were brought at a commercially viable level.

The first of a series of remote sensing satellites developed by the French Space Agency, called *SPOT*, was launched in 1986. This satellite provided the first proper stereoscopic capability, by being able to point to the right and to the left of its ground track during subsequent orbital passes, instead of pointing only directly downwards. Like the new generation of Landsats, SPOT is operated on a commercial basis (Kelly 1991; Leinders et al. 1989). Now, in the year 2003, there are numerous satellites orbiting the earth and plans for launching sophisticated, non-American remote sensing satellites in the coming years show that the virtual monopoly of the USA in this field is now coming to an end.

A spectrometer is an apparatus in which radiation (mostly light) can be split into spectral lines. Modern satellite technology deals with the degree of spectral resolution in the development of 'imaging spectrometers', providing data across the reflected spectrum with a spectral resolution of about 10–20 nm. *Imaging spectrometry*, both in the field and via satellites, is a promising technique for detecting 'hotspots' of soil and sediment pollution, such as occur with heavy metals accumulation in river floodplains, for example. Once the relation is known between soil composition (sand, silt, organic matter) and the adsorbed sediment pollution (heavy metal concentration), the degree of spectral resolution of the various soil types recorded by the imaging spectrometer can provide a quick scan of the distribution of contaminating heavy metal concentrations. This method has indeed been applied by environmental scientists in studying the spatial variation of heavy metal pollution in floodplain sediments and its consequences for accumulative effects in the food chain (Kooistra et al. 2002).

Depending on the field of view of the imaging system and the altitude at which it is deployed, the area covered by an image ranges from a few hundred metres to thousands of kilometres across (see also Section 10.6: 'Integrated indicators for environmental quality assessment'). An *image* is a synopsis of conditions captured over several minutes at most and in some cases nearly instantaneously. In practice, the spatial resolution of the information content of an image is governed by the means used to present the imagery data. Images are made up of *rectangular picture elements* (*pixels*) analogous to the dots used in newspaper photographs. Because different parts of the electromagnetic spectrum can be imaged in this way, information from these various regions can be combined in threes in an analogous way to the red, green and blue dots that make up colour TV pictures.

The advantage of digital images made up of pixels is that they can be manipulated by computers in numerous ways. Such image processing can serve a variety of purposes, for example (Leinders et al. 1989):

- Images from different imaging systems on board different vehicles and captured at different times can be combined with one another and with conventional map projections.
- Image defects hampering visual interpretation can to some extent be removed.
- Image contrast and focus can be enhanced as aids to better visual interpretation by highlighting subtle features present in the digital data.
- Various procedures can be applied that rely on a computer's ability to group together pixels with similar 'signatures', detect boundaries and shapes, and identify areas that have changed over a given period of time.

Developments in geology, oceanography, ecology, agriculture and many other fields of science all show a long history of local and regional exploration of resources, using expensive methodologies geared to establishing 'ground truth', i.e. collection of data out in the field. Large parts of the earth's continents and oceans, mainly in (sub-)tropical and (sub-)arctic regions, are known only in a sketchy and sometimes outdated manner, despite the attention of many generations of scientists over several centuries. We know a lot about many small areas of the earth. The new, comprehensive maps and inventories accruing from remote sensing can serve as a database with which to link the

---

Box 10.6 GIS and remote sensing in river management

Leuven et al. (2002) provide useful examples of the application of GIS and remote sensing. Data collection is always a combination of aerial photographs and satellite images on the one hand and ground truth (i.e. field data) on the other. This text box describes the methodology's application in three different areas: river geomorphology (Middelkoop 2002), ecology (Kooistra et al. 2002) and land use and river management planning (Farthing and Teeuw 2002).

Middelkoop (2002) applied remote sensing and GIS-based techniques to analyse floodplain sedimentation in the lower Rhine. Using Landsat TM images, the downstream changes in sediment concentrations and spatial patterns of sediment depositions within the floodplain were analysed after the extreme discharges of the Rhine in December 1993. Ground truth data comprising sediment trap measurements were interpolated in remote sensing images and converted to sedimentation maps in a GIS. These data formed the basis for development of a GIS-based mathematical model for simulating floodplain sedimentation. As already mentioned, floodplain soils along the Rhine show great spatial variability in terms of pollutant concentrations.

Kooistra et al. (2002) used GIS to quantify the spatial variability of pollutants in relation to soil characteristics, based on the definition of homogeneous landscape units that incorporated the foraging behaviour of exposed terrestrial organisms. Using a probabilistic exposure model the exposure to heavy metal concentrations of a typical floodplain food chain was estimated: from floodplain soil via earthworms, shrews and voles up to the Little owl (*Athena noctuca*) as top predator. To estimate exposure risks to cadmium for the Little owl, site-specific risk maps of the floodplain were constructed, showing high-risk areas for the bird along the river. Based on these maps, recommendations for future management of the floodplain were made.

Farthing and Teeuw (2002) used aerial photography and Landsat imagery to produce detailed maps of Thames region floodplains. Data integration was performed in a GIS environment using hydrological, geomorphological, and ecological data, together with information on flood defence, fisheries, physical planning schemes and archaeology. Among the output of the programme are river catchment management plans, 'state of the environment' reports and action plans.

small, disconnected parcels of existing understanding generated by traditional means of study into many-sided concepts. These concepts are no longer static but more truly reflect the dynamics of nature and can often discriminate between natural phenomena and anthropogenic changes to the natural environment. Today remote sensing has numerous applications (see for instance Box 10.6) and to that extent should be considered an important methodology in the environmental sciences (Leuven et al. 2002).

10.6 Integrated Indicators for Environmental Quality Assessment

Many of the most meaningful questions in environmental science deal, in one way or the other, with environmental 'quality'. As already mentioned in the Introduction, *environmental quality* comprises the structural and functional properties of the environment in the context of human appreciation thereof, either positive or negative (Boersema et al. 1991: 24). In this section, a number of integrated (holistic) indicators for the assessment of environmental quality are discussed. The aim here is not completeness and there are certainly other indices that might be mentioned, particularly those combining natural science and social aspects (see Chapter 11). The indicators discussed range from relatively 'simple' metric compounds to very complicated ecological quality assessments methodologies.

10.6.1 Index of Biotic Integrity – IBI

The *Index of Biotic Integrity (IBI)*, originally developed for fish communities in streams in several US states, incorporates a dozen biological attributes into a single index (Karr 1991). The strength of the index is that the biological attributes used are expressed in numerical values permitting objective comparison with future situations: total number of native fish, number and identity of sunfish species, proportion of individuals that are carnivores, etc. Fish and invertebrate metrics can also be combined into a single index. The values of these simple or combined metrics can be compared to the values expected for a relatively undisturbed stream of similar size in a similar geographic region (*reference image*). Many state agencies in the USA have adopted the biotic integrity approach for use in water management. Today field data from biological monitoring programmes for fish and macro-invertebrates are used to calculate IBIs for specific streams and rivers all over the world. This has become a significant tool for water managers, whose task it is to describe the condition of rivers, diagnose the causes of any degradation and develop rehabilitation plans where necessary.

The IBI concept has proved so successful that international conferences have been devoted to this way of assessing the ecological integrity of running waters. '*Biotic integrity*' and '*ecological integrity*' are often used synonymously. Both approaches are concerned with assessing environmental quality, in this case the condition of running waters. Concepts such as 'integrity' (unimpaired, or in original condition) and 'health' (shorthand for in good condition; see Section 10.6.3: 'Ecosystem Health') appear to be relevant and worth imitating because they are grounded in science and yet appeal to citizens.

10.6.2 The TRIAD Method

Another type of environmental monitoring is embodied in the 'TRIAD' approach, which integrates methods from several sciences, combining chemical and ecological data with bioassay results to assess the quality of a terrestrial or aquatic compartment or habitat. To ensure results are consistent, the use of sophisticated laboratory equipment as well as properly conditioned experimental set-ups is required.

As an example, let us consider how the TRIAD method might be applied to evaluate the quality of estuarine underwater sediment. For this purpose the *Sediment Quality TRIAD* has been developed, a synoptic measure comprising three dimensions: sediment contamination, toxicity and infaunal (fauna living in sediments) community composition. These quantities are measured, respectively, by means of (1) chemical analysis, (2) bioassays and (3) field observation. The chemical analyses cover all relevant toxic substances thought to be present in the sediment, such as heavy metals and organic micropollutants, which are expressed in terms of weight per kg dry sediment. The bioassays yield absolute mortality figures per unit time for selected benthic invertebrates exposed to contaminated sediments. The ecological field data are combined to yield a composite index of biological diversity, i.e. a ratio comprising the number of species and the number of individuals per species of selected benthic invertebrates.

The measured values obtained in (1), (2) and (3) are divided by values from a (clean) reference location to give the so-called ratio-to-reference (RTR) values. These RTR values are averaged per group of variables, yielding as a final result a set of three figures per locality, for (1), (2) and (3). These figures are then plotted onto the three axes of a TRIAD diagram (see Fig. 10.11). The larger the triangle connecting the three data points, the more sediment quality deviates from the reference values. What the TRIAD diagram illustrates are merely correlations, rather than causal chains, and it is ultimately the subjective choices and assumptions made by the researcher that are the decisive factor for how the quality of the underwater sediment habitat is assessed. Particularly important are the choice of chemical and biological variables and the test organisms used in the bioassay experiments. Another prerequisite in this methodology is a 'clean' reference location. Given current levels of anthropogenic disturbance, however, it is almost impossible to find the kind of location required for comparison while still excluding geographical or ecological differences unrelated to pollution effects (Chapman et al. 1987).

10.6.3 Ecosystem Health

To understand and manage complex ecosystems, we need to have some measure of the system's overall performance (or 'health') in which both structure and functioning are represented. Over the past few decades a considerable number of attempts have been made to evaluate and monitor ecosystem health (for a review, see e.g. Rapport et al.

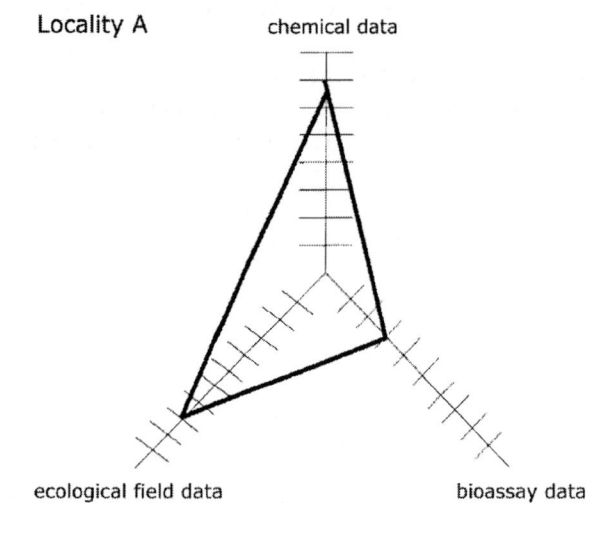

Fig. 10.11 The Sediment Quality Triad at locality A
Source: Derived from Chapman et al. (1987)

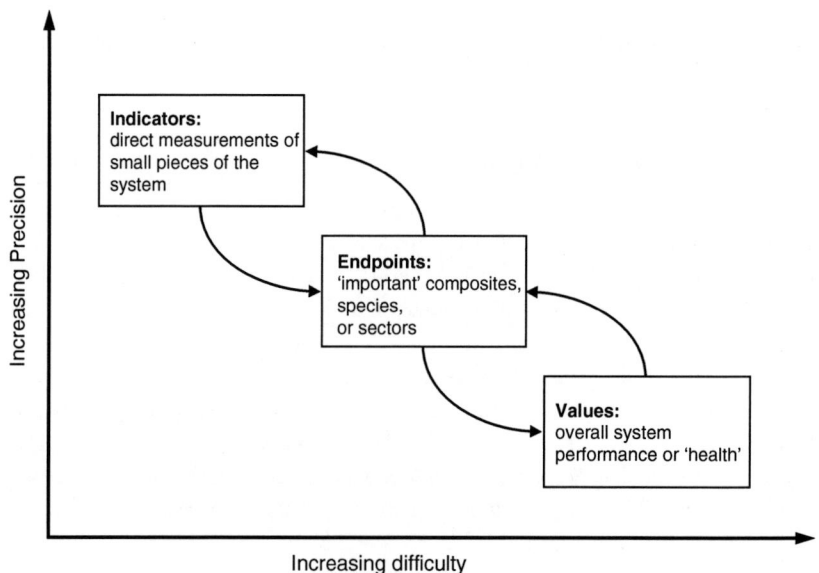

Fig. 10.12 Relationship between indicators, endpoints and values
Source: After Costanza et al. (1992)

1995). Costanza and Mageau (1999) define *ecosystem health* as a comprehensive, multi-scale, dynamic, hierarchical measure of system vigour, organisation and resilience. These concepts are also embodied in the term '*sustainability*', referring to a system's ability to maintain its structure (*organisation*) and function (*vigour*) over time in the face of external stress (*resilience*). In the simplest terms, health is a measure of the overall performance of a complex system built up from the behaviour of its parts. Such measures of system health imply a weighted summation (or some more complex manipulation) of results for the component parts, with the weighting factors embodying an assessment of the relative importance of each part to the functioning of the whole. Such an assessment of relative importance incorporates 'values', which can range from the subjective and qualitative towards the (more) objective and quantitative as more knowledge is gained about the system under study.

Figure 10.12 shows the progression from directly measured 'indicators' of the status of a system component, via 'endpoints' that are composites of these indicators, to overall system performance (health) employing 'values' based on subjective, normative choices. Measures of health are inherently more difficult to quantify as well as more comprehensive, require more modelling and synthesis, and involve greater uncertainty as well as less precision. To encapsulate the functioning of an entire ecosystem, however, they hold far greater relevance than direct measurements of individual facets of the ecosystem, such as a biochemical or physiological anomaly in some vital process of a particular animal.

Health is also a scale-dependent characteristic. We may reasonably expect a cell in an organism to have a relatively short lifespan, the organism itself to have a longer life, the species an even longer one and the planet a still longer lifespan. A *healthy and sustainable*

Table 10.3 A tabular model of ecological succession: trends to be expected in the development of ecosystems

Ecosystem attributes	**Developmental stages**	**Mature stages**
Community energetics		
1 Gross production/community respiration (P/R ratio)	Greater or less than 1	Approaches 1
2 Gross production/standing crop biomass (P/B ratio)	High	Low
3 Biomass supported/unit energy flow (B/E ratio)	Low	High
4 Net community production (yield)	High	Low
5 Food chains	Linear, predominantly grazing	Web-like, predominantly detritus
Community structure		
6 Total organic matter	Small	Large
7 Inorganic nutrients	Extrabiotic	Intrabiotic
8 Species diversity – variety component	Low	High
9 Species diversity – equitability component	Low	High
10 Biochemical diversity	Low	High
11 Stratification and spatial heterogeneity (pattern diversity)	Poorly organized	Well-organized
Life history		
12 Niche specialisation	Broad	Narrow
13 Size of organism	Small	Large
14 Life cycles	Short, simple	Long, complex
Nutrient cycling		
15 Mineral cycles	Open	Closed
16 Nutrient exchange rate, between organisms and environment	Rapid	Slow
17 Role of detritus in nutrient regeneration	Unimportant	Important
Selection pressure		
18 Growth form	For rapid growth ('r-selection')	For feedback control ('K-selection')
19 Production	Quantity	Quality
Overall homeostasis		
20 Internal symbiosis	Undeveloped	Developed
21 Nutrient conservation	Poor	Good
22 Stability (resistance to external perturbations)	Poor	Good
23 Entropy	High	Low
24 Information	Low	High

Source: Odum (1969)

system in this context is one that attains its full expected lifespan. Since ecosystems undergo 'succession' under the impetus of their own development, today possibly modified by climate change or other such factors, they too have a limited, albeit fairly long lifespan. The key here is to differentiate between changes due to the normal limits of the system's lifespan and changes that cut it short as a result of anthropogenic impact or disturbance. When aquatic ecosystems become eutrophied under human influence, for example, it is their very nature that becomes radically changed: the lifespan of the oligotrophic system is brought to a end, and that of a more eutrophic system initiated. By the earlier definition, we should call this process 'unhealthy', since the lifespan of the first system was cut 'unnaturally' short. The water basin may have eventually eutrophied of its own accord, but the anthropogenic stress caused the transition to occur prematurely.

What methods are available for measuring ecosystem health? Ecosystem health can be said to comprise three components: vigour, organisation and resilience. The classical dichotomy in ecology is between the functioning and structure of biological 'units' (organism, species, community, ecosystem). *Vigour* stands for the *functioning* of a system, and *organisation* for its *structure*.

Ecosystem health basically uses the attributes formulated by Odum (1969) in his famous paper on the strategy of ecosystem development (Table 10.3). In this paper Odum presented an understanding of ecological succession, the principles of which have a fundamental bearing on the relationships between man and nature. Indeed, the framework of succession theory needs to be conscientiously examined as the very basis for resolving our present environmental crisis. The table distinguishes 24 attributes of ecological systems under six headings and highlights trends in each as the ecosystem develops from young to mature. The attributes grouped under community energetics and nutrient cycling correspond more or less to our 'vigour', the traits of community structure and life history our 'organisation'. Selection pressure (with '*r* selection' predominating in the early stages of colonisation and '*K selection*' prevailing as more and more species attempt to colonise as the ecosystem matures) and overall homeostasis in terms of resistance to external perturbations come close to our 'resilience'.

According to Odum (1969) the overall 'strategy' of ecosystem development is directed towards achieving as large and diverse an organic structure as possible within the limits set by the available energy input and prevailing physical conditions. Eugene P. Odum (1977) was an environmental scientist '*avant la lettre*' by stating that "ecology must combine holism with reductionism if applications are to benefit society. To achieve a truly holistic approach, not only ecology, but other disciplines in the natural, social and political sciences as well must emerge to new hitherto unrecognised and un-researched levels of thinking and action". Against this background let us now consider the aspects of ecosystem health in more detail:

1. The *vigour* of an ecosystem is simply a measure of its activity, metabolism or primary productivity. Examples of such measures include gross or net primary production in ecological systems, and gross national product in agricultural and other economic systems. It has been hypothesised that a system's ability to recover from stress, or cope with it, is related to its overall metabolism, i.e. the energy flowing through it (Odum 1969), or its 'scope for growth' (Bayne 1987), i.e. the difference between the energy required for system maintenance and that available to the system's activity patterns, mainly respiration.

2. The *organisation* of an ecosystem refers to the number and diversity of interactions between the component elements of the system. Indicators used to measure it should reflect the diversity of species and communities comprised by the ecosystems as well as the number of pathways of material exchange between these individual elements. A system comprising species feeding on only one or two specific prey items and in turn serving as prey for only one or two predators will exhibit a less degree of organisation than a system comprising the same number of generalist feeders connected by multiple pathways of exchange. Organisation thus extends traditional measures of diversity by also considering the patterns of exchange between system components.

3. The *resilience* of an ecosystem refers to its ability to maintain its structure and behavioural patterns in the face of stress (Holling 1986). In the context of this chapter, it may refer to the system's ability to maintain its vigour and organisation in the face of *stress (disturbance)*. A healthy system is one that has enough resilience to survive a wide variety of small-scale perturbations. The concept of ecosystem resilience has two main aspects. Most com-

monly it refers to the time it takes for a system to recover from stress. It may also refer, however, to the *amount* of stress from which a system can satisfactorily recover, or the system's specific thresholds for absorbing particular forms of stress.

How, then, can ecosystem vigour, organisation and resilience be quantified?

Vigour is the most straightforward of the three to measure. In most ecosystems it can be measured directly and relatively straightforwardly using standard ecological methods and can be expressed in such terms as gross primary production (GPP), net primary production (NPP), gross ecosystem production (GEP) or metabolic rate. It is debatable whether economic measurement of gross national product (GNP) is to be classed among the valid parameters for expressing ecological vigour.

Organisation is more difficult to quantify, as it involves both qualitative and quantitative aspects of biological diversity as well as the exchange pathways between species and individuals. Diversity indices and multi-species indices fail to incorporate exchange pathways connecting system components. Network analysis is a potential approach to solving the problem of measuring organisation. This involves quantitative analysis of interconnections between ecosystem components (the species) and their interconnections within the larger system (their biotic environment), i.e. the total number of potential pathways of material exchange between system components (see Section 10.4.3: 'Ecosystem structure and function', and in particular: *connectance*).

Measuring the *resilience* of an ecosystem is also difficult, because it implies an ability to predict the dynamics of the system under conditions of stress. Predicting ecosystem impacts over time generally requires use of dynamic simulation models. One measure of resilience is recovery time: the time it takes for an ecosystem to recover, in terms of vigour and organisation, from a wide variety of stresses to some previous steady state. When calculating this measure of ecosystem resilience the choice of indicators to be tracked over time is very important. The population of a single species would be easiest to track, but this would tell us least about the overall system's response to stress. Increasingly complicated measures such as those suggested for vigour, organisation and their combination (*ascendancy*; Ulanowicz 1980) will tell us more about system response, but these parameters are far more difficult to measure and lack precision and reliability (Costanza and Mageau 1999).

In contrast to definitions of ecosystem health based solely on ecological criteria (see Costanza and Mageau 1999), judgements of ecosystem health must also include *human values*, *goods* (such as food) and *services* (such as waste assimilation) derived from the system. Boulton (1999) elaborates on this theme for the 'health' of rivers. His concept of '*river health*' is based on a blend of *ecological values* and human values. Boulton (1999) uses *ecological integrity* as an overarching concept for ecological values, roughly covering Costanza and Mageau's (1999) and our 'vigour' and 'organisation' (compare the limited meaning of 'ecological integrity' in Section 10.6.1: 'Index of Biotic Integrity').

In environmental economics the ecological values of an ecosystem are referred to as *natural capital stocks*. Human values, on the other hand, represent the benefits derived by human populations, directly or indirectly, from these ecological values. The goods and services provided by ecosystems and the natural capital stocks from which they are generated are critical to the functioning of the earth's life-support system. A great number of such ecological goods and services can be identified: water supply, food production, raw materials (goods), climate regulation, erosion control, sediment retention, waste treatment (services) and also recreational and cultural services (Costanza et al. 1997). The search for integrated indicators for ecosystem health will certainly help to define and underpin 'sustainability', one of the key concepts in environmental science today.

Rapidly increasing human population and consumption, and concomitant demand for food and energy, are making society increasingly dependent on services provided by remaining natural and managed ecosystems. The question is how changes in organisation (in this context biodiversity) of an ecosystem affect the functioning of ecosystems and their ability to provide goods and services for humans. Biodiversity experiments have shown that greater numbers of plant species lead to a greater production of biomass. The long-term stability of an ecosystem service – the annual production of biomass and thus of potential biofuels and livestock fodder – also depends on biodiversity. Biodiversity can therefore be an important element for the reliable and sustainable provisioning of ecosystem services (Sala 2001; Tilman et al. 2006).

10.6.4 The AMOEBA Method

Environmental quality is a pivotal notion in the environmental sciences. It encompasses not only 'objective' measurements using methods from the natural sciences but also 'subjective' choices and assumptions, as emerged in Section 10.6.3: 'Ecosystem Health'. Environmental quality indicators have spatial as well as temporal characteristics. As we move up the sequence: microhabitat (where the individual lives) – habitat (where part of the population lives) – community – ecosystem – landscape (e.g. sector of a river) – geographic region (e.g. floodplain) – river drainage basin, there is growing ecological complexity. Table 10.4 lists key characteristics of the extremes in this sequence, the micro-habitat and the landscape, on the spatial scale from 'small' (= simple) to 'large' (= complex). Most characteristics come from the natural science domain, but some are oriented towards human society (policy relevance, management practice).

Table 10.4 shows the contrasts between the high specificity of indicator values and the large quantity of scientific information at lower levels of the hierarchy, and the uncertainty, unpredictability and lack of management guidelines at higher levels. In keeping with its interdisciplinary mission, environmental science should pursue environmental problem analyses and solutions having as much policy relevance as possible, thus ranking high in the biogeographical hierarchy (ecosystem or landscape level). In scientific terms there is substantial knowledge about lower levels of ecological integration. The more complex the system, the higher its policy relevance to society, but – and herein lies the twist – the less scope for defining good management practice and the longer it takes for the system to recover. In other words, if some relatively small fraction of a plant or animal population is disturbed, ecological restoration is likely to be successful, provided we are knowledgeable about the species in question and its habitat is still undisturbed. If the entire floodplain of a river were annihilated, though, with its numerous plant and animal populations, it would take drastic and as yet possibly unknown measures and a very long recovery period to restore the system. Although the restoration or preservation of ecosystems and landscapes is the most ambitious aim of nature conservation policy, it is often at the population level that measures are most successful.

Table 10.4 Characteristics on the spatial scale from 'small', i.e., 'simple', to 'large', i.e., 'complex'

Characteristic	Microhabitat	to Landscape
Complexity of system	Low	High
Precision of measurement	High	Low
Specificity of measurement	High	Low
Sensitivity of system	High	Low
Policy relevance	Low	High
Ecological relevance	Low	high
Recovery time	Short	Long
Manageability	High	Low
Management practice	High	Low
Policy indicator value	Low	High

Source: Modified after data of Costanza et al. (1992), Frissell et al. (1986)

The *General Method for Ecological Description*, generally referred to as the *AMOEBA method* (Ten Brink et al. 1991), is an example of a semi-quantitative method for describing and assessing the quality of aquatic ecosystems. The AMOEBA method comprises a number of indicators, called *target variables*: keystone plant and animal species, biological communities or habitats, whose recovery is sought through implementation of dedicated management measures in a disturbed ecosystem. These target variables have been chosen for policy reasons rather than on the basis of scientific argument, employing three main criteria: the species or community should serve as an indicator for the overall health of the ecosystem, i.e. it should (1) be related to water quality, (2) be readily recognisable by non-experts, and (3) be quantifiable (countable, measurable, etc.). This does not necessarily mean these target species are the most important in terms of optimum ecosystem functioning. A hidden fungus living on the roots of a shrub (mycorrhiza) may be more important for the survival of shrub stands than any other factor because it is indispensable for metabolism. In policy terms the shrub is the target species, then, although functionally the fungus is the most important. The shrub cannot survive without the fungus, while the fungus can without the shrub.

Figure 10.13 shows the AMOEBA for the coastal waters and estuaries of the Netherlands along the southern North Sea for the year 1995 (Baptist and Jagtman 1997). The circle of this so-called *radar diagram* represents the *target image* formulated for the long term (2010–2020). The slices of this pie-like radar diagram represent the *target variables*, i.e. the *indicator species* or communities. As can be seen, the Sandwich tern stands at 61% of its target. Marine mammals, several

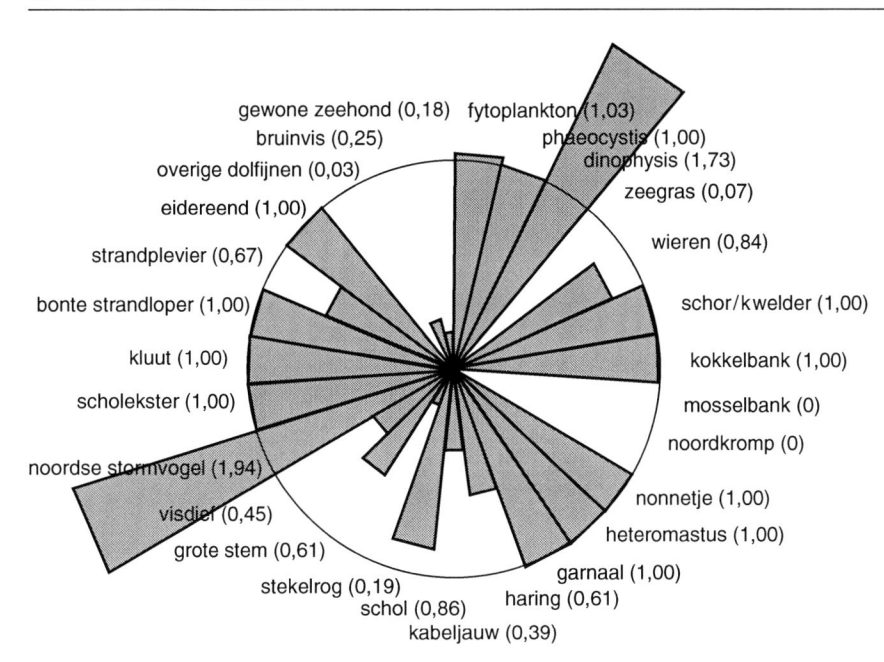

Fig. 10.13 AMOEBA-radar diagram for the coastal waters and the estuaries of the Netherlands, for the year 1995 (Baptist and Jagtman 1997). The AMOEBE is a Dutch method; the names of the target species and target communities (pieces of the pie) are given in Dutch. The index (0.1 etc.) indicates how much (in numbers or area) of the target (circle) has been reached (0.1 = 10%). Grote stern = Sandwich tern; Noordse stormvogel = Fulmar; Bruinvis = Porpoise; etc. For further explanation see text

Table 10.5 The ecological rehabilitation of the Sandwich tern population (target variable) along the coasts of the Netherlands, governed by four steering variables. Best professional judgement is used to indicate the expectations for short-term ecological restoration: L = low; M = mean; H = high

Steering variables	PCB-load	Fishery quota	Conservation of nesting sites	Turbidity of the water
Complexity of system	M	H	H	H
Precision of measurement	H	L	M	M
Specificity of measurement	H	M	L	M
Sensitivity of system	H	M	M	M
Policy relevance	H	H	H	H
Ecological relevance	H	H	H	H
Recovery time	M	M	M	M
Manageability	H	L	L	M
Management practice	H	M	M	M
Policy indicator value	H	M	M	M

bird and fish species and several biological communities (seagrass and natural mussel beds) are far below the desired numbers or area. A few species are above target. In the case of *Dinophysis acuminata*, an undesirable, toxic micro-alga, the causal factors (including eutrophication) leading to algal bloom need to be reduced. In the case of the Fulmar (*Fulmaris glacialis*) the 'surplus' numbers are due to natural fluctuations in population density.

The Sandwich tern (*Sterna sandvicensis*) thus serves as one of the target variables in projects to restore Dutch coastal ecosystems, the objective being to increase overall numbers from 10,000 to 40,000. The *reference image* is 50,000 individuals of this species, the approximate number recorded along the Dutch coast prior to the Second World War. The following *steering variables* are employed to achieve this objective: (1) river loads of PCBs, consumed by the birds with their food, thereby reducing fertility; (2) fishery quota, competing with the terns for food; (3) conservation of specific nesting sites, which are now scarce along the coastline; (4) water turbidity, which needs reducing so the terns can accurately localise their prey. Table 10.5 reviews expectations (low, medium, high) for short-term ecological recovery of the Sandwich tern population along the Dutch coast with respect to these four steering variables, employing the same criteria as in Table 10.4.

The results can be summarised as follows. (1) PCB loads, which disrupt the reproductive cycle of Sandwich terns, score high in virtually all respects: the problem was and still is very important, can ultimately be properly managed (a ban on PCBs) and can be easily monitored from any well-equipped laboratory. (2) Fishery quota are certainly of major policy relevance. However, the complexity of the ecological system is also high, being governed by weather and climatic conditions and by the reproductive cycles of the various contributing fish species. (3) Conservation of breeding sites is difficult, for many habitats have simply disappeared or have been permanently disturbed. Traditional forms of resource use (harvest of eggs) and management (building of small nesting sites in wetlands) have been abandoned and localities used as nesting sites are now used for other purposes. (4) The turbidity of the seawater is also a complex problem, related to algal blooms and hence to the eutrophication status of coastal waters and river nutrient loads.

Assume now that PCBs have been completely banned and are no longer present in the environment, that fish stocks have recovered and that water turbidity has been reduced to zero. Even then, if there are still no suitable breeding sites for the Sandwich tern, the objective of 40,000 birds in 2010 will never be achieved. In other words, there will generally be a limiting factor, which should be properly identified. Similar considerations will obviously hold for the other AMOEBA target variables.

Environmental quality indicators, though preferably formulated at a high hierarchical level (ecosystem, landscape), are most realistic as well as useful for managers and policy-makers if they are at the level of (parts of) populations. Even at that level, however, as the example of the Sandwich tern illustrates, if there is no compensation for lost habitat the species will not recover.

10.7 Summarising Conclusions

In this chapter, the 'environment' has been considered as the physical, chemical and biological (i.e. living and non-living) environment and as part of human society (its 'support base'), with which it stands in mutual relation. This implies that a basic knowledge of the principles of physics, chemistry and biology are essential for understanding the nature, characteristics, scope and dimensions of environmental problems. Every environmental scientist should thus have some basic understanding of the 'laws' of nature. This chapter has therefore endeavoured to present the 'fundamentals' of the natural sciences, to the extent that they are relevant to the analysis and solution of environmental problems. The principles of physics are exemplified in mass and energy balances and mass transport processes. The principles of chemistry are reflected in chemical kinetics and thermodynamics and in equilibrium processes. The principles of biology are illustrated with reference to the variability and dynamics to be found at all levels of biological organisation, from genetic variability via population dynamics to ecosystem structure and functioning. Essential biological notions underlying a wide variety of environmental problems are illustrated with reference to the flow of energy and the cycling of materials through biological systems.

In selecting an appropriate scientific method for tackling a particular environmental problem, it is of crucial importance to properly identify the spatial and temporal dimensions of that problem. Most physical, chemical and biological indicators provide direct measurements of small parts of the ecosystem, often with great precision. When it comes to measuring overall environmental quality, however, these kind of indicators are of little use, as they each provide only a partial picture of reality and a relatively arbitrary one at that. Ideally, environmental quality indicators should be comprehensive in scope, as relevant as possible for policy-makers and easy to manage. Precisely because of their comprehensiveness, though, such indicators are unfortunately not the easiest to measure. This outlines the basic dilemma of 'scaling', exemplified in the spatial and temporal dimensions of chemical distribution patterns and biodiversity units in the environment. A number of methods for specifically dealing with the scaling problem have been described: mapping, monitoring Geographical Information Systems and remote sensing. Finally, a number of integrated indicators for assessing overall environmental quality have been discussed.

The choices made have been subjective ones, and other integrated methods might just as well have been cited. The four integrated methodologies considered here range from the relatively simple to the highly complex. The first, the Index of Biotic Integrity, is a compound measure incorporating several numerical values. The second, the TRIAD method, comprises a combination of chemical and (experimental) biological criteria. The third, ecosystem health, is based predominantly on a comprehensive series of ecological criteria. The fourth and most complex, the AMOEBA methodology, establish target objectives and steering variables, based mainly on non-scientific policy arguments. All these concepts refer to ecosystems in a stable physicochemical environment. No doubt, the indicators presented here have serious limitations when presented against the background of climate change.

As a young and evolving science, environmental science has no 'standard set' of methodologies. As already said in the Introduction, the criterion used in selecting the methods from the natural sciences, explained in this chapter, is their usefulness for analysing and resolving environmental problems. Priority has thereby been given to methods that are particularly useful in multi- and interdisciplinary research and consequently of greatest relevance to society as a whole. A proper comprehensive approach to environmental problems can only be reached in combining selected methods from the natural sciences, the social sciences and the humanities.

References

Auer, M. T., Penn, M. R., & Mihelcic, J. R. (1999). Biology. In J. R. Mihelcic (Ed.), *Fundamentals of environmental engineering* (pp. 209–323). New York: Wiley.
Baird, C., & Cann, M. (2004). *Environmental chemistry* (3rd ed.). New York: W.H. Freeman.
Bakker, K., Mook, J. H., Nienhuis, P. H., Van Rhijn, J. G., & Woldendorp, J. W. (1995). *Oecologie*. Houten (in Dutch): Bohn Stafleu Van Loghum.
Baptist, H. J. M., & Jagtman, E. (Eds.). (1997). *De AMOEBES van de zoute wateren*. Watersysteemverkenningen 1996. Rapport RIKZ-98.027. Den Haag (in Dutch).
Bayne, B. L. (1987). *The effects of stress and pollution on marine animals*. New York: Praeger.
Boeker, E., & Van Grondelle, R. (1999). *Environmental physics* (2nd ed.). Chichester: Wiley.
Boeker, E., & Van Grondelle, R. (2001). *Environmental science, physical principles and applications*. Chichester: Wiley.
Boersema, J. J., Copius Peerboom, J. W., & De Groot, W. T. (Eds.). (1991). *Basisboek Milieukunde*. Meppel (in Dutch): Boom.
Boulton, A. J. (1999). An overview of river health assessment: philosophies, practice, problems and prognosis. *Freshwater Biology, 41*, 469–479.
Chapman, P. M., Dexter, R. N., & Long, E. R. (1987). Synoptic measures of sediment contamination, toxicity and infaunal community composition (the Sediment Quality Triad) in San Francisco Bay. *Marine Ecology Progress Series, 37*, 75–96.
Connell, J. H. (1978). Diversity in tropical rain forests and coral reefs. *Science, 199*, 1302–1310.
Connell, D. W. (2005). *Basic concepts of environmental chemistry* (2nd ed.). Boca Raton/New York: Lewis.
Costanza, R., & Mageau, M. (1999). What is a healthy Ecosystem? *Aquatic Ecology, 33*, 105–115.
Costanza, R., Norton, B. G., & Haskell, B. D. (Eds.). (1992). *Ecosystem health. New goals for environmental management*. Washington: Island Press.
Costanza, R., d'Arge, R., De Groot, R., Farber, S., Grasso, M., Hannon, B., et al. (1997). The values of the world's ecosystem services and natural capital. *Nature, 387*, 253–260.
DeMers, M. N. (2002). *Fundamentals of geographic information systems* (3rd ed.). Chichester: Wiley.
Devinny, J. S., Everett, L. G., Lu, J. C. S., & Stollar, R. L. (1990). *Subsurface migration of hazardous wastes*. New York: Van Nostrand Reinhold.
Farthing, K., & Teeuw, R. M. (2002). Use of geographical information systems and remote sensing for river management: Thames region, UK Environment Agency. In R. S. E. W. Leuven, I. Poudevigne, & R. M. Teeuw (Eds.), *Application of geographic information systems and remote sensing in river studies* (pp. 75–94). Leiden: Backhuys.
Frissell, C. A., Liss, W. J., Warren, C. E., & Hurley, M. D. (1986). A hierarchical framework for stream habitat classification: viewing streams in a watershed context. *Journal of Environmental Management, 10*, 199–214.
Golley, F. B. (1993). *A history of the ecosystem concept in ecology*. New Haven: Yale University Press.
Gordon, D. C. Jr., Boudreau, P. R., Mann, K. H., Ong, J.-E., Silvert, W. L., Smith, S. V., et al. (1996). LOICZ biogeochemical modelling guidelines, *LOICZ Reports and Studies* No. 5, Texel.
Grime, J. P. (1973). Control of species density in herbaceous vegetation. *Journal of Environmental Management, 1*,151–167.
Gumbricht, T. (1996). Application of GIS in training for environmental management. *Journal of Environmental Management, 46*,17–30.
Harremoës, P. (1997). The challenge of managing water and material balances in relation to eutrophication. In R. Roijackers, R. H. Aalderink, & G. Blom (Eds.), *Eutrophication research state of the art*. Wageningen: Wageningen Agricultural University.
Holling, C. S. (Ed.). (1978). *Adaptive environmental assessment and management*. Chichester: Wiley.
Holling, C. S. (1986). The resilience of terrestrial ecosystems: local surprise and global change. In W. C. Clark & R. E. Munn (Eds.), *Sustainable development of the biosphere* (pp. 292–320). Cambridge: Cambridge University Press.
Honrath, R. E., & Mihelcic, J. R. (1999). Physical processes. In J. R. Mihelcic (Ed.), *Fundamentals of environmental engineering* (pp. 139–207). New York: Wiley.
Houghton, J. J., Meiro Filho, L. G., Callander, B. A., Harris, N., Kattenberg, A., & Maskell, K. (Eds.). (1995). *The science of climate change*. Contribution of Working Group I to the Second Assessment Report of the Intergovernmental Panel on Climate Change. New York: Cambridge University Press..
Huston, M. (1979). A general hypothesis of species diversity. *American Naturalist, 113*, 81–101.
Hutter, K., Blatter, H., & Ohmura, A. (1990). *Climate changes, ice sheet dynamics and sea level variations*. ETH: Zürich.
Jackson, A. R. W., & Jackson, J. M. (2000). *Environmental science. The natural environment and human impact* (2nd ed.). Harlow: Longman.
Jones, A., Duck, R., Reed, R., & Weyers, J. (2000). *Practical skills in environmental science*. Harlow: Prentice Hall, Pearson Education.

Karr, J. R. (1991). Biological integrity: A long-neglected aspect of water resource management. *Ecological Applications, 1,* 66–84.

Kaye, G. W. C., & Laby, T. H. (1995). *Tables of physical and chemical constants* (16th ed.). Harlow/London: Longman.

Kelly, M. G. (1991). Remote sensing of the near-shore benthic environment. In A. C. Mathieson & P. H. Nienhuis (Eds.), *Intertidal and littoral ecosystems of the world* (Vol. 24). Amsterdam: Elsevier.

Kooistra, L., Leuven, R. S. E. W., Wehrens, R., Buydens, L. M. C., & Nienhuis, P. H. (2002). GIS as a tool for incorporating the spatial variability of pollutants in ecological risk assessments of Dutch floodplains. In R. S. E. W. Leuven, I. Poudevigne, & R. M. Teeuw (Eds.), *Application of geographic information systems and remote sensing in river studies* (pp. 133–146). Leiden: Backhuys.

Leinders, J. J. M., Drury, S. A., Rothery, D. A., Kirschner, P. A., Beyderwellen, W., & Smit, E. O. (1989). *Remote sensing, course book.* Heerlen: Open University.

Leuven, R. S. E. W., Poudevigne, I., & Teeuw, R. M. (Eds.). (2002). *Application of geographic information systems and remote sensing in river studies.* Leiden: Backhuys.

Lide, D. R. (Ed.). (1996). *Handbook of chemistry and physics* (77th ed. with yearly updates). Boca Raton, FL: CRC Press.

Middelkoop, H. (2002). Application of remote sensing and GIS-based modelling in the analysis of floodplain sedimentation. In R. S. E. W. Leuven, I. Poudevigne, & R. M. Teeuw (Eds.), *Application of geographic information systems and remote sensing in river studies* (pp. 95–118). Leiden: Backhuys.

Mihelcic, J. R. (Ed.). (1999). *Fundamentals of environmental engineering.* New York: Wiley.

Mihelcic, J. R., Urban, N. R., Perlinger, J. A., & Hand, D. W. (1999). Chemistry. In J. R. Mihelcic (Ed.), *Fundamentals of environmental engineering* (pp. 43–138). New York: Wiley.

Odum, E. P. (1969). The strategy of ecosystem development. *Science, 164,* 262–270.

Odum, E. P. (1977). The emergence of ecology as a new integrative discipline. *Science, 195,* 1289–1293

Rapport, D. J., Caudet, C. L., & Calow, P. (Eds.). (1995). Evaluating and monitoring the health of large-scale ecosystems. *NATO ASI series* (Vol. 128), Berlin: Springer.

Sala, O. E. (2001). Price put on biodiversity. *Nature, 412,* 34–36.

Schwarzenbach, R. P., Geschwend, P. M., & Imboden, D. M. (1993). *Environmental organic chemistry.* New York: Wiley.

Ten Brink, B. J. E., Hosper, S. H., & Colijn, F. (1991). A quantitative method for description and assessment of ecosystems: The AMOEBA-approach. *Marine Pollution Bulletin, 23,* 265–270.

Tilman, D., Reich, P. B., & Knops, J. M. H. (2006). Biodiversity and ecosystem stability in a decade-long grassland experiment. *Nature, 441,* 629–632.

Ulanowicz, R. E. (1980). A hypothesis on the development of natural communities. *Journal of Theoretical Biology, 85,* 223–245.

Van de Meent, D., & Huijbregts, M. A. J. (2005). Evaluating ecotoxicological effect factors based on the Potential Affected Fraction of species. *Environmental Toxicology and Chemistry, 24,* 1573–1578.

Valiela, I. (1995). *Marine ecological processes.* New York: Springer.

Wadsworth, R., & Treweek, J. (1999). *GIS for ecology: An introduction.* London: Longman.

Wania, F., & MacKay, D. (1996). Tracking the distribution of persistent organic pollutants. *Environmental Science & Technol., 30,* 390A–396A.

Ward, J. V., Tockner, K., & Schiemer, F. (1999). Biodiversity of floodplain river ecosystems: ecotones and connectivity. *Regulated Rivers: Research & Management, 15,* 125–139.

Chapter 11
Analytical Tools for the Environment-Economy Interaction

Paul Ekins

Contents

11.1 Introduction 177
11.2 Input-Output Analysis 178
  11.2.1 Introduction 178
  11.2.2 Description 178
  11.2.3 Using Matrix Notation 180
  11.2.4 Extension to Energy and the Environment 182
11.3 Resource and Environmental Accounting 185
  11.3.1 1Introduction 185
  11.3.2 Basic Issues in National Environmental Accounting 186
  11.3.3 On Differing Approaches to National Environmental Accounting 187
  11.3.4 Ways Forward for National Environmental Accounting 189
  11.3.5 Corporate Environmental Accounting 191
11.4 Costs and Benefits in Environmental Decision-Making 192
  11.4.1 General Issues and Considerations 192
  11.4.2 Cost-Effectiveness Analysis 195
  11.4.3 Cost-Benefit Analysis 198
11.5 Multi-criteria Analysis 201
References 204

11.1 Introduction

This chapter is concerned with relating the flows of matter and energy through the social system to the flows of money through the economy. Clearly there is a close connection between the two types of flows. Most resources flowing through the social system have been mobilised by the economy, and these resources inevitably become wastes of production and consumption, and are emitted into the environment, in due course. The extraction of resources and the emission of wastes are important causes of environmental damage and degradation.

The central role of the economy in generating environmental damage raises two sets of questions:

1. Which sectors of the economy produce which environmental impacts, and what is the relationship between economic value added and environmental impact in these sectors? The answers to this set of questions would seem to be important in order to be able to gain insights into which economic sectors are relatively damaging per unit of their wealth creation. The analytical tools which seek to provide answers to these questions are input-output analysis and, building on this, resource and environmental accounting.

2. What are the socially desirable levels of different kinds of environmental quality? The economic approach to this set of questions recognises that the environment contributes in a number of ways to human welfare, i.e. it provides benefits. At the same time, protecting the environment, so that it may continue to provide these benefits, can involve costs (e.g. to reduce emissions to the environment) or require that certain economic activities, which themselves yield benefits, are scaled down or forgone completely. Analytical tools which focus on the costs and benefits of environmental protection are cost-effectiveness analysis (CEA) and cost-benefit analysis (CBA). An analytical tool which seeks to explore many of the same issues, but outside a cost-benefit framework, is multi-criteria analysis (MCA).

These tools will now be discussed in turn.

P. Ekins
Sustainable Development at the University of Westminster
Policy Studies Institute, London, UK.
e-mail: p.ekins@psi.org.uk

11.2 Input-Output Analysis

11.2.1 Introduction

In a modern industrial economy the economic activity of firms is highly inter-connected. Input-output tables represent this interdependence between industries, and input-output analysis, which is also called inter-industry analysis, is the technique developed to analyse this interdependence and the implications of changes in economic structure or demands.

Although the technique can be traced back to François Quesnay, who in 1758 published a table showing how expenditures can be systematically tracked through an economy, its modern development is credited to work by the economist Wassily Leontief from the late 1930s, for which he received the economics Nobel Prize in 1973. In their widely used textbook, Miller and Blair (1985: 1) state that today 'input-output analysis is one of the most widely applied methods in economics.'

11.2.2 Description

In order to produce their goods and services most firms buy goods and services from other firms. Conversely, most firms sell at least some of their products to other firms, to be used in further production, rather than to final consumers such as households. Some of the goods bought by firms from other firms are relatively long-lived (for example, buildings and machines) and are classified as investment. The rest of the goods, and all the services, bought by firms from other firms are used up by the purchasing firms in their production process in the course of the accounting period, and are classified as 'intermediate' products, or are said to satisfy 'intermediate demand'. In contrast, goods and services sold to 'final' consumers, as well as those sold for investment, are called 'final' products, and are said to satisfy 'final demand'. Final consumers comprise private households and local and national government.

It is clear that everything produced by the economy, its total output, must go to intermediate demand, final demand or exports. This leads to the equation:

$$X = X_I + Y + E \quad (11.1)$$

where X = the total output of the economy
X_I = intermediate demand (i.e. goods and services sold to firms)
Y = final demand (i.e. goods and services sold to final consumers, plus investment)
E = exports

and all the quantities are expressed in money terms.

Sometimes it is desirable to split final demand into its components, such that:

$$Y = C + I + G$$

where C = goods and services sold to households
I = investment goods
G = goods and services sold to government

For the purpose of analysing economic and industrial structure and change, it is convenient to group firms involved in similar activities into 'sectors'. These sectors may be more or less disaggregated. At the most aggregate level, only three sectors may be distinguished: primary (including agriculture, mining, primary energy industries), manufacturing and services. At the most disaggregated level, sectors may be identified according to particular products. The US National Input-Output Tables reproduced by Miller and Blair (1985) were prepared according to 85-, 365- and 496-sector aggregations, which they aggregate to the 7- and 23-sector classification given in Table 11.1.

Output tables may be constructed by collecting data grouped as in Equation (11.1) for each sector. Thus, for a 7-sector aggregation:

$$X_1 = X_{1I} + Y_1 + E_1 \quad (11.2a)$$

$$X_2 = X_{2I} + Y_2 + E_2 \quad (11.2b)$$

$$\ldots$$

$$X_7 = X_{7I} + Y_7 + E_7 \quad (11.2g)$$

where X_1 = total output from sector 1
X_{1I} = intermediate demand for sector 1's products
Y_1 = (domestic) final demand for sector 1's products and
E_1 = exports of sector 1's products etc.

These equations state that all of each sector's output may be accounted for as sales to intermediate demand, (domestic) final demand or exports. These equations may be expressed in tabular form as in Table 11.2.

In Table 11.2, X_{1I} has been disaggregated into its seven component parts, such that X_{11} is the use of

11 Analytical Tools for the Environment-Economy Interaction

Table 11.1 7 and 23-sector aggregations of industry sectors

7-Sector aggregation	23-Sector aggregation
1 Agriculture	1 Agriculture, forestry and fishing
2 Mining	2 Metal mining
	3 Petroleum and natural gas mining
	4 Other mining
3 Construction	5 Construction
4 Manufacturing	6 Food, feed and tobacco products
	7 Textile products and apparel
	8 Wood products and furniture
	9 Paper, printing and publishing
	10 Chemicals and chemical products
	11 Petroleum and coal products
	12 Rubber, plastics and leather
	13 Stone, clay and glass products
	14 Primary and fabricated metals
	15 Machinery, except electrical
	16 Electrical equipment and supplies
	17 Transport equipment and ordnance
	18 Other manufacturing
5 Transportation and trade	19 Transportation and trade
6 Services	20 Electric, gas and sanitary services
	21 Other services
7 Other	22 Government enterprises
	23 Scrap and second hand goods

Source: Miller and Blair 1985: 406

Table 11.2 Sectoral output, divided into inter-industry purchases, final demand and exports

	1	2	3	4	5	6	7	Y	E	X
1	X_{11}	X_{12}	X_{13}	etc.				Y_1	E_1	X_1
2	X_{21}	X_{22}	X_{23}	etc.				Y_2	E_2	X_2
3	X_{31}	X_{32}	X_{33}	etc.				etc.	etc.	etc.
4	etc.	etc.	etc.							
5										
6										
7	X_{71}	X_{72}	X_{73}	etc.				Y_7	E_7	X_7

sector 1's output in its own production process, X_{12} is the use of sector 1's output in sector 2's production process, etc. In general, X_{ij} is the amount of sector i's product used by sector j in its production process. In monetary input-output tables, all entries are expressed in money values.

Rows 1–7 in Table 11.2 give the relation between each sector's total output (X) and its constituent parts. It may also be seen that columns 1–7 give the inputs from different sectors to a particular sector. Thus X_{11}, X_{21}, X_{31} etc. are the sectoral inputs into sector 1, X_{12}, X_{22}, X_{32} etc. are the sectoral inputs into sector 2, and so on.

However, the input columns in Table 11.2 are incomplete, because in addition to intermediate inputs from other sectors, each sector must buy inputs from labour (i.e. pay wages, L), and pay for all other value added (N, for example interest charges on financial capital employed), as well as buy imported inputs. Incorporating these items leads to the full input-output table of Table 11.3.

Each entry in the final row of Table 11.3 is the sum of the column above it. For the industry sectors this final row entry represents the sum of each sector's payments for its inputs. This is the same as the value of the sector's total output. The final demand entries in the final row are the totals for each category.

Each entry in the final column of Table 11.3 is the sum of the row entries to its left. For the industry sectors these are the same as in Table 11.2, expressing the fact that each sector's output must go to intermediate or final demand, or to exports. These entries represent the total receipts of each sector, from other sectors, from (domestic) final demand and from foreign demand (exports). The receipts add up to the total output of each sector, which is the same as the sector's total payment for inputs, shown in the corresponding column. It is the equality between payments and receipts that gives the inter-industry part of Table 11.3 its symmetry.

Also in the final column of Table 11.3, M is total imports (with M_C, M_I and M_G being imports bought by households, by firms for investment, and by government, and M_E being imports bought for re-export), L is total payments for labour and N is total payments for all other value added.

Table 11.3 may therefore be seen to represent a number of fundamental relationships in the economy:

1. The inter-industry relationships (the X_{ij} elements in the table)
2. The division of industry output, by sector, between intermediate and final demand (the rows)
3. The inputs into each industry sector from other sectors and by factors of production (the columns)
4. The macroeconomic aggregates of output and income

The macroeconomic aggregates emerge from the final row and column of Table 11.3.

From the final row:

$$X = X_1 + \ldots + X_7 + C + I + G + E \qquad (11.3a)$$

Table 11.3 Full input-output table for an economy disaggregated into seven sectors

		Industry sectors							Final demand (Y) and exports (E)				Industry output (receipts)	
		1	2	3	4	5	6	7	C	I	G	E	X
Industry sectors	1	X_{11}	X_{12}	X_{13}	etc.				C_1	I_1	G_1	E_1	X_1
	2	X_{21}	X_{22}	X_{23}	etc.				C_2	I_2	G_2	E_2	X_2
	3	X_{31}	X_{32}	X_{33}	etc.				C_3	I_3	G_3	E_3	X_3
	4	etc.	etc.	etc.					etc.	etc.	etc.	etc.	etc.
	5												
	6												
	7	X_{71}	X_{72}	X_{73}	etc.				C_7	I_7	G_7	E_7	X_7
Imports		M_1	M_2	M_3	etc.				M_C	M_I	M_G	M_E	M
Value added		L_1	L_2	L_3	etc.								L
		N_1	N_2	N_3	etc.								N
Industry inputs (payments)		X_1	X_2	X_3	etc.				C	I	G	E	X

From the final column:

$$X = X_1 + \ldots + X_7 + M + L + N \quad (11.3b)$$

$$M + L + N = C + I + G + E$$

$$L + N = C + I + G + (E - M) \quad (11.4)$$

Equation (11.4) is a fundamental equation for the national accounts. It states that factor incomes, or value added (L + N) in an accounting period are equal to gross domestic product, GDP (final demand plus net exports). GDP may therefore be computed by summing the value added in all the industry sectors (L + N) or by summing the components of final demand plus net exports,

i.e. GDP = L + N = C + I + G + (E - M)

It should be noted that the individual inter-industry transactions are excluded from GDP. This is because these transactions are accounted for in final demand and net exports, and to include them again would amount to double counting.

11.2.3 Using Matrix Notation

Table 11.3 represents all the basic inter-industry relationships in the economy, but its form is not convenient for carrying out analysis of the implications of changes in those relationships, or in components of final demand, which is the principal purpose of input-output analysis. Far more useful for this purpose is a matrix representation of the input-output relationships. A detailed explanation of matrix algebra and manipulation is beyond the scope of this book, so the basic facts will simply be stated here. A far more detailed exposition may be found in Miller and Blair (1985) and Proops et al. (1993).

The first stage in moving towards a matrix representation of input-output relationships is to define the 'technical coefficients', a, which determine how much one sector needs of another's production for its own production. Let

$$a_{ij} = \frac{X_{ij}}{X_j} \quad (11.5)$$

where X_{ij} is the amount of sector i's output which sector j needs to produce output X_j. Then a_{ij} is the amount of sector i's output required to produce one unit of sector j's output. Equations (11.2a)–(11.2g), for a 7-sector economy, may then be re-written as follows:

$$X_1 = a_{11}X_1 + a_{12}X_2 + a_{13}X_3 + \ldots + a_{17}X_7 + Y_1 \quad (11.6a)$$

$$X_2 = a_{21}X_1 + a_{22}X_2 + a_{23}X_3 + \ldots + a_{27}X_7 + Y_2 \quad (11.6b)$$

...

$$X_7 = a_{71}X_1 + a_{72}X_2 + a_{73}X_3 + \ldots + a_{77}X_7 + Y_7 \quad (11.6g)$$

11 Analytical Tools for the Environment-Economy Interaction

$$X_1 - a_{11}X_1 - a_{12}X_2 - a_{13}X_3 - \ldots - a_{17}X_7 = Y_1 \quad (11.7a)$$

$$X_2 - a_{21}X_1 - a_{22}X_2 - a_{23}X_3 - \ldots - a_{27}X_7 = Y_2 \quad (11.7b)$$

...

$$X_7 - a_{71}X_1 - a_{72}X_2 - a_{73}X_3 - \ldots - a_{77}X_7 = Y_7 \quad (11.7g)$$

Now define the following matrices (in what follows, letters representing matrices or vectors are in **bold type**):

$$\mathbf{A} = \begin{bmatrix} a_{11} & a_{12} & \ldots & a_{17} \\ a_{21} & a_{12} & \ldots & a_{17} \\ \ldots & \ldots & \ldots & \ldots \\ a_{71} & a_{72} & \ldots & a_{77} \end{bmatrix} \quad \mathbf{X} = \begin{bmatrix} X_1 \\ X_2 \\ \ldots \\ X_7 \end{bmatrix} \quad \mathbf{Y} = \begin{bmatrix} Y_1 \\ Y_2 \\ \ldots \\ Y_7 \end{bmatrix}$$

$$\mathbf{I} = \begin{bmatrix} 1 & 0 & \ldots & 0 \\ 0 & 1 & \ldots & 0 \\ \ldots & \ldots & \ldots & \ldots \\ 0 & 0 & \ldots & 1 \end{bmatrix}$$

Then Equations (11.7a)–(11.7g) may be expressed as

$$(\mathbf{I} - \mathbf{A})\mathbf{X} = \mathbf{Y}$$

$$\Rightarrow \mathbf{X} = (\mathbf{I} - \mathbf{A})^{-1}\mathbf{Y} \quad (11.8)$$

where $(\mathbf{I} - \mathbf{A})^{-1}$ is the inverse matrix of $(\mathbf{I} - \mathbf{A})$.

Equation (11.8) is the fundamental tool of input-output analysis. Through the $(\mathbf{I} - \mathbf{A})^{-1}$ matrix, called the *Leontief inverse*, it permits the change in total output resulting from a change in final demand to be calculated. The two assumptions that are necessary for this purpose are that there are no economies of scale (i.e. a proportionate change in inputs produces a proportionate change in output), and fixed technical coefficients (i.e. the inputs to each production sector are required in fixed proportions, there is no possible substitution between them). While these are important assumptions, in the short term and for relatively small changes in output they seem not unreasonable.

The following simple numerical example shows how the equation can be used, for a simple two-sector economy with no imports.

Consider the input-output Table 11.4. From the definition in Equation (11.5) it may be seen that the technical (a) coefficients for the input-output table in Table 11.4 are:

$$a_{11} = 150/1000 = 0.15; \quad a_{12} = 500/2000 = 0.4;$$
$$a_{21} = 200/1000 = 0.2; \quad a_{22} = 100/2000 = 0.05.$$

Table 11.4 An input-output table for a two-sector economy

From industry sectors	To industry sectors 1	2	Final demand Y_i	Total output X_i
1	150	500	350	1,000
2	200	100	1,700	2,000
Value added	650	1,400		2,050
Total inputs	1,000	2,000	2,050	5,050

Source: Miller and Blair 1985: 15

The **A** matrix for this input-output table is therefore:

$$\mathbf{A} = \begin{bmatrix} 0.15 & 0.4 \\ 0.2 & 0.05 \end{bmatrix}$$

From this the $(\mathbf{I} - \mathbf{A})^{-1}$ matrix may be computed as:

$$(\mathbf{I} - \mathbf{A})^{-1} = \begin{bmatrix} 1.2541 & 0.3300 \\ 0.2640 & 1.1221 \end{bmatrix}$$

From this it is easy to confirm that Equation (11.8) holds:

$$\begin{bmatrix} 1000 \\ 2000 \end{bmatrix} = \begin{bmatrix} 1.2541 & 0.3300 \\ 0.2640 & 1.1221 \end{bmatrix} \begin{bmatrix} 350 \\ 1700 \end{bmatrix}$$

Now it is possible to calculate the effect on total output of a change in final demand. For example, suppose final demand changes from $\begin{bmatrix} 350 \\ 1700 \end{bmatrix}$ to $\begin{bmatrix} 600 \\ 1500 \end{bmatrix}$

Then the new total output may be calculated from Equation (11.8) as:

$$\begin{bmatrix} X_1 \\ X_2 \end{bmatrix} = \begin{bmatrix} 1.2541 & 0.3300 \\ 0.2640 & 1.1221 \end{bmatrix} \begin{bmatrix} 600 \\ 1500 \end{bmatrix} = \begin{bmatrix} 1247.46 \\ 1841.55 \end{bmatrix}$$

In words, because of the inter-industry connections given by the **A** matrix, it is necessary for sector 1 to produce output of 1,247 units, and sector 2 to produce output of 1,842 units, in order to produce a final demand for their products of 600 and 1,500 units respectively.

The significance of the $(\mathbf{I}-\mathbf{A})^{-1}$ matrix may be understood by considering the output required by final demand as the result of several rounds of requirements. In the first round it is obviously necessary that each sector produce at least as much as the final demand for its output. So, in the first round, $X_i \geq Y_i$.

But, to produce Y_i for each sector, it is known from the **A** matrix that at least **AY** must be produced, because of the inter-industry connections. This is the second round requirement to produce a final demand of Y, and X for each sector must be at least as large as the sum of the two rounds. So $X \geq Y + AY$.

But, as the third round requirement, AY requires $A(AY)$, or A^2Y of output because of the inter-industry connections, so $X \geq Y + AY + A^2Y$.

The argument may be extended to an infinite number of rounds, leading to the equation:

$$X = Y + AY + A^2Y + A^3Y + \ldots$$

⇨ $X = (I + A + A^2 + A^3 + \ldots) Y$ (11.9)

Because each element in the **A** matrix is less than 1, the higher powers of the matrix are diminishing and will eventually become negligibly small.

Comparing Equations (11.8) and (11.9) it may be seen that

⇨ $(I - A)^{-1} = (I + A + A^2 + A^3 + \ldots)$

The elements of the Leontief inverse actually capture the effects of an infinite number of rounds of tracing back the production requirements, due to inter-industry connections, for ever smaller quantities of intermediate output.

The basic input-output model expressed in Equation (11.8) has been extended in many directions and applied to many different situations, which are outside the scope of this book (see Miller and Blair 1985 for details). Here the only extensions which will be considered are the use of input-output analysis for issues of energy and the environment.

11.2.4 Extension to Energy and the Environment

So far all the entries in the input-output tables have been in value (money) terms, as a convenient way to express transactions involving enormously different goods and services in a common unit. For energy and environmental variables, however, the flows through the economy are of interest in physical as well as monetary terms. Indeed, many environmental flows take place outside the money economy altogether.

Section 11.3 discusses ways of accounting for energy and environmental flows that are consistent with the national accounting framework, and it is no surprise that it has become common to use an input-output format, which provides the basis for the national accounts, for environmental data as well. The next two sub-sections discuss briefly how input-output analysis can be applied to these increasingly important areas.

11.2.4.1 Energy

Energy inputs into the economy come in many forms (coal, oil, gas etc.), but they can all be measured in a common unit (kWh, therms etc.). This provides the basis for analysing energy flows through the economy, and the implications of changes in economic structure or final demand.

Two common methods of input-output energy analysis have been developed, one based on matrices with hybrid units, the other working with coefficients of energy intensity. They will be briefly described in turn.

11.2.4.2 Hybrid Matrices

This method proceeds by distinguishing between the energy-producing sectors in an economy, and all other sectors. The output of the former is expressed in energy units, while the output of the latter remains in monetary units.

Just as, in the original analysis of inter-industry flows, the total output of a sector was the sum of the intermediate and final demands, so, for the energy-producing sectors, the total energy produced by a sector is equal to the sum of the demands from other sectors and from final demand. Thus, analogously to Equations (11.2a)–(11.2g), but this time in physical energy units:

$$F_i = E_{i1} + E_{i2} + \ldots + E_{iY} \quad (11.10)$$

where F_i = the energy produced by sector i
E_{ij} = the demands for sector i's energy by other sectors
E_{iY} = the final demand for sector i's energy

Hybrid matrices are constructed for each of the standard matrices used in input-analysis. (X_{ij}, Y, X) and A and $(I - A)^{-1}$ matrices are constructed using these hybrid matrices. Expressions can then be derived for the coefficient matrices both of the direct energy use of final demand, and the total energy use of final demand (Miller and Blair 1985: 203–4). Using these coefficient matrices, the impacts on direct and total energy use of changes in final demand can be calculated.

11.2.4.3 Energy Intensity

Just as technical coefficients, a_{ij}, were defined to express the 'intensity' of use of sector i's product in the production process of sector j (see Equation (11.5)), so energy coefficients may be defined as follows:

$$e_{ij} = \frac{E_{ij}}{X_j} \quad (11.11)$$

where e_{ij} = the intensity of use of energy from sector i by sector j
E_{ij} = the physical amount of energy from sector i used by sector j
X_j = the monetary output of sector j.

Thus e_{ij} has units of energy/money. E_{ij} and so e_{ij} will be 0 unless i is an energy-producing sector.

Then, if **E** is the matrix of the elements e_{ij}, from Equation (11.10):

$$\mathbf{F} = \mathbf{E}\,\mathbf{X} + \mathbf{E}_Y \quad (11.12)$$

where **F** = total energy use (all the elements in **F** corresponding to non-energy sectors will be 0)
E X = the indirect energy use due to intermediate demand
$\mathbf{E}_Y$ = the direct energy use by final demand.

Using the basic input-output relation of Equation (11.8), Equation (11.12) may be re-written:

$$\mathbf{F} = [\mathbf{E}\,(\mathbf{I} - \mathbf{A})^{-1} + \mathbf{Q}]\,\mathbf{Y} \quad (11.13)$$

where **Q** is a matrix of direct energy intensities of final demand.

Using either of these two methods as appropriate (Miller and Blair 1985 discusses differences between them which are beyond the scope of this book), the implications for energy use of changes in final demand may be determined.

The methods have found a wide range of applications in:

- Net energy analysis (comparing the outputs from an energy sector with the energy inputs into it).
- Calculating the energy cost of goods and services (i.e. the total energy used during their production).

- Assessing the impacts of new energy technologies (by estimating new sets of energy and technical coefficients corresponding to the new technologies and using them in the input-output tables).
- Forecasting the effects of an energy tax (by seeing how the tax is passed through the economy).

11.2.4.4 Environment

As with energy, there are a number of ways in which environmental impacts can be incorporated into an input-output framework.

11.2.4.5 Sectoral Pollution Coefficients

The simplest way is to construct a matrix of sectoral pollution coefficients (say **V**), each element of which is the physical quantity of a pollutant emitted by a monetary unit of production of a given sector. Then the total amount of pollution is given by:

$$V = \mathbf{V}\,\mathbf{X} \quad (11.14)$$

Or, using Equation (11.8):

$$V = [\mathbf{V}\,(\mathbf{I} - \mathbf{A})^{-1}]\,\mathbf{Y} \quad (11.15)$$

Equation (11.15) enables changes in pollution to be derived from changes in final demand. The difference between Equations (11.14) and (11.15) makes clear

Table 11.5 CO_2 emissions for UK producing sectors, 1993 (million tonnes)

	Sector	Total output	Final demand	% Change
1	Agriculture, forestry, fishing	5,371	4,799	−11
2	Electricity	167,041	74,552	−55
3	Mining and quarrying	176,541	169,593	−4
4	Chemical products	16,881	17,457	3
5	Basic industry	29,976	18,384	−39
6	Manufacturing	7,866	29,508	275
7	Household products	19,749	36,296	84
8	Construction	3,085	19,534	533
9	Transport services	58,320	29,940	−49
10	Services	53,403	138,168	159
	Total	**538,232**	**538,232**	

that the increase in pollution from a given increase in a sector's total output may be expected to be different from that deriving from an equal increase in that sector's final demand. It can make a great difference whether responsibility for pollution is accorded to sectors on the basis of the pollution deriving from their total output, or on the basis of the pollution associated with their final demand, derived from input-output analysis of the inter-industry connections, as the following example shows.

Table 11.5, derived using the UK input-output tables, provides an example of the difference a focus on total output or final demand can make. It shows carbon dioxide (CO_2) emissions from ten UK economic sectors. The Total Output column records the emissions from each sector's direct use of fossil fuels. The Final Demand column consists of emissions from the direct use of fossil fuels by the portion of the sector's output that goes to final demand, plus indirect emissions from other sectors' fossil fuel use that provide intermediate inputs.

While the overall total of emissions is the same in each case, it can be seen from Table 11.5 that agriculture etc., electricity, basic industry and transport services have much lower emissions when they are calculated on a final demand basis. This is because a substantial portion of those sectors' output goes as intermediate inputs to other sectors, to which the corresponding CO_2 emissions are allocated. Conversely, manufacturing, household products, construction and services have far higher CO_2 emissions on a final demand basis, because the production of their output consumes relatively little fossil fuel directly, but uses as intermediate inputs the products of sectors which do.

Other applications of this form of environmental input-output analysis include a comparison of the different total CO_2 emissions from different forms of electricity generation (Proops et al. 1996), and an assessment of the CO_2 implications of deep cuts in US defence spending, depending on how the saved government revenues are redeployed (Heyes and Liston-Heyes 1993).

11.2.4.6 Augmented Leontief Model

Another way of incorporating environmental variables into the input-output framework is by augmenting the $(I - A)$ matrix by adding a row and column for each variable. For pollution, X_P, this is done by calculating the pollution coefficient, a_P, for each of the sectors, analogously to the technical coefficients, such that:

$$a_{pj} = \frac{X_{pj}}{X_j} \quad (11.16)$$

a_{Pj} is then the pollution emitted per unit of output of sector j.

Where X_P is total pollution, for a two-sector economy it may be given by the equation:

$$X_P = a_{P1}X_1 + a_{P2}X_2 \quad (11.17)$$

$$a_{P1}X_1 + a_{P2}X_2 - X_P = 0 \quad (11.18)$$

This allows the pollution coefficients to be incorporated into the 2-sector input-output framework as follows:

$$\begin{bmatrix} I-A & | & 0 \\ \cdots & | & 0 \\ -a_{P1} & -a_{P2} & | & 1 \end{bmatrix} \begin{bmatrix} X_1 \\ X_2 \\ X_P \end{bmatrix} = \begin{bmatrix} Y_1 \\ Y_2 \\ 0 \end{bmatrix}$$

$$\begin{bmatrix} X_1 \\ X_2 \\ X_P \end{bmatrix} = \begin{bmatrix} I-A & | & 0 \\ \cdots & | & 0 \\ -a_{P1} & -a_{P2} & | & 1 \end{bmatrix}^{-1} \begin{bmatrix} Y_1 \\ Y_2 \\ 0 \end{bmatrix}$$

This allows the change in pollution, X_P, from a change in final demand to be calculated.

11.2.4.7 Economy-Environment Commodity Flows

A third way of treating the environment in an input-output framework is to represent all the resource flows into the economy as extra rows (inputs) in the input-output table, and all pollution as extra columns (outputs) from the relevant industry sectors. Table 11.6 is a representation of a 3-sector economy, which has inputs of water and land (in physical units), and produces emissions of sulphur dioxide (SO_2) and hydrocarbons (HC) (also in physical units). From this representation of resources and pollution as commodity flows it is possible to calculate each of the resource inputs required to produce one unit of each sector's final demand, and the corresponding output of each pollutant (see Miller and Blair 1985: 243, for the calculation).

Input-output tables such as Table 11.6 provide the basis for far more ambitious attempts at integrated economy-environment modelling. Table 11.7 gives the

Table 11.6 Resource inputs and pollution outputs in an input-output table

Inter-industry transactions		Consuming sectors			Final demand	Total output	SO_2	HC
		Agriculture	Mining	Manufacturing				
Producing sectors	Agriculture	1	3	5	3	12	0	1
	Mining	0	2	10	0	12	0	2
	Manufacturing	0	2	6	16	24	4	3
Ecological commodity inputs	Water	5	4	8				
	Land	10	10	1				

Source: Miller and Blair 1985: 242

Table 11.7 Basic structure of economy-environment models

	Industries	Ecological processes
Industries	Flows between economic sectors	Flows from industry to the ecosystem
Ecological processes	Flows from the ecosystem to industry	Flows within the ecosystem

Source: Miller and Blair 1985: 252

basic structure of such models, the detailed construction and operation of which are beyond the scope of this book. However, it should always be borne in mind that the further into the future such modelling is taken, the more unrealistic become the two fundamental input-output assumptions of fixed technical coefficients and no economies of scale.

Economy-environment models require large quantities of environmental data in a form that is consistent with the accounting conventions that structure economic data. It is to the topic of resource and environmental accounting, dealing with issues of structuring, interpreting and using environmental data in the framework of the national accounts, that the next section is devoted.

11.3 Resource and Environmental Accounting

11.3.1 Introduction

This section explores ways in which efforts have been made to incorporate environmental considerations into the accounting systems which both guide decision-makers and inform them about the economic development that has been achieved. It is mainly focused on the system of national accounts, but also briefly discusses corporate environmental accounting.

The internationally accepted system of national accounts (SNA) is a framework for the systematic ordering of quantitative information about a national economy, so that the economy can be managed and its performance measured and compared with that in previous periods and with those of other countries.

When the national accounts were being systematised in the 1940s, environmental issues had a low perceived importance, and the accounting structure adopted simply ignored environmental issues. The national accounts have now matured into the fundamental instrument of macroeconomic management and, more controversially, means of indicating economic progress. The 1992 Earth Summit in Rio de Janeiro committed countries to producing national strategies for sustainable development, but it is most unlikely that these will carry economic weight unless they are integrally related to the national accounting system.

There are three basic shortcomings of the national accounts with regard to the environment. Firstly, they neglect the scarcities of natural resources that can pose a serious threat to sustained economic productivity. Secondly, they neglect the effects of economic activity on environmental quality. When such activity results in environmental deterioration (as is often the case), this amounts to neglecting costs of production. When environmental protection expenditures are then incurred to mitigate or remedy such damages, they appear as additions to national product (and therefore economic welfare), when in fact they are simply serving to compensate for unaccounted costs. Thirdly, they pay only limited attention to the effects of environmental quality on human health and welfare. Thus current treatment of the environment in the national accounts is economically misleading. A number of studies (see Ekins 2000, Chapter 5 for a brief survey) have now shown that GNP growth rates and net capital formation can be greatly overstated by failing to account properly for natural resources and environmental damage.

Over the past ten years it has gradually been accepted by both policy makers and national statistical offices that environmental impacts due to economic activities need to be incorporated in some way into the national accounts which provide a systematic way of reporting on and monitoring those activities. One expression of this consensus may be found in the Handbook on Integrated Environmental and Economic Accounting produced by the United Nations Statistical Division (UNSD 1993).

However, the consensus that, in principle, the national accounts should include environmental elements has not been carried over to the practical means by which this should be achieved. Since the mid-1980s a wide range of methods to incorporate the environment in the national accounts have been put forward (see, among others, Ahmad et al. 1989, Lutz 1993 and UNSD 1993). The methods proceed from a number of different theoretical positions, which derive in turn from some basic differences in perception. On the basis of a brief discussion of some of these differences, the sections below present a summary of the various possible adjustments and supplements to the national accounts, and its aggregates, which might be part of a green national accounting framework. First, however, some core issues of green national accounting are addressed.

11.3.2 Basic Issues in National Environmental Accounting

One issue in national environmental accounting which has won general agreement is that the accounting for ecological capital should be carried out in a 'satellite', or ancillary, account rather than in the System of National Accounts (SNA) proper. The advantage of proceeding initially with such an account is that it brings previously excluded areas within the frame of reference of the SNA, while permitting further experimentation and redefinition in the light of experience, without disturbing existing data series.

The first required component of an environmental satellite account is a systematic description of natural resources and the state of the environment in quantitative physical terms, and of changes in these quantities over time. Many countries now have extensive experience with such physical environmental and resource accounts. The second required component is a monetary valuation of the natural resources and environmental qualities, and of changes in them, calculated in such a way that the figures generated are directly comparable with the quantities in the conventional national accounts.

A country's system of national accounts has two main categories: *stocks* of assets which produce *flows* of goods and services. Another name for the stocks is *capital*. Both stocks and flows are measured in monetary terms. The purpose of the national accounts is to measure the stock of a country's capital (its wealth) and the flows of goods and services to which this wealth gives rise (its Gross Domestic Product (GDP), or Gross National Product (GNP), the difference between which is not material here). GNP contains a substantial element of consumption of the stock of manufactured capital, but the national accounts specifically deduct an estimate of this to arrive at the figure of Net National Product (NNP) or National Income, which is consistent with Hicks' (1946) definition of income as consumption net of changes in the capital stock.

For the environment to be fully integrated into the national accounting framework, the same procedure must be followed in the environmental satellite account. The flows in this case correspond to environmental depletion and degradation (capital consumption) or environmental regeneration (capital formation); the stock comprises land (including subsoil deposits), air and water, and the plants and animals (biota) they support. An environmental balance sheet, together with the steps for calculating from opening to closing stocks in both physical and monetary units, is set out in Table 11.8.

Rows 2,3,4 in Table 11.8 are changes in assets that correspond to flows and which would need to be accounted for in any adjustment to NNP. New findings of subsoil deposits are not included as flows, although they add to stocks, because they do not result from the production process, and are accounted for under Row 6. Row 7 accounts for the effects of natural disasters (e.g. floods, fires, earthquakes etc.)

While the ultimate aim in environmental accounting is to produce a complete balance sheet for each kind of environmental asset, of more immediate importance for the assessment of sustainability of current production are the flow items 'depletion' and 'net degradation'. These represent consumption of environmental assets and, ideally, their value must be subtracted from NNP to arrive at an income figure which is environmentally sustainable in terms of Hicks' definition given earlier.

11.3.3 On Differing Approaches to National Environmental Accounting

11.3.3.1 Production, Income and Welfare in the National Accounts

The purpose of economic activity is to increase human welfare. NNP is a measure of production (or, equivalently throughout the discussion below, income) which, other things being equal, may be assumed to contribute positively to human welfare. However, there continues to be profound disagreement as to whether NNP should be treated not just as a *contributor to* human welfare, but as an *indicator of* human welfare.

There is no shortage of pronouncements from economists and statisticians on the theoretical problems and heroic assumptions involved in using the national accounts aggregates as welfare indicators. Despite these, there is a continuing robust academic tradition of advocating the practical use of NNP as a welfare indicator, with the proviso, arising from current concerns about the environment, that it is suitably adjusted for environmental effects. This has led to two distinct approaches to the attempted adjustment of the national accounts to take account of environmental (or other) issues:

1. One approach treats the national accounting aggregates as indicators of production and not of welfare. Any adjustment to these aggregates is then justified on the grounds that it is necessary to improve GNP and NNP as indicators of production, and not of welfare. If it is desired to derive a welfare indicator that takes account of the contribution of income to welfare, then NNP may be combined with indicators of other perceived contributors to welfare to arrive at a composite welfare index.
2. The other approach treats NNP (or one of its components, such as consumer expenditure) as an imperfect welfare indicator, and then seeks to add to or subtract from it various elements that will improve it in this role. Attempts to calculate such a welfare measure include Daly and Cobb (1989)'s Index of Socio-Economic Welfare (ISEW), which has, however, been criticised for the arbitrariness of its adjustments to NNP (see Ekins 2000, Chapter 5, for a discussion). Such calculations also depend on methods of valuing environmental effects in money terms which are also very controversial. These are discussed below.

It is possible to relate environmental effects to the national accounts framework by focusing on the stock or flow elements. If it is the flows that are being considered, the loss (gain) of environmental goods and services during the period could be subtracted from (added to) GNP to arrive at an aggregate figure for economic goods and services adjusted for any change in environmental goods and services. If it is the stocks that are being considered, then the depletion and degradation of environmental resources could be regarded as loss of natural capital and, after appropriate valuation, subtracted from GNP together with the consumption of manufactured capital to arrive at an environmentally-adjusted NNP, ENP (sometimes also called Green GNP). Such an approach would be fully theoretically consistent with the way GNP and NNP are computed, and interpreted as indicators of gross and net production. But its practical realisation depends on the appropriate valuation of the environmental depletion and degradation being feasible. This is the next issue to be addressed.

Table 11.8 Schema for an environmental balance sheet

	Physical units	Multiplied by price	Monetary units
1	Opening stocks	× opening price	= Value of opening stocks
2	+ Natural growth		
3	– Offtake/depletion		
4	– Net degradation		
5	2 + 3 + 4 = change in stocks due to economic activity	× opening price	= Value of change in ecological capital
6	± Changes due to economic re-appraisal (e.g. discoveries)		
7	± Changes due to non-economic events (e.g. earthquakes)		
8	= Closing stocks	× closing price (revaluation)	= Value of closing stocks

11.3.3.2 Valuation of Environmental Change

Depletion

Resources used in economic activity often have a market price, which facilitates the valuation of depletion, which, for both renewable and non-renewable resources, may be valued according to all or part of the economic rent of the depleted resource. Various techniques have been proposed for the valuation of natural resource depletion, with no consensus yet being reached on the correct one. However, each technique is feasible and the calculations they yield, while different, are robust.

Environmental Degradation

Environmental degradation, mainly comprising pollution, is not normally valued in the market. Two main ways of imputing values to environmental degradation have been proposed:

1. According to the cost of the damage incurred. Where the damage is to marketed goods (e.g. crops, fish), then the economic damage can be valued in principle as the loss of production and income resulting from the damage. Where the damage is to non-market goods or to non-economic (environmental) functions of natural assets, the valuation may be carried out by various techniques, including those involving surrogate markets or contingent valuation. These techniques can be controversial and can yield wide ranges of valuations. At its present stage of development contingent valuation does not command the necessary consensus among statisticians to be used by national statistical offices for this purpose.
2. According to the abatement/avoidance/restoration cost that would be required to achieve a certain standard of emissions, or maintain or restore certain standards of environmental quality. Sometimes collectively called 'maintenance costs' (UNSD 1993: 105 ff.), these costs are hypothetical, in that they represent expenditures that would have to be made to achieve or maintain certain standards of environmental quality. Where these costs are substantial, then there is little doubt that actually incurring them would fundamentally affect prices throughout the economy. Maintenance costs have been divided into five types: reduction in or abstention from economic activities; changes in economic outputs or consumption patterns; changes in economic inputs or processes; prevention of environmental deterioration without changing products or processes (e.g. by end-of-pipe abatement); environmental restoration or protection from harmful effects.

These two ways of attaching money values to environmental degradation yield cost figures which have quite different meaning and significance in relation to economic activities. The next section discusses this difference, in particular with regard to the macroeconomic aggregates with which the cost figures may be compared.

On Environmental Costs and National Accounting Aggregates

The basic equation of environmental adjustment, that is incorporated into UNSD 1993, is:

   GDP − Consumption of fixed capital = NDP
   NDP − (Imputed) environmental costs = EDP

where EDP = Eco, or Environmentally-adjusted, Domestic Product. (The validity of the equation is unchanged by substituting GNP, NNP and ENP for the equivalent domestic quantities.)

In this equation the 'imputed environmental costs' are the costs of environmental depletion and degradation. Unfortunately neither of the methods of calculating the environmental costs outlined above can be used in practice to calculate EDP: damage costs cannot be estimated sufficiently such that they are robust enough (quantitatively or methodologically) to be combined with national accounting figures; and maintenance costs, as noted earlier, are a quite different kind of hypothetical number to the macroeconomic aggregates which reflect the actual equilibrium prices in the economy.

To conclude, there would seem to be no way to arrive at a practical and robust estimate of EDP, or 'Green GDP', as it is sometimes called, simply by subtracting cost figures from the relevant macroeconomic aggregate. While such a procedure may be theoretically sound, whether NNP is regarded as an indicator of welfare or net production, in the former case there is no way to arrive at credible estimates of the damage costs, and in the latter case it is not valid to deduct the maintenance costs, which might be credible, from NNP. 'Green GNP', as originally conceived in terms of making deductions from NNP, is not an operational concept. The ways forward for green accounting from this conclusion are laid out in the next section.

11.3.4 Ways Forward for National Environmental Accounting

11.3.4.1 NAMEA (National Accounting Matrix Including Environmental Accounts)

As noted earlier, the foundation of any environmental accounting system, whether or not it is intended to relate it to the national economic accounts, is current and reliable physical data about the environment. The NAMEA approach to environmental accounting, developed in the Netherlands, seeks to present environmental data in a module that mirrors the national accounting structure so as 'to provide a complete account of all linkages between changes in the environment and the transactions recorded in the national accounts' (De Haan et al. 1993: 1).

A NAMEA consists of a conventional input-output table, structured according to the Standard Industrial Classification, with extra rows at the bottom giving environmental inputs into the economy, and extra columns giving environmental emissions from the economy, as discussed in the previous section. In this way it is very easy to identify the resource-using and emission-producing sectors. Environmental imports and exports from and to the rest of the world can also be accommodated within this framework.

The NAMEA approach has won wide acceptance among national accountants and statisticians. Although the initial NAMEA work focused largely on air emissions, there is now substantial international activity to extend it to cover emissions to land and water, and the use of natural resources as well.

11.3.4.2 Calculating Net and 'Genuine' Savings

The characterisation of environmental losses as the consumption of natural capital may be used to deduct the figure for such losses from the net investment (savings less depreciation) of different countries to compute a figure for 'net savings'. Hamilton (1994) further adjusted net savings by adding education expenditures (as investment in human capital), to arrive at a 'genuine savings' indicator. On the basis that a fundamental precondition of sustainability is the maintenance of the capital stock, figures of net or genuine savings rates that are near zero or negative would indicate clearly that development is not sustainable.

Calculations by Pearce and Atkinson (1992) of net savings figures for 21 different countries indicate that eight are unsustainable according to the net savings rule. Surprisingly, developing countries that have experienced and are experiencing great environmental degradation (Brazil, Costa Rica and Zimbabwe), as well as a number of industrial countries, are all shown as sustainable by this rule. In Hamilton (1994: 16), OECD countries are shown to have positive genuine savings rates. A factor which improved the net savings indicator for oil-producing countries during the 1980s was the fall in the oil price, which reduced the estimates of UK depletion costs.

These results show the limitations, as well as the uses, of such indicators. Environmental data do not lend support to the hypothesis that industrial economies are environmentally sustainable. Moreover the influence of the oil price on the sustainability indicator, irrespective of physical depletion rates, emphasises that what is being measured is economic, and not environmental, sustainability. Such considerations suggest that adjusted savings indicators should at most be used as indicators of environmental unsustainability, when they are negative, but positive net or genuine savings do not necessarily indicate the reverse. Indications of environmental sustainability must be derived first and foremost from the physical conditions by which it is defined. One possible way of carrying this out is described in the next section.

11.3.4.3 Estimating the Sustainability Gap

Sustainable development has become one of the core organising concepts of environmental policy. One element of this concept is environmental sustainability, which has been defined as the maintenance of important environmental functions into the indefinite future (Ekins 2000: 80). From this definition of environmental sustainability, sustainability principles can be derived and, from them in turn, sustainability standards for each individual environmental impact may also be derived. It seems desirable that the sustainability standards are formulated as far as possible on the basis of objective considerations deriving from environmental science concerning the maintenance of important environmental functions, rather than being influenced by considerations of cost or political feasibility.

The difference between the current level of environmental impact from a particular source, and the sustainable level of impact according to the sustainability standard, may be termed the 'sustainability gap' (SGAP).

By combining the SGAP figures with calculations of the costs that would be required to reduce the SGAP to zero, a monetary SGAP indicator may be derived which shows the extent to which economic performance is environmentally unsustainable. This technique is discussed in more detail in Ekins and Simon 1998.

11.3.4.4 Modelling Environmentally Sustainable National Income

GNP is calculated from prices and quantities in the economy that take account of economic interactions and interconnectedness. Using the maintenance cost approach it is possible to define the least cost way of achieving environmental standards that are deemed to be compatible with environmental sustainability. As noted earlier, because these are hypothetical, static cost calculations, rather than being derived in the context of full economic interaction, they cannot simply be deducted from GNP. Rather, any calculation of an Environmentally Sustainable National Income (ESNI), using the costs of technologies that would have to be implemented to bring environmental performance within sustainability standards, requires the implementation of these technologies to be simulated in the context of an economic model that contains these interactions and which projects the economic implications of progressively meeting the sustainability standards over time. Once the standards had been met, the model's economic output could be regarded as the ESNI, namely the level and composition of economic output that, for each accounting period being considered, was compatible, on the basis of current or projected future technologies, with the achievement of environmental sustainability.

Such modelling is a major undertaking, requiring integrated economy-energy-environment modelling that is disaggregated enough to enable the various technologies and environmental effects to be adequately represented.

11.3.4.5 Calculating Defensive Expenditures

Environmental defensive expenditures are defined in UNSD (1993: 42) as expenditures which prevent environmental damage; restore environmental damage; avoid damage (e.g. to humans, buildings) from an environment which has been damaged; and repair damage (e.g. to humans, buildings) from a damaged environment. The first of these categories has now been systematised in detail in the Environmental Protection Expenditure Account (EPEA) of the European System for the Collection of Economic Information on the Environment (SERIEE, EUROSTAT 1994).

There is continuing debate as to how defensive expenditures should be dealt with in the System of National Accounts (SNA). It is clear that defensive expenditures are beneficial in themselves, they add to welfare, and it might be thought that they should therefore be included in GNP and NNP. However, this is only part of the necessary argument. The 'real' output of the total economy-environment system comprises both economic and environmental goods and services. If the production of economic goods and services reduces the output of environmental goods and services (which is what 'a negative impact on the environment' means in economic terms), then 'real' output from the economic activity should be accounted net of this reduction. Alternatively, the economic goods that only serve to offset this reduction, to maintain the output of environmental goods at their level before the damaging economic activity took place, should not be included in the net output of the economy during the accounting period.

This characterisation of the relationship between GNP and defensive expenditures illustrates an important point that is valid when considering any adjustment to the macroeconomic aggregates: the adjustment must be consistent with the interpretation being given to the aggregate and with the use to which it is being put. Thus if GNP is being used to represent the money flows through the economy, then clearly defensive expenditures should be included in GNP. If, on the other hand, GNP is intended to be interpreted as the net output of the economy during the accounting period, then defensive expenditures should be deducted from it, as, of course, should capital depreciation. Similarly defensive expenditures should be excluded from any figures of net output that are being considered as a contribution to economic welfare.

11.3.5 Corporate Environmental Accounting

11.3.5.1 Physical Flows

The national accounts are derived from business surveys and reporting, so that it is not surprising that, parallel with concerns about environmental accounting at the national level, there have been attempts at the corporate level to incorporate, and make more transparent in firms' accounting systems, both the environmental impacts associated with business activity and the expenses associated with managing those impacts. An incentive for these attempts, which is absent at the national level, is that proper accounting for environmental impacts and expenses can also reduce business costs, as will be seen. As with national accounting, corporate environmental accounting may be associated with either physical or financial flows, or both.

For physical flows, a common technique is the derivation of an ecobalance, the primary focus of which is 'the flow of energy, materials and wastes associated with a company, site, process or product' (Jasch 1999: 151). The physical flows are tracked using a form of input-output analysis, such that all the inputs to the company, site, process or product are followed through various stages of use or transportation, and then accounted for as a balancing quantity of outputs.

The focus on the physical flows provided by an ecobalance can often lead to more efficient resource use and waste management. Because many of the inputs considered by an ecobalance are purchased, and because many of the outputs incur waste management expenses, this can lead to cost savings. For example, Rauberger and Wagner (1999: 180) report that use of the ecobalance technique by the company Kunert AG enabled it to make a number of cost-effective environmental improvements. Over 5 years an investment of US$550,000 in ecobalance accounting and 20 resulting environmental projects produced savings of US$1.55 million.

As noted above, the physical outputs from a company's activities may cause or contribute to environmental problems, and this may also be measured in a manner similar to the physical 'sustainability gap' methodology described above for the national accounts, but called the 'distance to target' method in this application. An example is the Environmental Policy Indicator (EPI) used by Unox (Van der Werf 1999), a food-producing subsidiary of Unilever in the Netherlands. Unox groups its environmental impacts into the same eight environmental themes as the Dutch National Environmental Policy Plan (NEPP). Using a target for each of these themes derived from the NEPP, the individual theme indicators may be combined into an overall indicator of environmental performance. If the targets may be identified as the environmentally sustainable levels of environmental impact, then these theme indicators may be regarded as the corporate sustainability gap, and the EPI as the firm's overall physical distance from sustainability.

11.3.5.2 Financial Flows

Most companies already incur financial costs for environment-related business activities, whether as the costs of resource inputs, the disposal of waste outputs, environmental management during the production process or payments for permits or insurance against possible accidents. For some companies these costs may be substantial, amounting, for example, to around 20% of non-feedstock operating costs at an Amoco refinery, and similar percentages of manufacturing costs for a Dupont agricultural pesticide and of non-raw-material costs for a Ciba Geigy product (Ditz et al. 1995: 48, 97, 131). A number of studies have shown that 'in many companies, the main part of environment-related costs such as energy, water, waste disposal and the salaries of environmental staff are likely to be included in overheads' (Bennett and James 1998: 46–47). The result is that managers are not adequately aware of these costs, nor can projects which would reduce them be accurately evaluated. This introduces a bias against pollution prevention.

A confusing number of terms are used to describe rather similar techniques which have been developed to address this problem, including Activity Based Costing (ABC), Total Cost Accounting (TCA), or full-cost environmental accounting. All of them take a similar approach to the following definition of ABC, which 'tries to create more meaningful cost information by tracking costs to products and processes on the basis of the underlying 'drivers' which cause those costs to be created in the first place' (Bennett and James 1998: 47).

An example of the use of these accounting techniques is the Green Activity Matrix developed by ATandT (Bennett and James 1998: 38), the first dimension of which corresponded to general ledger codes (for example, people, materials and supplies, services and consulting fees, depreciation or equipment, energy and utilities etc.), while the second dimension listed some thirty to forty types of environment-related activity (for example obtaining permits, treating on-site waste, environment-related training, waste disposal).

As with ecobalance analysis ABC/TCA can result in better financial decisions which can save money. Table 11.9 shows one application of TCA to an investment that converted a solvent/heavy metal to an aqueous/heavy metal-free coating at a paper coating company. The company analysis column shows how the company's conventional accounting system assessed the project's costs and benefits. The TCA column includes costs and benefits that were accounted in the company analysis under headings which obscured their relation to the project. These hidden, or indirect, costs and benefits included costs of waste management, utilities (energy, water, sewerage), pollution control/solvent recovery and regulatory compliance. It can be seen that the project under TCA was substantially more profitable than with the conventional company analysis.

Finally, it is possible to develop a monetary 'sustainability gap' at the corporate level analogous to that described above for the national accounts. Howes 1999 describes a project carried out for the Interface carpet company, which calculates the abatement, avoidance or restoration costs which the company would need to expend in order to reduce its environmental impacts to sustainable levels. These costs may then be subtracted from the company's reported profits to give its environmentally sustainable profits, namely the profits which would have remained had the necessary investments for environmental sustainability actually been made.

There is much current interest in corporate environmental accounting, and in corporate environmental reporting more generally, among both business and government. Already government is putting pressure on businesses or, in some cases (e.g. Denmark, see Rikhardsson 1999), requiring firms to give general environmental reports. It seems likely to be only a matter of time before detailed environmental accounts are as normal, and mandatory, a part of corporate accounting as the financial accounts that now appear in companies' Annual Reports.

11.4 Costs and Benefits in Environmental Decision-Making

11.4.1 General Issues and Considerations

For the economist it is natural to adopt a framework for environmental policy analysis which seeks to elucidate the various costs and benefits which arise from a particular use, or change of use, of the environment, in order to recommend that course of action which has the highest benefit/cost ratio greater than 1. This is the purpose of cost-benefit analysis, which is discussed further below.

First, however, it is worth noting that there is an inescapable philosophical element in the valuation of environmental benefits, which derives from the need

Table 11.9 Financial data for a project comparing conventional company analysis with a Total Cost Assessment (TCA)

	Company analysis	TCA	Difference
Total capital costs	$623,809	$653,809	6%
Annual savings (BIT)[a]	$118,112	$216,874	84%
Net present value, years 1–10	($98,829)	$232,817	336%
Net present value, years 1–15	$13,932	$428,040	2,972%
IRR, years 1–10	12%	24%	12%
IRR, years 1–15	16%	27%	11%
Simple payback (years)	5.3	3.0	−43%

[a]Before interest and taxes.
Source: Dorfman et al. 1993: 203

to identify the ethical or philosophical basis of value, to articulate what constitutes the source of the value in question. As will be seen below, use of cost-benefit analysis is based on the presumption that utilitarianism provides the appropriate philosophical framework for assessing such values in situations where it is to be used.

Alternative philosophical approaches to valuing environmental benefits have been suggested by De Groot (1992: 1), and include:

- *Sentimentalism*, 'stressing the need to achieve a certain harmony between man and nature'
- *Ethical argument*, 'based on the perception that all living beings have an equal right to exist'
- *Educative argument*, based on the view that nature is necessary to humans' discovery of their identity and their place in the universe
- *Survival argument*, 'that the continued functioning of natural processes is essential to human existence on earth'

The ethical argument may be extended to the formulation, common in definitions of sustainable development and in the justification of an objective of environmental sustainability, that no group of people has the right, in their provision for themselves, to degrade the environment such that the capacities of other people, in the present or future, to make similar provision are reduced. It may be noted that of the four arguments above only the second is not anthropocentric (i.e. not oriented towards human well-being) but all go beyond the utilitarian focus on individual preferences and none are amenable to techniques of economic valuation.

Economists are inclined to assume that, of the different philosophical approaches to the valuation of environmental benefits, only utilitarianism is 'practical'. In fact, there are numerous examples of human institutions that uphold values on a basis other than utilitarianism. Perhaps most obvious among these are systems of justice, which are not based on cost-benefit analysis. The robber is not exonerated of theft however much he may show that he can put wealth to greater social advantage than his victim. This is simply not a relevant consideration with regard to justice, and it may well be argued, for some environmental issues, that justice is a better analytical framework within which to decide mitigation imperatives than economic rationality.

Zerbe and Dively (1994) have helpfully distinguished between the tasks of setting a policy objective (for example, the target level of environmental quality) and taking account of the costs of achieving it. They have distinguished between four models that can be used in environmental decision-making:

- *Cost-oblivious model*. This involves taking a decision (for example, with regard to the appropriate environmental target) without any regard to the costs involved. Such an approach might be justifiable in situations in which economic logic is generally agreed to be wholly subject to other considerations or simply inapplicable. Issues of morality or justice are obviously relevant here. For example, environmental regulations for safeguarding human health, especially from life-threatening effects, are often formulated with regard to maximum permissible health impacts. Companies in breach of these regulations cannot absolve themselves of guilt by showing that the benefits derived from doing so are greater than the costs of damage incurred, however great the margin between cost and benefit. Similarly it could be argued in relation to global warming that, whatever the benefits to be derived from burning fossil fuels, this does not justify the threat to the integrity of the biosphere that global warming represents. This model is often invoked on the basis of an intuition that the issues under consideration, such as justice, morality, or life and death, are in some sense 'beyond price'.
- *Cost-effectiveness model*. Here, as in the previous model, the interaction between costs and benefits is not thought to be sufficiently important to influence the setting of the objective under consideration. However, the model prescribes that, once set, the objective should be achieved at least cost. This model is discussed further below.
- *Cost-sensitive model*. This model envisages that the relation between costs and benefits has some influence in determining the objective to be set as well as how it will be achieved. This approach is exemplified, for example, in the prescription of BATNEEC (Best Available Technology Not Entailing Excessive Cost) in UK environmental policy. The key to the implementation of such a concept, of course, is in the interpretation of the word 'excessive', which usually reflects the regulator's judgments about the availability, affordability and environmental effec-

tiveness of the technology concerned, rather than being based on a rigorous assessment of the costs and benefits involved.
- *Strict cost-benefit analysis (CBA).* Because strict CBA assumes that *all* costs and benefits can be and, in its computations have been, taken into account, the outcome, whether in terms of benefit-cost ratio or optimality, is more a directive for decision-making than a guide. If all relevant considerations of a number of options have been converted into money values and included in the analysis, then the result unequivocally indicates whether any given course of action is the best possible. The attraction, and power, of this technique is that it offers unambiguous recommendations based on quantitative results. What is not always kept in mind is that the seemingly unambiguous outcome is entirely dependent on the assumptions and methodologies on which CBA is based, and on the inclusiveness of the analysis as a whole. These issues are discussed further below.

In reality, these four models represent not so much discrete alternatives as different points on a spectrum, any position on which is largely determined by the weight that is to be attached to economic logic and calculation. In strict CBA such calculations determine the optimum, and therefore in rational decision-making they fully decide the objective, as well as its means of achievement. Moving along the spectrum, more and more weight is accorded to non-economic principles and considerations, until at the other end economic calculations are irrelevant to the decision to be taken.

In general, any point on the spectrum is subject to critique. Starting from the end of cost-oblivion it may be observed that resources, after all, *are* scarce, and that where the administration of even the grandest principles, such as justice, requires resources, this cannot in practice be exempt from economic argument. Moreover, there are few principles so sacrosanct that they justify adherence regardless of cost. Moving towards cost-effectiveness, where principle is still predominant, it is not difficult to imagine situations where even the most cost-effective implementation of a worthy principle involves costs so high that such implementation may be brought into question.

From the other end, it will be seen below that there are certain situations in which the application of the strict cost-benefit metric seems fraught with unresolvable difficulties. Abandoning this metric in such cases in favour of cost-sensitivity may not resolve the problems, for it immediately raises the question of who is to be sensitive to whose costs. Pure economic logic has little purchase on distributional issues.

It is sometimes argued that, again because of the inherent scarcity of resources, any decision that involves their allocation has an implicit price. By this argument, whatever the formal decision-making mechanism, the social value of lives can be deduced by the amount that governments will spend on improving traffic safety or health services. Where the values differ markedly between deductions, more lives could be saved if resources were shifted from the higher to the lower value use. But there is a huge philosophical and practical difference between using such *ex post* valuations for the more cost-effective allocation of public money, to using them to decide *a priori* whether certain activities should be permitted because they will create more wealth than the 'value' of the loss of life they may cause.

Similarly, it is sometimes argued that, if rich countries care enough about the lives of poor country citizens in the future to undertake substantial abatement of global warming now, they could actually save the lives of more such citizens now by transferring a comparable quantity of resources in the form of development aid (Schelling 1992: 7). Even if this were true (and the uncertainties connected with both damage and abatement costs of global warming mean that this cannot be determined), this contention misses the point that the arguments for development aid are quite different from those relating to whether rich country lifestyles should cost poor country lives.

These contentions and arguments reveal the full complexity of human decision-making in which economic and non-economic logics, principles and imperatives jostle with each other demanding compromises which can never be evaluated according to a common numeraire. Undoubtedly the addition of the environment to the already crowded decision arena has made decision-making more complex still. No analytic tool for decision-making can give the 'right' answer as to the relative weight to be given to economic development and environmental protection, where these objectives conflict. The most important consideration in the use of any tool in this area would seem to be to make transparent the value judgments and assumptions on which its methodology and application are based.

11.4.2 Cost-Effectiveness Analysis

As noted above, cost-effectiveness analysis (CEA) is a tool for determining how to meet a given objective at least cost. As such, its use is relatively uncontroversial, for it is not easy to argue in favour of excessive costs *per se*.

The first use of CEA was in relation to health objectives. The objective may be in relation to a general objective (e.g. to save lives), when the cost-effectiveness of different options may be expressed as a ratio giving the average cost per life saved for each option; or in relation to a target (e.g. to save a certain number of lives), when the total cost will depend on the aggregation of the minimum marginal costs required to reach the target.

When applied to the environment (and often in other applications), the objective to be met, and how such an objective should be derived, will itself be the subject of debate. In environmental policy a concept that is being increasingly used in such debates is the concept of *environmental sustainability*, which is one half of the composite idea of sustainable development, which now commands such widespread political support in speeches and, increasingly, in practical policy.

The basic meaning of sustainability is the capacity for continuance more or less indefinitely into the future. Its new prominence on the policy agenda reflects the fact that the past 20 years have seen a substantial accumulation of evidence that, in aggregate, current human ways of life do not possess that capacity. This is either because they are destroying the environmental conditions necessary for their continuance, or because their environmental effects will cause unacceptable social disruption and damage to human health.

The key contribution of the environment to the human economy and to human life in general can be perceived to be through the operation of a wide range of 'environmental functions' (De Groot 1992) or ecological or ecosystem services (Daily 1997), which at the most general level may be thought of as the generation by the environment of resources for human activities, its absorption of wastes from those activities and its provision of other 'services' such as climate stability, and opportunities for amenity and recreation. Using this idea, environmental sustainability can be defined as the maintenance of important environmental functions or ecological services.

Environmental sustainability derives its legitimacy as a fundamental objective of public policy from a number of sources. Partly it has an ethical basis: some consider it wrong to diminish the environmental options of future generations below those available today; others consider wrong the impacts on other life-forms that unsustainability implies. Partly the concern derives from perceived self-interest. Unsustainability threatens costs and disruption to ways of life in the future that are or may be greater than those incurred by moving voluntarily towards sustainability now. The self-interested concern about unsustainability will obviously become stronger as the time-scale within which the costs and disruption will be experienced is perceived to shorten. This is probably the reason for the current strengthening of public anxiety about the environment.

With regard to the cost models, the adoption of sustainability as the underlying principle of environmental policy is consistent with all the models except that of strict cost-benefit analysis. As already noted, any principle is superfluous to CBA because an optimal outcome is attainable through the consideration of costs and benefits. For cost-sensitivity sustainability could be the underlying principle of policy, but one that was occasionally honoured in the breach rather than the observance when the costs of observance seemed excessive. Perhaps this would be most likely to apply in practice by the lengthening of the time scale during which sustainability would be approached. In the cost-oblivious model, sustainability could play the major role in suggesting action and behavioural change. For cost-effectiveness sustainability would be the objective, and the purpose of economic analysis would be to show how the costs of achieving it could be minimised.

Given an objective of environmental sustainability (or any other general objective), quantitative targets across all relevant environmental areas need to be derived. For environmental sustainability, such targets might include the rates of depletion and recycling of non-renewable resources; the minimum biologically acceptable stock, and maximum sustainable rate of harvest, of renewable resources; the maximum emissions and concentrations of pollutants that are consistent with the maintenance of human health; the critical loads of pollutants for ecosystems; and the avoidance of risks with potentially very high costs.

Once these targets have been agreed, CEA would then analyse all the various ways in which they could be met, whether by reducing production and consumption, by technical change, by product substitution or by structural change in the economy, and the various instruments which might be used to bring about these changes. In environmental policy, the whole discussion about the best choice of policy instrument, as between regulations, taxes and negotiated agreements, is highly influenced by questions of cost-effectiveness.

In seeking to determine cost effectiveness it is necessary to distinguish between different types of cost, depending on the level at which the cost effectiveness is being analysed. For the costs of individual projects of environmental protection, or the costs relating to the environmental protection efforts of single firms, it is the *direct financial costs* of different ways of achieving the given environmental objective that need to be calculated and compared.

Where a given environmental improvement (for example, the reduction of carbon dioxide emissions) can be achieved using different technologies, then a cost-reduction curve may be built up by ranking the different technologies, and the amount of improvement they can achieve, in an order of increasing marginal cost (the cost of one additional unit of environmental improvement).

Thus, in Fig. 11.1, the cheapest technology has a marginal cost of emissions reduction of $\$a_1$ per tonne (t) and can achieve an emission reduction of b_1 tonnes. The next cheapest technology has a marginal cost of $\$a_2$ per tonne (t) and can achieve an emission reduction of $b_2 - b_1$ tonnes, and so on. In the cost curve shown, the cheapest technologies are shown to be able to reduce emissions at negative net cost, implying that these technologies save more money, through making more efficient use of resources, than they cost, as well as reducing emissions. The existence and extent of such 'win-win' technical opportunities for environmental improvement is controversial, but some estimates have put the potential for cost-free energy saving, for example, as high as 20% of total energy use.

The cost-effective approach to environmental improvement in these cases is to implement the cheapest technologies first, up to the target level of improvement that has been set, shown in Fig. 11.1 as an emission reduction of S. Where this target is identified as consistent with the overall goal of environmental sustainability, then it may be termed a 'sustainability standard' for that environmental impact.

At the next level of complexity, the direct financial costs of particular abatement measures may be applied to a whole economic sector, using a sectoral economic model which incorporates such assumptions as the underlying rate of growth of the sector, the costs of other production inputs, and the price elasticity of demand for the sector's products. The costs of the environmental improvements, again on the basis of the implementation of the cheapest technologies first, will then feed through the model into price increases of the sector's output and reduction in the demand for it. It is

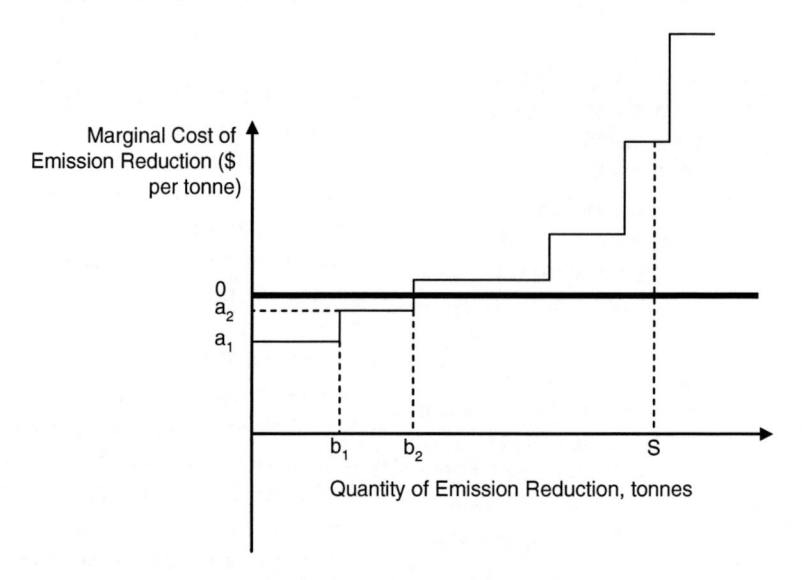

Fig. 11.1 A hypothetical cost-reduction curve for reducing carbon emissions

this reduction in demand, and therefore in sectoral output, that is then the measure of the *sectoral economic cost* of the environmental improvement. In CEA this sectoral cost may then be compared with achieving the same level of environmental improvement through measures in other sectors.

For example, with regard to climate change, carbon emissions are produced by the power generation, manufacturing, service, commercial, transport and household sectors, and may be sequestered from the atmosphere by the forestry sector. Using appropriate models for each sector, CEA would seek to determine the least-cost mix of carbon reduction from each sector that would reach the overall carbon reduction target that had been set.

Treating sectors as isolated in this way is called partial equilibrium analysis. In a real economy, of course, the sectors are connected to each other, and what happens in one sector affects the others and the economy as a whole. Modelling the sectors as part of an integrated model of the whole economy can capture these inter-sectoral connections, permitting general equilibrium analysis of the cost of environmental improvements. The macroeconomic cost of these improvements is then the reduction in the total output of the economy (its Gross Domestic Product, or GDP), which may be the result of achieving them. Where the improvements may be achieved in different ways, with different contributions from different sectors, CEA seeks to determine which combination of sectoral contributions reduces the economy's GDP by the least amount. It may be noted that, in general, because of the sectoral interactions, general equilibrium (GE) analysis will result in a different sectoral contribution to least-cost environmental improvement than partial equilibrium (PE) analysis. Moreover, the macroeconomic cost, derived from GE analysis, of achieving a certain environmental target will be less than the sum of the different sectoral economic costs derived from PE analysis. This is because a GE or macroeconomic model permits the different sectors to adjust to changes in costs in order to minimise the effects of those costs on the economy as a whole.

In addition to modelling different sectoral contributions to environmental improvement, GE and macro-economic models can be used to compare the costs of different environmental policy instruments, such as regulations or environmental taxes. CEA in such situations will seek to discover the instrument or combination of instruments that produces the least reduction in GDP to attain a given environmental target.

It may also be noted that the ultimate purpose of CEA is to find the minimum welfare cost of achieving a given level of environmental improvement. Even the calculations of macroeconomic cost in terms of a reduction in GDP only yield an approximation to this. For example, it is consumption rather than GDP which is normally associated with welfare, and changes in GDP may conceal shifts in the relative proportions of consumption and investment, both of which are components of GDP. Moreover, even consumption is an imperfect indicator of welfare, because welfare may not increase linearly with consumption, and may be sensitive to the distribution of consumption, as well as its level, which is not indicated in aggregate consumption figures. However, it is difficult, both theoretically and empirically to pass from GDP (or macroeconomic) costs to welfare costs, and most CEA analysis does not attempt to do so.

Finally it must be stressed that the economy, and the economy-environment, are dynamic systems, such that interventions may have different economic and environmental implications in the long-term than in the short-term. Two examples of these dynamic effects that feature prominently in discussions of environmental policy are the Porter hypothesis and the rebound effect, both of which may be illustrated in relation to Fig. 11.1.

The Porter hypothesis, probably most succinctly expressed in Porter and Van der Linde (1995), suggests that, far from reducing competitiveness and economic output, environmental (and other) policy can stimulate innovation and long-term competitive advantage, especially when such policy comes to be followed by other countries. In the terms of Fig. 11.1, this could be interpreted as saying that, although achieving the sustainability standard S might incur an immediate marginal cost of a_s (and total costs given by the area between the marginal cost curve and the horizontal axis), this may in due course generate net economic benefits if it results in the development of less environmentally damaging technologies and ways of doing business which become widely adopted by other countries. The general validity of the Porter hypothesis is still the subject of much debate (see Ekins 2000: 172 ff. for a discussion).

The rebound effect is most often hypothesised to occur in relation to energy efficiency and energy use.

The hypothesis is that cost-effective investments in energy efficiency (in relation to Fig. 11.1, where this refers to energy use, these would be investments in energy saving up to b_2) will save money for firms or households, and that this money will then be spent on other goods or services which also use energy. At the economy-wide level, more efficient ways of using energy may open up whole new applications of energy use that become important economic sectors. The implications of the rebound effect are that, at the least, the savings from the energy efficiency investments will not be as high as calculated beforehand, and in theory the extra energy use could even exceed the amount of energy saved. While the existence of the rebound effect is generally recognised, its magnitude has been and is still much debated (see Greening et al. 2000 for a survey).

11.4.3 Cost-Benefit Analysis

Cost-Benefit Analysis (CBA) is 'a procedure for: (1) measuring the gains and losses to individuals, using money as the measuring rod of those gains and losses; (2) aggregating the money valuations of the gains and losses of individuals and expressing them as net social gains and losses' (Pearce 1983: 3). Pearce (1983: 7) goes on clearly to spell out the value judgments on which CBA is based: (1) the individual preferences (of everyone with such preferences about the issue in question) should count; (2) those preferences should be weighted by the existing distribution of income.

As will be seen, the second of these judgments may be altered without undermining the foundation of CBA. The first judgment, however, is fundamental, and is what distinguishes CBA from decision-support processes that are driven by experts, authoritarians or particularly powerful interest groups.

In a situation where the marginal social cost is above its marginal social benefit (with both the social costs and benefits calculated to include non-market effects, so-called 'externalities') CBA seeks to identify the point at which abatement of the environmental damage contributing to the social cost would be such that the marginal abatement cost is equal to the marginal social cost. This is the 'optimal' point, below which further abatement could increase welfare, while above it further abatement would cost more than it is worth. In many environmental situations, it is not possible to identify the marginal social cost and benefit curves with sufficient accuracy to determine this optimal point, so that the CBA result is expressed in terms of a comparison between the total costs and benefits of different situations. For example, this was the approach chosen by the Stern Review in its widely quoted conclusion:

'If we don't act, the overall costs and risks of climate change will be equivalent to losing at least 5% of global GDP, now and forever. If a wider range of risks and impacts is taken into account, the estimates of damage could rise to 20% of GDP or more.

In contrast, the costs of action – reducing greenhouse gas emissions to avoid the worst impacts of climate change – can be limited to around 1% of global GDP each year.' (Stern 2007: xv)

11.4.3.1 Short History

There is now an extensive academic literature relating CBA to welfare economics (Johansson 1993), but the practice of CBA and CEA emerged in the 1930s unrelated to economic theory, driven by the need of public policy makers to make choices which could be justified in economic and social terms. The first practical formulation of the CBA approach was in the US Flood Control Act of 1936, which stated that the Federal Government 'should improve or participate in the improvement of navigable waters … for flood control purposes, if the benefits to whomsoever they accrue are in excess of the estimated costs.' (quoted in Pearce 1983: 14).

Where the basis of the valuation of costs and benefits is straightforward and generally agreed, CBA is an uncontroversial technique, and its application increased over the years to the 1970s. Then its use in situations of social and environmental complexity (for example, concerning the siting of a new airport), and its incorporation of valuations of non-market impacts (including valuations of human life), attracted both criticism and outrage. The next two sections explore some of the issues that have made CBA controversial.

11.4.3.2 Valuation Issues

The relationship of people with the natural environment is complex, and they value it for different reasons. These reasons have been classified according to the *use* values (deriving from current uses) of the environment, or some part of it, *option* values (deriving from its potential use), and *existence* values (deriving from appreciation of its existence). Even more fundamental is the value deriving from the environmental processes that sustain life on Earth, about which processes even scientists have a lot to learn.

The application of CBA to environmental issues requires that all these values are expressed in terms of money. For marketed environmental features and resources, this presents no problem in principle, although it is not always the case that market prices will capture all the values of the feature or resource in question. For non-market environmental goods and services a range of techniques has been developed (which cannot be discussed in detail here), some of which infer values from markets in which the environmental goods and services are implicated (for example, the component of house prices that derives from some environmental attribute of its situation). Others impute value to the environmental goods and services by placing them in a hypothetical market and asking people how much they would be prepared to pay for the good or service if in fact it was to be bought (or what compensation they would need to receive in order to consent to its loss).

Contingent valuation, as this latter technique has come to be called, has progressed significantly in recent years in terms of finding answers to many of the obvious issues that arise in connection with questions about hypothetical markets. But there remain two fundamental theoretical issues, and one practical one, that continue to call its results into question.

The theoretical issues concern the difference between hypothetical and actual payments and market values; and the willingness and ability of people to assign market values to non-market situations and decisions. Adams (1980) raised many of the relevant points related to the second issue, which have still not been resolved. The practical issue relates to the wide range of valuations which have emerged from applying contingent valuation to some environmental issues, which has both limited its usefulness and damaged its credibility in some decision-making situations.

A further issue which makes contingent valuation both more difficult and controversial relates to what is called 'benefits transfer'. Strictly, because environmental costs and benefits are felt in particular places at particular times, they should always be evaluated in relation to those places and times. However, contingent valuation is a costly and time-consuming exercise, which would be much facilitated if valuations made in one context could be applied to others, if the valuations could be 'transferred' between them. After some discussion of the various issues around and criteria for the validity of benefits transfer (Eftec 2006: 40) concluded: 'Benefits transfer is not a substitute for original research. In fact, as environmental conditions change, the definitions of the 'current situation' used in the existing studies become outdated requiring new studies to fill the gap.'

One of the most well-known controversial practical applications of environmental valuation was carried out by Costanza et al. (1997), who sought to calculate the total monetary value of the ecological services 'of the entire biosphere'. The issues raised are discussed in detail in Costanza and Farber (2002) and the journal Special Issue to which this is an Introduction.

11.4.3.3 Weighting Issues

Time Discounting

Many environmental issues involve costs and benefits over an extended time period. In order for costs and benefits occurring at different points in time to be compared, standard economic practice requires that they be converted to present values by use of a discount rate. Where the quantities of money concerned are private financial flows, discounting at the cost of capital (the interest rate) is relatively uncontroversial, because this reflects the real intertemporal costs involved.

Where the quantities of money relate to environmental costs and benefits, which affect different generations of people, there is general agreement that that 'social discount rate' consists of two parts, a 'pure rate of time preference', relating to how current generations should, or do, value the welfare of future generations, and a term relating to marginal utility of income (it is generally perceived that richer people value a marginal unit of income less than poorer people, so that if future generations are richer than current generations, they will value given costs and benefits less so that the discount rate applied by current generations should be higher). Stern (2007: 46ff.) provides an accessible introduction to the rather complex ethical and economic arguments involved.

Table 11.10 Discount factors for conversion to present values, for different discount rates over various periods of time

Years in future	Discount rate		
	3%	6%	8%
10	1.34	1.74	2.16
20	1.81	3.21	4.66
30	2.43	5.74	10.06
50	4.38	18.42	46.90
100	19.22	339	2,200

Despite agreement on the general approach to be applied, there is still dispute within the economics profession and elsewhere, as to the appropriate discount rate to be applied. Yet this choice can make an enormous difference to the present value of future costs and benefits, as can be seen in Table 11.10. This shows the discount factors (the amount which the cost or benefit in the future needs to be divided by to convert it to a present value) that are derived from different discount rates applied over different periods of time. Comparing the 3% and 8% discount rates, both of which have been recommended in relation to such issues as climate change, it can be seen that the discount factor varies by more than a factor of two over 20 years, 4 over 30 years, 10 over 50 years and 100 over 100 years. The choice of discount rate can therefore effectively determine the outcome of a CBA, yet, as noted above, there is no generally agreed way of making that choice. Some experts even consider that, in some economic and environmental situations, the correct discount rate is zero, which would mean that any indefinite stream or costs or benefits would be of infinite size.

Risk Aversion

Some environmental changes (for example, climate change, biodiversity loss) may result in very large costs. Because of uncertainties, it is often not clear what probabilities should be attached to such eventualities (indeed, it may not even be clear what the full set of possible outcomes is – see Stirling 2003 for a discussion of this issue), which adds an extra dimension of difficulty to the valuation issues discussed earlier. But even where the valuation can satisfactorily incorporate the uncertainty, it is unlikely to be appropriate to give the same weight of consideration to moderate costs as to those arising from potentially catastrophic situations.

Societies are normally perceived to be risk averse, that is, they give relatively more weight to potentially very large costs than to moderate costs. Indeed, this was recognised at the Rio Summit of 1992 by the inclusion of the Precautionary Principle in the conclusions of the conference. Unfortunately there is no agreement as to what weighting factors should be applied to potentially large costs to take account of risk aversion, but, as with discounting, the choice of weights may have an important effect on the CBA result.

Distributional Effects

Environmental costs and benefits differ in their distribution within the current generation as well as between generations. The victims of environmental damage may be different from both those who caused the damage and those who benefited from the activities which caused it. It may also impact differently on different income groups. Even if it does not, richer groups are likely to be better able to cope with a given cost than poorer groups (i.e. they will suffer a lower loss of utility, or welfare).

For all these reasons it may be desirable to weight differently the costs and benefits of environmental effects to take account of their distribution. Unfortunately, again, while it is often agreed that there should be such weighting, there is no agreement as to what the weights should be in different situations.

The choices made over discount rates, and weights to take account of risk aversion and distributional effects, are very likely to be the cause of controversy in situations where they make a significant difference to the result of a CBA, yet there are no criteria to decide which choices are 'right'. This introduces a disturbing element of arbitrariness into CBAs, which can seriously compromise their usefulness to decision-makers.

Conclusion

Where the basis of valuation of the costs and benefits of an environmental effect is generally agreed, and it is perceived that there are no problematic issues of discounting, risk aversion and distribution to take into account, CBA can be a powerful decision-support tool, giving detailed quantitative advice and, sometimes, a recommendation of an optimal decision.

However, the above conditions often do not apply to environmental issues, and the controversies over CBA have arisen when the techniques used for valuation have been disputed, or when issues of discounting, risk aversion and distribution are perceived not to have been given sufficient, or the correct, weight. Pearce (1983: 21) has written that one of the reasons for the widespread use of CBA in decision support has been 'a fundamental attraction of reducing a complex problem to something less complex and more manageable'. Where valuation and weighting issues can be resolved to the satisfaction of all parties, this is undoubtedly true. Where such resolution is not possible, the attempted use of CBA may inflame debate and make issues less manageable, rather than more so. Under such circumstances it may be preferable to maintain an explicit distinction between the different elements of an issue, making explicit the different criteria against which the issue is being assessed, rather than trying to convert them all at an early stage into money values. This is the aim of Multi-Criteria Analysis (MCA), which is the subject of the next section.

11.5 Multi-criteria Analysis

The preceding discussion was differentiated according to the influence that was accorded to the economic dimension in setting policy objectives, from situations in which cost was given no role to those in which the objectives were entirely determined through cost-benefit analysis.

It will often be the case, with environmental policy making and with other complex areas of policy, that the policy is expected to contribute to the achievement of a number of different policy objectives or, at least, will be assessed across a number of different dimensions. It is therefore necessary for policy makers to have the means of comparing the performance, or likely performance, of possible policies across these different dimensions. The purpose of the comparison may be to make the best possible policy choice (optimisation), to ensure that the policy is satisfactory across the dimensions of concern (satisficing), or simply to be able to justify the policy choice after the event (what Nijkamp and Voogd 1985: 63 call 'justificing').

Where the policy performance can, for all relevant concerns, be expressed in money values, then cost-benefit analysis can provide powerful guidance for policy choice and will often recommend policy decisions that appear optimal under the analytical assumptions that have been made. The problem is that the assumptions in question are, as was seen earlier, far from trivial. Stirling (1997: 187–8) lists them as follows:

- The possibility of characterising the decision-situation in question, and the outcomes of any policy choice, with the necessary comprehensiveness and certainty.
- 'The validity and acceptability of the fundamental ethical underpinning' of assigning money values to the different decisions.
- The credibility and robustness of the various techniques that are required for 'the prioritisation and reduction to a single metric of indices applied across a disparate range of incommensurable criteria.'

Stirling (1997: 189) also identifies a number of even more fundamental assumptions which must be satisfied for any quest for a single 'rational' or 'optimal' policy choice to have a valid outcome:

- Decision-makers (or the analysts) must know in advance 'all relevant details and their respective degrees of importance' concerning the situation in question.
- 'There exists a single definitive set of considerations' that is relevant to any individual decision, and 'there can be only one rational chain of inference from any single set of propositions'.
- Where diverse perspectives about the situation exist, it is possible to reconcile and integrate these into 'a single coherent structure of social preferences'.
- 'The consequences of any policy action are in principle predictable.'

Stirling (1997: 191) believes that environmental policy choices very often have six characteristics which comprehensively invalidate these assumptions, and therefore the techniques which are based on them, namely:

- They are path-dependent, with many possible channels of development.
- They are multi-dimensional, with the appraisal criteria for the different dimensions being mutually incommensurable and often unquantifiable.
- Appraisal of their outcomes, and of the issues with which they deal, is perspective-dependent, such that no generally valid appraisal metric exists or can be constructed.

- They present themselves in situations of fundamental uncertainty and ignorance.
- They are processes, requiring continuous re-appraisal in changing circumstances, rather than discrete acts.
- They have widespread social implications and embody broad social concerns, and so require the participation of a range of social actors, with implications for the transparency, style and content of the discourse.

In order to take into account these characteristics, Stirling has developed a technique of Multiple Criteria Mapping (MCM), which attempts to provide a quantitative map of the issue in question and to rank the different policy options according to the priorities attached to various criteria derived from different perspectives on the issue on question. Figure 11.2 gives a schematic illustration of Stirling's MCM technique.

MCM can provide two sets of insights of use to policy makers. At the most general level it can give policy makers, as its name suggests, an overall policy map of the relevant issue, identifying the different perspectives and the priorities and criteria that are regarded as important. At the very least, policy making using such a map should ensure that points are not overlooked due to ignorance.

Beyond this, MCM can sometimes reveal surprising points of agreement between what at first sight might seem very different perspectives. This is illustrated by Stirling in two applications of MCM, a hypothetical application to electricity supply options (Stirling 1997: 200 ff.) and a pilot study of alternative views about genetically modified crops (Stirling and Mayer 1999). In the former case it was shown that the very small proportion of renewable energy (about 3%) in the current UK electricity supply mix is only justified when a very low importance is assigned to both external environmental costs and the diversity of sources of electricity supply. The latter study showed widespread dissatisfaction with conventional agricultural systems across all perspectives, and a generally positive attitude towards organic agricultural systems.

MCM is one type of a broad number of techniques going under such names as multi-criteria evaluation, appraisal, decision analysis or decision support (see Stagl 2007 for a review of some of these techniques). All the techniques have the objective of evaluation, where this is defined as 'a set of activities to classify and conveniently arrange the information needed for a choice in order that the various participants in the choice process are enabled to make this choice as balanced as possible' (Nijkamp et al. 1990: 15). All these techniques also involve the identification of different criteria of relevance to the decision in question, and some of them assign weights to these criteria in order to arrive at a score or ranking for the various decision options. MCM only does this for individual perspectives which may be presumed to share a world view and fundamental underlying assumptions about the issue, leaving policy makers to find other ways

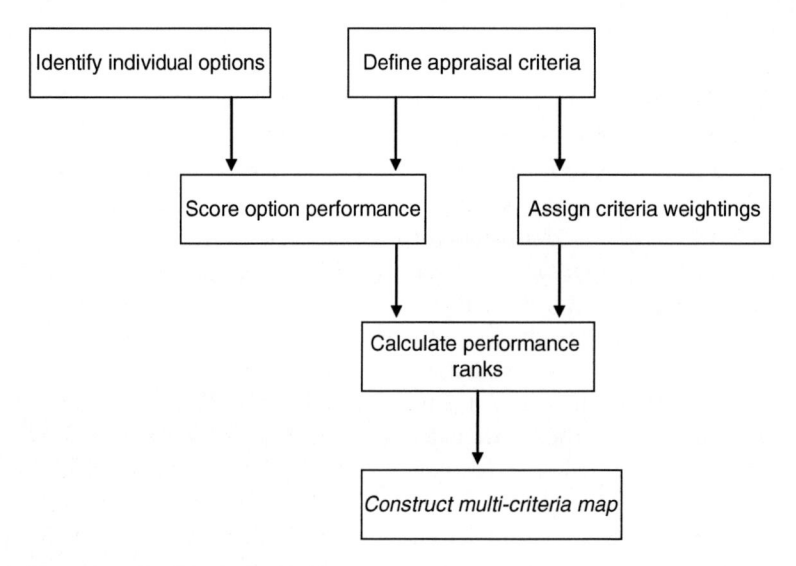

Fig. 11.2 Schematic illustration of multi-criteria mapping
Source: Stirling 1997: 197

of judging and deciding between fundamentally different perspectives. Other techniques of multi-criteria appraisal seek, implicitly or explicitly, to extend the scoring across the different perspectives as well.

Multi-criteria (or multiple criteria) analysis (MCA) emerged during the 1960s (i.e. when cost-benefit analysis was already well established), as a response to the increasing complexity in environmental, and other, policy making, a desire to include intangible and incommensurable effects in policy evaluation, the increasingly conflictual nature of environmental policy choices, and the need to make the assumptions behind and the implications of the different choices more transparent both to the policy makers and the general public.

An important first step in any kind of policy evaluation is an analysis of all the possible impacts of the various policy options, and of the no-policy baseline. It is from this impact analysis that the various general criteria of concern are identified, as well as the extent and importance of the impact which allows each criterion to be given a score. In CBA the score is, of course, a money value, which enables all the impacts to be compared directly with each other and with the financial costs and benefits of the policy options. With MCA the scores across different criteria are first recorded in their natural units. Weights are then attached to the different criteria to indicate their relative importance, either according to expert judgment (where it is perceived that a unique expert judgment may be formulated) or by the policy maker (who is then implicitly representing the preference structure of society at large) or, with MCM, by representatives of different, defined perspectives. There are a number of different ways of assigning weights to the criteria, some involving the direct estimation of the weights, others requiring them to be estimated from preference statements. Some techniques require that the criterion scores be normalised. Sometimes there may be minimum requirements or thresholds for some or all of the scores (which may enable the set of acceptable alternatives to be reduced). Sometimes the criteria may have some element of lexicographic ordering. Nijkamp et al. (1990: 39 ff.) provide an overview of weighting methods.

Nijkamp et al. (1990: 65 ff.) also distinguish between discrete multi-criteria methods, where the decision choice is between a number of discrete options, and continuous methods (101 ff.), where the decision variables may take a potentially infinite number of values. Where both the impact weights and scores involve cardinal numbers, the simplest of the discrete methods simply involves the summation of the products of the scores (normalised if necessary) and the weights of the different options. This amounts to an assumption of a linear utility function. Where such an assumption appears undesirable, a more complex technique called concordance analysis may be used to compare the alternatives pairwise. Indices of concordance or dissonance are computed depending on the relative performance of the different alternatives over the different criteria, and the criteria weights. Sometimes further assumptions are necessary to derive a definitive ranking of alternatives through this method. Other techniques have to be used when either the scores or weights or both are only expressed in ordinal form. A Mixed Evaluation matrix (EVAMIX) may be used when part of the information is cardinal and part is ordinal.

Where the policy options are continuous, rather than discrete, different methods have to be adopted. For example, the pairwise comparison on which a number of the discrete methods rely cannot be made. One solution is to choose discrete options from among the continuous possibilities, and then apply discrete choice methods, but this has the disadvantage of potentially excluding attractive choices from consideration. A common initial approach in continuous evaluation methods is to limit the options under consideration to those that are efficient, i.e. those for which no alternative performs better across all criteria. The problem then becomes one of choosing between efficient options. This can sometimes be done using algorithms, although the computing requirements can soon become substantial, or through interaction with the decision makers. This involves specifying the maximum and minimum attainable levels of each objective and formulating a compromise solution which is feasible, in accordance with the decision maker's previously stated preferences. The decision maker is then able to specify a desired improvement in performance with respect to one criterion, and an alternative efficient set which contains this (along with reduced performance with respect to one or more other criterion) can be computed. This process can be iterated until a set is generated which the decision maker finds satisfactory.

Examples of MCA require too much detail for their adequate explanation to be included here, but the technique has been extensively used in real-life decision making. Nijkamp et al. (1990) give examples of MCA applied to the location of power stations, industrial islands and gravel pits in the Netherlands, which well illustrate how the theoretical considerations in the methods adopted are applied in practice to particular situations.

In conclusion it may be observed that the methods of multi-criteria mapping (MCM), through other techniques of multi-criteria analysis (MCA) to cost-benefit analysis (CBA) form a continuum of analytical techniques, which are appropriate in different situations. MCM is suited to highly contested situations characterised by strongly divergent perspectives and world views. It enables these perspectives to be fully articulated in their own terms, and for their implications for the options under consideration to be fully explored. Sometimes unexpected points of agreement may emerge between the different perspectives, but even where this does not occur, MCM fully illuminates the context in which a decision must be taken. However, beyond identifying the points of most agreement (or least disagreement), MCM gives little guidance to decision makers on which choice to make between the different options presented.

MCA provides the guidance lacking in MCM by formulating and applying a set of weights that are deemed to be generally socially applicable, rather than just pertaining to particular perspectives. These weights may derive from expert judgment, or from the decision maker in his role as responsible social representative. To the extent that the expert views are generally accepted, or the decision maker is accepted as representative, this may be a perfectly acceptable way to proceed. However, MCA techniques often generate satisfactory rather than optimal choices, and the units of choice are arbitrary, so that the actual net gains from the chosen option are difficult to comprehend.

CBA is the technique which promises to deliver not only the optimal choice but also one in which the net gains, expressed in monetary units, may be readily compared with economic situations generally. However, it achieves this not only by imposing a single weighting system across social preferences, as with MCA, but also by assuming that all these preferences can be accommodated within an economic analytical framework and assigned money values. It has been noted earlier that there are many environmental decision situations in which such assumptions do not seem to be justified, with the scope of validity of CBA being correspondingly circumscribed.

There are no hard and fast rules for determining which of these broad approaches to policy evaluation should be chosen. Rather, as Nijkamp et al. (1990: 128) note: 'Policy analysis is essentially an art, the quality of which depends strongly on human and institutional factors'. Consideration of these factors should be strongly present both prior to the choice of analytical approach and during its implementation if the decision process is to be adequately informed.

References

Adams, J. (1980). ....And how much for your grandmother?. In: S. E. Rhoads (Ed.), *Valuing life: Public policy dilemmas*. Boulder, CO: Westview Press.

Ahmad, Y., El Serafy, S., & Lutz, E. (Eds.). (1989). *Environmental accounting for sustainable development*. Washington, DC: World Bank.

Bennett, M., & James, P. (1998). The green bottom line. In: M. Bennett & P. James (Eds.), *The green bottom line: Environmental accounting for management* (pp. 30–60). Sheffield: Greenleaf Publishing.

Costanza, R., & Farber, S. (2002). Introduction to the special issue on the dynamics and value of ecosystem services: integrating economic and ecological perspectives. *Ecological Economics, 41*(3), 367–373.

Costanza, R., d'Arge, R., De Groot, R., Farber, S., Grasso, M., Hannon, B., et al. (1997). The value of the world's ecosystem services and natural capital. *Nature, 387*, 253–260

Daily, G. (Ed.). (1997). *Nature's services: Societal dependence on natural ecosystems* (pp. 23–47). Washington, DC: Island Press.

Daly, H. E., & Cobb, J. (1989). *For the common good: Redirecting the economy towards community, the environment and a sustainable future*. Boston, MA: Beacon Press (UK edition 1990 Green Print, London: Merlin Press).

De Groot, R. S. (1992). *Functions of nature*. Groningen: Wolters-Noordhoff.

De Haan, M., Keuning, S., & Bosch, P. (1993). *Integrating indicators in a national accounting matrix including environmental accounts (NAMEA)*, National Accounts Occasional Paper Nr. NA-060. Voorburg: Central Bureau of Statistics.

Ditz, D., Ranganathan, J., & Banks, R. D. (1995). *Green ledgers: Case studies in corporate environmental accounting*. Washington, DC: World Resources Institute.

Dorfman, M., White, A., Becker, M., & Jackson, T. (1993). Profiting from pollution prevention. In: T. Jackson (Ed.), *Clean production strategies: Developing preventive environmental management in the industrial economy* (pp. 189–206). Boca Raton, FL/London: Lewis.

Eftec. (2006). *Valuing our natural environment*. Final Report NR0103, for Department for Environment, Food and Rural Affairs (DEFRA), March, www2.defra.gov.uk/research/project_data/

Ekins, P. (2000). *Economic growth and environmental sustainability: The prospects for green growth*. London/New York: Routledge.

Ekins, P., & Simon, S. (1998). Determining the sustainability gap: National accounting for environmental sustainability,

In: P. Vaze (Ed.), *UK environmental accounts: Theory, data and application*. London: Office for National Statistics.

EUROSTAT (1994). *SERIEE*. Luxembourg: EUROSTAT.

Greening, L., Greene, D.L., & Difiglio, C. (2000). Energy efficiency and consumption – the rebound effect – A Survey. *Energy Policy, 28* (a Special Issue on the rebound effect), 389–401.

Hamilton, K. (1994). Green adjustments to GDP. *Resources Policy, 20*(3), 155–168.

Heyes, A. G., & Liston-Heyes, C. (1993). US demilitarization and global warming: An empirical investigation of the environmental peace dividend. *Energy Policy*, December, 1217–1224.

Hicks, J. (1946). *Value and capital* (2nd ed.). Oxford: Oxford University Press.

Howes, R. (1999). Accounting for environmentally sustainable profits. *Management Accounting, 77*(1), January, 32–33. London: Chartered Institute of Management Accounting (CIMA).

Jasch, C. (1999). Ecobalancing in Austria: Its use in SMEs and for benchmarking. In: M. Bennett & P. James (Eds.), *Sustainability measures: Evaluation and reporting of environmental and social performance* (pp. 151–169). Sheffield: Greenleaf Publishing.

Johansson, P.-O. (1993). *Cost-benefit analysis of environmental change*. Cambridge: Cambridge University Press.

Lutz, E. (1993). Toward improved accounting for the environment: An Overview. In: E. Lutz (Ed.), 1993 *toward improved accounting for the environment* (pp. 22–45). Washington, DC: World Bank.

Miller, R. E., & Blair, P. D. (1985). *Input-output analysis: Foundations and extensions*. Englewood Cliffs, NJ: Prentice-Hall.

Nijkamp, P., & Voogd, H. (1985). An Informal Introduction to Multi-Criteria Evaluation. In: G. Fandel & J. Spronk (Eds.), *Multiple criteria decision methods and applications*. Berlin: Springer.

Nijkamp, P., Rietveld, P., & Voogd, H. (1990). *Multi-criteria evaluation in physical planning*. Amsterdam: North-Holland.

Pearce, D. (1983). *Cost-benefit analysis* (2nd ed., 1st ed. 1971). London: Macmillan.

Pearce, D., & Atkinson, G. (1992). *Are national economies sustainable?: Measuring sustainable development*, CSERGE Discussion Paper GEC 92–11, University College London.

Proops, J., Faber, M., & Wagenhals, G. (1993). *Reducing CO_2 emissions: A comparative input-output study for Germany and the UK*. Berlin: Springer.

Proops, J., Gay, P., Speck, S., & Schröder, T. (1996). The lifetime pollution implications of electricity generation, *Energy Policy, 24*(3), 229–237.

Porter, M., & Van Der Linde, C. (1995). Toward a new conception of the environment-competitiveness relationship. *Journal of Economic Perspectives, 9*(4), 97–118.

Rauberger, R., & Wagner, B. (1999). Ecobalance analysis as a managerial tool in Kunert AG. In: M. Bennett & P. James (Eds.), *Sustainability measures: Evaluation and reporting of environmental and social performance* (pp. 170–184). Sheffield: Greenleaf Publishing.

Rikhardsson, P. (1999). Statutory environmental reporting in Denmark: Status and challenges. In: M. Bennett & P. James (Eds.), *Sustainability measures: Evaluation and reporting of environmental and social performance* (pp. 344–352). Sheffield: Greenleaf Publishing.

Schelling, T. (1992). Some economics of global warming. *American Economic Review, 82*(1), 1–14.

Stagl, S. (2007). *SDRN Rapid Research and Evidence Review on Emerging Methods for Sustainability Valuation and Appraisal*. Sustainable Development Research Network (SDRN): London.

Stern, N. (2007). *The economics of climate change: The stern review*. Cambridge: Cambridge University Press.

Stirling, A. (1997). Multi-criteria mapping: Mitigating the problems of environmental valuation?. In: J. Foster (Ed.), *Valuing nature: Economics, ethics and environment*. Routledge: London/New York.

Stirling, A. (2003). Risk, uncertainty and precaution: Some instrumental implications from the social sciences. In: F. Berkhout, M. Leach, & I. Scoones (Eds.), *Negotiating environmental change* (pp. 33–76). Cheltenham: Edward Elgar

Stirling, A., & Mayer, S. (1999). *Rethinking risk: A multi-criteria mapping of a genetically modified crop in agricultural systems in the UK*, August, Science Policy Research Unit (SPRU). University of Sussex: Brighton.

UNSD (United Nations) (1993). *SNA draft handbook on integrated environmental and economic accounting*. New York: Provisional Version, UN Statistical Office.

Van der Werf, W. (1999). A weighted environmental indicator at Unox: An advance towards sustainable development. In: M. Bennett & P. James (Eds.), *Sustainability measures: Evaluation and reporting of environmental and social performance* (pp. 246–252). Sheffield: Greenleaf Publishing.

Zerbe, R. O., & Dively, D. D. (1994). *Benefit-cost analysis in theory and practice*. New York: Harper Collins.

Chapter 12
Analysis of Physical Interactions Between the Economy and the Environment

Helias A. Udo de Haes and Reinout Heijungs

Contents

12.1 General Introduction	207
12.2 Definitions and Terminology	209
12.3 **Environmental Risk Assessment: A Tool for Analysing Risks Due to Activities and Substances**	211
12.3.1 Short History	211
12.3.2 Risk Assessment and Risk Perception	211
12.3.3 Risk Assessment, Risk Evaluation and Risk Management	213
12.3.4 Types of Risk Assessment	214
12.3.5 Environmental and Human Risk Assessment for Chemicals: Technical Framework	215
12.4 **Material Flow Analysis: A Tool for Analysing the Metabolism of Materials and Substances**	219
12.4.1 Short History	219
12.4.2 Definition and Applications	220
12.4.3 Technical Framework	220
12.4.4 Goal and System Definition	220
12.4.5 Inventory and Modelling	222
12.4.6 Interpretation	224
12.5 **Life Cycle Assessment: A Tool for the Analysis of Product Systems**	224
12.5.1 Short History	224
12.5.2 Definition and Applications	225
12.5.3 Technical Framework	226
12.5.4 Goal and Scope Definition	226
12.5.5 Inventory Analysis	227
12.5.6 Impact Assessment	231
12.5.7 Interpretation	232
12.6 **Evaluation and Prospects**	232
12.6.1 Limits to the Usefulness of Analytical Tools	232
12.6.2 The Choice Between Physical and Monetary Models	232
12.6.3 The Location of Category Indicators in the Environmental Causal Chain	233
12.6.4 Scope and Level of Detail	233
12.6.5 How to Deal with Uncertainty?	234
12.6.6 The Contribution to Industrial Ecology	235
References	236

H.A. Udo de Haes (✉)
Institute of Environmental Sciences, Leiden University (CML), The Netherlands
e-mail: udodehaes@cml.leidenuniv.nl

12.1 General Introduction

In this chapter methods for analysing the physical interactions between the economy and the environment will be discussed. The historic roots of such methods lie in the 19th century and go back to Karl Marx and Friedrich Engels, who used the term 'metabolism' (*Stoffwechsel*) to imply a relationship of mutual material exchange between man and nature, an interdependence beyond the widespread notion of man simply 'utilising nature'. Like many of his contemporary economists in the mid-19th century, John Stuart Mill linked this concept of metabolism to the idea of a 'stationary state', a form of economic development with no physical growth. This first phrasing of 'sustainable development' was then forgotten for some time.

It was not until the 1960 of the 20th century that the physical interactions between the economy and the environment again formed a basis for scientific thought, induced by the upcoming acknowledgement of the side effects of economic growth. Thus the economist Kenneth Boulding was worried that a 'cowboy economy' might not be compatible with 'Spaceship Earth', and outlined a coming change to a 'spaceman economy' that was suitably cautious in its dealings with finite resources. At the end of the 1960s, the physicist Robert Ayres and economist Allen Kneese laid the basis for a physical model, for the United States, of the material and energy flows between the economy and the environment, proposing to view environmental pollution as a mass balance problem for the entire economy.

Analysis of these physical interactions also forms the roots of what is now called Life Cycle Assessment. Starting in the early 1970s, under various names and in different countries (in particular Sweden, the UK, Switzerland and the USA), consumer and other products were subjected to an integrated analysis of resource and energy requirements and wastes arising, thereby including all the processes associated with the product 'from the cradle to the grave'. This may be regarded as an extension of the energy analysis that gained popularity following the publication of the Club of Rome's 'Limits to Growth' report in 1972 and the energy crisis of 1973.

A third particular contribution was made by the economist Nicholas Georgescu-Roegen, who around 1970 published on the significance of the entropy law for economics. His central thesis was that the creation of order in the form of useful products created from raw materials is necessarily accompanied by creation of disorder in the form of waste products and pollutants. Later research has focused on clarifying the relationship between thermodynamic concepts and evolutionary-economic concepts, most notably growth.

One core characteristic of almost all these types of analysis is their reliance on systems analysis. In systems analysis the object of analytical interest is considered to consist of a system of discrete elements assumed to interact according to specified relationships. In this way a macroscopic picture of a system as a whole is obtained through knowledge of the behaviour and interdependence of its constituent elements. A terminological word of warning is in order here: in systems analysis one speaks of a system and 'its' environment, which is assumed to act as an infinite source and sink of materials and energy. In environmental systems analysis, we should not confuse 'its' environment (in a systems-analysis sense) with 'the' environment (in an ecological sense). One of the prime aims of environmental systems analysis is to study interactions between the economy and the natural environment from a systems-analysis perspective. A primary motive is that 'the' environment is *not* an infinite source and sink of materials and energy. Thus, the physical existence of environmental problems implies that there may be shortages of nutrients, resources and so on in the environment and excesses of pollutants, wastes and so on. From a systems analysis perspective it would therefore be more appropriate that 'the' environment is not the environment of a system (i.e., the economy), but rather a subsystem of a larger system comprising both the economy and the environment. The environment of this larger system is then the alleged immutable source and sink, including the perennial flux of solar energy.

Together with the analysis in monetary terms discussed in Chapter 11, the analysis in physical terms treated in this chapter forms the core of the body of knowledge comprising the field of industrial ecology (Ayres and Ayres 2002), the systematic analysis of the interactions between the economy and the environment at the macroscopic level. The different types of analysis, or analytical tools, to be discussed here are in fact attuned to answering different types of questions and are used in different kinds of application. Fields of application of physical analysis include environmental impact assessment (EIA), industrial licensing, environmental management of companies, eco-design of products ('design for the environment' or DfE), with green procurement as its mirror activity, life cycle management and associated liability issues, strategic policy development and development of business strategies. The analyses may be either qualitative or quantitative, with applicability again depending on the type of question and required level of detail. This does not preclude cultural perspectives or differences between schools of scientific thought playing a role, however, thus possibly giving rise to competition in the application of the various methods.

Most of the analytical tools to be discussed are being developed within distinct scientific communities, many of them with their own scientific journals. Increasingly, though, these communities are exchanging knowledge and experience and coming to interact in other ways too. In addition, a number of methods are presently being standardised by an authoritative body. Of particular interest are the activities of the (private) International Organization for Standardization (ISO), with its series of standards on environmental management (14,000 ff.), environmental audit (14,010 ff.), environmental labelling (14,020 ff.), environmental performance indicators (14,030 ff.), Life Cycle Assessment (14,040 ff.), and integration of environmental aspects in product development (14,062). Where relevant, these will be referred to in the text.

This chapter starts with a section on terminology, which is also relevant for the discussion of monetary analysis (Section 12.2). This is particularly important here, as the information to be presented derives from a number of different scientific communities, each at least in part with its own terminology. We shall also define terms such as 'concepts' and 'tools'. Next three tools, or in fact families of tools, will be discussed: Environmental Risk Assessment (Section 12.3), Material Flow Accounting together with Substance Flow Analysis (Section 12.4) and Life Cycle Assessment (Section 12.5). The chapter concludes with a section analysing the limits and prospects of the various tools, which also includes a comparison with monetary tools (Section 12.6). Integration of the information on these tools within the framework of a case study is the subject of Chapter 20 of this book.

12.2 Definitions and Terminology

The first key distinction to be made is that between concepts, tools and policy instruments. The term 'concepts' refers to general principles, such as cleaner technology, design for the environment, industrial ecology, industrial transformation, ecological modernisation, life cycle thinking and total quality management.

'Tools' are operational methods for the analysis and management of relationships between the economy and the environment. An important further distinction can be made between the analytical and the procedural elements of such tools. The former comprises all aspects relating to problem content and, more specifically, the availability of data, models and rules for computing such results as impacts on the environment or sustainability indicators. Procedural elements, in contrast, cover a wide variety of organisational, legal and institutional aspects. Although almost every tool has both analytical and procedural elements, some tools have a stronger analytical focus, while others emphasise the procedural element, at least at present. A distinction has consequently developed between, on the one hand, *analytical tools*, such as Environmental Risk Assessment (ERA), Substance Flow Analysis (SFA) and Life Cycle Assessment (LCA), as well as the monetary tools of Chapter 11, such as Cost-Benefit Analysis (CBA) and Input-Output Analysis (IOA); and, on the other hand, *procedural tools*, such as Environmental Impact Assessment (EIA) and the Environmental Audit (EA).

'Policy instruments', finally, are measures implemented specifically to reduce or solve environmental problems. Examples include regulations, licences, taxes, agreements and so forth.

A further distinction can be made between analytical tools as such and the *technical elements* they comprise, which they may share with other tools. Examples of technical elements, into which appropriate data are fed, include mass balances, environmental dispersion models, dose-response relationships or, in the monetary sphere, cost functions. An overview of these terms, with examples, is presented in Fig. 12.1. The focus of this chapter will be on analytical tools, although in relation with other items.

Another point on which terminological clarification is useful is the distinction between three different levels in the environmental cause-effect network. Here the so-called DPSIR framework of the European Environmental Agency may well be taken as basis (see Fig. 12.2). In this framework:

Concepts for environmental sustainability
- Life Cycle Thinking
- Life Cycle Management
- Design for the Environment
- Cleaner Technology
- Total quality Environmental Management
- Dematerialization
- Eco-efficiency
- Industrial Ecology
- Extended Producer Responsibility
- End of Life Management
- Others

Decision process → Towards implementation

Analytical tools for environmental aspects
- Life Cycle Assessment (LCA)
- Material Flow Accounting (MFA)
- Substance Flow Analysis (SFA)
- Material Intensity per Service unit (MIPS)
- Cumulative Energy Requirements Analysis (CERA)
- Extended Input/Output Analysis (IOA)
- Environmental Risk Assessment (ERA)
- Checklists for Eco-design, eco-audit
- Others

Procedural tools
- Environmental Management System
- Environmental Audit
- Environmental Performance Evaluation
- Environmental labelling
- Eco-design
- Environmental Impact Assessment
- Green procurement
- Voluntary agreements
- Quality management system
- Others

Analytical tools for technology, social & economic aspects
- Market analysis
- Socio-economic assessment
- Life Cycle Costing (LCC)
- Total Cost Accounting (TCA)
- Cost Benefit Analysis (CBA)
- Input/Output Analysis
- Partial equilibrium model
- Optimization models
- Applied general equilibrium models
- Technology Assessment
- Others

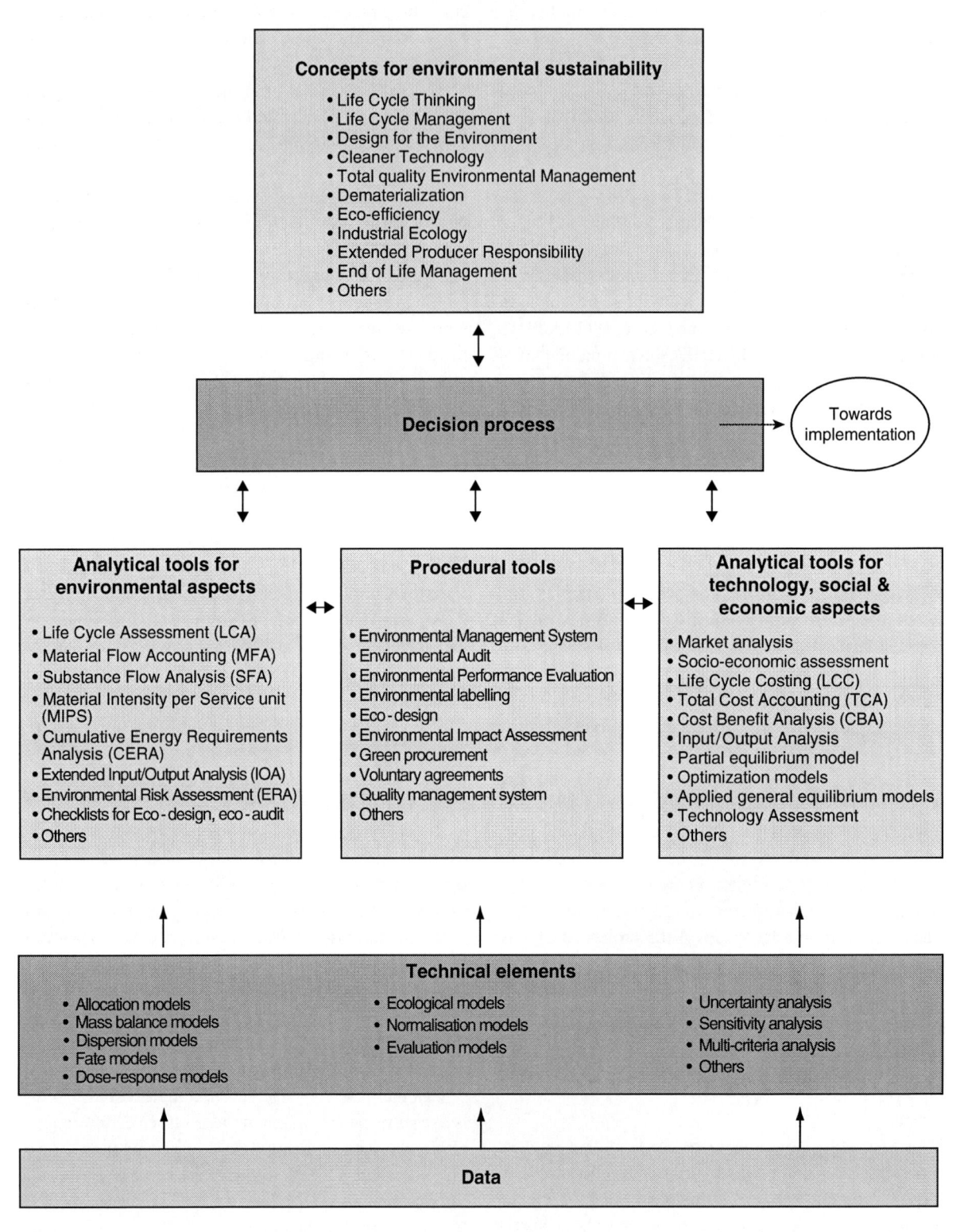

Technical elements
- Allocation models
- Mass balance models
- Dispersion models
- Fate models
- Dose-response models
- Ecological models
- Normalisation models
- Evaluation models
- Uncertainty analysis
- Sensitivity analysis
- Multi-criteria analysis
- Others

Data

Fig. 12.1 Overview of general concepts, procedural tools, analytical tools and technical elements of tools in the field of environmental management

Source: Adapted from Wrisberg and Udo de Haes 2002. Decision-making is supported by the tools, with guidance from the concepts

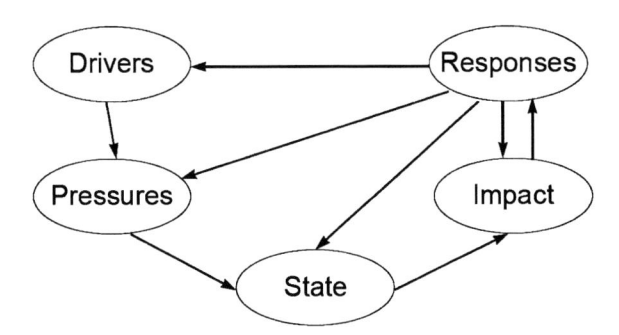

Fig. 12.2 Overview of the DPSIR framework
Source: Smeets and Weterings (1999: 6)

- D stands for driving forces within human society, like population growth, economic growth or technology development.
- P stands for pressures on the environment, like extractions of resources, emissions of hazardous substances and changes in land use.
- S stands for state of the environment, like concentrations of various substances or groundwater table.
- I stands for impacts on human health, on ecosystems or on the economy (also often called the "damage level").
- R stands for responses from society to avoid or reduce undesirable impacts.

12.3 Environmental Risk Assessment: A Tool for Analysing Risks Due to Activities and Substances

12.3.1 Short History

Environmental Risk Assessment, or ERA, is concerned with assessing the environmental risks posed by a single activity or substance. As yet there is still no generally agreed, formal definition of risk or risk assessment, under the auspices of the ISO, for instance. In principle, in a natural science or technological context risks are interpreted as hazards with probabilistic consequences for man and/or the environment. Pierre Simon Laplace was the first to give a mathematical definition of risk; he defined it as the sum of the products of the positive or negative consequences associated with a given activity and their probabilities. He also made a statistical analysis of the life expectancy of people with and without smallpox. Later in the 19th century risk analysis focused mainly on the economic costs and benefits associated with economic investments. Since the beginning of the 20th century broader applications have been envisaged, including technological risks and industrial risks, and since World War II environmental risks have also emerged in the public focus. Over the last decade organisations like OECD, WHO, EU and US-EPA have devoted considerable efforts to formalising and quantifying environmental risks, in particular those of chemicals.

In contrast to Laplace's approach, today the term 'risk' is applied solely to adverse impacts, not to possible benefits. Box 12.1 presents different definitions of risk, all of which start from a probabilistic conception of risk. Note that some of these definitions (numbers 4 and 5) go beyond a purely science-based approach, as they include weighting in terms of prohibitive limits. Although it is clear that there is no unique, objective definition of risk, in the following we shall consider elucidation of a given type of risk assessment as a scientific endeavour.

12.3.2 Risk Assessment and Risk Perception

There is an important difference between risk assessment as a science-based activity aiming at identifying

---

Box 12.1 Different definitions of "risk", starting from a probabilistic concept of risk

1. Risk = probability times specific type of negative impact, possibly environmental effects only.
2. Risk = sum of the products of probability and effect associated with a given activity, for negative effects only.
3. Risk = sum of the products of probability and effect associated with a given activity, for both negative and positive effects.
4. Risk = probability to power a, times effect to power b, expressing increasingly prohibitive limits in either probability or effect.
5. As one of the above, but specifically aiming at a comparison between alternatives.
6. As above, but in addition specifying the beneficiaries and/or victims of the different types of effect.

and quantifying risks, and risk perception by the public. The latter is discussed in Chapter 8 (Section 7) and Chapter 9 (Section 4.2). In Table 9.3 the dimensions of perceived risk have been outlined. In Fig. 12.3 below results from risk perception exercises are plotted against the results of science-based risk assessment. There appears to be a pattern in the differences. In general, we see that the incidence of low-frequency categories of mortality such as food poisoning is overestimated, while that of high-frequency categories such as diabetes or stroke is underestimated.

Personal characteristics like education, culture and religion are also important in both the perception and acceptance of risks, but even these prove to be to a certain extent predictable. Thus, Starr concluded that the public will accept risks from voluntary activities (such as skiing) that are roughly 1,000 times as great as would be tolerated from involuntary hazards (such as consumption of food additives). In this respect the difference in risk acceptance between active and passive smoking is also familiar. Particularly in the case of voluntary risks, people tend to introduce some form of weighting to represent personal benefits.

The above comparisons between perceived risks and science-based risks, assume that the latter are more correct. However, science-based risk assessments may also deviate substantially from reality and public perception may, in contrast, prove more correct. Such was the case when institutional science failed to correctly assess the potential ecological and human health impacts of agricultural pesticides. The damage to birds of prey caused by persistent organochlorine compounds was first revealed in the early 1960s by concerned field biologists like Rachel Carson, not by experts in the field of risk assessment. In the UK the Pesticides Advisory Council (PAC) repeatedly stated that for a given group of herbicides there was no scientific evidence, based on laboratory studies, for their causing harm to humans, despite a wealth of anecdotal evidence to the contrary provided by farm workers themselves. The PAC corrected its statements by adding that the herbicides should be produced and used under appropriate conditions. Thus, assessment of 'absolute' risks proved to be highly dependent on management conditions. Funtowicz and Ravetz (1990) extensively discuss the cases of the Chernobyl disaster and the Challenger explosion to demonstrate the problems involved in piling assumption on assumption, without a sound analysis of the quality of the information thus yielded.

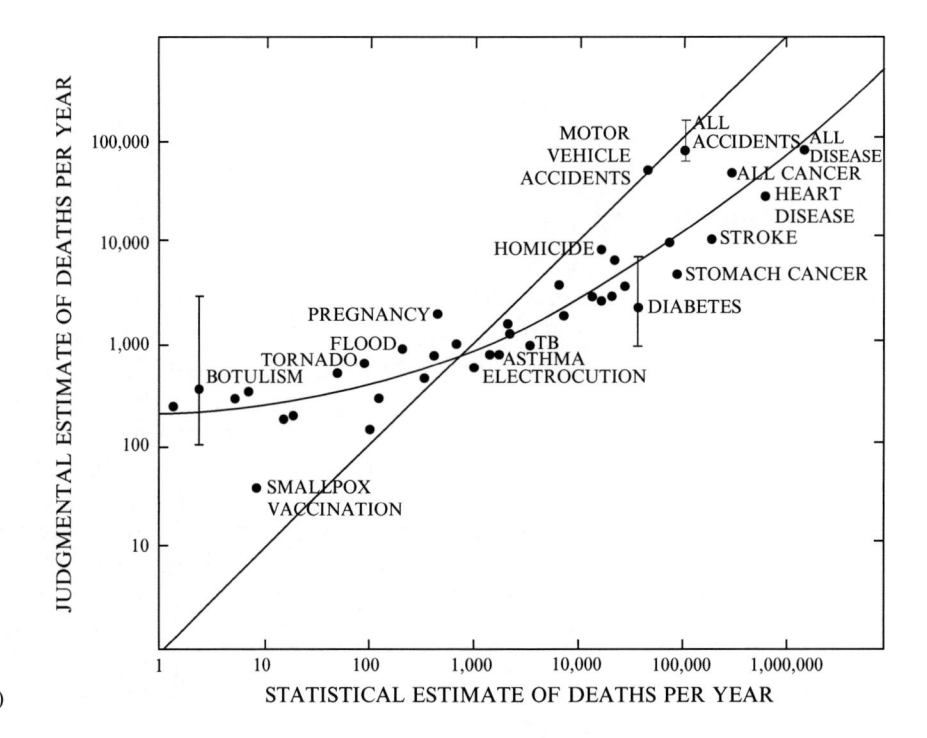

Fig. 12.3 Relationship between actual and estimated incidence of mortality categories
Source: Fischhoff et al. (1981)

12.3.3 Risk Assessment, Risk Evaluation and Risk Management

Risk assessment is, in principle, a science-based activity to analyse the risk of a given event or state of affairs and its estimated consequences. *Risk evaluation* is a structured procedure for determining whether risk reduction is required, by comparing the costs and benefits of such a reduction. *Risk management* is the selection and implementation of measures to reduce risks. (Note that 'risk classification' is sometimes distinguished as a separate step, but is included here in risk evaluation; in addition, the term 'risk management' is sometimes used to cover the entire process, including assessment and evaluation steps). In practice, however, the boundaries between these three consecutive steps are not that sharp, in part because the assessment step is not based entirely on science. There are two major reasons why it may be crucial to involve the public in the process from the very start. Firstly, this may extend the information basis and thus improve the quality and relevance of the risk assessment, for instance by distinguishing between different groups of affected people. Secondly, it may improve the quality of the dispute-resolving mechanism, by involving the various stakeholders more actively in the process. Particularly in the USA and Canada, this process of 'environmental mediation' is becoming a valuable means of reaching consensus.

To a large extent risk evaluation is based on subjective appraisal rather than natural science. The factors presented in Table 12.1 can play a role on both the benefit and the cost side. A relatively simple procedure is to perform a cost-benefit analysis to establish the economic efficiency of introducing particular risk reduction measures. However, as pointed out in Chapter 11, the successful application thereof presupposes that there are no problematic issues of discounting, risk aversion and risk distribution to take into account. Also structured decision rules can be used to support such evaluation. Such rules include application of best available technology (BAT), best practicable means (BPM), best available technologies not entailing excessive costs (BATNEEC) and aiming for risks as low as reasonably achievable (ALARA). These procedures have as a common element that they include a crude cost-benefit comparison, but they may also incorporate more general measures aimed at risk prevention and appropriate action in the case of accidents. The above decision rules can be seen as an initial step towards concrete implementation of the precautionary principle, as discussed in Chapter 8.

Table 12.1 Annual fatality rates per 100,000 persons at risk

Risk	Rate
Motorcycling	2,000
Aerial acrobatics (planes)	500
Smoking (all causes)	300
Sport parachuting	200
Smoking (cancer)	120
Firefighting	80
Hang gliding	80
Coal mining	63
Farming	36
Motor vehicles	24
Police work (non clerical)	22
Boating	5
Rodeo performer	3
Hunting	3
Fires	2.8
1 diet drink/day (saccharin)	1.0
4 T peanut butter/day (aflatoxin)	0.8
Floods	0.06
Lightning	0.05
Meteorite	0.000006

Source: Slovic 1991 (in Mayo and Hollander 1991: 53)

Decision rules for risk management often take the simple form of a binary distinction between two risk levels: a maximum permissible level and a negligible level. If the measured or predicted level of risk is:

- Above the maximum permissible level, risks are regarded as unacceptable and appropriate action is required.
- Between the two levels, further information is required, including cost-benefit evaluation of preventive measures.
- Below the negligible level, risks are regarded as acceptable and no action is required.

It should be noted that the notion of 'acceptable risk' is a formalised policy term. What is deemed officially 'acceptable' may well deviate from what is considered tolerable by a given individual. It should also be mentioned that the term 'acceptable' is sometimes used (as in 'acceptable dose') to refer to more or less science-based situations, as when ecosystems are exposed below harmful levels, and that the term 'accepted' is then used (as in 'accepted dose') to refer to how the societal context modifies a limit value to include aspects of feasibility and political will.

12.3.4 Types of Risk Assessment

We shall now focus further on the scientific element of environmental risk assessment. Three major types of analysis can be distinguished: risk assessment of single activities, comparative risk assessment of a set of activities, and chemicals risk assessment. Risk assessment of single activities is designed to estimate the probability of rare, high-consequence accidents in complex systems such as nuclear reactors, liquefied natural gas (LNG) plants or chemical factories. It consists of constructing process trees, each branch of these trees with its probability of failure and ensuing consequences, as well as procedures for dealing with those consequences. Conversely, it may start from a calamity, analysing error trees from which such an event may potentially arise.

The best known example of the latter is the so-called Rasmussen report published by the US Nuclear Regulatory Commission (US-NRC) in 1975. This report addressed the following main questions. Under what conditions might a core melt accident occur, and what is the chance of occurrence of such an event? What features are provided in reactors to cope with such an accident? How might a loss-of-coolant accident (LOCA) lead to a core melt? How might a reactor transient (that is, events requiring the reactor to shut down) lead to a core melt? How likely is a core melt accident? The outcome of this extensive study was a calculated total probability of one core melt in 20,000 reactor-years. With 100 reactors on stream in the US, this would be one core melt-down every 200 years. The report added, however, that in the 200 reactor-years of commercial operation (up to 1974), there had been no fuel melt accidents. Cost calculations were made for a bottom-line core melt accident, based on an area 45° wide downwind 40 km long.

These calculations did not take into account the accidents that had occurred in the plutonium works at Majak, in the southern Ural. In 1957 a waste reservoir exploded there, causing the death of some 1,000 to 4,000 factory-workers and military personnel in the immediate aftermath and cancer in some 10,000 employees over the next 40 years. Up to the year 1993 another ten smaller accidents occurred in Majak. The world was similarly unprepared for the accident that took place on March 28, 1979 at Three Mile Island, 16 km south-east of Harrisburg, Pennsylvania, in the US. This accident involved losses of cooling agent and gas, and developed over the next few days, through a coincidence of technical failures, incorrect prescriptions, misplaced corrective actions and ignorance, into the most severe crisis in the history of the American nuclear industry. The world was even less prepared for the melt-down of the Chernobyl nuclear power plant in April, 1986. The resultant radioactive fall-out stretched to the far north of Norway, seriously disrupting reindeer herding by the native Sami. Although it may be true that circumstances in Majak and Chernobyl were different from those in countries like the US, this does not hold for Three Mile Island, and the public message of the Rasmussen report implicitly concerned nuclear power in its entirety, and not just in the US. Following the latter accident the attitude of the US-NRC was severely criticised. More importantly, the image of nuclear power had been severely damaged.

The second type of risk assessment is concerned with comparing the risk of different activities, which are rendered comparable through suitable definition of the type of impact they produce: the incidence of human casualties, for example. The data used for this purpose are generally based on annual statistics and do not imply in-depth analysis of the underlying processes. A fairly basic statistic is the annual fatality rate associated with different activities. This is illustrated in Table 12.1.

These mortality rates do not capture the fact that some hazards (for instance, pregnancy or motor cycle accidents) cause death at a much earlier age than do others (for instance, lung cancer due to smoking). One way to cope with this difference is to calculate the average loss of life expectancy due to exposure to the hazard, based on the distribution of death as a function of age. Several examples are shown in Table 12.2.

Yet another way to gain a perspective on risk was devised by Wilson (1979), who presented a set of activities all estimated to increase the risk of dying during any particular year by one in a million (not corrected for differences in age this time) (see Table 12.3).

In all the above examples the only effect variable identified is death. Alternatively, a more sophisticated approach may be adopted. In a well-known study

Table 12.2 Estimated loss of life expectancy due to various causes

Cause	Days
Cigarette smoking (male)	2,250
Heart disease	2,100
Being 30% overweight	1,300
Being a coal miner	1,100
Cancer	980
Stroke	520
Army in Vietnam	400
Dangerous jobs, accidents	300
Motor vehicle accidents	207
Pneumonia, influenza	141
Accidents in home	95
Suicide	95
Diabetes	95
Being murdered (homicide)	90
Drowning	41
Job with radiant exposure	40
Falls	39
Natural radiation (beir)	8
Medical x-rays	6
Coffee	6
All catastrophes combined	3.5
Reactor accidents (UCS)	2[a]
Radiation from nuclear industry	0.02[a]

[a] These items assume that all U.S. power is nuclear. UCS is the Union of Concerned Scientists, the most prominent group of critics on nuclear energy.
Source: Cohen and Lee 1979 (in Mayo and Hollander 1991: 54)

Table 12.3 Risks estimated to increase chance of death in any year by 0.000001 (1 part in 1 million)

Activity	Cause of death
Smoking 1.4 cigarettes	Cancer, heart disease
Spending 1 h in a coal mine	Black lung disease
Living 2 days in New York or Boston	Air pollution
Travelling 10 miles by bicycle	Accident
Flying 1,000 miles by jet	Accident
Living 2 months in Denver on vacation from New York	Cancer caused by cosmic radiation
One chest x-ray taken in a good hospital	Cancer caused by radiation
Eating 40 T of peanut butter	Cancer caused by aflatoxin B
Drinking 30, 12-oz cans of diet soda	Cancer caused by saccharin
Drinking 1,000, 24-oz soft drinks from recently banned plastic bottles	Cancer from acrylonitrile monomer
Living 150 years within 20 miles of a nuclear power plant	Cancer caused by radiation
Risk of accident by living within 5 miles of a nuclear reactor for 50 years	Cancer caused by radiation

Source: Wilson 1979 in Mayo and Hollander (1991: 55)

published in the mid-nineties, 'The global burden of disease', Murray and Lopez (1996) analyse human health impacts in terms of years of life lost (YLL; this is the measure used in Table 12.2), years of life disabled (YLD, in which different types of health impairment are weighted against YLL) and a combination of the two, Disability Adjusted Life Years (DALY). Parallel to the DALY concept there is also its mirror, QALY: 'Quality Adjusted Life Years'. The latter two measures include two more steps. In the first place future health impacts are discounted, at 3% per year; and second there is age weighting, distinguishing between the alleged importance of a year of life at different stages of maturity, with greatest weight being assigned to a year in the life of the young and middle-aged compared to a year in the life of young children and the elderly. Although there is room for debate on the moral desirability of these two further steps, the widening of the YLL concept to include different types of disability is an important step forward. Table 12.4 shows the results for ten major risk factors, among which environmental management relates to the first (malnutrition), second (poor water supply) and last (air pollution).

The third type of risk assessment is concerned with the risks posed by everyday use of hazardous chemicals. These involve not a multiplication of risk of occurrence and effect, but the product of regular exposure and effect, exposure being the sum total of processes from the release of the chemicals into the environment to the affected impact (or damage) level. This type of risk assessment, Environmental Risk Assessment and Human Risk Assessment, together abbreviated as HERA, are discussed in more detail in the following section.

12.3.5 Environmental and Human Risk Assessment for Chemicals: Technical Framework

Environmental and Human Risk Assessment for the purpose of regulating hazardous chemicals is based on the so-called 'risk quotients approach'. For Environmental Risk Assessment this concerns the

Table 12.4 Global burden of disease and injury attributable to selected risk factors, 1990

Risk factor	Deaths (1,000s)	As % of total deaths	YLLs (1,000s)	As % of total YLLs	YLDs (1,000s)	As % of total YLDs	DALYs (1,000s)	As % of total DALYs
Malnutrition	5,881	11.7	199,486	22.0	20,089	4.2	219,575	15.9
Poor water supply, Sanitation and Personal and Domestic hygiene	2,668	5.3	85,520	9.4	7,872	1.7	93,392	6.8
Unsafe sex	1,095	2.2	27,602	3.0	21,100	4.5	48,702	3.5
Tobacco	3,038	6.0	26,217	2.9	9,965	2.1	36,182	2.6
Alcohol	774	1.5	19,287	2.1	28,400	6.0	47,687	3.5
Occupation	1,129	2.2	22,493	2.5	15,394	3.3	37,887	2.7
Hypertension	2,918	5.8	17,665	1.9	1,411	0.3	19,076	1.4
Physical Unactivity	1,991	3.9	11,353	1.3	2,300	0.5	13,653	1.0
Illicit drugs	100	0.2	2,634	0.3	5,834	1.2	8,467	0.6
Air pollution	568	1.1	5,625	0.6	1,630	0.3	7,254	0.5

Source: Murray and Lopez (1996: 311)

calculation of the quotient of the predicted environmental concentration of a substance and the no-effect level of that substance (the PEC/PNEC quotient). For Human Risk Assessment this concerns the quotient of the predicted daily intake and the acceptable daily intake of a given substance (the PDI/ADI quotient). A number of organisations are involved in such assessment activities, including the OECD, the European Plant Protection Organisation (EPPO) and the US-EPA. Although these organisations all adopt approximately the same approach to risk assessment, distinguishing an exposure route and an effect route that come together in risk characterisation, there is as yet no standardised technical framework for the process as a whole. We propose the framework presented in Fig. 12.4, the various steps of which are discussed in the following sections. This formal framework for the risk assessment process comprises four steps: hazard identification, exposure assessment, effects assessment and risk characterisation. These steps may subsequently be followed by risk evaluation and risk management.

12.3.5.1 Hazard Identification

The aim of the hazard identification step is to establish whether a risk is present. To this end data are gathered on exposure conditions and inherent effect characteristics. Sources vary and may include available test data, exposure models, the incidence of accidents, field measurements or monitoring data, and possibly measures taken in the workplace. If in this preliminary analysis a potential risk is identified, further steps are to be undertaken. In fact, this is a procedural step, indicating that Risk Assessment can be an iterative process, starting with a screening step.

12.3.5.2 Exposure Assessment

'Exposure' covers all the steps from emissions through intake up to damage to human health and to ecosystems (the impact or damage level identified in Section 12.2). In particular, it includes the emission, transport, dilution, accumulation, transformation and intake of chemical and radioactive substances. For ecological impacts exposure generally ends with the predicted environmental concentration (PEC) as endpoint, and for human health with the predicted daily intake (sometimes referred to as 'PDI') integrated over the various possible intake routes (food and water intake, breathing and skin contact).

With respect to emissions, two forms of risk assessment can be distinguished. In the first, the risks associated with emissions of one or more substances by one particular activity are assessed. These emissions then constitute the foreground and other emissions of the same substance(s) from other activities the background level. Alternatively, the focus may be on the risks associated with a particular pollutant in a given region. In that case all emissions are taken into equal account. In many cases there will be no empirical data available on actual emissions. These must then be estimated using 'emission factors' for specific processes, groups of processes or entire economic sectors.

Fig. 12.4 Technical framework for environmental risk assessment, risk evaluation and risk management

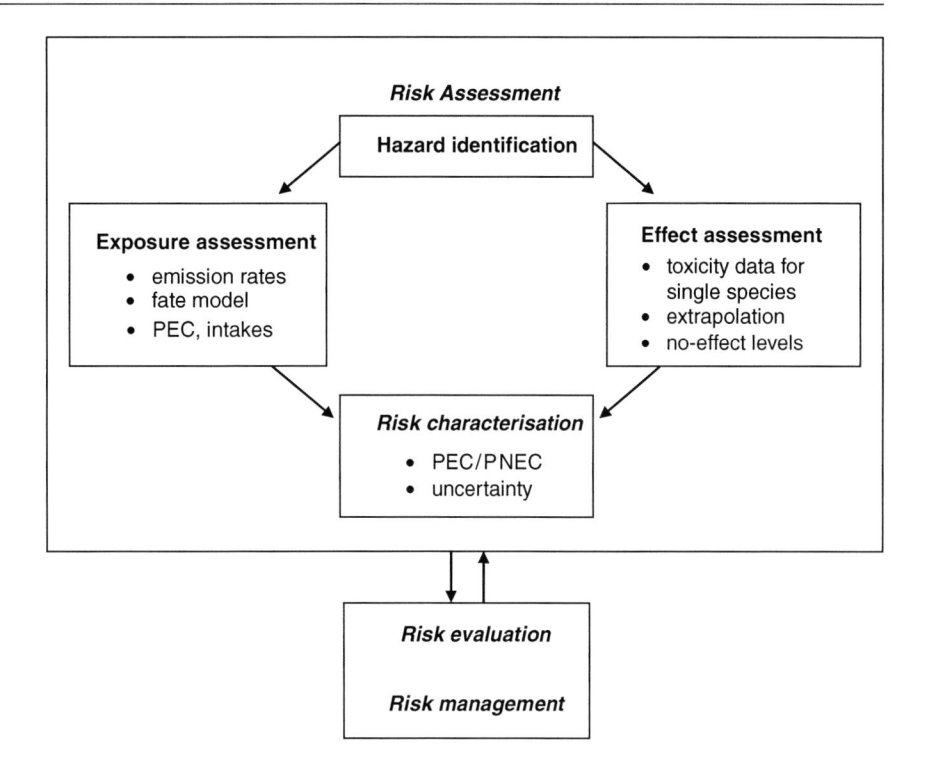

Risk Assessment

Hazard identification

Exposure assessment
- emission rates
- fate model
- PEC, intakes

Effect assessment
- toxicity data for single species
- extrapolation
- no-effect levels

Risk characterisation
- PEC/PNEC
- uncertainty

Risk evaluation

Risk management

The core element of exposure assessment comprises analysis of the processes of physicochemical and biological transport, accumulation and transformation. In modelling these processes, an important choice is definition of the time aspect: steady state or dynamic, the latter yielding a time series of exposure levels. When analysing some kind of current situation, steady-state models are generally used. The best of these are the 'multimedia models' developed by Mackay and coworkers (Mackay 2001), which use thermodynamic principles to determine the equilibrium partitioning of substances between the various environmental media. Its 'multimedia' aspect is then that it covers air, water, soil, sediment and possibly other media, too; see Table 12.5. The principal abiotic transformation processes in exposure models are hydrolysis, oxidation, reduction and photochemical degradation. Biotic transformation processes mainly take the form of enzymatic reactions and generally involve specific groups of compounds. Bioaccumulation (within a single organism) and biomagnification (within a food chain) are now receiving more attention than they did in the past. Until now these processes have been addressed by using uncertainty factors. In the case of human exposure these steps are followed by an assessment of uptake via the different intake routes.

12.3.5.3 Effects Assessment

Effects assessment, or dose-response assessment, is concerned with establishing critical levels of exposure and is based on a number of standard tests. In the case of human health these tests focus mainly on acute and chronic toxicity, carcinogenicity, teratogenicity (deformities arising during foetal development) and mutagenicity and impacts on fertility (damage to offspring or impacts on the number of offspring) as endpoints. The major test animal is the laboratory rat, but other tests are also used (guinea pigs, tissue cultures, etc.).

Tests for ecotoxicity differ from tests for human health, because they must be able to handle extensive taxonomic diversity, within a generally shorter time frame and on a smaller spatial scale, but at the same time with the potentially greater role of biomagnification. Although most tests focus on mortality or immobilisation of individuals of a species, they may also be directed towards damage to organs or tissues, or to any number of physiological processes. The results are expressed in terms of such metrics as the EC50 (the median concentration causing 50% response), the LC50 (the median lethal concentration), or the NEC (no effect concentration), NOEC (no observed effect

Table 12.5 Multimedia model showing different compartments

Level	Type	Information needed	Outcome
I	Equilibrium, conservative[a]	• Physicochemical properties • Model environment parameters • Amount of chemical in the system	• Distribution of the chemical between the compartments
II	Equilibrium, non-conservative	• Level-I + • Overall discharge rate • Transformation and advection rates in different compartments	• Distribution between compartments • Environmental life time
III	Steady state, non-conservative	• Level-II + • Compartment specific discharge rates • Intermedia transfer rates	More accurate estimation of: • Life time • Chemical quantities and concentrations in different compartments
IV	Non-steady state, non-conservative	• Similar to Level-III	• Time needed before steady state is achieved or • Time required to disappear when discharge ends

[a] Conservative or non-conservative in the sense that degradation of the chemical is (or is not) possible.
Source: Van de Meent et al. (in Van Leeuwen and Hermens 1995: 123)

concentration) and NOAEC (the same but then focused on adverse effects).

Various groups of tests are available for the different environmental compartments, in part standardised by such bodies as the OECD. For aquatic toxicity these include microbial tests, tests with plants like algae or Duckweed, life cycle tests with invertebrates (like Daphnia) and so on. For sediments the best known test uses larvae of the mosquito *Chironomus riparius*, while for terrestrial toxicity there are tests with litterbags to measure toxic impacts on bacteria and fungi, for example, and tests with invertebrates like earthworms. The honey-bee is another invertebrate that is rather often employed in toxicity testing. For ecotoxicity testing, either aquatic or terrestrial, vertebrate test animals are rarely used.

Given the enormous number of potentially toxic substances in use and coming onto the market and restrictions of time and money, toxicity tests can be performed for a limited number of substances only. And even for these, data may continue to change due to ongoing further research. There is therefore growing focus on the potential for predicting these impacts on the basis of the chemical characteristics of the substances involved. Thus, much research is now focused on the elaboration of 'quantitative structure-activity relationships' (QSARs), which link aspects of chemical structure to specific types of activities (e.g. bioconcentration) or effects (e.g. no-effect level).

Toxicity test results are used to derive the 'predicted no-effect concentrations' (or PNECs) for ecosystems, thus broadening the NEC concept for single species to the PNEC for multiple species. As indicated above, for humans similar critical levels are known as the 'acceptable daily intake' (ADI). In extrapolating experimental laboratory data to man or to aquatic or terrestrial ecosystems, 'uncertainty factors' or 'extrapolation factors' are employed to introduce a safety margin reflecting the degrees of uncertainty in such an exercise. These factors usually lie in the range 10 to 10,000. Without such factors large sections of the human population would remain unprotected, because of specific uptake routes (in the workplace, for example), differences between chronic and acute exposure and/or differences in individual sensitivity. Large portions of ecosystems would likewise remain unprotected, because of taxonomic diversity and the existence of highly specific transport and transformation processes. Perhaps the best known set of extrapolation factors in this context is that of US-EPA.

12.3.5.4 Risk Characterisation

Risk characterisation is the last step of Environmental Risk Assessment. It consists of calculating the quotient of the predicted concentration or dose and the no-effect concentration or dose, yielding PEC/PNEC values for aquatic, terrestrial and possibly sediment ecosystems and PDI/ADI values for man, for instance. These are dimensionless ratios, which should not exceed unity (i.e., 1) from a toxicological perspective.

12.3.5.5 Risk Evaluation and Risk Management

Two additional steps relate to the social response to the outcome of the risk assessment step. As already stated, the aim of risk evaluation is to decide whether risk reduction is required. In line with Fig. 12.4 a 'maximum permissible level' and a 'negligible level' are identified, both in terms of PEC/PNEC or PDI/ADI, determining, in principle, whether abatement measures are required or, alternatively, additional information. Risk Management is concerned with the selection and implementation of measures to reduce risk. In practice, however, the procedure is not always that straightforward and we often see the PEC/PNEC ratio simply being taken as a measure to characterise the present situation, possibly at values (well) above unity, as is frequently the case with soil contamination. Consequently, the upper limit may become rather elastic. Another key problem is the dilemma posed by the enormous number of substances which are now routinely emitted to the environment. Comprehensive risk assessment has been performed for between 50 and 250 substances only (depending on the number of test species involved). Effect data are available for several thousand substances: 2,500 are documented at the European Chemical Bureau (Ispra, Italy), while US-EPA has drawn up ordinal 'PBT-scores' on persistence (P), bioaccumulative potency (B) and toxicity (T) for some 3,000 substances. This is still a very small number, however, in comparison with the approximately 100,000 substances registered at the European Inventory of Existing Commercial Chemical Substances (EINECS), however. In 2006 the EU adopted a new directive on de Registration, Evaluation, Authorisation and Registration of Chemicals (REACH), which aims to register all high volume chemicals produced or imported in the EU, together with all additional chemicals which are considered of very high concern. Substances with a market volume of less than 1 t per producer or importer are free from registration. In total this will amount to the testing of about 30,000 chemicals. Only 3% of these are as yet tested, and of 11% base set data are available.

Assessments are made of some 3.9 million vertebrate test animals which will be needed for this process. Discussions are going on how this number can be reduced by making use of all existing information, use of QSAR information (see above), data on in vitro tests, grouping of substances and waiving of substances. Still, there will remain a need for a precautionary approach, partly because of the various remaining uncertainties, but also because even below PNEC thresholds toxic chemicals still fill 'environmental space', potentially giving rise to negative effects in combination with other substances, and may therefore be undesirable (see Section 8.7).

12.4 Material Flow Analysis: A Tool for Analysing the Metabolism of Materials and Substances

12.4.1 Short History

Material Flow Analysis (or Accounting) or MFA is concerned with analysing the flows of materials and substances within an economy. Environmental flows are also sometimes included in the analysis, and the system under study then consists of two subsystems: the economy and the environment.

In contrast to HERA, MFA focuses on the entire set of processes in the economy involving the particular material or substance of interest. The first people in recent history to advocate tracing the physical transformation of materials as they passed through the economy were Robert Ayres and Allen Kneese, as mentioned in the introducing section of this chapter. Since the late 1970s a number of important MFA case studies have been performed and since the early 1980s descriptive material balances have come to be included ever more commonly in environmental statistical reports. In recent years, there has been greater focus on developing more sophisticated analytical tools for investigating material flows in the economy. Leading developments take place in Germany, Austria, Switzerland, the Netherlands, Sweden, the US and Japan.

The MFA tools currently in use are many and varied, ranging in their object of analysis from companies to countries, and from watersheds and continents up to the global level, as well as in the type of material flows under study, from modelling of individual substances through to aggregated (e.g. national) mass flows. So far, no standardisation has been attempted of either the purpose or methodology of material flow studies. This is well understandable, because until now the various tools have been used mainly for descriptive and analytical purposes, rather than for comparative purposes aimed at influencing the market, as is the case with Life Cycle Assessment (LCA). As the field becomes more structured, however, there is an increasing need to compare the results of different studies and this now calls for a harmonisation of the methodology. The work implied concerns mainly terminology, the technical framework and guidance as to which type of method should be used for which type of application. (Actual methods should not be prescribed, at least not at the present stage.) The following review can be seen as a step in this direction.

12.4.2 Definition and Applications

MFA seeks to analyse the physical flows of materials and substances, or groups of such, through the economy and possibly also the environment, and identify any accumulations therein. The analysis requires that a system first be defined that is bounded geographically (e.g., Tokyo, Canada or the Rhine basin) as well as temporally (e.g. 1 January until 31 December 1998). MFA comprises a whole family of tools, suitable for analysing different types of materials. Bulk-material flows are studied using Bulk-Material Flow Analysis (bulk-MFA); (chemical) substances or groups of related substances are studied by means of Substance Flow Analysis (SFA). These tools can be used for a variety of purposes:

- To analyse trends and their underlying causes and identify major problem flows to the environment, tracing these back step by step to their origins in society. Thus, hidden 'leaks' from processes in society can be traced back, and an assessment can be made of the degree of closure of material cycles.
- To identify, and assess the effectiveness of, potential pollution abatement measures and help prioritise them. Modelling can also provide insight into possible problem-shifting, as may occur after redirection of substance flows, for example. Scenario studies are a more complex application of this type of modelling.
- To serve as a screening tool, to identify issues for further investigation by means of other analytical tools.
- To provide support for data acquisition, thereby also functioning as an error check procedure, as inconsistencies in aggregated numbers can often be traced back to errors in separate contributing figures. Missing flows can likewise be identified.

12.4.3 Technical Framework

To structure the analysis, MFA should be performed within a well-defined technical framework comprising a number of distinct phases. If the technical framework in place for LCA (ISO 14040, see Section 12.5.3) is also essentially adopted here, the following basic phases can be distinguished: goal and system definition, inventory and modelling, and interpretation. In Sections 12.4.4 to 12.4.6 the content of these three phases will be briefly discussed.

12.4.4 Goal and System Definition

This first phase is concerned with establishing the aim and scope of the study, i.e. the (type of) material, substance or substance group of interest, the goal of the study (one of the applications described in Section 12.4.2) and the precise nature of the system under study. The system definition should cover the geographical and temporal system boundaries, any subsystems the system may contain, the inputs and outputs of both system and subsystems, and any processes within the system and subsystems, i.e. any flows and transformations related to the metabolism of the chosen substance(s) within the defined boundaries.

The definition of the system boundaries depends on the goal of the study: political boundaries in the case of a study relating to administrative units, for example, or ecological boundaries when analysing a watershed. Other types of geographical boundaries can be envisaged. An MFA study may thus focus on a single industrial sector or firm, or may analyse all the consumption-related flows of a given substance in a given region in a given year. In the latter case it may also be decided to include flows outside that region in the system, if these are related to the consumption in question (the so-called 'ecological rucksacks'), thereby shifting from a regional to a functional approach. The analysis then in fact follows the structure of LCA, described below.

As already mentioned, the economy (or socio-economic system) and the environment are usually distinguished as two subsystems of a larger system. The focus is then generally on analysing the processes within the economic subsystem and, more precisely, the material basis of the economy, i.e. in contrast to a financial-economic analysis. The physical flows in the economic subsystem can be broken down further into flows in particular economic sectors and these in turn into the final unit of analysis, so-called unit processes like electricity production or waste incineration. If the environment is also included as a separate subsystem, it can comparably be subdivided into the lithosphere, pedosphere (soil), hydrosphere, atmosphere and biosphere. The lithosphere is that part of the environment, which does not or hardly interact with the other media for the material or substance(s) under study. It is sometimes regarded as a separate, third subsystem.

The inputs and outputs of the system as a whole and of the different subsystems can be defined in analogous fashion. Thus, the economic and the environmental subsystem will both have their own inflows and outflows, their exchanges with the immobile stocks in the lithosphere and their interactions in the form of emissions, extractions and forms of land use. It then remains to define the processes taking place within the subsystems, the flows between these processes and the accumulations occurring at this process level. In the processes the flows are redistributed, either into other flows or into stocks.

The above issues can be put together into a single substance flow diagram. Taking the substance flow of nitrogen compounds within the EU as an example, the result is shown in Fig. 12.5 (after Van der Voet et al. 1996). The two core boxes are the economic subsystem (A) and the environmental subsystem (B). Both have a number of input and output flows. For the given example, EU N-metabolism, we already observe at this

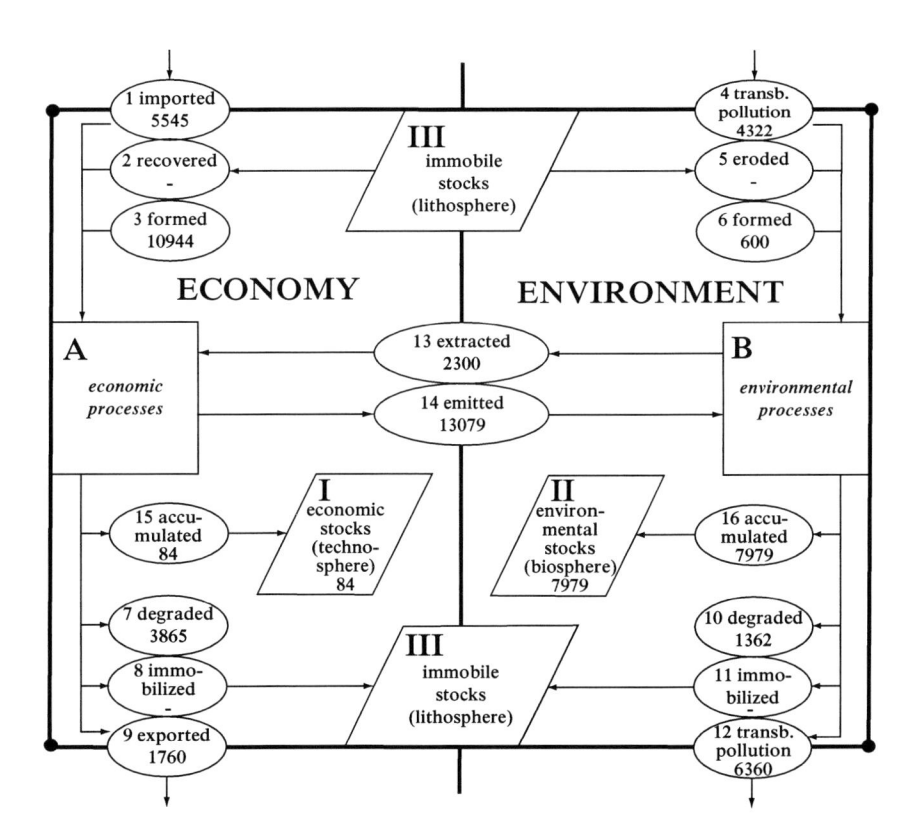

Fig. 12.5 Substance flow diagram for nitrogen compounds in the EU in 1988 Source: From Van der Voet et al. (1996). Numbers represent kilotonnes/year

overview level that total emissions to the environment far exceed total extractions from the environment; that total accumulation within the economy is far smaller than total accumulation within the environment; that production in the economic subsystem is about twice as high as import into the economy; that the environmental subsystem is a net exporter of nitrogen; and that the mass load generated by society dominates the processes in the environment. These observations already provide quite a good characterisation of the nitrogen problem. Depending on the aim of the MFA study, certain processes of interest within the two subsystems are obviously still to be analysed and presented in greater detail.

12.4.5 Inventory and Modelling

This phase is concerned with computing the flows and stocks of the material or substance under study for the year of interest. This can be done on the basis of a so-called bookkeeping system or by using a steady-state or dynamic model. These options will now be discussed in turn.

In a bookkeeping system a flowchart is first developed showing all relevant stocks, flows and processes in both the economic and environmental subsystems. Then, for the given year, empirical data are collected and duly attributed to the stocks, flows and processes. For the economic subsystem this means that, particularly in SFA studies, statistics on production, consumption, waste management and trade are linked to data on the concentration of the substance under study in relevant products and materials. In the case of the environmental subsystem it is predominantly empirical data that will be used, from environmental monitoring networks, for example. The unit processes serve merely as points for the redistribution and increase or decrease of flows. For conservative substances, like metals, the inflows and outflows of each individual process are balanced, unless there is accumulation. For nitrogen no such balance can be made, because the focus is on nitrogen containing compounds, generally leaving free nitrogen gas out of consideration. Overviews thus obtained can be used as an error check, to find missing data or identify present or future problem flows, by identifying major accumulations in the economy or in a specific environmental medium, for example.

The second option is to set up a static (steady-state) model, in which processes are formalised in such a way that outputs can be formally computed from inputs, or, conversely, that the inputs can be computed that satisfy a given set of output values. The model describes a steady-state situation. In principle, empirical data from the bookkeeping overview can be used to calibrate the mathematical equations of the processes. It is preferable, however, to use data on the characteristics of the processes themselves, for instance on the basis of physical, chemical or biological knowledge, to exclude coincidental factors from the equations as far as possible. The core aim here is to develop a consistent mathematical structure specifying the relations between the different flows and stocks within the system.

One of the principal applications of this type of model is assessment of the consequences of changes in for instance household demand, in imports or in technology by calculating the new flows and stocks in the system. The ultimate origins of specific problem flows can also be analysed. Conversely, the effectiveness of potential policy measures can be estimated, which is not possible with the bookkeeping approach.

For instance, for the nitrogen flows presented in Fig. 12.5, it could be shown that for atmospheric deposition and groundwater pollution the agricultural sector is the main source, and for eutrophication of the North Sea this is the households. But the main ultimate origin lies in the production and import of ammonia for fertilizer production. Another example of origin analysis concerns the case of cadmium. The main result of this analysis has been to reveal the inelastic supply of cadmium, which is not extracted independently as an ore but always produced as a by-product of other metals, especially zinc. Thus the analysis revealed that, as long as zinc ore is extracted and processed, a certain amount of cadmium will find its way to the marketplace, in the form of batteries, for example. Post-consumer intake of batteries and recovery of cadmium will consequently lead to an accumulation of cadmium in the economy, probably leading to a lowering of the price of the metal and, ceteris paribus, leading to an increase of usage. Policy measures to tackle this problem should therefore focus on reducing the use of virgin zinc and on safe storage of cadmium waste, rather than only on cadmium recycling. A rather similar situation applies to mercury, which, besides continued metal extraction in countries such as Russia and Spain, is also produced and sold as a by-product of natural gas.

12 Analysis of Physical Interactions

Finally, an MFA study can be based on a dynamic model in which time is also included as a variable in the process equations. In this way not only can the long-term steady state of a certain regime be calculated, but also the path towards this steady state and the time required to reach it. For a true scenario analysis this option is the most suitable. It permits detailed analysis of time paths resulting from changes in patterns of consumption and technological innovation or due to the time lags between production, use and waste management. However, this option also has the greatest data requirements: a complete overview of stocks and flows for the initial year, and time-specific relations between the different flows and stocks. Given the difficulty of obtaining reliable data on all these points, the resulting projections may be of only limited accuracy. A dynamic approach is therefore not automatically preferable to a static approach. At the same time, however, it should be borne in mind that in steady-state models, too, there will be uncertainties in the mathematical relationships and their coefficients.

An example of dynamic modelling is presented in Fig. 12.6, which shows the global stock of copper products in the economy, with a distinction between short-lived and long-lived products. The stock of the short-lived products seems to level of at about 200 million metric tonnes, the stock of the long-lived products does not seem to have reached a steady state with 3 billion metric tonnes by the end of this century. These figures do get more significance if we compare them with data on the copper reserves. In 1999 the total world (economically

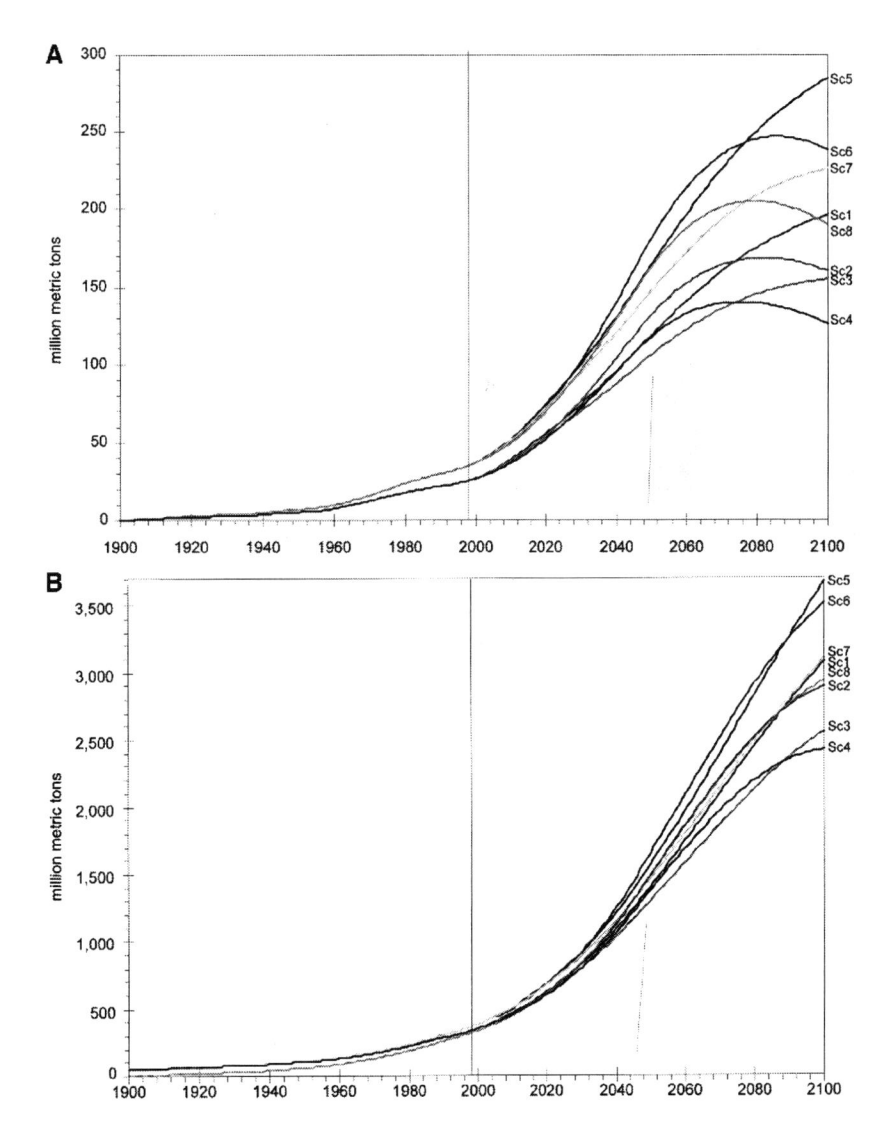

Fig. 12.6 Global stock of (A) short-lived and (B) long-lived copper products, 1900–1998. Eight scenarios
Source: After Ayres et al. (2003)

extractable) reserve was estimated at 340 million metric tonnes, and the total reserve base at 650 metric tonnes. The presently sharply increasing price of copper is a foretoken of future scarcity and should work as a driver for replacement and conservation.

MFA has its core applications in sustainable metals management. Typical issues include for instance: the depletion of reserves, like that of copper or platinum, but also of rare metals like that of indium, bismuth or ruthenium; the limits of metals recycling; the increase of metal stocks in the economy, and their impending risks; dematerialization of metals turnover; and prospects for more sustainable metals industries (Von Gleich et al. 2006).

12.4.6 Interpretation

For interpreting the results of an MFA study different types of indicator can be distinguished. The first type measures the overall efficiency of processes within the economy. A wide variety of indicators are at hand here, all of which can be subsumed under the term 'driver indicators'. They include total resource inputs, fraction of secondary materials used, recycling rate, process efficiency (i.e. ratio between process output and process inputs), degree of substance dissipation in the economy, degree of accumulation, or the volume of waste generated. Underlying drivers, like size of population or level of consumption, are outside the scope of analysis of MFA.

The second type of indicator measures the exchanges with the environment in terms of the environmental extraction of resources as well as emission of hazardous substances. These are known as 'pressure indicators'. The third type, the 'response indicators' measure the environmental consequences due to the extractions of emissions (either at state or impact level of the DPSIR framework). These can be based on the contributions of the given substances to a set of environmental problems or environmental impact categories, such as climate change, ozone depletion, acidification or toxicity (cf. Section 12.5.6).

As an example, let us consider the results of an SFA study on chlorinated hydrocarbons in the Netherlands, one of the aims of which was to assess the contribution of the various compounds or groups of compounds to ambient ecotoxicity (based on so-called toxicity equivalency factors; cf. Section 12.5.6) (Kleijn et al. 1997). Comparison of the various stages of the life cycle of the given compounds (here: production, use, transport and waste treatment) shows that by far the greatest toxic impacts occur during product use, where emissions are predominantly from the intentional use of chlorine-containing pesticides, responsible for about 10% of total ecotoxic impact in the Netherlands. In comparison the chemical industry appears to be a relatively minor polluter, as far as its production processes are concerned, although this does not, of course, relieve the industry of responsibility for use of its products after they have left the factory gate. Another point of interest is that these results are based on about 99% of chlorine flows in the Netherlands. The results of this study have stimulated some researchers to focus further research on the last 1%, which pose a potentially high risk, particularly in terms of ecotoxicity and human toxicity, in the production stage as well as elsewhere.

At the moment a wide variety of driver, pressure and response indicators are in use within different institutional settings, and defined in different ways. Interpretation of MFA results will be facilitated by harmonisation of the indicators used. Thus, a set of driver indicators may consist of the total level of use of a material or substance, the total extraction of virgin material, the fraction of secondary material used, the recycling rate, the efficiency of processes, the accumulation in the economy, the dissipation in the economy, total wastes and emissions to the environment, and the pollution exported outside the boundary of the system concerned. Pressure indicators will focus on specific extractions from the environment and emissions into the environment. And finally, a set of response indicators may consist of the environmental accumulation, the environmental concentration in different media, the daily intake and the different types of damage.

12.5 Life Cycle Assessment: A Tool for the Analysis of Product Systems

12.5.1 Short History

Life Cycle Assessment, or LCA, analyses the environmental impacts associated with products, or more precisely, as we will explain later, product systems. LCA has its origins in the early 1970s, when LCA-type studies were performed in a number of countries, in particular Sweden, the UK, Switzerland and the USA.

The method had its roots in energy and waste management, and the products given greatest attention in this initial period were beverage containers and diapers, topics that predominated in LCA discussions for a long time. During the 1970s and 1980s numerous life cycle studies were performed, using different methods and without a common theoretical framework. The consequences were largely negative, LCA being applied directly in practice by companies keen to substantiate their market claims. Even with the same objects of study there was often marked variation in the results obtained, which has hampered development of LCA into a more widely accepted analytical tool.

Since around 1990, there has been a growing exchange of knowledge and experience among LCA experts. Under the auspices of the Society of Environmental Toxicology and Chemistry (SETAC) efforts were undertaken to harmonise methodology. Since 1994 the International Organization for Standardization (ISO) has also played an important role. While the work of SETAC is primarily scientific, focusing on methodological development, ISO is formally handling the task of methodology standardisation, as now laid down in the ISO 14040 series of standards. And since about 2000 UNEP has also played its part, mainly in stimulating the application of life cycle approaches in practice.

As LCA methods become more sophisticated, dedicated software packages and databases are also being developed. For results to be acceptable to potential users, though, standard procedures are also an essential prerequisite. Part and parcel of this development is that there has grown a major need for input by key stakeholders in the LCA process, and for an independent peer review of the results of an LCA study. In the ISO standards these procedural elements are included as mandatory for public comparative studies, i.e. LCA studies concerned with comparison of several competing products and the results of which are made available to the public. This section describes the state of the art in LCA methodology as well as ongoing developments and future perspectives. An illustration of the LCA technique is included in Chapter 17.

12.5.2 Definition and Applications

The ISO 14040 standard defines LCA as follows: 'LCA is a technique for assessing the environmental aspects and potential impacts associated with a product by compiling an inventory of relevant inputs and outputs of a system; evaluating the potential environmental impacts associated with those inputs and outputs; and interpreting the results of the inventory and impact phases in relation to the objectives of the study'. The term 'product' should be understood here as referring to either goods or services providing a given function. LCA may also have other types of object of study, such as an entire industrial sector or the functioning of a firm as a whole. In the following, however, we shall refer simply to products, as a *pars pro toto* for all objects of LCA studies, unless specified otherwise.

More specifically, the reference point of an LCA study is the *function* provided by the product(s) of interest, and it is this function that serves as a basis for inter-product comparison. In LCA all the environmental impacts occurring during the 'life cycle' of a product are analysed and quantitatively related to this function. The life cycle is formed by the entire chain of industrial and household processes associated with the product, 'from cradle to grave', i.e. from resource extraction through to final waste disposal. This set of unit processes, interlinked by material, energy, product, waste and/or service flows is referred to as the 'product system'. It forms part of the wider, economic subsystem that comprises the following main activities: mining of raw materials; production of materials, products and energy; use and maintenance of products; waste treatment and processing of discarded products; and transport. Wherever possible the product system is studied using quantitative, formal mathematical methods.

LCA can be used in different contexts. It can be applied in an *operational* context for product improvement, product design or product comparison. Examples of the latter include the underpinning of environmental labelling programmes, like eco-labels for consumers or Environmental Product Declarations (EPDs) for business-to-business agreements. These programmes can also support green procurement by governmental agencies. LCA can also be applied at a *strategic* level, by private companies or public authorities. In the corporate context LCA may help guide business strategies, including decisions on what types of products to develop, what types of resources to purchase and what types of investments to make in waste management. Public authorities may use LCA for comparing and contrasting environmental policy options in such areas as waste management, energy policy, transportation or the building sector. Typical questions addressed include the following. What are the environmental gains of composting organic waste compared with fermentation or incineration with energy recovery? What is the preferred application of thinning wood: electrical power generation or paper production? How do biomass-for-energy programmes affect climate change? What mode of freight transport is environmentally preferable: road or rail, distinguishing different levels of technological improvement? What requirements should be set on sustainable building activities, in terms of both the energy characteristics of the building(s) in question and the construction materials used?

Use of LCA in public comparative studies makes the highest demands on methodology and data and on procedural provisions. One example is the use of LCA in the consumer oriented eco-labelling programmes mentioned earlier. These programmes are often associated with studies that meet with public criticism; they may also lead to disputes among commercial firms. This is far less the case with the business-to-business EPDs, or with in-house company applications aimed at product improvement or new product design, or studies that seek to improve a company's overall environmental strategy.

12.5.3 Technical Framework

In order to make LCA a workable tool for comparative purposes, the first step required is to establish a standard technical framework and terminology. As mentioned above, it is this task that ISO has taken upon itself. The ISO framework consists of the following phases: Goal and scope definition, Inventory analysis, Impact assessment, and Interpretation (see Fig. 12.7). As can be seen from the figure, LCA is not a linear process that simply starts with the first and ends with the last phase. Instead it often follows an iterative procedure, at increasing levels of detail. The individual phases of an LCA will be described below. The standardisation of these phases is laid down in the new ISO 14044 standard.

12.5.4 Goal and Scope Definition

The Goal and scope definition phase of an LCA includes the following elements: specification of the purpose and scope of the study; definition of the functional unit (as explained below); establishment of the level of detail required for the application at hand; and elaboration of a procedure for ensuring due quality of the study. Specification of the purpose and scope of the study includes a choice as to which products will and which will be not included in the analysis, a point that may lead to serious debate. In addition, the 'functional unit' must be defined: the quantified function provided by the product system(s) under study, for use as a reference value throughout the LCA study. This definition should be established with great care. It would be incorrect, for example, to

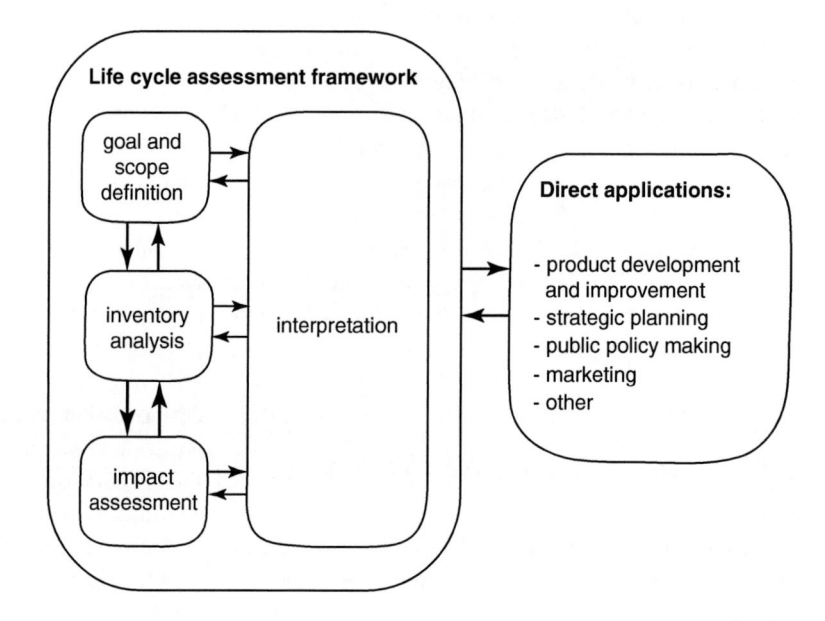

Fig. 12.7 Technical framework for Life Cycle Assessment
Source: ISO 14040

compare one milk bottle with one milk carton, as the former is used many more times than the latter. 'The packaging of 1,000 l of milk' may be a better functional unit. Still better may be 'the packaging required for the consumption of 1,000 l of milk', which also takes losses from breakage into account. In comparing different types of paint the unit of '1 litre of paint' will clearly be inadequate. 'The paint required for the painting of 1 m^2 of wall' is already a better functional unit; still more precise is 'the paint required for 1 m^2 of the wall over a period of 10 years', as this also takes the durability of the painted surface into due account. The definition of the functional unit also determines what type of alternatives are to be included in the analysis. '1 km of transport' may possibly include all transportation modes, '1 km of car transport' clearly not. The functional unit will generally have no significance in absolute terms, serving solely as a reference value for analysis purposes. In public comparative applications, all stakeholders should be able to participate in the process of specifying the exact product function under study, identifying which products adequately serve this function and defining the functional unit. All the processes to be investigated in an LCA study (usually referred to as unit processes) are quantitatively related to this functional unit in the Inventory analysis (see below).

As already mentioned, the reliability of an LCA study depends not only on its level of detail but also on procedural aspects, including in particular those relating to quality assurance. The latter may be provided for internally, but may also be performed by an external, independent panel of experts or panel of stakeholders. For public comparative assertions the latter is a strict ISO requirement. All decisions on the degree and form of participation and critical review are made in the Goal and scope definition phase.

12.5.5 Inventory Analysis

The Inventory analysis phase is the most objective and generally also the most time-consuming part of an LCA study. It comprises the following steps: drawing a flow chart of all the unit processes involved; defining the boundaries between the product system and the environment; specifying the unit processes and associated data gathering; allocating multiple processes; and compiling the so-called inventory table. If a study is performed up to Inventory analysis only, it is referred to as an LCI study, i.e. a Life Cycle Inventory study.

The *flow chart* identifies all the relevant and interlinked processes of the product system with their associated material and energy requirements. In drawing up the chart, product use forms the basic point of departure; from here, the processes ramify 'upstream' to product production and to processing and extraction of the various constituent raw materials, and 'downstream' to various forms of waste management. An example of a flow chart is given in Fig. 12.8.

In preparing the flow chart, system boundaries must be defined between the product system (as part of the economy) and the environment. This means that the flows across these boundaries have to be defined. Definition of these so-called environmental pressures is not as easy as one might think, however. Is wood a resource brought into the economy from the environment, or is a forest already part of the economy? Similarly, is a landfill to be regarded as part of the environment or as part of the economy? In the first case all landfilled materials must be regarded as emissions into the environment; in the latter this will hold only for emissions from the landfill to air and groundwater.

Having established the system boundaries with the environment, the next step is *specification of the unit processes* and *data retrieval* for these processes. This is the most time-consuming activity. It involves collecting quantitative data on such diverse issues as electricity use for steel production, carbon dioxide emissions of heavy trucks and toxic releases by paper producers. To keep data collection within manageable proportions, there is a need for a cut-off procedure determining which of the unit processes associated with the product system can be omitted if the efforts required to retrieve relevant data are inordinately large compared with the relative contribution. When analysing the impacts of car use, for instance, it might be decided to ignore the effects of a car radio. A more recent development is the use of Input Output Analysis (see also Chapter 11) for this purpose, translating data on monetary flows in specific economic sectors into a set of key environmental indices, thus enabling to avoid the above mentioned cut-off and to in principle include all connected processes (Suh and Huppes 2005).

Fig. 12.8 Example of a flow diagram in an LCA study on cars. The dashed line is the system boundary, boxes represent processes, arrows within the system boundary represent flows of products, materials, energy and waste, and arrows outside the system boundary represent flows to and from the environment and to other product systems

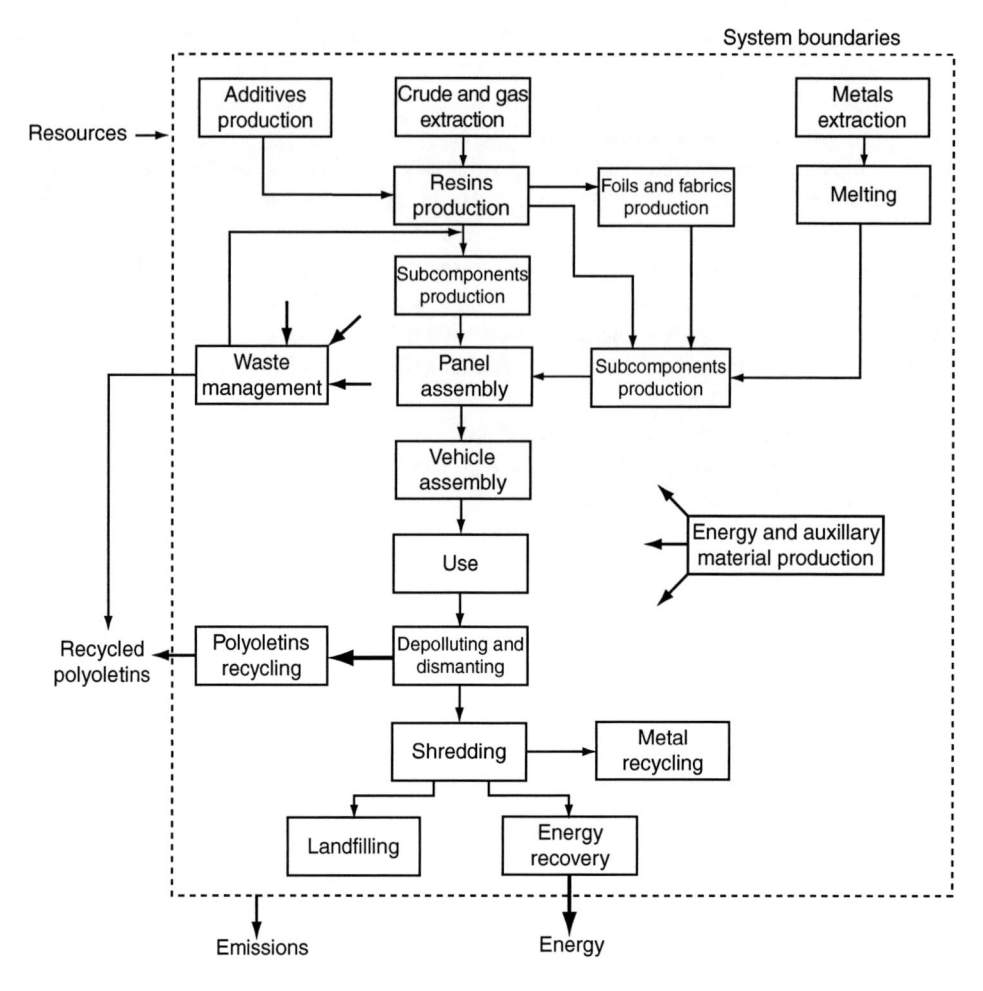

Another problem faced in data retrieval is that data are often obsolete, highly variable or secret. There are literally thousands of different technological specifications in existence for producing electricity. When the product system of interest involves electrical power supply, it seems logical to work with a weighted average for the technologies in current operation according to their share in the national grid. This partly solves the variability problem. To a growing extent the problem of limited data availability is being resolved through the establishment of large databases containing process data for many basic materials and processes. For more specific processes such databases are of less help, however, also due to the secrecy issue.

In LCA, one must quantitatively relate the different processes to one another, quantifying them each in relation to the magnitude of the defined functional unit as reference. It should be noted that the processes modelled in the Inventory phase may be sequentially linked or, alternatively, looped ('the electricity required for producing the steel required for electricity production'). The basic calculation procedure employed in LCA is fundamentally different from that of MFA and HERA. While LCA includes processes to the extent that they contribute to the defined functional unit, in HERA and MFA, in contrast, the processes are always included in their entirety. As mentioned earlier, a functional approach may also sometimes be adopted in MFA (i.e. in the so-called 'rucksack' concept), but the analysis then essentially follows the structure of LCA. In fact, we may reinterpret functional MFA as LCA, although not entirely along ISO-lines.

One specific issue in Inventory analysis concerns the procedure adopted for allocating the environmental pressures associated with so-called 'multiple processes', i.e., processes providing more than one function. This is known as an *allocation procedure*. Consider the example of a refinery producing a whole array of outputs. If petrol is the only product required by a given product system, how should the environmental impacts of the refinery processes be allocated to this petrol? This is a typical, and difficult, question for LCA that is not encountered in ERA or SFA. There are, typically, three types of multiple processes for which allocation is necessary. The first is co-production, as in the above example of a refinery. The second is combined waste treatment, when waste flows generated by different product systems are processed in one and the same waste management process. The third is recycling, when waste flows from one product system are used as a secondary material in another product system. The question is always: how to properly attribute the environmental pressures associated with the multiple processes to the different product systems involved.

A single, coherent allocation framework should be developed for dealing with all multiple processes. This is a key issue in ISO standard 14044, regarding inventory methodology. The prescribed procedure does not imply a specific choice on the part of ISO but rather a prioritising of different options, to be assessed sequentially for applicability:

1. Check whether allocation can be avoided by dividing processes into sub-processes
2. Check whether allocation can be avoided by expanding the boundaries of the system, that is, by including more than one function
3. Apply principles of physical causality for allocation of the burdens and finally
4. Apply other principles of causality, for instance economic value

Although the general thrust of this guidance is clear, the ramifications of practical application are still being debated and refined with particular focus on appropriate criteria for deciding whether or not a given option is acceptable.

The Inventory analysis concludes with *compilation of the inventory table*, an aggregated list of all the environmental pressures (extractions, emissions and types of land use) associated with the product systems under investigation. Apart from quantitative entries, the inventory table may also include issues that cannot be dealt with quantitatively but must still be considered in the final interpretation of results.

12.5.6 Impact Assessment

In ISO 14044 life cycle impact assessment is defined as the 'phase of life cycle assessment aimed at understanding and evaluating the magnitude and significance of the potential environmental impacts of a product system'. In this phase the environmental pressures of the product system listed in the inventory table are translated by means of simple conversion factors into terms of the various types of environmental impacts they cause. For practical reasons these impact data are aggregated, because a list of 50 or 100 or so items is too long for the decision-making context. The wide diversity of emissions and other environmental pressures are thus assigned to a smaller number of 'impact categories', such as climate change, acidification and human toxicity.

In the 1970s impact assessment was performed in an implicit fashion, by defining a number of broad, inventory-based parameters considered indicative of the overall spectrum of impacts. These parameters included net energy consumption, total resource input and total solid waste output. These approaches are time-efficient and can yield robust results. However, various types of impact may not be adequately covered by such a small number of inventory-based indicators.

Since the mid-1980s a variety of methods has been developed for aggregating the diverse environmental impacts of chemical and other substances into a manageable number of impact categories. Guidance on this point is given by ISO standard 14044, which provides a stepwise procedure intended to separate as far as possible the scientific and the normative (i.e. value-based) steps involved in Impact assessment. In the *selection and definition* step a set of key impact categories is selected and defined, together with the underlying characterisation models to be used for relating the environmental interventions from the previous phase to specific impact categories, thus permitting aggregation of results within each category. Box 12.2 presents a fairly standard list of impact categories, which can be used both as a framework for analysis and as a checklist for completeness. It is worth noting that several familiar impact categories are not included in the table, in particular energy consumption, resource use, solid waste production and land use. The first three of these categories involve flows of substances and materials that are not followed up to the product system boundary and which should, if possible, be translated into environmental pressures related to the categories of Box 12.2. If this is not possible, these categories can be taken along as supplementary categories. The latter category, land use, is highly relevant, but can only to a limited extent be included in the LCA format of analysis.

A key element of this step of LCA is the selection of appropriate characterisation models for the respective impact categories. For the output-related ('pollution') categories this modelling should encompass both the fate and effect of the substances emitted. Fate modelling maps pollutant distribution across the various environmental media and physicochemical and/or biological persistence in each. Once again, steady-state multimedia models often form the basic point of departure. The models can be used to derive the effects of the emission of one additional unit of pollutant or of the extraction of one additional unit of resource. The derived numbers, referred to as characterisation factors, are tabulated in several guidebooks and software for LCA.

> **Box 12.2 Impact categories for life cycle impact assessment**
>
> **A. Input-related categories ('resource depletion')**
>
> 1. Extraction of abiotic resources
>   - Deposits like fossil fuels and mineral ores
>   - Funds like groundwater, sand and clay
>   - Flows like solar energy, wind and surface water
> 2. Extraction of biotic resources (funds)
> 3. Land use
>   - Increase of land competition
>   - Degradation of life support functions
>   - Biodiversity degradation due to land use
>
> **B. Output-related categories ('pollution')**
>
> 4. Climate change
> 5. Stratospheric ozone depletion
> 6. Human toxicity (incl. radiation and fine dust)
> 7. Ecotoxicity
> 8. Photo-oxidant formation
> 9. Acidification
> 10. Nutrification (incl. BOD and heat)
>
> Source: After Udo de Haes et al. (1999)

In the *classification* step, data on extractions, emissions and land use are assigned qualitatively to relevant impact categories. Thus, SO_2 is assigned to at least acidification and human toxicity, and possibly also to climate change, due to the limiting impact of sulphate aerosols on radiative forcing.

In the *characterisation* step, the inventory data, having been assigned qualitatively to impact categories, are multiplied by the characterisation factors selected in the selection step to yield a quantitative figure. With these factors the interventions associated with each of the substances in the inventory table can be expressed in terms of the common units of the category indicator. Examples include so-called global warming potentials (GWP) for climate change and ozone depletion potentials (ODP) for stratospheric ozone depletion. Although being the general aim, characterisation factors cannot be defined on the basis of scientific knowledge alone. They also have to be based on value judgment, to a greater or lesser extent. This is even the case for the widely accepted GWP values, based as these are on possibly debatable scenarios or background levels (ambient levels) of different greenhouse gases, and on the choice how far into the future these scenarios are taken. A pragmatic approach to this problem is to employ characterisation factors authorised by an international body such as UNEP and to ensure that these factors are updated at regular intervals.

The next step is *normalisation,* whereby the results for each of the impact categories are expressed relative to a well-defined reference value, for instance the total magnitude of the environmental pressures in a certain area in a certain year for that category. In this way the relative contributions of individual product systems to a particular impact category can be calculated and compared.

In the *weighting* step the (normalised) indicator results for each impact category are assigned numerical factors according to their deemed relative importance, multiplied by these factors and possibly aggregated. Such weighting can be done on a case-by-case basis, or by using a generic set of weighting factors. Although weighting is based as far as possible on formalised, quantitative procedures, the choice of weighting principles and factors is obviously a delicate and highly subjective area. Among the principles currently used are: translation to monetary units, (using Willingness-to-pay- or Collectively Revealed Preference methods), comparison with environmental standards (using 'Distance-to-Target' methods), and use of so-called social panels.

The resulting aggregate environmental indices are very practical, particularly for the eco-design of products. They permit rapid comparison of materials which have all their environmental characteristics expressed in one single number. According to the ISO guidelines, these weighted indices are not to be used for the so-called public comparative assertions, because of the value choices involved.

Figure 12.9 presents an example of an LCA study on UMTS mobile communication. Over the full life cycle of UMTS nearly 50% of the total environmental impact is caused by the mobile phones; nearly 25% by the base station; and the rest by the antenna, the switching system, the connection network and the administration. These results are being presented in terms of the so-called EcoIndicator, a special form of weighting including the use of social panels. In Fig. 12.9a the impacts of the UMTS mobile phones are compared with those from a GSM mobile phone. For both types of mobile phones the main impacts are in the production phase, those of UMTS being clearly larger. However, strictly speaking both devices do not provide the same function; per transferred amount of data UMTS is ecologically preferable. Figure 12.9b shows the impacts of the base stations of the two devices. For both, the use phase is dominant due to electricity use, with a larger difference between the two devices.

12.5.7 Interpretation

The Interpretation phase is designed to help the LCA practitioner or user formulate conclusions on the outcome of the LCA study. To this end the results are compared with the initially defined goal of the study. Is additional analysis required, or does the goal perhaps need to be redefined? More specifically, the interpretation phase deals with evaluating the assumptions

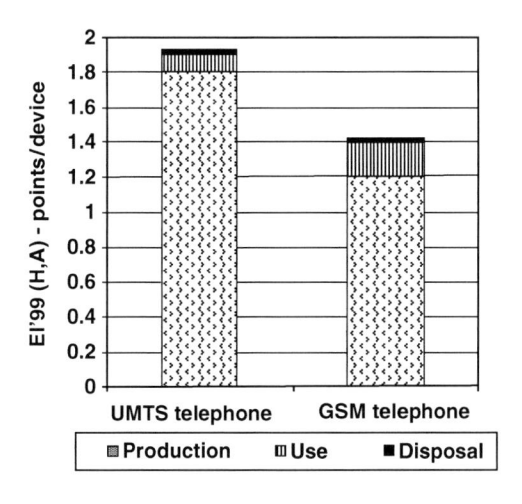

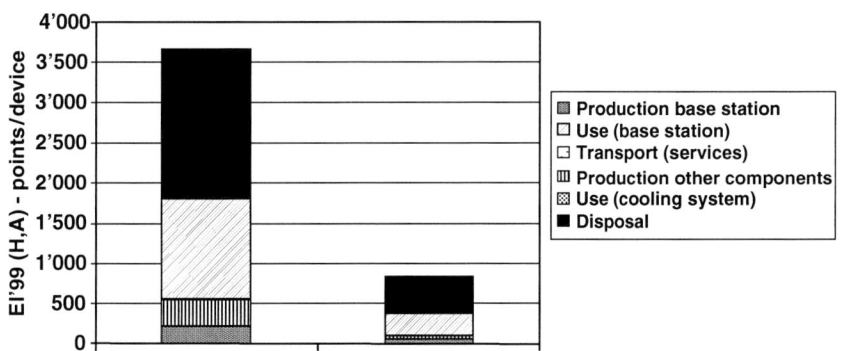

Fig. 12.9 Attribution of EcoIndicator points for a UMTS and a GSM phone (A) base station (B) to the different life cycle phases (production, use in Switzerland, disposal) Source: Faist Emmenegger et al. (2006: 273)

that have been made as well as the quality of the data and models used. How robust are the results? What has been ignored? Might a more sophisticated LCA study yield contrasting results? Interpretation also includes dealing with the uncertainty of the results (see for this Section 12.6.5 below). The Interpretation phase may also prompt a new round of data collection if the predefined goal does not appear to have been satisfactorily achieved.

12.6 Evaluation and Prospects

Having considered several tools for environmental analysis, some more closely than others, let us now evaluate them with reference to the following five questions: (1) What are the limits to the usefulness of analytical tools? (2) When should models based on physical parameters be used, and when are monetary models (cf. Chapter 11) preferable? (3) How far along the 'environmental mechanism' should category indicators be defined? (4) What scope and level of detail should be aimed for? (5) What procedures should be adopted for dealing with uncertainty in data? And, finally: (6) What are the prospects of the tool contributing to the developing field of industrial ecology?

12.6.1 Limits to the Usefulness of Analytical Tools

The ultimate aim of all these analytical tools is to support decision-makers, in both the public and private sectors. But before going into the questions dealing with the application of the different tools, there may be more fundamental problems at stake. Analytical tools may involve paradigmatic assumptions that are not shared by all the stakeholders in the process or activity in question. For example, the general public often rejects a risk assessment approach to nuclear power facilities, because for this technology there is a large divergence with data on perceived risks. The same holds true for resistance to genetically modified organisms. Likewise, in the case of the chlorine industry and the controversies about its future, LCA has failed to move the public debate forward in either Sweden or the Netherlands (Tukker 1998). Although technical uncertainties will certainly have played a role, the technology under scrutiny met so much inherent resistance that a rational, analytic approach could not be of any avail. An important question is then whether a new phrasing of the questions, together with changes in the tools to be used and with a planning process with more public involvement, will be able to solve the point. For the moment we must conclude that the use of quantitative, analytical tools presupposes an agreement on certain paradigmatic starting points. Having said this, we shall now discuss several rather more technical issues.

12.6.2 The Choice Between Physical and Monetary Models

The choice between modelling in monetary as opposed to physical terms has been an issue of debate since the 1970s. This holds true for the modelling of both unit processes (within the economy) and environmental processes. In the tools discussed in the present chapter, the flows between economic processes have been expressed in terms of physical parameters. In contrast, in tools such as Input-Output Analysis (IOA) or Cost-Benefit Analysis (CBA) these are expressed in monetary terms (see Chapter 11). Physical terms have the advantage of being amenable to presentation in the form of a mass balance and also of relating directly to environmental pressures. Monetary terms have the advantage of working in a single unit, which permits easy aggregation, and high data availability, although only for flows with a price tag attached. Even waste flows, just flows with negative product values, cannot be easily incorporated in these models.

Similar reasoning holds for environmental flows. Environmental impacts can also be described in either physical or monetary terms. Modelling in terms of physical parameters stays closer to physical reality, and avoids the assumption that everything of human value can be expressed in market terms. Thus, expressing human life in monetary terms is often perceived as a degradation of the value of that life. On the other hand, translation into monetary units puts the results in terms very close to the practice of decision-making. If this can be feasibly achieved, the relevance of the results will generally increase. In addition, use of monetary values can be seen as the ultimate step in aggregation, expressing all impacts into a single index.

A number of the methods for monetarising environmental impacts have been discussed in Chapter 11. Here we wish to make two additional remarks. The first is that some impact categories are more amenable to monetarization than others. For damage to crops or buildings this approach may well be appropriate, while for damage to human health it is under debate. Damage to ecosystems is most often left out of consideration altogether, given the general lack of clarity as to an ecosystem's economic values and how these should be measured. The second point is that not all – or perhaps even: only a few – decisions require aggregation of the information into a single index. Often, there is an explicit need for information about individual categories of environmental impact. In such situations monetarisation is not appropriate, as it will conceal the nature and scale of impacts as well as underlying modelling assumptions.

12.6.3 The Location of Category Indicators in the Environmental Causal Chain

Our third point of general evaluation concerns the question how far along the environmental causal chain (the 'environmental mechanism' according to ISO-LCA) to choose the category indicators used to express modelling output. Examples of modelling early in the causal chain concern the changes in climate forcing due to greenhouse gases or the release of protons due to acidifying substances. Alternatively, modelling outputs can be defined at the so-called impact or damage level, such as damage to human health, ecosystems, crops, man-made materials or works of art. In fact, there is a fundamental choice at stake here. Either modelling is taken up to the pressure variables themselves or to some midpoint level. The results are then nearly science based, will be relatively certain and will involve a broad range of impacts. But these results will relate to *potential* impacts only, in the sense that it remains obscure whether impacts will in fact occur further down the 'environmental mechanism', at the impact or damage level of the DPSIR framework. In addition, aggregation (when desirable) can only be performed by experts, conversant with the full range of potential impacts. The second option is to continue modelling up to the impact or damage level. The results will then be less certain and will generally be limited to certain selected items for which such modelling is best feasible. Where such results are available, though, they will be better understandable. Despite inherent limitations, as discussed in Chapter 11, in some cases conversion to monetary terms is then performed, using market prices, contingent validation (that is, "willingness-to-pay") methods, or preferences revealed by government expenditures. Current research efforts are focusing increasingly on modelling to the impact or damage level for all categories, in order to make both options possible.

12.6.4 Scope and Level of Detail

LCA aims to be 'all-encompassing' with respect to the types of impact it endeavours to analyse. In practice, however, there is always a bias towards emissions and resource extractions. Although land use is by far the most important factor affecting biodiversity, it has proved difficult to relate this pressure variable and its consequences for biodiversity to a functional unit. But apart from these methodological limitations, many policy studies just take CO_2 as the main indicator for environmental burden, ignoring other types of impact.

As to the desirable level of detail, economic analysis acknowledges the useful and widely accepted difference between macro-, meso- and micro-economics. Macro-level studies are concerned with processes occurring in society as a whole, possibly distinguishing between individual economic sectors; meso-level studies with interactions among these sectors, or the consequences of changes in consumer preferences or of certain external price developments; and micro-level studies with the activities of single firms or other actors, with their respective aims and motives. In environmental analysis there is an urgent need for a similar division into levels, but no such division is yet generally accepted. Thus, any study performed at the allegedly macro- or meso-level is likely to be subject to the criticism that impacts at some localised level are still not sufficiently well understood, and that the study cannot therefore provide a sound basis for decision-making.

12.6.5 How to Deal with Uncertainty?

As already mentioned, the use of analytical tools will generally be accompanied by various areas of uncertainty, under the main headings of data uncertainties, methodological assumptions, scenario assumptions and value choices. Let us consider a few examples. In SFA, there is technical uncertainty with regard to the empirically established coefficients for a number of trace substances such as dioxins from combustion processes. Apart from data uncertainty, there may also be gaps in knowledge. Thus, on more than one occasion allegedly safe chemicals like CFCs have turned out to have major environmental impacts. With respect to methodological uncertainty, a key assumption in ERA is that extrapolations are to be performed in moving from the laboratory to field conditions. Another methodological assumption is that most if not all of the modelling relationships employed in the various tools are assumed to be linear, while relationships in the real world are often within a relatively small range, tending to deviate from linearity at a certain point. Scenario uncertainties occur when, in predicting future environmental effects, we must employ projections of, say, the world population or the number of cars in a certain country. Value choices, finally, have to be made in LCA, with respect to the selection and prioritisation of impact categories, most explicitly in the weighting step of this tool.

There are a number of options available for dealing with these various types of uncertainty, which will be described in the following sections. However, it should be borne in mind at all times that analysis is never simply a question of mechanical application of rules, just as justice is not to be found in the mechanical application of laws. In environmental analysis, as elsewhere, human judgement and debate remain essential ingredients. The function of analytical tools is to support decisions, not to make them.

12.6.5.1 New Measurements

Faced with uncertain indicator values, the first option is to make new measurements. This may be done by performing new dose-response experiments in the laboratory, through field validation of laboratory-to-field extrapolations of multi-media dispersion model results, or by carrying out new field experiments on, say, ecosystem impacts. This may be termed the high road of uncertainty abatement. However, it is timely and costly and will be not be a realistic option for an individual case study.

12.6.5.2 The Choice of Robust Indicators

A second possibility is to choose indicators that are more robust, i.e. relatively insensitive to (re-evaluation of) data, modelling assumptions and model design, methods and value choices. Increasing modelling detail will improve the accuracy of results, but at the same time generally increase uncertainty margins. Conversely, use of more robust indicators will be at the expense of accuracy. For example, the effectiveness of policies on chlorine chemicals can be assessed in terms of the practical leverage they provide on impact category indicators, such as those used in LCA. These have quite a high power of resolution, but at the same time many of them are relatively uncertain. Alternatively, these policies can be assessed in terms of the total kilograms of organic chlorine emitted, as is generally the case in MFA studies. Such a metric is very robust. On the other hand, this approach obscures very important differences between the chlorine compounds emitted. Going one step further, one may even leave quantification behind altogether and opt for qualitative indicators such as 'made from recyclable material' or 'biodegradable'. There are, however, important scientific objections to such crude rules of thumb.

12.6.5.3 Uncertainty Analysis

Given a set of indicators, the uncertainties they comprise can be assessed by calculating standard errors, although in practice this approach is rarely adopted. Such errors may arise in any of numerous links in the chain of processes underlying the indicator in question and may, furthermore, relate to uncertainties in data, to methodological assumptions and/or to value choices regarding any of these different links. The results of uncertainty analyses consequently soon become extremely complex and may well pile uncertainty on uncertainty. A more sophisticated approach is Monte Carlo simulation. For every element in the uncertainty of a given indicator the probability of different values is estimated. A series of computational runs is subsequently made, in which the various elements of uncertainty are fixed independently, each according to their own probability distribution. The final result will show a more realistic range of outcomes that avoids artificial accumulation of uncertainties, Unfortunately, however, realistic data on probability distributions are rarely available. The applicability of this approach is restricted by the fact that information on the uncertainty of data items is very hard to obtain; indeed, one must often be content with merely finding the data items themselves available.

An entirely different approach to uncertainty analysis is to use so-called safety factors. These are standard factors, such as those used for assessing the 'predicted no effect concentration' (PNEC) of toxic substances, for example. The smaller the number of species from which the PNEC is derived, the higher the factor by which the result must be divided.

12.6.5.4 Sensitivity Analysis

If no data errors can be found and if no corrections can be made for suspected elements of uncertainty, a sensitivity analysis can be performed by experimentally modifying the modelling set-up in various ways. Thus, changes can be introduced in the input data, the methodological assumptions or the value choices made in individual methodological steps. The effect of such changes on the final result can then be evaluated. This procedure is adopted fairly frequently, as it makes only modest requirements on study resources and still provides important insight into the sensitivity of final results.

12.6.5.5 Scenario Analysis

While sensitivity analysis generally addresses the data, methods and value choices associated with individual parameters, in scenario analyses choices on any or each of these are amalgamated into consistent packages. Thus a worst-case, best-case or most-likely scenario can be constructed. Scenario analyses help structure the results of sensitivity analyses, making them more readily comprehensible for decision-making purposes.

12.6.5.6 International Standardisation

International standardisation in the field of analytical tools is focused predominantly on terminology, technical frameworks, general methodological guidance and procedural requirements. It may go a step further, however, to establish 'best available data' or 'best available practice'. The Intergovernmental Panel on Climate Change (IPCC), under UNEP authority, thus establishes the 'best available knowledge' regarding the climate change potentials of various greenhouse gases, while the World Meteorological Organisation (WMO) does the same for the ozone depletion potentials of relevant gases. Although considerable uncertainties as well as value choices are involved in such standardisation, this approach provides due guidance in practical applications and helps avoid arbitrariness in selecting appropriate data or models.

12.6.5.7 Procedural Checks

The above options for dealing with uncertainties all relate to technical (i.e. modelling) characteristics. A quite different approach starts from the other side, as it were, in the decision-making procedure in which the ensuing analytical results are to be used. For instance, the results may be reviewed by an independent panel of experts. If the results pass such a review procedure this may well contribute more to their credibility than any of the aforementioned technical procedures. For this reason, there is considerable focus at the moment on the scope for explicitly incorporating analytical tools in decision procedures in which independent experts have a clearly defined role. These procedural scripts also often define a role for stakeholders, for industry, government, environmental NGOs and so on. Although this kind of interactive process will indeed generally lead to better societal acceptance of the results, a starting point should be that stakeholder participation will focus on the value part of the analysis; 'science by voting' cannot be seen as an improvement from the scientific point of view.

12.6.6 The Contribution to Industrial Ecology

The assemblage of analytical tools, both monetary and physical, and together with the specialised knowledge embodied in the technical elements they comprise and the information on their specific applicability, constitutes a growing body of knowledge of increasingly scientific quality and consistency. Although the term 'toolbox' is often used in this context, it is somewhat misleading in that it suggests the ready availability of a collection of tools specifying how to go about environmental management. This is clearly not the case. The tools are all still under development and will indeed never achieve a 'finished' state. Assumptions and approximations will always have to be made, and data will never be complete. And there will always be paradigmatic differences between stakeholders, limiting the applicability of quantitative tools altogether. Within these constraints, however, a set of tools is being developed with which to rationalise environmental decision-making and render it more consistent. As information from a variety of scientific disciplines becomes increasingly integrated, a core body of knowledge is being formed for the new multidisciplinary field of industrial ecology.

References

Ayres, R. U., & Ayres, L.W. (2002). *A handbook of industrial ecology*. Cheltenham: Edward Elgar.

Ayres, R. U., Ayres, L.W., & Rade, I. (2003). *The life cycle of copper, its co-products and byproducts*. Dordrecht: Kluwer.

Cohen, B., & Lee, I. (1979). A catalog of risks. *Health physics, 36*, 707–722.

Faist Emmenegger, M., Frischknecht, R., Stutz, M., Guggisberg, M., Witschi, R., & Otto, T. (2006). Life cycle assessment of the mobile communication system UMTS – towards eco-efficient systems. *International Journal of Life Cycle Assessment, 11*(4), 265–276.

Fischoff, B., Lichtenstein, S., Slovic, P., Derby, S. L., & Keeney, R. L. (1981). *Acceptable risk*. Cambridge: Cambridge University Press.

Funtowicz, S. O., & Ravetz, J. R. (1990). *Uncertainty and quality in science for policy*. Dordrecht: Kluwer.

Kleijn, R., Tukker, A., & van der Voet, E. (1997). Chlorine in The Netherlands, Part I: an overview. *Journal of Industrial Ecology, 1*(1), 95–116.

Mackay, D. (2001). *Multimedia environmental models. The fugacity approach* (2nd ed.). Boca Raton, FL: Lewis.

Mayo, D. G., & Hollander, R. D. (Eds.). (1991). *Acceptable evidence, science and values in risk management*. Oxford/New York: Oxford University Press.

Murray, C. J. L., & Lopez, A. D. (Eds.). (1996). *The global burden of disease*. Volume I of global burden of disease and injury series, WHO/Harvard School of Public Health/World Bank. Boston, MA: Harvard University Press.

Slovic, P. (1991). Beyond numbers: a broader perspective on risk perception and risk communication. In D. G. Mayo & R. D. Hollander (Eds.), *Acceptable evidence, science and values in risk management* (pp. 48 – 65). Oxford/New York: Oxford University Press.

Smeets, E., & Weterings, R. (1999). *Environmental indicators. Typology and overview*. Copenhagen: European Environmental Agency.

Suh S., & Huppes, G. (2005). Methods for life cycle inventory of a product. *Journal of Cleaner Production, 13*(7), 687–697.

Tukker, A. (1998). *Frames in the toxicity controversy, based on the Dutch chlorine debate and the Swedish PVC debate*. Ph.D. thesis, Tilburg, The Netherlands.

Udo de Haes, H. A., Jolliet, O., Finnveden, G., Hauschild, M., Krewitt, W., & Müller-Wenk, R. (1999). Best available practice regarding impact categories and category indicators in life cycle impact assessment. *International Journal of Life Cycle Assessment, 4*(2), 66–74 and (3), 167–174.

Van de Meent, D., de Bruijn, J. H. M., de Leeuw, F. A. A. M., de Nijs, A. C. M., Jager, D. T., & Vermeire, T. G. (1995). Exposure modelling. In C. J. van Leeuwen, & J. L. M. Hermens (Eds.), *Risk assessment of chemicals: An introduction* (Chapter 4, pp. 103–145). Dordrecht/Boston/London: Kluwer.

Van der Voet, E., Kleijn, R., & Udo de Haes, H. A. (1996). Nitrogen pollution in the European Union, origins and proposed solutions. *Environmental Conservation, 23*(2), 120–132.

Von Gleich, A., Ayres, R. U., & Gössling-Reisemann, S. (Eds.). (2006). *Sustainable metals management; securing our future – steps towards a closed loop economy*. Dordrecht: Springer.

Wilson, R. (1979). Analysing the daily risks of life. *Technology review, 81*(4), 40–46.

Wrisberg, N., & Udo de Haes, H. A. (2002). *Analytical tools for environmental design and management in a systems perspective*. Dordrecht: Kluwer.

Further Readings

ERA

Crouch, E. A. C., & Wilson, R. (1982). *Risk/benefit analysis*. Cambridge, MA: Ballinger.

Leeuwen, C. J., van, & Hermens, J. L. M. (Eds.). (1995). *Risk assessment of chemicals: an introduction*. Dordrecht/Boston/London: Kluwer.

OECD (1989). *Report of the OECD workshop on ecological effects assessment*. Paris. OECD Monographs no. 26.

MFA and SFA

Adriaanse, A., Bringezu, S., Hammond, A., Moriguchi, Y., Rodenburg, E., Rogich, D., & Schütz, H. (1997). *Resource flows: The material basis of industrial economies.*, Washington, DC/Wuppertal/The Hague/Tsukuba, Japan: World Resources Institute/Wuppertal Institute/Netherlands Ministry of Housing, Spatial Planning and Environment/National Institute for Environmental Studies.

Ayres, R., & Kneese, A. V. (1969). Production, consumption and externalities. *American economic review, 59*(3), 282 – 297.

Brunner, P. H., & Baccini, P. (1992). Regional materials management and environmental protection. *Waste Management and Research, 10*, 203–212.

Fischer-Kowalski, M. (1998). Society's metabolism: the intellectual history of materials flow analysis, part I: 1860–1970. *Journal of Industrial Ecology, 2*(1), 6178. And: Fischer-Kowalski, M., & Hüttler, W. (1999). Part II, 1970–1998. *Journal of Industrial Ecology, 2*(4), 107–136.

Matthews (2000). *The weight of nations. Material outflows from industrial economies*. Washington, DC: World Resources Institute.

Von Gleich A., Ayres, R.U., & Gössling-Reisemann, S. (Eds.). (2006). *Sustainable metals management – securing our future – steps towards a closed loop economy: Eco-efficiency in industry and science* (Vol. 19). Dordrecht: Springer.

LCA

Curran, M. A. (Ed.). (1996). *Environmental life-cycle assessment*. New York: McGraw-Hill.

Guinée, J. B., Gorrée, M., Heijungs, R., Huppes, G., Kleijn, R., de Koning, A., et al.. (2002). *Life cycle assessment. An operational guide to the ISO standards. I: LCA in perspective. IIa: Guide. IIb: Operational annex. III: Scientific background*. Dordrecht: Kluwer.

International Organization for Standardization: Environmental management – Life cycle assessment – Principles and framework (ISO 14040, 1997); Environmental management – Life cycle assessment – requirements and guidelines (ISO 14044, 2006).

Goedkoop, M., & Spriensma, R. (1999). *The Eco-indicator 99. A damage oriented method for life cycle impact assessment*. The Hague, The Netherlands: DGM.

Heijungs, R., & Suh, S. (2002). *The computational structure of life cycle assessment*. Dordrecht: Kluwer.

Pedersen Weidema, B. (1997). *Environmental assessment of products*. Helsinki: Finnish Association of graduate engineers TEK.

Reijnders, L. (1996). *Environmentally improved production processes and products. An introduction*. Dordrecht: Kluwer.

Van den Berg, N. W., Dutilh, C.E., & Huppes, G. (1995). *Beginning LCA, a guide into environmental life cycle assessment*. The Netherlands: NOVEM, Utrecht.

Wenzel, H., Hauschild, M., & Alting, L. (1998). *Environmental assessment of products. Volume 1. Methodology, tools and case studies in product development*. London: Chapman & Hall. And: Hauschild, M., & Wenzel, H. (1998). *Environmental assessment of products. Volume 2. Scientific background*. London: Chapman & Hall.

Chapter 13
Environmental Policy Instruments

Gjalt Huppes and Udo E. Simonis

Contents

13.1 Introduction 239
13.1.1 Approaches to Governance 239
13.1.2 Outline of the Chapter 241

13.2 Policy Instruments in a Long-Term Perspective 244
13.2.1 The Future Context of Instruments 244
13.2.2 General Tendencies in Society 244
13.2.3 Globalisation Tendencies 247
13.2.4 General Tendencies in Environmental Problems 248
13.2.5 Prospects for Instruments 250
13.2.6 Strategic Instrument Choices Ahead 251

13.3 Policy Instruments: What Are They Good for? 254
13.3.1 Policy Instruments: What Are They? ... 254
13.3.2 Why Have Policy Instruments? 255
13.3.3 A Framework for Analysing Policy Instruments 255
13.3.4 Policy Instruments in Context 257
13.3.5 Environmental Problems: Causes and Solutions 259
13.3.6 Evaluation Criteria for Policy Instruments 261

13.4 Design, Analysis and Evaluation of Policy Instruments 264
13.4.1 A Framework for Design and Analysis 264
13.4.2 A Thousand Instruments Defined, with Examples 271
13.4.3 Choice of Instruments in Policy Design 273

References 279

G. Huppes (✉)
Institute of Environmental Sciences at Leiden University (CML), The Netherlands
e-mail: huppes@cml.leidenuniv.nl

13.1 Introduction

13.1.1 Approaches to Governance

What is the role of environmental policy instruments? In simplified terms, environmental policy instruments can be said to link policy development and decision-making to policy implementation. Starting from policy development, the policy problem is translated into operational goals, the appropriate instruments are chosen, and their implementation achieves the goals. This picture of policy as a linear, stepwise activity is an over-simplification. For instance, the definition of the policy problem is often already based on instrument choice, resulting in a circularity that disrupts the seemingly rational linear picture. Permits for the operation of installations tend to define problems in terms of the effects the installation has on its environment, for instance in terms of noise, stench and eutrophication of a nearby lake. Having defined the problem in this way, the choice of policy instrumentation is then more or less limited to variants of the permit, as emission taxes and liability rules simply do not fit the problem definition. Such more abstract instruments require a more aggregate view of the problem, involving for instance groups of installations or activities (see Huppes 1993).

It is also possible to take a more complex view of the policy process, starting with the social processes involved. Policy implementation presupposes a particular context in which the instruments function, involving public and private organisations and many individuals, each with their own roles, knowledge and ideas. In some countries, installation permits are specified and become legally binding without having much effect on behaviour. Actual behaviour may deviate from the permit regulations because monitoring is lacking, officials agree to allow deviations, or sanctions cannot be effectively applied. In such quasi-illegal, seemingly ineffective regulatory situations, however, private organisations may still decide to introduce environmental improvements, on a variety of grounds, and may communicate and coordinate their actions with officials. These may then sanction the actual behaviour, either informally or by ultimately issuing a formal permit befitting this behaviour. Such an informal style of regulation has been dominant in England and the Netherlands, where officials communicate with firms and agree on actual operations without necessarily issuing a permit. As a Dutch environment officer once expressed it: 'As long as there is no permit, I can influence my firms. As soon as the permit is issued officially, my influence is gone.' In such situations, there is no clear distinction between the setting of policy goals, instrument choice and policy implementation, let alone that they occur in this precise sequence. A context in which such officials have all sorts of, preferably broad and unchecked, powers allows them to negotiate. It is these operational practices in society, embedded in culture and institutions, which may largely determine policy development and policy instrumentation. In a more legalistic culture, as in the US, such informal procedures may play a limited role.

These two views of the environmental policy process relate to two basic views on the way society can be influenced, one vertical and the other horizontal. The vertical view links a formal democracy which sets policy goals with a Weberian ideal type of hierarchical bureaucracy implementing these goals (see Weber 1947). The horizontal view is associated with direct democracy and social integration through meso-level bodies, like trade associations. Covenants between all public and private parties involved set environmental goals and leave their implementation to those most directly concerned. Even if one does not specifically opt for one of these views, one still has to deal with these diverging aspects, as ignoring social aspects may well make enacted policy ineffective. In the Netherlands, three decades of formal regulations on the manure problem have been to no avail, as local governments refused to effectively implement the successive laws enacted (see Huppes and Kagan 1989). Pressure by the EU in this field, with heavy penalties being imposed on governments for not meeting quality standards for air quality and eutrophication, seems to create an impulse towards effective policy implementation. The two views also meet at a more strategic level. In formalised bureaucratic societies, like the United States, the rigidity of regulations and their increasing complexity exerts a pressure towards more informal procedures. Conversely, the lack of effective regulation in more informal societies, as England and the Netherlands used to be, has led to pressures towards more formal procedures, as have the formal rules introduced by the EU. The integration of process aspects and instrumentation requires a more abstract approach to the policy development and implementation process. German authors (especially Mayntz 1978; Luhmann 1989) have been highly influential in this respect, also in the US, linking the Weberian bureaucracy theory with political process theory into modern administrative sciences.

How does this chapter deal with such fundamental differences of view when discussing the subject of policy instruments? We have decided to apply both. For some instrument types, this chapter uses the hierarchical vertical model, mitigated by the knowledge that policy development and implementation is a cyclic process and that context, in terms of culture and institutions, is a decisive factor in the functioning of such instruments, especially in situations where horizontal government is the dominant mode. At the other end of the scale, there are policy-making processes in which consultations between those involved create a consensus between them which hardly needs formalising. A covenant, without any sanctions for non-compliance, may be the only externally visible instrument in this situation. In between, there is a range of instruments which guide private choices without ever specifying the behaviour of the actors involved in any detail. Emission taxes and environmental information are examples of situations in which very limited public actions have a far-reaching influence on many decisions. Taxes on petrol in Europe and Japan have led to much lower gasoline consumption and related emissions than in the US, while subsidies on energy in the former communist countries, still not fully being broken down after the collapse of communism, have led to extremely high energy consumption per Euro of national income. Such mechanisms work regardless of the intentions with which the subsidies and taxes were created; it is through anonymous markets that they exert their influence. The middle European countries now having become members of the EU have adapted to world market prices and Western European taxing systems and are transforming rapidly to the Western European energy consumption levels per Euro of national income.

Furthermore, this chapter covers not only the vertical and horizontal approaches to public policy, but also purely private developments. An example is that of the international standards for environmental performance specified in the ISO 14000 Series. ISO (International Organisation for Standardisation) is a private organisation. These standards, in combination with cultural values and liability rules, are having a profound influence on corporate behaviour in particular. The inclusion of such non-public policy instruments in the analysis results in a broad view on instruments.

Finally, we had to decide which environmental subjects to cover. In principle, this chapter includes all subjects for which policy formation has developed in such a way that clearly specifiable domains of social activities are linked to states of the environment with real or potential undesirable effects, either directly on the environment or through the environment on society. Hence, all environmental problems relating to emissions and a-biotic extractions are extensively covered. To some extent, the chapter also covers biotic extractions, such as fishery rights in common waters and quality control on sustainable forestry. Other areas are not extensively covered in this chapter. Policy development for land-use related subjects has not yet developed much beyond zoning laws. Globalisation has fundamental ecological and evolutionary consequences, for instance through the spreading of species to all ecosystems in our globalising world, from viruses and bacteria to plants, fungi and animals. Its effects on biodiversity will be substantial, while health risks may result from recombination of viruses and bacteria. However, no preventive instruments are available other than border controls, which are steadily diminishing. Similarly, amenity values of landscapes, including aspects of cultural history, cannot yet be broadly and systematically covered by instruments, as building regulations can only partly affect such amenity aspects. Clearly, then, there are domains of environmental policy in which steering has not yet become sufficiently routine and formalised to allow institutionalised instruments to be used.

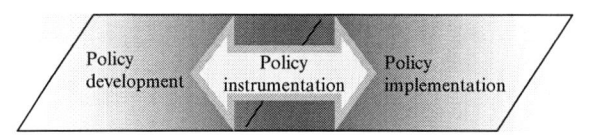

Fig. 13.1 The place of policy instruments in the policy process

It is now widely accepted that human society has to find a sustainable way of dealing with its environment. The basic ethical position taken in sustainability is one of distributive justice within and between generations. The implications for empirical analysis and social action are enormous, and in terms of concrete actions obviously depend on the more detailed ways in which the sustainability concept may be specified. Somehow, society has to incorporate and maintain feedback mechanisms which redirect the actions of many, a process which Luhmann (1989) calls *autopoiesis*. The development of policy instruments is part of the institutional development of society towards the kind of self-steering which is required for long-term human welfare for all, and ultimately even for human survival. See Fig. 13.1 for a main conceptual framework. The institutional development required, often referred to as ecological modernisation, centres around the development of new policy instruments, going from curative to preventive, from centralised to decentralised, and from over-regulated to stimulating (see van Tatenhove and Leroy 2000).

13.1.2 Outline of the Chapter

The chapter moves from the general to the specific, in that it starts with broad societal developments at a global level, then goes into the social contexts of instrument functioning, and finally discusses the detailed specification of instruments, analysing their strengths and weakness. Section 13.2, entitled 'Policy instruments in a long-term perspective', investigates the global social context from a long-term strategic perspective. As choices on policy instrumentation bind society for a long period, long-term changes in context have to be taken into account, in terms of structural, cultural, economic and political developments. Major structural developments include globalisation, which changes all organisations in business, science and technology. Small countries have little decisive influence on technologies which have been developed elsewhere. A major cultural trend is towards individualisation, with intermediate organisations like churches, trade unions and trade organisations loosening their traditional grip on individuals. A key economic characteristic is the continuing growth of income, based on structural developments at a global level like better education, technological progress through ongoing investments in science and technology and increased labour participation not yet offset by shorter working hours. Such fundamental developments are first surveyed, with a view to finding out how instrument choices can take such major contextual developments into account. A number of strategic criteria for instrument design and evaluation emerge from this analysis, which takes for granted certain rough ideas about what policy instruments are and why we need them.

Section 13.3, entitled 'Policy instruments: what are they good for', discusses basic analytics, in a bottom-up approach. Why do we need instruments for environmental policy? Could integrated policies, without specific instruments, not cover all problems? As instruments are placed in their administrative setting, while emphasising 'horizontal governance' – in which public and private partners together design solutions to common problems – the question why we need them crops up again. The answer given here is that they are needed for environmental effectiveness, for the simplification of the policy process, and for building into the fabric of society the safeguards for long-term sustainability. After these hurdles have been taken, the analysis is constructed around the theme of the nature of environmental problems and the general mechanisms that cause undesirable environmental consequences. Concepts like 'external effects', 'collective goods' and 'free rider problems' are surveyed, as these constitute the particular context in which instruments should provide solutions. A final introductory theme is that of evaluating alternative instruments for environmental policy. A distinction is made here between first-order criteria like effectiveness and costs; second-order criteria like requirements on administrative capacity and effects on technology development, which are mostly difficult to quantify; and strategic third-order criteria relating to instruments, such as fitting them into overall regulatory and broader institutional developments, which are never quantified.

Finally, Section 13.4, entitled 'Design, analysis and evaluation of policy instruments', specifies the main dimensions for defining instruments. It shows that it is not at all clear how policy instruments can be classified and described. Nor is it clear how a consistent evaluation of policy instruments can be set up. Still, as some ordering is necessary for instrument development and instrument choice, an analytic framework is developed, covering not only the traditional regulator-regulatee relations but also instruments structuring the relations between various governmental organisations, and instruments structuring relations between private actors, both individuals and organisations. Some practical guidelines are given for policy development at a case level.

Before discussing first strategies, then the contextual framework and next the actual instrument analysis, let us first try to visualise this complex subject as a whole, to gain some perspective (see Fig. 13.2). The starting point is some sort of policy goal that is to be achieved. In designing instruments, one first has to decide on the actors involved, like regulators and regulatees, but also the broader groups involved, which together form the stakeholders to the policy. Next, the goal of the policy has to be defined at an operational level, like an emission volume, an immission concentration, an immission load or an allowable health risk. The next element required is something that can be regulated, something that is the object of regulation, like an installation, a product or a behaviour. Finally, something is needed that sets the policy in motion, such as an operational influencing mechanism. Some positive or negative incentive has to be created, like a prohibition, a subsidy or the obligation to display information, to act as the instrument's influencing mechanism. What happens next, empirically speaking, is set in motion, that is, caused, by the instrument.

This then leads on to an empirical part of the analysis, involving all sorts of mechanisms in society which may have direct effects, involving simple causal chains in the short term; indirect effects, taking into account longer effect chains; or secondary effects, involving broader effect mechanisms. The direct effect of an energy tax, for instance, is a change in energy prices and consumption, while its indirect effect is increased spending on non-energy expenditures and a shift towards more energy-efficient products, and its secondary effects may involve adjustments to the goals of research and

Fig. 13.2 Design, empirical analysis and evaluation of policy instruments

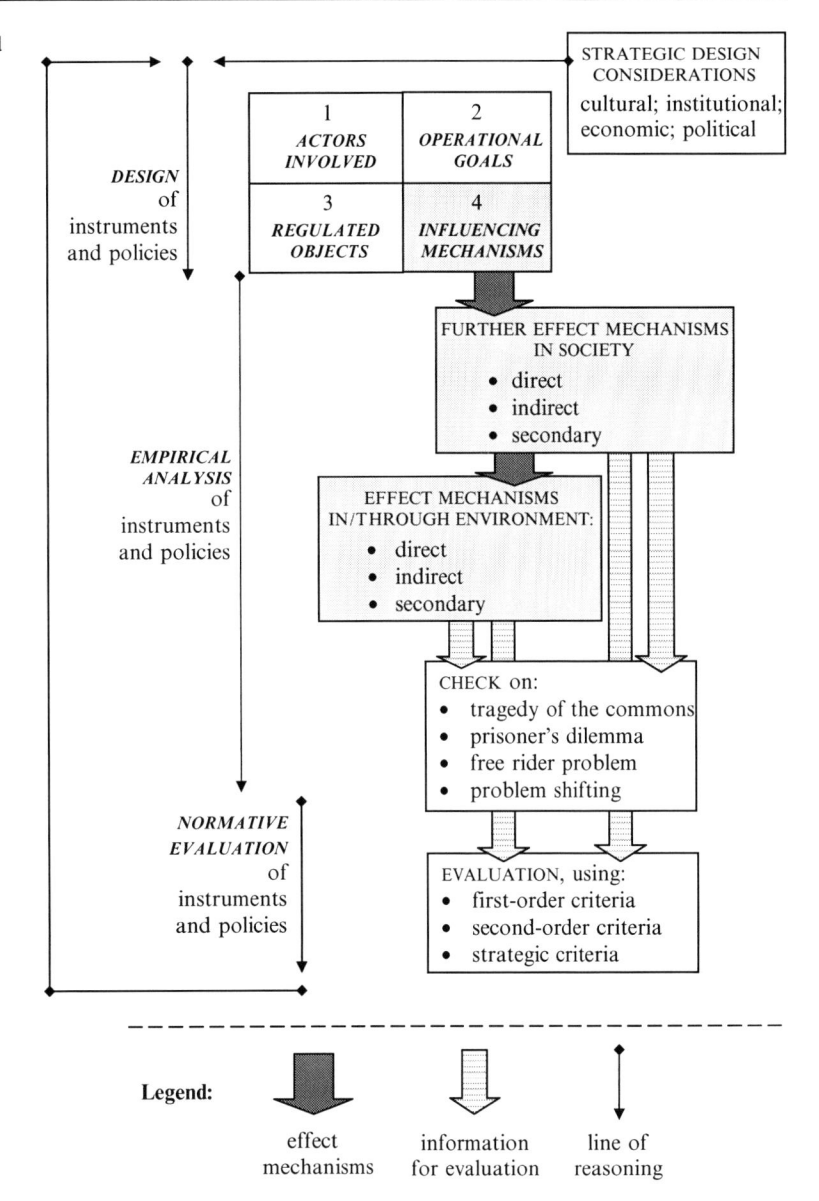

development in large sections of society. This empirical analysis of instrument application in society is then followed by an empirical analysis of the resulting environmental effects, with feedback to society.

A third element in the analysis is evaluative: is it worthwhile to use the instrument, reckoning with its implementation costs and effects? This question is related on the one hand to the empirical functioning of the instrument, involving evaluation criteria like environmental effectiveness, cost and regulatory effort. On the other hand, an important set of evaluation criteria is of a more abstract nature. How do subsidies for not polluting relate to liability law, where it is an accepted principle that those who cause damage have to pay for it? The highest level of strategic considerations takes into account the expected developments in our world, relating to trends like globalisation. In a world where international trade becomes more pervasive, national regulations increasingly lead to loss of competitiveness. Instruments differ in this respect, with emission taxes having major effects and consumption changes minor ones, and instruments may be designed in such a way as to prevent this unintended consequence. The empirical and normative analyses feed back to instrument choice and policy design. The empirical analysis is of course not based on actual empirical performance, which is not yet there, but on expected consequences. This completes the policy instrumentation cycle (see the arrow on the left in Fig. 13.2).

13.2 Policy Instruments in a Long-Term Perspective

13.2.1 The Future Context of Instruments

Policy instruments tend to bind society for a long period of time, as changing them often requires complex administrative procedures and a decision to adopt a particular instrument in one context influences instrument application in other contexts. Given this long-term character of instrument choices, the context of their functioning is not the situation in which they are being introduced but more that of the future. Although the future is hard to predict and also shaped by deliberate choices, some general developments may be specified, such as structural trends. Such structural changes, both in society and the environment, lead to shifting conditions, resulting in a different functioning of existing instruments and emerging options for new instruments. At the core of an instrument, there may be simple social mechanisms which were rightly taken for granted when the instrument was introduced, but are not self-evident in the long run. For instance, the workhorse of environmental policy in England and the Netherlands is the permit for individual installations. It was assumed that decisions on how to operate these could be discussed with those responsible before the installations were designed and built, and long before a permit was due. The firm's planning was geared to that of the permitting procedure, with quite some procedural flexibility. In many sectors, the periods for planning and implementation technologies have now been reduced to months rather than years, and involve integrated decisions on technologies implemented worldwide. Discussions with the local environmental authority will only relate to some of the details, but not to the technology itself. Nor are consultations allowed to address factors involving real cost shifts, as expected costs have already been incorporated in the decision-making process in the network of firms involved in the technology. Thus, the role of local authorities in the strategic and tactical aspects of technology development has been reduced to virtually zero in most permitting situations. Only additional instruments like large subsidies can exert some influence in exceptional cases. For instance, highly efficient and environmentally benign coal gasification plants for power generation are only built with huge subsidies now, and no more than a few dozens of such installations have been built in the world so far. Therefore, developing the instruments for environmental policy, as an operational set, requires such long-term developments in society to be taken into account, not only in the local firms but also in the broader societal settings.

A number of these developments are addressed below, with some indications of their potential significance for the instrumentation of environmental policy. Relevant and significant developments are taking place in the overall structure of society, in general cultural developments, in developments in the economy and specifically in industrial relations, in the changing role of government and, last but not least, in the changing nature of environmental problems. We will also discuss in some more detail the consequences of globalisation. The next aspect to be addressed is that of the prospects for various policy instruments in a changing world. These prospects do not automatically lead to clear indications as to which instrument is to be used, when and how. In this respect, policy instrumentation as a societal process is itself also partly developing through causal mechanisms of a social nature, and is partly based on explicit strategic considerations and decisions. Some major strategic choices in policy instrumentation are identified in the final section of the chapter, which approaches the arena of political discussion.

13.2.2 General Tendencies in Society

The social structure of Western countries is showing a trend towards a decreasing role of intermediate organisations like churches, trade unions, parties and clubs. Institutional integration is weakened and its indirect control over individuals is reduced. This structural development reflects a double cultural development. As early as the 1950s, Riesman (1950) described a change from *inner directed* to *other directed* control of people's behaviour. This means that it is not internalised norms and values that guide specific choices, but the notions of others about this subject as determining his choices. Moral precepts are replaced by self-centred considerations of expediency, based on how others view an action and react to it. General considerations on what is right and wrong lose their place in guiding actions. The second tendency, stated even longer ago by Tönnies (1887), relates to the question who these others may be. The reference group for one's norms and values is shifting from a closely knit small group of partners for life, one's 'Gemeinschaft', to larger groups of more shifting acquaintances, the 'Gesellschaft'. These basic developments provide opportunities for new forms of integration. The differences between national cultures are diminishing, as a global culture is developing through shared TV programmes, advertisements, books and movies, through the nearly universal marketing of products and through internationalisation of contacts via tourism and the Internet. Although some debate is possible on new modes of small group integration, the tendencies specified above all lead in the same direction, in which normative control on individuals in their roles as consumers and producers is decreasing. The legitimacy of measures founded in morality has been reduced in the process, and no longer plays its invisible (quasi-automatic) role the way it used to. Therefore, the assumption that rules will be followed more or less automatically may increasingly come to be questioned. Although cultural controls on behaviour are shifting to higher-level collectivities, the subjective view at a global level is ultimately that of increasing individualism, with people deciding on ever more subjects for themselves, arguing from their individual views of the world. This theme of individualisation has been extensively discussed by Beck and is summarised in Beck and Beck-Gernsheim (2002). In a world dominated by relatively superficial types of information, the reference for individual decision-making may be quite poor and simple, relating to individual experiences with simple generalisations. This de-socialising tendency (with integration levels becoming less binding and content-poor, compare Durkheims *anomie*) may be countered by other structural mechanisms, like the globalisation of science and literature. It is difficult, however, to link these global activities to new types of normative and broader cultural integration.

The global economy is going through an extensive structural change, based on new technologies and new forms of communication. The amount of specialised knowledge embodied in any given product is expanding and the technological complexity in or behind most products is increasing. The innovative capacity of firms is increasingly based on functionally differentiated, more or less independent innovation generating organisations. At the same time, the organisation of successful firms is more open to external innovation options. A large firm like Shell has incorporated its main research capacity in an independent organisation, called Global Solutions, which has Shell as its main client, but operates on the world market. The market for innovations in industrial production has increasingly gone global, including and relying on the information and service industries. It is here that future technologies with their specific environmental consequences are born and start to diffuse throughout the world. Technological innovation, viewed by Schumpeter (1942/1976) half a century ago as the capacity of owners of firms to innovate their own activities, has now become a more or less independent capacity at the service of all other firms. It is an open question whether 'Schumpeterian dynamics' and the Rio imperative of sustainability can be made compatible, based on the development of *zero emission* or *Factor Ten* technologies which will not arise spontaneously. If firms want to invest in new installations, they will often be faced by a stark choice: here and now, or not here at all. Delays due to environmental permitting procedures are becoming increasingly unacceptable to a firm which may just have acquired new and superior technology from the specialised technology developers abroad, with only a few weeks or months of head start on its competitors. In this new situation, traditional regulatory controls on technologies clearly have to be redefined, as they can no longer be based on insights by the regulators into the firms they are regulating.

Another structural change in the economy is the shift *from product to service*. This development also affects traditional hardware, such as cars that are not bought but increasingly being leased. In addition, it is embodied in the emerging information and communication technologies, whose hardware is mainly owned by providers, while clients pay for services only. This development is taken one step further when firms do not provide services but act as service-providing organisations, leaving the actual physical activities related to service provision to smaller units. Franchising is one example that has been in existence for some time, but in industrial production, designers and marketers now externalise production to a much higher degree than they used to. A fast growing firm like Nokia is an example, in which suppliers of all major parts of mobile phones and physical networks are being chosen flexibly every few months. The actual service provision is by competitive network operators, which ensure that the phones function for the consumers who are paying them. And if services are not externalised outside the firm, larger organisations set up business units, which in many respects function as independent firms. They sell and deliver their goods and services to other business units in the organisation but also to the outside markets, while other business units are not obliged to buy their inputs from business units within the same firm. No doubt, this downsizing of organisations, combined with increased international competition for more or less standardised activities, makes environmental control on such standardised mass production activities very hard for individual countries. The non-standardised, creative and strategic aspects of the innovation process may hardly have discernable environmental effects, and are not influenced by binding instruments or market instruments, leaving the scene to 'softer' informational instruments and to structural instruments with a limited environmental scope.

A shift in a similar direction is the changing nature of coordination in the economy. There is a marked shift from hierarchical control to coordination by contracts and markets, quite in line with the tendency towards individualisation. Contracts and markets are not fully anonymous but involve flexible relations, increasingly based on global communication networks. Anything can be bought anywhere in the world at short notice. The main fixed points are the locations of consumers and most employees and some bulk resources, while most of the other aspects of the physical economic activities are variable in terms of location, due also to decreasing transport costs. Formerly the domain of large multinationals, now small and medium sized enterprises are also turning to international trade, integrating in international networks. As a consequence of these developments, traditional national regulators must feel their powers vanishing.

Will there be an end to the increase in global production and consumption, easing the pressure on the environment? Long-term developments in this respect are pointing in the opposite direction. Economic growth is fairly constant in the long run, with real growth in the last decade being of the order of 3% per year, doubling consumption every 23 years, while international trade is increasing by 7% per year. The implied growth in labour productivity is not matched by a proportional decrease in total labour time. On the contrary, there are pressures towards longer working hours per worker. This means that total production will rise, as will the physical activities related to negative environmental effects. The shifting nature and composition of consumption do not indicate an automatic solution to environmental problems. For instance, services may seem attractive in that they are free of matter. However, the system activities required for growing services like mobile phoning and international holidays are highly energy-intensive and require substantial material inputs of a very specialised nature.

Tendencies in government reflect these broader developments in structure, culture and economy. The ideal of planning the future is dead. Although it is still reflected in names like the Central Planning Bureau in the Netherlands, planning, even the sort of indicative planning that existed in France, has disappeared both from government and from business. (Mintzberg, the 'guru' of strategic planning in the firm, has named one of his books 'The end of strategic planning', Mintzberg 1994). Of course, targets are still being set, such as the reduction targets for greenhouse gas emissions in the Kyoto protocol, and the targets in national policy plans. However, these targets are not part of a broad vision on societal development, and the link between the quantitative goal and implementation activities is decidedly weak. In fact, global implementation mechanisms in the form of policy instruments are largely lacking. Although some of the vocabulary is thus still there, integrated strategic planning and control are fading. In particular, the strong control that used to be provided by the informal modes of regulation in England and the Netherlands has first lost its glamour and then its effectiveness. Ultimately, the informal flexibility used to be backed up by the power of officials to implement what they liked, even if it was unreasonable. Negotiating 'in the shadow of the law' is increasingly difficult and the nature of the instruments involved has changed. It seems such countries are all shifting towards the American, more formalised and litigative procedures, which is not very helpful in the new situation of global competition for technologies and products.

New developments also include avoiding the complexities and costs of formalised regulation by concentrating on consensus building. Consensual processes are major vehicles for change, often with horizontal government as a guiding principle. Building on available integration in society, the domain of application of this approach will tend to become more local. In local affairs, those involved know each other already, and horizontal government is not so much of a change. For higher level problems, as environmental problems increasingly are, this is a change relative to existing practice. It is a tendency that builds on corporate government ideas of involving all major parties involved in a negotiating procedure, in which win-win situations are created, to the advantage of all. In corporate government, it was the top echelons of the socio-economic institutions which made the deals; now it is 'those involved' in general, that is, the stakeholders, who together decide on some problem or action. On the one hand, this tendency reflects the decreasing power of national governments. On the other hand, horizontal government is also an impetus for less active and less binding types of regulation. The consequence for environmental policy could be that for most problems, the old habit of setting standards more stringently than those involved think reasonable is no longer an option; only information, stimulation and financial incentives may remain available if this tendency continues.

13.2.3 Globalisation Tendencies

One recurring element in the above survey of tendencies was globalisation. Environmental problems are increasingly becoming transboundary or global; economic production processes integrate at a global level for a global consumer market; and a global culture is emerging, at least in consumption. International political integration in blocks is losing momentum in favour of, as yet limited, integration at the global level. Examples of the latter are the strength of the World Trade Organisation (WTO) and the vehemence of the political discussion on its further expansion vis-à-vis social and environmental interests. Together, these developments pose severe problems for national environmental policy formation and instrumentation.

The consequence in terms of instruments is that binding instruments based on command and control at a technological level are becoming harder to apply. Any instrument with real effectiveness leads to costs. The idea that economic-environmental win-win situations will emerge spontaneously is attractive but highly improbable, as the structural causes of environmental problems remain and environmental pressures increase because of population growth and economic growth. It seems that win-win situations are related to weak sustainability, in which innovations are attractive, environmentally and economically, per unit of product. Since, at the same time, economic growth is implied, the overall effects at the macro level will usually be detrimental to environmental quality. We therefore assume that, for the foreseeable future, environmental policy will not be expendable at all. Quite the opposite, in fact. Its nature, however, will have to change.

Therefore, effective national policies in the global context require international co-ordination. This co-ordination should not only be effected at the level of setting aims, as is now increasingly the case, but to a certain extent, the design of instrumentation also has to be agreed upon internationally, if nationally effective policies are to be achieved. Some early steps have been taken in climate policy and biodiversity policy, where international instruments between countries have been worked out for joint implementation, in the form of a political-administrative instrument for emission trading and a clean development mechanism. For instance, countries may now trade their emission reduction obligations on a bilateral basis to improve overall efficiency in emission reduction. European countries are investing in new technologies in the former communist countries to reduce emissions there relative to the assumed autonomous growth of emissions in these countries. However, any simple trading system based on future emissions of a substantially fictitious nature may easily erode collective efforts and cannot easily be brought into line with the systematic development of regulatory instruments. Imagine that a country can claim credit for helping to close down an old power plant or helping to plant a forest in another country. This is now possible under joint implementation. The net influence of these supporting activities may become negligible if these developments would have taken place anyway. There is then no actual result, only a clear result on paper. Political deals with some financial gains to the main parties involved, may then become more important than real efforts for emission reduction.

Tradable emission permits might also be implemented as a regulatory instrument. The question is then who would receive the emission rights being traded and what the initial distribution between countries would be. Whatever the initial distribution, after some time real emitters would have to have the emission rights corresponding to their emission volumes. While such a system would be more transparent, it would have serious implications in the normative set-up of instrumentation. Achieving global efficiency in climate policy would mean that the trade in private emission rights would have to be preferred, on the same efficiency grounds now used to defend joint implementation. Broader ethical considerations embodied in the normative-legal structure of most countries would favour instruments that are in line with the polluter-pays principle (see OECD 1995). In this case, emission taxes are to be preferred. In the case of carbon dioxide emissions in particular, such taxes could in principle be set up with relative ease. The main choices in working out such taxes relate to who receives the proceeds. If it is the national governments implementing them, this would create an incentive for effective implementation. However, the carbon resource owning countries would effectively pay a large part of the bill as a result of reduced prices for gas, oil and coal. Ideas on legal frameworks as being discussed in the WTO tend towards giving these countries a right to compensation. Alternatively, the emerging global political community, potentially in the form of a *World Environment and Development Organisation* (see Simonis 1998) needs financing, and the carbon tax proceeds would seem an ideologically acceptable source.

13.2.4 General Tendencies in Environmental Problems

The structural economic developments described above have changed the nature of our relations with the environment (see also Box 13.1). Five basic trends may be discerned, each of which is important in its own right. Together, they are creating a challenge for policy instrumentation which cannot yet be met.

Depletion problems of materials supply, never really high on the political agenda, would lead to a greater emphasis on all forms of recycling. It seems that the real environmental drive behind this emphasis is not so much depletion but avoiding the detrimental effects of primary production and final waste disposal.

Traditional environmental problems of final wastes and emissions tend to become *more global* in nature. The remaining 'old' toxicity problems are of a more global nature, being related to substances that are not broken down and circulate in the environment for a long time, like heavy metals and pseudo-estrogens. Acidification and eutrophication now also tend to involve larger areas, like subcontinents and coastal seas. Some of the newer problems, like ozone depletion and climate change, are essentially of a global nature.

Appropriation of nature (Fischer-Kowalski et al. 1997; Ingold 1986; Imhoff et al. 2004, going back to Marx's *Das Kapital* in a slightly different meaning), in terms of the share of biomass production controlled, harvested and consumed by humans, is growing. Room for natural type ecosystems is vanishing fast and the remaining ecosystems are controlled for their human oriented biomass production. Tendencies towards biomass production for energy applications will give further impetus to this development. This will decrease the actual volume of non-edible algae and plants and of all creatures living off them – animals, fungi and one-celled organisms.

The *share of non-man made ecosystems* is decreasing, with full human control emerging except for protected nature reserves. This leaves ever smaller patches of ecosystems, which in principle cannot be stable in terms of the number of species they can support, and hence will become more dynamic.

The combined effects of the appropriation of nature and the increasing spatial scale of human control reduce the ecological basis of humanity not only in terms of volume but also in terms of quality, as many ecosystems require large spaces for their long-term existence.

The *homogenising of ecosystems*, closely related to the globalisation of production and the emergence of global recreation, may well become a major new focus

> **Box 13.1** Policy instruments in a long-term perspective
>
> **Culture**
> - Reduced role of internalised norms and values
> - Reduced role of local groups
> - Increasing role of international mass media
> - Some cultural elements globalised
>
> **Social stucture**
> - Fewer intermediate organisations
> - Expanded liability
> - More private and national ownership of biotic resources
>
> **Government**
> - Formalised and litigative tendencies in binding instruments
> - Decreasing role of binding instruments
> - Increasing role of horizontal mechanisms, especially locally
> - Uncertain role of purely private mechanisms
>
> **Population**
> - Continuing growth over the next decades
> - High growth rates in many poor countries
> - No further growth in industrialised countries
> - Higher average age in industrialised countries
>
> **Prospects for instruments**
> - Dynamic effectiveness of instruments: halving environmental effects per unit of income required every 25 years
> - Internationally co-ordinated instrument choices essential
> - Instruments for safeguarding ecological resources to be developed
> - Decreasing overall effectiveness of instruments
> - Financial instruments essential for high level of trade-off between economy and environment
>
> **Economy**
> - Increasing technological complexity
> - Functional differentiation of innovation
> - Shift from product to service
> - Shift from service provider to service organiser
> - Co-ordination: shift from hierarchies to markets and condtracts
> - More flexible co-ordination in networks
>
> **Environment**
> - Decreasing ecological resources
> - Local problems dealt with reasonably
> - Emphasis shifting to continental and global problems
> - A-biotic depletion only in very far future
> - Increasing appropriation of nature and biotic depletion
> - Continuing climate change
> - Continuing loss of biodiversity
> - Natural areas diminishing fast
> - Threats of ecosystem instability
>
> **Strategic instrument choices ahead**
> - Tradable right to environmental damage versus collective right to unspoiled environmental quality
> - Equal right to environmental use space versus equal efforts for damage reduction
> - Global equity versus global efficiency
> - Means-directed technology-specific instruments versus goal-directed environmental incentives
> - Normative integration of policies with broad internalisation in society versus political-administrative debate per item of choice

for policy. As species from different ecosystems mix, the diversity of ecosystems is reduced. As new species are introduced, from viruses and bacteria to fungi, plants and animals, it will be the generalists that have the greatest chance of survival, out-competing the niche specialists. In addition, competition between niche specialists will occur, reducing the numbers of their species. As a result, new ecosystems will evolve, with new selective pressures towards new species. Since new species of higher organisms evolve over a longer time scale than smaller and simpler organisms, it is especially new variants of viruses and bacteria and other mono-cellular organisms that will arise in the short term, posing threats to humans and crops, as well as to other species.

Overall, the pressures of emissions, combined with the three other types of structurally increasing influence on the biotic environment will lead to rapidly decreasing biodiversity, instability in the genomes of smaller creatures and an influence on the life support functions of natural ecosystems that is more difficult to specify. Policy options may focus on prevention of these mechanisms, or on the control of undesirable consequences. Examples of the first are bans on the transport of species, which are already in place in many countries for plants and plant material, for agricultural reasons. Solving problems after they have occurred often seems difficult. Disease control by vaccination is one activity that is obviously to be stimulated in terms of research and instruments for their operational application to large parts of the population. This level of problem analysis cannot yet be generally related to policy instruments, thus allowing the problem to grow. This is an undesirable state of affairs for broad sustainability reasons. Ultimately, the influence of human activities may reach a point where viable ecosystems can no longer exist unless fostered by specific human activity. Such a total control of our biotic environment, and hence also of our a-biotic environment, is not feasible at a scientific level, let alone that policy instruments could already be envisaged.

13.2.5 Prospects for Instruments

As explained above, the role of technology binding instruments is expected to decrease further. Only in industries not exposed to international competition, like building and infrastructure, and in the context of some internationally binding agreements on prohibiting emissions like the Montreal Protocol for most ozone depleting substances, may instruments binding private addressees remain dominant (see e.g. Castells Cabré 1999). International coordination, for instance by technology guidelines or Best Available Technology rules as in the EU IPPC Directive (Integrated Pollution Prevention and Control), may help to leave some space for technology binding instruments, especially in matured industries with relatively little technology development. In other fields, different types of instrument will take over or at least become more important. Private choices in production and consumption are increasingly being guided by cultural and informational instruments like life cycle assessment (LCA) and Environmental Audits, both standardised by ISO.

Such analysis instruments may have a broad influence especially if the information in the instruments is complemented with normative information about the relative importance of different types of effect. However, they create a limited incentive only, due to the collective nature of most of the environmental effects involved. Where real choices are to be made, with substantial costs involved in environmental improvements, information and normative statements will not suffice. If binding instruments lose their importance in limiting options, only financial instruments and liability instruments can substantially correct the pay-offs for those making choices, be they governments, business and environmental associations, or private producers and consumers. Softer cultural instruments are important to support prime movers and to generate political support.

Thus, a major problem for future environmental policy relates to methods compensating for the diminishing role of technology-specific binding regulations. With few exceptions, the role of financial instruments has so far been limited. Broader use will be based on a number of conditions. To begin with, a clear choice is required between issuing emission permits, against the polluter-pays principle, or creating emission taxes, in line with it. Secondly, the operational applicability at the level of emissions and resource use (e.g. CO_2 taxes), as opposed to application at product and technology levels, has to be improved. Thirdly, international coordination in terms of the design and levels of taxes is needed to avoid unfair and overall costly shifts in competitiveness between firms in different countries. Fourthly, better integration of various environmental aspects is required for cultural instruments, as real actions always involve virtually all environmental problems that exist. This is not only a matter of information but of clearer normative guidance as to the relative importance of various environmental interventions, based on their potential consequences. Lack of conceptual clarity is a major problem here. Can we realistically specify the ultimate consequences of CO_2 emissions at an 'endpoint' level? Or should we evaluate at the 'midpoint' level of global warming potentials? Or are potential instability and uncertainty about possible catastrophes the prime motives for reducing greenhouse gas emissions? Such questions need to be answered in a very practical manner if a trade-off between, for instance, energy use, land use and a diverse array of emissions is at stake, as is the case in most practical decisions in production and consumption. Such normative guidance, which is only partly based on empirical evidence, is not only a prerequisite for practical decisions on specific technologies, but also for more aggregate developments, as in steering technology development and creating more sustainable lifestyles.

What is ultimately needed is not only a view on the relations between the various environmental effects involved, and the trade-offs between environmental interventions implied in our actions. The trade-off between economy and the sum total of environmental effects also needs to be stated as a clear principle, guiding the quantification of all types of instruments in a uniform way. Without such guidance, equal treatment of similar cases cannot be achieved, leading to both injustice and substantial static and especially dynamic inefficiencies.

Finally, structural instruments, such as changes in institutions, constitute an option for new policy instrumentation that has not yet been fully exploited. For example, cases against the tobacco industry in the US have shown how large the payments can be in a specific judicial setting. In all cases, creating the right incentives should be accompanied by removing the wrong incentives. The situation in many countries is such that there are substantial subsidies on energy use, as in Eastern European countries, and similarly by tax exemptions as for aircraft fuel on international flights; that the cost of infrastructure is not reflected in prices, as with un-priced roads; that new technologies are difficult to implement due to complexities of regulation; and that a global perspective on environmental policy development is prevented, as implied in emerging WTO regulations. If a broad shift in policy implementation were to emerge, the result would be a more balanced internalisation of environmental consequences, both in public and private actions.

Instrumentation for newly emerging environmental problems is usually lacking. The *appropriation of nature*, that is, a decreasing share of nature in the total biomass production, is accompanied by an even faster reduction in natural biomass breakdown, which is the source of food for all fungi and animals. This is one of the new environmental problems, defined here at 'midpoint' level, relating to biodiversity, the life support function of ecology and the quality of nature. Instruments that can work at a global level hardly seem to be available. For some aspects, such as the protection of available genetic information, the road towards structural-institutional instruments has already been taken, with governments or private organisations owning the species on their territories and their genes, whether naturally found in organisms or constructed. This may help to protect and regulate the existence of gene pools. It does, however, not necessarily help to create healthy ecological conditions. The preponderance of single protected genes, as in agricultural production, may even lead to monocultures of such economically dominant genes, with probably negative overall ecological effects through reduced biodiversity.

Two lines of instrumentation are presently available for saving and creating ecological values. One is the creation of nature reserves, as an option-creating instrument. However, increased spatial needs for food production to feed the growing and more affluent populations and increased mono-cultural production of biomass for energy purposes literally leave less room for this instrument, given the largely fixed area of land on earth (which even will decrease due to rising sea levels). The other, 'integrational', line is to create more ecological quality in areas primarily used for agriculture, recreation, infrastructure, production and housing. Again, the options have hardly been investigated and instrumentation is largely lacking. Apart from emission reducing instruments, there is a clear lack of instruments safeguarding ecological richness in diversity and volume, not only in nature reserves but also in human-dominated ecosystems. Furthermore, ecology-oriented instruments still lack theoretical foundations and operational development.

13.2.6 Strategic Instrument Choices Ahead

Our first assumption here is that it is not possible to avoid choices regarding environmental policy instruments by doing nothing. Economic growth, population growth and globalisation tendencies make internationally coordinated policy development inevitable. At the same time, it is impossible to make independent choices on policy instrumentation for each individual case. A well-argued strategy has to guide serious development of policy instruments. If the polluter-pays principle is accepted, all instruments have to take it into account. Basic choices should preferably be made consistently, according to generally recognised principles. Some major lines of development are discussed here, relating to basic liability rules, environmental ethics, the importance of efficiency and equity, means-directed or goal-directed types of instruments and, finally, principles of policy integration.

13.2.6.1 Liability Rules

Liability rules have traditionally been set up to prevent active infringements on the goods or rights owned by others, either individually or collectively. The 1960s saw the start of a debate in economic circles on the other option, that is, giving everybody the right to infringement of the goods or rights of others, especially in the environmental domain. The discussion was opened by Coase (1960), who showed that there was no difference between these options for the outcome in real terms, if transaction costs needed to arrive at these outcomes could be neglected in both cases. In the debate following his paper, the latter restriction has been somewhat relaxed in that Coase's conclusion also holds if transaction costs are similar. Transaction costs result from the necessary contacts, primarily the labour time of all involved, including costs of negotiation, control and litigation.

This point of view has had a major influence on environmental policy, where the same period had seen broad acceptance of the polluter-pays principle. Different versions of this principle exist. A basic element is that polluters have to pay for the environmental damage they cause, thus internalising environmental aspects in their decision-making. Since Coase showed that the principle is not required for cost-effective policies, policy-makers may reason towards policy instruments on the basis of net costs in terms of real outcomes including transaction costs, and do 'what is best', on a case-by-case basis. This has opened up the path to tradable emission permits, which give the owners a direct right to pollute.

The conflict between the polluter-pays principle and the pragmatic do-what-is-best in each case principle somehow needs to be clarified and resolved to allow basic innovation in national, and particularly in international policy instrumentation. In-between options are possible, but not necessarily more attractive (see Tisdell 1998). One option is to market emission permits with a very limited term of validity. Effectively, this means that the right to emit has to be bought again each time to operate an installation. If the number of permits brought on the market by governments is set so as to achieve a predetermined price level, the difference with an emission tax of a similar level is really very small in terms of effects, and may even be similar in its administrative operation.

At a practical level, the liability instrument works by linking cost to environmental interventions. The US has taken a practical approach in the field of waste management, by making firms liable for the sanitation cost of illegal landfill with chemicals. This programme (Superfund Programme 1986) has conferred substantial cost on polluters, based on ongoing sanitation activities pre-financed by government, see e.g. EPA (1999). Similar programmes have been developed in the Netherlands. In the US, these programmes function against a background of already existing extended liability ('chained and several liability'), where larger firms in the chain are liable, and if the damages result from emissions of several firms, each one is liable. In Europe, developments towards extended environmental liability are ongoing, based on the Environmental Liability Directive.

13.2.6.2 Ethical Norms

In the international setting of a globalising world, basic discussions relate to the ethical principles which provide guidance in handling distributional effects. Though sustainability involves an accepted distributional principle, concrete ethical norms have not been agreed upon. Should all citizens of the world have an equal share in the environmental use space? Is this share tradable? Does every citizen have the same right to a certain minimum environmental quality? Are the costs of environmental improvements to be distributed equally per head of the population? Or is an equal percentage of income to be spent on environmental protection?

The answers to such questions have a direct bearing on instrument choice. Internationally, the use of emission rights as a political-administrative instrument, with initial rights distributed to countries according to their share in world population, would be in line with equal environmental use space. Emission permits inversely related to average income give more per capita emission use space to the poor. Equal emission taxes worldwide would roughly lead to equal shares of income being paid for environmental protection. If such principles are to be more than a guise for tactical interest protection, i.e. if they have real meaning, the implications for policy instrumentation will be quite direct. Still, there is not one principle that can force the decision.

13.2.6.3 Efficiency and Equity

A further strategic choice concerns the relation between *global efficiency* and *global equity*. For the sake of efficiency, marginal costs of environmental protection or improvement should be equal across the entire world, which means that environmental improvements should be achieved in all choices, up to a particular level of cost per unit of improvement. If this rule is not satisfied, with some doing less and others doing more, the world could benefit from a shift in effort, with those still having cheap options for improvement doing more and those with high costs of improvement doing less. A real Pareto improvement is then possible, with everybody being better off if those who reduce efforts compensate for those increasing their efforts for environmental improvements.

The broad emphasis on efficiency as a guiding principle for trade relations indicates that this principle would also have prime importance in environmental policy instrumentation. Internationally tradable private emission permits and emission taxes that are the same the world over would be prime instruments. Who is receiving the 'grandfathering' rewards of initial permit distribution and who is receiving the proceeds from the emission taxes is not relevant to efficiency considerations. This indicates that there is some room for combining efficiency with equity, by redistributing proceeds. Full emphasis on equity will lead to other instruments, however. The justice principle (as embodied in the polluter-pays principle) would shift the choice from emission permits to taxes on adverse environmental impacts.

13.2.6.4 Goal-Oriented and Means-Oriented Instruments

A further major choice concerns the aim that has been made operational in the environmental policy instruments. For instance, an emphasis on easy implementation means that regulators have a clear grip on technology development and makes policy integration an aspect of policy development. Dynamic efficiency, the most important cost aspect in the long run, remains a problem in this approach. This means-oriented approach is contrasted with the goal-oriented approach, in which policy instruments have to internalise sustainability goals as fully and directly as possible, allowing for decentralised technology choices with incentives for environmental improvements.

It is clear that efficiency considerations also imply a choice to adopt the goal-oriented option. In the liberal ideal, the choice is clearly to adopt goal-directed instrumentation, while in social-democratic and socialist circles, preferences might lean more towards means-directed instrumentation. However, ideas in European social-democratic parties indicate that the broad integration of environmental considerations in private decision-making is to be preferred over the option of having governments decide on technology choices on a case-by-case basis. Thus, there is a broadly supported political tendency favouring a shift towards goal-oriented instruments, though means-oriented instruments remain the main vehicle for environmental policy, also because the controls for regulators seem more easy to implement.

13.2.6.5 Principles of Policy Integration

In means-oriented policy instrumentation, the integration of the various environmental aspects involved is implicit. One may assume, optimistically, that a single policy-maker will be consistent in the way he or she uses the implicit trade-offs between various environmental aspects, as well as the trade-offs against social and economic aspects. If multiple policy-makers are involved, in both public and private organisations, consistency cannot be expected to come about automatically. In goal-oriented policies, there is a logic to making an explicit statement on the relative importance of various environmental aspects related to activities, at the operational level of emissions, extractions and disturbances, that is, trade-offs are to be stated more explicitly, allowing for equal trade-offs in different situations. An example is the equivalence between 14 t of SO_2 and 1 t of CO_2 which is being used informally in Dutch environmental policy. Worldwide, however, explicit statements on such trade-offs are lacking, while decentralised environmental decision-making cannot do without them. Where tradable emission permits come up, the comparative reference may be the price paid for the permit. This may be the case in the European Union for climate changing emissions (in the Emissions

Trading Scheme) if it develops to a mature system, and in the US for acidifying emissions. These markets are very partial and not yet global however.

Reasoned choices in this respect require explicit statements about the reasons for choosing particular trade-offs and the links between these environmental aspects and concrete economic actions. Such relations depend on evolving normative ideas about what is important, on the state and development of the environment and society, and on the way these relations can be modelled. Consistency, including consistency over time, can only come about on the basis of an explicit and comprehensive debate. This ideological superstructure to operational policy is poorly developed. In Dutch environmental policy, the 'themes approach' was developed as a conceptual framework over a decade ago, including themes like eutrophication, acidification and ozone depletion. This approach has been followed in various ways by other countries. The European Union, for instance, has defined a large number of Preferential European Environmental Problems, not as a systematic treatment of the subject but based on political consensus, including quite incommensurate items ranging from waste prevention to biodiversity preservation. Waste prevention, however, is not an environmental aim but a means to achieve environmental aims. This means that the means-oriented approach is reintroduced through the backdoor. The explicit and general normative integration of environmental policy aims, as opposed to the implicit choices sufficing for technology binding instruments, is a clear task ahead.

13.3 Policy Instruments: What Are They Good for?

13.3.1 Policy Instruments: What Are They?

There are many instruments that might be relevant to environmental policy, including analysis tools, checklists and plans. More generally, environmental policy instruments can be seen as the means of implementing such policies. Here, a slightly more restrictive definition is used:

Environmental policy instruments are structured activities aimed at changing other activities in society to achieve environmental goals in a particular time schedule.

The main focus here will be on public policy, but the definition is not limited to this. The internal tradable permit system developed by Shell (see Box 13.4 in Section 13.4) is a perfect example of an environmental policy instrument. It has been superseded by public carbon trading systems like in Norway and the EU, see Hoffman (2006). Of course, not all policy instruments are intended for environmental policy. Other instruments for public policy, like those on energy and transport, may include environmental policy goals in addition to the prime non-environmental goal, or instruments may be multi-goal oriented, or the goals are not explicit, as with excises on petrol and alcohol. Multiple goals are implied in integrated policies. The division between environmental and non-environmental instruments is thus not strict. This is not a real problem, however, as policy instruments for non-environmental goals may be analysed in a very similar way.

Not all policies are structured, in the sense of being institutionalised in terms of instruments. Setting up the high-speed rail link to reduce air traffic between Paris and Lyon did indeed reduce air traffic at first, and reduced its growth afterwards. Green politicians may exhort the people in public speeches to leave their cars at home for at least 1 day a week, with some success. Such incidental activities towards policy goals, however, are not seen as instruments, as instruments would need to have an element of repetition in their application and operation. Hence, if high-speed railroads are built consistently on trajectories with rising air traffic, this provision of infrastructure can be regarded as an option-creating type of policy instrument. If a politician's speeches are part of a series set up for public education, they too may be seen as part of a communicative instrument. The dividing line is not strict, but, again, this is not really a problem.

As a final distinction in the definition, it may not be clear what exactly the environmental policy goals of some instruments are, and if these are really environmental goals. Raising prices for dumping waste in landfill sites may have non-environmental aims, for instance to increase the availability of easily accessible landfill sites, or to provide an incentive for increased use of under-utilised incineration plants. Or it may be seen as a means of reducing primary materials production, reducing resource depletion and the environmental effects related to ores processing. What exactly constitutes the 'real' prime motive is often difficult to establish, but also not very relevant. Such borderline instruments may still be analysed as instruments for environmental policy, with environmental effectiveness as one aspect of their assessment. The actual environmental effectiveness of instruments may not always be a distinguishing criterion. In certain circumstances, subsidies on environmental improvements, for instance, may have adverse effects, by delaying structural change which otherwise would have taken place. These subsidies are then environmental policy instruments that are not adequate for the particular situation, but they still are environmental policy instruments.

13.3.2 Why Have Policy Instruments?

Why bother about environmental policy instruments when actual policies based on integrated assessments can integrate environmental and other consequences in everyday actions? The main reason is that the complexity of all empirical relations and that of assessment are so great, and the information requirements so vast, that this option is not really available beyond a limited number of relatively simple occasions. Instruments work by simplifying a complex reality. They can be studied and assessed at a general level to decide what conditions can be identified for their sensible application. This may reduce the complexity of policy-making. At the receiving end, in society, most policies have their effects not in terms of directly correcting current activities of regulatees but, to a large extent, by guiding the planning of and decision-making on future activities. Having instruments whose nature is known from the relevant literature and past experience will make policies more predictable and facilitate adjustments. These adaptive mechanisms in society, if structured in stable patterns, can be regarded as a part of policy instrumentation as such, and may be more important than specific policy applications themselves. Without such applications, however, the general adaptive mechanism would cease to exist.

13.3.3 A Framework for Analysing Policy Instruments

Various policy instruments may be characterised in a common framework, with an empirical part explaining how they work and an evaluative part providing criteria on their value and adequacy. Such an empirical and evaluative analysis may be part of the policy cycle, in which the effects of policy, combined with other developments, feeds back into new policy preparation. The evaluative part is worked out in Section 13.3.6 below. The framework for the empirical part of the analysis has four main components: regulators, regulatees, society and the environment. These four components, as applied to one particular country or region, are mirrored by the same entities abroad (see Fig. 13.3). The framework defines the basic structure for modelling the functioning of environmental policy instruments. In a very basic mechanistic model, there is a single causal chain from the regulator's actions to the environmental effects. This limited framework already opens up a world with a rich variety of instruments and a high level of complexity of mechanisms in society.

The starting point in the model is some public regulation, which may be collective, as when a public body sets an emission standard in the metals plating industry, or private, as when the management board of a firm sets an environmental performance target. As a first step in the causal chain, there is the technical adjustment enforced among regulatees as the subjects of instrument application, see step 1 in Fig. 13.3. A second step usually centres around economic mechanisms and

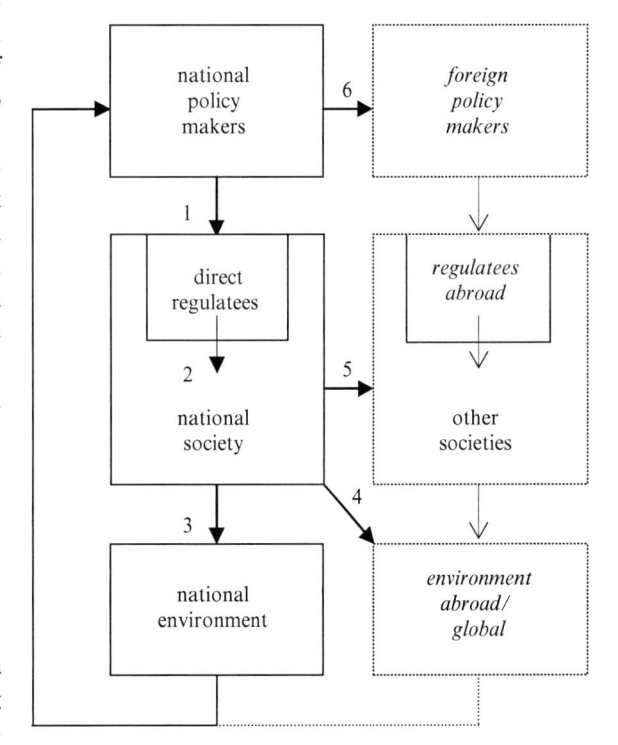

Fig. 13.3 Regulation: a simple model without feedback loops

relates to the costs imposed on regulatees, that is step 2 in Fig. 13.3. The degree to which such secondary effects are taken into account may vary. Effects on markets and other technologies will usually be part of the analysis, and will depend on the specific circumstances in these markets. Stricter emission standards in a small, open economy with a few large internationally operating firms may lead to emission reduction by shifting production to locations abroad, without necessarily changing technologies, and without reducing emissions at a global level. Conversely, technology adjustments in a large country with many small firms producing for the national market will be more pronounced, with only limited changes in the volumes produced. For a given set of national technical effects and volume changes, the net resulting environmental interventions can be defined and linked to effects on the national environment (3) as well as on the environment abroad and ultimately the global environment (4). As most markets are now international, national policies will induce economic changes abroad (5), which also have certain environmental effects. Finally, policies in one country may directly influence policies abroad (6). Dutch excises on petrol, for instance, are limited by the German excises on petrol, as too large a difference will lead to the closing of petrol stations in the border regions. Similarly, Californian regulations on 'emission-free cars' have set in motion regulatory activities and technology development in Japan and Europe.

The model with one-way causalities does not, however, correspond fully to reality, where feedback mechanisms, always dynamic, abound. Such feedback mechanisms may be quite complex. If regulatory capacity is limited, as in some way it always is, using regulatory power to solve one problem precludes its application for solving others. Also, using one type of instrument for one problem, like covenants to achieve the best available practice for energy saving in industry, will render the later introduction of emission taxes on CO_2 and NO_x rather unacceptable to industry. Negotiations on a covenant depend on what the industry sees as an alternative to the covenant: maybe emissions taxes or maybe costly actions. Hence, such negotiations necessarily take place 'in the shadow of the law', as Galanter (1981) and Scharpf (1991) phrased it. Indirect effects in society, in terms of the economic and environmental developments they induce, result from further feedback mechanisms. A very common mechanism is that by which regulations induce costs and hence lead to market changes and technology adaptations. For instance, costly measures to reduce emissions in the metals plating industry have induced a shift to high quality coatings, with other types of emissions resulting and with other policies being required. On the other hand, inducing changes in an industry may mean that cost-saving innovations that were already available come to be introduced faster and on a wider scale.

The ultimate feedback, of course, is that relating to environmental quality. The poor air quality in Mexico City not only raises direct costs of production by requiring air filters in many processes, it also reduces the legitimacy of the government and makes it difficult for firms to attract managers and specialists from abroad. Visible actions, in terms of standards and regulations, are most suitable to remedy these negative effects in the short term, by showing that 'something is being done'. Less visible actions, like changes in liability rules and market structure, might, however, be more effective in the long run. Hence, various feedback loops can influence instrument choice. Feedback mechanisms by their very nature are dynamic, with past choices determining future ones. Once a particular policy for some environmental problem has been designed, for instance one issuing emission rights, it becomes very difficult socially and even juridically to change over to policy instruments that are more in line with the polluter-pays principle, where emitters have to pay for their infringement of the right of others not to be polluted.

Why is the question how instruments function so important for their characterisation? The answer is that in reality, instruments are not independent given 'things'; their definition and description, and the analysis of their functioning, are intertwined and closely related. Most instrument definitions focus on their functioning, mostly on only one aspect of it. Thus, technical prescriptions focus on technologies applied in industry; tradable emission permits focus on the equalisation of marginal emission reduction costs between firms, industries, or countries; liability rules focus on specific enforcement procedures and actual compensation and covenants focus on the procedure in the policy formation process. None of these descriptions take all steps in the framework into account, let alone the feedback loops that usually exist.

Limited description may easily lead to simple assumptions about the other steps in the causal chains required for environmental effectiveness. For instance, many believe that emission permits may not be ideal in terms of minimising costs, but that at least they are a sure means of reaching specified results. In most countries, however, this belief is not well founded (Bardach and Kagan 1982; Bonus 1998; Hawkins 1984; Vogel 1986). Rules often exist on paper only and are not necessarily linked to actual practice, though the tendency towards formalisation as taking place in Europe may lead to closer correspondence. While environmental standards and regulations in the former Soviet Union were among the most stringent in the world, environmental quality was worse than in most other countries. This means that defining rules and regulations, or at least viewing their functioning in a broader framework, helps to avoid the myopia of partial views, making the context of their functioning more important.

13.3.4 Policy Instruments in Context

Most people would agree that policy instruments should be placed in the broader framework of their functioning. However, if this functioning were the basis of their definition, this could lead to counterintuitive results. Specifying this framework may show that what is referred to as one instrument is actually something different in different contexts. And the implementation of one and the same instrument may be very different in different prevailing circumstances. In litigious societies with limited general legitimacy, technology-binding legislation may be implemented effectively only by various types of lawsuit, with years of delay, while in highly integrated and less formalistic countries, legislation and implementation may be nearly synchronous, seemingly originating from informal consensual communication (see Vogel 1986). Conversely, for regulatees, technology-binding legislation may lead to adverse reactions or to internalisation of the rules enacted. Are these two separate instruments or are they context-related applications of the same instrument?

The broader effects of policy instruments in society depend greatly on previously established institutions. Although several communist countries have enacted emission taxes, they have had little effect. As increases in production volume were the primary aim of state-owned firms, with prices fixed and with the state reaping the profits and shouldering the losses, emission taxes were simply entered on the balance sheet, with no behavioural consequences in the firm itself (see Cole and Clark 1998) (Box 13.2). By contrast, similar taxes in capitalist countries with competitive markets may induce far-reaching behavioural changes. For instance, Dutch wastewater taxes enacted in the 1970s were followed by overall decreases in effluent volume by a factor of 20, mainly through process-integrated technology changes (Bressers 1988; Huppes and Kagan 1989). Endres (1997) has described the contextual requirements for an effective application of market-based instruments.

In some Western countries, like England and the Netherlands, policy development and implementation have been linked in a less recognisable way. It was broadly accepted that firms functioned without the obligatory permits (see Vogel 1986). In such a 'slightly illegal' situation, regulators might actually have greater influence on developments than if they had issued seemingly strict permits which tended to petrify the past. Tendencies towards a more formalised and litigious type of society, as in the US, have made this style of regulation more difficult. In the old situation, there

---

Box 13.2 Main lines of argumentation in Section 13.3

- The prime role of policy instruments is in *reducing social complexity* to manageable proportions.
- Instruments as institutional arrangements may not only be seen as tools used by governments to influence private behaviour, but also as a means of guiding behavioural relations between public bodies, and between individuals and private organisations.
- Environmental problems mainly result from the *external effects* economic activities have on *collective goods*.
- Environmental policy instruments help to avoid the *tragedy of the commons* by solving *the prisoner's dilemma* and preventing the *free rider problem*.
- The evaluation of instruments for environmental policy is not only based on *first-order criteria* for the evaluation of effects, like eco-efficiency and distributive justice, but also involves *second-order* criteria, like effects on competitiveness and influence on technology development, and *third-order strategic criteria*, like fitting in with general institutional, cultural and economic developments.

was bargaining in the shadow of the (possibly unreasonable) law, with a permit as the eventual short-term fixation of a situation. In the new, more formalised situation, it is not so clear how administrators can have a flexible influence. The covenant has taken over the bargaining step, while the shadow has not been clearly defined. Thus, the precise definition of an instrument depends on a more accurate assessment of its functioning. In the Dutch and English contexts as they were, permits were actually not the 'real' instruments for environmental policy at all; rather, they formed the background for negotiations, with mostly informal deals between regulators and regulatees achieving the actual environmental improvements (Box 13.3).

Recognition of the contextual specificity of policy instruments seems to be detrimental to the basic aim of distinguishing policy instruments, which is to simplify reality and make the behaviour of all concerned more predictable. If the instruments in official use hide what is actually happening, they increase complexity and may more usefully be removed. In administrative science and in sociology of law, the consequence has been that the idea of standard instruments has more or less been abandoned. In an ideal horizontal government, all stakeholders participate, in principle on an equal footing, resulting in deals that are best suited to the situation (see von Benda-Beckmann and Hoekema 1987). Again, why then bother about instruments? The answer is not straightforward. Governments may increasingly stimulate decentralised use of analytical tools like LCA to guide the outcomes of negotiating procedures in the right environmental direction. Such sensible developments should not be denied when discussing the role of instruments; they can instead be

---

Box 13.3 Sectoral emission reduction covenant

A **covenant** between the Dutch central government and NOGEPA (Netherlands Oil and Gas Exploration and Production Association) specifies goals for emissions reduction and the procedures to be used to achieve them. These goals relate to the performance of the firms involved, not individually, but as a group. The means used to meet the emissions reduction goals, like technical measures, are not specified in the Covenant. The procedures to meet the goals are specified, however, including procedures to redefine the goals when circumstances require this, for instance because of technical and economic criteria. For the actual implementation, all firms involved are to specify an Environmental Business Plan (BMP). Collectively, the result of the efforts indicated constitutes the Industry Environmental Plan, leading to the goals specified.

There is no strict relation between the overall emission reduction targets and the contribution different firms have to make to this target, which is at least partly a subject for collective consideration. The Covenant stipulates that specific measures are to be selected based on their cost-effectiveness, or eco-efficiency, with specific rules to quantify this efficiency. The usual private costing methods are applied for the economic analysis. The environmental analysis starts with the life cycle environmental interventions for each technical measure, based in principle on LCA inventory modelling of the activity with and without the measure.

The further effects on the environment are specified in terms of policy themes (see Guinée et al.). Eco-efficiency is established by first aggregating these theme scores into one eco-indicator score, using covenant-specific weights established by all those concerned in the covenant (see Huppes et al. 1997). This is one of the few examples worldwide, in a regulatory context, where trade-offs between environmental aspects are specified explicitly. Combining it with cost allows the environmental improvements per Euro of expenditure, that is, the cost-effectiveness, to be established for each emission-reducing technique in its specific context. The firms involved can make an inventory of possible techniques and rank them according to this cost-effectiveness. The most attractive techniques in this respect are included in the four-yearly Environmental Business Plan. Plotting this against the cost per unit of environmental improvement results in a flat curve, which at a certain point starts to rise steeply. It is in the interest of the firms involved to develop technologies in the flat range, to avoid high costs in the next Environmental Business Plan.

For a general description of the covenant, see www.nogepa.nl, section Environment. The full text of the Covenant (in Dutch only) is available at the same website (Last visited: August 2007).

made part of the development of environmental policy instruments. Therefore, instruments can also cover situations where governments may be quite invisible or even absent.

Thus, the question is not really why we need environmental policy instruments but which ones we need, or are able to have. There are several reasons why the decentralised horizontal instruments do not suffice. Firstly, there are limits to horizontal government, in terms of the human resources and knowledge required for adequate negotiations. This limitation exists on the side of government but also on that of other stakeholders. Most firms hate continuous negotiation because it drains their management capacity and so endangers their current and future functioning. Sustainability requires continuous adjustment of the behaviour of all firms and all consumers, which is now mainly guided by market considerations. Influencing this behaviour is clearly beyond the scope of the negotiating government. Hence, corrections to the outcomes of in-firm decision-making, including technology development and product design, and of market processes can be the subject of negotiating governance only in special situations, within the capacity limits of regulatory bodies. In addition, there is a more positive reason for having environmental policy instruments. Institutional development in society somehow has to cope with sustainability in a structural manner. Leaving a key value like sustainability to day-to-day negotiations by private persons and lower level officials would be unwise or even immoral. Somehow, quasi-automatic mechanisms, such as institutions, have to be shaped to safeguard the sustainability of operational, tactical and strategic decisions. Environmental policy instruments that do not depend on horizontal negotiations will play an indispensable part in these mechanisms. In between, there are the horizontal negotiations to decide which instruments to apply and how to apply them in practical policies. If such negotiations are to succeed, other instruments must be available as well; these are the 'harsh' instruments, serving as more or less ready options for governments to base their negotiating position on. They constitute the 'shadow of the law' in which governments can safeguard the sustainability of the outcome of negotiations in the networks involved.

There remains a domain where instruments at first sight may not seem to be relevant, as with some single big issues. For instance, should we not just curb the further growth of passenger air transport with its noise, stench and emissions? One option here would be to limit the growth of airports, which would not require specific instruments for environmental policy, as long as building airports requires public decision making. Another option, however, would be to use the price mechanism for environmental purposes. Taxing the emissions of carbon dioxide, nitrogen oxides and noise at sufficient rates will reduce these emissions, not only through technology adjustment but also through a reduction of the numbers of passengers transported. The growth of airports would then be reduced as a consequence of environmental measures, not as an indirect means of fostering the environmental aim of emission reduction. Using instruments for environmental policy in this way may prove to be more useful than seemingly simple measures like preventing airports from growing.

13.3.5 Environmental Problems: Causes and Solutions

Understanding the working mechanisms of instruments in solving environmental problems requires some insight into the causative mechanisms resulting in such environmental problems. In virtually all causal models for societal actions, some rational actor models play a key role. It is in such rational actor models that many of the causes of environmental problems can be identified and specified, at least for a start.

Common to all environmental problems is the causal mechanism by which direct private advantages of some actions outweigh the adverse environmental effects for the persons (or organisations) deciding on that action, while at the same time this adverse effect may affect others. To this extent, the adverse effect is external to these private considerations: it is an *external effect* of private actions on some large collectivity. If the single owner of an island cuts down his forest to create a garden, he simply prefers the garden to the forest, and there is as yet no environmental problem involved. It is only if others are bothered by the disappearance of the forest or the consequences of its disappearance that there is an environmental problem. This shows the *collective good* nature of environmental quality, and is a necessary mechanism for environmental problems to occur. In economic jargon, it is external effects of private actions that are detrimental to a collective good.

An additional mechanism in problem development is that the detrimental effects usually result not from the actions of a single person but from the combined actions of many persons. Though not strictly necessary in a logical sense, this is the typical situation for nearly all environmental problems. It is the *tragedy of the commons*. Single actions may contribute very little to the problem, but it is the multitude that leads to overall undesirable effects, and ultimately to the breakdown of the ecological system.

Together, these two mechanisms have a power that is hard to break. If an individual producer or consumer corrects his behaviour, his action may have negative effects on himself, in terms of costs and efforts, even if only in terms of the burden of nuisance. At the same time, the positive environmental effects of the adjusted behaviour may well be negligible. Such cases result in the *prisoner's dilemma*. Rational actors will only choose the behaviour with the preferred outcome if they can expect most others to act in the same way. In the 'wrong' situation, actual behaviour by others proves that this expectation is not justified and rational actors will choose the sub-optimal behaviour of contributing their share to the environmental problem. If corrective action is taken by all but a few, the environmental problem is largely solved, benefiting these few while they do not bear their share of the costs. This is the *free rider problem*. If the free riders are visible, social norms on collective action may easily become eroded, and with it the collective environmental good. If the free riders are invisible, the flesh is weak. It then takes highly internalised values for most people to remain on the right track. Environmental policy instruments somehow have to break this circle, either by effectively influencing all (or almost all) people, or at least by creating the confidence that most others will act accordingly.

A criticism of this model has been that the rational actor model underlying it is based on too simple a view of the real motives of real people. In reality, many actors indeed often behave altruistically, because they like doing so, and groups of actors often have an explicit or tacit mutual understanding to avoid 'bad' behaviour (see Sen 1977). At least some people bother about separate collection of waste streams, even in instances where others cannot see what exactly they are doing. They have internalised environmental norms to some extent. However, even if the restrictions on rationality are relaxed to include such social aspects of human behaviour, the unpleasant situation remains that detrimental environmental effects do occur and are re-created all the time by people who know the problems exist. So, even after taking into account the social nature of behaviour, the wrong choices are still made so often that environmental problems result.

On the basis of these theoretical considerations, sometimes named the 'field model', it is now possible to specify what environmental policy instruments should do. They should:

1. Avoid external effects on the environment and thus save collective goods, in this case related to environmental quality
2. Avoid the tragedy of the commons
3. Solve the prisoner's dilemma
4. Prevent the free rider problem

The model also allows us to indicate the mechanisms that policy instruments may be aimed at. The behaviour of an individual actor can be corrected in a number of ways (see Bressers and Klok 1988 for a fuller treatment).

1. The set of available alternatives can be changed.

This can be done by offering new alternatives, like separate collection facilities; by removing alternatives physically, as by fencing conservation areas; or by improving knowledge of existing but as yet unknown alternatives, as in nature education programmes.

2. The consequences of alternatives can be changed.

This may be done positively, as by subsidies on unleaded petrol, or negatively, as by threatening those who dump toxic wastes with jail sentences or other penalties.

3. The evaluation of the consequences of alternatives can be changed.

This can be done by changing the value system of actors, through educational processes, or by improving their active knowledge of the consequences of available alternatives, as by eco-labelling schemes.

These three types of mechanism in a policy instrument may avoid the *external effects* on the collective good, as well as the *prisoner's dilemma* and the *free rider* problem. In many situations, however, this is possible to a limited extent only and the situation may still be that of a prisoner's dilemma, with the free riding option lurking in the background. If only heavy industries were induced to reduce their CO_2 emissions substantially, using a large number of specific measures, the prisoner's dilemma would remain for all other actions not regulated, for which free riding may still remain the norm. In addition, the non-heavy industries would still be riding free. Not only would they continue to emit, but lower energy prices because of reduced demand by heavy industry may even allow them to increase their emissions. Thus, there is another role of policy instruments in avoiding the *tragedy of the commons*, by solving the prisoner's dilemma through creating the trust that everybody will indeed take part in the creation of the collective good.

4. *Justified trust in everybody's positive and due contribution to our common good can be created.*

This would make free riding virtually impossible. In this situation, the individual may seem to be deciding alone. In fact, however, he is making his decisions as if he were the collectivity, deciding for all people simultaneously. This co-operative solution is a very direct option for solving environmental problems, with collective values being internalised in individual decision-making. Although this option seems highly idealistic, it is a normal solution to many problems, at least in small communities. Most people take the trouble to store their waste in waste bins, even in larger communities. Only in situations of reduced awareness of norms, as when alcohol is used collectively, do the norms on littering break down. Tasks for the common good are executed, that is, behaviour is adjusted, because one expects everybody to do so. Still, this ideal is not always achieved, even if only a small number of people are involved, as can be seen in some families where children (or indeed parents) may try to avoid the daily dish-washing duties, always with good excuses at hand.

What are the requirements for this type of co-operative behaviour? One element would be that free riders undergo negative sanctions for free riding when they are caught. This would mean that the bad behaviour is forbidden at the level of the individual, and hence the co-operative approach is absent. However, control and sanctions may be more informal, not involving police and administration but friends and family, or the next-door neighbours. Another prerequisite is that the behavioural norm is clear and non-compliance is visible.

13.3.6 Evaluation Criteria for Policy Instruments

The analysis of how instruments for environmental policy work is one part of policy instrumentation hat is indispensable for any evaluation. The question is, however, what criteria to use in judging policy instruments. Somehow, the framework for the evaluation of instruments, and instrument-related policies, is to mirror the empirical analysis, which ultimately has to indicate effects in terms of these evaluation criteria. This adds a layer of analysis of a normative and political nature. As it is the consequences of instruments which are used as the basis of evaluation, the approach is that of consequentialism, not in the narrow sense of a utilitarian type of economism, but in the broad sense given to this term by Sen (2000). In this broad view, consequences obviously encompass the preferences of individuals, as is the case with utilitarianism exclusively. However, collective aspects like 'sustainability' may well also be covered by this broader consequentialist approach. Most of the criteria for evaluation specified below belong to this second approach (see Table 13.1). Three groups of criteria are distinguished. First-order criteria are related to more or less direct operational consequences of the application of the instrument, with environmental effectiveness as the first criterion. Second-order criteria relate to broader aspects of administration and economy, like administrative capacity and technology development. Strategic criteria, the most general category, link the instrument to the broader culture and institutions in society.

Sustainability may be an agreed general goal behind all specific environmental policies, as in many countries it already is. Its specification is normative and political. While the environmental effectiveness of instruments is an indisputable part of their evaluation, 'sustainability' and 'environment' are to some extent catch-all concepts, without precise underlying definitions.

The question is in which terms and with what level of detail this environmental effectiveness is to be established empirically, taking into account the mechanisms and time horizon? Is there a right to some minimum quality everywhere, for which permits might be the most suitable instruments, or is some overall level of emission reduction enough, for which emission taxes might be an adequate instrument? Are

Table 13.1 Criteria for evaluating policy instruments

First-order criteria	Second-order criteria	Strategic criteria

• Effectiveness (st, lt)[a]
• Social costs (st, lt)
• Eco-efficiency
• Distributive justice:
  – Intragenerational
  – Intergenerational
  – Justice as fairness
• Generative equality
	• Social and political acceptability
	• Within administrative capacities
	• Limited changes in competitiveness
	• Incentive for sustainable technology development
		• Fitting in with the broader conceptual framework for public policy
		• Fitting in with the broader institutional framework of society
		• Fitting in with general cultural developments
		• Fitting in with general economic developments

[a]st: short-term; lt: long-term.

cost-effectiveness and efficiency important parameters for assessment, as costs will be important for most people? Or are distributional effects more relevant for instrument choice? Are economic and environmental effects abroad to be taken into account, or only national ones? Do other aspects of justice, like the right to pollute versus the right not to be polluted, play a role in instrument choice? Is freedom of choice on the part of producers and consumers an independent criterion? Even if one were to refrain from normative choices in these respects, the same questions return as empirical ones in developing and implementing policies. Other people *will* mind distributional effects, people abroad *do* mind being polluted, and a broadly defined polluter-pays principle is generally accepted, implying that he who pollutes should be held responsible for the consequences of his actions. If people perceive policies as running counter to their values, the legitimacy of these instruments is reduced, as is their effectiveness. No doubt, the normative acceptability of instruments is one major empirical factor in both their political relevancy and their environmental effectiveness. Thus, the normative questions return through the backdoor.

When unleaded petrol came on the market – at a slightly higher price than leaded petrol, accompanied by a government campaign to 'buy green petrol' – many people's reaction was that if they were to buy green, they would be part of a minority shouldering the costs, while the main problem would remain unsolved. Thus, buying green would have a limited or even negative effectiveness, combined with an unfair sharing of burdens. On the basis of their normative appraisal, many regulatees decided not to co-operate, forcing the government to use other policy instruments. While straightforward product rules might have been the preferred option, the Dutch government chose to solve this collective action dilemma differently, by making leaded petrol more expensive through a tax measure, equivalent to taxing lead. Thus, leaded petrol was effectively forced off the market, and everybody paid the higher price of unleaded petrol. In this solution, burdens for environmental improvements are shared equally, in the sense that everybody pays the same price per litre. This is in line with one of several justice criteria, which states that the effort required for a certain amount of environmental improvement should be the same for everybody, at the margin. It is not an equal effort per head, as those who drive more and drive bigger cars also pay more. This criterion happens to be nearly equivalent to that of economic efficiency, in its static variant (Baumol and Oates 1988). This efficiency goal is central in economists approaches to instrument design and evaluation, see e.g. the survey paper by Bohm and Russell (1985).

It would, of course, be strange if criteria for judging environmental policies were different from those valid for other policies. Hence, the criteria are related to general views on the tasks of government. The sum of neo-liberal and social-democratic views roughly encompasses the entire political spectrum, with different political groups emphasising different aspects, but by and large involving the same ingredients. Giddens has compared the new consensus on public policy tasks with more traditional views. These new views are very much related to structural developments in the economy, with global markets and international networks replacing command and control in firms and fixed contractual relations between them. The emphasis in policy is also shifting from 'control' to 'generative policies', which allow 'individuals and groups to make things happen, rather than have things happen to them, in the context of overall social concerns and goals' (Giddens 1994: 15). The value of equality shifts from distributional equality, in terms of disposable income, to generative equality, in terms of security, self-respect and self-realisation (Giddens 1994: 191). These broader societal developments inevitably involve developments in environmental policy instruments.

What criteria can be used for the practical evaluation of instruments? In assessing environmental policy, the first criterion is probably environmental effectiveness. In the case of integrated environmental policies, however, effectiveness cannot easily be established separately from other values. How important is a toxic effect on child development compared to cancer risks at a later age and on a much longer time scale, and compared to the loss of plant species that might have had pharmaceutical value? Hence, environmental effectiveness can be established only on the basis of broader, non-environmental judgements. One such judgement relates to time. The time scale of effects requires choices in terms of the relative importance of future effects. The specific location of effects not only influences their type and magnitude, but involves different social groups as well. The spatial distribution also relates to the way effects abroad should be taken into account at home. Should national policies also aim at effects abroad as a part of their overall effectiveness? WTO regulations run counter to such considerations, unless the rules apply equally to home products and imported products. For imported products not produced at home, environmental rules will mostly be seen as undue trade limitations. Another problem is how to deal with low-probability, high-impact effects, where evaluating effectiveness is based on the degree to which risk avoidance or precaution is deemed important.

In addition to environmental effectiveness, itself value-based, there are also other values. One broadly accepted value is that of costs, or rather, welfare effects in terms of production losses required for environmental improvements. Instruments which help stimulate environmental technology development, like market-oriented, pricing-based, economic instruments, will entail lower costs in the long run (for the theoretical aspects, see Baumol and Oates 1988, for existing instruments see Opschoor and Vos 1989 and for empirical aspects, see Hemmelskamp 1997). A clear distinction is thus to be made in the cost criterion, between short-term costs (st) and long-term costs (lt). In multi-purpose instruments, the environmental cost-effectiveness (or 'eco-efficiency') can only be established by attributing one part of the cost to environmental goals and other parts to each of the other objectives contributed to. Other values relate to ethical categories of justice and equality, covering traditional distributional justice within and between generations, justice as fairness, and the newer generative equality (on these ethical issues, see Rawls 1972; Giddens 1994, 1998). Intergenerational justice has been made operational in an environmental context as part of 'sustainability' in the 1987 Brundtland Report. It is on a par with the discounting practice economists use, in which future effects count less than current ones.

However broadly the effects of environmental policy instruments are modelled, there will always remain relevant aspects beyond modelling that have to be taken into account and have to be specified as second-order criteria or as strategic criteria (see Table 13.1). The government has to operate with some legitimacy, which means that, on average, some minimum level of social and political acceptability and support is required in instrument application. Quantifying this aspect is difficult, but any environmental policy that forgets about such aspects will falter in the long run. Furthermore, instruments have to fit in to some extent with the capacities of the existing administration. Emission taxes are easier to implement in countries with a tradition of effective direct taxation. At a different level, major changes in sectoral competitiveness may create social instability and should generally be avoided. Another element, which is lacking in most quantified models, is how instruments influence technology development. Although these, partly overlapping, aspects are hardly quantifiable, they may be essential for a well-founded judgement on environmental policy instruments, and for the long-term effectiveness of environmental policy.

Since instrument choice may bind society for years or even decades to come, such choices are to be placed in a strategic context, not only taking account of relations as they are now, but also of developments taking place in this longer time perspective. Four main strategic areas can be distinguished, relating to politics, social structure, economy and culture:

1. Instruments are to fit in with the broader conceptual framework for public policy, e.g., along the lines sketched by Giddens (1998).
2. They should be in line with the broader institutional framework of society, e.g., in terms of increased mobility, functional specialisation of organisations and internationalisation of organisations.
3. They should take into consideration the general cultural developments, such as individualisation, mass culture and other-directedness, as outlined by many sociological studies.
4. Finally, instruments have to be adapted to general economic developments, such as the globalisation of markets, shifts from hierarchical co-ordination to network co-ordination, and shifts from the production of commodities to the production of services, as outlined by Castells (1996).

Some people, especially economists, simplify the evaluation by reducing it to an economic analysis. In principle, such an evaluation may cover all environmental effects; it is based on a specific discount rate; it uses a risk avoidance factor of zero; it uses an equal weight for every euro or dollar, thus disregarding income distribution; it takes only current private preferences into account; and it assumes these preferences to be independent, meaning that nobody cares about the welfare of anybody else. In this situation, each emission or environmental intervention may indeed have an environmental price tag in terms of a (negative) net present value. Of course, this also assumes that empirical consequences can be fully specified in terms of items relevant to such hypothetical individuals. Environmental policy instruments can then be evaluated in one unit: money. This overall score is the sum of the environmental effects, transformed into a net present value as outlined above, and the direct economic (market-related) effects. The single euro or dollar figure resulting then indicates which instrument to use in which situation.

In reality, of course, this hypothetical situation does not exist. Where price tags can be put on emissions, these prices relate to partial effects and will mostly be based on less than realistic assumptions. Several aspects of justice, such as equality and fairness, are omitted or included only superficially. Moreover, second-order and strategic criteria are mostly not suitable for economic quantification. Therefore, this option is too narrow to be the sole basis of a convincing instrument evaluation, although costs of course do play a role in such an evaluation.

13.4 Design, Analysis and Evaluation of Policy Instruments

Having described the role of environmental policy instruments in a long-term perspective, and having indicated their role in policy and the criteria for their evaluation, the final question is how to design, analyse and evaluate them at a more operational level. It should be clear by now that there is no one unique way of doing so. Many dimensions have their due place and not all of them can be included at the same level of aggregation. Four main dimensions can be distinguished, which together create a framework for design and analysis. The framework may be used to generate a large number of distinct policy instruments, well over a thousand, opening up options for well-argued choices on instrumentation. This section presents some existing examples of policy instruments, using the framework for their specification. The ultimate evaluation of designs and of implemented instruments is based on their actual functioning. Evaluation for the purpose of the design, or revision, of instruments has two main types of input (see Fig. 13.2 in the introduction to this chapter), one covering strategic points of view going beyond the level of the case, the other referring to the analysis and evaluation of the functioning of the instrument in its specific domain of application, i.e. for particular cases. The discussion of these two types of input completes the chapter on instruments for environmental policy.

13.4.1 A Framework for Design and Analysis

Instruments as societal structuring mechanisms bring order to the relations between actors and help to guide their behaviour. What is common to all policy instruments, and hence also to environmental instruments, is that they have to bring about a change in behaviour relative to the behaviour without application of the instrument. In specifying instruments, we distinguish four main types of dimension. In question form, these are:

1. Who influences whom?
2. What is the influencing mechanism?
3. What object is being influenced?
4. What is the operational goal?

These four empirical dimensions are quite general; in principle, they are the same all over the world, regardless of cultural differences. In addition, and again in principle, these four dimensions can be analysed more or less independently, while other instrument dimensions seem more closely tied to specific cultures and institutions. Juridical status, for instance, is often used as a defining characteristic, but juridical categories are linked to the specificity of judicial systems. For instance, an EU regulation does not have a counterpart in most other countries, while Anglo-Saxon common law is not used in most European or former communist countries. The systems converge however, as new legislation in Anglo-Saxon countries is statutory, while the courts in French law based European countries have increasing interpretational power.

Using these four dimensions, can we now specify in more detail what makes instruments into environmental policy instruments? The relations between actors, i.e., who is influencing whom, might give some clue. One could argue that instruments issued or used by a ministry or department of the environment are environmental policy instruments, using the first dimension. This would mean that an administrative reorganisation, for instance shifting policy-making in this area to the department of agriculture, could change their status. It would also mean that social instruments could never be environmental instruments. The second dimension, the influencing mechanism, is not suitable to define the environmental nature of a policy instrument either. Permits, taxes and excises and prohibitions have a general nature in all regulatory contexts. The third dimension, the object influenced has no specific environmental status either, as there are many rules on products and installations not related to the environment. It is the fourth dimension, the operational goal, indicating the direction of behavioural adjustment, that decides whether an instrument is an environmental instrument: its goal has to be an environmental one, or at least its ultimate aim should be. This is not always a clear-cut criterion. If, for example, a government stimulates hydropower with the intention of becoming less dependent on imported fossil fuels, this also helps to reduce CO_2 emissions. Hence, the operational goal of stimulating hydropower need not be based on environmental considerations. Such borderline cases will certainly exist in practice. Even if none of the participants sees a particular instrument as an environmental policy instrument, it could still be categorised and analysed as such, relating the instrument to this goal and ultimate aim.

Although the general tasks of environmental policy instruments – avoiding the *tragedy of the commons*, solving the *prisoner's dilemma* and preventing the *free rider problem* – could easily have been made into defining characteristics, they have not been included here because of their rather abstract and strategic nature. However, they do still play a role in instrument design and instrument evaluation, be it in general or at a case level. The normative evaluation criteria specified in the previous section have not been included here either, thus creating a distinction between empirical-descriptive elements of policy instruments and their normative evaluation. Of course, there has to be a link between descriptive elements and evaluation, since it is ultimately the evaluation that counts. The evaluation is based not only on the direct effects of the instruments; decisions should be guided by the 'ultimate' effects. There is a tendency to include some standardised effect mechanisms into policy instruments, like including global warming potential (GWP) in national and international climate policy. Thus, some mechanism may play a role in the goal specified in the instrument. Most social and environmental mechanisms, however, will be independent of the instrument, which means that their analysis must be included as a separate step in the evaluation of policy instruments, not in their definition.

The four main dimensions chosen now need to be defined in greater detail. It should be clear that there is not one general truth at this level either. For instance, 'actors' can be described in many dimensions, for instance as individuals or firms, while firms can be described as small or medium sized firms, large national firms or multi-nationals, etc. What would be the guiding principle for such further choices? In the end, the question that has to be answered is how policy instruments can fulfil their function of simplifying the complexities of reality to allow effective and concerted actions towards environmental goals, and distinctions should serve this ultimate purpose. However, the purpose of this section is also very practical, namely to decide how real instruments can be created and how their expected functioning can be evaluated. In approaching this task, there is a tendency to introduce further distinctions relevant to the situation. As our four main dimensions already lead to over a thousand instrument categories, further systematic detailing, however relevant it may be, should be used sparsely if at all. Such relevant additional aspects are more suitable for the evaluation of, in this sense, more sparsely defined instruments.

13.4.1.1 Who Influences Whom?

In deciding who is influencing whom, a major distinction can be made between governments on the one hand and non-governmental private actors, like individuals, firms and organisations, on the other. These two types of actors involved in instrument application lead to a further categorisation of instruments, involving a distinction between three types of actor relations (see Table 13.2).

This section uses the three main types of actor relations to categorise the instruments as political-administrative instruments, regulatory instruments and social instruments.

Regulatory instruments are the most common type. An environmental permit is a major instrument for governments to influence private actors (including publicly owned firms), as are emission taxes, such as SO_2 taxes creating a market incentive for reducing SO_2 emissions. Political-administrative instruments may work at an international or national level. An international treaty like the Montreal Protocol on substances that deplete the ozone layer is an instrument between governments, while an EU environmental regulation binds national governments in the EU. The ISO 14001 industry standard is an example of a social instrument. By specifying rules, it guarantees that environmental audits have a degree of generality and reliability in describing the environmental performance of firms, creating an incentive to take environmental aspects seriously. Other examples are private certification systems, such as those used for food in supermarkets, influencing the behaviour of food producers and creating options for environmentally oriented choices by consumers.

Table 13.2 Actor relations based instrument specifications

Actor relations	Name of instrument	Examples
Governments influencing governments	Political-administrative instruments	• Montreal Protocol • EU regulations
Governments influencing private actors	Regulatory instruments	• Environmental permits • SO_2 emission charges
Private actors influencing private actors	Social instruments	• ISO 14000 Series • Private certification systems

13.4.1.2 What Is the Influencing Mechanism?

The influencing mechanism specifies how one actor influences the other. We follow the main categories distinguished by Bressers and Klok as explained in the previous section, with some further differentiation as found in regulatory practice, see Table 13.3 for a survey. The first mechanism involves changing the set of available alternatives, for instance by their prohibition or prescription, or by the creation of new options. These define the first three instrument categories. Secondly, the consequences of alternatives can be changed, for instance by influencing market prices and market volumes. This is the category of economic instruments. The third major mechanism is the change in the evaluation of consequences, achieved for instance through information on the effects of actions and through normative guidance. These are the cultural instruments. In addition to these actor theory based categories, we distinguish the influence exerted by changes in the institutional structure of society, as structural instruments influencing actions. Liability rules are a major example. Finally, there are the procedural influences, functioning as procedural instruments. They are introduced separately here, as their use is widespread in specifically environmental contexts, as in covenants and audit systems.

The influencing mechanism is to be clearly distinguished from mechanisms further down the causal chains. Such further effects may well involve the same

Table 13.3 A typology of instrument mechanisms

Mechanism-based instrument specifications	Examples
Prohibiting instruments	• No cadmium stabiliser allowed in PVC as a building material
Prescriptive instruments	• Legal obligation for separate waste collection
Option-creating instruments	• Multiple waste containers for separate collection
Economic instruments	• Volume: auctioned car ownership rights • Price: energy tax, SO_2 tax
Cultural instruments	• Normative: ecolabel on products • Relate organic solvents to summer smog
Structural instruments	• Liability rules • Public decision-making safeguards
Procedural instruments	• Obligatory environmental officer in firm • ISO 14001 audit

mechanisms, but do not form the basis for instrument categorisation. For instance, a change in liability rules will ultimately involve market mechanisms and the creation of options with lower liability risks. The influencing mechanism, however, is that of a structural instrument. The terminology used in practice varies somewhat. For instance, the prohibition and prescription of options is also referred to as direct instruments or as juridical or legal instruments. This terminology seems awkward, however, as option creation can be seen as a direct instrument as well, and financial instruments like emission taxes also have a distinct legal status. The combination of prohibiting and prescriptive instruments is also referred to as binding instruments.

13.4.1.3 What Object Is Being Influenced?

A further dimension is the nature of the object being affected by the influencing mechanism. A first basic distinction in objects being regulated is that between single objects and classes of objects. This relates to the applicability of the instrument and to the options for implementation. A second distinction is that between products, which are mostly mobile physical entities; installations, which are mostly immobile physical entities; and activities, focussing on more general behavioural options. Such behavioural options will usually be linked to some physical object, like speed limits for cars in a city. These objects may be defined in very abstract terms, however, like environmental auditing rules for firms, relating to all their installations. In the US, most regulated physical objects are called installations. Since no example of a single activity being regulated comes to mind, this class has been omitted. As the classes of objects are necessarily restricted as to spatial scale level, a further distinction is made based on classes of objects being regional or global. This results in eight types of object (see Fig. 13.4 and Table 13.4). While no specific instrument names are linked to these classes, the term *product policy* is becoming popular in EU policy and in a number of countries, referring to classes of products in the EU.

Regulating material objects, as 'things', is not done because of their inherent properties. Ultimately, it is only the processes in which they function, as activities, which influence the environment by causing environmental interventions. Environmental policy instruments try to influence these interventions, not the techniques

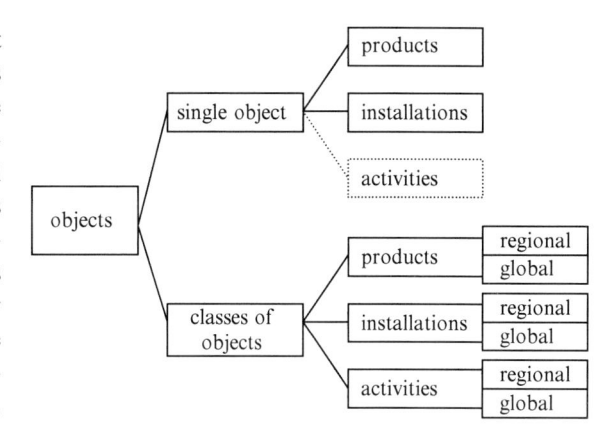

Fig. 13.4 A classification scheme of regulated objects

Table 13.4 A typology of influenced objects

Object influenced	Examples
Single product	• Single aircraft flying permit
Single installation/firm	• Testing whether a building harmonises with the landscape
	• Permit requiring safety valve on some pressure vessel
Product classes, regional	• EU rule on compulsory three-way catalytic converters in cars
	• Rules on NO_x concentrations from household boilers
Product classes, global	• WTO rules on non-discrimination
Installation classes, regional	• German rules on allowable SO_x emissions per kWh at power generation
Installation classes, global	• IAEA rules on safety requirements for nuclear installations
Classes of activities, regional	• Local speed limits for passenger cars
Classes of activities, global	• ISO 14001 requirements on environmental planning in firms

as such. However, regulating things may be easier than regulating activities, as most things can be inspected at any time, while effective inspection of behavioural aspects is much more complicated. Measuring the concentration of NOx in a boiler outlet requires real-time measurements, so it is easier to prescribe a burner type. Cadmium in PVC stabilisers in building materials, however, can be measured at any time. For the sake of regulatory efficiency, requirements are often specified in terms of the technical composition of the product or installation, assuming that these will result in an emission reduction. In a considerable proportion of cases, however, this relation is not fixed. Bypassing a flue gas purification installation, for instance, saves costs. Illegal bypassing can be stopped only if the inspector calls at the right time.

Classes of activities may be defined narrowly, as with denied access to National Parks in Germany after sunset, or broadly, as in the environmental auditing of all firms in the world. A good deal of complexity can be introduced in defining classes of activities. LCA (Life Cycle Assessment) refers to all processes (= activities) involved in having the function of a product delivered. In environmental audits of firms, it is becoming common to incorporate the supply chains involved. This is similar to LCA, as the object is also a system of linked activities. Such larger entities are attractive, as a policy instrument may influence all of them in a balanced way. Since some arbitrariness is involved in defining such systems, binding measures may be more difficult to apply.

13.4.1.4 What Is the Operational Goal?

Operational goals specify what the instrument states as the objective to be approached or achieved. 'Cadmium in PVC window frames' may be forbidden, as in the US, or a 'weighted maximum of emissions' may be allowed, as in a number of Dutch environmental covenants. We can categorise the goals in relation to how close they are to ultimate environmental aims, see Table 13.5 for a survey. Most traditional binding environmental policy instruments, like the operating permit, have relied heavily on easily verifiable rules for the composition of a product or installation, like banning the application of PCBs, or prescribing a filter on air outlets, or banning some mercury-based technology for chlorine production. Since the 1980s, the emphasis on efficiency and problem prevention has induced a shift towards goal specification closer to the ultimate environmental goals. Goal-oriented permits specify the maximum emissions of the installation, or the maximum emissions per unit of product produced, as in SO_x per kWh of electricity produced in Germany. Policy integration, as a central element in efficiency, has led to further steps, through integration in terms of policy themes like global warming and acidification. The Kyoto Protocol includes steps to integrate various substances in terms of their contribution to the policy theme of climate change, based on their Global Warming Potential. Estimates by the World Bank indicate that such integrated policies, as opposed to the focus on CO_2 emission by energy systems, can reduce the cost of climate policy by 60%. Further integration, across various policy themes, has been occurring over the last decade, like the use of some eco-indicator as an integration of similar effect types, or even integrating all environmental effects in a normatively based overall environmental score. An overall evaluation of all environmental interventions makes the relative importance of various environmental effect categories explicit. By allowing a shift between them, based on the explicit trade-off, it is their combined effect which is regulated. The NOGEPA (Netherlands Oil and Gas Producers Association) covenant in the Netherlands specifies such an integration. It forms the basis for one further integration step, combining environmental and economic goals, in order to select the most eco-efficient measures for environmental improvement.

The closer the instrument goals are to environmental aims, the more efficient the instrument can be, by avoiding unnecessary technology fixation. If a policy instrument uses the mass of a car as a proxy for its emissions,

Table 13.5 Typology of operational environmental goals in instruments

Operational goals	Examples
Composition of product or installation	• Cadmium stabiliser in PVC as a building material
	• Percentage of post-consumer paper waste in paper
Technology characteristic	• Double skin in oil tankers
	• Inherent safety in nuclear installations
	• Take-back legislation (several European countries)
Single environmental intervention	• Noise-based landing fees in airportsCO_2 tax
	• Ambient air quality standards
Set of environmental interventions	• Set of allowable emissions as a goal permit
Single theme score	• Emission reduction targets (Kyoto Protocol)
	• Environmental policy plans
Set of theme scores	• LCA-based rules for waste prevention (Germany, Netherlands)
Effect-oriented eco-indicator scores	• Limited set of eco-indicators in building regulations
Single integrated environmental score	• Best Available Technology specification (EU)
	• EPS based design rules in industry
	• Theme-based weighting (NOGEPA covenant)
Combined environmental and economic score	• Eco-efficiency as a selection criterion for emission-reducing technical measures (NOGEPA covenant)

the regulation does not influence the type of engine characteristics or the driving style. If only the mass indicator is influenced, cars will get lighter but not necessarily cleaner at the same rate. If emissions are taxed or bound to a maximum, these other characteristics will primarily be influenced, with probably a limited effect on car mass. Measurement techniques for such instruments are not yet available but may well be developed in the near future.

As regards integrated policy goals in instruments, there is a gap between what modelling can more or less realistically achieve (with adequate validity and reliability), and what is needed for integration in the single environmental score. Somehow, the modelled multitude of environmental interventions and other effects have to be transformed into an overall judgement. There are several methods for this purpose, which do the undoable. Economists derive overall evaluations in monetary terms on the basis of past behaviour and the stated preferences of individuals. In a very different approach, impact assessment in LCA first integrates, on the basis of modelling, into several environmental policy themes and then through a weighting procedure merges these into an overall evaluation. In whatever way this integration is achieved, for reasons of policy consistency it would be necessary to use the same trick every time. There is a modest requirement for the overall rationality of environmental policy, which is that the trade-offs between various effects are equal for each policy. This simple requirement can be transformed into a conditional statement (see von Neumann and Morgenstern 1953; Sen 1982): if policies are rational, there is a single set of weights on their effects which can 'explain' all policy choices made. If this condition is not met, this means that it would have been possible to reach the policy aims in a less costly, that is, more eco-efficient way. Given the theoretical and practical limitations of modelling, policy integration can only be reached through practical choices, based on incompletely developed arguments. Making a start here is better than accepting the even less attractive current state of affairs.

When making this start, the best one can do is to strive for a consistent and transparent solution, on the one hand taking into account the as yet only partly known real mechanisms, and on the other hand specifying the normative background of the evaluation. This problem area, it seems, has not yet been under serious scientific scrutiny. Some practical solutions are available, like using a panel of officials and experts (e.g., NOGEPA covenant, see Huppes et al. 1997); using policy aims for weighing emissions into one score (e.g., Swiss or Norwegian Ecopoints); using a mixture of partial economic valuations or some equivalency factors (e.g. ExternE and EPS); and applying some preferences or value types, see Guinée et al. (2002) for a survey.

All these practical models have been developed in limited domains, in hardly peer-reviewed studies, without a broad public discussion. Important questions are only touched upon and not answered. How can we differentiate between reversible effects, like ecosystem degradation, and irreversible effects, like species extinction? How can we differentiate between low-probability-high-risk effects, such as possible runaway effects in climate change, and more probable slow-change scenarios? Assuming that uncertainties can be specified in terms of risk or at least subjective probability, how can we evaluate options with different probabilities? Even in the extremely well researched area of climate change, surprises and outright disasters have not yet been incorporated in policy models, though the first frameworks are being developed (see Schneider and Kuntz-Duriseti 2002). Given the uncertainties involved for each policy theme, how can we make a comparative evaluation of climate change effects, which can hardly be specified in economic (i.e., welfare) terms, as against effects of acidification resulting in reduced crop yields and increased corrosion, which can quite easily be specified in terms of economic losses? Such fundamental problems have yet to be solved before integrated goals can be used more widely in policy instruments and in policy.

13.4.1.5 Combining Actor Relations, Mechanisms, Objects and Goals

Combining the four main dimensions, those of actor relations (3); influencing mechanisms (7); objects influenced (8); and goals (9), results in a large number of instrument types: $3 \times 7 \times 8 \times 9 = 1,512$. Not all combinations may be relevant, however, or even possible. A technology specification, as a goal, cannot easily be applied to 'classes of activities', as an instrument object. The combination of more binding instrument mechanisms with more encompassing objects of regulation and with more integrated policy goals makes policy development and implementation increasingly complex. However, the problems involved in this integration at the instrument level are mainly the same as those at a policy level using several instruments. Criteria like consistency are very visible at the instrument level and not at the policy level. The aim would be to have consistency at the latter level as well, to avoid unnecessary costs in achieving environmental aims.

The aim of consistency in environmental policy, and the related aim of eco-efficiency, may be achieved through adapted markets, as by emission taxes, tradable emission permit systems and liability rules. Assuming some degree of competitiveness in the markets, the normative information on policy aims and the empirical relations in markets and technologies are combined in the price structure faced by individual decision-makers in society, and taken into account in their decisions. Such broadly applicable 'macro-instruments' (see Huppes et al. 1992; Huppes 1993) may, however, not be ready for widespread application, for administrative-technical as well as political reasons. The market-based information system, as Adam Smith's invisible hand, is therefore mainly lacking, guiding economic decisions in the wrong directions from an environmental point of view. It is to be created by environmental policy, as efficiency is a generally accepted central goal. All the more complex object types and goals types have been set up to create this integrated view on environmental regulation. Attempts to avoid problem shifting to other emissions, to other policy themes and to other times, which are preconditions for efficiency, constitute a key element in a tool like LCA, which is beginning to be incorporated in policy instruments.

In using the more integrated *classes of activities* as objects, their definition as a system with internal relations has become a research subject in itself, that of environmental systems analysis. Major tools for this purpose are SFA (substance flow analysis), LCA (life cycle assessment) and E-IOA (environmentally extended input-output analysis). They are mainly simple models, based on linear relations and constant technological relations. However, the fact that they are simple, with a limited validity and reliability, means they can be made operational. Where decisions on technologies and markets are made at a decentralised level without much direct government influence, as is increasingly the case in a globalising world, we need a decentralised tool that comprehensively indicates the ultimate effects of such decisions, incorporated in a policy instrument creating the incentive to use it. Mostly, such tools are incorporated in social instruments or in weak regulatory instruments. Thus, the potentially most efficient instruments can only play a very limited role at present, while the not-so-efficient regulation of technologies by prohibiting and prescriptive instruments still carries the main burden of environmental policy implementation.

When specifying integrated goals for complex objects in policy instruments, it should be clear that instruments cannot comprise the full extent of all real effect mechanisms. The creation of direct effects inevitably leads to a whole range of direct and secondary effects on society and the environment, not all of which can be incorporated. It is not even the most sophisticated models available which play a role in instruments, as in covenants in Germany and the Netherlands. The simplified standardised modelling in these instruments is an approximation, which should not be confused with state of the art modelling of real effects. By being operational at a decentralised level, the approximation may represent an improvement on other ways of regulating technologies and products, which may not take the simple effect mechanisms into account at all.

Fortunately, things in real world situations are sometimes less complicated than they are theoretically perceived to be. In such cases, environmental policy instruments can be simpler as well. For example, when banning a toxic and persistent agrochemical for which slightly more expensive alternatives are available, the real effect route in the economy hardly has primary or secondary effects in the chain, nor is there much complexity in the environmental pathways towards valued endpoints in terms of human and eco-toxic effects. There is then no reason to complicate the instrument and burden it with complex effect mechanisms and evaluations. A simple prohibition of the agrochemical will do, after a relatively simple analysis of effect chains in the policy formation process, including a check on the availability of not too costly alternatives. Such easy solutions have mostly already been implemented, however. After more than 30 years of active environmental policy, it seems that most simple end-of-pipe (add-on) measures and simple product prohibitions have already been enacted. Such policies may now even start to hamper environmental progress by fixing old technologies. The remaining problems are more complicated and may well require a more sophisticated instrumentation.

13.4.2 A Thousand Instruments Defined, with Examples

A policy instrument may be defined by combining elements from the four basic instrument dimensions discussed above, in principle leading to well over a thousand major instrument types. These dimensions are more or less independent, so they may be used as a framework for instrument development, as an *instrument generator*. Any combination defines the main lines of an instrument. Figure 13.5, 'the instrument generator', gives some examples. Take, for instance, 'social instrument' from the *actor relations* column; use 'economic instruments', used here in the sense of pricing, from the set of *instrument mechanisms*; take 'product classes, global' from the set of *objects influenced*; and take 'single environmental intervention' from the set of *operational goals*. This instrument then can be further specified in terms of the product classes, for instance using 'aircraft', with different noise levels, measured in a specified way, as the fully operational goal in pricing. The result is the 'noise related airport landing fees', which is a fairly common social instrument used by airports near larger cities. The motivation behind a social instrument may be another policy instrument, especially a regulatory one. In the example given here, the motivation-creating regulatory instrument may be the operating permit of the airport, stating maximum noise levels in surrounding residential areas, which is a prescriptive regulatory instrument for a single installation, involving both technology characteristics and sets of environmental interventions as its goals. A second motivation step, behind this regulatory instrument, may be a political-administrative instrument, such as an EU directive on permissible noise levels in residential areas. Another example, from the US, is the extended liability which has been achieved for toxic wastes that have not been treated properly, to avoid damages to the environment. It involves a structural mechanism and pertains to all waste sites with toxic wastes in the US. All highly hazardous substances are involved, as potential environmental interventions. Another example is the Kyoto protocol,

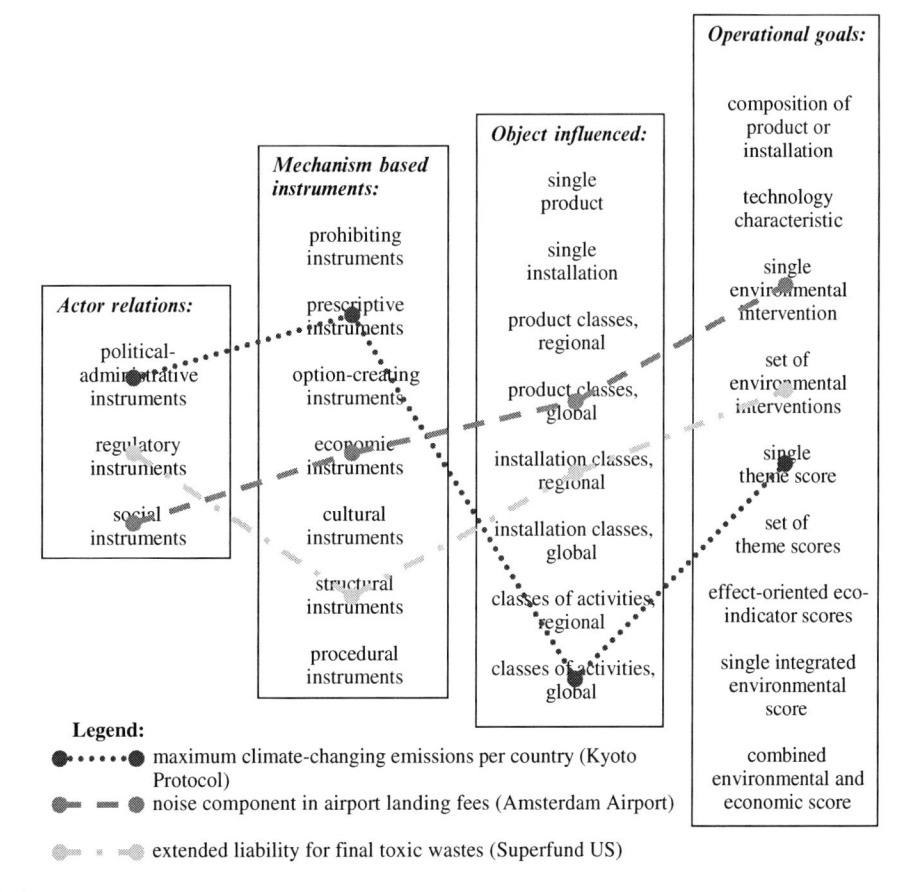

Fig. 13.5 The instrument generator

which limits the total amount of climate changing emissions in terms of their global warming potential. These limits are set for countries, involving all emitting activities in each country, as one level of geographic region.

Using only the first two dimensions, for a start, results in 27 instrument categories. An example of each is given below.

I Political-administrative instruments (guiding relations between public bodies)

The meaning of the word government is restricted here to the regulatory part of governments, engaged in planning, developing and implementing policies, and using policy instruments in such policies (in this case environmental policies). Other operative public tasks, like building and maintaining roads, canals and dikes, maintaining an army and distributing electricity, are productive or consumptive activities, to be regulated like any other economic activity. The relations between governments as regulators, and private persons and organisations as regulatees, are in most cases hierarchical. The relations between governments may be hierarchical as well, as when the EU binds the policies of countries with Directives, and national governments prescribe policies to regional and local governments. However, in the international context, most relations between environmental policies are horizontal, as in bilateral and multinational treaties. Some hierarchy is implied when international public bodies are involved. In addition, where bilateral relations seem to be involved, there may still be a hierarchy. For instance, while 'joint implementation' is dealt with at the interstate level, the rules for joint implementation are dealt with in the Kyoto Protocol, and the possible future extensions to that protocol might be designed hierarchically under UN leadership.

The prohibiting and prescriptive instruments are combined here as *Binding instruments*. In principle, the seven main implementation mechanisms discussed above can be applied.

1. Binding instruments

- International treaties and conventions including binding elements, like the Montreal Protocol, the Kyoto Protocol and the Biosafety Protocol
- EU environmental directives for member states

2. Option-creating instruments

- Clean development mechanism, as under the Kyoto Protocol
- Multilateral Ozone Fund under the Montreal Protocol

These options seem to stretch the concept a bit. However, their basic function is to allow states to develop regulatory activities which would not be possible or at least rather unlikely without the explicit development of the option.

3. Economic instruments

- Internationally tradable emission reduction obligations for climate changing emissions, as might be based on the Kyoto protocol

4. Cultural instruments

- International guidelines, as by OECD, and in the EU IPPC/BAT (Integrated Pollution Prevention and Control/Best Available Technology) rules
- Rio Declaration, AGENDA 21
- National indicative guidelines for local zoning laws
- ILO conventions regarding environment-related labour standards

5. Structural instruments

- WTO rules on environmental considerations in restrictions on trade

6 Procedural instruments

- International Criminal Court (ICC) (no environmental example available)

II Regulatory instruments (guiding public regulator – private regulatee relations)

Regulatory instruments guide regulator – regulatee relations; they are the traditional environmental policy instruments.

1. Prohibiting and prescribing (= binding) instruments

- Binding instruments can be either prohibiting or prescriptive. Prohibiting instruments are usually conditional, in that something is forbidden unless some requirements are fulfilled. In symmetrical situations, as with speed limits, the difference between prohibiting and prescribing instruments is small: it is forbidden to drive faster than the limit or prescribed to drive more slowly.
- Allowable coolants in household refrigerators
- Operating permits for installations, the workhorse of environmental policy
- Land use regulations and zoning laws

2. *Option-creating instruments*

  • Separate waste collection facilities, in all Western countries

Option creation can be direct, as in providing separate waste collection facilities to households that may voluntarily separate their wastes. No prescription or prohibition is involved.

3. *Economic instruments: price-changing (financial) instruments*

  • Emission taxes (or: charges, levies, excises) on CO_2 (e.g. Norway) or on SO_2, formerly in Japan, now in China
  • Road pricing (most Western countries)

4. *Economic instruments: market volume instruments ('things' only)*

  • Tradable emission permits, like SO_x permits in the US
  • Tradable production rights, like fishery rights
  • Tradable product ownership permits, like car permits in Singapore

5. *Cultural instruments: non-compulsory structured information*

  • Public eco-labelling schemes
  • Public certification of firms, as for refrigerator repair firms in the Netherlands

6. *Structural/institutional instruments*

  • Extended liability
  • Good housekeeping ownership rules
  • Educational system, Copernicus charter, etc.

7. *Procedural instruments*

  • Covenants, voluntary agreements 'in the shadow of the law'
  • Environmental Impact Assessment rules
  • Obligatory information disclosure, as in the US Toxic Releases Inventory (TRI) Act

III Social instruments (guiding relations between private actors)

These instruments are similar to political-administrative instruments in that they may reflect horizontal relations between equals, or have a hierarchical element in them, as is often the case in environmental supply chain management. Again, the six main implementation mechanisms apply.

1. *Binding instruments*

  • Contractually specified rules for waste management, as when firms commit themselves to delivering a certain amount of waste over a longer period of time (Netherlands)

2. *Option-creating instruments*

  • Battery take-back facilities in supermarkets

3. *Market instruments*

  • Noise-related landing fees at airports
  • Deposit-refund system on cadmium-containing rechargeable batteries for household appliances, on a voluntary basis
  • In-firm tradable emission permits (see Box 13.4)

4. *Cultural instruments*

  • Green marketing (all Western countries)
  • Green accounting (ISO)
  • Ecolabelling rules (ISO)

5. *Structural instruments*

  • Standard contracts specifying adherence to environmental standards, set up for instance by a branche organisation

6. *Procedural instruments*

  • ISO 9000 Series, on quality control
  • ISO 14000 Series, on environmental performance measures

The international standard on environmental auditing, ISO 14001, for instance, is a procedural instrument, requiring firms to take due notice of environmental aspects in their operations, in the sense of having an environmental policy plan, having officials responsible for checking its progress, etc. If rules were incorporated on how to further specify environmental performance, the instrument would become a cultural instrument.

13.4.3 Choice of Instruments in Policy Design

Using the framework for environmental policy instrumentation and the evaluation criteria indicated above would seem to suffice for a rational development of

Box 13.4 Multinational company tradable emission permit

Some major oil companies, including Shell, have introduced emission trading between the firms comprising these multinational companies. The emission trading focuses on climate-changing emissions, like carbon dioxide and methane. Each independent business unit within Shell has an number of emissions rights, which may be sold to other Shell units. There is an accounting system which establishes the actual emissions of each unit. Emissions without a permit are not allowed, resulting in a company-internal cost penalty. If a unit has more permits than it needs, it will try to sell them to other units. If it wants to expand, it may acquire permits on the internal but global company market. The total amount of emissions permitted is being reduced slowly, based on the company's environmental plans, by reducing the allowable emission volume per permit each year. The combination of business expansion and the reduction of the overall emission volume permitted results in upward pressure on permit prices, while environmentally oriented technological development leads to a downward pressure. What will the effects of this instrument be?

The effects in terms of company emissions are quite clear: the goals of Shell's environmental policy plan are met, while leaving the business units their technological freedom. The emission reduction is achieved in the most efficient way, as each business unit reduces its emissions to the level where cost reductions are (roughly) equal to the costs of having the permit. A major problem in implementing such a system is choosing the system boundaries. How can firms partly owned by Shell and partly by other companies participate in the scheme? What happens to the total number of Shell permits if Shell sells some of its activities, or acquires others?

What the net environmental effects in the global society will be, in terms of reduced climate-changing emissions, is less clear, which in this case is due to quite complex indirect effects. For activities where a company has competitors with less stringent policies, its costs will show a relative rise. Hence, in the course of time, there will be a shift to firms not participating in this (or a similar) emission permit trading scheme. Also, questions arise as to how company environmental policy relates to public environmental policies in the various countries where the company operates. If more stringent policies are introduced in some countries, the permit system will no longer have effects there, as induced costs of emission reduction are higher than the permit costs. In countries with emissions taxes, the firms involved will have a greater incentive to reduce emissions than other firms in the company, reducing the overall efficiency within the company. In this sense, companies using such a scheme will create an argument against more stringent national policies. Conversely, if public policies are less stringent than the company scheme, they become superfluous. In this situation, multinationals like Shell create an incentive for national governments to implement more stringent policies. The overall effect will be that public policies will tend to be harmonised at a global level towards the level of emission reduction indicated by the large multinationals. If most multinationals were to come up with similar and equally stringent schemes, there would be a clear drive towards uniform policies, at the level of stringency chosen by those firms, and not by governments. It should be relatively easy to extend the tradable permit system to trade between firms. The choice of their policy instrument will influence policy implementation by governments as well, making it very difficult for instance to implement emission taxing schemes on top of the company tradable emission permit scheme. Shell chairman Moody-Stuart has called upon governments to implement similar market-based mechanisms to achieve their Kyoto targets. For further information on the Shell tradable emission permit system (STEPS), see www.shell.com/climate.

policy instruments for the environmental problems facing us. The question may be asked, however, how the analytical approach depicted here relates to process aspects in instrument and policy design at a case level. This also involves preliminary choices that may start and guide the process, possibly leading to different results. In this final part of the chapter we comment on two such preliminary considerations, those underlying the definition of the case, and some case-independent considerations guiding instrument choice (Box 13.5).

> **Box 13.5** Main lines of argumentation in Section 13.4
>
> - Policy instruments are not given entities to be examined; they are social constructions with many degrees of freedom.
> - Four main dimensions are central to the definition of specific instruments, though probably not enough for a full specification of operational instruments. They are: the nature of actor relations; the instrument mechanism in implementation; the objects influenced; and the operational environmental goals embodied in the instruments.
> - After specification, these four dimensions create an 'instrument space'. Criteria, ultimately evaluation criteria, signpost the route through this instrument design space for relevant instrument choices.
> - Instruments are building blocks in the process of policy formulation and policy implementation; they are not the policy itself.
> - In actual policies, public and private, consensual acts are at the core of behavioural adjustments. This should not obscure the fact that power and interests play key roles in such processes and that power is very much based on the availability of operational policy instruments.
> - Given some maximum regulatory effort for environmental policy, there is a limit to the overall effectiveness of such policy. Focussing on social procedures in regulatory instruments may enhance the effectiveness of specific policies, but implicitly excludes other policies from being developed and implemented.
> - Structural instruments like liability rules and emission taxes may exert their influence at low transaction costs and with potentially high environmental effectiveness, but for the time being only on a limited number of environmental effects.
> - In design and evaluation of policy instruments and policies, one part of the analysis is empirical while the other is normative.
> - The empirical analysis is partially subjective and concerns direct, indirect and, as far as possible, secondary effects. No broadly accepted models are available, leaving much room for complex debate.
> - The criteria for instrument and policy evaluation refer to direct expected effects, but also include second-order criteria and strategic criteria, viewed from a long-term perspective of the development of environmental policy instrumentation.
> - Basic choices on the desired nature of regulatory instruments, in view of overall institutional developments in global society, would have a direct bearing on instrument choices in specific cases.

13.4.3.1 Cases

Environmental policy instruments can best be chosen and designed in greater detail by specifying the characteristics most relevant to the case at hand. However, it is not always easy to say what exactly is 'the case', as the case is defined both by the problem context and by preliminary choices made before a specific instrument is designed. These reflexive or interactive relations make instrument design a much less rational and mechanical activity than might seem possible at first sight. If, for instance, one first defines the ecological effects of eutrophication as a manure problem, and then the manure problem as a consequence of too many animals per hectare, the choice to regulate the number of animals per farm seems logical. If one already had the traditional operating permit in mind, one would probably define the problem this way. By contrast, if the problem is defined in terms of a lack of oligotrophic areas, or as a lack of differentiation in nutrient concentrations, a regional regulatory scope will be more logical. If one already had a different instrument option in mind, like a substance deposit (see Huppes 1988), the scale of the problem could well be defined at a regional level. The use of individual permits then no longer seems so obvious. Evidently, it is not only the empirical context that indicates choices; normative considerations on how to regulate may well already play a role in the problem or goals definition phase. The polluter-pays principle reflects the normative principle that he who pollutes has to pay for the consequences of his action – and for the costs of preventing them, as is the case in liability law in all Western countries. Tradable permits lead to prevention costs being borne by the polluter, but not the damages. In this respect, an emission tax is more in line with general social and legal considerations than a tradable permit. If it is specified as a regional substance deposit, the focus on individual emitters vanishes more or less completely. With such caveats in mind, we now turn to design choices in instrumentation, within the framework developed.

13.4.3.2 Process Aspects of Design

A possible objection to the above procedure might be that in this top-down approach, the uniqueness of actual problems and options for their solution will be lost. This is a valid objection. The answer is that for any *institutionalised* policies, the knowledge of the concrete cases is so far removed from the regulators that real knowledge of 'the case' will usually not be available. Another objection might be that, as there are so many relevant evaluation criteria, these at least should be reflected in the design space developed here. Such further criteria may indeed help to specify actual instrument design in a relevant way. Another objection may be that the top-down approach is a hierarchical, technocratic approach, even an anti-democratic one. Analytic hierarchy, however, is not necessarily linked to social top-down approaches. On the contrary, the quality of the democratic process may well be improved by reducing the complexity of the subject of regulation while at the same time giving a broader and more systematic perspective.

Policy design, in terms of selecting and applying instruments, is not a mechanical procedure with results independent of the wider social context and independent of the qualities of the actors involved. There have been simple descriptions of the policy process which assume that the legislator enacts what is best, after which the regulations are implemented by law-abiding officers, leading to the intended effects, provided of course that the technical preparations for legislation had been done properly. Political scientists have long since shown (see e.g. Easton 1965) how at a systems level, policy-making is related to political support, limiting options for politicians and making the expected outcome of regulations only one aspect in the process. Sociologists of law have shown that similar laws work out differently depending on the administrative and social context in which they are functioning. A major difference, for instance, is that between the litigative American style of regulation, where laws are often fiercely debated and enacted after lengthy litigative procedures, and the more horizontal policy process in England and the Netherlands, with officials influencing private decisions through discussion and information, and only ultimately through threats of harsh regulatory actions against non-cooperative regulatees (cf. Jänicke and Jorgens 1998; Vogel 1986). In such systems, implementation may take place 'in the shadow of the law', without any new laws or permits being enacted, or in a private context, with contracts or covenants being signed.

In administrative science, this has led to greater emphasis on the process of policy formulation and implementation, reducing the emphasis on the more formal characteristics of policies in terms of instrumentation. Policy-making is then easily regarded as a discursive process between all those involved, with outcomes in terms of their environmental actions based on the power, interests, resources and the shrewdness of the actors involved. The present authors prefer what they regard as a balanced view in this respect, indicating the role of policy instruments both in terms of structuring discussions and as indispensable means to wield power and shape both society and the environment. Of course, this does not deny the fact that politics play an essential part in policy development, nor that social processes are fundamental in terms of both policy-making, including instrumentation, and policy implementation, using instruments.

Distinguishing between the political-administrative process of policy and instrument development, and the policy instruments being used in the policy implementation process is sometimes straightforward. The US, for instance, has enacted laws on tradable emission permits for SO_2 emissions, after lengthy research on how this instrument could function and lengthy political discussions on its advantages and disadvantages in terms of efficiency, effectiveness and ease of implementation. Implementation is a largely administrative process, upheld by checks and balances in which self-regulation plays a key role. Since nobody wants their competitors to have a free ride, all trading parties support officials checking the outcomes of emission trade, especially if focussed on their competitors. The instrument is clearly differentiated from its broader social and political context. With other instruments, however, the distinction is not so clear. Covenants between governments and groups of firms may be viewed from different angles. In some instances, they create the discussion platform leading to concrete actions, as in the Dutch packaging covenant. In this sense, it is not an instrument but a procedure that may lead to instrumental use, if needed. Or covenants may already specify concrete actions for specific parties, as is also the case in the Dutch packaging covenant. The 'shadow of the law' is very explicitly present in this covenant, which states that it replaces direct regulation and, if not successful, will be followed by more direct regulation. The threat, of course, only works if regulatees – in this case the private partners to the covenant contract – expect government to be able to come up with this legislation if deemed necessary.

Voluntary approaches (including voluntary agreements) cover procedural variants of regulatory instruments, as well as most social instruments. They are to be distinguished from the social procedures followed in the preparation of other types of regulatory instruments, like permits or emission taxes, which themselves are not voluntary. The difference is not always clear though, as in the example of 'permit preparation', which was formerly a major instrument for environmental policy in England and the Netherlands.

13.4.3.3 Instrument Analysis for Evaluation

Evaluating specific policy instruments requires a combination of normative and empirical analysis. The normative analysis guides the empirical analysis, as only results that are relevant in normative terms are relevant to the evaluation. As usual, however, things are not as simple in practice as they are in principle. In empirical terms, two types of mechanism are involved in the effect route towards environmental policy aims or, using a broader definition, sustainability aims. The first group includes mechanisms in society, the second those in the environment. For both types of mechanisms one may distinguish between primary effect mechanisms, essentially reducing causalities to one single chain, and secondary mechanisms, involving feedback loops, modelled in simple or more advanced ways. In normative terms, there is no well-structured set of values that can be linked to environmental policy. There is a possible classification, however, into major value fields related to human health, economic prosperity and the quality of nature. To these may be added amenity aspects, distributional aspects, the nature of our relation with the biotic and abiotic environment and other normative aspects. The evaluation criteria specified in a previous section may serve as a guide.

Since empirical analysis is often very scanty in environmental matters, it may be necessary to fall back on not fully proven assumptions. A striking conclusion from research on the effectiveness of voluntary approaches is that little is known about the functioning of such approaches (cf. Harrison 1999). Their effectiveness has not been thoroughly studied, and where it is assumed, this seems largely a matter of belief, similar to the old belief that binding instruments would automatically lead to the effects specified. This may have been the case to some extent in the US, but in most European countries, there is a well-known gap between legislation and implementation. In addition, legislation may enact what would have happened anyway. This safeguards effectiveness in terms of being in line with legislation, while effectiveness of policy in terms of a behavioural adjustment for environmental improvement may be more or less lacking. Similarly, instrument debates often state that tradable emission permits lead to definite environmental effects, unlike emission taxes. Let us take CO_2 emissions from electricity production as an example. If, due to unusually hot and cold weather periods, electricity consumption and related CO_2 emissions have eaten up the available permits, it can hardly be imagined that electricity production will therefore be shut down for the last half of December. Seemingly rigid instruments are not applied to the full, while soft instruments like taxes may have very predictable results at an aggregate level.

This state of affairs in empirical analysis for evaluation may be discomforting but should not lead to inertia. In real life, as opposed to science, a best guess is better than none at all, and defective but comprehensive evaluation schemes are to be preferred to doing nothing, or to focusing policy on some partial effects because other things have not been fully proven. If a balance is to be struck, one at least needs to know what is not fully proven, and to see where evaluation problems reside. The scheme of Fig. 13.2 at least indicates a number of relevant types of information. An analysis and evaluation which may be faulty in some respects is better than none at all, and policy instruments should preferably be set up in a way which best reflects available knowledge, limited though it may be.

13.4.3.4 Strategic Considerations Above the Case Level

We now return to strategic aspects of instrument choice and evaluation going beyond the design and evaluation of specific problem-and-instrument combinations. One question is whether a specific problem can best be solved by one or a few instruments, as opposed to a broad mixture of instruments. Another question is whether instrument choice is to focus on a limited set of instruments for environmental policy, as opposed to a set that is as broad as possible in order to fit all purposes. Piecemeal improvement at the micro level of individual instruments may well be sub-optimal at the macro level of societal environmental policy as a whole. Instrument choice in the sense of adding new instruments may not simply be a matter of finding the best options for the environmental problem case at hand. Both of these general strategic questions relate back to instrument choice at the operational level as discussed above.

The first question actually asks whether there is a minimum, optimum and maximum number of instruments; in other words, it asks how sparsely instruments should be used and which level of variety is necessary and wise. In the context of macro-economic policy, there is a clear preference for using Occam's razor: the minimum number of instruments required is also the maximum number to be used. There is a basic logic in mathematical models that says that the number of independent variables, that is, the instruments, has to be equal to the number of dependent variables, the goals or aims (see Tinbergen 1967). This holds for a wide range of model types, but most clearly for linear models. The reasoning is different for the two types of possible deviations: fewer or more. A smaller number of instruments will not allow the goals to be achieved, and non-effectiveness is obviously a mortal sin in instrument design. If there are more instruments than goals, as is often advised, the system has an infinite number of solutions. The value of the instrument variables can only be derived by arbitrarily reducing the number of instrument variables to the point where the remaining number once more equals that of the goals. This could hardly be called a well-founded procedure. It would probably be more rational and adequate if, based on considerations external to the model, a selection of instruments could be made beforehand.

The discussion on policy integration is closely related to this subject. If trade-offs between different emission types are made, they define a number of equations (usually of the type $ax = y$). An example is the global warming effect as empirically modelled for various substances, for instance with 20 units of carbon dioxide (ax) equalling 1 unit of Global Warming Potential for methane (y). By eliminating the specific substances and their equations, the number of goals is reduced to one: global warming (more precisely: time-integrated climate forcing). This integration can be pursued further, on a normative basis, for instance by integrating the overall toxicity scores of all toxic emissions with this total global warming effect of all climate-changing substances, reducing them to one evaluation-based denominator. The level of aggregation defines the problem, in that this discussion on numbers of instruments and goals may hold for a given problem level.

Before accepting these quite stringent conclusions on instrument numbers equalling the number of goals, however, it should be examined if this line of reasoning is really applicable. There are several reasons why this does not seem to be the case. Some are based on the nature of goals in environmental policy, others relate to the somewhat loose nature of instrument and goal specification in relation to the underlying dirty and fuzzy models of the world. It may also be questioned whether the goals are really discrete values or are of the nature of variables to be maximised, for which overshooting the mark is no problem at all, merely yielding a slightly higher environmental quality than originally expected or intended. If this is true, the problem should be seen as a maximisation (or in this case minimisation) problem with boundary conditions, rather than as a solution to arrive at a discrete value. This implies that the requirement of equality between numbers of instruments and numbers of goals vanishes, as is generally accepted in administrative sciences (see e.g. Gunningham and Sinclair 1999). Different instruments may each take care of particular improvements, possibly supporting each other in their functioning, in that for instance an informational instrument may be conducive to the active development of permits. Furthermore, the relation between instruments and goals may not always be so clear as to imply a one-to-one link. In most real life situations, there is no clear model with instruments put in at one end and goal achievement pouring out at the other. The relation between instruments and goals is not so straightforward. Ultimately, each instrument is at least to some extent linked to all environmental problems, albeit in widely different degrees and with very large uncertainties. Also, instruments may be mutually supportive. The underlying models for these instruments then differ somewhat in scope. A more practical approach would be to see them as parallel applications of several instruments.

Finally, some types of instruments belong together and may be mutually supportive, while others are conflicting. In environmental policy, there has always been a tension between the control-oriented permitting system and technology-binding and behaviour-prescriptive approaches on the one hand, and liability and taxing approaches on the other. A great deal of detailed legal sophistication has developed to reconcile these conflicting approaches in regulatory practice. Introducing the tradable emission permit, which is already operational in the US on sulphur dioxide emissions and is now being considered for climate changing emissions in Europe, leads to a substantial shift in this balance, for one thing because carbon dioxide emissions are intimately linked to all economic activities. The second approach, following the pure version of the polluter-pays principle that was worked out by the OECD three decades ago, is compromised by the advent of tradable emission permits. It will become quite impossible to shift to such a taxing system once the tradable permit system has been established, whereas such a shift is still possible whilst technologies are being regulated with permits. Such basic institutional choices, which need to be made explicitly, would have to be based on a rethinking of environmental regulation in particular, of public regulation in general, and of the even more general principles of social organisation. The outcome of this broader societal analysis and decision-making process could help guide all choices on instruments for environmental policy.

References

Bardach, E., & Kagan, R. A. (1982). *Going by the book; the problem of regulatory unreasonableness*. Philadelphia: Temple University Press.
Baumol, H., & Oates, W. E. (1988). *The theory of environmental policy*. Cambridge: Cambridge University Press.
Beck, U., & Beck-Gernsheim, E. (2002). *Individualization: Institutionalised individualism and its social and political consequences*. London: Sage.
Bohm, P., & Russell, C. S. (1985). Comparative analysis of alternative policy instruments. In A. V. Kneese & J. L. Sweeney (Eds.), *Handbook of natural resource and energy economics*. Amsterdam: North Holland.
Bonus, H. (Ed.). (1998). *Umweltzertifikate. Der steinige weg zur marktwirtschaft (sonderheft 9/1998 der zeitschrift für angewandte umweltforschung)*. Berlin: Analytica.
Bressers, J. T. A. (1988). A comparison of the effectiveness of incentives and directives: The case of the Dutch water quality policy. *Policy Studies Review, 7*(3), 500–518.
Bressers, J. T. A., & Klok, P. J. (1988). Fundamentals for a theory of policy instruments. *International Journal of Social Economics, 15*(3/4), 22–41.
Brundtland Report: World Commission on Environment and Development. (1987). *Our common future*. Oxford: Oxford University Press.
Castells, M. (1996). *The rise of the network society*. Oxford: Blackwell.
Castells Cabré, N. (1999). *International environmental agreements. Institutional innovation in European transboundary air pollution policies*. Amsterdam: Free University Press (doctoral dissertation).
Coase, R. H. (1960). The problem of social cost. *Journal of Law and Economics, 3*(1), 1–44.
Cole, D. H., & Clark, J. (1998). Poland's environmental transformation: An introduction. In J. Clark & D. H. Cole (Eds.), *Environmental protection in transition. Economic, legal and socio-political perspectives on Poland*. Aldershot: Ashgate.
Easton, D. (1965). *A systems analysis of political life*. Englewood Cliffs, NJ: Prentice-Hall.
Endres, A. (1997). Incentive-based instruments in environmental policy: Conceptual aspects and recent developments. *Konjunkturpolitik, 43*(4), 298–343.
EPA. (1999). Description of 9 proposed sites and 10 final sites added to the national priorities list in October 1999. *Intermittent Bulletin, 2*(5) [online].
Fischer-Kowalski, M. et al. (1997). *Gesellschaftlicher Stoffwechsel und Kolonisierung von Natur*. Amsterdam: Verlag Fakultas.
Galanter, H. (1981). Justice in many rooms: Courts, private ordering and indigenous law. *Journal of Legal Pluralism, 19*, 1–47.
Giddens, A. (1994). *Beyond left and right* (p. 15, 191). Cambridge: Polity Press.
Giddens, A. (1998). *The third way: The renewal of social democracy*. Cambridge: Polity Press.
Guinée, J. B., Gorrée, M., Heijungs, R., Huppes, G., Kleijn, R., de Koning, A., et al. (2002). *Handbook on life cycle assessment. Operational guide to the ISO standards*. Dordrecht: Kluwer (now Springer).
Gunningham, N., & Sinclair, D. (1999). Regulatory pluralism: Designing policy mixes for environmental protection. *Law and Policy, 21*(1), 49–76.
Harrison, K. (1999). Talking with the donkey: Cooperative approaches to environmental protection. *Journal of Industrial Ecology, 2*(3), 51–72.
Hawkins, K. (1984). *Environment and enforcement. Regulation and the social definition of pollution*. Oxford: Clarendon Press.
Hemmelskamp, J. (1997). Environmental policy instruments and their effects on innovation. *European Planning Studies, 5*(2), 177–194.
Hoffman, A. J. (2006). *Getting ahead of the curve: Corporate strategies that address climate change. PEW centre on climate change. See chapter on shell*. Ann Arbor, MI: University of Michigan.

Huppes, G. (1988). New instruments for environmental policy: A perspective. *International Journal of Social Economics, 15*(3/4), 42–50.

Huppes, G. (1993). *Macro-environmental policy, principles and design.* Amsterdam: Elsevier.

Huppes, G., & Kagan, B. (1989). Market-oriented regulation of environmental problems in the Netherlands. *Law and Policy, 11*(2), 215–239.

Huppes, G., van der Voet, E., van der Naald, W. G. H., Vonkeman, G., & Maxson, P. (1992). *New market-oriented instruments for environmental policies.* London: Graham & Trotman.

Huppes, G., Sas, H., de Haan, E., & Kuyper, J. (1997). Efficiënte milieuinvesteringen (efficient environmental investments). *Milieu, 12*(3), 126–133.

Imhoff, M. L., Bounoua, L., Ricketts, T., Loucks, C., Harriss, R., & Lawrence, W. T. (2004). Global patterns in human consumption of net primary production. *Nature, 429*, 870–873.

Ingold, T. (1986). *The appropriation of nature: Essays on human ecology and social relations.* Manchester: Manchester University Press.

Jänicke, M., & Jorgens, H. (1998). National environmental policy planning in OECD countries: Preliminary lessons from cross-national comparisons. *Environmental Politics, 7*(2), 27–54.

Luhmann, N. (1989). *Ecological communication.* Cambridge: Polity Press.

Mayntz, R. (1978). *Volzugsprobleme der umweltpolitik (application problems in environmental policy).* Stuttgart: Kohlhammer.

Mintzberg, H. (1994). *The rise and fall of strategic planning.* New York: Free Press.

OECD (Organisation for Economic Co-operation and Development). (1995). *Climate change. Economic instruments and income distribution.* Paris: OECD.

Opschoor, J. B., & Vos, H. B. (1989). *The application of economic instruments for environmental protection.* Paris: OECD.

Rawls, J. (1972). *A theory of justice.* London: Oxford University Press.

Riesman, D. (1950). *The lonely crowd: A study of the changing American character.* New Haven, CT: Yale University Press.

Scharpf, F. W. (1991). Die handlungsfähigkeit des staates am ende des zwanzigsten jahrhunderts. *Politische Vierteljahresschrift, 32*(4), 621–634.

Schneider, S. H., & Kuntz-Duriseti, K. (2002). Uncertainty and climate change policy. In S. H. Scheider & A. Rosencranz (Eds.), *Climate change policy: A survey.* Washington, DC/London: Island Press.

Schumpeter, J. A. (1976). *Capitalism, socialism and democracy.* London: George Allen & Unwin.

Sen, A. K. (1977). Rational fools. A critique of the behavioral foundations of economics. *Philosophy and Public Affairs, 6*, 317–344.

Sen, A. K. (1982). *Choice, welfare and measurement.* Cambridge, MA/London: Harvard University Press.

Sen, A. K. (2000). Consequential evaluation and practical reason. *Journal of Philosophy, 97*(9), 477–503.

Simonis, E. U. (1998). Internationally tradable emission certificates: Efficiency and equity in linking environmental protection with economic development. In H.-J. Schellnhuber & V. Wenzel (Eds.), *Earth system analysis: Integrating science for sustainability: Completed results of a symposium organized by the Potsdam Institute.* Berlin: Springer.

Superfund Programme. (1986). This programme has been adopted in two steps. CERCLA (Comprehensive Environmental Response, Compensation and Liability Act) in 1980 and SARA (Superfund Amendments and Reauthorisation Act) in 1986.

van Tatenhove, J., & Leroy, P. (2000). The institutionalisation of environmental politics. In J. van Tatenhove, B. Arts, & P. Leroy (Eds.), *Political modernisation and the environment: The renewal of environmental policy arrangements.* Dordrecht: Kluwer.

von Benda-Beckmann, K., & Hoekema, A. J. (1987). Hoe horizontaal bestuurt de overheid (how horizontal is government?). *Recht der Werkelijkheid, 8*(2), 3–12.

von Neumann, J., & Morgenstern, O. (1953). *Theory of games and economic behavior.* Princeton, NJ: Princeton University Press.

Tinbergen, J. (1967). *Economic policy: Principles and design: See Section 3.3.* Amsterdam: North-Holland.

Tisdell, C. (1998). Economic policy instruments and environmental sustainability: Another look at environmental-use permits. *Sustainable Development, 6*, 18–22.

Tönnies, F. (1887). *Gemeinschaft und gesellschaft: Abhundlung des communismus und des socialismus als empirischer culturformen.* Leipzig: Fues's Verlag.

Vogel, D. (1986). *National styles of regulation: Environmental policies in Great Britain and the United States.* London: Cornell University Press.

Weber, M. (1947). *The theory of economic and social organisation.* New York: Oxford University Press.

Chapter 14
Environmental Institutions and Learning: Perspectives from the Policy Sciences

Matthijs Hisschemöller, Jan Eberg, Anita Engels, and Konrad von Moltke[†]

Contents

14.1 Political Institutions, Knowledge and Power 281
14.2 Environmental Policy Analysis and Learning 284
  14.2.1 Institutionalism and Constructivism 284
  14.2.2 Conceptualizing the Policy Process 284
  14.2.3 The Advocacy Coalition Framework 285
  14.2.4 Policy-Oriented Learning 286
  14.2.5 The ACF and Environmental Policy 287
14.3 Institutionalisation of Policy-Science Interfaces and Environmental Policy 288
  14.3.1 Different National Traditions 289
  14.3.2 International Assessment 291
14.4 Biases with Respect to Stakeholder Participation in Environmental Policy Institutions 292
  14.4.1 Policy as Rule 294
  14.4.2 Policy as Negotiation 295
  14.4.3 Policy as Accommodation 295
  14.4.4 Policy as Learning 295
14.5 Methods to Enhance Stakeholder Participation 296
14.6 Summary and Conclusions 300
References 302

M. Hisschemöller (✉)
VU University Amsterdam, The Netherlands
e-mail: Matthijs.hisschemöller@ivm.vu.nl

[†] Konrad von Moltke passed away in 2005

14.1 Political Institutions, Knowledge and Power

This chapter on environmental policy institutions and learning opens with a brief introduction on the problems and dilemmas political institutions face in dealing with knowledge and power. The reason is twofold: on the one hand, such an introduction provides a framework which, in our view, is critical for an adequate understanding of the challenges for environmental institutions and the policy science's concepts, theories and methods to investigate and advise with respect to these institutions. On the other hand, the study of environmental institutions and the way they may adapt to the needs of society may provide valuable lessons for the current discussions on political institutions in general.

In this chapter, the concept *institution* is defined in a somewhat different way than may happen in everyday speech, where the concept is often used as identical to *organisations*. In the social sciences, *institutions* refers primarily to formal and informal 'rules of the game' that shape individual and group behaviour. Defined in a broad sense, this may include organisations as well, but institutions may exist with and without organisations. Examples of political institutions are visible organisations such as national states, the European Union or the United Nations. But equally important are less visible though influential institutions such as markets, formal and informal policy-science interfaces and laws.

One of the major problems for political philosophers and practitioners, working on the design and evaluation of political institutions, has been the question how such institutions can do justice to the will of the people and to scientific knowledge. For those, who did not find themselves bound by democratic principles or even rejected democracy as mediocracy that would render

Jan J. Boersema and Lucas Reijnders (eds.), *Principles of Environmental Sciences*,
© Springer Science+Business Media B.V. 2009

wise decisions impossible, this problem did not exist. Plato suggested to give power to a group of people who had qualified for ruling through a thorough philosophical training. The 19th century philosopher Saint-Simon proposed governance by three chambers consisting of scientists, artists, engineers and captains of industry. In his view, effective political institutions should be run by representatives of these groups, as they are qualified to contribute what is relevant for good governance: knowledge, creativeness, the capacity to carry out big works and the capacity to make money. The American political scientist Dahl (1985) cites these and similar political theories, which all share this basic characteristic that rulers rule by virtue of certain qualifications instead of rule by 'the people'. He refers to this kind of antidemocratic theories as 'guardianship', because they argue for political institutions led by 'guardians' who know what is good for the people.

For democrats, however, the tension between what is right and majority will (knowledge and power) has been an issue from the early days of democratic theory onwards which by times evolved into a huge dilemma. The French philosopher De Condorcet, a founding father of game theory, expressed this dilemma most clearly. For the years after the French revolution, he was both chair of the Academy of sciences and member of the Assemblee National in which capacities he was one of the first to stand up for colonial independence and full racial equality. Condorcet argued for a type of democracy with institutional guarantees for science based decision making. Elected civil servants had to be trained experts at the same time, capable of an objective and impartial judgment. As a scientist, Condorcet tried to calculate the probability for a majority decision to be 'right', which probability he expected to increase to the extent the decision makers would have a better training (Hisschemöller 1993).

Although it maybe argued that De Condorcet was more of an enlightened guardian than a democrat, 'real' democrats too struggled with the issue of knowledge and power. Many democrats have phrased the issue of knowledge and power in an indirect way, i.e. through the articulation of the conflict between private interest and the public good or, as we would say today, as a social dilemma. Rousseau, who rejected any subjection of people by the state, argued for a direct democracy that would impose the 'general will' (volonté general) through public deliberation. As for other supporters of direct democracy, Rousseau faced the problem that citizens, in developing the general will, might show a bias in favour of particular interests. In order to reduce this risk, Rousseau and some liberal democrats from the 19th century were actually opposed to the formation of interest group organisations and political parties, which they considered a threat to free deliberation on the public cause. A convinced democrat like John Stuart Mill was reluctant to endorse general franchise, because he was worried that the 'less intelligent majority' would in elections outweigh the educated minority. Therefore, he proposed an arrangement where every adult would have the right to vote, but to the extent people were educated they would have two, three or four votes to compensate for the unequal distribution of knowledge in society (Mill 1861/1973). Hence, democratic philosophers by that time shared the view that training and education could bring the ideal of participatory democracy closer to realisation, as the educated public would become qualified to enter into debate on the public good.

Also in political philosophies which argue for a representative system, a view already strong in the US that became dominant in the 20th century, considerations with respect to the role of knowledge in political decision making are critical. Ezrahi (1990) argues that science provides liberal democracies with a powerful means to render rule by persons, elected because of their views, programmes and promises, into a set of depersonalised and objectified acts. In this view, science and expertise are not enemies of the democratic process, but indispensable for the legitimacy of democratic political institutions.

To what extent are these insights relevant for the study of environmental policy institutions today? Two views can be distinguished in this respect: on the one hand, political scientists, like Dahl (1985) and Fischer (1990, 2001) have pointed out that notions of guardianship, although theoretically opposed to principles of democracy, can be found alive in present environmental policies around the world. In analysing the short history of environmental policy and science, Fischer (2001) refers to the growing power of experts and what he refers to as a technocratic tendency in environmental policy. From the early foundation of environmental policy institutions in the 1960s and 1970s, environmental policy has indeed been indebted to science and to independent scientists who struggled for bringing environmental issues to national and international agendas. However, environmental issues have also been issues for participation by stakeholders, such as environmental NGOs, local citizen groups and individuals. Over time, these groups and individuals may have become distinguished experts in their fields, but they have often started from outside the vested policy-science networks. Hence, the increased impact of expertise on environmental policy has contributed to a significant decrease of citizen participation in environmental policy and thereby to a loss of vision and knowledge.

On the other hand, it has been argued that the trend toward participatory policy making has not weakened at all but that instead this trend has weakened traditional political institutions, such as elected government and parliamentary control (Ezrahi 1990). The underlying driving force behind the weakening of political institutions is not so much a technocratisation of policy making but a changing view on science in society. It has been argued that academia has lost its monopoly of knowledge production, and a new mode of knowledge production gains prominence (Gibbons et al. 1994; Nowotny et al. 2001; Funtowicz and Ravetz 1993; Ravetz and Funtowicz 1999). In this new mode, scientific knowledge has to compete and share position with non-scientific views and knowledge claims, referred to as *practical knowledge*, *lay knowledge* or *tacit knowledge*. Hence, subjective stakeholder knowledge has gained importance at the expense of 'objective' science. From this perspective, the weakening of political institutions in liberal democratic societies has been caused by the fact that science to a much lesser extent performs the function of legitimising public policies.

Environmental policy offers a fertile ground to explore and investigate into both contradictory views on the evolution of environmental institutions. And we may not be surprised if in the study of environmental institutions, we will observe both developments at the same time: on the one hand a technocratisation of environmental policy institutions and on the other a growing awareness of the contested character of knowledge and attempts, at national and international levels, to involve stakeholders in the strategic production and evaluation of knowledge for policy.

However, what constitutes the main relevance of reflecting upon the history of political philosophy for current trends in environmental institutions is given by the critical assumption which both views have in common despite their differences: whereas classical political philosophy, i.e. pleas for guardianship as well as for democracy, starts from the assumption that the common good is knowable through the application of standardised and objective scientific methodology, current reflections show that this is not the case. The common good is conceived of as a social construct that maybe negotiated by the policy and science communities but not as a given waiting to be discovered. Hence, there is a growing tendency toward reaching consensus on knowledge as a basis for political interventions. A central concept here is learning by networks of policy makers, stakeholders and scientists with different, often conflicting views on the nature of the environmental problems to be addressed.

The role of science in these processes is subject of an ongoing debate (Hisschemöller et al. 2001a). Traditionally, policy and science are viewed as different institutions with clearly demarcated boundaries. But in processes of stakeholder involvement and learning traditional boundaries between science and policy have become diffuse or even invisible (Halffman 2003). Scientists as professionals are confronted with manifold questions and new expectations related to a growing claim for participatory approaches, such as new forms of quality control. These new expectations can be felt very strongly in the field of environmental science. In the context of these changing expectations of science, successful scientific assessments need to appropriately deal with questions of policy relevance, scientific quality and legitimacy at the same time. It will be shown that for shaping the interaction between stakeholders and scientists a wide range of institutional forms is available, depending on power relations, national traditions and the character or 'structure' of the environmental problems at hand.

This chapter explores various perspectives on the study of environmental institutions from both the angles from policy and science. Section 14.2 presents an analytical perspective from which to deal with environmental policy as a dynamic and refractory policy field, which focuses on policy change and policy-oriented learning. Section 14.3 focuses on how environmental institutions have been shaped in the light of different (inter)national historical contexts. Section 14.4 focuses on the bias that environmental institutions show with respect to the construction of environmental problems and the implications for modelling policy-science interactions. Section 14.5 discusses methods to structure and facilitate open dialogues between policy-makers, stakeholders and scientists aimed at escaping impairment caused by institutional bias. Section 14.6 wraps up and draws some conclusions with respect to the design and evaluation of environmental institutions and learning, as well as the implications of the environmental experience for the future of political institutions in general.

14.2 Environmental Policy Analysis and Learning

14.2.1 Institutionalism and Constructivism

Social relations is all about action and order. It would be ideal to study every social situation (phenomena, processes) as a *dynamic* interplay of *actors* and *structures*. Yet, in order to understand (often complex) policy processes, we might just as well try to approximate this and apply one or more eclectic, or rather comprehensive, theoretical frameworks. These will be the products of the vast and cumulative 'body of theory' in social and political science.

Policy-making is an interactive, creative, yet often capricious process, circumscribed by either man-made rules and patterns, or other (partly unpredictable) contextual and situational conditions. Two aspects are of special interest: social constraints and social constructs, studied in the traditions of institutionalism and constructivism. In the light of the premise that policy processes are inherently dynamic, these aspects will represent the dimensions of 'structure' and 'actors' in a framework of policy change. Institutionalism studies institutions: rules, norms, values, procedures, also organisations that (each specifically) produce and propagate rules and norms; it is about the limitation, steering and organisation of social action. Constructivism studies social constructs: knowledge and meanings as a result of social interaction; this is about perceptions from which people and organisations act. Institutionalism and constructivism are linked. Social action always results from intentions and limitations, an interplay between actors and structure.

Relevant institutions and social constructs, respectively, represent long term and short term policy changes. Put differently: patterns and structures are more durable than plans and projects, and *mores* outlive trends. Environmental policy should be approached from both short term and long term policy changes. In the short run, problems are reacted upon, and policy is developed and formulated. In the long run, policy goals will change, as will the policy context, the regulatory structure, the administrative culture, and the fundamental social values. Long term and short term policy changes are connected. The long term political culture and structure forms the institutional 'background' for short term policy development. In addition, sustainable dominant or successful policy (in short term) will institutionalise (in the long run).

14.2.2 Conceptualizing the Policy Process

Policy processes are neither uniform nor predictable. For every policy problem, policy-making and implementation, as well as the corresponding policy objectives, strategies, and instruments, will differ. A policy process is dynamic in two ways: it is a process of formation and implementation, and it will be continually adjusted according to the momentary situation, until it will be terminated. Policy dynamics have been modelled in various ways.

In policy research, often so-called stages heuristic models are used, such as the sequential phases model, the policy cycle model, the (structure approach) systems model, and the (actor approach) barrier model. These models describe the policy process as a temporal sequence or a cycle of different stages starting with agenda setting and passing through policy formulation, implementation and evaluation. None of these images will reflect practice. The two most important criticisms of these models are the premise of a sequential order of processes, and a legalistic top-down approach with a dominant role for government policy actors. This neglects both the whimsicality of the policy process as well as the roles of other actors in the policy area. Policy-making may take all kinds of forms, and during the process there may be different occasions and motives for policy change.

Today's politics are 'on the move', not only in terms of administrative decentralisation and internationalisation, but also in terms of 'sub-politicisation of society' (Beck 1992), that is the shift of political action from the traditional power centres to uninstitutionalised actors (e.g. consumers, marketers). Policy processes result from an interplay between policy problems, policy actors, and several factors influencing the policy area. Processes of policy change are not only influenced by power struggles, conflicting interests, and political pressures, but also by strategic choices, shifting frames of meaning, and the improvement and exchange of knowledge and beliefs. Also learning processes are involved. In his comparative study on pollution control strategies, Weale (1992) discusses 'modes of politics' to deal with the current waste issue. He favours a combination of institutional analysis and policy discourse approaches: the institution of discourse. As he states: 'Without understanding the meaning and function of the competing discourses we are unlikely to be able to understand the changing politics of pollution' (217). 'None the less [..] the force of ideas in practice is likely to be severely conditioned by the institutional factors of political acceptability, administrative feasibility and straightforward bureaucratic politics' (220). '[..] and hence those factors must themselves be understood' (222). According to Weale, this can best be done within a framework of social learning theory.

Environmental policy-making will be better understood when, besides its development of contents, outputs, and impacts, also its roots are studied. Knowing more about the underlying policy styles and political cultures helps to explain the differences between the several environmental policy belief systems involved. This requires a general theoretical framework of short-term and long-term policy changes. A major contribution to political theories of policy change and learning is made by Paul Sabatier, who developed a general framework, applicable to a wide variety of policy domains and political systems.

14.2.3 The Advocacy Coalition Framework

Mainly drawing on insights of Lakatos (1971), Heclo (1974), and implementation literature, the American political scientist Paul Sabatier developed a General Model of Policy Evolution Focusing on Competing Advocacy Coalitions within Policy Subsystems (Sabatier and Jenkins-Smith 1993, 1999) as an alternative to the stages heuristic of public policy.

According to this *Advocacy Coalition Framework* (ACF), policy change over time is a function of three sets of processes:

1. The interaction of competing *advocacy coalitions* within a policy subsystem. An advocacy coalition consists of actors from a variety of positions and institutions (elected and agency officials, interest group leaders, researchers, etc.) who share a particular belief system, and who show a nontrivial degree of parallel action over time. Coalition actors seek to translate their beliefs into public policies throughout the governmental system. The concept of an advocacy coalition assumes that it is shared beliefs that provide the principal 'glue' of politics.
2. *Changes external to the subsystem* in socio-economic conditions, public opinion, system-wide governing coalitions, and decisions from other policy subsystems.
3. The effects of changes in *relatively stable system parameters*: the basic attributes of the problem area, the basic distribution of natural resources, fundamental socio-cultural values and social structure, and the basic constitutional structure. For both the cause and solution of environmental issues, these parameters maybe critical.

[T]he challenge confronting the ACF has been to explain an exceedingly complex situation involving hundreds of actors from dozens of organisations seeking to influence the overall policy process over periods of a decade or more in situations where relatively technical information concerning problem severity and causes cannot be neglected. (Sabatier 1998: 101–102)

The time perspective of a decade or more is necessary to understand the process of policy change (and the role of technical information therein). This argument for an extended time period comes from findings concerning the importance of the 'enlightenment function' of policy research. This knowledge utilisation model of Weiss (1977) argued persuasively that a focus on short-term decision making will underestimate the influence of policy analysis because such research is used primarily to alter the belief systems of policy makers over time (Sabatier and Jenkins-Smith 1999: 118). Weiss' argument also supports the policy-science interface, more specifically the influence of science to policy: not the instrumental use of knowledge but the conceptual use, i.e. background theories, ideas, frames of thinking. The conceptual contribution of research changes the vocabulary of policy makers and thus influences the way they think about problems and consider possible solutions.

Figure 14.1 shows an overview of the ACF. Within the subsystem, it is assumed that actors can be aggregated into a number of advocacy coalitions. At any particular point of time, each coalition adopts a strategy envisaging one or more institutional innovations that members feel will further policy objectives. Conflicting strategies from various coalitions are normally mediated by a third group of actors, here termed 'policy brokers', whose principal concern is to find some reasonable compromise that will reduce intense conflict. The end result is one or more governmental programmes, which in turn produce outputs at the operational level (e.g. emission standards for waste incinerators). These outputs result in a variety of impacts on targeted problem parameters (e.g. operating costs, environmental quality) as well as in various side effects.

The ACF proposes that the belief system of an advocacy coalition is structured into three categories, arranged in order of decreasing resistance to change:

- A Deep Core of fundamental normative and ontological axioms that define an actor's underlying personal philosophy
- A Near Policy Core of basic strategies and policy positions for achieving deep core beliefs in the policy area or subsystem in question
- A set of Secondary Aspects comprising a multitude of instrumental decisions and information searches necessary to implement the policy core in the specific policy area

14.2.4 Policy-Oriented Learning

Many aspects of a coalition's belief system are susceptible to change on the basis of scientific and technical analysis. For example, science has dramatically increased our understanding of environmental problems after

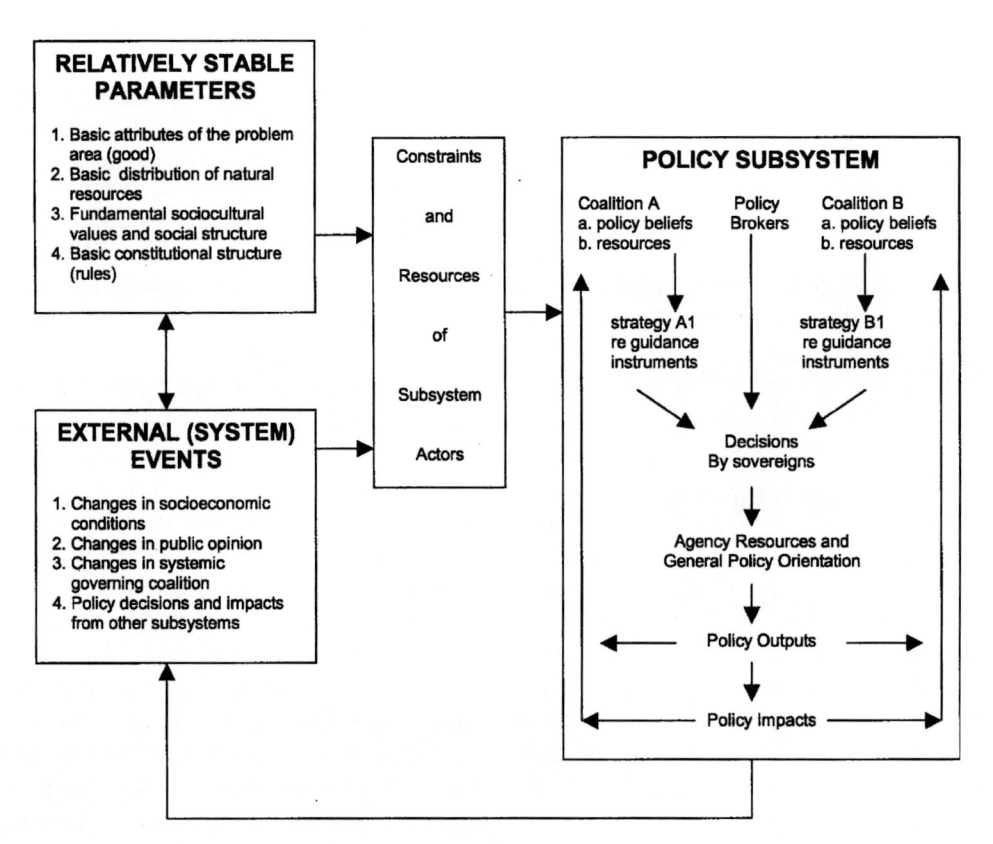

Fig. 14.1 The Advocacy Coalition Framework of policy change
Source: Adapted from Sabatier and Jenkins-Smith (1993: 224)

three decades of debates concerning the seriousness of the problems and the appropriate policy instruments to be used in addressing the issue. Within the process of policy change, the ACF focuses on these processes of *policy-oriented learning*. Following Heclo, the ACF defines policy-oriented learning as relatively enduring alterations of thought or behavioural intentions that result from experience and are concerned with the attainment (or revision) of public policy. Policy-oriented learning involves the internal feedback loops depicted in Fig. 14.1, perceptions concerning external dynamics, and increased knowledge of problem parameters and the factors affecting them. The framework assumes that such learning is instrumental, that is, that members of various coalitions seek to better understand the world in order to further their policy objectives. They will resist information suggesting that their basic beliefs may be invalid or unattainable, and they will use formal policy analyses primarily to buttress and elaborate those beliefs. Although learning from experience is very difficult in a world where performance is difficult to measure and well-developed causal theories are often lacking, those who can most effectively marshal and apply political resources and credible arguments are more likely to win in the long run.

Policy-oriented learning is a form of information and argumentation management. The ACF distinguishes two types of policy-oriented learning: *within* a coalition's belief system, and *across* the belief systems of different coalitions. The first type of learning means that members of an advocacy coalition are seeking to improve their understanding of variable states and causal relationships consistent with their policy core ('puzzle-solving'). The second type of learning refers to a productive analytical debate between members of different advocacy coalitions. One or more coalitions are led to alter policy core aspects of their belief system (or at least very important secondary aspects) as a result of an observed dialogue rather than a change in external conditions. The ACF hypothesises that learning across coalitions is facilitated by a moderate level of conflict, an issue that is analytically tractable (i.e., has widely accepted theories and quantitative indicators), and the presence of a professionalised forum in which experts from competing coalitions must justify their claims.

A clear example of policy-oriented learning in the environmental policies of various western nations is the way they have changed from sector policy to integrated policy and from source oriented policy to effect oriented policy. This indicates that the objectives of environmental policy have altered because the original assumptions about the nature of environmental problems and how to attack them no longer appeared to be tenable. In that way, a coalition of government actors learned, partly on its own, but also from a coalition of environmental organisations arguing on the basis of policy effectiveness and sustainability. An environmental lobby and the disclosure of a variety of environmental scandals supported this.

14.2.5 The ACF and Environmental Policy

Analysing processes of policy change and learning according to the advocacy coalition approach enables one to structure both institutional aspects ('left-hand side' of the model) and constructivist aspects of the policy process ('right-hand side' of the model). This is especially appropriate to a policy field as dynamical as environmental management.

Employing the ACF, policy change within a subsystem can thus be understood as the product of two processes. First, policy-oriented learning. Advocacy coalitions within the subsystem attempt to translate the policy cores and the secondary aspects of their belief systems into policy programmes. Although most programmes will involve some compromise among coalitions, there will usually be a dominant coalition and one or more minority coalitions. Each will seek to realise its objectives over time through increasing its political resources and through policy-oriented learning. Second, external perturbation, that is, the effects of system-wide events on the resources and constraints of subsystem actors. The ACF argues (and hypothesises) that, although policy-oriented learning is an important aspect of policy change and can often alter secondary aspects of a coalition's belief system, changes in the core aspects of a policy will usually result from the influence of these external events.

The major arguments of the ACF generally seem to be supported by the case study researches in the 1993 volume of *Policy Change and Learning*, edited by Sabatier and Jenkins-Smith. Between 1988 and 1998, the ACF has been applied to at least 34 cases, see Table 14.1 (Sabatier and Jenkins-Smith 1999: 125).

Table 14.1 Environmental policy related research cases applying the ACF 1987–1999

Year(s)	Researcher(s)	Case(s)
1987/1989/1990/1993	Sabatier et al.	Environmental policy at Lake Tahoe
1988	Lester, Hamilton	Ocean waste disposal (U.S.)
1988	Heintz	Outer Continental Shelf (OCS) leasing policy (U.S.)
1988/1990	Jenkins-Smith	U.S. Energy policy
1988	Weyent	U.S. natural gas policy
1991	Jenkins-Smith	Nuclear waste and weapons (U.S.)
1991/1993	Jenkins-Smith, St. Clair, Woods	OCS leasing policy
1993	Munro	California water supply policy
1995	Grant	Auto pollution control in California
1995/1997/1998/1999	Sabatier, Zafonte	San Francisco Bay/Delta water policy
1996	Lertzman et al.	Forestry policy in British Columbia
1996	Sewell	Climate change in U.S. and Netherlands
1996	Wellstead	Forestry policy in Ontario and Alberta
1997	Duffy	Nuclear power (U.S.)
1997	Eberg	Waste management in Netherlands and Bavaria
1997	Freudenburg, Gramling	OCS leasing (U.S.)
1998	Andersson	Environmental policy in Poland
1998	Elliot	Forestry policy in Indonesia, Canada, and Sweden
1999	Burnett, Davis	U.S. forest policy
1999	Herron et al.	Nuclear waste and weapons (U.S.)
1999	Leschine et al.	Water pollution in Puget Sound (U.S.)
1999	Loeber, Grin	Dutch Water quality
1999	Sabatier, Zafonte, Gjerde	U.S. auto pollution control

Source: Based on Sabatier and Jenkins-Smith (1999: 126)

Sabatier and Jenkins-Smith continue to refine the ACF on the basis of their own research and that of others. On the basis of studies critically applying the ACF, several review essays, and his own reflections, Sabatier (1998) reaches a number of conclusions regarding the ACF's strengths, weaknesses, and needed revisions. The existence of advocacy coalitions proved to be largely beyond dispute; the model of the individual was clarified, as well as its implications for coalition stability and coalition behaviour; the defining characteristics of policy core beliefs were improved, and the set of topics covered by the policy core was extended; criteria for the existence of a mature policy subsystem were formulated; the argument for the contribution of policy-oriented learning to policy change was found to be confirmed; and finally, whereas one should be able to distinguish major policy changes quite easily, also the conditions under which these changes occur should be acknowledged. These reflections have basically supported the value of the ACF as a framework for analyzing long-term policy change (see also Sabatier 1999; Sabatier et al. 2005).

14.3 Institutionalisation of Policy-Science Interfaces and Environmental Policy

Scientific research could proceed without any policy-science interface. Indeed, many scientists feel that involvement with policy is nothing but a distraction from their real work. In consequence the institutionalised policy-science interface is always driven by the needs of policy-makers rather than scientists. These typically seek answers to highly deterministic questions such as whether to control a substance or not, and if so how rigorously to go about this, and what the costs and benefits of such actions are likely to be. These are not the questions that scientists seek to answer, so the policy-science interface represents a process to translate science into conclusions that can inform specific questions of policy.

The core criterion for successful institutionalisation of the policy-science interface is the *policy relevance* of the outcome. This explains why significant differences exist in the institutionalisation of this interface depending on the nature of the policy process that is being informed. The scientific dimension of the policy-science interface is reflected in the need for *high quality scientific results*. The criteria for determining the quality of the research are largely science-based and not policy-dependent. They are also largely independent on the policy environment, even though there is often a preference for 'indigenous' research, that is work undertaken by scientists within the jurisdiction in question. Despite the need for policy relevance, effective policy-science interfaces must still respond to the demand for *scientific legitimacy* of the advice. This is generally achieved in reference to the scientific credentials of the participants in the process and the process by which policy advice is generated. In many instances transparency of the process and the publication of results constitute additional structures to promote legitimacy.

An international comparison of environmental policy-science interfaces shows that in some countries institutions are based on the idea that a clear demarcation between policy and science, knowledge and power, is desirable although in practice not always feasible. In these cases, the soundness of scientific quality and scientific integrity are highly valued. In other countries, the institutions are much more based on pragmatic considerations with respect to the usability of scientific advice. Here, scientific relevance appears to be valued most, while scientific quality and integrity are not supposed to suffer from political pragmatism.

14.3.1 Different National Traditions

The three variables above, policy relevance, quality of science and scientific legitimaticy, are reflected in the many examples that exist for the institutionalisation of the policy-science interface. In fact national traditions for the institutionalisation of the policy-science interface are very diverse, reflecting differences in political and administrative culture as well as different institutions of scientific research. It is worth discussing some of the more important examples at the national and international level.

In *Germany*, the debate about 'useful' science goes back to the early 19th century. When establishing the university in Berlin that was subsequently named after Alexander von Humboldt, there was discussion as to whether the model should be the research-based University of Göttingen or whether the new university should be required to produce research that was useful for industry and commerce. The former view prevailed and firmly established the German tradition of science for its own sake, a tradition that had a powerful impact on universities world-wide.

German academia has maintained a difficult relationship with policy makers – as exemplified by Max Weber's famous articles on science as a profession and politics as a vocation (Weber 1946, 1992), and by its tortured relationship to the excesses of the Nazi regime. The situation was rendered even more complex when the vagaries of occupation politics caused the Humboldt University and the Academy of Sciences in Berlin to lie in the Russian sector. West Germany created new organisations to replace them, organisations that have persisted beyond unification of Germany.

As a result, there are no general purpose science/policy institutions in Germany but rather specialised advisory bodies, for example five major economic policy research institutes and the Council of Experts for the Environment (Sachverständigenrat für Umweltfragen: www.umweltrat.de) and an Advisory Council on Global Change (Wissenschaftlicher Beirat der Bundesregierung Globale Umweltveränderungen – WBGU: www.wbgu.de), composed exclusively of university professors. Germany has also evolved a highly developed structure of public research agencies for health, worker safety, the environment and conservation. These agencies channel assessments into the policy process. All of these bodies subscribe to the assumption that science and policy should be clearly separated even though they will typically engage in negotiations with policy makers in the definition of the problems that they address in their advice.

Finally, Germany has invented the remarkable tool of the commission of enquiry of the German Bundestag, whose membership is composed of an equal number of experts and members of parliament. Two such commissions have produced seminal reports on ozone depletion and climate change (Deutscher Bundestag 1991). A further commission is currently looking at globalisation and its consequences. These commissions, being constituted of both scientific experts and policy makers, are in many ways an exemplar of a more pragmatic approach which values policy relevance more than clear policy-science boundaries.

Science based policy decisions emerge from the dynamics of this rich institutional landscape between government and academia. It is characterised by continuous interaction and by a remarkable stability of personnel and functions, attributable in large measure to the fact that many researchers are civil servants, or enjoy an equivalent status under a negotiated union contract. Policy relevance is achieved by the creation of interface institutions specialised in certain areas of policy, frequently augmented by close organisational relationships between advisory bodies and the relevant ministries or agencies. Nevertheless most German advisory bodies exercise a good deal of discretion in determining the topics they will address. Few of these German bodies undertake primary research of their own. They draw on the work of the relevant German research institutions, augmented by research that is recognised internationally. The issue of scientific legitimacy is largely solved by drawing heavily, or even exclusively, from the ranks of senior university researchers.

In the middle of the 19th century, a debate about useful research also occurred in the *United States*, but with a different outcome. It led to the creation of land grant universities, public institutions of research and teaching with an obligation to provide free advice to commerce and agriculture (Cross 1999). At about the same time as it established land grant universities, the US Congress authorised the creation of a National Academy of Sciences, on condition that its members be available to provide advice to policy makers (www.nationalacademies.org/arc.nsf). This led to formation of the National Research Council (NRC), the largest and most potent science-based institution for policy advice in the world (www.nas.edu/nrc). The NRC has played an important role on numerous environmental policy issues (www.nas.edu/subjectindex/env.html). Its reports typically circulate widely in the United States and beyond. The work of the NRC is funded by the interested agencies of the US government and the questions to be considered consequently are formulated by these agencies, presumably in negotiation with the NRC. This reflects the balance of scientific and policy interests that are to be expected in any science assessment process.

The NRC works through committees composed of members of the National Academy of Sciences and invited researchers. These in turn establish, and oversee, the panels that do the actual work. Even though membership of panels reflects a wide variety of backgrounds, the discourse within the panels is rigorously circumscribed by the practices of science and issues that are submitted to panels are typically designed to elicit unambiguous scientifically based responses.

The existence of the NRC has tended to overshadow other institutions in the policy-science interface. None matches the NRC in terms of its legitimacy and its acceptance by agencies that seek advice, and are required to pay for it. The policy relevance of NRC reports is rarely in doubt. The quality of science is supported by access to research throughout the United States and a tradition of drawing on the best research available anywhere, helped of course by the fact that the working language of the NRC is also the language of the most widely accepted scientific publications worldwide. The issue of scientific legitimacy is more complex, since the membership of the panels is typically drawn so as to reflect some sense of balance between public research institutions, private universities, industry-funded institutions, and often even not for profit institutions. This additional selection criterion overlays the purely scientific criteria. Committees, and consequently panels, are served by long term scientifically qualified staff and reports are frequently subject to a process of peer review, drawing on the membership of the Academy, which oversees the entire institution.

Several European countries have adopted the 'academy of sciences' model of assessment. The academies of *Sweden* (www.kva.se), *Norway* (www.sol.no/dnva) and the *United Kingdom* have all played a role in environmental policy, albeit with differing degrees of enthusiasm. These academies typically date back to the 18th century, a period when universities were places of doctrinal theological orthodoxy and hostile to scientific enquiry. With deep roots in the history of science, the national academy model adheres quite rigidly to the ideal of strict separation between science and policy.

In some countries, notably those of *Central Europe* and in *France*, the most advanced research has traditionally been separate from the universities, located in academies or similar institutions that were actually major research enterprises in their own right. How policy makers derive science-based advice from these institutions differs from one country to another. In smaller countries – with the exception of the *Netherlands* with its strong tradition of consensus-based decision-making – there will be fewer group processes, and individual scientists can play a dominant role. In larger countries, the relationship between researchers and government will be significantly affected by the funding patterns. In France, with a tradition of strong administrative control over education and research, the assessment process also tends to be under the control of policy makers. In the United Kingdom, with a tradition of independence for scientific organisations, there is significant resistance on the part of researchers to being used for policy purposes. Consequently, there is a larger amount of government-funded, policy-oriented research, the results of which can have major and quite direct impact on policy.

What is striking is the degree of variation between countries in addressing the core issues of policy relevance, quality of science, and scientific legitimacy of advice. In each country the solutions that have been adopted will reflect the traditions of both policy-making and the scientific establishments, as well as research funding patterns. The willingness to draw on research undertaken in other countries is typically tempered by the likelihood that questions submitted for advice will impact the local environment – in which case the relevant research is more likely to be local as well – or significant interests within the political system – in which case these interests can be expected to marshall locally whatever scientific advice supports their position. In some instances a language barrier may limit the ability of scientists from outside a country to participate effectively in the deliberative processes, although smaller countries are more likely to use the English language on a continuing basis.

14.3.2 International Assessment

There is no tradition of *international assessment*. In some instances, the existence of an international regime has given rise to 'epistemic communities' – groups of like-minded researchers (Haas 1991). These are not, however, engaged in attempting broadly based assessments. They are generally scientists and expert administrators whose concern is directly affected by the international regime. As often as not, they feel marginalised in the domestic context. In practice, such epistemic communities are a phenomenon of 'international civil society', individuals and groups whose allegiances are increasingly divided between domestic constituencies and international peer groups (Newell 2000).

Some international environmental regimes have generated their own forms of assessment, from the review conferences organised under the Vienna Convention that contributed to the Montreal Protocol, to the scientific committees of the International Whaling Commission, to the elaborate processes of the Convention on International Trade in Endangered Species (CITES) when determining whether certain species are to be listed on certain Annexes of the Convention (UNEP 1998). Scientists employed or nominated by member governments typically staff these institutions – thus their work is conditioned by the concerns of the governments to which they are responsible.

The Intergovernmental Panel on Climate Change (IPCC) is by far the most important international science assessment process (www.ipcc.ch). It is probably the most elaborate science assessment process ever mounted at any level, by now outstripping the various attempts by governments to validate nuclear energy for sceptical electorates. This reflects the importance of the issues, the complexity of the science and the fact that few governments are likely to act on the assessment of other governments on an issue of such significance. An international assessment becomes a necessary condition for international action. Time and again it has been demonstrated that governments armed with domestic assessments that are entirely convincing to them and their constituents prove unable to convince other governments to act. The IPCC process has also highlighted the extent to which assessment is conditioned by political and administrative traditions of the participants. On issues such as climate change, it is important to ensure that adequate research occurs in a sufficiently large number of jurisdictions to provide assessment results that are credible across a wide spectrum of countries.

Even when there has been no institution comparable to the IPCC, assessment has played a significant role in international environmental regimes. The Convention on Long Range Transboundary Air Pollution (LRTAP), whose parties include all the countries of Europe, plus Canada and the United States, has seen a remarkable evolution since it was signed in 1979 (Wettestad 1999; see also Chapter 18). Since its entry into force in 1983, it has been extended by eight protocols. From the outset, there was significant disagreement between countries about the significance of the evidence concerning acid rain. There is no formal process of science assessment within the LRTAP structure itself. However, events in two countries influenced the regime. In Germany a changing assessment of the health of the country's forests – based on work by one of the governmental research-based agencies – led to a dramatic shift in position that changed the dynamics of the entire regime. In the United Kingdom, deep governmental scepticism about the seriousness of acid rain was first nurtured by government scientists but overcome, at least in part, by a joint assessment undertaken by the British Royal Society and the Norwegian and Swedish academies of sciences.

Once the LRTAP regime began to evolve in a constructive manner, it was capable of taking up issues that were more complex than acid rain. The first such issue concerned nitrogen oxide (NO_x) emissions, which were implicated in both acid rain and tropospheric ozone formation, resulting in the 1988 Sofia Protocol concerning the Control of Emissions of Nitrogen Oxides or their Transboundary Fluxes (NO_x Protocol). The preparatory process for the NO_x Protocol involved a series of meetings of researchers from many member countries organised by Norway and Sweden (Nilsson 1986). These meetings, while not formal assessments because they were structured by the organisers with certain outcomes in mind, produced the key concept of 'critical loads'. This moved the LRTAP regime from the control of emissions to the environmentally much more sound – but much more complex – issue of acceptable levels of deposition of pollutants. This concept depends on the affected ecosystem, since critical loads will differ depending on the nature of the receiving environment. The critical loads concept was integrated into the NO_x Protocol and is in many ways the first international operationalisation of the precautionary principle since it turns the burden of proof around, requiring the identification of levels of deposition that do not cause harmful effects in the ecosystem.

The core issues of policy relevance, quality of science, and scientific legitimacy exist at the international level no less than at other levels of governance. Yet the institutional constraints are such that governments play a much more central role in all international science assessments, sometimes influenced by particular actors or groups of actors, depending on the underlying issues and interests. In practice, all international science assessments are policy-driven rather than science-derived. In many instances it is the participants in national science assessment processes who will also play the central role at the international level. The result is an elision of the different views with respect to policy-science boundaries. Participants in international assessments generally have experience in domestic institutions and consequently bring backgrounds that are rooted in the typically national practices. At the international level, the difficulty of problem identification is such that it requires a joint effort on the part of experts and policy makers alike. This moves the process closer to a policy-science interface where both sides are dependent on one another rather than operate in distinct networks. This is driven though by pragmatic needs rather than by any theoretical concept of improved assessment.

14.4 Biases with Respect to Stakeholder Participation in Environmental Policy Institutions

The rules of the game in policy vary according to the (perceived) structure of the policy issue in question. Problem structure is defined as the relationship between contents and process (Hisschemöller and Hoppe 2001). Theories on problem structure focus attention on institutional biases. They do this by focusing policy and stakeholder assumptions (often implicit or taken for granted assumptions) on the problem at hand. These assumptions may have to do with national interest, such as: 'The Kyoto protocol is not in America's interest.' This claim articulates messages at different levels: at the level of facts and figures or in the terminology of the Advocacy Coalition Framework (Section 14.2.3) the 'secondary aspects', it refers to economic calculations indicating that the US economy might be negatively affected by implementing the Kyoto protocol. At the same time it denies the relevance of figures that indicate otherwise. At a more substantial level (policy core) it says something about how the problem of climate change is perceived, i.e. as a 'global' rather than an 'American' problem. Finally, the claim may be interpreted as an implicit however very important message related to the power relations among the actors involved in the international negotiations.

The assumptions with respect to problem contents may not only relate to the perception of the (dominant) actor itself, but also to expectations with respect to other actors' views and behaviour. An example is the assumption widely shared among public officials that 'industry is only interested in Greenhouse Gas reduction options in as far they are cost-effective on the short term.' The inference of this claim is that the range of options seriously considered becomes significantly reduced: solar photovoltaic energy is not expected to have some impact before 2040 and beyond, while CO_2 removal and underground storage are expected feasible and cost-effective already. What is left out of consideration is the possibility that for industry short-term cost-effectiveness is not the only relevant criterion in evaluating the feasibility of reduction options. They might for example also be interested in options that generate a high innovation potential, which makes them more interesting on the long term.

Problem assumptions have far reaching process implications. They imply which stakeholders become involved in the policy process and who are excluded from participation. Taking the example of Kyoto being not in the US interest, it is well-known that the US are not so fond of multi-lateral treaties, because countries can form coalitions to oppose US proposals; in order to safeguard their interest the US prefer bilateral agreements with separate countries. They also point to the (implicit) rules of the game that shape the interaction among the parties involved, e.g. what they talk about and what is left out of the discussion. In the above example of industry being only interested in cost-effective (cheap) options, the energy options considered relevant are the options perceived as cheap on a reasonable short term rather than, for example, options that have a large innovation perspective on the long term. In fact, what happens is that the assumption about cost-effectiveness introduces an unwarranted bias in the discourse on options for reducing greenhouse gas emissions.

In general terms, two institutional biases can be distinguished that are critical for the evaluation of environmental policies:

- A *technical bias* to a problem implies that the relevant knowledge (including skills and methods) is available and that those who are in possession of this knowledge are qualified to address the problem. Participation by citizens and interest groups is largely restricted, because they are considered incompetent. The information these actors may provide is considered irrelevant, as it might be a disguise for emotional response, resistance to change or narrow self-interest. A clear illustration of this bias is the claim that 'the doctor is always right.' The technical bias reflects the notions of guardianship (Section 14.1) that are still paramount in environmental policy. The doctor is not supposed to act in a self-interested way; he cares for his patients. And he is qualified, so he knows what is good for them in spite of their own thoughts.

- The *market bias* to a problem underlies the assumption that all relevant values are known or agreed upon and that settlement of the issue requires a certain (re)distribution of costs and benefits over affected stakeholders. This may happen through expert calculations or through negotiation between public authorities and participating stakeholders. In the case of negotiation, the parties involved are bound to this basic rule of the game that they may discuss only a given set of values or options (emission reduction to the lowest costs). Once they try to extend or narrow the scope of the debate ('Let's discuss options for innovation.'), they run the risk of 'putting themselves outside the discussion' and may loose their legitimacy vis-à-vis the other participating actors ('We are discussing cost-effective reduction options here'). Hence, this bias may lead to the exclusion of information without excluding actors from participation.

For the analyst, it is important to know that the technical and market bias only hold under specific conditions. The technical bias holds as long as the experts draw one line vis-a-vis the 'outside world'. Once opposition against the expert view gets support from within the expert community itself, the technical bias might collapse. The anti-nuclear opposition in the 1980s could only succeed because leading experts in the field openly expressed doubts as regards nuclear safety and the issue of nuclear waste. Disasters such as in Harrisburgh or Tsjernobyl certainly accelerated this process but were insufficient to initiate policy change.

The market bias is tenable unless a significant number of stakeholders (significant in the meaning of either large or influential) concludes that a specific vulnerable interest maybe harmed by the outcome of the negotiations. A clear example is the issue of mining or drilling under protected natural reserves. Even if risk is considered low and benefits may outweigh costs, the policy process may end up in deadlock. Historical examples of siting of hazardous waste show that policy makers may feel an inclination to offer financial compensation in order to overcome resistance. This strategy may fail dramatically, as the opposition takes the offer as to confirm that the decision makers in charge do not take their arguments seriously.

Figure 14.2 presents a meta-theory, which distinguishes different types of policy problems according to their *structure*. The typology summarises how the technical and market bias may affect the policy process. It also alludes in what way the biases reflected in policy process shape and limit the possibilities for scientists to make a contribution to policy. The main message of this typology is that learning among policy-makers, stakeholders and scientists requires an open dialogue and a willingness to discuss conflicting perspectives. Given the constructivist perspective (see Section 14.2.1), it should be noticed that the distinction between knowledge and values is ideal-typical, as in actual policies knowledge and values are always articulated in a specific way. The typology is explicitly meant to assist the analyst in the search for (hidden) biases and to anticipate what may happen if these biases are avoided. The question to be asked for any given case is: does it make sense to assume consensus or certainty on relevant knowledge and values, or is there reason to believe that observations indicate otherwise?[1]

14.4.1 Policy as Rule

Policy as Rule (see Fig. 14.2) applies to structured problems, i.e. problems characterised by consensus on what knowledge is relevant and what values are at stake. In situations where consensus is real, the problems can be considered technical and often (but not necessarily) routine. Policy allows for a heavy reliance on scientific and technical experts. This is not to imply that scientists get formal decision-making status. To the contrary, the traditional boundaries between policy and scientific advice are kept in tact, science advises policy. Scientific advise is *de facto* binding for policy decisions. The decision maker in this type of policy is usually one monolithic actor. The advisors are part of a closed policy-science network. The role of science is problem solving. Policy as Rule assumes scientific consensus. The most interesting cases arise when scientists get divided. The common reaction from the dominant policy-science interface will be to defend the status quo, including

Consensus on relevant knowledge? \ Consensus on relevant values?	NO	YES
NO	(A) UNSTRUCTURED PROBLEM Policy as learning Science as problem finding	(B) MODERATELY STRUCTURED PROBLEM Policy as negotiation Science as advocate
YES	(C) BADLY STRUCTURED PROBLEM Policy as accommodation Science as mediator	(D) STRUCTURED PROBLEM Policy as ruling Science as problem solver

Fig. 14.2 Four types of policy problems and policy styles and their bearing on the role of science in public policy

Source: Hisschemöller et al. (2001b)

---

[1] Obviously, matters of relevance are not free from subjective judgment. It must be stressed that the typology helps the analyst to formulate an argued judgment about how a problem is addressed at a certain point in time and how its contents may have changed over time. Given the social-constructivist perspective, one cannot be entirely sure as to whether a problem has been adequately structured, as there may be a person somewhere with a deviating position. However, see also the discussion in Section 14.5 on this matter.

its privileged position. Most likely, it will put forward that critics are driven by political rather than by scientific motives. The historic experience with the nuclear energy debate and, more recently, the debate on GMOs, illustrates what may happen if policymakers and scientists hide behind a would-be consensus. It will take quite some conflict before critical information is recognised as science. Once this happens, the problem is no longer treated as structured and shifts to a more pluralist policy setting.

14.4.2 Policy as Negotiation

The moderately structured problem is characterised by consensus on the values at stake, i.e. some public good that needs protection, but uncertainty and conflict as regards the best way to realise common ends. Like in the case of a structured problem, consensus can be imposed. Even if some actors do not really believe in the imposed consensus (such as climate change or reduction of fish-stocks) they play the game according to its rules in order to maximise gain and minimise losses. In this policy type, research-and-analysis becomes an intellectual ammunition in the pluralist interest group struggle. Processes of partisan mutual adjustment (Lindblom 1965) work like a selection device for scientific arguments for previously determined policy stands. Each and every interest will mobilise its own science-based expertise to bolster its case. In this system, policy analysts are like lawyers, and their business is advocacy (Hisschemöller et al. 2001b). In the adversarial model, separate actors defend or strengthen their respective positions in the short run, while in the long run policy-oriented learning may result (Lindblom and Cohen 1979; Sabatier and Jenkins-Smith 1993; Eberg 1997, see Section 14.2). Needless to say, this context seriously limits the opportunities for scientists to take a nuanced position which may provide a 'third way' out of the conflict. Most likely, their work is either misused or ignored. Although conflict may resolve over time, there may be deadlock when powerful minorities, like farmers unions, mining companies or (local) citizen groups, contest the legitimacy of the policy outcome. In that case, the issue shifts to another policy type which, again, brings other opportunities for science involvement.

14.4.3 Policy as Accommodation

The badly structured problem can be best understood as a conflict between irreconcilable values, a dilemma without a solution perspective. The best one can hope for is a compromise which keeps the main conflicting parties on board. The strategy for working out a compromise is characterised by depoliticisation of the value conflict, in other words by transforming the social and political issue into a technical one. The first step in this direction is to move away from a specific problem situation to a problem at a more general level, abstract and long-term. Politics and research concentrate on the invention and internalisation of concepts rather than specific interventions, such as 'sustainability' or 'ecological modernisation' rather than 'emission reductions' (Hajer 1995). The politics of compromise are highly symbolic but may enhance a process of policy learning, which may in the end result in specific interventions. Science in a mediatory role may flourish under this kind of policy; there is a need for multidisciplinary research and, to support the process of accommodation, interaction between scientists and policy makers. But still, the borders between science and policy are intact, as well as most of the disciplinary boundaries. In order to make the mediation between conflicting policy views succeed, scientific consensus is a must. This type of policy-science interaction can frequently be seen in national environmental policy, especially in countries with a strong consensus tradition. But it can be observed even more clearly at the level of international environmental regimes that need to accommodate states with divergent conceptions of their national interests. International relations scholars share the view that scientific consensus, or the existence of so-called epistemic communities (Haas 1991; see Section 14.3.2) is a vital condition for the success of any environmental regime. The Intergovernmental Panel on Climate Change (IPCC) is frequently cited as a successful example.

14.4.4 Policy as Learning

Situations where there is uncertainty about what knowledge is relevant as well as dissent on the relevant values at stake are characterised by the unstructured nature of the underlying problem. Once this situation has been recognised, it may be possible to engage in a process of problem structuring, i.e. to identify, confront, compare and, where possible, integrate different views (Hisschemöller and Hoppe 2001). Policy-oriented learning supposes that parties reconsider their (vested) interests, which makes learning both a cognitive and an emotional experience. What policy learning produces, is one or several consistent though competing problem alternatives. This policy type is rare but not unknown. Co-production occurs, especially in case of concrete problems that come up at a local or regional scale (Botts et al. 2001). Science has traditionally played a significant role in the signalling and agenda setting of environmental problems. However, the more complex these problems are, the greater the need for scientists to work in an interdisciplinary manner, which implies the identification, confrontation, and where possible integration of different scientific perspectives. The boundaries between science and practical knowledge get diffuse and may even become irrelevant, as stakeholders have insights to contribute that scientists by themselves cannot make (Schön 1983). Therefore, experts are bound to cooperate in their very core business of knowledge production with non-experts or, to put it differently, experts in other fields, which requires specific qualities and costs a great deal of time. The difficulty with unstructured problems is that policymakers usually try to avoid them, because they are so hard to manage and the outcome of the policy is far from certain.

14.5 Methods to Enhance Stakeholder Participation

The last decades have witnessed the rise of an approach aimed at involving stakeholders and scientists in an open dialogue to assess options for environmental policy. Participatory technology assessment (Hoppe and Grin 1995) and participatory integrated environmental assessment (IEA) (Hisschemöller et al. 2001c) provide quite some examples. Examples from IEA include the Canadian Mackenzie Basin Impact Study (Cohen 1997), which focuses on a regional risk assessment and the European ULYSSES (Urban LifestYles, SuStainability and integrated Environmental aSsessment) project, which inter alia focused on interaction between modellers and local citizens in several European cities to assess policy options for addressing climate change (Kasemir et al. 2003; Van der Sluijs 2001). The Dutch COOL (Climate OptiOns for the Long term) project focused on the identification of strategies for Dutch long term climate policy (Van de Kerkhof 2004). In this study, the realisation of −80% geenhouse gas emissions by the year 2050, was taken as a starting point for the Netherlands. From there, stakeholders from industry, environmental NGOs, agriculture, transport, consumer NGOs, Housing and energy companies discussed the possibilities and pitfalls for emission reduction options such as efficiency, wind, solar, biomass, hydrogen, CO_2 removal and underground storage and sinks.

This section first discusses some of the methods that are used to enhance an open dialogue. Then, it evaluates the pros and cons of participatory assessments, as comes forward from the science and policy communities.

Participatory methods have in common that they prescribe a more or less precisely defined process for realising a more or less precisely defined outcome, which can be distinguished from other (social science) methods in that groups of people are brought together at a specific location (which may also be at the www) in order to make some sort of assessment. A dialogue is not automatically open, even if peoples' intentions are good. We can learn from the former sections that learning processes normally take a decade or more, which is the period for new knowledge to get settled on policy agendas (Section 14.2). Depending on culture and tradition, policy stakeholders and scientists may have difficulties to communicate with one another (Section 14.3). Then, there are the institutional biases that relate to problem structure (Section 14.4). Before a dialogue can actually start we must ask ourselves at least five questions:

1. What is the aim of the dialogue? What kind of results do we expect?
2. Who are the stakeholders to be involved?
3. To what extent do the stakeholders think differently about the subject matter for discussion and what do they expect to get from it (jointly as well as for themselves)?
4. What is the knowledge to be involved?
5. Are we able to get knowledge from different sources (disciplines) and to what extent can we expect the information provided by different experts to be contradictory or even conflicting?

It is very important to anticipate as to whether we want to 'educate' a group of stakeholders or to encourage a debate on conflicting views and perspectives including knowledge. Mayer (1997: 251) distinguishes seven forms of stakeholder dialogue-participation:

1. *Information/education.* The primary function of stakeholder involvement is to make them aware of scientific findings and to explore the usability of the information offered.
2. *Consultation.* Stakeholders are asked what they know about the problem and what should be done about it.
3. *The anticipation of future developments.* Often used in IEA. Forecasting and backcasting are methodologies which fit in with this approach.
4. *Mediation.* Here, the question is: What do participants know about mutual values? What level of consensus can they reach?
5. *Co-ordination.* Addresses questions such as: what interdisciplinary knowledge should participants generate? What is the relation with other policy issues or sectors?
6. *Co-production.* A concept introduced by Susskind and Elliot (1984), relates to joint problem solving. The main question is: what shared responsibility can participants achieve?
7. *Learning.* This kind of participatory activity enhances a change in core knowledge and attitudes. Participants are asked to explore new styles and strategies for policy-making.

A closer look at the distinction between participatory forms reveals that they in fact reflect a scale, which ranges from a lower to a higher level of actual participation. Methods cannot be used to serve several or all of these objectives. Dependant on their scope and focus, they shape interactive processes in a particular way. We may distinguish methods according to their goal and underlying assumptions and briefly discuss as to whether they are fit to be used in an open dialogue meant to facilitate learning among stakeholders, scientists and policy makers.

Brainstorming has its roots in the past. The concept has become part of every-day language, but as a method it has been developed in the 1930s and 1940s by Osborn (1954). *Brainstorming* is meant to identify options for creative problem solving in the field of marketing. This method is however not meant to structure a discussion. Participants must listen but are not allowed to discuss the options mentioned. That people are doing this exercise together is because they are supposed to mutually stimulate creative association. This requires a high tempo and, as Osborn clearly points out, a well defined problem. If not, the focus of the group gets lost. Also *Focus Group Methodology* has been originally developed in the field of marketing, although the name of the method is sometimes used to allude to all sorts of interactive exercises (scenario workshops, consensus workshops etc.). Its aim differs somewhat from brainstorming. In its authentic form, focus group methodology is meant to articulate the opinions and underlying dimensions that people (consumers) use to evaluate a product or policy. The procedure is quite open in that a small group of people talks about a certain issue for a few hours. The input maybe newspaper articles or pictures. Basically, the facilitator does not intervene except for adding new information to the discussion. The tool has been used to assess the popularity of politicians or policies but also to help developing questionnaires for surveys. More recently, focus group approaches have been developed for participatory sustainability assessments (Kasemir et al. 2003).

Methods known as *Simulation and Gaming* originate from the military and for the training and design of complex systems such as nuclear power plants and aircraft and are aimed at eliminating 'group-think' among for example risk managers (Parson 1996). *Simulation and gaming*, too, is a label that covers a wide range of procedures and claims to realise many objectives, including the identification of policy options as well as learning about conflicting insights and views. In this method, distance between the participant and the subject matter is essential. As Parson explains:

[T]hese methods involve displacing participants from their real and immediate tasks, roles, identities and decision contexts. They pose decision situations that are future, or counter factual, to move participants outside their normal habits and positions, and encourage creative thinking, new ideas, and insights. (Parson 1996: 12)

So, in a game, European participants maybe invited to negotiate a climate change agreement in 2025 on behalf of China.

Controversies on issues related to science and technology in the 1970s and 1980s gave rise to methods that aim at communicating scientific controversy to a wider audience. Methods such as *scientific mediation* (Abrams and Primack 1980) or *Citizens' Juries*

(Seley 1983) are meant to enable policy makers, stakeholders and citizens to make a political judgement on complex issues taking into account contradictory scientific information. They have been used to support decision making with respect to hazardous waste facility siting and for the underground storage of nuclear waste. A rather new development is the *Consensus Conference* (Joss and Durant 1995). This approach is also meant to involve a wider public in decision making on complex technological issues. The procedure followed may vary, but the basic idea is that expert advisement is complemented by a citizen's report assisted by a multidisciplinary team of experts. This report may be used in decision making. The Consensus Conference may not necessarily lead to consensus or even to a shared understanding among experts and non experts. This is illustrated by an observation on a Dutch conference on the risks related to GMOs. Asked as to whether this conference had produced consensus about the major topics involved, the experts answered that this was indeed the case, while many of the non experts had the feeling that most of their arguments were not heard (Jelsma 2001).

Policy delphi (Linstone and Turoff 1975; Dunn 2003) and *backcasting* (Dreborg 1996; Weaver et al. 2000; Carmichael et al. 2004) have their roots in future research and technology assessment. They were originally meant to assist experts in analyzing possible future developments, but, given the intersubjective character of this work, they have been developed to serve as interactive tools. *Policy delphi* is meant to explore conflicting views with respect to future trends and to identify the assumptions that explain for these differences. *Interactive backcasting* is considered the opposite of *forecasting*, as it does not extrapolate current trends and developments as in *forecasting*, but takes a possible future image as a starting point for analysis from which stakeholders 'reason backwards' to the present situation, identifying milestones, opportunities and obstacles that occur along the way. Interactive backcasting thus shapes the diversity between a possible future and the problematic present and helps to identify opportunities for change. Furthermore, the possibility to start the analysis from various future states enables interactive backcasting to shape the diversity of views and interests within a heterogeneous stakeholder group. In The Netherlands, interactive backcasting has been applied in the COOL project (Climate OptiOns for the Long term): a stakeholder dialogue to explore the feasibility of 80% reduction of greenhouse gas emissions by the year 2050 (Van de Kerkhof 2004).

Social scientists, who participated in the democratic wave of the 1960s and 1970s, developed what has become known as deliberative methods, such as *Planungszelle* and *Citizens Forum* in Germany (Renn et al. 1995; Renn 2004). This approach shares characteristics with the Citizen's Jury, the Consensus Conference and the Policy delphi. It is meant to provide policy makers with a well-articulated view on specific policy issues, using well-defined procedures and social science methods to structure and facilitate the debate. Renn et al. (1995) have given much attention to the conditions for an open dialogue process. In their view, an open dialogue must meet two requirements, *fairness* and *competence*. Fairness relates to a transparent process which gives participants equal opportunity to make their claims and discuss the claims made by others. Competence relates to the availability of information for all stakeholders involved and their willingness and capacity to use the information in their discussions.

The *Dialectical Method*, developed from the 1960s onward by Mason and Mitroff (1980) in the US is meant to assist stakeholders in making strategic decisions in public and company policies. The method is based on the concept of problem structuring, i.e. the articulation, confrontation, comparison and, where possible, integration of as many contradictory arguments as possible (Hisschemöller and Hoppe 2001). This method takes a more radical approach than others in that it is based on the assumption that the articulation and exploration of conflict is a necessary vehicle to enhance good decisions.

So far the discussion of methods that are used to enhance an open dialogue and stakeholder learning. The discussion indicates that not all of these methods are equally well defined, but also that their scope and focus varies as well as the form of participation they may be applied to. They have been developed in quite separate fields, such as marketing, future research, sociology of science and technology and management science. Van de Kerkhof (2004) makes the observation that quite some well known methods have a limited value for a dialogue that is aimed at articulating and discussing conflicting views among policy makers, stakeholders and scientists. Some methods (brainstorming in particular) do not allow for any discussion at all, while others (gaming and simulation in particular) create an artificial distance (in time and place) between the stakeholders and the issues for discussion. It may also be concluded that a most important question relates to the evaluation and quality assessment of stakeholder dialogues and the methods used to facilitate them. There is a growing attention in the literature for this issue (Renn et al. 1995; Van de Kerkhof 2004).

It must be noticed that participatory integrated environmental assessment is not an undisputed approach. Both within the science and policy communities arguments pro and con have been put forward. From a policy maker's point of view, the main arguments in favour of participation are (1) an extension of social awareness for environmental problems and social support for policy-making, (2) mutual understanding and creation of trust among stakeholders and (3) an increased potential for social and policy *learning*. However, it has also been argued that participation may undermine the transparency of the democratic process. Once business representatives, scientists, senior civil servants and citizens have reached consensus, it is almost impossible for elected agencies such as national, regional and local parliaments and councils to decide differently. However, there is legitimate concern that participants will not choose in favor of sustainable solutions but may incline to postpone the 'hot potatoes' in order to avoid an escalation of conflict. If government denies the outcomes of participatory processes, it may jeopardize the legitimacy of the policy process and the guiding institutions themselves.

Also on the part of scientists, objections exist. The case of the evaluation of the European Forum for Integrated Environmental Assessment – EFIEA – (Engels 2002) provides an overview of the pros and cons as identified within the integrated assessment community itself. The major argument against stakeholder involvement in the production and evaluation of science is that it may harm scientific integrity, as science becomes a battle-field for competing private and public interests. An other argument is that stakeholders lack the knowledge to seriously consider science, especially where scope and focus of scientific methods are concerned. This view on participatory assessment can be labeled Sound Scientific Expertise. In this view, the production and the use of knowledge occur in institutionally separate spheres. The interface between science and policy is used to generate policy questions and to communicate scientific answers to these questions.

To safeguard the integrity and credibility of scientific advice, however, the scientific knowledge production as the core of the assessment is kept apart from the influences that control the decision making process. The internal mechanisms of scientific quality control – methodological rigor, peer review – are used to produce sound scientific expertise, with the underlying hope that this can be fed into the policy process and can be used there as a source of information that helps to clarify policy options.

Arguments pro participatory assessment stress that it might increase the relevance of science, as stakeholder involvement does justice to the complex and unstructured nature of many environmental problems. In this view, labeled as 'Mutual Learning', the research process is more open to the broader public or to various types of decision-makers and stakeholders. Different mechanisms of quality control – participatory science, extended peer review – are used not only to improve the quality but also to increase the relevance and legitimacy of scientific advice. The social preconditions for the validity of knowledge are taken into consideration. Interactions between science and policy already occur in the early stages of the problem definition and throughout the assessment. Policy-science interfaces are seen as a recursive relationship between tightly coupled spheres that mutually inform and influence one another.

The debate on the pros and cons of participatory assessment has led to identify critical questions as regards the conditions that are favorable for a participatory process. Box 14.1 lists the critical issues that come out of the EFIEA evaluation.

It is interesting to note that there are many assumptions and hypotheses with respect to effectiveness of participation, but that empirical evidence still has to become available. This being so, there is support for the assertion that the added value of participation is dependant on three main factors. First, it must be very clear to all what the ambitions and expectations of the assessment are, on the side of the project team as well as on the side of the respective stakeholders. Second, there must be clarity with respect to how the assessment results will be used in policy-making. This is one of the toughest issues, as policy makers often consider participatory assessment as interesting intellectual exercises but ignore their potential bearing on policy. And third, there must be consensus on the settings of the process, including tasks and responsibilities of the actors involved, the rules of the game and process

> **Box 14.1** Checklist for a better understanding of the policy context
>
> - What exactly is the policy issue that is addressed? At what level of detail should it be addressed?
> - Which phase of the policy process is actually involved (agenda setting, policy formulation, target definition, implementation)?
> - Which are the relevant regulatory bodies? How many of them are involved? How do they relate to one another?
> - What does the current legal framework look like?
> - What is the structure of the policy problem?
> - What kinds of conflicts does the respective policy issue entail? Can a polarised debate be expected?
> - Who are the relevant stakeholders? In what sense are they relevant for the respective policy issue?
> - How are these stakeholders organised? How are they represented in the process of decision-making?
> - What is the goal of the specific policy-science interface (awareness creation, scientific warning, ex-ante impact assessment of policy options, ex-post evaluation of actual policy choices, monitoring of ongoing implementation processes, etc.)?
> - Or, in other words, what role should science play in the policy-science interface (problem solving, policy advocacy, mediation, problem finding)?

design and management. Participants must have a good understanding of what they are supposed to contribute, how and why. The more these basic conditions are being met, the more one may expect an inspiring (instead of irritating) process with surprising (novel) outcomes.

14.6 Summary and Conclusions

The classical dilemma between knowledge and power that has occupied many philosophers and practitioners working on the design and evaluation of political institutions, is paramount in environmental policy and policy analysis. After all, the environmental policy field is dependant on social support as well as sound scientific information. We have argued that environmental policy is actually confronted with changing expectations among stakeholders and the broader public, which has also a considerable impact on the production and use of scientific knowledge. These changing expectations entail interesting challenges for environmental policy institutions. Difficulties arise where the institutional logic of policy and decision-making easily dominates and determines the rules of the game, whereas a successful environmental policy needs an open assessment as to identify and explore information that meets with the requirements of policy relevance, scientific quality and legitimacy at the same time.

Hence, improvement of environmental institutions requires careful attention to stakeholder involvement as well as to policy-science interfaces. This chapter addressed these issues from four different perspectives developed in political science and sociology.

Section 14.2.3 discussed the Advocacy Coalition Framework (ACF). The ACF has proven to be a useful model to study environmental policy change in two ways. First, as an ordering instrument for policy analysis in general and second, more specifically, through linking the concept of *policy belief systems* to that of *advocacy coalitions*. In addition to this, the merit of the ACF is also expressed by its emphasis on learning as one of the forces affecting policy change over time. Especially in environmental policy, research and political debate play a prominent role. This focuses attention on policy-oriented learning as a form of information and argumentation management, and as an important problem structuring strategy. Analyses in the field of environmental policy have shown clear examples of policy-oriented learning within and across the belief systems of advocacy coalitions, both at an instrumental level and at a conceptual level. Finally, the ACF has proven to be a useful theoretical framework to study policy change and learning within and across countries.

In Section 14.3 we compared the institutionalisation of specific policy-science interfaces and their impact on national and international environmental institutions. Each environmental institution handles the problems of policy relevance, scientific quality and legitimacy of research in its own way. Actual experience in European countries and the US suggests that some countries prefer an institutional model where strict science-policy boundaries are being maintained and others tend towards the weakening of boundaries in order to enhance mutual learning. The transfer of a particular form of policy-science interface from one jurisdiction to another needs to overcome these differences and is, therefore, not very easy to implement – an issue of importance in the context of the European Union. The huge diversity of institutional arrangements in which policy-science interfaces unfold needs to be recognised as a determining – and difficult – environment for developing new and more adequate assessment procedures where policy makers and stakeholders can meet with scientific experts. How assessments are translated into policy is again subject to significant differences, depending on the institutions and traditions of governance in each jurisdiction. Some governments can move directly from science assessment to policy formation. By and large, the precautionary principle covers this activity – the determination by governments that the accumulated scientific evidence or the lack thereof justifies a certain action.

Section 14.4 focused on institutional bias in the context of the structure of policy problems, which concept was defined as the relation between policy contents and process. Two biases are considered particularly important in that they shape opportunities for stakeholder involvement in environmental policies. The technical bias reflects notions on guardianship that are still strong in environmental institutions within democratic polities. It builds upon the assumption that non-experts should abstain from participation and leave the decision making to experts. Where the line is drawn between experts and non-experts may differ from case to case. The market bias reflects strong notions on economic rationality in environmental policy. It builds upon the assumption that in decision making the sets of alternative courses of action (values) are given. Stakeholder participation maybe allowed in as far the stakeholders negotiate their goals and interests within this set of values. The technical and market biases are found in different combinations in environmental policy, each of which faces specific opportunities for action but also specific obstacles for an open stakeholder dialogue. Policy learning is considered possible once all relevant actors recognise the unstructured character of the issue at hand and are committed to engage in problem structuring, i.e. to identify, confront, compare and where possible integrate as many different views and knowledge claims with respect to the issue at stake as reasonably possible.

Section 14.5 discussed methods used for enhancing stakeholder dialogues with scientists and policy makers in the context of participatory integrated environmental assessment and participatory technology assessment and discussed their potential. Stakeholder dialogues can serve quite different objectives and may take on many different forms, such as education (most passive form of participation), consultation, anticipating future developments, mediation and learning. If one wants to use a dialogue as a means to improve a scientific assessment, a decision-making process, or both, there are some fundamental questions one should address in advance, related to the objectives and kind of outcomes envisaged, the stakeholders to be involved and their specific views and questions, as well as the information available from different sources. It is equally important to note that not all participatory methods are fit to enhance an open dialogue process in the same way. Caution is needed where some methods are better fit to deal with contradictory information than others. Among policy makers and scientists there are different views with respect to the value of participatory assessments. The improvement of quality control seems a task for the (social) sciences community in the first place, as the social sciences, political science, sociology and psychology in particular, focus on the study of social interaction and conflict.

As regards the learning potential of environmental institutions there is also a huge challenge for policy in this respect. Guardianship does not work, as it tends to neglect an open process of choice among policy options. Environmental institutions today still reflect biases that prevent open dialogue between policy makers, scientists and stakeholders. The characteristics of today's environmental problems are such that science may deliver information not sufficiently relevant or even contradictory. Stakeholder participation may be helpful to assess the feasibility of options for policy. At the same time, political choices remain necessary. Traditional mechanisms to deal with the problems of knowledge and power are to be replaced by others that facilitate the articulation and evaluation of policy alternatives. Strict boundaries between policy and science have become an obstacle for innovation. The same holds for decision making that is guided by the inclination to keep everyone satisfied and therefore tends to avoid hard choices. Hence, the observations from environmental policy provide a challenge for the (re)design of environmental institutions and political institutions in general.

References

Abrams, N. E., & Primack, J. R. (1980). Helping the public decide, the case of radioactive waste management. *Environment, 22*(3), 14–40.

Beck, U. (1992). *Risk society. Towards a new modernity.* London: Sage.

Botts, L., Muldoon, P., Botts, P., & von Moltke, K. (2001). The Great Lakes water quality agreement. In M. Hisschemöller, R. Hoppe, W. N. Dunn, & J. Ravetz (Eds.), *Knowledge, power and participation in environmental policy analysis. Policy studies review annual* (Vol. 12, pp. 121–144). New Brunswick, NJ: Transaction.

Carmichael, J., Tansey, J., & Robinson, J. (2004). An integrated assessment modeling tool. *Global Environmental Change, 14*, 171–183.

Cohen, S. J. (1997). Scientist-stakeholder collaboration in integrated assessment of climate change: Lessons from a case study of Northwest Canada. *Environmental Modelling and Assessment, 2*, 281–293.

Cross, C. F. (1999). *Justin Smith Morrill: Father of the land-grant colleges.* East Lansing, MI: Michigan State University Press.

Dahl, R. A. (1985). *Nuclear weapons: Democracy versus guardianship.* Syracuse, NY: SUNY Press.

Deutscher Bundestag. (1991). *Protecting the Earth. A Status Report with Recommendations for a New Energy Policy.* Third Report of the Enquete Commission of the 11th German Bundestag Preventive Measures to Protect the Earth's Atmosphere. Deutscher Bundestag: Bonn.

Dreborg, K. H. (1996). The essence of backcasting. *Futures, 9*(28), 813–828.

Dunn, W. N. (2003). *Public policy analysis* (3rd ed.). New York: Prentice-Hall.

Eberg, J. (1997). *Waste policy and learning. Policy dynamics of waste management and waste incineration in the Netherlands and Bavaria.* Delft (The Netherlands): Eburon.

Engels, A. (2002). *Evaluation of four science-policy interface workshops.* Amsterdam: EFIEA.

Ezrahi, Y. (1990). *The descent of icarus. Science and the transformation of modern democracy.* Cambridge, MA: Harvard University Press.

Fischer, F. (1990). *Technocracy and the politics of expertise.* Newbury Park, CA: Sage.

Fischer, F. (2001). Beyond technocratic environmentalism: Citizen inquiry in sustainable development. In M. Hisschemöller, W. N. Dunn, R. Hoppe, & J. Ravetz (Eds.), *Knowledge, power and participation in environmental policy analysis. Policy studies review annual* (Vol. 12, pp. 29–46). New Brunswick, NJ: Transaction.

Funtowicz, S., & Ravetz, J. (1993). Science for the post-normal age. *Futures, 25*(7), 739–755.

Gibbons, M., Limoges, C., Nowotny, H., Schwartzman, S., Scott, P., & Trow, M. (1994). *The new production of knowledge.* London: Sage.

Haas, P. (1991). *Saving the Mediterranean. The politics of environmental cooperation.* Cambridge, MA: MIT Press.

Hajer, M. A. (1995). *The politics of environmental discourse: Ecological modernization and the policy process.* New York: Oxford University Press.

Halffman, W. (2003). *Boundaries of regulatory science: Eco/toxicology and aquatic hazards of chemicals in the US, England and the Netherlands, 1970–1995.* University of Amsterdam: thesis.

Heclo, H. (1974). *Social policy in Britain and Sweden.* New Haven, CT: Yale University Press.

Hisschemöller, M. (1993). *De democratie van problemen. De relatie tussen de inhoud van beleidsproblemen en methoden van politieke besluitvorming.* Amsterdam: Free University Press.

Hisschemöller, M., & Hoppe, R. (2001). Coping with intractable controversies. The case for problem structuring in policy design and analysis. In M. Hisschemöller, R. Hoppe, W. N. Dunn, & J. Ravetz (Eds.), *Knowledge, power and participation environmental policy analysis. Policy studies review annual* (Vol. 12, pp. 47–72). New Brunswick, NJ: Transaction.

Hisschemöller, M., Hoppe, R., Dunn, W. N., & Ravetz, J. (2001a). Knowledge, power and participation in environmental policy analysis: An introduction. In M. Hisschemöller, R. Hoppe, W. N. Dunn, & J. Ravetz (Eds.), *Knowledge, power and participation in environmental policy analysis. Policy Studies Review Annual* (Vol. 12, pp. 1–28). New Brunswick, NJ: Transaction.

Hisschemöller, M., Hoppe, R., Groenewegen, P., & Midden, C. J. H. (2001b). Knowledge use and political choice in Dutch environmental policy: A problem structuring perspective on real life experiments in extended peer review. In M. Hisschemöller, R. Hoppe, W. N. Dunn, & J. Ravetz (Eds.), *Knowledge, power and participation in environmental policy analysis. Policy Studies Review Annual* (Vol. 12, pp. 437–470). New Brunswick, NJ: Transaction.

Hisschemöller, M., Tol, R. S. J., & Vellinga, P. (2001c). The relevance of participatory approaches in integrated environmental assessment. *Integrated Assessment, 2*(2), 57–72.

Hoppe, R., & Grin, J. (Eds.). (1995). Interactive strategies in technology assessment [Special issue]. *Industrial and Environmental Crisis Quarterly, 9*(1).

Jelsma, J. (2001). Frame-reflective policy analysis in practice: Co-evolution of a policy regime and an intractable controversy in biotechnology. In M. Hisschemöller, R. Hoppe, W. N. Dunn, & J. Ravetz (Eds.), *Knowledge, power and participation in environmental policy analysis. Policy Studies Review Annual* (Vol. 12, pp. 201–227). New Brunswick, NJ: Transaction.

Joss S., & Durant, J. (Eds.). (1995). *Public participation in science. The role of consensus conferences in Europe.* London: British Science Museum.

Kasemir, B., Jaeger, J., Jaeger, C. C., & Gardner, M. T. (2003). *Public participation in sustainability science. A handbook.* Cambridge: Cambridge University Press.

Lakatos, I. (1971). History of science and its rational reconstruction. *Boston Studies in the Philosophy of Science, 8*, 42–134.

Lindblom, C. E. (1965). *The intelligence of democracy.* New York: Free Press.

Lindblom, C. E., & Cohen, D. K. (1979). *Usable knowledge: Social science and social problem solving.* New Haven, CT: Yale University Press.

Linstone, H. A., & Turoff, M. (Eds.). (1975). *The delphi method, technics and applications.* Reading, MA: Addison-Wesley.

Mason, R. O., & Mitroff, I. I. (1980). *Challenging strategic planning assumptions. Theory, cases, and techniques.* New York: Wiley.

Mayer, I. (1997). *Debating technologies. A methodological contribution to the design and evaluation of participatory policy analysis.* Tilburg, The Netherlands: Tilburg University Press.

Mill, J. S. (1861/1973). Considerations on representative government. In J. M. Robson (Ed.), *Collected works of John Stuart Mill.* Toronto: University of Toronto Press.

Newell, P. (2000). *Climate for change. Non-state actors and the global politics of the Greenhouse.* Cambridge: Cambridge University Press.

Nilsson, J. (Ed.). (1986). *Critical loads for nitrogen and sulphur.* Report from a Nordic Working Group (no 1986:11). Copenhagen: Nordisk ministerrd.

Nowotny, H., Scott, P., & Gibbons, M. (2001). *Re-thinking science. Knowledge and the public in an age of uncertainty.* Cambridge: Polity Press.

Osborn, A. (1954). *Your creative power. How to use imagination.* New York: Charles Scriber.

Parson, E. A. (1996). Three dilemmas in the integrated assessment of climate change. An editorial comment. *Climatic Change, 34,* 315–326.

Ravetz, J., & Funtowicz, S. (1999). Post-normal science – An insight now maturing. *Futures, 31,* 641–646.

Renn, O. (2004). The challenge of integrating deliberation and expertise. Participation and discourse in risk management. In T. L. McDaniels & M. Small (Eds.), *Risk analysis and society. An interdisciplinary characterization of the field.* Cambridge: Cambridge University Press.

Renn, O., Webler, T., & Wiedeman, P. (1995). *Fairness and competence in citizen participation. Evaluating models for environmental discourse.* Dordrecht: Kluwer.

Sabatier, P. A. (1998). The advocacy coalition framework: Revisions and relevance for Europe. *Journal of European Public Policy, 5*(1 March), 98–130.

Sabatier, P. A. (Ed.). (1999). *An advocacy coalition lens on environmental policy.* Cambridge: MIT Press.

Sabatier, P. A., & Jenkins-Smith, H. (Eds.). (1993). *Policy change and learning: An advocacy coalition approach.* Boulder, CO: Westview Press.

Sabatier, P. A., & Jenkins Smith, H. (1999). The advocacy coalition framework: An assessment. In P. A. Sabatier (Ed.), *Theories of the policy process.* Boulder, CO: Westview Press.

Sabatier, P. A., Focht, W., Lubell, M., Trachtenberg, Z., Vedlitz, A., & Matlock, M. (2005). *Swimming upstream: Collaborative approaches to watershed management.* Cambridge: MIT Press.

Schön, D. A. (1983). *The reflective practitioner. How professionals think in action.* New York: Basic Books.

Seley, J. F. (1983). *The politics of public facility planning.* Lexington, MA: Lexington Books.

Susskind, L., & Elliot, M. (Eds.). (1984). *Paternalism, conflict and coproduction. Learning from citizen action and citizen participation in Western Europe.* New York: Plenum.

United Nations Environment Programme (UNEP). (1998). *Policy effectiveness and multilateral environmental agreements* (Environment and Trade 17). Geneva: UNEP.

Van de Kerkhof, M. F. (2004). *Debating climate change. A study on stakeholder participation in an integrated assessment of long-term climate change in the Netherlands.* Utrecht, The Netherlands: Lemma.

Van der Sluijs, J. (2001). Integrated assessment modelling and the participatory challenge. In M. Hisschemöller, R. Hoppe, W. N. Dunn, & J. Ravetz (Eds.), *Knowledge, power and participation in environmental policy analysis. Policy studies review annual* (Vol. 12, pp. 317–347). New Brunswick, NJ: Transaction.

Weale, A. (1992). *The new politics of pollution.* Manchester: Manchester University Press.

Weaver, P., Jansen, L., van Grootveld, G., van Spiegel, E., & Vergragt, P. (2000). *Sustainable technology development.* Sheffield: Greenleaf.

Weber, M. (1946). Politics as a vocation. In H. H. Gerth & C. Wright Mills (Eds.), *From Max Weber. Essays in sociology.* New York: Oxford University Press.

Weber, M. (1992). *Wissenschaft als beruf 1917/1919 (science as profession) (gesamtausgabe I/17, wolfgang mommsened.).* Stuttgart: Mohr/Siebeck.

Weiss, C. (1977). Research for policy's sake: The enlightenment function of social research. *Policy Analysis, 3*(Fall), 531–545.

Wettestad, J. (1999). More 'discursive diplomacy' than 'dashing design'? The convention on long range transboundary air pollution (LRTAP). In J. Wettestad (Ed.), *Designing environmental regimes. The key conditions* (pp. 85–123). Cheltenham: Edward Elgar.

Chapter 15
Technology for Environmental Problems

Karel Mulder

Contents

15.1	Introduction	305
15.2	The Role of Technology: Culprit, Saviour or Both?	306
15.2.1	A Short History of Technology and Society	306
15.2.2	On the Inseparability of Technology, Culture and Structure	309
15.3	A Brief Taxonomy of Environmental Technologies	312
15.4	Classic Environmental Technologies: Triple D	314
15.4.1	Modern Triple D	315
15.5	Good Housekeeping Technologies: Triple M	316
15.6	End-of-Pipe Technologies	317
15.6.1	Gas Separation and Treatment Technologies	318
15.6.2	Water Separation and Treatment Technologies	319
15.6.3	Solid Waste Treatment	321
15.6.4	Other End-of-Pipe Solutions	322
15.6.5	Environmental Restoration	322
15.7	Process Adaptation and Damage Prevention	324
15.7.1	Changing the Primary Energy Base	324
15.7.2	Changing the Raw Materials Base	325
15.7.3	Avoiding By-products and Emissions	326
15.8	Towards Sustainable Production and Consumption	327
15.9	Sustainable Energy Technology	330
15.9.1	Storage	331
15.9.2	Ecology	331
15.9.3	Risks	331
15.9.4	Space	331
15.10	Closing Remarks	331
References		332

K. Kulder (✉)
Technology Dynamics & Sustainable Development at the Faculty of Technology,
Policy & Management of Delft University of Technology,
The Netherlands
e-mail: k.f.mulder@tudelft.nl

15.1 Introduction

This chapter examines the potential role to be played by technology in resolving today's environmental problems. First, though, we consider some of the positions that can be taken on the extent to which technology can itself be held responsible for these problems and how this bears on the extent to which technological innovation can help solve them. To provide some background for this key issue, we briefly explore the history of western technology. We then move on to consider the range of environmental technologies in use today as well as several principles that can be applied to design inherently more benign technologies. The chapter closes with some thoughts on sustainable development and technology.

Current technology is neither the best possible, given present scientific understanding, nor the result solely of socio-political choice.[1] In the large-scale technological systems of today, social institutions and technological hardware form a seamless web and any distinction between the 'social' and 'technological' dimensions of these systems becomes futile. Particularly when systems fail, attempts are made to blame casualties on either 'human' or 'technological' factors. Such attempts are doomed to failure, though, for it is in fact impossible to distinguish the human and technological factors in any given technological system: is it the hardware that is not properly adapted to the humans operating, administering or maintaining it, or are the humans not

---

[1] Socio-political choice is reflected in technology. However, the resulting technology often outlives the socio-political conditions that created it and can even influence choices on new technologies much later. Roman carriages defined the width of British roads. These in turn set the standards for the width of British railroads, which became the standard in most industrialised nations.

functioning in accordance with the demands set by the hardware they are dealing with? This question cannot be answered empirically.

New technologies arise out of the specificmomentary conditions pertaining in technological systems as well as changes in the environment of those systems. At the same time, though, the technological options available for the future are shaped by the specific path by which previous technologies have been developed (reflected in standards, methods and equipment) and the knowledge, rules and beliefs of the technological community.

Technological changes are often slow and can easily lag behind the more rapid pace of change in society as a whole. However, the creativity of technologists also leads to new products and systems that revolutionise social life, like cellular phones and computers. This is not to say that every new revolutionary technology is accepted by society. Indeed, many new technologies were not accepted, and are hardly remembered. Civil aircraft for vertical take-off and landing or soluble tablets to replace toothpaste are just some of the vast array of technologies that have been rejected. In the case of nuclear power, the issue of acceptance has still not been settled.

It might therefore be argued that, in so far as technology can be distinguished from its social environment, the relationship between them is a (co-)evolutionary one: they adapt to one another, but there are many mismatches. Such mismatches occur especially in times of rapid change, due either to massive breakthrough of new technologies, as in the case of IT in the 1980s, or to rapid changes in society's preferences.

The technologies applied to solve environmental problems are products of the present and the past, of society's norms and values and of technologists' experiences and paradigms. They shape our common future while reflecting contemporary standards and interests. To develop the technologies that mankind needs for sustainable development is therefore not only a formidable challenge to technologists but to our entire technology-dependent civilisation.

New technologies always entail social change. The successful introduction of a new technology is therefore always a matter of socio-technical change. In the following sections, a range of socio-technical solutions for environmental problems are discussed. These technologies have been categorised according to their degree of 'radicalism', i.e. the degree to which they affect current technological systems:

1. Pre-industrial technologies
2. Classic environmental technologies
3. Good housekeeping technologies
4. End-of-pipe technologies
5. Process adaptation and damage prevention
6. Sustainable technologies

Most industries today use type 2 technologies or better. When confronted with the environmental impact of their activities, operators usually adopt type 3 technologies, with the next category only being applied if the problem remains unsolved. Simply substituting a technology from a higher category will not render the original technology superfluous, however; on the contrary, it will often need to be adapted and improved. Technologies in a higher category often require larger investments, thereby creating higher production costs. Even if they do not increase production costs, they require greater changes to the production system. The greater the change in the production system, the greater the risk of unforeseen events disrupting the continuity of production. The greater the changes required, moreover, the greater the likelihood that the conserving forces of organisational or technological tradition will be mobilised.

15.2 The Role of Technology: Culprit, Saviour or Both?

15.2.1 A Short History of Technology and Society

Many technology critics, e.g. the philosophers Habermas and Marcuse, have argued that technology is the root cause of the lack of sustainability in society. More recently, the book of Brauni (1995) is a good example. There are many, though, who hold that technological change must be at the very heart of any move towards sustainable development. To explore the role of technology in society, this section reviews some of the forces that drive technological change. Although the industrial revolution has roots dating back partly to antiquity, the most profound changes in the relationship between humanity and the natural environment have taken place over the past 500 years and it is this period we shall focus on here.

It is doubtful whether there has ever been an enduring stable human society living in complete harmony with its natural surroundings. Even before antiquity, the human species drove other species to extinction. The early classical civilisations, in Mesopotamia, China or India, for example, all disrupted their natural environment. These disruptions sometimes threatened human society, as when productive farmland was ruined by salinisation and production systems had to be adapted to prevent social collapse. For many cultures the limitations encountered in their dealings with nature became part of their normative framework. Modern, industrial consumer society appears to have a different dynamic. For a start, it is expanding far faster than any previous society. Because of its vastly greater size and intensity of material consumption, moreover, it is threatening not just local and regional ecosystems but the global ecosystem itself. Today the traditional limitations arising in the natural world are being ever more replaced by what might be termed the purely expansionist rationality of industrial consumer society.

The roots of this expansionist thinking can be traced back to the Middle Ages and even further. In the Middle Ages technological 'progress', such as it was, diffused only very slowly. The case of papermaking may serve as an illustration. By the 8th century, Chinese papermaking technology had reached Samarkand in Central Asia. This technology was established in the Muslim world in the 10th century and in the 11th and 12th in Christian Europe, where it was initially denounced as a technology of the Muslim foe. This attitude remained essentially unchanged until introduction of the printing press in the 15th century, when paper finally became accepted in Western Europe. This example shows that innovations that would today reach every corner of the world within a few years were only very slow to spread, if at all, in medieval times.

The resistance to papermaking illustrates another feature of medieval society. The slow pace of technological change was not due simply to a lack of individual inventiveness. In a society that sought its solace beyond this earthly life, technological change was, rather, considered to be relatively unimportant. Medieval thinking on how nature and society were ordered stressed the importance of objects and peoples each remaining in their 'allotted place'. This implied that (financial) prosperity through technological innovation was not encouraged. As long as new technology could be absorbed within existing institutions, it was more or less tolerated, being sought actively only in times of sheer despair: starvation, conquest, flooding and so on. Competition was stifled, moreover, by the institutional structure of economic production, i.e. agricultural production by a peasantry dependent on the nobility and production of a narrow range of goods by craftsmen organised in a guild system. The general result was to foster tradition rather than spur innovation for a 'better' future.

Still, technological change did occur, but mainly as a result of outside influence and in those societies where the old 'divine order' was challenged most. Thus, it was Crusaders who probably brought the windmill to Europe in the early 13th century. Although it was of great advantage in lowlands lacking waterpower, the technology diffused only slowly because of its clear impact on social structure, upsetting the nobility's traditional milling rights. In the Northwest European lowlands, however, the windmill became an increasingly popular and effective means of pumping water. With sea levels rising and agricultural peat lands setting after centuries of use, these countries were in dire need of pumping power. Windmill-driven pumps were consequently not introduced primarily to expand economic activity, but to protect agriculture from deteriorating natural conditions. Later, population growth prompted a need for agricultural expansion and, with most arable land already in use, windmills opened the way for 'reclaiming' new land.

The changes that took place during the Renaissance (14th–16th century) influenced the social climate for technological change tremendously. The emphasis in society was now on the individual's struggle for power, property and status. Intellectuals were striving for individual expression, which included the active design of new machines, as we know from the extensive draught work of Leonardo da Vinci, for example. The changes also fuelled a drive to discover the world, facilitated in particular by new, superior techniques of shipbuilding and navigation. Although this led to a series of world-famed discoveries, however, the vast majority of the population felt the effects of technological change only very gradually, if at all.

The Enlightenment (18th century) can be seen as the closing phase of this transformation. Its basic philosophy of rationalising society implied that technology should be freed of its guild-based rules and traditions of craftsmanship. The newly emerging nation states established engineering corps to create national transportation systems, strengthen military power and facilitate colonial exploitation. These engineers were not to work according to local circumstances (as the mediaeval craftsman did), but to apply the universal rules of science throughout nations or empires. It is important to realise that until the 18th century, science and technology were almost entirely separate realms of activity: academic science was practised out of curiosity and was not geared to improving technology, which was designed and maintained by craftsmen. A very early example of the application of science to technology can be seen in Galileo's work, around the turn of the 17th century. Galileo was the first to show that, if air resistance is disregarded, the acceleration of a falling body is independent of its mass, allowing him to calculate the parabolic orbits of projectiles. The resulting tables proved to be a powerful tool for gunners, enabling them to accurately determine the firing angle needed to hit any target at a given range.

In that era, it was rare for the results of scientific experiments to be commercially exploited in the form of new or improved products or processes. Although Dutch windmill-builders improved their designs during the 17th century, for example, their efforts were not based on scientific principles. It was not until 1759 that John Smeaton presented the results of the first scientific experiments on windmill sails in his *On the construction and effects of windmill sails*, thereby corroborating the value of 17th century changes in windmill design. Indeed, quite a number of key technologies were created long before the underlying scientific principles were formulated. For example, the invention of the first steam engines (by Newcomen 1712; and Watt 1770) preceded the formulation of the thermodynamic principles explaining their operation by more than a century (by Carnot 1824; and Joule 1845). As L. J. Henderson penetratingly wrote in 1917: '*Science is infinitely more indebted to the steam engine, than is the steam engine to science*' (Barber 1970).

In this era, the process by which technical innovation came about was generally one of trial and error. As failures were often dramatic, there was a strong reluctance to change technologies. Churches, for example, were constructed according to rules of thumb that had remained unchanged for centuries. However, there are even modern techniques, such as those employed in bridge construction, that are still in part trial and error, sometimes leading to unforeseeable failure. In 1996, for example, the city of Rotterdam had to close its new Erasmus Bridge because of oscillations set up by the wind in the suspension cables. Similarly, in 2000 pedestrian-induced oscillation of the London Millennium Bridge necessitated closure and substantial reconstruction.

The substitution of tradition by rationality promoted in the 19th century led to a need for a new profession: the engineer. The engineer was not a traditionally trained craftsman but someone scientifically trained to design and operate technologies according to rationalist principles. By 1847 the founders of the Dutch Royal Institute of Engineers were explicitly requiring their members to be trained scientifically.

In the 19th century, especially, statesmen were convinced they needed scientifically trained engineers to replace the traditional craftsmen trained over many years of apprenticeship. There were important tasks for these new professionals. The nation states that took shape in the 18th and 19th century transformed the European economies, shifting their basis from local autarchy to national or even imperial autarchy. For this transformation, a variety of civil works were needed. Colonies supplied the surpluses required for investments and the engineers drew up the designs. The engineers had to bring 'Progress' by constructing railroads, canals, roads, bridges and water control and management systems. They rationalised agricultural production and designed machinery and factories to efficiently supply large quantities of goods to the masses. The technologies they designed made the national state a real entity.

By the end of the 19th century, the 'social issue' dominated the public debate: industrial production had created an urban proletariat and sharp social conflict. Many engineers were sympathetic to political visions in which the state was assigned a leading role in advancing social justice. For a profession that had gained its status by rationalising infrastructure and production, rationalist planning of economic activities was a highly attractive proposition. Generally speaking, one could say that the engineering profession of the first half of the 20th century was very committed to solving society's problems by means of rationalised planning.

In the large-scale 'social market' economies of the second half of the century, new problems emerged. With society now effectively rationalised, many people felt alienated by the complexity of the technological systems they were part of. Rationalised spatial planning had led to heavy-handed uniformity of the built environment. As the scale of production grew, environmental pollution, ecological destruction and resource depletion became more than simply local problems. Critics of modern technology blamed it for creating the arms race, nuclear and chemical risks, pollution, threats to privacy and, indeed, threats to mankind's short and long-term survival. In the debates of the day, this was often referred to as the 'price' to be paid for the unprecedented affluence in which people lived, an affluence created by technological 'progress'.

Although at first specific technological products were criticised for their detrimental impact, 'technology' itself was soon under attack. 'Technological thinking' was accused of being based on a principle of domination over both nature and people. This technocratic attitude was condemned as undemocratic, in that it sought to control people rather than interact with them, and as short-sighted, in that intensive and exponentially growing use of natural resources and ditto production of waste and emissions were bound to end in catastrophe.

As engineers and their technologies were blamed for creating these very substantial problems, engineers became far less committed to social issues. Having grown used to the criticism of environmentalists, engineers did not jump on the issue of sustainability when it was introduced in 1987. Although sustainable development might equally well have been have been interpreted as a new challenge to the engineering community, many engineers regarded the notion as suspicious, concerned that it might mean a new attack on their profession.

Over the past decade, there have been some very promising initiatives to once again harness engineers and their skills in addressing the challenges facing society, among which sustainable development is arguably the principal today.

However, might technological innovation really be the key to Sustainable Development, or will it make things worse by increasing the technocratic domination of nature and people? If so, is there a new and different way of designing technology that can help solve existing problems without introducing new ones? In the next section we examine this question by briefly reviewing some of the positions people have taken on the problematical relationship between technology and society.

15.2.2 On the Inseparability of Technology, Culture and Structure

One of the most outspoken critics of modern technology was the French philosopher Jacques Ellul. Adopting a historical perspective, Ellul sought to understand how medieval technologies differed from the technologies spawned by the rationalist attitudes of 19th century engineers. In his view, they did so in a number of salient respects: they were:

- Restricted in their sphere of application: technologies were often based on specific local resources and therefore rarely transferable.
- Dependent on limited resources and on well-developed skills, such as making and repairing tools, or judging weather conditions and tides.
- Local in character: technological solutions to specific problems were embedded in local culture and tradition.

Together, this meant that individuals and local communities were able to influence and shape the technologies they applied. There was, in other words, a degree of technological choice. Modern technology, in Ellul's analysis, has quite a different set of characteristics:

- Automatism: there is just one 'best' way to solve each particular problem, and the accompanying technology appears to be compelling, across the planet.
- Self-perpetuation: new technologies reinforce the growth of other technologies, leading to exponential growth.
- Indivisibility: the technological way of life must be accepted completely, with all its up- and downsides.
- Cohesion: technologies used in disparate areas have much in common.
- Universalism: technologies are omnipresent, both geographically and qualitatively.

For Ellul, this implied that modern technology destroys human freedom. In his view the future of mankind was very bleak, and there was no way back.

Some of Ellul's basic lines of argument can be recognised in the 'Unabomber manifesto'. In the 1980s and early 1990s, Kaczynski, a Californian professor of mathematics, launched a series of attacks on US airlines and research institutes. In his Unabomber Manifesto, he explained that it was only possible for an individual to participate in modern society if he or she accepted its technologies. Although these technologies were often legitimised as creating more freedom, the Unabomber explained, they in fact took away more and more of our freedom because they proved to be increasingly compelling for the individual. In the vision of the Unabomber, too, this would ultimately lead to the destruction of human freedom. And in his view, too, there was no alternative and no way back.

However, in their analysis, Ellul and the Unabomber fail to recognise the tremendous advantages that technologies have brought us. The number of people living a relatively good life on planet Earth, i.e. well nourished and healthy, is historically unprecedented. Criticising the alienation created by technology, as Ellul and the Unabomber did, is one-sided, to say the least, and therefore leads to unjustified conclusions.

At heart, the reasoning of many technological optimists is rather similar, though: like Ellul and the Unabomber, they often paint a picture of a high-tech future as if this future were not man-made, but an inevitable reality. The *technological determinism*[2] inherent in this line of reasoning is highly questionable. It assumes that the relationship between science, technology and society is unidirectional, i.e. that technology itself results from some mechanism of 'technological self-perpetuation' (see above) and from the ever-growing pool of (scientific) knowledge available to the technologist.

As the example of the steam engine shows, however, technology results not merely from the simple application of scientific principles. Various historical analyses show, moreover, that the creation of new technologies is not a process with a single ('optimum') outcome, as Ellul holds, but is the result of a process of choice, a characteristic claimed only for traditional technology.

In the view of *constructivists*, technological change is a process in which social actors with varying perceptions of a given technology make choices and interact, thereby determining the function and shape of new technological artefacts and systems. This choice cannot be made arbitrarily, however, for technological artefacts possess a certain stability. Changes in technology will not, therefore, coincide with changes in society but rather lag behind. As technological change can affect social and economic conditions, a dynamic pattern emerges.

Between these two rather extreme positions, conceptions have meanwhile been developed that aim to integrate the determinist's main assertion, that technology is self-propelling, with that of the constructivist, that technology is the outcome of socially determined choice. These new conceptions seek to account for the conditions under which technological change is propelled predominantly by social forces and the conditions under which social forces can scarcely influence that process.

For technological determinists, attempts to influence science and technology for commercial, ethical or political reasons are futile, producing no more than minor ripples in the pond. For the constructivist, the stability of technological artefacts is explained by the stability of the different environments in which those artefacts are embedded:

- The socio-economic environment: an artefact must meet the demands imposed on it by all the relevant social actors in its environment (for a car: price, status, comfort, safety, appeal, speed, etc.). New technologies must, in other words, solve the problems that actors think can be solved by the artefact in question.
- The physical environment: every artefact is adapted to other artefacts, to technological infrastructure, to maintenance systems, energy sources and so on. For example, cars must fit on roadways, use available, standardised fuels, be fitted with familiar steering mechanisms and meet various performance standards before they are approved. New technologies must be compatible with these existing conditions.
- The technological knowledge base: technologies are based on the know-how, rules and heuristics accepted within a particular technological community. New technological artefacts arise from the state of knowledge and the shared beliefs regarding possible improvements within the community of practitioners designing and constructing them. New artefacts will consequently accord with the practices and shared beliefs of these practitioners as to what is possible and worthwhile. More radical technological change therefore implies changing mainstream ideas within technological communities or

---

[2] This is the view that the state of technological development determines the nature of a society, rather than vice versa. The technological deterministic worldview is reflected in such historical characterisations as 'the Stone age', 'the Bronze age' or 'the age of the Automobile'.

breaking their power by creating an alternative technological community to take its place.

These mechanisms of stability and change lead to an evolutionary pattern of development. Since major upheavals in all three realms are rarely simultaneous, radical innovations are likewise rare. For example, the technological system in which the automobile operates (refineries, gas stations, garages and safety regulations) makes it very hard to change to other sources of energy than oil-based fuels. In the technological regime of automobile propulsion the main focus was for a long time on increasing the efficiency of the internal combustion process, with almost no expertise available on all the other technologies that might be used to propel cars. It is only recently that efforts have been undertaken to develop fuel cells for motor vehicle propulsion. The current socio-economic environment favours development of larger cars – preferably with a green image, too.

Because of the various forces favouring technological stability, technological artefacts generally change only incrementally, over long periods of time. As in biological evolution, however, a technology may become extinct, or split into several species adapted to specific niches and circumstances.

Technologies adapt by means of small steps to changes in their environment. At some point in its development, a technology might be unable to adapt any further. Under these circumstances, especially, a technological 'revolution' may occur as new technologies supersede the existing ones. This comparison of processes of technological change with biological evolution has its limitations, though. In standard evolution theory, variations occur at random and the natural environment selects survivors. Selection and variation are thus strictly separate processes. However, the creators of new technologies often actively seek to change the 'selection' environment, to create the right conditions for their new 'baby'; inventors actively seek to create a demand for their invention or try to remove barriers (legal restrictions on genetic modification, for example). Social actors sometimes actively seek to influence the 'variation' process in order to generate the specific artefacts they favour. For example, the US Air Force had vast programmes in the 1950s to stimulate the chemical industry's R&D efforts on heat-resistant materials, which were needed for the new generation of jet aircraft.

Today, the deterministic view of technology has hardly any adherents among students of technological change. However, technological determinism is still common in popular reasoning, whether in the form of a bright future being foreseen for mankind, soon to start out on the conquest of the final frontier of space, or in the form of a society deprived of human values, which will ultimately collapse for lack of social cohesion or through unbridled consumption. Not only is this vision empirically untenable, it is also morally questionable, as it encourages lethargy instead of action when it comes to the problems currently faced by society.

The implication of these insights into the dynamics of technology is that a new technology should never be introduced just to solve a strictly physically defined problem, as these problems are always social in nature, too.

Decision-making with respect to our technological future is probably more important for the long-term outlook of society than any other political issue today. Very often, however, new technologies are not recognised as political issues requiring due debate but viewed, rather, as inescapable necessities. We should therefore seek to uncover the political nature of processes of technological change and develop mechanisms for democratising those processes.

The 1970s saw the first attempt to help improve politicians' understanding of the impacts of new technologies, an exercise then referred to as Technology Assessment. Initially, the issues raised by the US Congress Office of Technology Assessment, and later by its European counterparts,[3] were not generally high on the politicians' agendas. The efforts of these TA institutions were criticised for being technocratic, moreover, because they did not aim to increase public participation. European TA institutions, in particular, began to change their course. They started emphasising the need to establish a dialogue on technological issues with stakeholders, rather than studying the potential impacts of new technologies. This form of TA has become known as *Constructive (or Interactive) Technology Assessment*. By encouraging debate in the early stages of technological development, harsh controversy can be avoided at a later date. This can steer

---
[3] Every EU member country has some kind of Parliamentary Technology Assessment facility. Moreover, the European Parliament has its STOA in Luxembourg.

technology and technological R&D in a different direction and lead to greater social acceptance of the resulting technologies without curbing the innovative capacity of the economy. We need this innovative capacity to be able to make our societies truly sustainable. However, it is still questionable in how far these procedures to democratise technological change can truly influence the course taken by technology in a profit-driven market economy.

15.3 A Brief Taxonomy of Environmental Technologies

Environmental problems come in all shapes and sizes. Ever since the Neolithic agricultural revolution, mankind has been accustomed to adapting his natural environment for the purpose of subsistence. These adaptations were often necessitated by the impossibility of moving to new lands. However, the methods used by human societies to shape their natural environment are not generally referred to as 'environmental technologies', this term being reserved for technologies designed to protect natural systems and cycles from damage or disruption. This kind of impact emerged at a new level and intensity as the Industrial Revolution got underway in the 18th and 19th century.

New technologies enabled entrepreneurs to start tapping into the enormous reserves of fossil fuels and ores under the earth's surface. Although the technologies were rather inefficient, the sheer abundance of resources opened the way for unprecedented economic growth. This combination of low resource efficiency and rapid growth caused widespread pollution and disruption of natural ecosystems. Gradually, environmental technologies began to be developed to eliminate some of the worst impacts, especially those affecting or endangering human health.

It should be noted that technologies to protect human society from the hazards of nature, such as surge barriers and vaccination, are not termed environmental technologies. Likewise excluded are technologies used for measurement and analysis. Although of great importance, these technologies are not generally specific to environmental problems.

Restoration technologies (for soil remediation and so on) are included in the definition, however, as are so-called 'end-of-pipe technologies', and it is to this kind of environmental technology that we now turn, thereby following the categories of Table 15.1.

1. Historically, the first environmental problems were of a local nature. The easiest way to tackle these problems was by using classic, so-called *triple D* technologies:

Table 15.1 Environmental classification of technologies

	Technology type	Function	Resources predominantly used	Resource efficiency	Emissions	Impact on natural systems
	Pre-industrial technologies	To provide	Renewable	Low to high	Low	Some over-exploitation, often compensated by low density of population
1	Classic environmental technologies	To prevent harm by pollution	Non-renewable	Low	High	Ecological destruction outside settlements
2	Good housekeeping technologies	To mitigate pollution	Non-renewable	Medium to low	Medium to low	Mitigation of ecological destruction
3	End-of-pipe technologies	To prevent pollution after process	Non-renewable	Medium to low	Low	Less pollution, at the expense of extra resource use
4	Process adaptation/ damage prevention	To prevent pollution arising in process	Non-renewable and renewable	High	Low	Less pollution and less resource use
5	Sustainable technologies	To provide within the limits of the Earths' carrying capacity	Renewable	High	None	Balance between humanity and the natural environment

- Dumping (of waste in pits, etc.)
- Displacement (of pollution by sewerage, smoke stacks, etc.)
- Dilution (of gaseous and liquid waste)

In this category of environmental technologies, the pollution is not chemically or biologically transformed.

2. As pollution became problematical, people began to think about pollution prevention. The first initiatives in this direction are always the easiest: options to minimise pollution are sought within the constraints of existing modes of production. This generally comes down to basic precautionary measures or what is nowadays called *good housekeeping,* or *triple M*:

- Monitoring
- Management
- Maintenance

3. Reduction of the remaining waste can be achieved by means of '*end-of-pipe technologies*', including in particular:

- Incineration
- Pyrolysis
- Separation
- Fermentation
- Chemical transformation
- Catalytic reduction
- Shielding (radiation, noise)

Recycling and re-use technologies that feed waste (product) back into production processes can be termed 'end-of-pipe' if the waste in question is from a different process loop, as in the case of waste being used as a fuel, for example. If waste is re-used in the same production loop without requiring much additional energy or generating much pollution, however, recycling may be sustainable (as in the case of reprocessed metals).

End-of-pipe technologies are often denounced as providing no real solution to the problem of pollution. In many cases these technologies create new problems, as the pollution prevented from being emitted must be stored, treated or discharged in some other way. For the short-term future we cannot do without them, however, as the introduction of alternatives will often require tremendous efforts.

Restoration technology is a specific kind of end-of-pipe technology. We have an obligation, at the very least, to clear up the worst pollution of the past and insulate polluted sites from their unpolluted environment. Areas where restoration is required include the following:

- Polluted soils
- Polluted lake and river bottoms
- Space debris
- Plastic wastes in the oceans
- Nuclear waste
- Non-indigenous species introduced into ecosystems

4. In many cases, however, the preferred option is to reduce the environmental burden by *changing the production process*. In this way, further reductions of pollution and resource consumption can be achieved. Complete redesign of production processes can lead to both environmental gains and cost reduction. Various tools are available for this purpose:

- Industrial ecology: integral design of industrial systems so as to minimise resource consumption and waste production by intelligent combinations of facilities (see Section 8.6)
- Life Cycle Assessment: analysing the overall production chain and identifying the main target areas for environmental and resource improvement
- Pinch technology: minimising resource consumption in production processes by minimising process redundancies

5. Ultimately, we must develop technologies for *sustainable production and consumption*, for none of the above technologies will suffice to solve the environmental problems we face. *Sustainable technologies* go beyond environmental technologies. While the latter are concerned with producing goods and services with minimal pollution, sustainable technologies have a far broader aim: to enable us to fulfil the needs of the whole humanity without:

- Exhausting the earth's non-renewable resources
- Exceeding its ecological recovery capacity
- Consolidating or promoting inequity

These technologies must enable humanity to survive in the longer term, that is to say, sustainable technologies are a necessary condition for the continuity of human civilisation. In and of itself, however, development of such technologies is not sufficient for sustainable development.

A further treatment of these environmental technology types will be found in Sections 15.4 to 15.9.

15.4 Classic Environmental Technologies: Triple D

When the first nomads settled they were faced with new problems, in particular that of human waste. As sites became more permanently inhabited, (organic) waste accumulated. Still, the quantities remained relatively small, with human excrement and other waste often being put to productive use, as animal feed, fertiliser or even raw material. Such practices continued until the Middle Ages, with human waste being used in industrial-scale processes for papermaking, dying and saltpetre production, for example. As human settlements increased in size, however, the rotting organic waste accumulating in public spaces became an intolerable burden for inhabitants, especially when the weather was warm. It was not only the stench that became a contentious nuisance; the rotting waste formed a breeding ground for contagious diseases and their insect vectors. At first, the preferred 'solution' was to dump the waste into rivers and canals, stockpile it at special sites or bury it in pits, generally outside the city gates. Such operations did not require complex technology.

Disposing of waste in rivers was a reasonable solution as long as there was sufficient flow to carry it downstream. Rivers not only displaced the pollution, but also diluted it. In many major European cities, river capacity proved inadequate, however. In 1404, Parisian authorities ordered butchers and tanners to move their businesses downstream to reduce pollution of the Seine. In the summer of 1859, the British Parliament was suspended because of the stench coming from the Thames as the river began to 'seethe and ferment under a burning sun'. Similar occurrences were frequent in cities centred on canals or slow-flowing rivers. The Dutch city of Delft, for example, had several canals that were used as open sewers as well as a source of freshwater for the inhabitants and their 16 breweries. In the 16th century, the city council had a special pumping windmill installed to improve water circulation, supplemented by a system of sluices to remove the accumulating dirt and refuse.

Dumping organic waste in private pits was a good solution as long as waste streams were not too large and the population not too dense. Archaeologists still use these pits to study everyday life in the distant past.

As cities grew, many of them set up waste collection systems in the course of the 19th century, removing waste to sites at city perimeters, where it could often be re-used, as animal feed or fertiliser, for example. In this era, it was also gradually recognised that contamination of drinking water with faeces was the cause of a number of contagious diseases. As a result, both drinking water supply and faecal disposal were improved in the latter half of the 19th century. Sewerage systems to carry away human excrement and street dirt were designed and implemented. Sewerage systems were in themselves nothing new, having already been used in antiquity. The Romans had their Cloaca Maxima, a 600-m drain discharging into the Tiber through a semicircular archway, which is still visible today.

However, sewerage as such was not enough. In 1780, Parisian authorities had forbidden their citizens to throw urine, faeces and other refuse out of the window. People needed in-house facilities to deliver their faeces to the sewerage system. During the 19th century, water closets were gradually introduced. This was not without public resistance, however, for water closets meant that faeces were flushed away and could no longer be collected and put to agricultural or industrial use. An alternative system, named after its inventor Liernur, collected the excrement. Valves connected toilets and the system was regularly emptied by steam-engine-driven vacuum pumps. The faeces were sold as fertiliser. However, the Liernur system could not handle large volumes of water, and as water closets were introduced on a wider scale the technology was abandoned. Water closets, for their part, had gained in popularity precisely because of the availability of large volumes of water transported through urban piping systems.

Because the first modern sewerage systems were designed particularly to collect street dirt (as all dirt ended up in the street), the system had to be able to handle large volumes, the dirt being flushed away only after rainfall. These large volumes of polluted water generally ended up in the downstream reaches of a river, in lakes or in the sea. At the sewerage outfall there were often problems. In 1990, for example, Uruguay's capital Montevideo had to lengthen its outfall into the Rio de la Plata estuary because the city beaches were severely polluted by human excrement.

One solution is to treat the sewage water. However, this is hampered by the joint removal of dirt and rain: in times of high rainfall, the storage capacity of the sewerage system is often insufficient and sewage must therefore be discharged untreated. The solution to this problem is to design a dual system for removal of rainwater and sewage water. New smart sewage technologies can also separate the clean rainwater from the system and infiltrate it into the soil, especially attractive in drought-prone regions. The main barriers to the introduction of these sewage technologies are the costs involved in adapting existing sewage systems.

With the growth of the 'consumer society', the autarchy of towns and villages diminished. Industrially produced and packaged products created mounting waste streams containing large amounts of non-degradable materials that could no longer be disposed of by burial in the ground. In large, densely populated cities, there was little scope for waste burial, if any at all. Still, in most industrialised countries, it was triple D technologies that were predominantly adopted to deal with waste until the end of the 20th century. Pits and tips were the preferred solution for the problem of solid waste, and sewage systems for that of liquid waste. As the new millennium approached, however, contamination with chemical waste frequently proved to have turned tips into environmental problems in their own right, while sewer outfalls were creating major problems for receiving bodies of water. If the sewage was treated, though, the residual sludge often had to be stored because of growing chemical contamination.

15.4.1 Modern Triple D

The main characteristic of most early environmental technologies is that pollution is not treated or prevented, but dumped, displaced or diluted as a form of nuisance control. Besides the sewerage systems in use worldwide, landfills for waste disposal are another important, and often criticised, example of such triple D technologies. Industries still use such tips to dispose of waste that cannot be treated. The zinc industry, for example, produces large amounts of jarosite, an iron-ammonia complex, for which there is no economic use. It is dumped in landfills, sometimes under tight controls, but often with none at all. Mining and related industries produce vast quantities of solid waste like the 'red mud' resulting from alumina production. The chemical industry often neutralises dilute acids to form gypsum, which is sometimes re-used as a building material but is often dumped. Nuclear waste is a particularly controversial category of end waste and various technologies have consequently been developed to store it deep underground in salt mines, salt pillars and other geological formations. As geologists cannot fully guarantee the stability of salt layers (especially under conditions of prolonged radiation), storage technologies have been developed that allow the waste to be retrieved if geological stability is affected.

Smokestacks transport gaseous waste to a height that generally prevents direct harm to the local environment. Particles are diluted before they fall to earth. No matter how high smokestacks are, however, the solids will come down somewhere, something that came home to Shell in the 1970s, when it heightened its refinery smokestack near Rotterdam to over 200m. While the conflict with the village adjacent to the refinery diminished, a new one emerged with market gardeners further downwind.

Landfills still constitute an important waste 'treatment' technology in large areas of the world, often being used to dispose of both domestic and industrial waste, including toxic materials. At modern landfills, however, a number of precautions are usually taken to prevent local pollution. The sites are generally provided with several impermeable layers that prevent toxins moving down to the water table. Rainwater percolating through the dumped waste is controlled for toxins and treated. When the landfill is full it is sealed. At some sites, the methane produced by the rotting of organic matter is siphoned off and used as a fuel. If this landfill gas is properly treated, removing CO_2 and halogenated compounds, it can be fed into the natural gas grid (see Table 15.2). However, it is more efficient to find a

Table 15.2 Composition of landfill gas

Landfill gas	Constituent gases (average)
Methane (CH_4)	50%
Carbon Dioxide (CO_2)	45%
Nitrogen (N_2)	5%
Oxygen (O_2)	<1%
Hydrogen sulphide (H_2S)	21 ppmv[a]
Halides	132 ppmv

[a] ppmv = parts per million by volume.
Source: EIA (1997)

specific local use for the untreated methane. An additional advantage of gas removal is that it stimulates the fermentation processes within the landfill, thus stabilising it and freeing the site up sooner for other uses. For this reason, the methane is sometimes flared. This is still to be preferred to non-treatment, as methane is a far more powerful greenhouse gas than CO_2.

Releases of waste heat form a special category of emissions. Electrical power plants, in particular, discharge large amounts of (hot) cooling water to the environment. As heat is valuable, it may be profitable for industrial facilities to recover it. Very often the waste heat can be used for pre-heating feedstock or for various forms of heating. It may even be attractive to implement such activities as greenhouse horticulture specifically to use the heat. If such use is inconvenient, though, because the heat is released irregularly, say, it may be dispersed to the air via cooling towers. By diluting heated effluent before release into open waters its impact on wildlife can be reduced. Even so, ecosystems may still suffer from thermal releases, particularly as a result of outbreaks of botulism, a bacterial disease affecting ducks, during periods of hot weather.

In many cases, triple D technologies are harmful to the environment because they merely solve problems locally, by transferring them elsewhere at the expense of extra energy consumption. In some cases there may be no alternatives, however, as in the case of nuclear waste. This is one of the main arguments against the nuclear industry. There are also cases where triple D technologies are unproblematic: some substances can harm the local environment but cause no harm if suitably removed. This holds, for example, for salt water (which until the 1980s was discharged into the Rhine by the French mining industry), which would cause no harm if discharged directly to sea. It may also be argued that similar considerations hold for organic wastewater, which must be treated to avoid local or regional health problems and control the eutrophication and stench it causes. Water treatment requires considerable energy, for stirring and pumping and so on. In diluted form and reasonable quantities, however, organic wastewater can be discharged into the environment without serious problem, as microbes will break it down.

Another argument against triple D technologies originates from the global character of the problems that a number of compounds create. This naturally holds for the gases that are damaging the ozone layer and contributing to climate change, but other compounds like heavy metals (which are non-degradable toxins) and persistent organo-chlorine compounds also constitute an enormous problem. Certain organo-chlorine compounds affect the endocrine system of aquatic and amphibian species, to an as yet unknown extent, thereby threatening successful reproduction. Organo-chlorine compounds are found in every corner of the Earth, including the fat layers of penguins and the breast milk of British women. The World Wide Fund for Nature (WWF 1999) claimed recently that more than 350 manmade pollutants had been identified in the breast milk of women in the UK, including 87 dioxins.

15.5 Good Housekeeping Technologies: Triple M

When companies and other organisations first start tackling environmental issues, their initial response is to assess the environmental damage they are causing and implement relatively easy measures to control it. Programmes of environmental care are then generally set up, based on the 'triple M' of monitoring, management and maintenance. As this form of response is often only partly technological in nature, we shall examine it only briefly here.

In many cases the first form of environmental action undertaken by corporate management was to study the environmental impacts of the organisation's operations and target those impacts that were easiest to control, i.e. those requiring little or no expenditure, investment, organisational change or changes to the production process. Impacts due to individual neglect on the part of staff (such as forgetting to switch off lights and computers or close valves and so on) or necessitating minor investment or even saving money (recycling of paper waste, replacing incandescent light bulbs by low-energy alternatives, repairing leaks, insulating buildings, etc.) could soon be implemented.

Monitoring of valves and pipes should be undertaken at regular (e.g. monthly) intervals, as effectiveness increases considerably with frequency. Some leakage maintenance and repair can be carried out on equipment in service, as in the case of bolt-tightening or packing lubrication. Often, however, plant processes must be interrupted and work is therefore often postponed to coincide with a scheduled shutdown, as emissions from leaks are often far lower than those occurring during process shutdown.

These measures were generally referred to as 'good housekeeping'. Today, this type of environmental management is supported by a variety of technologies:

- Time switches (to shut off water taps or lights or put monitors on 'sleep')
- Fuel-efficient vehicles
- Organic Vapour Analysers (OVA) to detect excessive hydrocarbon levels, a key element of leak detection and repair, especially in chemical and petroleum processing
- Geiger counters to monitor radioactive pollution, similar to OVA
- Acoustic equipment to locate leaks in underground pipelines

Since a prime aim of this monitoring is to motivate staff to get involved in improving the company's environmental performance, feedback is very important. Software to register waste production and energy consumption is important for monitoring the organisation's performance and to motivate personnel. It can give due warning if performance slackens.

The organisational culture is a key element in maintaining the environmental performance of a production system. Pollution often results from poorly maintained plants and equipment. It is often forgotten that proper maintenance not only prevents leakage (or worse) but also minimises interruptions to production.

Industrial production processes are often considerably oversized. Moreover, intermediate products are often stored. These redundancies allow production to continue when specific parts of the process are interrupted. However, excess capacity and intermediate storage often give rise to extra environmental costs, generally in the form of energy, space and materials (used in equipment). Proper maintenance and optimised production monitoring could make these storage facilities and extra capacities redundant, with benefits to both the environment and corporate returns.

A corporation that sees environmental performance as an important aspect of its operations will also report on it. Environmental reports have a double function: Not only do they allow external stakeholders to scrutinise the company's environmental performance. To be able to create an acceptable 'green' image, a company needs to adequately monitor its performance and improve it continuously.

In 1996 an international certification procedure was introduced for Environmental Management Systems (EMS): ISO 14001, comprising the following six key elements:

- *Environmental policy*

Establishing an environmental policy and the requirement to pursue this policy via objectives, targets and environmental programmes.

- *Planning*

Analysis of environmental aspects of the organisation (including its processes, products and services as well as the goods and services it itself uses).

- *Implementation and operation*

Implementation and organisation of processes to control and improve operational activities that are critical from an environmental perspective (including both the products and services supplied by the organisation).

- *Checking and corrective action*

In particular, monitoring, measurement and recording of parameters and activities having a potentially significant impact on the environment.

- *Management review*

Review of the EMS by the organisation's top management to ensure its continuing suitability, adequacy and effectiveness.

- *Continual improvement*

This is a key component of the EMS; it completes the cyclical process of 'plan, implement, check and review', which it continually improves.

When first confronted with environmental issues, most corporations saw them as a burden. Since the early 1990s, however, more and more corporations have come to realise that they can only survive as long as they effectively fulfil the needs of society and respect the limits set by that society.

15.6 End-of-Pipe Technologies

End-of-pipe treatment is, by definition, not pollution prevention. End-of-pipe technologies scarcely affect production processes or the organisation of the corporations operating those processes, as end-of-pipe technologies are always appended to existing processes. This makes them easy to apply, in two respects:

- They require no changes to the core of the production process (which would involve risks to product quality and continuity of production).
- Their cost and operation are predictable and controllable, because the equipment can be ordered 'off the shelf' and operations possibly contracted out.

The required investments may be considerable, however. End-of-pipe technologies can, moreover, set their own limits on the magnitude of treatable waste streams (toxin volume and concentration in biological wastewater treatment, for example). It is therefore often financially attractive to minimise waste streams before applying end-of-pipe technologies and/or build facilities for temporary storage to ensure their efficient operation.

End-of-pipe technologies are not a 'technological category' in their own right. In industrial production, filters as well as other end-of-pipe technologies are also used to obtain material flows of a required composition, for example.

Because they often require considerable investments, end-of-pipe technologies may actually hamper environmental improvements, as any decrease in waste volumes creates redundancy in waste treatment capacities, or in other words deficits on the balance sheet.

End-of-pipe technologies can be divided into four categories:

- Treatment technologies like incineration, fermentation and catalytic reduction that eliminate or transform harmful substances and materials.
- Separation technologies to separate harmful from harmless substances and materials.
- Shielding (for noise, radiation and light).
- Compensation (planting forests to absorb CO_2, etc.); it might also be argued that these are restoration technologies.

End-of-pipe technologies are generally categorised according to the carrying medium of the waste: gas, liquid, solid.

15.6.1 Gas Separation and Treatment Technologies

Gas can be contaminated either by other gases or by solid or liquid particulate matter (aerosols). The smallest of these particles (less than $1\,\mu m$) are produced mainly by combustion and condensation processes, those between 1 and $10\,\mu m$ are often process dust (especially coal/peat/biomass/diesel burning), while most of the largest originate in physical operations like grinding, spraying and wind erosion.

There are a variety of technologies available for removing particulate matter from industrial gas streams:

- *Gas separators*

A simple and efficient means of removing larger particles from gases is inertial separation. Centrifugal forces are exerted on the gas stream in a *cyclone*, separating out the heavier solid or liquid particles. The gas stream can also be deflected around a suitable barrier, causing the heavier particles to collide and fall into a collecting chamber. This is the principle of so-called *impingement separators*.

- *Wet scrubbing*

In wet scrubbers, particulate matter is removed from the gas stream by wetting the particles. In the simplest design, the gas stream enters a tower near the bottom and leaves at the top, passing through a spray of water as it moves up. The water droplets collect particles (and gases) on their way down, which can then be collected at the bottom. Depending on the particulate matter removed, the water can be recirculated or treated. There are a number of more elaborate variations on this basic design. Wet scrubbers can be effective for particulates upward of $0.5\,\mu m$ in diameter.

One possible disadvantage of a wet scrubber as an end-of-pipe gas-cleaning device is that it creates a condensate plume (mist), thereby suggesting pollution. If water is used as a scrubber liquid, there is generally substantial corrosion. Considerable quantities of scrubber liquid may have to be used or (if recirculated) treated. Power consumption may also be high. Wet scrubbers are robust, in the sense that they can deal with a wide variety of gas stream contaminants, although the cleaning of hot gases is inefficient, as the heat tends to vaporise the liquid particles.

- *Electrostatic precipitators*

In this case the gas stream is passed over a discharge electrode (usually wires), causing the particles to become electrically charged. The charged particles are attracted by an electrode of opposite charge (in the form of a plate). Dust can be removed in dry form or using liquids. Electrostatic precipitators are very efficient and can be operated at high temperatures. The equipment is large and expensive, however, and requires safety measures due to its high voltage (40,000–70,000 V).

- *Filters*

Particulate matter can be separated from gases using fabric filters, which can remove particles down to a size of 0.3 μm with over 99% efficiency. Fabric filters usually consist of tubes of woven or felted fabric held in a housing structure. As the captured particles accumulate, however, filters needs to be regularly cleaned. There are three methods available: shaking, reversing the gas flow and high-pressure pulsation. By these means the dust-cake on the filter cloths can be made to crack and can then be collected.

By applying new, robust materials, filtering can now also be carried out at relatively high temperatures. This is essential for filtering stack gases at electrical power plants, especially if low-sulphur coal is burned, making it hard to filter the fly ash using traditional electrostatic methods.

These are the technologies most commonly used to remove particulate matter from gas streams. For removing gaseous contaminants, a variety of other methods are available:

- *Absorption*

There are two basic types of absorption:

– Chemical: the contaminant gas is reacted with a liquid solvent.
– Physical: the contaminant gas is merely dissolved in the liquid solvent.

Absorption is commonly used to remove SO_x, H_2S, HCl, Cl, NH_3 and NO_x. In one of the processes most commonly used, a limestone solution is sprayed into a SO_x-containing gas stream (comparable to wet scrubbing), the lime reacting with the SO_x to form gypsum.

- *Condensation*

Vapour-phase contaminants can be removed by cooling the gas stream until the vapour forms a liquid, which can then be readily collected and disposed of. This type of gas cleaning can even be made to occur of its own accord if the residual heat of a waste stream is re-used by feeding it through a heat exchanger.

- *Flaring*

Some flammable waste gases can be flared off as an open flame. Naturally, this option should be used only if the emissions are irregular (as in the case of emergencies), for regular emissions may represent a valuable source of energy. Generally speaking, flare systems must be designed to handle large volumes of gas and a seal is therefore required to prevent flame flashback when emissions are low (for this reason the gas is commonly fed through a water tank).

- *Incineration*

Incinerators, or afterburners, are used to burn low concentrations of hazardous combustible gases. The extent to which hazardous gases are eliminated by this treatment may well be inadequate (see Fig. 15.1).

15.6.2 Water Separation and Treatment Technologies

Water pollution differs from air pollution in two important respects:

- Biological contaminants often play a key role.
- Thermal pollution is also often a major factor.

Wastewater can be subdivided into domestic and industrial wastewater. Although the flow of domestic wastewater may vary over time, on average its content is fairly stable, consisting mainly of various species of organic matter. Inorganic contaminants of contemporary household wastewater include ammonia, phosphates, chlorides and heavy metals.

Historically speaking, biological pollution was the driving force behind the construction of sewer systems. In general, public health problems could be dealt with fairly adequately by discharging wastewater into rivers or seas away from populated areas, as discussed above. There comes a point, however, when oxidation of the organic matter in the receiving waters leads to serious deoxygenation, killing fish and damaging or even devastating local ecosystems. The same holds for organic industrial wastewater, such as effluents from the food industry, dairy industry, distilleries and pulp and paper industries.

Although the discharge of massive quantities of organic waste can create problems at the local level, discharge of untreated organics is not a problem *per se*; indeed, it has always been the way to dispose of excrement and other waste. It is only through accumulation that it becomes a problem. As aerobic wastewater treatment requires energy, it may justifiably be argued that

Fig. 15.1 Air pollution control system of the Municipal Solid Waste incinerator of the city of Amsterdam. The flue gases are first quenched and washed (using NaOH and NH_3 as chemical absorbers) in the left tower (B), they are further washed (using NaOH) in the central tower (1). Afterwards, the gas flue is treated by a wet electrostatic process in the right tower. The remaining droplets are separated from the gas stream by a radial flow system (D). The wastewater that is collected in the three towers is cleared and recycled (E). The sludge is further filtered to turn out clean water (G) and dust-cake

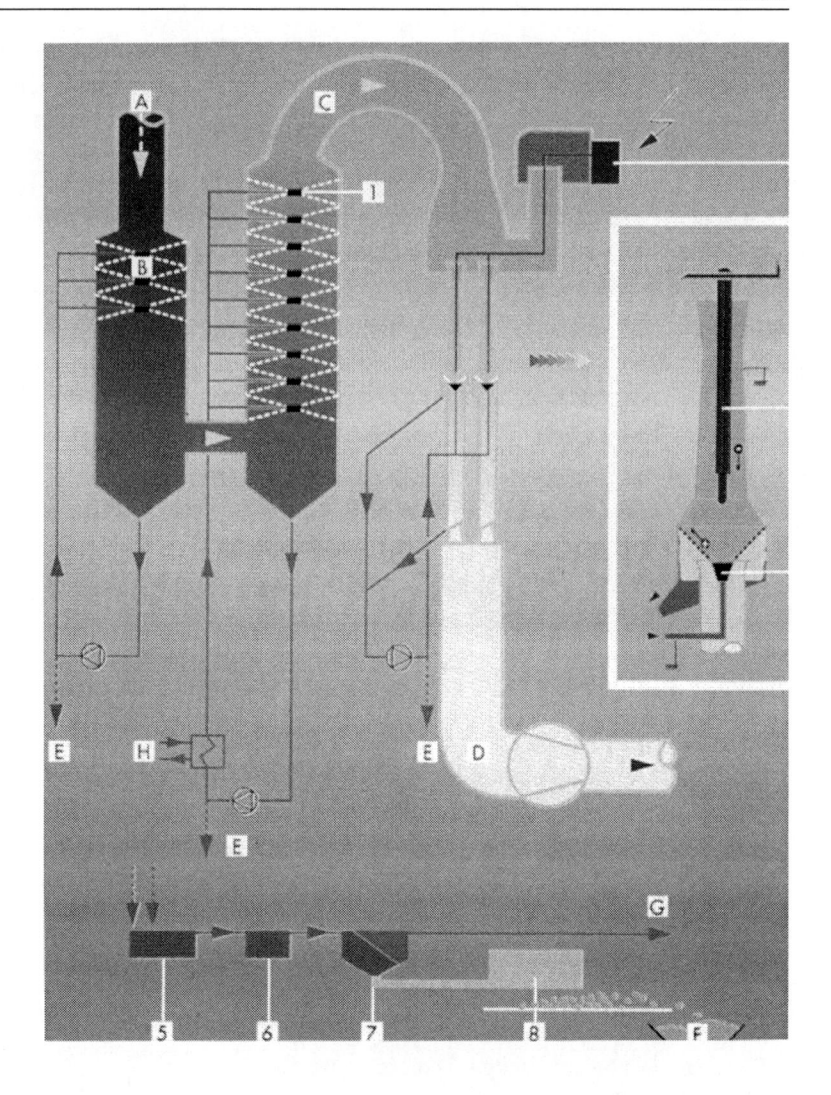

untreated discharge is preferable to aerobic treatment, as long as local ecosystems are not affected.

It is important that the wastewater flow does not exceed treatment capacity, for otherwise untreated wastewater will have to be discharged.

Wastewater treatment often starts with the removal of solids like sand, rock and debris, using various types of *screens*. The solids and particles are generally trapped by making use of their greater specific gravity in a curved flow. Finer particles that pass these screens can be trapped by *sedimentation*, i.e. trapping them on the bottom of ponds in which the water is stored. *Flotation processes* can be used to separate substances floating on the surface of the wastewater, like oil and grease. Flotation can be stimulated by aerating the water with fine gas bubbles, which attach to the suspended matter.

Aerobic biological treatment is generally the main step in the domestic wastewater treatment process. Oxygen is mixed with wastewater in an aeration basin and the suspended or soluble organic waste is oxidised by microorganisms to form carbon dioxide and water. In a second step (the clarifier) the microorganisms are separated from the treated water and returned, in part, to the aeration basin.

In *trickling filters* wastewater is trickled over a microbial surface on a solid medium, with organic substances again being oxidised. A mechanism must be provided to guarantee a minimum flow of wastewater, to keep the microorganisms alive.

Besides aerobic biological treatment of wastewater there is also *anaerobic biological treatment*. This is currently used to clean up effluents from industries handling large volumes of organic substances such as the food industry. In anaerobic biological treatment methane is generated, which can be used in energy generation, giving anaerobic treatment an energetic advantage over aerobic treatment.

Filtration is frequently used as a means of separation. A variety of media can be used as porous barriers, including sand, paper, filter-cloth and polymer membranes. Membrane filtration through a porous polymeric sheet can be very effective, even for filtering salt ions out of seawater to produce fresh water. However, the pores often become quickly clogged and these filters must therefore be cleaned very regularly by reversing the flow. Membranes can be expensive, especially compared with the low cost of many other water treatment systems. For energy-intensive purification processes like distillation, however, they may be serious competitors.

Chemical treatment, using chemicals to bind contaminants in insoluble form, and *disinfection* (using ozone, for example) are also used for water purification, but mainly to produce drinking water.

15.6.3 Solid Waste Treatment

Solid waste is, by its very nature, generally less diluted than liquid or gaseous emissions. Recycling is therefore almost always a solution to consider very seriously. However, solid waste must often be pre-treated before it can be recycled or converted into more desirable forms. Municipal solid waste (MSW) consists of a wide variety of materials of varying and irregular shape and size. MSW is therefore often shredded into more uniform particles. If the constituent waste streams, or fractions, are not collected separately, initial separation can be achieved on the basis of weight and electromagnetic properties. Organics and plastics constitute the lighter fractions, with sand and glass the heaviest. Metals are separated on the basis of their magnetic properties (ferrous) and conductivity (non-ferrous). Separated metals are often recycled commercially.

Glass can be readily recycled, especially if it is collected separately. Recycling glass saves energy and avoids the need for landfills. However, its success is often exaggerated: glass is hardly voluminous and therefore little of a landfill problem and waste glass collection uses up a considerable part of the energy saved by recycling.

Plastic products are very amenable to re-use and recycling. Polycarbonate milk bottles, for example, can be cleaned and re-used about 40 times before their quality falls below statutory limits. At that point, the polycarbonate is recycled to make new bottles.

However, plastics discarded along with other MSW are difficult to recycle. The number of commercially available plastic compounds is so large that sorting is extremely difficult. Moreover, it is virtually impossible to recycle plastics waste originating from MSW without risk of biological contamination of the recycled material. For this reason, it is not permitted to package food and drugs in recycled plastic. For plastics collected from MSW, incineration may be the preferred solution, in order to avoid landfill and recuperate the energy content of the waste.

For the organic fraction of the waste, the following treatment options are available:

- Incineration, not generally preferable as MSW is generally fairly moist.
- Fermentation, yielding 'biogas'.
- Composting, a form of re-use, yielding compost.
- Gasification (by pyrolysis), yielding a mix of gases that can be used as a fuel.

Much of the waste produced in transport and trade (pallets, foils, etc.) can be recycled more easily, as it is often rather uniform in composition and the quantities make it commercially attractive. The recycling technologies are the same as for MSW.

Some industries produce large quantities of waste; this is especially true of the mining and base metals industries. Red mud, jarosite and gypsum have already been mentioned, while slag heaps can still often be seen in old industrial areas. These wastes are usually contaminated with hazardous substances that may leach into the environment. Although there are technologies available for extracting and immobilising most of these toxins, they are expensive. Government regulation is the main driving force behind technologies for dealing with industrial wastes.

Power plants are also major producers of waste. Removing dust from the stack gases leaves the power plants with large quantities of fly ash. Although this ash is presently be used as a secondary road-building material, it is still being debated whether it creates an environmental risk, through leaching. Using it in buildings is too risky, however, as the chemicals might be released during drilling of walls. The drive to desulphurise power plant stack gases also created an enormous flow of gypsum waste. Happily, though, this proved to be a rather pure form of gypsum, which is now commercially marketed.

Combustible solid wastes that are particularly toxic can be burned under controlled circumstances in specially designed incinerators. In general, in all forms of waste incineration, control of the temperature and humidity is very important to prevent (highly) toxic compounds forming in the stack gases.

There are still several forms of waste (highly toxic compounds, radioactive materials) that cannot be adequately treated and that must therefore be stored in special facilities. Such substances may be vitrified in glass or packed in special canisters and stored in mines. This is no true solution to the problem, however, as it may leave the real clean-up to future generations.

15.6.4 Other End-of-Pipe Solutions

The term 'end-of-pipe solutions' can be applied more widely than merely to waste streams. We might also call the culling of selected species in order to restore the balance in an ecosystem an end-of-pipe solution, or even zoo breeding programmes that seek to protect certain animal species from extinction.

Noise and radiation are often hard to control once they have been generated and end-of-pipe solutions like noise and radiation barriers may therefore be the only effective way to limit their harmful effects. A high-tech means of controlling noise more actively might be to deaden it, by adding waves of the same frequency but of opposite phase. This solution is very expensive, though, and warranted only under rare circumstances.

15.6.5 Environmental Restoration

Many environmental problems are so enormous that we can scarcely conceive of where to begin in addressing them. It is difficult to imagine an effective policy for a problem on the scale of today's sharply declining biodiversity, for example. The same applies to CFCs: we can stop using them, but we cannot realistically prevent the millions of tonnes already present in the atmosphere from damaging the ozone layer far into the 21st century (see Fig. 15.2 and Fig. 6.7B).

In the case of space debris, i.e. satellite and spacecraft waste orbiting the earth, there are no known technologies capable of tackling the problem. The problem is therefore becoming steadily worse (Fig. 15.3).

The problem of climate change seems even further away from a solution, as consensus among scientists on

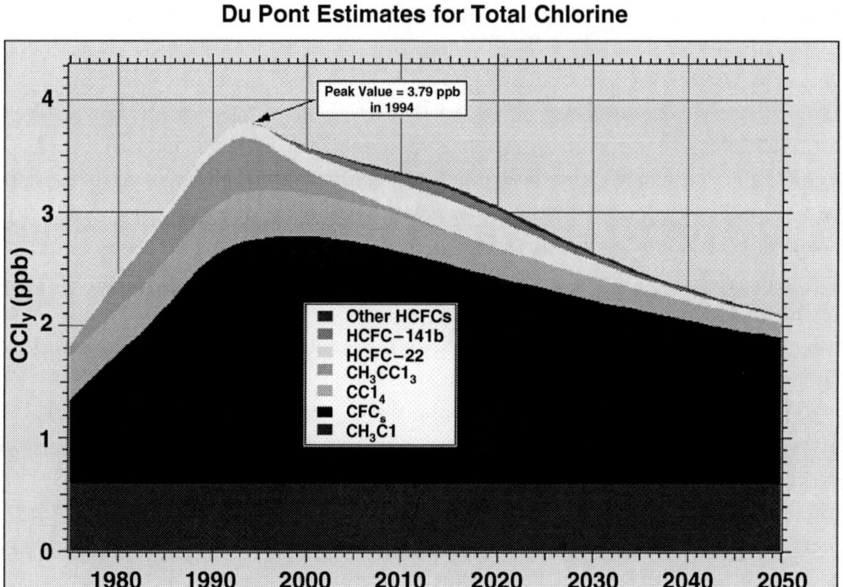

Fig. 15.2 Du Pont estimates for total chlorine gases damaging the ozone layer

Fig. 15.3 Multiplication of space debris orbits, 3 months, 1 and 4 years after a spacecraft explosion

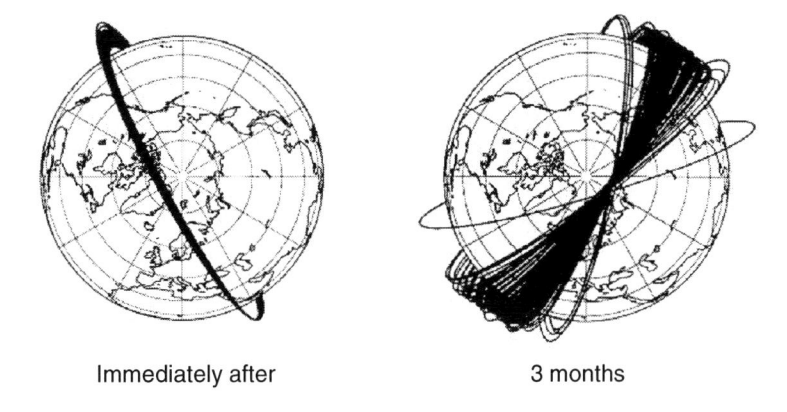

Immediately after   3 months

Time after the explosion

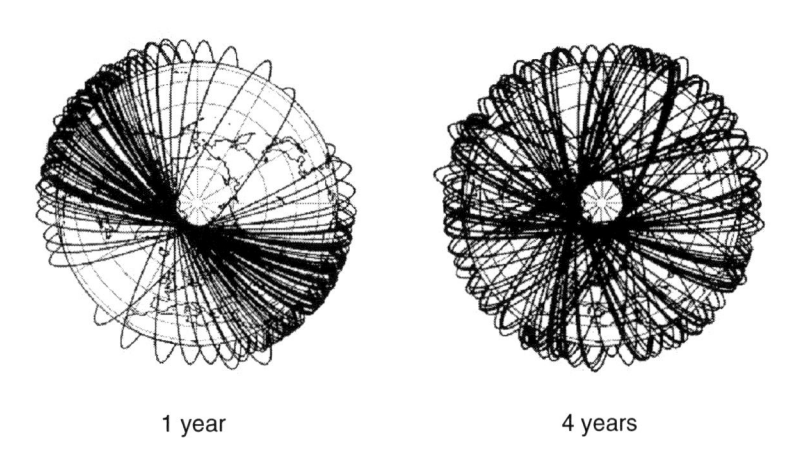

1 year   4 years

the cause and magnitude of the problem has only recently been established in international scientific fora. Dissident opinion among scientists is still undermining the willingness of politicians to take action, however.

There are several studies that suggest CO_2 might be stored in the oceans. In the autumn of 2000, the German research vessel 'Polarstern' carried out experiments in which algal bloom was stimulated by adding an iron solution to the ocean. As they die and sink, algae transport carbon to the ocean floor. In arctic waters, particularly, far greater algal growth was achieved. It has also been suggested that CO_2 might be dissolved directly in oceans. However, due caution is in order here. Such interventions might disturb the oceanic ecosystem dramatically, as the acidity of the oceans might gradually increase, affecting underwater life and especially coral reefs. It may well also be the case that injected CO_2 is transported to the ocean surface, to be vented in the future back into the atmosphere.

Problems that are more local in character are easier to solve technically. Polluted soil is generally first separated from clean soil by drilling steel walls into the ground or by laying geofoils. In the early days of environmental policy, polluted soil was usually treated by removing the most toxic soil and dumping it in a landfill or cleaning it by burning off the combustible pollutants. This solution is only cost-effective for dealing with fairly concentrated pollutants, however. Biological (i.e. bacterial) treatment may be useful for removing synthetic chemical contaminants like petroleum derivatives when present at lower concentrations. Heavy metals can be removed from the topsoil by plants, especially tobacco. The plants take up the metals from the soil and can then be burned to concentrate the pollutants in the ashes. This form of soil treatment, called 'phytoremediation', may take many years, but it is suitable for large areas that would be too expensive to clean any other way.

Cleaning up soil pollution is generally expensive and in the case of some hydrocarbon contaminants, especially, the best option may be to simply leave the soil to be regenerated by natural processes. Removal of the topsoil, combined with confinement of the polluted soil to prevent diffusion, would then suffice to clean the soil. This might take considerable time.

Bio-invasive species, i.e. species intruding into and disturbing the original ecosystem, can sometimes be successfully eradicated, as with the muskrat in England. Successful attempts at eradication are scarce, however. As far as possible, the influx of new invaders must be terminated at source. In many regions, this is scarcely possible, though, as with migration of marine life between the Mediterranean and the Red Sea through the Suez Canal.

Wildlife can be brought back to preferred areas by removing or adapting roads, canals, hydroelectric dams and other civil works. Polders can sometimes be returned to the sea and waterways allowed to meander naturally to restore the hydrological balance. Reintroduction of locally extinct indigenous species (wolves or beavers in Western Europe, for example) may contribute to restoring the balance of disturbed ecosystems.

There are also problems to solve indoors: many houses still contain asbestos, for example. As long as this material is not worked on or demolished, it forms no problem, but when it is, great care should be taken to prevent the spread of asbestos particles. On rare occasions in very old houses, lead piping is still used. It should be replaced, as it contaminates the water.

15.7 Process Adaptation and Damage Prevention

As end-of-pipe technologies have little or no *overall* impact on any specific industrial production system it is generally the preferred solution of industry when it comes to tackling the environmental crisis. From the perspective of the overall production system, however, such solutions are usually sub-optimal. In some cases, moreover, problems may be impossible to solve once they have been created. For example, even though emissions of lead by motor vehicles were acknowledged to be a major public health hazard, it proved unfeasible to remove the lead from car exhausts. The only solution was therefore not to create these fumes in the first place, by ceasing to add lead to petrol and finding a replacement, redesigning car engines, opening a separate supply line for unleaded petrol and waiting 20 years until the vehicle fleet was replaced.

Therefore, adapting the overall production process may often represent a more effective solution. Three basic forms of process adaptation can be distinguished:

- Changing the primary energy base, and improving the energy efficiency of the production system
- Changing the raw materials base and re-using discarded end-products
- Avoidance of by-products or captive re-use

15.7.1 Changing the Primary Energy Base

The coming decades are likely to see major changes in the sources of primary energy used in the industrialised nations. This will be discussed further in 15.9 here the problem is dealt with from the viewpoint of an individual energy-consuming company. Naturally, some companies may be in a position to generate renewable energy themselves by using solar cells, wind turbines or geothermal heating. In general, though, they will be dependent largely on energy supplied by others.

The noxious emissions of industry and power stations need to be reduced and one way to achieve that aim is to switch from coal- or oil-based fuels to natural gas.

Production processes can be adapted to use electricity as a primary energy source. In some regions, a considerable number of vehicles are electrically powered. Some critics call these 'emission elsewhere vehicles', rather than 'zero emission vehicles'. Still, switching to electricity as a primary energy source may have environmental benefits:

- Power plants burn fuel much more efficiently than other industrial processes.
- In the future it may be easier to convert a limited number of power plants to sustainable energy operations than countless small fuel users.
- Power plants pollute, too, but the pollution can be better controlled because it is far less diffuse.

- Energy users often pollute the air in (e.g. urban) areas where it is more harmful, while power plants can be sited to cause least damage.
- A large, robust and well-integrated power grid with a high-voltage backbone can absorb the irregular supply of various sustainable energy sources: wind power can be readily transported over long distances from windy to windless regions, for example, and hydropower dams can be used for energy storage, supplying power when all other sustainable sources fail.

For some purposes, however, conversion to electrical power as a primary energy source is not to be recommended. For example, high-efficiency boilers for domestic heating consume far less primary fuel than electric heaters do indirectly.

Energy efficiency can be increased enormously by:

- Applying thermal insulation
- Using heat exchangers to pre-heat raw materials and fuels and simultaneously cool emissions
- Redesigning industrial processes to prevent heating and cooling processes working in opposition

15.7.2 Changing the Raw Materials Base

Changing the primary raw materials base of today's industrial production processes could help reduce environmental pollution and the depletion of non-renewable resources significantly. Industry produces various wastes that are, in principle, re-usable but that are not re-used for technical/economic reasons (no cost-effective process available). Naturally, the processes that are available might become cost-effective by changing if there were a change in legislation or in the market prices of raw materials. Non-availability is therefore seldom to be regarded as a pure fact of nature, but usually as a social fact.

Re-use and recycling can be carried out at various places in the material cycle (see Fig. 15.8, plastics cycle):

- As a consumer product
- As a part
- As a material
- As a compound
- As energy

Consumer products that end up in the trash bin are generally landfilled or incinerated, although a substantial proportion could be recycled or re-used as a raw material. Before these materials can be efficiently re-used, however, sorting and cleaning technologies must be further developed.

A new line of thinking on industrial production is 'Industrial ecology', which is based on the notion that industry should follow nature's example, i.e. produce with no net waste. Perhaps the best-known example of industrial ecology in practice is the Kalundborg complex of industrial facilities, in Denmark, which minimises its collective use of energy and raw materials through mutual exchange of surplus energy and useful wastes, thereby also minimising overall waste production and emission levels (see Fig. 15.4). With respect

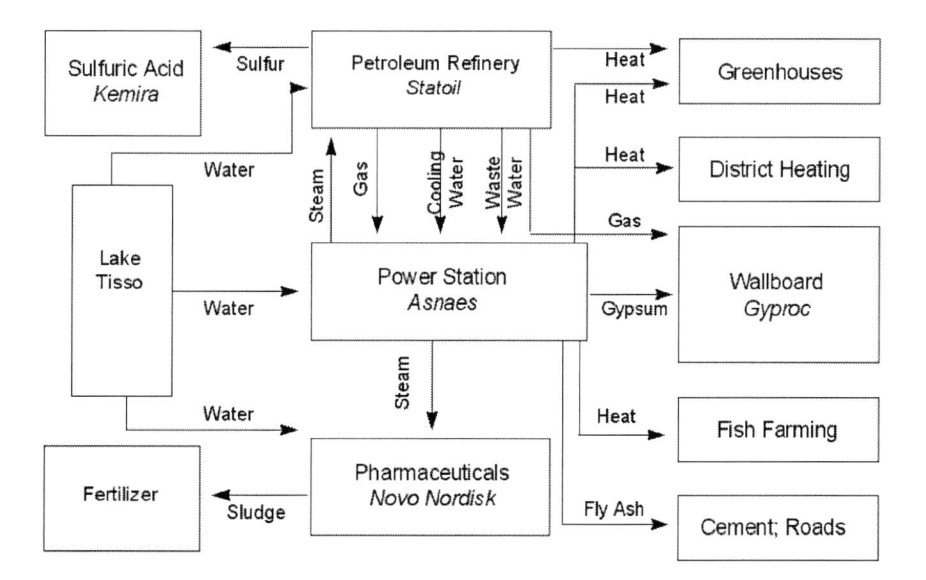

Fig. 15.4 Industrial ecology at Kalundborg, Denmark
Source: www.indigodev.com/Kal.html

to the recycling of household waste fractions, efforts thus far have focused predominantly on metals, glass and paper, with plastics being recycled in a few countries only. Although plastics could well be recycled, it hardly makes sense: as long as we burn vast quantities of crude oil in power plants, why not burn plastics too, for the energy cost of plastics recycling is considerable, and the health risks in various applications (like food and pharmaceuticals packaging) are serious. Plastics recycling does make sense, but only if we embark on a course that leads to a termination of fossil fuel consumption.

15.7.3 Avoiding By-products and Emissions

In a variety of chemical processes the product is yielded in several isomeric forms, i.e. compounds having the same atomic composition but differing in their spatial structure. Very often, only one of the isomers is the 'useful product', and the others must be dealt with as waste, or be reprocessed or perhaps marketed at a cheaper price. Para-phenylene-diamine (PPD), for example, is produced in considerable quantities as an intermediate for high-performance aramid fibres, along with an equal volume of its isomer ortho-phenylene-diamine (OPD). Aramid fibre producer AKZO-Nobel had no use for the latter compound and as the company needed large quantities of PPD, the development of a selective process to produce only this isomer became a key research target (although mainly for economic rather than environmental reasons).

Traditional oil refineries had little scope for controlling the relative fractions of petrol, heavy oils and tar they produced. As there was little or no demand for the latter fractions (by-products), which contained high concentrations of sulphur, these had to be sold at very low prices. In this case, too, there was therefore a strong economic motive to control the refinery process. For refinery operators, catalytic conversion of tar-like and heavy oil fractions was a means of increasing profits. It also meant they could tackle the high sulphur content of these fractions during conversion to petrol and related products. These improved process control technologies therefore permit more efficient conversion of scarce raw materials into useful end products.

Economic incentives are not always sufficient. In the early 1970s, for example, the polyvinylchloride (PVC) industry emitted considerable quantities of vinyl chloride feedstock to the air, until this was found to be a carcinogen (affecting mainly workers). Under government pressure processes had to be tightly sealed and maintenance schemes adapted to bring emissions down to acceptable levels (see Fig. 15.5).

In many production processes, a range of solvents, cleaning agents, filtering materials and other so-called 'auxiliaries' are used that do not show up in the end product. Although these substances are sometimes recycled or re-used, they all eventually end up as waste or emissions. As these substances are not part of the core production process, their use is often not given much thought. For example, the microelectronics

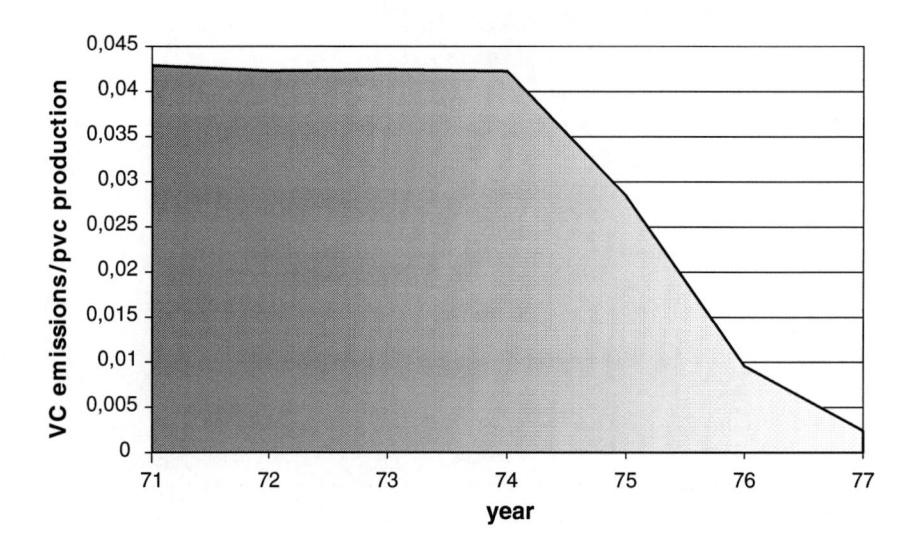

Fig. 15.5 Reduction of VC emissions in PVC production in Germany

industry had never considered an alternative for CFC 113, which they used for cleaning printed circuit boards. Despite its protests, however, the industry was forced by the 1987 Montreal Protocol to seek alternatives The effect was astonishing: the industry discovered that 70% of their CFC 113 use was in fact entirely superfluous and the rest could be avoided almost completely by cleaning the circuit boards with soap and water and then drying them. This solution was in fact rather counter-intuitive, as electronics engineers are generally trained not to use water, because it corrodes connectors.

Design modifications to roadways and waterworks and the construction of special fly-overs or tunnels to protect wildlife and their habitats can also be regarded as falling under the theme of this section: 'Process adaptation and damage prevention'. In the livestock industry, the limitations set on intensive meat production can be seen as similar in nature, to the extent that reducing ammonia emissions and manure production is the underlying motive.

15.8 Towards Sustainable Production and Consumption

Technologies that are truly sustainable must go beyond the principles discussed in the previous sections. A sustainable technology means more than merely producing goods without pollution, either at-source or at 'end of pipe' (or even after the pipe). Sustainable technology means fulfilling people's needs in such a way that the recovery capacity of the planet and local ecosystems is not exceeded. It therefore implies using no more fossil fuels than are formed in geological processes, which means using virtually none at all. It also implies closing materials cycles, not emitting substances that only occur artificially and limiting other emissions so as to no longer exceed the earth's recovery capacity.

The changes required to create truly sustainable technologies are enormous. This will be a long-term enterprise, in the course of which mistakes will undoubtedly be made. Some idea of the magnitude of the required changes can be derived from general principles and the so-called 'IPAT' equation (see Chapter 1 and Box 15.1), which can be summarised as follows:

Box 15.1 'IPAT' equation

$$I = P * A * T$$

I = Environment impact
P = Polulation
A = Affluence (consumption of services and products per capita
T = Technology (environmental burden per product or services unit))

Source: Ridken 1972

- The recovery capacity of the natural world is limited: we should not appropriate more of it for human needs but preferably half our use (I = 0.5–1.0).
- In the next 50 years the world's population will grow to about 9.5 billion (US Bureau of the Census 1999) or perhaps 9 billion (United Nations Population Division, World Population Prospects: The 2004 Revision). In the short run, this growth will hardly be affected by birth control measures, as it is mainly the result of the current demographic pyramid: there are relatively few older people who will die in the near future, and relatively many young people who are still to reproduce. It will take at least three, but more probably five decades before birth control measures lead to a stable world population (P = 1.5, see Fig. 15.6).
- Moderate growth in consumption in the North (1.5%) and accelerated growth in the South (5–6%, reaching about half the wealth of the North in 2050) will lead to a *factor* 4–8 growth in global average consumption levels by 2050 (A = 4–8).

Although various objections may be made to the simplifications of the IPAT formula, the basic conclusion is clear: T, the environmental pollution per unit product or service, must be cut back by at least a factor 6 by the year 2050 merely to hold the environmental burden at current levels. To achieve any reduction of the burden (say by half), or if there is greater economic growth (as is actually happening in the North, and desirable in the South) further reductions will be required. A factor 20 reduction in unit environmental impact might therefore be a more realistic target.

Against this background several publications have appeared stressing the potential for major environmental impact reductions associated with emerging technologies

Fig. 15.6 Predictions of world population
Source: United Nations Population Division, world population prospects: The 2004 revision

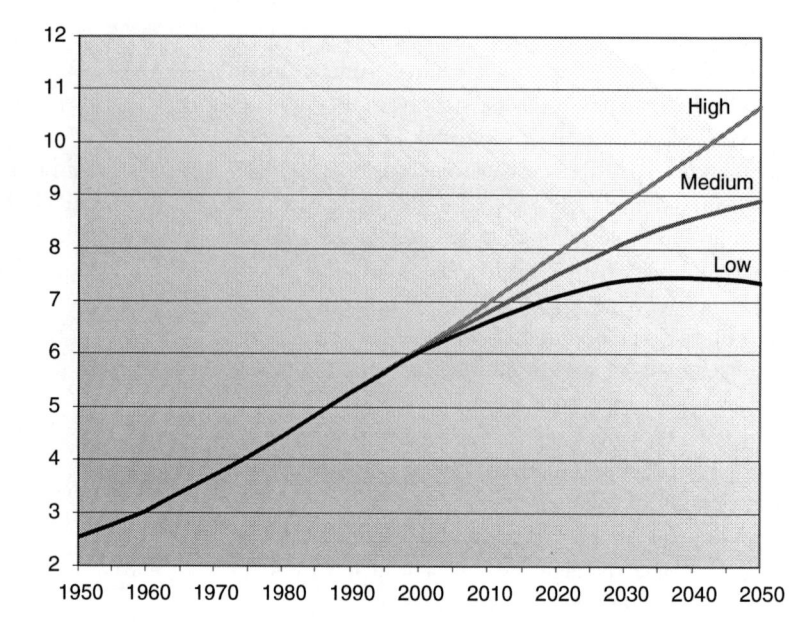

and/or technologies that are yet to be developed. The resulting discussion has been called the Factor X debate, as impact reductions by a factor 4 to 50 have been advocated. As used in practice, this factor X may relate to a product (such as an automobile), a service (e.g. transport over a certain distance at a certain speed), a need (e.g. clothing), a sector of the economy (e.g. transport) or the economy as a whole.

Key contributions to the Factor X debate have been made by von Weiszäcker et al. (1997); Schmidt-Bleek (1997); Hawken and Lovins (1999); and Weaver et al. (2000). All these contributions show considerable optimism about cutting environmental impacts in a big way through changes in technology. For the near future impact reductions in the order of a factor 4 are typically felt to be feasible, while in the longer term reductions by a factor 10 or 20 or even more are held to be achievable. The debate has not been purely academic, moreover. A factor 10 impact reduction goal has found its way into the policy objectives of several European countries and the company Pratt and Whitney has adopted 'factor 10 goals'.

Reductions of environmental impact by a factor 20 or similar can obviously not be achieved using the kind of technologies discussed so far. Although there have been considerable environmental gains as a result of management measures and logistical and organisational restructuring, especially in services, across the board these gains have generally been fairly modest. End-of-pipe technologies may control emissions, but in doing so they generally use more energy and create (other) waste or emissions. Fundamental changes to production processes are a more promising option than adding on end-of-pipe technologies, for they often allow emissions to be avoided entirely and energy consumption to be cut considerably. However, such changes should be judged not only on their direct impact on the production process and its organisation, because from the perspective of sustainability the entire supply chain is equally important. One principle here is that environmental gains achieved somewhere down the supply chain should not lead to new problems elsewhere in the chain. Because sustainability is about more than just the environment, due consideration should also be given to cultural and social conditions all the way along the supply chain: technologies shift the power relations among different groups in society, create new jobs but destroy existing ones and almost always introduce new risks of technological malfunction.

In nuclear power controversies between 1970 and 1985, for example, it was not only the direct environmental impacts of nuclear power plants (radiation and waste) that were at stake, but also such issues as: centralised versus decentralised power generation (empowering local communities or large utilities), risks of nuclear proliferation, risks of terrorism and risks of a 'war on terrorism' creating a police state. The importance of energy generation for continued economic growth and for preventing unemployment was also an important issue.

Generally speaking, these kinds of social and economic effects are only partially discussed in the public arena before technologies are introduced.

The environmental gains achieved using improved production technologies are often limited by the specifications of the consumer product in question. For example, to melt a metal will always require a minimum amount of energy, no matter how efficiently a process is designed. If further improvements are sought, high-efficiency recycling schemes may be introduced. For some applications of metals like zinc this is impossible, however, as these materials are dissipated in use. In the case of galvanised steel, for example, the zinc layer dissipates into the environment. Further progress by means of technological innovation would therefore seem to be out of the question… until it is realised that consumers do not necessarily want a galvanised steel product! What consumers want are durable products that render them a specific service. The first step should therefore be to analyse the *function* performed by the product. Can this function be carried out in a different way? If so, are these alternatives preferable in environmental, social and ethical terms'?

If this is carried out vigorously, then in general many alternatives may show up, originating in entirely different production sectors. Alternatives for galvanised metals, for example, might include (treated) wood, painted metals (non-corroding metals), plastics or even a completely redesigned product that renders all these options superfluous. It may even be the case that manufacture of the product can be superseded by providing a service, replacing a transport vehicle by a means for communication, for example.

Such radical innovations involve a new production system as well as new marketing channels and new forms of consumption. They are *systems innovations*. Systems innovations require huge investments and are always destructive for the system they supersede. Intercontinental flight destroyed the transatlantic luxury passenger liner transport system, the trains destroyed stagecoach transport and the computer destroyed the typewriter business. As a consequence, the interests of existing production systems are generally opposed to alternative systems. As there are always risks involved in radical technological change, there is always ample scope for opposition against what may be seen as 'cannibalising' innovations. However, as Sony founder Akio Morita once said, a company is better off cannibalising its own technologies than having somebody else do it for them.

We cannot predict the precise nature of the sustainable technologies of the future. Various unforeseen effects may occur. People tend to consume more of products that are claimed to be 'environmentally sound' or 'sustainable', for example. This so-called 'rebound effect' is also reflected in the example of people buying a second car because it is 'environmentally sound', cancelling out any gains that may be achieved through technological innovation. The extent to which such a rebound effect occurs in actual practice has been studied extensively in the context of improved energy efficiency. Here the conclusion was that, though rebound effects definitely occur, overall most of the energy savings are retained.

The following principles may contribute to create sustainable technologies:

- No damage to ecosystems beyond their recovery capacity
- No depletion of non-renewable resources
- Closed material loops
- Resource efficiency ('do more with less')

Consider, for example, the future of plastics, a class of product that can be said to be unsustainable in several respects:

- They deplete non-renewable (or, better: very slowly renewing) resources such as crude oil (formed over millions of years).
- They generate a variety of emissions.
- They create a waste stream of plastic end products.

At its most basic level, one could say that the problem with plastics is that we are very rapidly transforming crude oil into plastic garbage and CO_2 (See Fig. 15.7). However, plastics are light-weight materials and

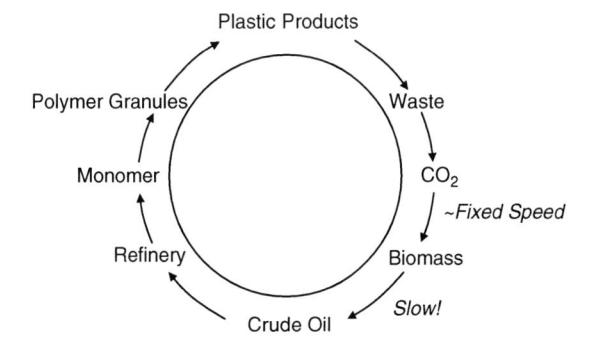

Fig. 15.7 The current plastics cycle

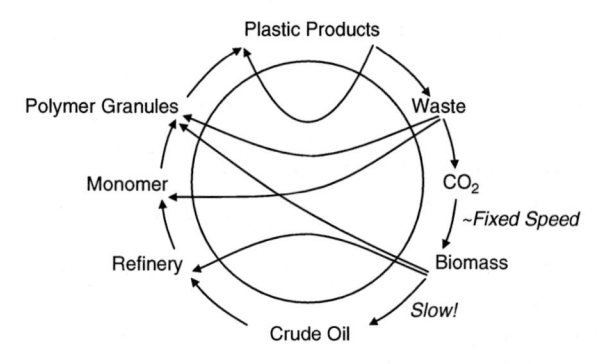

Fig. 15.8 A sustainable future for plastics?

therefore efficient in use, and they are not corrosive. If we could:

- Produce plastics solely from biomass
- Re-use plastics products as high up in the 'value chain' as possible
- Burn the remaining waste to recuperate its energy content

we could speak of a sustainable plastics 'cycle' (Fig. 15.8).

15.9 Sustainable Energy Technology

Our energy supply systems will be key elements on the road to sustainable development (see Table 15.3). If we are able to supply the world with energy generated by systems that utilise only renewable primary sources, do not endanger local or global ecosystems (during normal operation as well as incidents) and enable local communities to exert control over crucial elements of their societies, then many other production systems will move far closer towards sustainability.

The main renewable forms of energy are:

1. Solar energy
2. Wind energy
3. Biomass energy
4. Hydropower
5. Wave energy
6. Ocean heat (using the differences in temperature with upper layers of the atmosphere)
7. Geothermal energy
8. Tidal energy

Some people would like to add nuclear fission and nuclear fusion to this list of sustainable energy sources. However, it is rather doubtful whether nuclear fusion, using the vast amounts of deuterium available in the water on the planet, can ever become more than a scientific curiosity; after half a century of research, the odds are against it. Nuclear fission has left us with enormous amounts of radioactive waste and the Chernobyl catastrophe that will have a death toll for centuries to come.

Items 2–6 on the list are in fact all forms of solar energy. The sun supplies the Earth with about 1,500 times the power used by society today. There is therefore ample renewable energy available; the main problem is how to use it effectively. Further optimisation of technologies is needed to decrease costs and boost efficiency. Under favourable conditions, though, it is estimated that technologies might be developed that can generate electricity from these sources for $0.02–0.06 per kW-h.

Every form of energy conversion is associated with losses and costs. Heat (solar or geothermal) can be used most effectively for direct heating and wind for direct propulsion of shipping vessels, for example.

Given the dependence on natural circumstance, renewable sources of energy should be combined prudently in systems in order to use them effectively:

Table 15.3 Primary energy consumption by fuel, in million tonnes oil equivalent, 1999

	Oil	Natural Gas	Coal	Nuclear	Hydro-electric
USA	882.8	555.3	543.3	197.7	25.8
N America	1047.1	651.5	581.2	219.3	58.2
Europe	755.2	399.6	348	246.1	51.5
Asia Pac.	928.7	241.7	912.5	125.9	46.2
World	3462.4	2063.9	2129.5	650.8	226.8
OECD	2171.2	1128.5	1078.7	564.2	117.2

Source: www.nef1.org/ea/eastats.html

15.9.1 Storage

Solar, wind, tidal and wave energy cannot be tapped when and as we chose, and storage is therefore required. If solar energy is collected as heat, it can be stored as hot water. If electricity is to be generated, however, storage becomes a problem. This can be solved in two ways:

- By converting the electricity into other forms of energy, for instance by:
  - Pumping water to high reservoirs and using it later as hydropower
  - Making hydrogen, which can be used to generate electricity in fuel cells
  - Charging batteries
  - Increasing pressure behind pressure valves

All of these forms of storage are expensive and are associated with considerable losses, as double conversion is required.

- By integrating these sources of energy into a far larger electricity supply system. An individual country may sometimes be clouded, but never an entire continent. Likewise, strong winds in one country can compensate for calm weather elsewhere. Larger electricity systems may be able to use hydropower to compensate for temporary shortages of these non-controllable energy sources.

However, there are serious drawbacks to several renewable energy technologies.

15.9.2 Ecology

Hydropower often ruins local ecology. Valleys are turned into lakes and barrier dams also act as barriers to wildlife like salmon. Millions of people may have to be displaced, as in the case of the Three Gorges dam in China. Over time, the retaining lakes may become rather less effective, as they fill up with plants and sediment. They may also become a breeding ground for insects.

Windmills are often criticised for killing birds. However, if sites are chosen carefully, avoiding seasonal migration routes, the number of bird casualties can be kept very low. Windmill noise and shade may be serious problems, but only in the direct vicinity. In this respect prudent site selection is therefore also an important issue.

15.9.3 Risks

In the event of an earthquake or military conflict, the lakes required for hydropower represent an enormous risk, for if a dam were to collapse the enormous rush of water might wipe out millions. The same may hold for notions of utilising ocean heat: this could probably only be done by erecting enormous towers evaporating low-boiling liquids like ammonia and generating electricity by collecting the condensed liquid at the top of the tower. In the event of an earthquake or an aircraft collision an enormous disaster might ensue.

15.9.4 Space

Because it is readily stored, biomass is for many people the preferred energy source. However, the *traditional burning of biomass causes considerable health risks*. If the world's population increases to 9 to 10 billion, more agricultural land will be needed to produce sufficient food and materials. As the most productive land is already used for farming, a 50% rise in population might imply a far greater increase in agricultural land use, unless agricultural efficiency is improved substantially. Can more agricultural land be used for energy generation? This would undoubtedly imply further damage to ecosystems.

There are many promising technologies with which to create sustainable energy supply systems. The main challenge will be to create a reliable energy system based on renewable sources that can effectively overcome these drawbacks, while enabling local and regional communities to regain or maintain control of their energy systems.

15.10 Closing Remarks

In this chapter we first sketched the historical relationship between technology and the environment. Various possibilities were then described as to how technologies might contribute to solving environmental problems. Technologies can and will have a major role to play in sustainable development, but never as the 'Deus ex machina' that will relieve us of our problems at no cost and without requiring any further adaptation.

Solutions will always be 'socio-technical' in nature; they will, in other words, encompass technological as well as social transformation. Finally, two words of warnings are in order here:

Improved technology may be the enemy of truly sustainable technology. Studies of technological change teach us that the development of technology is a path-dependent process, i.e. we cannot choose freely but are bound by the choices of the past. Similarly, our choices today have consequences for the range of options left open to future generations. As a consequence, what today may appear to be an improvement may in fact mean embarking on a technological strategy that prevents us choosing an even better solution later, a phenomenon known as 'lock-in'. An example of an environmentally improved technology that would preclude real long-term sustainability is substitution of the electrical process used to produce aluminium by a blast furnace process like that used in steel-making. Although at first this would probably boost the energy efficiency of production, it might well block the path to a truly sustainable solution, as it simply increases dependence on coal. There is no known way to produce coal sustainable. At some time in the future, though, we may be able to produce aluminium using only renewable sources of electricity. We should therefore adopt a long-term perspective, although it is by no means always clear what a truly long-term sustainable solution might entail.

Striving only for truly sustainable technologies may stop us taking action now. Solving the problems we face today will hurt. Doubts about the viability of particular solutions often therefore serve merely as an excuse for postponing action. In practice the precautionary principle is frequently reversed: as long as the existence and gravity of the problem has not been proven beyond doubt, action is deferred. This is an attitude we can no longer afford to adopt, however. To develop new technologies takes time. Even development of a new car, involving no revolutionary new technology, takes about 10 years. Revolutionary changes will take decades of concerted action on the part of governments, consumers and industry – and particularly the vehicle and energy supply industries.

It is important for society to try and foresee the impact of technological change, so that the merits of such change can be discussed democratically. However, experience with parliamentary Technology Assessment during the last few decades of the 20th century have shown that we can gain no more than limited insight into the future impact of technologies, for that impact is intimately bound up with – or even indistinguishable from – wider, more general cultural changes. Unforeseen rebound effects and new social dilemmas are frequent. They cannot be prevented and it is therefore crucially important that technological strategies towards sustainability are flexible. Our challenge today is to learn from past mistakes and correct what we can with due haste. And start doing so now.

References

Barber, B. (Ed.). (1970). *L. J. Henderson on the social system: Selected writings*. Chicago: University of Chicago Press.

Brauni, E. (1995). *Futile progress technology's empty promise*. London: Earthscan.

EIA (1997). *Renewable energy annual 1996*. See www.eia.doe. gov/cneaf/solar.renewables/ renewable.energy.annual/chap10. html.

Hawken, P., Lovins, L. H. (1999). *Natural capitalism – creating the next industrial revolution*. Colorado, USA: Rocky Mountains Institute.

Ridken, R. G. (Ed.). (1972). *Population, resources and the environment*. Washington, DC: US Government Printing Office.

Schmidt-Bleek, F. (1997). *MIPS and factor 10 for a sustainable and profitable economy*. Wuppertal: Wuppertal Institut.

United Nations (2005). *World populations prospect. The 2004 revision*. New York: United Nations.

Weaver, P., Jansen, L., van Grootveld, G., van Spiegel, E., Vergragt, Ph. (2000). *Sustainable technology development*. Sheffield: Greenleaf.

Weiszäcker, E. U. von, Lovins, A., Lovins, L. H. (1997). *Factor four: Doubling wealth – halving resource use*. London: Earthscan.

WWF (1999). See: www.wwf.org.uk/news/n_0000000183.asp.

Further Readings

Allen, D. T., Sinclair, K. (1997). *Pollution prevention for chemical processes*. New York: Wiley.

Ausubel, J. H., Sladovich, H. E. (1987). *Technology and environment*. Washington, DC: National Academy Engineering, National Academy Press.

Corbitt, R. A. (1999). *Standard handbook of environmental engineering*. New York: McGraw-Hill.

Graedel, T. E., Allenby, B. R. (1995). *Industrial ecology*. Englewood, NJ: Prentice Hall.

Kemp, R. (1994). Technology and the transition to environmental sustainability, the problem of technological regime shifts. *Futures, 26*(10), 1023–1046.

Kirkwood, R. C., Longley, A. J. (1995). *Clean technology and the environment*. London: Blackie.

Leal Filho, W. (1998). *Environmental engineering: International perspectives*. Frankfurt: Peter Lang.

Moors, E. (2000). *Metal making in motion, technology choices for sustainable metals production*. Delft: Delft University Press.

Moors, E., Mulder, K. F. (2002). Industry in sustainable development, the contribution of regime changes to radical technical innovation in industry. *International Journal of Technology Policy and Management, 2*(4), 434–454.

Mulder, K. (1998). Sustainable consumption and production of plastics? *Technological Forecasting & Social Change, 58*(1&2), 105–124.

Nazaroff, W. W., Cohen, L. A. (2001). *Environmental engineering science*. New York: Wiley.

Schipper, L., Grubb, M. (2000). On the rebound: between energy intensities and energy use in IAEA countries. *Energy Policy, 28*, 367–388.

Tierny, R. J. (2002). Green by design: factor 10 goals at Pratt and Whithey. *Corporate Environmental Strategy, 9*, 52–61.

Turkenburg, W. C. (2000). *Renewable energy technologies, Chapter 7 of the world energy assessment*. New York: United Nations Development Programme.

Chapter 16
Integration

Jan J. Boersema

Contents

16.1 Introduction 335
16.2 Two Methods 335
16.3 Levels of Integration 338
    16.3.1 Data/Numerical Level 338
    16.3.2 Administrative/Legislative/
           Institutional Level 340
    16.3.3 Societal Level 341
    16.3.4 Combinations 342
References 343

16.1 Introduction

In the preceding chapters the term 'integration' was used many times, but what is in fact meant by *integration,* or *integrated*, as in 'integrated policy instruments' or 'integrated environmental management'? According to the dictionary, to integrate means 'to combine parts into a whole', 'to complete by the addition of parts', or 'to bring into equal participation in or membership of society', as for instance in 'Can these immigrants be integrated into our society?' Applied to policy, 'integrated' often tends to correspond with 'integral' and statements on environmental issues often suggest that the word 'integrated' means 'covering all aspects' – although this is in fact usually a hollow pretence. In practice, it is rarely possible to integrate *all* relevant aspects in a project or method, if only because time or manpower restrictions make choices inevitable. On closer inspection, then,

J.J. Boersema (✉)
Environmental science and word views at the Institute for Environmental studies (IVM) at the vu University Amstedam, the Nethlands
e-mail: Janboersema@ivm.nl

'integrated' approaches usually turn out not to be 'fully integrated' in the absolute sense of the phrase, but 'more integrated', usually signifying that the analysis includes a few more factors than earlier approaches. This process may repeat itself, leading to greater and greater 'integrative' scope, but the concept may well lose some of its meaning along the way, reminding one of the detergents that claim to make your wash 'ever whiter'. There would thus seem to be a need for a more precise definition of the concept of integration.

16.2 Two Methods

There are also theoretical grounds for developing a more precise definition. The concept of integration is also used in mathematics, and it might be useful to see whether this 'auxiliary discipline' can clarify matters. In mathematics, integration is the calculation of an infinite number of infinitesimal quantities. It is an elegant method for exact summation of non-discrete quantities, applied, for instance, to calculate surface areas and volumes limited by two- and three-dimensional curves. Box 16.1 provides a simple illustration of the approach. In analogy to this example, one might say that integration in environmental management is only useful if it involves *defined integration*. This means the limits of the system must be clearly defined, which in turn means the approach can no longer be regarded as 'all-encompassing'. Insofar as phrases like 'comprehensive policy instruments' and 'comprehensive environmental policy' claim to be all-encompassing and give no clear indication of the system limits involved, they are fundamentally ill-defined and in practical terms meaningless. In this case, it would be better to call them 'integration-oriented instruments' and 'integrated environmental policies'.

The latter phrase entails the assumption that the entities considered in such policies have indeed been integrated. But what does that actually mean? A yet closer examination of the terms 'integration'/'integrating' is required, and once again elementary mathematics can help. In the example of Box 16.1, the numerical quantities set out on the X- and Y-axes can be simply and fully combined (i.e. integrated) to yield a new quantity. In this situation, integration is possible without loss of information. In most cases, however, heterogeneous entities are involved that cannot simply be converted into one another but must nonetheless be integrated. In such cases, 'integrating' boils down to 'adding up apples and oranges', or in more general terms, combining heterogeneous entities to obtain a homogeneous, or at least more homogeneous, whole. Closer inspection shows that there are basically two methods of achieving this kind of homogeneous combination. In a mathematical analogy, these can be designated the Greatest Common Divisor (GCD) and the Least Common Multiple (LCM) methods. These concepts are explained in Box 16.2.

Integration of the GCD type means seeking a relevant quality common to the entities to be integrated. The relation between the entity and this quality is then expressed as a value that can be combined with the corresponding value obtained for the other entity to yield an integrated overall value. The main advantage of this type of integration is its numerical nature, which allows exact comparison of previously dissimilar alternatives. This makes it possible to integrate concentrations of SO_x and NO_x, for instance, via their common quality 'acidifying capacity', into *acid equivalents*, or add up quantities of CO_2 and CH_4 by converting both to their GWP (Global Warming Potential, the capacity to warm up the earth's atmosphere per unit substance). It is clear that this type of integration involves a loss of information, for by definition it ignores certain characteristics of the substances. The choice of the common denominator depends on the purpose of integration and often involves subjective aspects. Once the denominator has been chosen, the value of the numerator can usually be calculated by means of reliable and/or veri-

---

Box 16.1 Integration

By way of example, let us take the equation

$$y = x^2$$

The graphic representation of the equation is a parabola. In many situations, only the right-hand part of the graph (x > 0) is of interest, for instance if the X-axis represents time. The general form of the integral of the function is:

$$\int x^2 dx = \frac{1}{3}x^3 + c$$

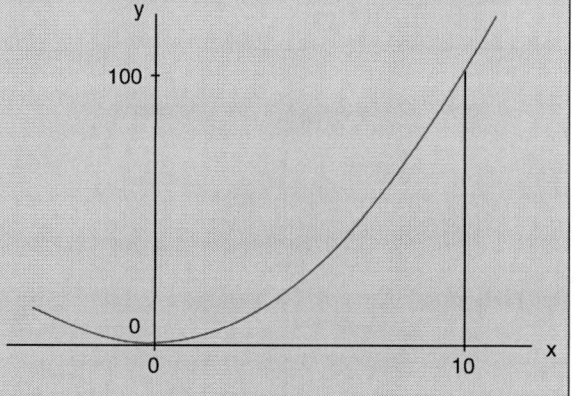

In which c is an indefinite number. This makes it an indefinite integral, and since no limits have been defined between which the process of integration is to take place, the integral cannot be numerically solved. If limits have been defined, the integral is called definite, and it can now be numerically solved. An example of a practical application of this is the following. Suppose that the wall of an oil tank develops a leak as a result of corrosion. The force of the oil leaking out widens the hole, and the amount leaking from the tank per day increases by a quadratic function, such as the one above. After ten days, the leak is plugged and the owners want to know how much oil has been spilled. If Y is the rate at which oil was leaking (in m³ per day), the integral of Y from day 1 to day 10 is the total amount of oil spilled. In the figure, the area under the curve represents this. The calculation goes as follows:

$$\int_0^{10} x^2 dx = \frac{1}{3}x^3 \Big|_0^{10} = \frac{1}{3}10^3 - \frac{1}{3}0^3 = 333\frac{1}{3}$$

It is only after clear system limits have been defined that an integral obtains practical, that is numerical, meaning.

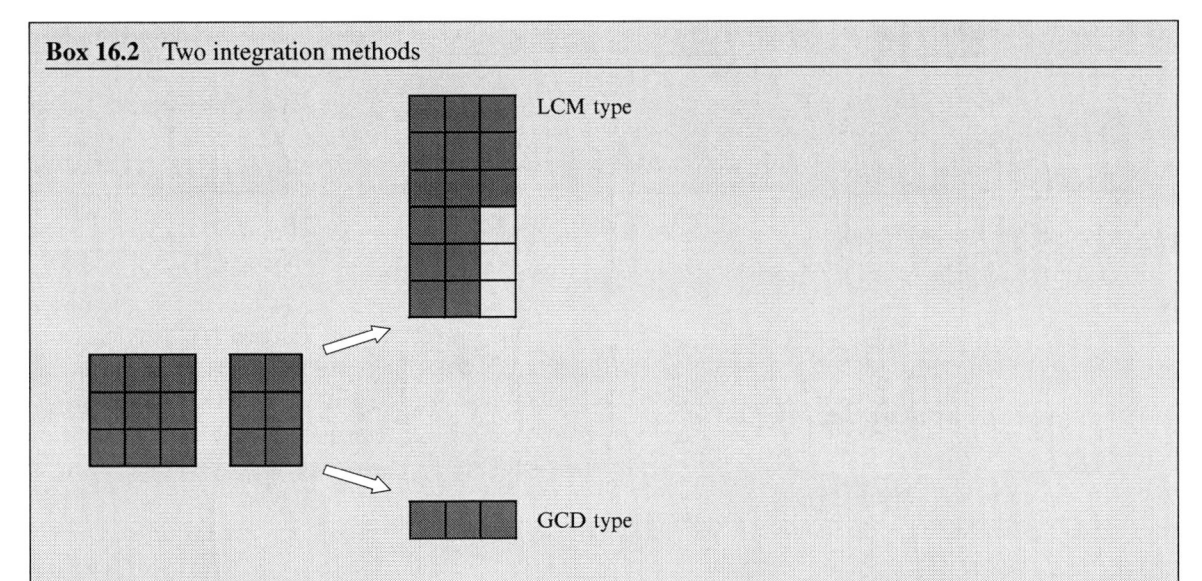

> **Box 16.2** Two integration methods
>
> LCM type
>
> GCD type
>
> The Greatest Common Divisor (GCD) is the largest integer by which two (or more) other integers can be divided. The numbers are reduced to the common denominator.
>
> The Least Common Multiple (LCM) is the smallest integer that can be divided by two (or more) other integers. It is, as it were, the smallest envelope that can fully contain them.

fiable methods. It may be noted that valuing entities using money as the common denominator is a special case of this type of integration. As this denominator changes over time, however, it may be contested whether monetary value is really a useful quality for integration.

Integration using the LCM method basically involves no loss of information, as the entities to be integrated are fully included in the 'envelope' or framework. Although it is the *least* common multiple that is being sought, there is even a certain degree of *redundancy*, for the envelope must to a certain extent be oversized in order to accommodate both entities. This type of integration is often of a qualitative nature. An example from environmental legislation would be the integration of two sectoral acts into one comprehensive framework act. The precise relation between the entities integrated into the larger framework is not always defined, but there are often rules and procedures that result in a more homogeneous whole. Even though the 'comprehensive environmental permit' (*integrale milieuvergunning*) introduced into Dutch environmental policies should not really be called 'comprehensive', it can serve a useful function as an integrative instrument. Models, concepts and the like can also fulfil such integrative functions. These are LCM-type entities (frameworks) that can sensibly combine certain heterogeneous entities to create a more homogeneous whole and that can do the same to new, as yet unknown entities. In terms of the apples and oranges, LCM type integration might involve using *fruit* as the integrative framework, allowing other types of fruit to be included in the integration process, too. In other words, the concept of fruit becomes the guiding principle of the integration process. In the remainder of this chapter, such concepts or frameworks, which both delimit and guide the integration process, will be called 'integrating concepts'.

Although these two types of integration can usually be clearly distinguished in environmental studies, the two are often combined. LCM-type integration often leads to – and sometimes actually implies – a GCD-type integration of components. For instance, so-called Multi Criteria Analysis (MCA; see Chapter 11) functions as an LCM-type integration that includes various components connected in parallel, each of which is integrated in a GCD-type integration. Similarly, several GCD-type integrations can be connected in series, with their outcomes then being combined to yield a single, overall, numerical environmental parameter, using weighting factors, for instance. Since environmental policy-making often involves choices between quantities that cannot readily be compared (being highly heterogeneous), there is a great demand for such integrated environmental parameters, particularly among policy-makers. How sensible, transparent and useful such parameters are depends partly on the level at which integration takes place and the purpose for which it is used. The next section therefore discusses the objectives of integration, the levels at which it occurs and the methods employed at the various levels.

16.3 Levels of Integration

For the purpose of the present chapter, the levels at which integration takes place can be divided into three groups, which can be designated as the data/numerical level, the administrative/legislative/institutional level and the societal level. Some of the main methods used at each of these levels are discussed below.

16.3.1 Data/Numerical Level

Several examples of integration at this level have already been given and can be found in Herendeen's excellent book on quantitative methods in ecology (1998). These methods generally involve integrating heterogeneous data into one overall parameter. This heterogeneity can take various forms and may relate to the following aspects.

16.3.1.1 Effects

In this case, identifying a common effect reduces heterogeneity. In the simplest cases, there is already reasonable consensus about the common effect, so that this can be used as the common denominator. Examples include parameters like 'acidification' and 'relative global warming potential', discussed above. When it comes to the impact of environmental factors on human health, it is becoming increasingly common to use the DALY (disability adjusted lost years) or DALE (disability adjusted life expectancy), a measure of years lived, adjusted for each year lived in incomplete health due to environmental pressures and other health-related factors (assessing complete health as 100%: Murray and Lopez 1996).

If no common effect can be found for certain substances, a denominator such as 'level of toxicity' can be constructed. There are two options for calculating such a level. In the first, the quantity of each substance is compared with, i.e. divided by, a toxicological reference such as the substance's No Adverse Effect Level (see Chapter 6). The second option also incorporates political decision-making about the risks society deems acceptable. In this case the quantity of each substance is divided by, say, the environmental quality or emission standard or TLV (threshold limit value) established for that substance. Risk studies sometimes use 'fatality equivalent' as a denominator, permitting integration of true fatalities and seriously wounded. Integration is problematic if it involves comparisons between data on different scales, i.e. on cardinal, ordinal or even nominal scales. A cardinal scale provides information on the distances between values; while an ordinal scale provides only a ranking order and a nominal scale no more than a typology. In such cases, it may be advisable not to use a single common denominator, but to give a choice of options. The most frequently used common denominator is probably money. Here, conversion might involve calculating the expenditures required to avoid various adverse effects and setting these off against anticipated revenues, for example.

Entities can also sometimes be integrated on the basis of particular characteristics or properties of relevance for their environmental impact. Two interesting aggregated measures have been developed in this context, one based on weight, the Material Intensity per Service unit (MIPS, Schmidt-Bleek 1993, 1998), the other based on area, the ecological footprint (Wackernagel and Rees 1996; WWF 2006).

Other integrative approaches are based on diagrams. The AMOEBA or RADAR is one example (Ten Brink et al. 1991). It is a diagramming device intended to show all indicators in a single format by comparing the real value to a ideal (sustainable) one. The method does not in fact generate a single value for the environmental impact but visualise the indicator values into a circular presentation. The actual shape of the AMOEBA depends on the value and the position of the indicators (see Fig. 10.13).

Finally, it should be noted that 'heterogeneity' may also relate to variation among those involved or affected, not only human beings (for instance unborn children versus adults) but also other organisms. If effects on people and wildlife are being compared, there will obviously be still greater variation. There is as yet no transparent system for including data on non-human organisms and abiotic factors in the comparison. A major and persistent point of debate is that of the 'valuation' of wildlife and non-human organisms. The often-advocated acknowledgement of nature's integrity and its intrinsic value offers few leads for the required weighting procedure. Many proposals have been made for such weighting, usually involving factors like numbers of individuals and 'emotional appeal'. It is this last factor, for instance, which explains why the nearly extinct California condor is given far more weight in policy decisions than some species of Hawaiian snail, even though the latter may also be on the verge of extinction.

16.3.1.2 Aspects of Scale

Environmental problems often involve different spatial or temporal scales. Both beneficial and adverse effects can be exported into other areas and/or into the future. In such cases, integration is controversial. In economics, for instance, it is common to 'discount' future risks by a certain percentage annually. If one objects to the export of risks to future generations, though, greater negative weight should be given to exported future risks than to current risks, although empirical research has shown that this conflicts with how people tend to view temporal aspects in evaluating environmental risks, with potential adverse consequences tending to be regarded as less important as these become more remote in time. Since all major environmental problems entail negative impacts that are relatively remote in this sense, solving this dilemma of temporality is a major task. 'Bringing forward' such remote adverse (as well as beneficial) effects would seem to be essential, but at the same time difficult to articulate (Boersema 2001). The first Dutch National Environmental Policy Plan (NEPP) drawn up by the Ministry of Housing, Spatial Planning and the Environment adopted an approach whereby certain policy alternatives are discarded as having effects beyond a certain horizon. Thus, given the government's position that environmental problems may not be passed on to our descendants, any alternatives leading to problems that cannot be solved within one generation (i.e. 25–30 years) are rejected.

16.3.1.3 Uncertainty

As we do not live in a deterministic world and our knowledge is limited, uncertainty is an intrinsic aspect of heterogeneity. This means that problems can often only be evaluated in terms of likelihood and bandwidth. In such evaluations, the upper end of the range is often considered more important than the average value. Governmental risk management philosophy may well attach greater weight to a very small risk of a major catastrophe than to a very high risk of a minor accident. Activities involving the risk of large numbers of casualties in a single event may even be regarded as unacceptable, while the same number of casualties spread over a longer period may well be deemed acceptable. In a sense, the 'fatality equivalent' approach is thus provided with a limit: a maximum number of casualties per event or unit of time. This tallies with the perceptions of the severity of environmental disasters that are common among the general public.

16.3.1.4 Examples of Numerical Integrative Instruments

Integrative instruments play an important role in environmental policies, as they can facilitate choices among alternatives. Homogeneous numerical quantities may have the advantage of permitting easy comparison not only among themselves, but, if properly chosen, also with standards. If well chosen, such quantities are useful not only in answering the question of which alternative is best for the environment, overall, but also in deciding whether the best alternative is compatible with the main standards and objectives of current environmental policies. The use of numerically integrated environmental parameters that can be used to compare products and services has been helpful in the rise of life cycle assessment (LCA). Both the statistical and dynamic variants of this technique have proved useful in deciding among alternative paths to the same goal (see Chapters 12 and 17). By varying the weight given to impacts associated with the life cycles of products or activities, different perceptions of such impacts can be accommodated. The LCA technique can be regarded as an example of 'bottom-up' integration, that is, integration based on a comprehensive analysis of most of the individual environmental aspects of a particular activity, service or product.

Another example of an integrative instrument is the total or aggregate environmental quality index that integrates different aspects of environmental quality. If the focus is on human health, the aforementioned DALY concept may be helpful for the purpose of integration. In this context there are also other approaches, though, one example being to divide actual pollution levels by environmental quality standards (e.g. Sol et al. 1995).

In an attempt to permit comparison across a more comprehensive range of impacts a number of researchers have proposed adopting *sustainability indicators* as standards (Bossel 1999; Hertin et al. 2001; Ekins and Simons 2001). To surmount just bookkeeping, indicators often imply thresholds or limit values and these are based on the idea that there is such a thing as 'environmental space', that is, the physical limitations of our human activities within the restrictions imposed by the sustainability principle. If these limitations to our exploitation of natural resources are graphically depicted as a function of time, the declining rate of utilisation means the graph tapers off. This can be regarded as a 'top-down' approach, as the environmental space defines the limits of all our activities. This integrative concept allows quantitative limits to be defined for specific activities: the sustainability indicators referred to above. An acceptable and consistent quantification and integration of 'green' environmental aspects like landscape and biodiversity still presents major problems in both these approaches, top-down as well as bottom-up. Although in theory it is simpler to quantify and integrate energy, raw materials and emissions, the complexity of the parameters constructed to that end may ultimately obscure essential differences. This is a more general problem that besets all integrative instruments to some extent or another. Criticisms focusing on this problem have in fact also been levelled at LCAs and indices for aggregate environmental impact. There is also the problem that collecting and processing the required data may require inordinate amounts of money and/or time.

A more fundamental question that arises is whether further quantification truly improves our understanding. This question is especially urgent in the type of approach characterised above as *bottom-up*, for the number of activities and products that can be evaluated with the help of such integrated environmental parameters is virtually limitless. Opting for more environmentally friendly activities and products does not automatically result in a sustainable society. This brings us to the next level of integration, in which qualitative elements play a greater part.

16.3.2 Administrative/Legislative/Institutional Level

One general characteristic of integration at this level is its combination of a framework encompassing a system of regulations and agreements, a combination described above as belonging to the LCM type. The framework, as well as the regulations and agreements, may be precisely defined, for instance in legislation, or less precisely, for instance in consultations among and between ministries and/or NGOs. Both the framework and the regulations aim to promote and steer the process of integration, but rarely prescribe the use of quantitative methods. Integration is thus procedural rather than analytical in nature. This is related to the kind of 'heterogeneity' that comes into play at this level, which is often synonymous with 'incoherence' and 'working at cross-purposes'. Environmental integration is a highly topical issue at this level, too, as is described in greater detail in Chapter 13 with respect to current government policies. For now, we limit ourselves to discussing a few methods characteristic of this level and describing some of the underlying tendencies.

Integration at the administrative level starts with a recognition of its necessity and an inventory of the elements that need to be integrated. This more or less defines the framework. At this stage, the method of integration is rarely precisely defined. The mere announcement of an integration process is often enough to spark off a number of activities which do indeed reduce the heterogeneity and result in a more homogeneous outcome. These activities might include simplifying consultative structures, exchanging information and improving coordination, say, in an attempt to reduce redundancy and incompatibilities and improve coordination in time and space.

Formal integration may be horizontal (between the activities of different ministerial departments or different ministries, for instance) or vertical (between various levels of government, i.e. local, district, state, multistate, global). It may take the form of integrated laws, covering a larger number of aspects of government activities, which is an example of LCM-type integration. It may also take the form of collaborative planning, as in the case of the Dutch National Environmental Policy Plan, a collaborative undertaking by four ministries. Environmental policy may itself in turn be integrated into a larger whole. This is indeed occurring in discussions about sustainability worldwide. Thus, for the Earth Summit in Johannesburg (2002) many countries produced integrated policy documents ranging in scope from the economic to the educational.

The integration of environmental care in business activities is another example of integration at the institutional level. Several types of integration can be distinguished here. Representing a relatively low level of integration is the firm with an environmental department that focuses on compliance with the law. A higher level of integration is achieved when management commits itself to 'eco-efficiency', for instance, seeking to continually improve its environmental performance to build up a 'green image'. In this case there will be greater involvement by other departments: environmental improvement will become a key research and development issue, the procurement department will be told to buy from suppliers meeting certain environmental standards, and the marketing department will focus on the 'greenness' of the firm's products. In such cases the group of perceived stakeholders may well be expanded to embrace environmental NGOs, say. The firm may also integrate its environmental department into a larger whole. Thus, a number of companies have integrated their (workers) health and safety department with their environmental department and in some cases even with the department for quality management.

LCM-type integration is clearly recognisable at the legislative level. Until recently, environmental legislation in many countries consisted of a range of sector-specific laws that took insufficient account of the complex and interrelated nature of environmental problems. The laws often overlapped and clear dividing lines were hard to establish. In addition, this sectoral legislation used to be, and in some instances still is, the responsibility of different ministries. Once again, good coordination is essential for the process of defining and implementing regulations and standards, which is why the laws have often been combined into a single system of environmental management regulations. In the Netherlands a comprehensive law of this kind, known as the Environmental Management Act (*Wet op het Milieubeheer*), has meanwhile been further extended to include a number of new regulations, such as those relating to Environmental Impact Assessment (*EIA*).

16.3.3 Societal Level

Integration involves data, legislation and institutions, but also people, both individually and as groups. Factors creating heterogeneity at the psychosocial level include character, knowledge and skills. If people with different characters collaborate in a particular project, they may have different views on matters like operating procedures, while differences in knowledge and skills may lead to disagreement about the problems to be examined, and even their definition. If such a project group is to cooperate fruitfully, it will have to come to some sort of agreement, which requires a basic willingness to enter into fruitful negotiations. This type of integration also needs to be adequately addressed in university courses on environmental science, for instance in multidisciplinary student projects. A fashionable way of improving integration processes at this level, in a team or company, is by means of *outward-bound* activities like 'survival' weekends. These aim to promote a sense of fellowship and solidarity among those having to work together in a team, and thus reduce psychosocial differences. In a survival weekend, people learn to solve problems together and encourage one another to persevere. They must depend on one another and learn to interact effectively, which may stand them in good stead in their work.

Societal heterogeneity involves other factors, too, such as interests and social status. This is especially relevant if the integration process includes other organisations. The methods used for this purpose, such as 'open planning' and 'target group consultations', have largely been derived from management theories. *Open environmental policy planning* or *interactive planning* attempts to coordinate the actions of various types of authorities and create a framework for future decisions. The plans themselves also seek to achieve greater policy coherence. Stakeholders are consulted in the early planning stages, which may lead to better mutual understanding. This process allows policies to be developed that enjoy wide support. Since the behaviour of the group(s) targeted by a policy is determined largely by the way the policy is implemented, interaction with the group(s) in question is essential for successful implementation. This is why the *target group consultation model* was developed in the Netherlands, a system of consultations between the Ministry of Housing, Spatial Planning and the Environment (VROM), the Ministry of Economic Affairs (EZ) and the business community with the aim of achieving the objectives set out in the consecutive NEPPs. This model seeks to give the business community greater responsibility for implementing environmental policies. The only parties in these consultations are VROM, EZ and the business community. While environmentalists have criticised the model for being too soft on industry, the business community feels due scope for negotiation makes for greater flexibility and allows adverse environmental impacts to be compensated by improvements elsewhere. The model allows the parties to come to detailed agreements, so they all know exactly where they stand.

The target group consultation model can be used in both the administrative and legislative contexts, to steer introduction of environmental care systems in companies, for instance. Both the above methods involve a certain amount of coercion on the part of government authorities, who at the end of the day are endeavouring to enforce something: introduction of certain rules and regulations, say. The aim of the consultations is then to facilitate negotiations and foster better understanding among the parties involved, by discussing such issues as bottlenecks and feasibility. Certain concessions may then be made to ensure the main objectives are ultimately achieved, often more simply and cheaply.

A more general elaboration of this method (or group of methods) is the *consensus* or *stakeholder approach* or *collaborative planning* dealt with by Hisschemöller in Chapter 14. This is a systematic method of dealing with conflicting opinions and interests. The aim is to achieve sensible, acceptable, sustainable solutions, as well as to improve relationships among the parties involved, or at least prevent them from deteriorating. One of the guiding principles is that anyone who feels involved in a problem bears some of the responsibility for its solution. This means that a solution can only be achieved if all those involved are engaged in the process and collaborate in finding a solution.

This method yields a set of alternative policy options, which can then become the subject of a public debate. The consensus approach will not necessarily produce the best possible solution, but it yields those solutions on which the greatest number of people agrees. The method can be used for any problem involving a conflict of interests, such as finding the right location for an industrial facility or setting appropriate conditions for development of conservation areas.

16.3.4 Combinations

The two integrative techniques discussed above, the analytical and the procedural can also be used in combination, yielding what may be called 'interactive integration'. In addition to 'classical' methods like multicriteria analysis, cost-benefit analysis (see Chapter 11) and environmental impact assessment, this includes *multi-attributive benefit analysis*. This method is applied if data cannot be expressed in monetary terms, by allocating a 'benefit value' to each. Allocating this kind of value is more complicated than allocating economic value, as benefits are ordinal while money is cardinal. Most analyses therefore operationalise benefits on interval scales. The allocation of benefits depends on the preferences of the allocators, and this is where the interactive element comes in. All benefit attributes allocated to each individual component must be combined into one overall benefit value. This analytical integration is usually achieved by attaching weights to the benefit attributes and then summing them to yield an overall benefit value. The option having the highest benefit value is then the preferred option.

Another popular method is *scenario analysis*, often adopted if the activities to be evaluated have many interrelated effects that cannot be predicted with sufficient certainty, generally long-term effects on social and natural systems. A scenario analysis includes descriptions of the initial situation, various possible future situations and the paths connecting them. Scenarios can be drawn up on the basis of the initial situation or on the basis of future situations, often with very different outcomes. Scenario analysis is a tool for developing long-term policies, yielding a range of options from which policy-makers can choose based on their own preferences. It is used above all in situations where time and uncertainty are major factors and in such cases is able to provide a better understanding of possible development paths.

A very popular tool for this purpose is *computer simulation,* particularly in the form of interactive models (Chapters 17 and 18). A model is an abstract representation of reality, involving a symbolic mapping of the observable elements of a natural system and the relationships between them in a formal mathematical system. Such models can be very inclusive, encompassing information from the natural sciences as well as social sciences. An important aspect of model construction is definition of the system limits. Models are frequently used to describe the distribution of substances over the various environmental compartments. This involves linking compartmental models to create an integrated environmental model, expanding the model's domain in space and time. An example of this is the RAINS model, which describes acidification in Europe (Chapter 18). The model itself can be used in two different ways. One is to run a simulation on the basis of a user-defined or model-defined scenario, while the other involves optimisation, using a predefined deposition pattern to calculate emission reductions. Models can thus be used to perform calculations on various options, to identify the best possible option for achieving a desired final outcome.

The RAINS model and the scientists applying it have played an important part in negotiations among decision-makers. The model allowed policy-makers' proposals to be simulated and the outcomes to be fed back to them. If outcomes were unacceptable, new proposals could be made, which could again be tested by the model. Environmental models are particularly useful in situations where space and time are important factors. This interactive method also provides the broader insights needed for policy decisions (Chapter 18).

New methods are still being developed, and the search for universal integrative techniques permitting truly comprehensive environmental assessment is particularly popular. For some scientists the aim is to develop a single yardstick by means of which all risks, effects, etcetera, can be measured and compared (see, for example, Bell and Morse 2003). Others, however, would prefer to see a set of indicators adopted.

References

Bell, S., & Morse, S. (2003). *Measuring sustainability. Learning from doing*. London: Earthscan.

Boersema, J. J. (2001). How to prepare for the unknown? On the significance of future generations and future studies in environmental policy. *Environmental Values, 10*, 35–58.

Bossel, H. (1999). *Indicators for sustainable development: Theory, method applications. A report to the Ballaton Group*. Winnipeg: IISD.

Ekins, P., & Simons, S. (2001). Estimating sustainability gaps: Method and preliminary applications for the UK and Netherlands. *Ecological Economics, 37*, 5–22.

Herendeen, R. A. (1998). *Ecological numeracy. Quantitative analysis of environmental issues*. New York: Wiley.

Hertin, J., Berkhout, F., Moll, S., & Schepelmann, P. (2001). *Indicators for monitoring integration of environment and sustainable development in enterprise policy*. Sussex: Science and Technology Policy Research, University of Sussex, UK.

Murray, C. J. L., & Lopez, A. D. (Eds.). (1996). *The global burden of diseases: A comprehensive assessment of mortality and disability from diseases, injuries and risk factors in 1990 and projected to 2020*. Cambridge, MA: Harvard School of Public Health.

Schmidt-Bleek, F. (1993). MIPS –A universal ecological measure? *Fresenius Environmental Bulletin, 2*(6), 306–311.

Schmidt-Bleek, F. (1998). *Das MIPS Konzept. Weniger Naturverbrauch – mehr Lebensqualität durch Faktor 10*. München: Droemer-Knaur.

Sol, V. M., Lammers, P. E. M., Aiking, H., De Boer, J., & Feenstra, J. F. (1995). Integrated environmental indexes for application in land use zoning. *Environmental Management, 19*(3), 457–467.

Ten Brink, B. J. E., Hosper, S. H., & Colijn, F. (1991). A quantitative method for the description and assessment of ecosystems: The AMOEBA approach. *Marine Pollution Bulletin, 23*, 265–270.

Wackernagel, M., & Rees, W. (1996). *Our ecological footprint*. Gabriola Island, B.C. Canada, New Society Publishers.

WWF (2006). *Living planet report 2006*. Gland: WWF International.

Chapter 17
Environmental Modelling

H. J. M. (Bert) de Vries

Contents

17.1	**Introduction and Scope**	345
17.2	**What Is a Model?**	345
17.3	**Science in a Complex World**	348
17.3.1	What Is Complexity?	348
17.3.2	Dealing with Complexity	350
17.3.3	Organizing Concepts	352
17.4	**Environmental Modelling**	353
17.4.1	Introduction	353
17.4.2	Resource Accounting	354
17.4.3	Differential-Integral Equations	354
17.4.4	Dynamic Simulation	360
17.4.5	Interactive Simulation and Policy Exercises	370
17.5	**Concluding Remarks**	370
References		371

17.1 Introduction and Scope

In the previous chapter, the concept of integration was introduced from a broad perspective. This chapter focuses on one of the major tools for understanding environmental system behaviour: models. It is not my intention to give a thorough and complete overview, but rather a feel for how models can be and are used, and some recent developments. I first discuss what a model is. Next, to suggest a model classification, the notion of complexity is discussed from an epistemological point of view. This is followed by an overview of some important developments in dynamic environmental modelling using illustrative examples and focusing on population-resource-environment models. In this manner, my aim is to prepare the way for the next chapter on integrated assessment models and their use.

17.2 What Is a Model?

Human knowledge consists of experiences encoded in a physical or mental representation (image, map). In a stricter sense, modelling is the process of coding observations and measurements into a formal system. The representation itself is a model (Fig. 17.1). Modelling is creating a relation between a 'natural' and a 'formal' system. To some extent, 'natural' (real-world) systems can be identified by means of 'observables', that is, qualities and quantities that can be observed and measured in a more or less controlled ('experimental') way. The social and 'interior' world can also be part of such a 'natural' system and can be investigated using participative and introspective methods (Wilber 1998). The 'formal' system can be any set of mental elements (objects, symbols, etc.) and connecting relations and rules that permit the generation of statements – as in language or logic. It is a set of conventions, and, as such, is solely a construct of the human mind. Its properties are determined by the definitions of its elements and the relations and rules between them. The rules can be used to generate new statements which then can be decoded in an attempt to explain and predict the 'natural' system. This is why they are called rules of inference (Fig. 17.1a).

Scientific method is associated with the 'knowledge spiral' of controlled experiments; the generation of laws and theories; and preparatory investigation leading to new experiments. In the first instance, observations and measurements gained from controlled experiments

H. J. M. (Bert) de Vries (✉)
Copernicus Institute Utrecht University, The Netherlands
e-mail: bert.devries@pbl.nl

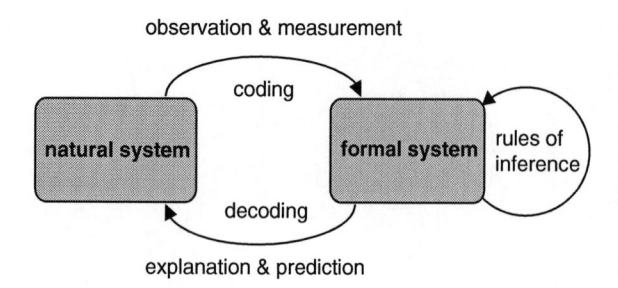

Fig. 17.1a Modelling as an iterative and interactive activity

generate experimental laws – an inductive process guided by pre-scientific intuitions and notions – and yield largely descriptive statements. At the next stage, the rules of the formal system become more important and are developed and used to generate theories – a deductive process during which the link with observations and measurements becomes less direct and the level of abstraction increases. In fact, the simultaneous development of experiments, measuring devices – which can be considered to be natural reference systems – and the formal systems of mathematics are at the root of the scientific revolution of the last couple of centuries. Here, I will use the word 'model' rather loosely to describe any formal representation of observations and measurements, ranging from empirical laws to universal theories.

This introduction to the philosophy of science is, of necessity, brief and incomplete, but to illustrate the modelling process in the context of scientific enquiry a simple example can be used. Let us observe an amount of gas in a cylinder. In classical thermodynamics, a description is provided on the basis of the following macroscopic observables: temperature (T), pressure (P) and volume (V). To quantify these observables, separate measuring devices are required, that is, reference systems (a thermometer, a manometer and a centimetre). Series of experiments in the 17th to 19th century carried out by Boyle, Gay-Lussac and other scientists led to the discovery of an experimental law: $PV = nRT$ in which n represents the number of moles, and R the gas constant. Since then, the observed data have been refined and so have the mathematical equations. For instance, Van der Waals proposed replacing V by $V - b$ to improve the descriptive adequateness and to correct for the non-zero volume of the gas particles. There are now a number of more advanced equations that describe the observations even better and across a larger domain of variables. However, they are just another formal description of empirical observations and do not add to our understanding. Only with the advent of statistical thermo-dynamics and quantum mechanics could the assumption of homogeneity in the intensive variables T and P be relaxed. Then a new operational model emerged incorporating, rather than invalidating, the previous model. This model used micro-level concepts, but with the same formal system of mathematics.

The natural sciences have benefited enormously from the use of formal mathematical systems. In combination with experiments, these systems have become the very hallmark of scientific and technical progress. In principle, the modeller has total freedom in choosing the observables in the natural system. In practice, however, this freedom is severely constrained by the measuring systems available. To give an example, the deterioration of the ozone layer could only be assessed after adequate measuring methods had been established. Much of the present scientific knowledge has been made available by extending the realm of measuring methods in various directions with the help of microscopes and telescopes, echoscopes, electromagnetic devices, isotope and pollen analysis, etc. – the reference systems referred to above.

The elements in a formal system can also, in principle, be chosen freely. In the natural sciences, a large set of formal systems have emanated from both theoretical and applied mathematics. Differential-integral equations, optimization algorithms and other techniques have become essential items in the scientific 'toolbox'. The mathematics is often quite advanced, and, for more complex problems, it does not yield a solution. In turn, this forces the researcher to design a simplified description or to use simulation. I will return to this observation. On the one hand, a well-developed set of formal systems (methods) is a condition for scientific progress, but, on the other hand, it may also become a prison – recall the dictum "If the only tool one has is a hammer, one treats everything like a nail". Another disadvantage is that the use of simulation tools can also lead to an erosion of modelling discipline and rigour, causing model calibration and validation to become a more tedious, if not impossible, venture.

Two different formal systems can describe sometimes one and the same natural system or, the reverse: two different natural systems can be encoded in the same formal system (Rosen 1985). This happens when some essential, though more abstract, features of two natural or formal systems are comparable – a homo- or isomorphism, mathematically speaking. In the evolution from empirical laws to universal theories, scientists construct ever more encompassing models. As long as there is no 'unified theory', two different descriptions may be accepted as complementary, as is the case with the wave and the particle description of light (Fig. 17.1b). Conversely, two different natural systems can be described with a single model. These systems are said to be analogues or, if the isomorphism is rather loose, metaphors (Fig. 17.1c). Such 'borrowing' of a formal system from, say, physics to describe observations in biology, for example, has led to fruitful heuristics, temporarily, at least. Famous examples are Hartley's hypothesis that the human heart works like a mechanical pump and LaPlace's identification of the planetary system with a clock. At the core of systems science is the recognition and use of models to describe different real-world systems at higher levels of abstraction.

Classical natural science owes its success to confining its domain of investigation to controlled experiments. This method has proved to be very successful for certain systems that are measurable at particular time and space scales – among them the artificial technical constructs ('machines') that are the very outcome of these scientific endeavours. The resulting 'strong' knowledge (Groenewold 1981) is based on extensive formal models that are thoroughly validated within their domain of observation and/or control.

Such a reduction to a description with a few observables that are then manipulated in controlled experiments is impossible or inadequate in many sciences, including most areas of environmental science. It is impossible to conduct on a large-scale controlled experiments on ecosystems or the atmosphere, on animals, on human social and psychological behaviour, or with government policies in an economy. Any attempt to do this usually generates highly restricted statements, as there are so many system variables and interdependencies that are still unknown. Models in biology, geology, meteorology and ecology, and even more so in archaeology, geography, economy, psychology and sociology are usually crude approximations of real-world events and processes. Crucial constraining assumptions of the natural sciences, such as the assumption of equilibrium and linearity, are invalid and inadequate. Scientific knowledge in these domains is 'weak' in the sense that the empirical 'laws' allow for divergent interpretations, often taking what is observable for what is relevant, and tacitly introducing disciplinary and methodological bias and values. The well-known *ceteris paribus* (everything else the same) in economic science is a polite acknowledgement of this situation.

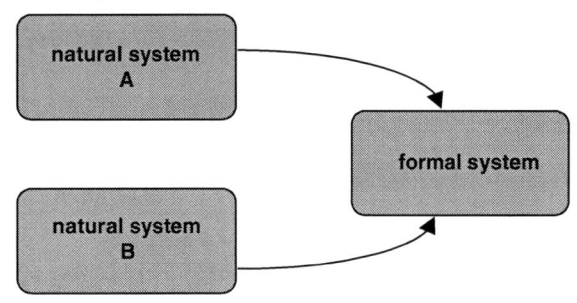

Of course, there is continuous progress in these 'weak' sciences too. One important avenue has been, and still is, the use of the above-mentioned analogues and metaphors – even though they sometimes also reflect misunderstanding and ignorance. Examples of a complex system that is understood in terms of [a model of] a simpler, better understood, system are the human brain as a computer; the human psyche as a steam engine; economic processes as analogues of chemical and mechanical, thermodynamical and chemical equilibria; and the earth as an organism (Gaia). Another is the fundamental equation of price–quantity equilibria in neo-classical economics: it is the formalisation of the quasi-static change of classical thermodynamics by describing physico-chemical systems as moving from one equilibrium to the next, irrespective of the path (de Vries 1989). Other recent examples are the use of biology in evolutionary

Fig. 17.1c Analogy: two natural systems represented by the same formal system (model). Usually, one system is relatively simple and well understood (system B), the other is relatively complex and not well understood (system A)

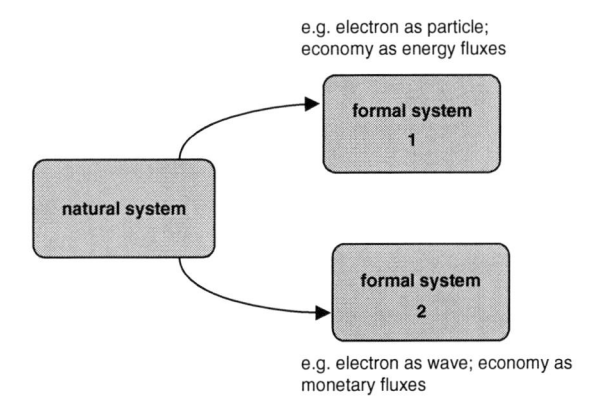

Fig. 17.1b Complementarity: two formal systems (models) represent the same natural system. Complementarity is often resolved as it becomes clear that it depends on observational filters, context and purpose (see interdisciplinarity)

economics (Boschma et al. 2002) and of immunology and linguistics in institutional economics (Janssen et al. 2002). Seen in this light, contemporary advances in economic science are the equivalent of the development of statistical mechanics as a microscopic theory of thermodynamics in late 19th and early 20th centuries.

It is important to realize that the evolution of scientific knowledge is itself a historical process. 'Scientific rationality' has become the hallmark of 'modernity'. However, as science enters into more complex fields of study, this form of rationality, rooted in the philosophy of Descartes and Newton and with a strong emphasis on 17th and 18th century natural science, can no longer claim universality. This insight is also at the root of post-modernist philosophy. For instance, Toulmin (1991) argues that its success has to be seen as a reflection of a desperate search for certainty during the chaos of the Thirty Year's War in the early 17th century. It caused the verbal word to be replaced by the written word, particular and local experience by universal and general laws, and contextual time by timelessness. It sought to provide, in the image of the solar system, legitimacy to a social order and a variety of forms of inequality and repression. By the second half of the 20th century, the shortcomings of this reductionism had become evident. With the growing interdependency of human society, ecology and biology increasingly provide the metaphors for a world which needs to recognise diversity and context.

In a similar vein, the notion of post-normal science has emerged (Funtowicz and Ravetz 1990) and the idea of distinctly different forms of rationality has been proposed (Thompson 1993). To support the rationality of the worldview to which one adheres, there is a tendency to ignore or politicize uncertainties and to fail to recognise the extent of our ignorance (Morecroft and Sterman 1992; Söderbaum 1999). Hence, different forms of knowledge, information gaps and ignorance will be dealt with, and, especially with regard to [environmental] policy-related research, the increased importance of notions of information quality and uncertainty management (van der Sluijs 1997). These developments also clearly reflect the ever-deeper penetration of science and technology into individual and collective life.[1]

This chapter looks at environmental modelling in an integrated population-resource-environmental framework. Environmental science is about complex issues: long-term, across several spatial and temporal scales, and including physico-chemical, biological, ecological and socio-economic aspects. It should, and does, involve many different scientific disciplines, some of them delivering 'strong' and others 'weak' knowledge (Müller 2003). For this reason, I consider the above distinction between 'strong' and 'weak' knowledge to be relevant in this chapter. Below, I will present concrete illustrative examples, but first, I will delve a bit deeper into the notion of complexity.

17.3 Science in a Complex World

17.3.1 What Is Complexity?

The increasing awareness of the inadequateness of prevailing [population-resource-environment] models coincides with new ideas about, and methods for dealing with complex systems. The notion of complexity is a difficult one, and an extensive discussion is beyond the scope of this text (see e.g. Lewin 1992; Allen 1994; Holland 1995). What is complexity? The everyday meaning is that a certain event, situation, machinery or process is considered to be difficult to understand. However, a car mechanic's judgment of the complexity of a car may be quite different from a layman's judgment. It is tempting to say: "Complexity is in the eye of the beholder". In other words: saying that something is complex gives it a contextual characteristic. It introduces the (human) observer. Dimensions of complexity such as identity and connection are not objective properties: they depend on what is distinguished by the observer. This has led Simon (1969: 3) to argue that *"The central task of natural science is to make the wonderful commonplace: to show that complexity, correctly viewed, is only a mask for simplicity; to find pattern hidden in apparent chaos"*.

This aspect is also evident in the ambiguity of its value-laden use. For instance, in constructing an apparatus or a set of procedural rules, low complexity is often appreciated, and, if it satisfies the objectives for which it is designed, it becomes associated with elegance and effectiveness. The word 'complicated' has acquired a pejorative connotation, precisely for this reason. Another illustration of the subjective aspect is that most people find it much easier to grasp systems on the social scale of the family or the spatial scale of the neighbourhood than the complex physical models of far simpler molecular structures. In other words, our intuitive acquaintance with everyday life events, as 'players', contrasts with our

---

[1] See Müller (2003) and Söderbaum (1999), among others, for a recent discussion of post-normal science, mode-2 science and the like, in the context of ecological economics.

awkwardness, as 'spectators', in the world of abstract mathematical models. This brings us to the question of what it is that determines a system's complexity: Is it its evolutionary or construction history, its behaviour over time, its openness to experimentation and use, or its diversity in terms of participants' descriptions and behaviour?

Obviously, complexity is a complex concept itself. It is a property of a real-world system which manifests itself in the inability of any one formalism to capture all its properties adequately. Complexity can be characterized by a lack of symmetry and it has two determining dimensions: distinction and connection. Neither of these are objective properties: they depend on what is perceived by the observer. It can be argued that one measure of complexity is the number of valid descriptions that observers can give to a system. In the narrow zone between order and chaos, where the highest information content is situated, complex systems may develop self-organizing patterns and thus exhibit surprising behaviour. In addition, during the evolution of these systems, emergent properties – those that cannot be defined or explained from the properties of their parts or their antecedent conditions – may also be acquired. 'Emergence', here, is an acknowledgement of failure.

Complexity can be framed in terms of systems, in other words, in relation to a set of material or non-material elements, interrelationships and procedures. Complexity is a feature of open systems consisting of elements that have diverse degrees of freedom and mutual interactions that interact with their environment in many different ways. A key feature is heterogeneity in time and space[2] – complex structures are positioned on a thin line between frozen order and chaotic anarchy. A complexity dimension can be associated with change along three interrelated axes:

- The [inter]action span in terms of the exchange of energy and matter – and thus information – with the environment of the system elements
- The[inter]connectedness, both in number and nature, of the system elements and
- The capacity of the system elements to make representations (models) of the past and future (memory, anticipation)

Low-complexity events, behaviour and structures are those of controlled experiments in physics and chemistry. Open physico-chemical systems and even more so the building blocks of biological evolution are of higher complexity. At this level, empirical observations can be made about physical reservoirs and flows of the biosphere and technosphere at various spatial and temporal levels. Model variables often have an explicit and formal correspondence with observable real-world phenomena and certain 'laws of nature' hold, such as the conservation of mass and energy. This becomes increasingly difficult with systems of higher complexity such as an ecosystem and its inhabitants with their genetic codes and intricate web of connections, which give rise to overall system behaviour that can only to be understood at the level of the system as a whole. Universally valid laws such as those of physics and chemistry have yet to be discovered.

As far as we know now, humans are at the highest levels of complexity in terms of individual and collective diversity of material and mental manifestations and interconnections. Here, behavioural and informational structures dominate, and these govern human impact on the underlying physical environment. The empirical basis is weaker and controversies abound, given the diversity and variety of behaviour that is, moreover, constantly evolving – a process of which these theories are themselves a part. What makes such socio-natural systems so complex is precisely the human actor who engages in processes of information-gathering, interpretation and processing. There is an increasing awareness of the complexity of these processes and their reduction to a few rules. For instance, rational optimizing behaviour seems unwarranted, as the notion of 'bounded rationality' indicates (see e.g. Morecroft and Sterman 1992).

Such an 'ascending order of complexity' is not a statement about reality *per se*, but about the ways in which humans can acquire [scientific] knowledge and understanding about [parts of] the world through interacting with it.[3] Figure 17.2 is an attempt to sketch this idea graphically. The aggregate complexity of a model

---

[2] This does not only refer to geographical space but also to 'behaviour space' and 'strategy space'.

[3] One might associate the levels of complexity dimension with an ascending order of intentionality, and consciousness (Vries and Goudsblom 2002). Whether this is read as an unfolding or push from below – as in the materialist-empiricist orientation – or as a teleological drift upwards – as in the metaphysical spiritual orientations – is as yet an open question for each individual human, as well as for human society at large. In fact, the existence of such an 'axis' appears to be part of 'perennial wisdom'. Teilhard de Chardin advanced complexity and interiorization as a third dimension alongside the infinitely large and the infinitesimally small; his concept of the noösphere is part of the process of unfolding. Aurobindo (1998) perceived an evolution towards an ever-higher consciousness; Jantsch (1980: 29) put forward the same vision in more scientific terminology. Many authors suggest a similar hierarchy from a more worldly orientation: a spectrum from 'low' to 'high' human needs and values, from worldly means to spiritual goals (see e.g. Harman 1993).

Fig. 17.2 Positioning sciences along the aggregate complexity dimension

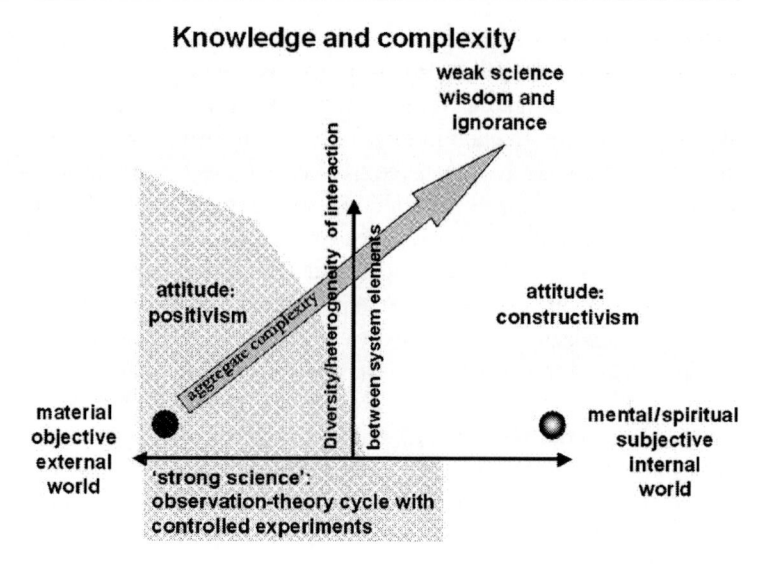

increases if the system under investigation changes from low-diversity, low-interaction systems in controlled experiments in classical mechanics to follow the direction of high-diversity, high-interaction systems on which economic and social sciences focus.[4] Philosophically, the domain of strong knowledge (lower left) tends to be associated with positivism; the domain of weak science (upper right) – and of literature, art and religion – will usually be considered from a constructivist point of view. This representation, which is actually a 'meta-model', indicates that the role of scientific reductionism is one in which more complex systems are simplified in order to acquire meaningful and relevant insights.[5]

As has been stated before, such a procedure has often been very helpful and may clarify key features of the system under investigation. At the same time, it has led to some persistently reduced representations and misunderstandings of real-world phenomena, particularly social and economic ones, which sometimes have an ideological function. Another consequence has been that the 'modern' scientific world view has increasingly separated the exterior world of science from the interior world of art and religion – the upper right parts in Fig. 17.2. As retrogressive movements towards pre-modern ways of life do not offer a way out, new avenues out of this painful situation will have to be found.

Figure 17.2 can be used as a framework for a model typology. Along the complexity dimension, a series of rather specific scientific methods and disciplines have evolved, all of them rooted in mathematics (see Table 17.1). The list, more or less mirroring the development of 'modern science', does not aim at completeness. In many ways, as previously indicated, the items are related to each other through the use of isomorphisms. The relevance of this list is to make us aware that the use of models in the quest for gaining a better understanding of [the evolution of] socio-natural systems will be faced with the characteristic ways of looking at reality implied in the choice of method and discipline. It can also help to bridge natural and social sciences in a constructive way.

17.3.2 Dealing with Complexity

Although this chapter focuses on environmental modelling, I will briefly discuss two broader, but related topics: the role of models in [environmental] management and planning and reason why 'meta-models' are organized. Models are constructed for a purpose, an objective. Knowledge to gain understanding increasingly becomes knowledge to gain control, particularly when it comes to human constructs and their impact on natural systems. The past decades have shown a

---

[4] This is distinct from algorithmic or deterministic complexity (e.g. Manson 2001).

[5] An interesting hypothesis in this context is that reducing the complexity of the environment is what civilization is all about (van der Leeuw 1998; de Vries and Goudsblom 2002).

Table 17.1 Methods and scientific disciplines along the complexity dimension

Methodology	[Environmental] modelling approaches and applications
Integral-differential equations	Physical and engineering sciences; pollutant dispersion
Optimization and control theory, linear/dynamic programming	Physical and engineering sciences; resource depletion; least-cost abatement strategies
Systems science, systems dynamics, Cybernetics	Resource and environmental economics; and Management
Network (graph, neural) theory	Economic input-output theory; social and information networks
Game theory	Common property resource management; social dilemmas
Cellular automata	Geography; land-use and land-cover dynamics
Multi-Agent Simulation (MAS)	Systems science and ecology; resource and ecosystem management
Complex Adaptive Systems (CAS)	Complexity science and ecology; archaeology; socio-natural system evolution
Scenario analysis	Systems science; connecting qualitative story-telling and quantitative modelling
Simulation gaming and policy exercises	Resource management

	Consensus on values	
	no	yes
Consensus on knowledge: no	unstructured problem: science as **problem recognizer**	moderately structured problem: science as **advocate**
Consensus on knowledge: yes	badly structured problem: science as **mediator**	structured problem: science as **problem solver**

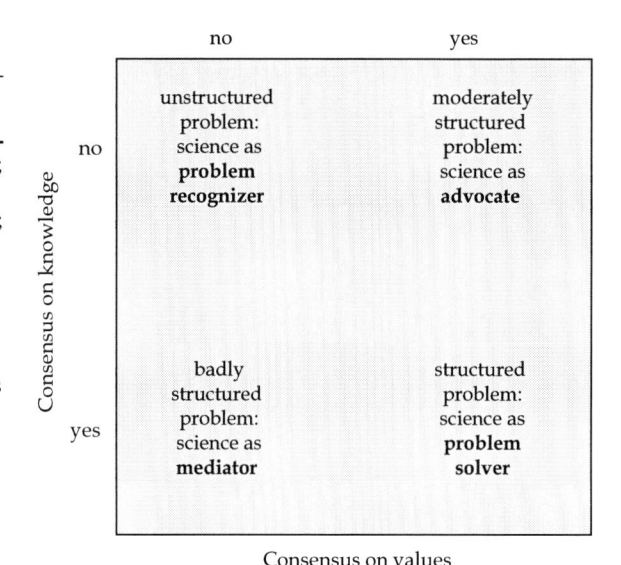

Fig. 17.3 The role of science in situations of low or high consensus on knowledge and values
Source: After Hisschemöller and Gupta (1999)

paradigm shift in this respect, which is also reflected in Table 17.1. Early, rather extreme, ideas about 'scientific management' applied methods from the physical and engineering sciences. With the acknowledgement of the complexity of socio-natural systems, there has been a shift to 'softer' forms of impact/control, such as adaptive management. In these more complex situations, science has to acknowledge uncertainty and ignorance. Hence management and planning, whether individual or collective (firms, governments), rely on value-laden interpretations of the situation and the scientific debate will be full of controversy and advocacy.

Figure 17.3 suggests a framework for discussion (Hisschemöller and Gupta 1999; de Vries 2006, see also Section 14.4). If there is a strong consensus on knowledge, especially on values, science can be invoked as a problem-solver. Many environmental models have served this purpose in the past – think of air pollutant dispersion or process models for water cleaning. In management science, this is called the rational zone. The buzzwords are planning, budgets, defined outcomes, and goal-seeking control. If no strong knowledge is [yet] available, the lack of knowledge consensus gives room for alternative models and scientists become advocates in the decision-making arena. Some elements of the debate on human-induced climate change are an example; for instance, the mechanisms of cloud formation or the technical potential of renewable energy sources. The expectation is that, with more research, such controversies will be solved. However, as long as the situation persists, individual scientists can play a major role and management science invites visionaries and missionaries to do so.

If there is consensus on knowledge, but none on values, the situation is one of mediation, offering rule systems and evaluation methods. 'Science' helps to find 'rational' ways of managing complexity and uncertainty in the form of decision support and value explication. This is one way of viewing methods such as resource accounting, cost-benefit analysis and [probabilistic] risk analysis – they are basically evaluation procedures and tools, rather than models in the above sense (see e.g. Pearce 2002). The various ways to bring about a reduction in greenhouse-gas emissions (den Elzen 2002; www.mnp.nl/fair/) is an example. For managers, this zone is political. Finally, there may be lack of consensus in both knowledge and values; from a management and planning perspective, the chaos zone. The primary role of science here is problem recognition. It appears that major global change issues such as the threats to, and consequences of changes in biodiversity and climate are in this quadrant of Fig. 17.3 (see e.g. Fiddaman 2002). This brief discussion shows that where there are unstructured problems related to complex systems, our knowledge will always be approximate and there will always be competing explanations of real-world observations which can be used to support one's behaviour and beliefs, and one's values and preferences.

17.3.3 Organizing Concepts

In dealing with ill-structured problems in complex socio-natural systems, there is a need for organizing concepts that frame the issues and express and communicate certain key ideas and values. I will refer to such mental maps as 'meta-models'. Although they express aspects of underlying scientific models, they are not models in the rather strict sense described above. The notion of ordered 'layers' of reality, as suggested in Fig. 17.2, is found in several approaches dealing with environmental and sustainability issues. The underlying idea is that physical 'nature' is the material substrate – or life-support system – which is the basis upon which derived demands for food, water, energy, etc. are based. The economic and infrastructure subsystems that relate the substrate to the higher levels of individual and collective behavioural and institutional structures are at this intermediate level. An example of such an imposed order is shown in Fig. 17.4.

Not all frameworks have this verticality. A rather widely used scheme in resource and environmental science and policy is the Driving forces-Pressure-State-Impact-Response (DPSIR) Chain or Cycle (see Chapters 1 and 18, Section 2.3), which helps to organize environmental issues within the broader context of causes and effects. It has been widely used to make lists of sustainable development indicators (CSD 2001). A problem with this scheme is the separation of natural and social systems and the rather mechanistic view taken on both of them.

A more sophisticated and integrative 'template' for structuring change processes is the *ecological cycle* proposed by Holling in the context of adaptive management and applied to socio-natural systems by Thompson (in: Gunderson et al. 1995; de Vries and Goudsblom 2002). In ecology, the process of succession ensures that an initially unstructured state – one huge niche filled with anarchic, opportunistic and competitive organisms (the r-strategists) – steadily becomes transformed into a climax community with more structured and stratified arrangement of diversified niches for the species that occupy them (the K-strategists) (Fig. 17.5). Social science has used this model as an analogue to argue that human societies experience a one-way transition from state A to state B – from, say, a simple (tribal) group to a complex (state) society.

Holling has argued that the climax community eventually makes itself so complex that it undermines its own stability, so that the inevitable consequence is

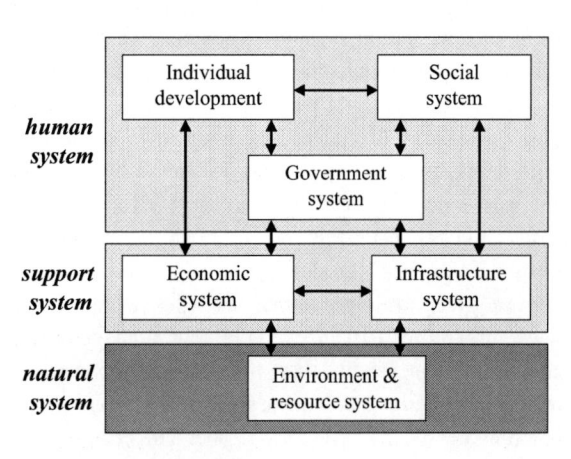

Fig. 17.4 Ordering scheme
Source: Adapted from Bossel (1998, 101)

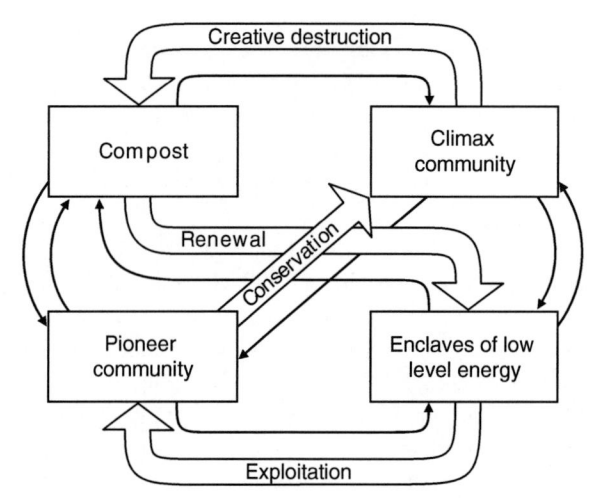

Fig. 17.5 The ecological cycle
Source: After Gunderson et al. (1995)

collapse.[6] At this catastrophic moment, all the energy that is tied up in all the niches and interdependencies of the climax community is released. Holling, well aware of the parallel with Schumpeter's (1942) theory of economic maturity, collapse and renewal, calls this transition from climax community to compost: 'creative destruction'. This is not the end of the road. The challenge is to harness the loose energy before soil leaching, for instance, causes it all to disappear. This is where the unspecialized and co-operative fence-builders – mostly micro-organisms – come into their own, gathering up the loose energy into small bundles that have no connections, as yet, with one another. Even this is not the end of the road, because the stage is now set for the unspecialized, but opportunistic, fast-breeding and highly competitive '*r*-selected' species. In exploiting and colonizing this environment, these generalized exploiters inevitably begin to push it into a rather more patterned and interconnected state, thereby making it less conducive to their way of doing things and more suited to the sort of energy-conserving strategies that characterize the '*K*-selected species'. These are the specialized, slower-breeding and often symbiotic plants and creatures that are the vanguard of the complex and increasingly ordered whole that constitute the climax community. This scheme was also suggested for economic and social systems, using divergent cultural perspectives as agents of change.

A recent attempt to structure complex phenomena at a high level of abstraction is the concept of *transitions*. A transition is defined as a series of simultaneous interrelated system changes. One of the first to be identified is demographic transition. Recently, the focus has been on health transition, comprising a change from environmental to welfare-related determinants of morbidity and mortality, and on technological transitions (Rotmans and de Vries 1997; Geels 2002). Sustainable development paths can be considered as transition trajectories in which society switches to high-efficiency, low-emission technological systems and associated economic institutions and social lifestyles. The idea of transitions thus puts more narrow and formal models of processes into a broader, integrated perspective. Let us now take a more down-to-earth perspective.

17.4 Environmental Modelling

17.4.1 Introduction

Numerous environmental models have been constructed in recent decades and quite a few typologies have been proposed (see e.g. Clark and Munn 1986; Wainwright and Mulligan 2004). Mathematically, a major distinction is between static and dynamic, and, within these categories, linear and non-linear. Another aspect is whether the model is purely deterministic or whether it uses stochastic variation or Markov probability chains. As is clear from the previous section, designing and using environmental models often corresponds with the choice of system boundary and the nature of the knowledge to be modelled. Material flows in environmental compartments are, for example, usually modelled in the form of [partial] differential equations, using factors such as cloud formation or rainfall as exogenous stochastic variables. Human behaviour, on the other hand, is often incorporated in black-box relationships derived from the statistical analysis of historical data, which, for instance, is the prevailing approach in econometrics. Simulation allows for a more explicit formulation of behavioural rules, in which case the reproduction of historical data can provide validation (van der Leeuw 1998). One of the forerunners of this approach was system dynamics. Finally, the choices of time scale (from short-term [seconds] to long-term [centuries]) and spatial scale (from local to global) matter and influence the type of modelling and method chosen.

Given the definition of environmental science (see Chapter 1), environmental modelling is the construction of a representation (model) of the interaction of human activities with their physical, biological and social environment. In line with the previous discussion on aggregate complexity, the following three classes of environmental modelling reflecting the arrow in Fig. 17.2 and the items in Table 17.1 can now be discussed:

---

[6] A related 'law' has been proposed by Tainter (2000) on the basis of investigations of the rise and decline of past civilizations. It states that societies are always faced with diminishing returns on complexity. In other words, if problems are addressed with increasing complexity, they ultimately fail to be solved.

- *Dynamic analytical*: differential-integral equations (Section 17.4.3)
- *Dynamic simulation*: systems dynamics and multi-agent modelling (Section 17.4.4) and
- *Interactive simulation*: simulation-gaming, policy exercises (Section 17.4.5)

They do not appear to be strictly separated, but are more of a continuum, with no claim to completeness. Although resource accounting methods are considered here as evaluative tools rather than models, a brief discussion of these is given first.

17.4.2 Resource Accounting

The first step to take here, as in most scientific endeavour, is to describe and organize one's observations. This process usually involves the construction of classes and concepts to relate the empirical observations to higher-level, (meta-) organization. Typical examples are the categorization of chemical elements into the periodic system; the classification of plants and animals; the construction of maps of human appropriation of biological fluxes (Fischer-Kowalski and Haberl 1998) and of ecological footprints (Vuuren and Smeets 2000; Wackernagel 2002). A well-known and widely used accounting framework is the input-output (I/O) analysis as developed in economics (see Chapter 11). This combines an empirical basis with mathematical tools and allows for detailed assessments of the indirect environmental consequences of production and consumption activities (Noorman and Schoot Uiterkamp 1998). The input-output framework can be connected to a neoclassical growth model; this is one way of exploring the consequences of exogenously introduced changes in (environmental) technology via the input-output coefficients (Duchin and Lange 1994). Such a dynamic input-output model is a promising step towards the integration of economic- and energy-environment models (Wilting et al. 2001).

For well-defined systems, the Resource Analysis, Energy Analysis and Life Cycle Assessment (LCA) methods describe the material fluxes for well-defined processes and products, taking into account the various, actual and anticipated, technologies. The results are presented in the form of indicators relevant for the problem at hand: for example, the amount of primary energy used for one unit of product X, or the greenhouse-gas emissions associated with a unit of activity Y. Such indicators are useful for comparative evaluation. Most of these resource accounting methods are static, but, of course, one can apply them in a dynamic context. An example of a Life Cycle Assessment is given in Appendix A of this chapter.

17.4.3 Differential-Integral Equations[7]

In the search to gain an understanding of human-environment interactions, many models of dynamic processes have been made. Initially, classical mechanics, thermodynamics and the engineering sciences very effectively provided the analogues and methods. A system with a well-defined boundary and a given initial state can be characterized by a set of state variables:

$$X(t)_1, X(t)_2, ... X(t)_n \text{ for time } t > t_0 \quad (17.1a)$$

which change according to a set of ordinary (simultaneous) differential equations:

$$\frac{dX(t)_i}{dt} = F_i(X(t)_1, X(t)_2 ... X(t)_n; G(t); t) \text{ for } i = 1 ... n \quad (17.1b)$$

with F some functional and G a set of constitutive and environmental parameters. For $dG/dt = 0$ this represents an autonomous system with one and only one trajectory through any point in phase space. Examples of state variables are the concentration of an element in a medium (soil, water), the population size of a biological species, or the size of a capital stock or a labour force.

Two ubiquitous examples of a linear models with a single state variable X are *exponential growth* (the solution for $dX/dt = aX$ is $X(t) = X_{at\ t=0}\ e^{at}$), and *exponential decline* (the solution for $dX/dt = a(A-X)$ is $X(t) = A - X_{at\ t=0}\ e^{-at}$) (Fig. 17.6). The former is a positive feedback that is well known from population growth,

---

[7] I use consistently differential equations, although most of what is written in this paragraph also holds for difference equations, which is the correct expression in the case of simulation models.

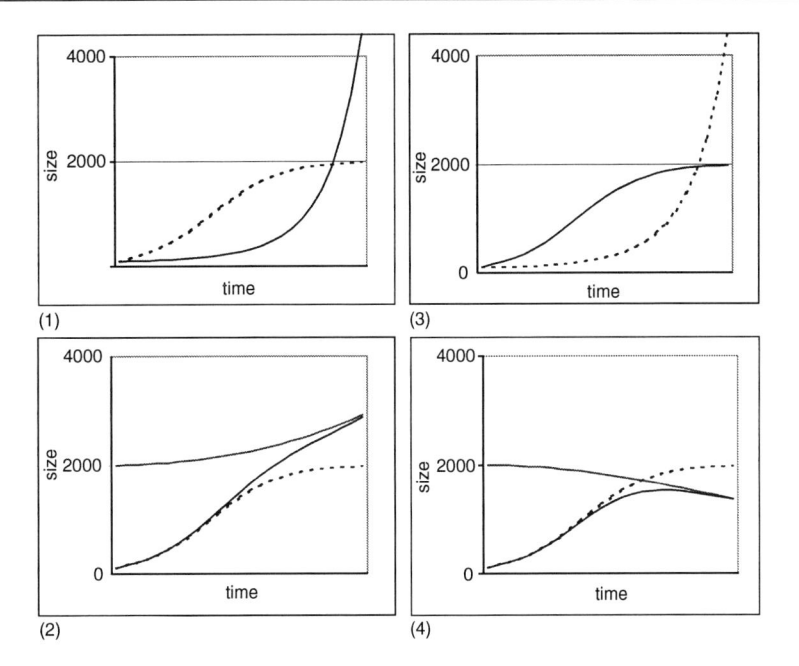

Fig. 17.6 Four ways of growth into a finite space: (1) exponential growth, (2) logistic growth (without overshoot-and-decline), (3) growth with upward pressure on the carrying-capacity and (4) growth with downward erosion of the carrying-capacity

and from capital- and interest-accumulation processes. The latter is a negative feedback; in other words, a balancing or stabilizing process where the state variable X tends towards an equilibrium state (or attractor). Textbook examples from engineering and environmental science are the flow of heat and substances across a temperature, pressure or chemical-concentration gradient. Other well-known examples are radioactive decay, the functioning of a host of technical devices of which the thermostat is the best known, and the equilibration of demand and supply in markets.

A linear model with two state variables X_1 and X_2 is often used to describe interaction. Let X and Y indicate two interacting population sizes:

$$\frac{dX_1}{dt} = F_1[X_1, X_2] = a_{11}X_1 + a_{12}X_2 + c_1$$

$$\frac{dX_2}{dt} = F_2[X_1, X_2] = a_{21}X_1 + a_{22}X_2 + c_2 \quad (17.1c)$$

This model arose from an early explanation of the arms race leading up to World War I, in which the state variables were the actual and perceived military budgets of two hostile countries (Richardson 1991). However, it also illustrates the possible dynamics between, for instance, tourists and pollutant flux or ecosystem degradation (Bossel 1992); and interactions, such as mutual cooperation and competition, parasitism and synergism between plant and/or animal populations (Boulding 1978). In a very special case, the set of (prey-predator) equations can be rewritten as a second-order differential equation, which is the classic model of a harmonic oscillator.

The great advantage of these models is that they benefit from the inference rules (see Fig. 17.1) of classic mathematical theory. One of the most interesting applications, not least for conceptualizing sustainable development, is stability analysis. The so-called Jacobian matrix $\underline{\mathbf{A}}$ with elements $A_{ij} = dF_i/dX_j$ is used to determine system behaviour: by calculating the Tr[ace] $\underline{\mathbf{A}} = a_{11} + b_{22}$ and the Det[erminant] $\underline{\mathbf{A}} = a_{11} a_{22} - a_{21} a_{12}$, the attractor[s] in the model – such as, a[n] [un]stable spiral, a saddle point or a stable oscillation – can be identified (see e.g. Edelstein-Keshet 1988; Bossel 1992) (Box 17.1).

Non-linear differential equations to describe change also have a long past. As linear models are usually far too simple, many more models have been proposed in which the functionals F_i (Equation (17.1b)) are non-linear. One of the best known is the equation to describe population growth in a finite resource environment:

> **Box 17.1** An example: water pollution
>
> An example of an environmental model is the pollution of a lake by some toxic material, for instance from mining or agriculture. Assuming that
>
> The pollutant influx *PolIn* (kg/day) is mixed immediately, resulting in a homogeneous concentration *PolCLake* (kg/m^3) equalling the ratio of total pollutant mass and lake volume.
>
> The pollutant is broken down by biological and chemical processes at a rate given by the clean-up coefficient *PolClUp* (1/day)
>
> Diffusion/sedimentation of pollutant between water and sediment on the lake bottom is governed according to Fick's law by a coefficient *DiffC* (m/day), and proportional to the difference in concentration in water and sediment *PolCLake-PolCSed* (kg/m^3)
>
> the dynamics is governed by a single differential equation:
>
> $$\frac{dPolCLake}{dt} = PolIn/V - PolCLake * PolClUp + \frac{PolCLake - PolCSed}{d} * DiffC \quad (17.2)$$
>
> with V the lake volume and d the lake average water depth.[8] This is a strongly simplified description of how a pollutant concentration in a lake can develop over time as a result of an emission influx.

$$dX/dt = \alpha * X * (K - X) \quad (17.3a)$$

This logistic or Verhulst equation indicates that a population tends to grow exponentially as long as it is far away from the carrying capacity indicated by the constant K (X << K), but that growth will decline to zero when the population approaches the carrying capacity (X ≅ K), (see Fig. 17.6). There are several other models for this kind of S-shaped growth trajectory.

This relatively simple model, which in its discrete form can exhibit chaotic behaviour for certain domains of α, has an interesting application in the field of technological dynamics (de Vries 1989). Long-term observations have shown that, during periods of rise and decline, new products or processes penetrate an economy according to:

$$\frac{d\log[f/(1-f)]}{dt} = k \quad (17.3b)$$

with *f* the market share of the product/process under consideration (Marchetti and Nakicenovic 1979; Grubler 2000). Using *d*log[*f*]/*dt* = (1/*f*)*df*/*dt*, this so-called logistic substitution model can be rewritten as *df*/*dt* = *kf*(1 - *f*) to show the isomorphism with the logistic growth equation.

Stability analysis is much more limited for nonlinear systems, but, under certain conditions, an analytical investigation of equilibrium points, or trajectories, is possible. An example from ecology is the limit cycle occurring in a system of two populations, one prey and one predator. One of the models proposed is a more complex model for prey-predator dynamics:

$$\frac{dN}{dt} = N * r * (1 - \frac{N}{K_d} - \kappa * \frac{P}{N+D}) \quad (17.3c)$$

$$\frac{dP}{dt} = P * s * (1 - \frac{hP}{N}) \quad (17.3d)$$

Here, *P* is the predator population e.g. fish, and *N* is the prey e.g. daphnia in a lake. The parameters *s* and *r* are the respective natural growth rates of the predator and prey populations; κ is the natural death rate of the prey population; K_d is the [fixed] prey-carrying capacity; 1/(N + D) is the catch effectiveness of the predator; and h/N is the food shortage of the predator. It can be shown that this system will show stable oscillatory behaviour under certain conditions, a so-called limit cycle. An interesting extension is the addition of one more prey population (Azar et al. 1995; Box 17.2). It can be shown that a one predator–two prey model such as this exhibits various stability domains, including periodic oscillations and chaos. Mathematical analysis also shows that, by harvesting the predator, its population is

---

[8] V and d are related through the surface between the water compartment and the soil compartment S: V = S * d. Dimension analysis is always an important way of checking the correctness of an equation.

Box 17.2 Modelling prey-predator dynamics

To model the growth of a population in an environment with a finite resource such as food, one often uses logistic growth into a 'space' which is characterized by a constant K, the so-called carrying capacity. An equation describing the growth of the population *Pop* can be:

$$\frac{dPop}{dt} = Gr * Pop * \left[1 - \frac{Pop}{D+K}\right] - Dr * Pop \quad (17.4a)$$

For $Dr = 0$ and $D = 0$, this is the well-known logistic Equation (17.3a). For $D \neq 0$, one gets the Michaelis-Menten equation and population growth becomes density dependent.

In many ecological and biological systems, there are more mutually interacting populations. Let us assume that two populations (suffix *s* for prey, suffix *f* for predator) interact according to the following set of equations:

$$\frac{dPop_s}{dt} = Gr_s * Pop_s * \left[1 - \frac{\lambda_s * Pop_s}{D_s + K_s} - \alpha * Pop_f\right] \quad (17.4b)$$

$$\frac{dPop_f}{dt} = Gr_f * Pop_f * \left[1 - \frac{\lambda_f * Pop_f}{D_f + K_f} - \beta * \frac{Pop_f}{Pop_s}\right] \quad (17.4c)$$

The cross-terms represent the species interaction: in the first equation it reflects that more predators increase the prey death rate (prey-per-predator relation), in the second equation that the amount of prey affects the predator death rate. On inspection, it can be seen that both equations return to the logistic growth (1) for $\alpha = 0$ and $\beta = 0$, i.e. when there is no interaction. Each species has its own carrying capacity. However, if α and β are non-zero, the two equations represent a prey-predator dynamic system. It shows for λ_i ($i = s, f$) and for certain parameter conditions that limit cycle behaviour: the state variables oscillate around a stable path attractor.

The graphs in Fig. 17.7 show such behaviour in the state variables Pop_s and Pop_f for various values of D_f and D_s. Clearly, depending on the density-dependence of ecosystem dynamics, the system tends towards

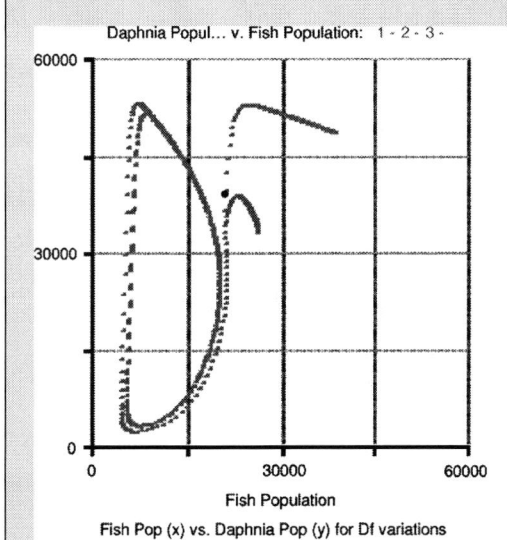

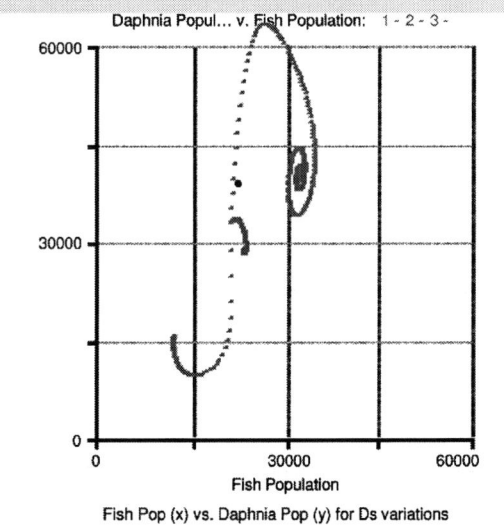

State space diagram (Kf=5000; Df=0, 20000 and 145000)   State space diagram (Ks=25000; Ds=25000, 60000 and 95000)

Fig. 17.7 Evolution in state space of a simple prey-predator system for different niche sizes D_f and D_s

(continued)

Box 17.2 (continued)

quite different equilibria. For instance, if the prey has a small niche (or 'ecological space'), the predator will rapidly over-exploit its food base and populations will oscillate (left graph, left curve). However, when predator expansion is severely constrained, the system will evolve from its initial state towards a higher population equilibrium (right graph, right curve). A well-known example of a real-world system which has exhibited these kinds of dynamics is the Kaibab plateau, which serves as an example of ecosystem behaviour and [mis]management (Roberts et al. 1983).

---

increased, and that the region of stable solutions is much larger for a constant harvesting effort ($H = hP$) than for constant harvest quota ($H = H_q$) (see e.g. Perman et al. 2003).

[Nonlinear] differential equations, such as those discussed in the box, can be quite instructive when exploring dynamic system characteristics, because they provide archetypical templates. Given the 'bounded rationality' mentioned earlier, people – among them, scientists – will tend to use a particular context for framing the issue and choosing method and focus. For instance, the carrying capacity K in Equation (17.3a) is not constant in real-world ecological and economic systems: it is often a function of the very state variable X. If this is the case, then one can either tell the story of human ingenuity ($dK/dX > 0$) or of human-induced erosion ($dK/dX < 0$), as shown in the two lower graphs of Fig. 17.6.

In these applications, there is often the additional aim of controlling the system and this has led to developments such as linear optimization and optimal control theory. Such applications are often borrowed from the engineering sciences, and this has given rise to the use of sophisticated tools in models that are [far] too simple. A famous example is the notion of sustainable yield. Let X indicate some capital stock (natural stock: fish, oil resources; economic stock: production plants, etc.). A general description of its use is then:

$$\frac{dX(t)}{dt} = F_i(X(t);G;t) - H(X) \quad (17.5a)$$

with H the exploitation rate (harvest/exploitation rate). For non-renewable resources, the term $F_i(X(t);G;t)$ is either considered zero or is used to introduce the process of exploration (de Vries et al. 2001). If X is a renewable resource, indicating e.g. tons of fish or biomass, then the simplest model is an extension of Equation (17.3a):

$$\frac{dX}{dt} = rX(K - X) - H \quad (17.5b)$$

with K a [finite] carrying capacity and H a constant. One may define sustainable yield as the maximum value of H for which the resource is still sustained, i.e., $dX/dt = 0$. This is the case for $H = rX(K-X)$ which implies the maximum H at $dH/dX = 0$, that is, at $X = ½K$. The situation is more complex if a more realistic harvesting function is chosen, e.g. $H = aX^aM^b$ with M an indicator of the harvesting effort such as the number of fishing boats.

In analogy with classical mechanics, optimal control theory has been applied in resource economics. Optimal use would imply the maximization of the discounted flow of net benefits:

$$\max \int_0^T e^{-rt}[U(R) - C(X,R,t)]dt \quad (17.6)$$

with $dX/dt = -R$, $X(0) = S$, $X(T) > 0$ and r the rate of time preference, T the planning horizon, $U(R)$ the utility derived from resource usage at a rate of R units per time period, X the amount of resource left in the ground, and $C(X, R, t)$ the cost of making the resource available for use. To benefit from advanced mathematics, the representation needs to be greatly simplified (Dasgupta and Heal 1979; de Vries 1989; Perman et al. 2003) (Box 17.3).

Linear and non-linear differential equations representing resource and environment systems have led to many insights and new research questions. Linear stability analysis has proven to be an important tool if one accepts the assumption of near-equilibrium and linearity for small perturbations. Examples can be found in this book and in many other texts on environmental modelling. Recently, a number of interesting research projects have focused on the non-linear responses of ecosys-

> **Box 17.3** Economic growth
>
> It is interesting to note a widely used isomorphism. The fundamental equation of the neo-classical economic growth model is analogous to that of the renewable resource harvest:
>
> $$\frac{dK}{dt} = I - mK = Y[K, L] - mK - C \qquad (17.7a)$$
>
> with K representing the capital stock, L labour, I the investment rate and m the depreciation rate, using $Y[K, L] = I + C$ with $Y[K, L]$ the production function and C the consumption rate (i.e. the equivalent of the exploitation rate H). Often, a simple Cobb-Douglas production function is used:
>
> $$Y[K, L] = Y_0 * (K / K_0)^a * (L / L_0)^b \qquad (17.7b)$$
>
> (usually with $a + b = 1$). More advanced: $Y[K, L, R] = Y_0 (K/K_0)^a (L/L_0)^b (R/R_0)^c$ and $dS/dt = -R$ with S representing a finite resource stock. Sustainable consumption can be calculated for a given capital stock as output Y minus depreciation mK. With $a + b = 1$ rewriting gives:
>
> $$\frac{dK}{dt} = Y_0 * (L / L_0) * (K / K_0)^a - mK - C \qquad (17.7c)$$
>
> which yields the recipe for maximum economic growth and 'sustainable consumption'.

tems that have been exposed to gradual changes in climate, nutrient loading, habitat fragmentation and biotic exploitation (Gunderson et al. 1995). Such catastrophic shifts have occurred in lakes when the pristine state of clear water and rich submerged vegetation suddenly became less transparent, with reduced levels of algae biomass and animal diversity (Scheffer et al. 2001). Experiments have shown that this happens if the nutrient loading passes a certain threshold so that the pristine state, with its self-stabilizing mechanisms, can only be restored by some kind of 'shock therapy' such as a strong reduction in fish biomass. Stochastic events often play an important role. Clearly, ecosystem management requires an adequate understanding of the complex underlying system. The concept of resilience, introduced by Holling in the context of ecological cycles (Fig. 17.5), is a most useful, but also difficult concept, in this respect.

Often in the form of a simplified or 'minimal' model, differential-integral equation models can be extremely helpful in improving understanding and management. They allow a thorough exploration of the conditions under which even very small disturbances or perturbations may cause multi-stable [eco]systems to exhibit catastrophic shifts. Nevertheless, such models quickly become analytically cumbersome or even intractable, if more real-world phenomena are added. In the next section, I will focus on some methodological innovations that try to overcome some of the shortcomings of the analytical approach. Bear in mind, however, that one should always question when and why it is actually necessary to add real-world detail, recalling Einstein's dictum: "The best explanation is as simple as possible ... but no simpler".

By way of introduction, listed below are three shortcomings in the kinds of model discussed above and ways in which they are being addressed:

1. Resource and environment dynamics exhibit important spatial interactions, so the assumption of system homogeneity in space has to be relaxed. As an aid, partial differential equations, as developed and applied in physics and chemistry, can and have been used. The method of cellular automata, discussed in Section 17.4.4.2, is a more recent approach to connect discrete objects and events. Besides relaxing the assumption of aggregation in space, one has to relax the assumption of average rate in time i.e. of events by using probabilistic methods (Allen 1994).

2. In real-world processes, there are many feedbacks, delays (or time-lags) and other forms of

information- and expectation-related interactions, such as, for instance, anticipatory behaviour (Rosen 1985). One of the approaches by which this aspect can be most explicitly addressed is to use system dynamics. This became known via the World3 Limits to Growth Model (Meadows et al. 1974) and, with the availability of supporting simulation software such as Stella (www.hps-inc.com) and Vensim (www.vensim.com), is being used increasingly. A detailed example is given in Section 17.4.4.1.

3 Whereas the two previous additions would make room for spatial and informational diversity in the models, it still does not allow for actor diversity. It is being acknowledged increasingly that, in order to understand the dynamics of socio-natural systems, for instance, as a basis for giving advice on policy, one should introduce actor heterogeneity – a broader and more diverse interaction of human beings with their physical and social environment. The development of object-oriented programming and multi-agent simulation and associated simulation tools such as CORMAS (http://cormas.cirad.fr) allow major steps forward in this direction. Recent investigations in the field of evolutionary ecology and economics indicate the prospect of major steps forward (Allen 1994). A few models of this type are discussed briefly in Section 17.4.4.3.

It should be noted that the differences between the previous models and those explored in the next section that use new methods, are gradual. For instance, formal approaches like network theory, catastrophe theory and multi-state and cohort/vintage approaches in demography/economy are all applied in both analytical and simulation contexts.

17.4.4 Dynamic Simulation

In essence, simulation is the numerical processing of the set of state equations for a sequence of values of the time variable (t) or of another variable with a finite distance ('time-step') between two subsequent values. How large this distance should be depends on the simulation objective and the available observations. Simulation only became a widespread activity with the advent of relatively fast computers.

Computers made it both possible and interesting to simulate systems, first in physics, chemistry and engineering, and later in management applications, for which no analytical solutions were available. It has led to an abundance of simulation tools and models in mathematics – and, more relevant here, in ecology, economy, geography and the environmental sciences. One important development has been the emergence of systems dynamics, which has deepened and spread the insights from systems science.[9] Rather basic new insights have emerged, such as inherent indeterminism ('chaos') in relatively simple non-linear equations like the logistic model (Equation (17.3a)). Such critical importance of initial state and path-dependence occurs if there are locally unstable equilibriums. A famous example are the Lorentz equations in meteorology; another example is the application of non-linear probability theory to explore path dependence and increasing returns in an economy (Arthur 1994).

In recent decades, a large number of simulation models have been constructed that are relevant to an integrative environmental science. Some are basic and disciplinary, providing universal building blocks (see Fig. 17.7); others are extensive, integrated and comprehensive enough to be used for the empirical validation of real-world processes. I will confine myself here to models that attempt to integrate various aspects of socionatural systems. Three methods are illustrated in the form of three modelling exercises: a systems dynamics model; a spatially explicit cellular automata model; and a multi-agent simulation model. For a more comprehensive treatment, the reader is referred to a number of overviews in the literature (Randers 1980; Roberts et al. 1983; Edelstein-Keshet 1988; Bossel 1992; Engelen et al. 1995; Ruth and Hannon 1997; Sterman 2000; Janssen 2002; de Vries and Goudsblom 2002).

17.4.4.1 An Exercise in [Un]sustainable Development: Lakeland

One way to explore the dynamic variety of strategies for sustainable development is to design *model worlds*. The tools for simulation have improved to such a

---

[9] See Richardson (1991) for an interesting account of the emergence of systems science and systems dynamics. Probably the most recent and complete textbook is Business Dynamics by Sterman (2000), but there are several others e.g. Roberts et al. (1983), Bossel (1992) and Ruth and Hannon (1997).

degree that rather complex model representations of [parts of] the real world can be explored both rapidly and interactively. Discussion will be confined here to integrated population-resource-environment models. The example of a rather simple country called Lakeland (see Box 17.4) is described below, which we think is representative of some past and present developments throughout the world (Jager et al. 2000).

Figure 17.8a represents our model world. It consists of two subsystems: the lake with its inhabitants (fish and their food – daphnia or shrimp, for instance) and water-sediment-pollutant, and the economy (people, gold). The two systems are connected: people catch fish for food, daphnia are affected by the water pollution resulting from gold mining.[10] The fish-daphnia-water ecosystem and the water-sediment-pollutant system function according to the equations described above (Equations (17.2) and (17.4b, c)), although with delays and smoothing processes to represent real-world inertia. Its representation in the stella-software is shown in Figs. 17.8b, c.

In a way, the economy can be considered as a predator one level higher in the hierarchy: fishermen and their boats and nets. In our model world, the capital stock of boats and nets, K, is described by the equation:

$$\frac{dK}{dt} = \left(\frac{DFC}{PFC} - 1 - \frac{1}{L}\right)K \qquad (17.8)$$

Box 17.4 Lakeland – a model world

Lakeland is an island with a large and beautiful lake which has been populated by fish as long as the Lakeland people can remember. The Lakelanders loved to eat the large fish which swam around so abundantly. The human population was in equilibrium with the ecosystem — a sustainable state as long as population growth remained small. However, one day, gold was discovered on the island, and it underwent 'development' in the form of a goldmine. The mining caused water pollution. Then another 'development' event took place: a group of Lakeland fishermen agreed with foreign businessmen to export fish, even though the island's government had set quotas on export fishing. After a few years, the fishermen started to notice that it was becoming harder and harder to catch fish. Fish on the local market was becoming more expensive, and, to their distress, the Lakelanders discovered that the water quality in the lake was declining. Not long after that, the gold resources also became depleted.

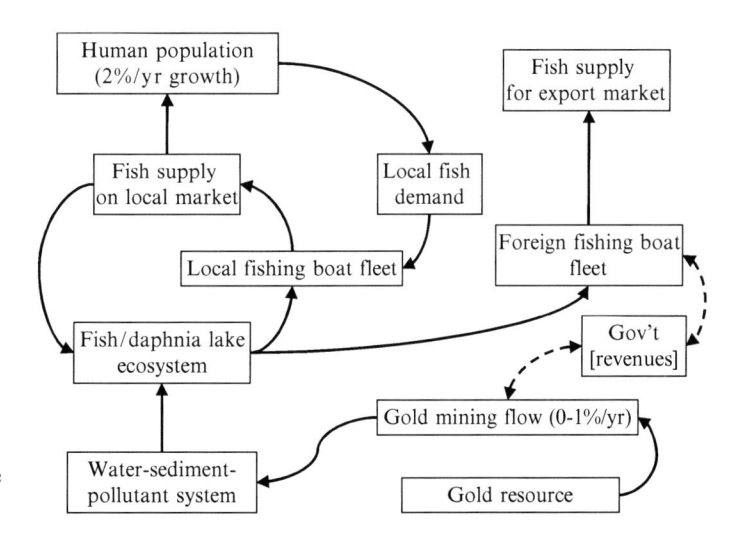

Fig. 17.8a Schematic representation of the Lakeland model-world

---

[10] I have also included some other elements that are absent in the differential equations and model results shown. The most important ones are: the less successful reproductive behaviour of fish below a certain density; slowdown in population growth if local fish supply falls short, unless the food import is activated; increasing boat productivity α in response to declining fish catches; rising cost and pollution load per unit of gold as mining proceeds. The model code is available on request (bert.devries@pbl.nl).

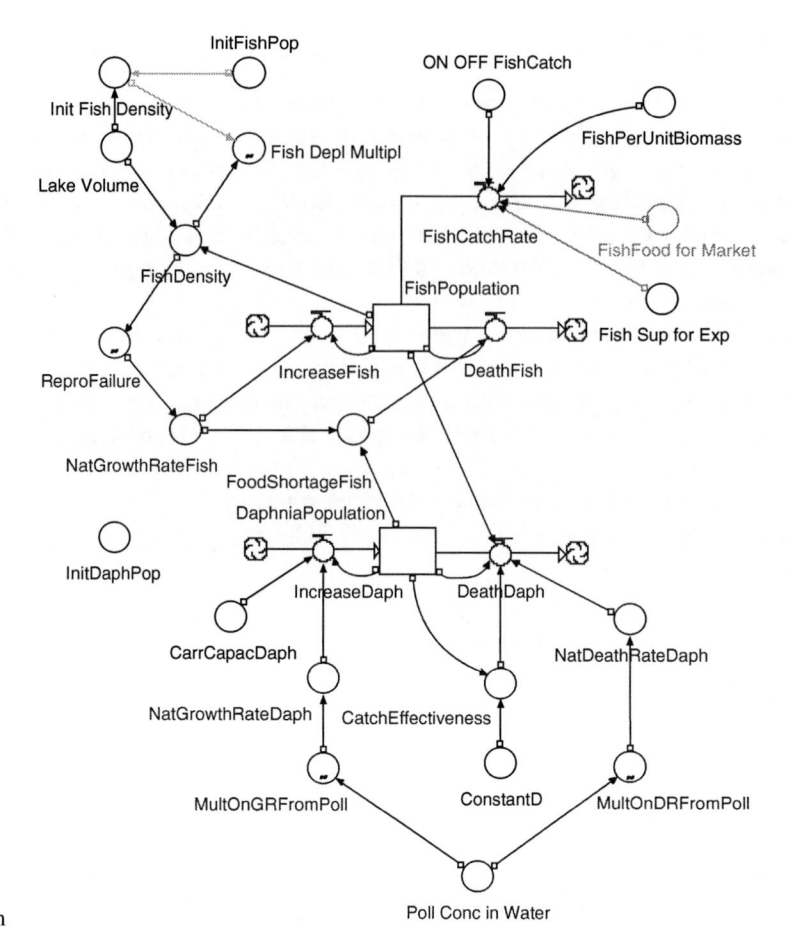

Fig. 17.8b Fish-daphnia system

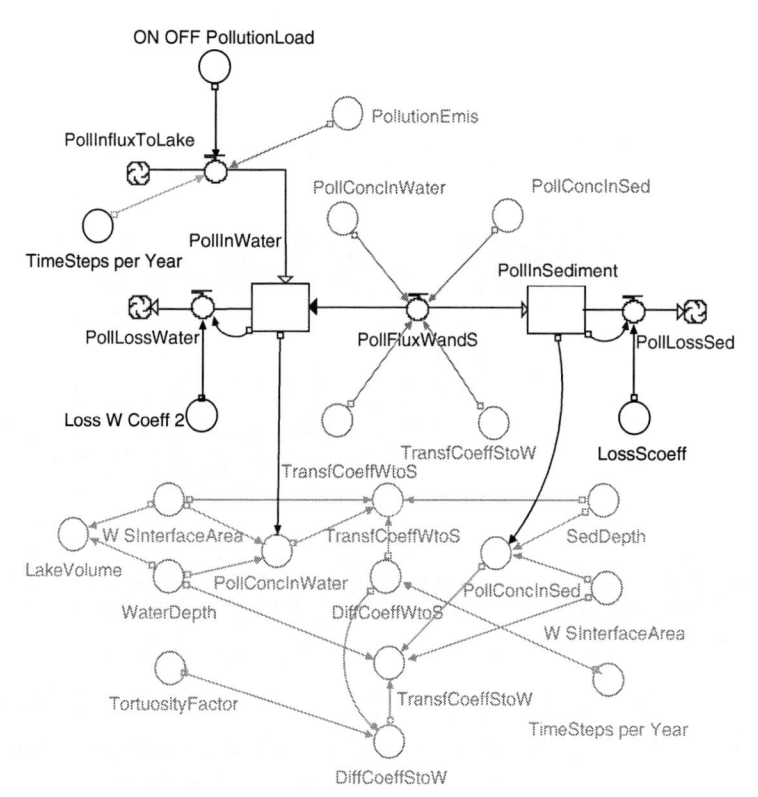

Fig. 17.8c Water-sediment-pollutant system

The capital stock to catch fish depreciates at a rate of one over the economic lifetime L. It grows when the desired catch *DFC* exceeds the potential catch *PFC*. The potential catch is related to the average catch per boat per year, α. The desired catch is determined by population, desired fish per person and export requirements. It is also assumed that demand for fish is influenced by price which equals cost with some fixed profit margin. Fish costs, in turn, are determined by the depreciation of the capital stock and its productivity α. The Lakelanders make their boats and nets from local materials and get their food, in the form of fish, in exchange for their labour. Fish is caught during the fishing season and delivered – with losses – to the marketplace and sold. The distribution of scarcity is partly a measure of higher prices and partly of food shortages.

The first, export-oriented sector in the economy is gold mining. If gold is mined and pollutants are allowed to flow into the lake, this affects the daphnia reproductive rate, which in turn affects fish and the fish catch. This is shown in Figs. 17.9a, b. The pollution per unit of gold mined is obviously crucial – but even at a five times lower value, it will bring down the fish catch within 12 years. One policy option is to tax mining revenues and spend part of it on pollution abatement. Figure 17.9 also shows that before pollution starts, the fish catch is sustainable, stabilizing the fish-daphnia system.

A second sector in the economy is the fishing-for-export. Its investments in fishing boats depend on the return-on-capital ratio: if it falls below 15%, new investments drop off rapidly. It is a top financial predator. Because this activity is also taxed, it is a second revenue source for the government. The entrepreneurs running the fish-export business have a rather strict profit criterium: if boat productivity α declines and the

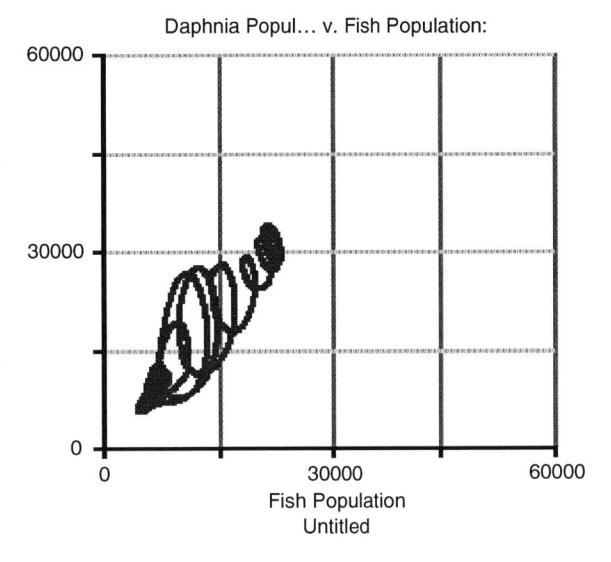

Fig. 17.9a State space diagram (1%/year of gold resource is mined) starting from upper right

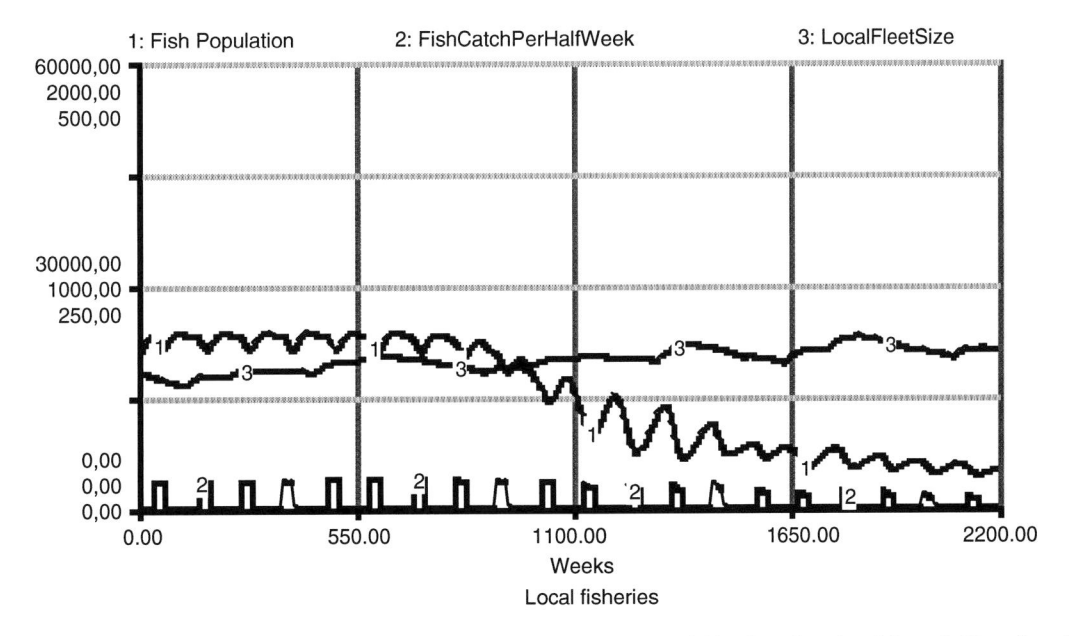

Fig. 17.9b Development of fish population, fish catch for local market and local fishing boat fleet (catch is per half-week period; 1 year is 110 units) with gold mining (1%/year) starting in year 4. Pollution affects daphnia, then fish and then catch

cost of catching fish increases, their income from fish sales at a world-market price will decline. This makes investing in new boats less attractive. Hence, the typical pattern is that the export-fishing fleet rapidly rises towards the maximum allowed by the government and then slowly becomes smaller again. As Figs. 17.10a, b shows, with such a large fleet, there is an initial tendency towards an over-exploitation of fish. However, with a maximum allowable foreign fleet of about three times the size of the local fleet, the system would recover, because foreign fishing boats would leave the area as profit targets could no longer be met. The combination of over-fishing and pollution from gold mining causes an almost irreversible collapse of the system.

From a policy point of view, the real trade-off is between a variety of assets, each with its own indicators: ecosystem health, gold resource, local food supply and supply security and, last but not least, government income from taxing gold mining and (foreign) fishing activities. If the institutional setting permits taxing and environmental policy, one can clearly define more and less sustainable development strategies.

This, purely illustrative, model world can be considered as a crystallized representation of generic knowledge about resource exploitation systems. It is also an attempt at integration: chemistry, ecology and economics. Quite a few models such as these have been constructed in the field of resource and environmental management. Many of them are disciplinary refinements on, for example, foodweb dynamics in the context of ecological stability (Neutel et al. 2002). Besides, there are many system-dynamics models, such as more sophisticated fishery models (Ruth and Hannon 1997), energy-economics models (de Vries et al. 2001), economy-climate models (Fiddaman 2002), and models of commodity systems such as shrimp, corn and forest (Sustainability Institute, www.sustainer.org).

The insights generated may seem rather 'strong' if only because of the mathematical formalism. However, both the participants in the system and scientific

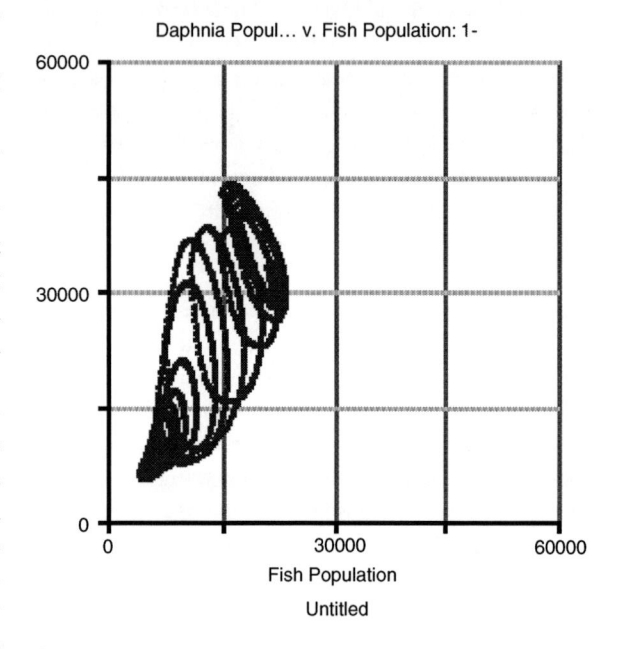

Fig. 17.10a State space diagram (foreign fleet of maximum 800 boats allowed and 1%/year of gold resource mined). Notice the initial movement to the upper left: overexploitation of fish

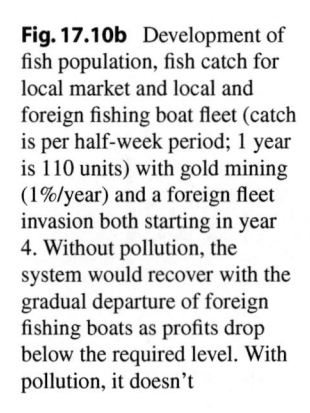

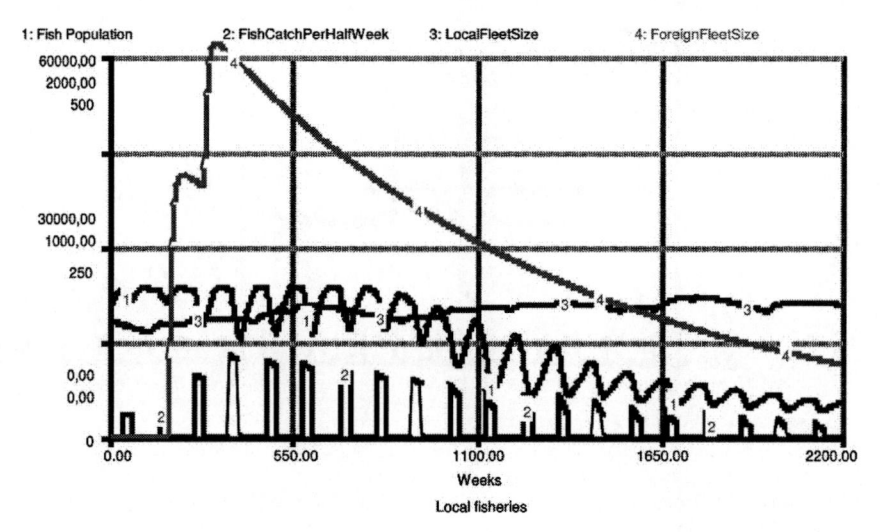

Fig. 17.10b Development of fish population, fish catch for local market and local and foreign fishing boat fleet (catch is per half-week period; 1 year is 110 units) with gold mining (1%/year) and a foreign fleet invasion both starting in year 4. Without pollution, the system would recover with the gradual departure of foreign fishing boats as profits drop below the required level. With pollution, it doesn't

experts would immediately point out the many deficiencies of this representation. One may base one's judgment on the dictum "all models are false, but some are useful". A step forward is to link the model to real-world observations in an attempt at model validation – and usually model refinement. Another is to look for improvements in its major shortcoming: homogeneity – in Lakeland, fish and people are measured as a single stock quantity. Let us first look at examples of spatial heterogeneity and then of actor heterogeneity.

17.4.4.2 Cellular Automata (CA): The Relevance of Spatial Interaction

Many population-resource-environment models are about land with inherent spatial characteristics – think of soil quality, climate and vegetation, hydrological features. The development of Geographical Information Systems (GIS) software, which started with spatial data representation, has allowed for the increasingly sophisticated inclusion of spatial aspects in dynamic models. An enormous stimulus has been provided by the vastly increased availability of satellite data.

Among the earliest attempts to model spatial relationships outside the framework of partial differential equations, as in gravitational and electromagnetic theory, and also in many disciplinary environmental models, was the cellular automata (CA) approach. A CA framework consists of a number N of spatially ordered cells. Each cell works as a kind of mini-computer: it is in a certain state $S(t)$ at time t and changes into a subsequent state $S(t + 1)$ at time $t + 1$. Among other factors, the transformation depends on the state of neighbouring cells. If the cells are ordered in a one-dimensional array, one has a sequence of interacting objects; if two-dimensional, it can be considered as (an isomorphism of) a spatial GIS representation in metric space.

Many CA-based models have been developed and applied in a policy context.[11] Early examples are a model of the population-economy-ecology interactions in Senegal and a model-based exploration of land degradation risks from tourism in Crete (Clark et al. 1995). A more systematic tool to explore the role of spatial dynamics is the Geonamica Dynamic Landuse Planning software (www.riks.nl). Against the background of spatial feature maps, it simulates, on the basis of CA, the dynamic interactions between various activities which are derived from a higher level dynamic scenario-generator. It has been used to explore a variety of environmental issues: the impact of climate change on the island of Santa Lucia, the impact of agricultural expansion on mangrove forests on the island of Sulawesi in Indonesia and a multi-criteria-based management of the Wadden Sea in the Netherlands.

Another example in the context of land use is the CLUE (Conversion of Land Use and its Effects) modelling framework at Wageningen University (Verburg et al. 2002) (www.cluemodel.nl). This is a CA model that simulates the transition in agricultural strategies as driven by biophysical and human factors. The model has particular relevance if one wants to simulate the agricultural processes in less-developed regions where traditional farming methods are gradually replaced and/or complemented by cash crops and other forms of commercial farming. For each land type, partly overlapping suitability classes are assigned to the (uniform) cells in a grid, and they may change over time as the result of feedback mechanisms such as land degradation. After establishing an infrastructure and initial land use, a dynamic process is simulated in which first the biophysical parameters are updated, then the wishes and requirements for regional land use are established, and finally, land use is allocated at grid level.

In the first step, regionally available food from the cells is calculated by incorporating the effects of past land use, pests or diseases, annual fluctuations with existing technology levels, economic values and 2 years of food/money reserves. In the second step, the regional land-use demand is calculated from population and some derived parameters (management level, urban expansion), from yield level and land-use strategy (both a function of [inter]national technology level) and related values and preferences, in particular those concerning cattle and the 'need' for natural vegetation. In the third step, a land-use allocation model translates regional demand into actual land-use changes. In this way, the model can be used to explore agricultural transitions, taking into account food security and technological adaptation. The model accommodates three kinds of land-use changes: land-use expansion (e.g. Amazonia), intensification (e.g. South-

---

[11] See some excellent introductions: Couclelis (1985), White and Engelen (1992), Engelen et al. (1995) and Uljee et al. (1996) and websites such www.riks.nl and www.iatp.org. Also the IMAGE model uses a CA formalism (see Chapter 18).

East Asia) and contraction (European Union). The results indicate that the causes of land-use changes are not controlled, but are only influenced by biophysical processes and that due consideration should be given to demographic and economic factors.

In the realm of agro-economics, there are also interesting modelling developments. For instance, Balmann (1997) has explored the regional structural change in agriculture, using the general idea of CA and rules founded in economic theory as well as on particular characteristics of the agricultural sector (such as the immobility of land). Farmers optimize their activities with respect to their objective function by considering their expectations, financial state, and existing assets within technical and economic restraints. Farms, with a given (initial state) location, have a number of action capabilities. They can close down, be founded, compete for and rent or give up land plots, invest and produce. Simulations of a sample region of 225,000 ha, with a plot size of 10 ha suggest, for different initializations and constraints, path dependence in land productivity and use evolution. As such, the model gives insights into the process of structural change, and, in particular, into the role of economies of scale.

Berger (2001), in a follow-up to Balmann's work, explores the dynamics of technology diffusion for a 650-km^2 area in Chili on a 2.5 ha resolution. Most conventional agro-economic models use econometric estimations for mathematical (linear/dynamic) programming, without explicit consideration of actor interaction. This implies, among other things, the absence of transaction and information costs and no explicit consideration of spatial dimension, thus neglecting internal transport costs and physical [im]mobility aspects. Besides a conventional 'smooth adjustment' scenario, Berger explores a scenario in which farmers are reluctant to innovate and refuse any technology adoption, thus reflecting some of the fierce resistance to the formation of an open Mercosur market region in South America. To this end, different implementations of thresholds and information dissemination are assumed. In practice, it meant a difference between two distinct communication networks: the campesino farms networks with small-scale family holdings of 2.5 to 12 ha, and a network of 'commercial' farms larger than 12 ha. Under 'ideal' conditions, farm incomes and on-farm labour intensity will increase. However, the inclusion of behavioural heterogeneity and bandwagon effects may cause this increase to vanish almost completely over the considered time horizon of 20 years. It suggests that efforts towards sustainable development, for instance in the form of water-saving technologies, should explicitly consider these dynamics. A similar approach has been used by Deffuant et al. (2002) to gain an understanding of the mechanisms and barriers for the penetration of organic farming.

Of course, these methods should be used with discrimination: Are they really giving new insights? Janssen et al. (2002) have explored the advantages of adding spatial heterogeneity to models of range-land management in Australia. Experimenting with various types of sheep behaviour using a relatively simple model – including fire and water points – it is found that non-uniform grazing due to different rules regarding biomass density, water point locations, and other sheep, may cause system behaviour to differ significantly from uniform grazing (as in non-spatial models). This could influence notions of ecosystem resilience and indicate the need for more prudent exploitation strategies. The cautious conclusion is that "understanding the implications of spatial processes is vital in the large open paddocks of range-lands; however, it is likely that these processes are also important, perhaps in a more subtle way, in other environments" (Janssen 2002: 123).

17.4.4.3 Multi-Agent Simulation (MAS): The Importance of Behavioural Variety

In most models, behaviour is simulated in the form of one aggregate actor described by one or a few differential equations, for example, resource use. However, we all experience heterogeneity in the behaviour of animals and people, and scientists are investigating it in more and more detail, using a variety of methods. To give a more realistic picture, a more comprehensive actor representation should be used, such as multi-agent simulation (MAS) modelling (sometimes also referred to as agent-based or multi-actor modelling). An agent, in this context, is a model entity that "can represent animals, people or organizations; can be reactive or proactive; may sensor the environment; communicate with other agents; learn, remember, move and have emotions" (Janssen 2002: 2).[12] Of

---

[12] To get a feel for the diversity in human beings beyond physical characteristics such as location, income, etc., one should have a look at recent classifications of world views and value orientations: at this level, heterogeneity is much larger and fuzzy than at the 'material' level; see NIPO (2002) and AGP. For an introductory text on multi-agent simulation, see Ferber (2000).

course, one may also read 'behaviour' in the differential equation descriptions discussed above. For instance, the interaction between fish and daphnia in Equation (17.4) is a condensed representation of random encounters in a homogeneous body of water, ascribing to the fish the behaviour of eating the daphnia. However, such formalization severely restricts an explicit range of agent strategies. Also, as seen above, the explicit inclusion of spatial heterogeneity often allows for, or even demands, more explicit agent formulation. Here, I confine myself to a brief discussion of a few examples.

The first example is the consumat project in which is investigated how the micro-level processes affect macro-level outcomes (Jager et al. 2000). People living in the Lakeland model world described above, are not just numbers, but are individuals who engage in cognitive processes such as social comparison, imitation and repetitive (habitual) behaviour. A small group (16) of individuals, called 'consumats', is placed in this micro world. There is access to two natural resources: a fish stock in the lake and a gold mine. The money earned from gold mining can be spent on fish imports and/or on non-fish consumption 'luxury' goods.

The consumats determine how they allocate their time to leisure, fishing and mining. They are equipped with certain abilities for fishing and mining and try to satisfy their four needs: leisure, identity, subsistence and freedom. These needs, which are supposed to play an important role in this case, are selected out of a larger set of needs, as described by Max-Neef (1991). The satisfaction of the need for leisure is related to the share of the time spent on leisure. The need for identity is satisfied by the relative amount of money the consumat owns, in comparison to consumats with similar abilities. The need for subsistence relates to the consumption of food, i.e. fish. The need for freedom is assumed to be related to the absolute amount of money the consumat owns, and on whatever assets the consumat prefers to spend them on.

All consumats have to satisfy their personal needs by fishing and/or mining, whereby they find themselves in a common dilemma when facing the risk of resource depletion. They can choose different strategies, all of which have been identified on the basis of psychological theory. Depending on the consumats' desired level of need satisfaction and behavioural control, which is linked to (un)certainty, they follow different cognitive processes: deliberation (reasoned, individual), social comparison (reasoned, socially determined), repetition (automated, individual) and imitation (automated, socially determined). If their performance falls below expectation, as recorded by memory, in a mental map, agents will switch strategy. Two archetypical consumats – *Homo economicus* and *Homo psychologicus* – have been created to study the effects of various assumptions. The former tends to favour deliberation, because he/she operates with high levels of need satisfaction and uncertainty reduction, whereas the latter tends to be quickly satisfied and uncertain, thereby inducing imitative behaviour.

One finding is that *Homo psychologicus* makes a transition from a fishing to a mining society that is very different from *Homo economicus,* who can only engage in the 'standard' mode of rational deliberation. The introduction of mining means that less time is spent on fishing. Because *Homo psychologicus* spends less time fishing than *Homo economicus* does, the latter depletes the fish stock to a greater extent. Yet *Homo psychologicus* depletes the fish stock at a higher rate than before the mining option came into existence. A major reason for this is that the slow, but persistent, move towards mining causes water pollution (Fig. 17.11a), which, in turn, negatively affects the fish stock by causing it to decrease. As the fish catch per hour decreases, this further accelerates the transition to mining. In turn, this leads to a different resource exploitation pattern (fish population, gold depletion, water pollution). A second finding is that, over time, the population and resource-use trajectory depends on the diversity of agents' abilities: for *Homo economicus,* it leads, on average, to a decrease in time spent working, but for *Homo psychologicus* it leads to an increase. These results, which have to be backed up with empirical research (fieldwork and laboratory experiments), indicate that modelling more diverse behavioural options, including motives and habits, can be more beneficial to gaining an understanding of the forces behind [un]sustainable development and resource exploitation than is most conventional [economic] modelling.

In another experiment, it is assumed that deliberation leads to fish-catching innovations and that mining is not possible. Not surprisingly, the fish resource is more rapidly over-exploited. However, the lower level of satisfaction and the more rapid diffusion through social comparison processes in a *Homo economicus* population causes a much faster collapse than in a *Homo psychologicus* community, where people are

Fig. 17.11a Pollutant concentration in the lake for both conditions for t = 1 to t = 50. The slower shift of *Homo psychologicus* to mining causes a slower rate of pollution than for *Homo economicus*
Source: After Jager et al. (2000)

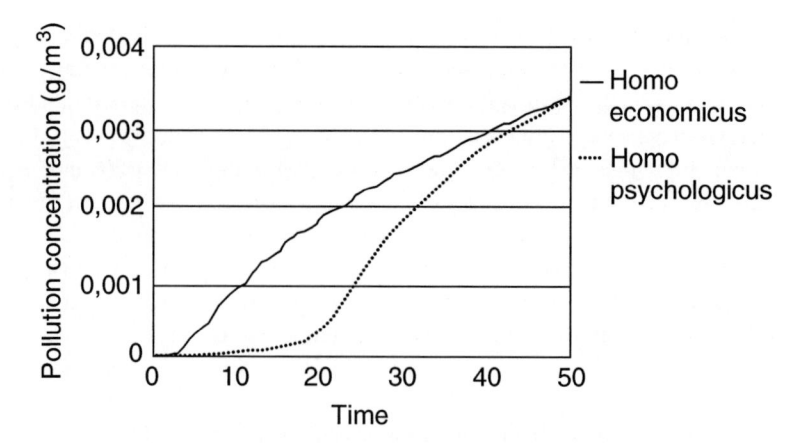

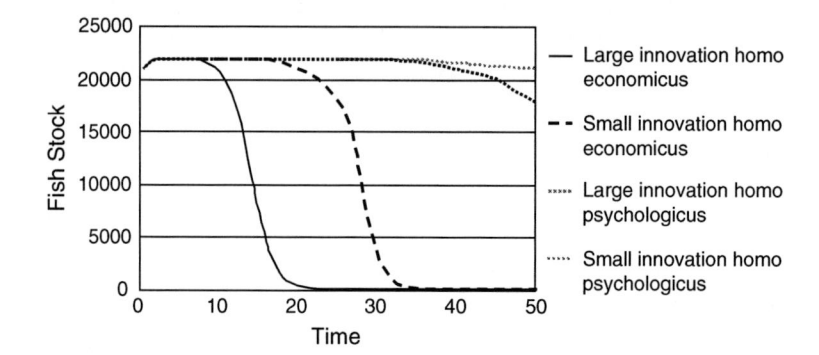

Fig. 17.11b Fish stocks in the lake under conditions of small and large technological innovations. Innovation, particularly for *Homo economicus*, accelerates fish stock depletion
Source: After Jager et al. (2000)

basically content and stuck in repetitive behaviour for a while, without noticing the innovation (Fig. 17.11b). This underlines the ambivalence of technological progress in issues of [un]sustainable patterns of resource use; it also suggests a possible dynamic in the confrontation of relatively advanced societies with traditional ones: living a satisfied life and having a low-risk alertness makes populations easy prey for more capable groups.

An interesting approach is the CORMAS framework of CIRAD Montpellier. This general framework aims to allow users to construct their own MAS model, starting from a set of sample models, mostly used for tutorial purposes. Some interesting environmental applications in less developed regions (Thailand, Senegal, Reunion) have been constructed and applied in an interactive way called 'companion modelling'.

Multi-agent simulation is being increasingly used in the social sciences (see e.g. Gilbert and Doran 1994; Tesfatsion 2001). The focus is often to explore and explain certain global regularities arising from repeated local interactions of autonomous agents, such as bottom-up decentralized market behaviour. Economic model worlds can now be populated with heterogeneous agents "who determine their interactions with other agents and with their environment on the basis of internalized social norms, internal behavioural rules, and data acquired on the basis of experience" (Tesfatsion 2001: 283). These developments may become part and parcel of the new branch of evolutionary and institutional economics modelling of natural selection processes and competitive and cooperative associations, instead of population-level laws of motion.

The work carried out on the role of trust amongst herdsmen in North Cameroon is a nice example, relevant to the present context. Nomadic herdsmen need to gain water-access rights from village leaders and access to pasture lands from farmers. Model experiments show "that the cost-priority model (based on ideas from transaction-cost economics, under which agents care only about minimizing their costs) and friend-priority models (based on ideas from institutional theory) of agent reasoning result in dramatically different experimental outcomes. In particular, the global efficiency of the cost-priority model is surprisingly low, relative to the friend-priority model, leading, in some cases, to the disappearance of herds" (Tesfatsion 2001: 288). This has a direct corollary with the resilience of socio-natural systems.

A key question is what should guide the selection of the model of the individual. This has been dealt with in some detail in connection with the important question of Common Pool Resource Management which can be seen as being archetypical for many environmental problems (Ostrom 1990). Some researchers stick closely to the perfectly rational individual with perfect foresight of a neoclassical economics model; others use a more inductive approach, based on insights from sociology and psychology, as in the consumat model above. As a branch of institutional economics, we now see the emergence of theoretical foundations, such as in the suggested assumptions to be made about individuals in MAS models (Ostrom 2000):

- *Preferences*, i.e. tastes or desires of individuals over objects and outcomes
- *Norms and values*, i.e. morally grounded internal valuations of particular types of actions
- *Information-processing capabilities*, i.e. the nature of, the access to, and the processing capacities with regard to certain information
- *Selection criteria*, i.e. what determines individual choices from among a number of alternatives and
- *Resources*, i.e. individuals' abilities to take action

Various model experiments about managing a common pool resource indicate the existence of multiple equilibriums, possible inefficiencies, path dependence and lock-in effects and show that "the presence of a belief structure in the agents [is] essential to the evolution of an effective group-level cooperative strategy in a potentially hostile environment" (Ostrom 2000: 158). One result is the importance of communication in cooperative strategies.

The above examples indicate that the incorporation of a micro-level perspective on human behaviour within integrated population-resource-environment models can contribute to a better understanding of micro-level dynamics, and hence to more adequate top-down representations (Table 17.2). The value orientation ascribed to human beings in their interaction with the environment and with each other may significantly affect the outcome. This points to the importance of collecting controversial interpretations about peoples' motives and means – cultural diversity – and incorporating them more explicitly in such models. Yet, here too, one has to remain critical. After all, even the more transparent and analytically tractable differential equation models have, albeit not very explicit, agent behaviour.

Table 17.2 Overview of modelling approaches mentioned in 17.4.4.2 and 17.4.4.3

Model name	Methodology	Source
Population-economy-ecology (crete)	Differential-integral equations and GIS	Clark et al. (1995
Climate-change impacts in the Caribbean; development in Sulawesi; Wadden Sea management	Three-layer simulation framework Geonamica: GIS, CA and system dynamics	www.riks.nl Uljee et al. (1996)
Land-use changes in less-developed regions (Costa Rica, China, etc.)	GIS-CA-based CLUE modelling framework	Verburg et al. (2002)
Structural change in agriculture; dynamics of technology diffusion in agriculture	GIS-CA with optimization rules for actors	Balmann (1997) Berger (2001)
Consumat model: *Homo economicus* and *Homo psychologicus*	MAS with behavioural diversity related to risk and satisfaction	Jager et al. (2000)
Land-use and market dynamics (France, Africa); barriers for organic farming	MAS framework	cormas.cirad.fr Deffuant et al. (2002)
Common Pool Resource management strategies	MAS and fieldwork/laboratory experiments	Ostrom (2000)

17.4.5 Interactive Simulation and Policy Exercises

The rapid development in computer technology and software applications increasingly allows for the interactive use of models, in other words, of constructing and applying models in close interaction with stakeholders – among them policy analysts and fellow scientists.[13] Several decades ago already, attempts were made to integrate scientific system insights with the real-world behaviour of people. Starting with war-games, simulation gaming also reached the realm of environmental sciences in the 1980s – as is evident from games such as 'Stratagem', 'Fish Banks', and others (Meadows 1990; de Vries et al. 1991; Meadows 1996; de Vries 1998). The related approach of policy exercises has also been used successfully in a variety of environmental contexts (de Vries et al. 1993; Toth 1994), (see Chapter 14, Section 14.5). The objective is usually a combination of teaching skills, conveying information, increasing motivation and interest, changing attitudes and inducing [self-] evaluation (Greenblat 1988). We are now entering the upper-right corner of Fig. 17.2.

These methods have been shown to provide a context for learning and decision making about environmental and, broader, sustainable development-oriented strategies, with a core of relatively 'strong' models in the background, and with support from other techniques such as scenario writing and back casting. It can help participants to understand the dynamics of the relevant processes, to formulate a shared problem perception and to design mutually beneficial strategies. Clearly, as with the previous approaches, one should not overstretch expectations.

In the context of environmental modelling, a most interesting use of models is as a tool in a policy exercise, where stakeholder involvement and simulation gaming are included. A recently published report describes an interesting example: the clarification and search for solutions to a conflict between sheep farmers, foresters and environmentalists in the Causse ('plateau') Méjac near Montpellier (Etienne et al. 2003). The perceived problem is that Scots pines and Austrian black pines are encroaching into a natural ecosystem of high biological diversity, affecting the development objectives of the stakeholders (sheep production, timber production, nature conservation) in different ways. A MAS model has been developed, using a 200 × 200 m² grid of 5,726 cells (4 ha each). Biological characteristics were introduced for the initialization, formulated in the form of 'spatial entities', i.e., management units making sense according to some specific perception of the ecosystem, such as already invaded ridges, farming and woodland areas. Then, three sets of actors were defined, each with their own strategy *vis-à-vs* the spatial entities: 37 sheep-farmer agents, two foresters (one dealing with afforestation, one with the exploitation of native woodlands), and a National Park ranger (the 'conservationist').

The model has been used in an interactive decision set-up, with an approach based on four concepts:

- *The point of view*, i.e. the specific way in which each stakeholder perceived resources and identified management entities
- *The viewpoint*, i.e. a spatial representation of a point of view, such as a vegetation map, a map of tree ages, a map of tree stands or a map of landscape units
- *Indicators*, i.e. a set of markers selected by the stakeholders to monitor the dynamics of the system and
- *Scenarios*, i.e. prospective management rules to tackle the problem at hand (such as pine encroachment of the Causse)

During these exercises, it became evident that two aspects were particularly effective in structuring the problem and looking for solutions: sharing dynamic processes and comparing contrasting scenarios. A major conclusion was that "it helped the participants to understand that all their views were legitimate, but also subjective and partial".

17.5 Concluding Remarks

Environmental science and modelling have experienced tremendous development and change over the past decades. Starting in the 1970s with monodisciplinary models of soil and water systems, air and soil pollution, ecosystems (e.g. vegetation, fisheries, forests) and resource depletion, there has been a steady increase in empirical data, theoretical concepts and computational tools. As a result, each of these research fields has deepened its understanding, developing at the same time

---

[13] Interactive model use is probably best known in education where it has not only penetrated science and engineering curricula, but also bridges the worlds of science and of computer-game fiction in a variety of ways. However, most of these applications are computer/model-oriented; a learning environment that goes beyond is rare and more difficult to design.

more sophisticated dynamic analytical and simulation models. Simultaneously, the first attempts at constructing larger, integrated, population-resource-environment models were pointing towards the need for a more systemic cross-disciplinary view. Since the 1990s, these developments have resulted in large and successful efforts to construct broad environmental, or maybe one should say sustainable, development-oriented models.

I have sketched this development as part of an evolution towards acquiring knowledge and understanding about more complex systems. In the natural sciences, there has been horizontal integration: earth, water, air and fire have become united. Biology, ecology and geography are showing the way towards vertical integration: the spatially explicit and diverse behaviour of animals and humans. Increasingly, the social sciences are contributing in ways that allow further integration in the form of conceptualizing agents or introducing real-world stakeholders, or both. The next chapter will present in more detail two large models, which reflect this development and use this evolution.

Acknowledgement The Author would like to thank Jodi de Greef for inspiring discussions and René Benders and Henk Moll for an early start.

References

Allen, P. (1994). Evolutionary complex systems: Models of technology change. In L. Leydesdorff & P. van den Besselaar (Eds.), *Evolutionary economics and chaos theory – New directions in technology studies*. London: Pinter.

Arthur, W. B. (1994). *Increasing returns and path dependence in the economy*. Ann Arbor: University of Michigan Press.

Aurobindo, S. (1998). *Le cycle humain*. Pondicherry: Sri Aurobindo Ashram.

Azar, C., Holmberg, J., & Lindgren, K. (1995). Stability analysis of harvesting in a predator-prey model. *Journal of Theoretical Biology, 174*, 13–19.

Balmann, A. (1997). Farm-based modelling of regional structural change: A cellular automata approach. *European Review of Agricultural Economics, 24*, 85–108.

Berger, T. (2001). Agent-based spatial models applied to agriculture: a simulation tool for technology diffusion, resource use changes and policy analysis. *Agricultural Economics, 25*, 245–260.

Boschma, R., Frenken, K., & Lambooy, J. (2002). *Evolutionaire economie – Een inleiding*. Bussum: Uitgeverij Coutinho.

Bossel, H. (1992). *Modellbildung und simulation – Konzepte, Verfahren und Modelle zum Verhalten dynamischer Systeme*. Braunschweig/Wiesbaden: Vieweg Verlag.

Boulding, K. (1978). *Ecodynamics – A new theory of societal evolution*. London: Sage Publications.

Clark, N., Perez-Trejo, F., & Allen, P. (1995). *Evolutionary dynamics and sustainable development – A systems approach*. Aldershot: Edward Elgar.

Clark, W., & Munn, R. (Ed.). (1986). *Sustainable development of the biosphere*. Cambridge/Laxenburg: Cambridge University Press/IIASA.

Couclelis, H. (1985). Cellular worlds: A framework for modelling micro-macro dynamics. *Environment and Planning A, 17*, 585–596.

CSD (2001). www.un.org/esa/sustdev/

Dasgupta, P., & Heal, G. (1979). *Economic theory and exhaustible resources*. Cambridge: Cambridge University Press.

Deffuant, G., Huet, S., Bousset, J.-P., Henriot, J., Amon, G., & Weisbuch, G. (2002). Agent-based simulation of organic farming conversion in Allier département. In Janssen M. (2002) pp 158–187.

den Elzen, M. (2002). Exploring climate regimes for differentiation of future commitments to stabilise greenhouse gas concentrations. *Integrated Assessment, 3(4)*, 343–359.

de Vries, B. (1998). SUSCLIME: A simulation/game on population and development in a climate-constrained world. *Simulation and Gaming, 29*(2), 216–237.

de Vries, B. (2006). Scenarios: guidance for an uncertain and complex world? In R. Costanza, L. Graumlich, & W. Steffen (Eds.), *Sustainability or collapse? An integrated history and future of people on earth (IHOPE)* – 96th Dahlem Workshop (Chapter 19, pp. 378–398). Cambridge MA: MIT Press.

de Vries, B., & Goudsblom, J. (Eds.). (2002). *Mappae mundi – Humans and their habitats in a long-term socio-ecological perspective. Myths, maps, and models*. Amsterdam: Amsterdam University Press.

de Vries, B., Dijk, D., & Benders, R. (1991). *PowerPlan – An interactive simulation model for electric power planning*. IVEM: University of Groningen.

de Vries, B., van Vuuren, D., den Elzen, M., & Janssen, M. (2001). *The targets image energy regional (TIMER) model technical documentation*. Bilthoven: RIVM: 188.

de Vries, H. J. M. (1989). *Sustainable resource use – An inquiry into modelling and planning*. Groningen: IVEM, University of Groningen.

de Vries, H. J. M., Fiddaman, T., & Janssen, R. (1993). *Outline for a global environmental strategic planning exercise (GESPE Project)*. Bilthoven, RIVM: Global Dynamics and Sustainable Development Programme.

Duchin, F., & Lange, G.-M. (1994). *The future of the environment*. Oxford: Oxford University Press.

Edelstein-Keshet, L. (1988). *Mathematical models in biology*. New York: Random House.

Engelen, G., White, R., Uljee, I., & Drazan, P. (1995). Using cellular automata for integrated modelling of socio-environmental systems. *Environmental Monitoring and Assessment, 34*, 203–214.

Etienne, M., Le Page, C., & Cohen, M. (2003). A Step-by-step approach to building land management scenarios based on multiple viewpoints on multi-agent system simulations. *Journal of Artificial Societies and Social Simulation* (JASSS) jasss.soc.surrey.ac.uk/6/2/2.html

Ferber, J. (2000). *Multi-agent systems – An introduction to distributed artificial intelligence*. London: Addison-Wesley.

Fiddaman, T. S. (2002). Exploring policy options with a behavioral climate-economy model. *System Dynamics Review, 18*(2), 243–268.

Fischer-Kowalski, M., & Haberl, H. (1998). Sustainable development: Socio-economic metabolism and colonization of nature. *International Social Science Journal, 4*(158), 573–587.

Funtowicz, S., & Ravetz, J. (1990). *Uncertainty and quality in science for policy*. Dordrecht: Kluwer.

Geels, F. (2002). *Understanding the dynamics of technological transitions – A co-evolutionary and socio-technical analysis* (p. 425). Enschede: Twente University.

Gilbert, N., & Doran, J. (Ed.). (1994). *Simulating societies – The computer simulation of social phenomena*. London: UCL Press.

Greenblat, C. (1988). *Designing games and simulations – An illustrated handbook*. London: Sage.

Groenewold, H. (1981). *Evolutie van kennis, waarden en macht*. Enschede: Twente University.

Grubler, A. (2000). *Technology and global change*. Cambridge: Cambridge University Press.

Gunderson, L., Holling, C., & Light, S. (Ed.). (1995). *Barriers and bridges to the renewal of ecosystems and institutions*. New York: Columbia University Press.

Harman, W. (1993). *Global mind change – The promise of the last years of the twentieth century*. San Francisco: Knowledge Systems/Institute of Noetic Sciences.

Hisschemöller, M., & Gupta, J. (1999). Problem-solving through international environmental agreements: The issue of regime effectiveness. *International Political Science Review, 20*(2), 151–174.

Holland, J. (1995). *Hidden order – How adaptation builds complexity*. New York: Helix.

Jager, W., Janssen, M., de Vries, B., de Greef, J., & Vlek, C. (2000). Behaviour in commons dilemmas: Homo Economicus and Homo Psychologicus in an ecological-economic model. *Ecological Economics, 35*(3), 357–379.

Janssen, M. (Ed.). (2002). *Complexity and ecosystem management – The theory and practice of multi-agent systems*. Cheltenham: Edward Elgar.

Janssen, M., Anderies, J., Stafford Smith, M., & Walker, B. (2002). Implications of spatial heterogeneity of grazing pressure on the resilience of rangelands. In M. Janssen (Ed.), *Complexity and ecosystem management – The theory and practice of multi-agent systems*. Cheltenham: Edward Elgar.

Jantsch, E. (1980). *The self-organizing universe – Scientific and human implications of the emerging paradigm of evolution*. Oxford: Pergamon Press.

Lewin, R. (1992). *Complexity – Life at the edge of chaos*. New York: Collier.

Manson, S. M. (2001). Simplifying complexity: A review of complexity theory. *Geoforum, 32*, 405–414.

Marchetti, C., & Nakicenovic, N. (1979). *The dynamics of energy systems and the logistic substitution model*. Laxenburg: IIASA.

Max-Neef, M. (1991). *Human scale development – Conception, application and further reflections*. New York/London: The Apex Press.

Meadows, D. (1990). *Stratagem-1 user's manual*. Durham: Institute for Policy and Social Science Research (IPSSR), University of New Hampshire.

Meadows, D. (1996). http://www.systemdynamicsapplications.com/models-from-literature/fish-banks.htm

Meadows, D. L., Behrens, W. W., Meadows, D. H., Naill, R. F., Randers, J., Zahn, E. K.O. (1974). *Dynamics of growth in a finite world*. Cambridge, MA: Wright-Allen Press.

Morecroft, J., (Ed.). (1992). Modelling for learning [Special issue]. *European Journal of Operations Research, 59*(1), 6–166.

Müller, A. (2003). A flower in full blossom – Ecological economics at the crossroads between normal and post-normal science. *Ecological Economics, 45*, 19–27.

Neutel, A.-M., Heesterbeek, J., & de Ruiter, P. (2002). Stability in real food webs: Weak links in long loops. *Science, 296*, 1120–1123.

NIPO (2002). *Het WIN-model: Waardensegmenten in Nederland*. Amsterdam, NIPO, 74.

Noorman, K.-J., & Schoot Uiterkamp, T. (Eds.). (1998). *Green households? Domestic consumers, environment, and sustainability*. London: Earthscan.

Ostrom, E. (1990). *Governing the commons – The evolution of institutions for collective action*. Cambridge: Cambridge University Press.

Ostrom, E. (2000). Collective action and the evolution of social norms. *Journal of Economic Perspectives, 14*(3), 137–158.

Pearce, D. (2002). An intellectual history of environmental economics. *Annual Review of Energy and the Environment, 27*, 57–81.

Perman, R., Yue Ma, J. McGilvray & Common, M. (2003). *Natural resource and environmental economics*. Essex, UK: Pearson.

Randers, J. (Ed.). (1980). *Elements of the systems dynamics method*. Cambridge, MA: Productivity Press.

Richardson, G. P. (1991). *Feedback thought in social science and systems theory*. Philadelphia: University of Pennsylvania Press.

Roberts, N., Andersen, D., Deal, R., Garet, M., & Shaffer, W. (1983). *Introduction to computer simulation – A system dynamics approach*. New York: Addison-Wesley.

Rosen, R. (1985). *Anticipatory systems – Philosophical, mathematical and methodological foundations*. New York: Pergamon Press.

Rotmans, J., & de Vries, B. (1997). *Perspectives on global change – The TARGETS approach*. Cambridge: Cambridge University Press.

Ruth, M., & Hannon, B. (1997). *Modeling dynamic economic systems*. New York: Springer.

Scheffer, M., Carpenter, S., Foley, J., Folke, C., & Walker, B. (2001). Catastrophic shifts in ecosystems. *Nature, 413*(11 October), 591–596.

Schumpeter, J. (1942/1975). *Capitalism, socialism and democracy*. New York: Harper.

Simon, H. A. (1969) *The sciences of the artificial*. Cambridge, MA: MIT Press.

Söderbaum, P. (1999). Values, ideology and politics in ecological economics. *Ecological Economics, 28*, 161–170.

Sterman, J. (2000). *Business dynamics – Systems thinking and modeling for a complex world*. Boston: Irwin McGraw-Hill.

Tainter, J. (2000). Problem solving: Complexity, history, sustainability. *Population and Environment, 22*(1), 3–41.

Tesfatsion, L. (2001). Introduction to the special issue on agent-based computational economics. *Journal of Economic Dynamics and Control, 25*, 281–293.

Thompson, M. (1993) Good science for public policy. *Journal of International Development, 5*(6), 669–679.

Toth, F. (1994). *Policy exercises. Simulation and gaming, 19*, 235–276.

Toulmin, S. (1991). *Cosmopolis: The hidden agenda of modernity*. New York: The Free Press.

Uljee, I., Engelen, G., & White, R. (1996). RamCo Demo Guide Version 1.0, RIKS, Maastricht: University of Maastricht.

van der Leeuw, S. (Ed.). (1998). *The Archaeomedes Project: Understanding the natural and anthropogenic causes of land degradation and desertification in the Mediterranean basin*. Brussels, Science Research Development European Commission Report EUR 18181 EN.

van der Sluijs, J. P. (1997). *Anchoring amid uncertainty: On the management of uncertainties in risk assessment of anthropogenic climate change*. Utrecht: University of Utrecht.

van Vuuren, D., & Smeets, E. (2000). Ecological footprint for Benin, Bhutan, Costa Rica and The Netherlands. *Ecological Economics, 34*(2000), 115–130.

Verburg, P., Veldkamp, W., Espaldon, R., & Mastura, S. (2002). Modeling the spatial dynamics of regional land use: The CLUE-S Model. *Environmental Management, 30*(3), 391–405.

Wackernagel, M. (2002). www.footprintnetwork.org

Wainwright, J., & Mulligan, M. (2004). *Environmental modelling – Finding simplicity in complexity*. London: Wiley.

Watts, D. J. (1999). *Small worlds – The dynamics of networks between order and randomness*. Princeton: Princeton University Press.

White, R., & Engelen, G. (1992). *Cellular dynamics and GIS: Modelling spatial complexity,* Research Institute for Knowledge Systems Maastricht: (RIKS).

Wilber, K. (1998). *The marriage of sense and soul – Integrating science and religion*. Boston: Gateway.

Wilting, H., Blom, W., Thomas, R., & Idenburg, A. (2001). *DIMITRI 1.0: Beschrijving en toepassing van een dynamisch input-output model*. Bilthoven, RIVM, 72.

Chapter 17A
An Illustration of the LCA Technique

Reinout Heijungs

Contents

17A.1 Introduction 375
17A.2 Goal and Scope Definition 376
17A.3 Inventory Analysis – The Basic Case 376
17A.4 Inventory Analysis – Some Complications ... 378
17A.5 Impact Assessment – Definition
of Impact Categories, Selection
of Characterization Models,
Classification and Characterization 381
17A.6 Impact Assessment –
Normalization and Weighting 382
17A.7 Interpretation 382
17A.8 Concluding Remarks 383
Reference 383

17A.1 Introduction

It has been mentioned in Chapter 12 that Life Cycle Assessment (or LCA) is one of the analytical instruments within the toolbox for analyzing the interface of economy and environment. This section provides a worked example of the LCA technique. It is a fairly elaborate demonstration of a quite simple hypothetical product system. No real case study and data were chosen, the reason being that the number of data items would be too large to be sorted out in an educational setting. All data are therefore fictional and no claims should be made with respect to the results of the exercise. With respect to the methods that are presented, a similar though weaker statement must be made. There

R. Heijungs (✉)
Institute of Environmental Sciences, Leiden University (CML), The Netherlands
e-mail: heijungs@cml.leidenuniv.nl

is not one unique interpretation and implementation of the principles of LCA, and any attempt to illustrate LCA by means of a concrete example should be understood as an illustration indeed. The purpose of this section is therefore to point out the main idea and methodological principles of LCA by means of a hypothetical example with fictitious data.

In doing so, we will concentrate on the computational content of LCA. Of course, the entire concept of LCA embraces (or: may embrace) many types of activities and aspects, including involvement of stakeholders, collecting of data, verification of data, manipulation of data, and carrying out peer reviews. To what extent stakeholders play a role and how this role should be embedded is something that not only is under discussion, but which also varies from situation to situation. The same applies to the process of peer review. Obtaining and verifying data are cumbersome activities which require a good deal of process-technological and ecological and toxicological knowledge. In contrast, the recipes that are used to manipulate the data are of an almost mechanical nature. There is not much disagreement (although at some places there is some). Strangely enough, almost all textbooks discuss those parts that are difficult and controversial, while the more or less rigorous basis is often left out of consideration. We feel, however, that a presentation of these basic mechanisms of the method for LCA is indispensable for a proper understanding of the meaning of the results of an LCA.

All calculations that are made in this chapter are simple enough to do by hand or with a spreadsheet. However, the use of dedicated software facilitates both procedure and calculation of LCA. The calculations presented below have been made with such a software tool, CMLCA, which is an abbreviation of Chain Management by Life Cycle Assessment. The software, as well as the file that contains the hypothetical data, can be downloaded, for educational purposes, free of charge, from the internet (www.leidenuniv.nl/cml/ssp).

17A.2 Goal and Scope Definition

The ISO standards for LCA are quite extensive with respect to the goal and scope definition, and many of the aspects that are mentioned, in fact duplicate many aspects of the inventory analysis and impact assessment. This is because carrying out the analysis is often an iterative process of trial and error: initial modeling assumptions are refined during the course of the analysis and it is only through a gradual process of improvement that the final results are reached. In this illustration, such a trial and error approach will not be followed. This then reduces the content of the goal and scope definition phase to a few aspects only. These steps relate to defining the exact question for which the LCA is carried out.

In this example we choose to analyze and compare two alternative methods of lighting a room: with the traditional incandescent bulb and with the modern energy saving fluorescent bulb. Of course, a comparison of one incandescent bulb and fluorescent bulb is not very useful. We choose to compare the product alternatives on the basis of the function 'having 10 hours of light of an intensity of 1000 lumen'. This has consequences for the power of the two product alternatives: an incandescent bulb of 60 W and a fluorescent bulb of 15 W. Both have different characteristics with respect to life span and material composition, but these aspects are addressed in the inventory analysis. The two alternatives differ in more respects, like frequency of the light emitted, warming-up time and so on. We decide in the goal definition of this example case to ignore these differences. In other cases, however, they may be felt to be insurmountable, so that the two products are not regarded as functionally equivalent, and a juxtaposition and comparison of the two is not a sensible idea.

17A.3 Inventory Analysis – The Basic Case

The main activity of the inventory analysis is collection and manipulation of process data. We will first do this for the incandescent bulb. The flow chart of unit processes is shown in Fig. 17A.1.

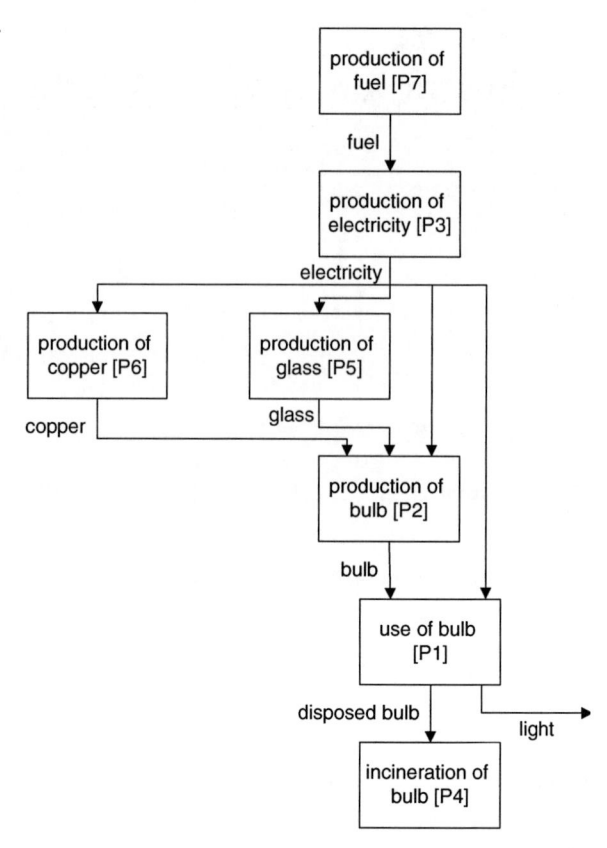

Fig. 17A.1 Flow chart of processes for one of the two example products: The incandescent bulb. Note that the word 'bulb' should be understood here as 'incandescent bulb' throughout

Every box represents a unit process, and every unit process needs to be represented with data. The data are tabulated in the process sheets of Table 17A.1.

For the fluorescent bulb the flow chart looks slightly different. For simplicity, however, it has been assumed that the process 'use of incandescent bulbs' is replaced by 'use of fluorescent bulbs', and that similar changes of incandescent into fluorescent are needed at other places, but that the structure of the flow chart is identical to that of Fig. 17A.1. The data for the unit processes that are not yet listed in Table 17A.1 are shown in Table 17A.2.

The basic calculation that takes place in the inventory analysis is to scale each unit process linearly with an appropriate scaling factor so as to satisfy a commodity balance for every product, material, service, or waste. This is referred to as the inventory problem. Let us consider the commodity balances for the case of the incandescent bulb. The external demand is '10 hr incandescent bulb light'. This means that we must find scaling factors

17 An illustration of the LCA technique

Table 17A.1 Data for the unit processes in the flow chart that constitutes the life cycle of the incandescent bulb

Process 1: Use of incandescent bulbs	
Economic inflows:	1 incandescent bulb
	10,000 MJ electricity
Economic outflows:	1 disposed incandescent bulb
	5,000 h incandescent bulb light
Process 2: Production of incandescent bulbs	
Economic inflows:	1,000 MJ electricity
	10 kg glass
	5 kg copper
Economic outflows:	1,000 incandescent bulbs
Process 3: Production of electricity	
Economic inflows:	500 kg fuel
Economic outflows:	1,000,000 MJ electricity
Environmental emissions:	1,000 kg CO_2 to air
	100 kg SO_2 to air
Process 4: Incineration of disposed incandescent bulbs	
Economic inflows:	100 disposed incandescent bulbs
Environmental emissions:	100 kg CO_2 to air
	2 kg copper to soil
Process 5: Production of glass	
Economic inflows:	100 MJ electricity
Economic outflows:	1,000 kg glass
Environmental resources:	1,000 kg sand
Process 6: Production of copper	
Economic inflows:	10,000 MJ electricity
Economic outflows:	100 kg copper
Environmental resources:	1,000 kg copper ore
Process 7: Production of fuel	
Economic outflows:	1,000 kg fuel
Environmental resources:	1,200 kg crude oil
Environmental emissions:	200 kg CO_2 to air
	5 kg SO_2 to air

Table 17A.2 Additional data for the unit processes in the flow chart that constitutes the life cycle of the fluorescent bulb

Process 8: Use of fluorescent bulbs	
Economic inflows:	5,000 MJ electricity
	1 fluorescent bulb
Economic outflows:	1 disposed fluorescent bulb
	25,000 h fluorescent bulb light
Process 9: Production of fluorescent bulbs	
Economic inflows:	3,000 MJ electricity
	20 kg glass
	150 kg copper
Economic outflows:	1,000 fluorescent bulbs
Process 10: Incineration of disposed fluorescent bulbs	
Economic inflows:	100 disposed fluorescent bulbs
Environmental emissions:	200 kg CO_2 to air
	4 kg copper to soil

for each unit process such that the scaled sum of outflows minus the scaled inflows of incandescent bulb light exactly yields a net outflow of 10 h incandescent bulb light. It is easily seen that the scaling factor for the unit process 'use of incandescent bulbs' is determined by this rule at a value of $10/5,000 = 0.002$. All other scaling factors are still undetermined. However, scaling of the use process brings about that the inflow of incandescent bulbs is set to 0.002. Now, we can construct the commodity balance for incandescent bulbs: the scaled sum of outflows minus the scaled inflows of incandescent bulbs should exactly balance to 0. The scaled inflow is, as was shown above, 0.002, and there is only one unit process ('production of incandescent bulbs') that produces this item, with an unscaled magnitude of 1,000. The scaling factor for this unit process is therefore determined to be $0.002/1,000 = 0.000002$. In this way, we can go through the entire flow chart of unit processes. Table 17A.3 presents the results of this calculation for both product alternatives.

Clearly, knowledge of scaling factors can not be the aim of the inventory analysis. The knowledge one desires is about emissions of pollutants and use of natural resources per functional unit of every product alternative considered. Scaling factors are only an intermediate step in obtaining these elementary flows. A scaling factor affects all flows of a unit process. Above, we have applied them only to the economic flows, like electricity and glass, not to the environmental flows, like CO_2 and sand. Therefore, we must apply the scaling factors to the environmental coefficients of each unit process as well, and calculate the emissions and resource uses of every unit process in relation to the reference flow of 10 h incandescent bulb light.

Table 17A.3 Scaling factors for the two product alternatives on the basis of 10 h of light

Unit process	Incandescent bulb	Fluorescent bulb
[P1] Use of incandescent bulb	0.002	0
[P2] Production of incandescent bulb	2 E–6	0
[P3] Production of electricity	2 E–5	2 E–6
[P4] Incineration of disposed incandescent bulbs	2 E–5	0
[P5] Production of glass	2 E–8	8 E–9
[P6] Production of copper	1 E–7	6 E–7
[P7] Production of fuel	1 E–5	1 E–6
[P8] Use of fluorescent bulbs	0	0.0004
[P9] Production of fluorescent bulbs	0	4 E–7
[P10] Incineration of disposed fluorescent bulbs	0	4 E–6

Table 17A.4 Inventory table, containing the emissions of pollutants (outputs) and the extractions of resources (inputs) for the two product systems investigated

Emission/resource	Incandescent bulb	Fluorescent bulb	Unit
Outputs			
CO_2 to air	0.024	0.003	kg
SO_2 to air	0.002	0.0002	kg
Copper to soil	4 E–5	1.6 E–5	kg
Inputs			
Sand	2 E–5	8 E–6	kg
Copper ore	0.0001	0.0006	kg
Crude oil	0.012	0.0012	kg

Finally, it is usual to aggregate the magnitude of each environmental flow over all unit processes into the inventory table, that lists the total emissions and resource uses that are associated to the reference flow. Table 17A.4 provides the inventory table of the incandescent bulb and the fluorescent bulb.

In this hypothetical example, fluorescent bulbs are better on all aspects, except for the extraction of copper ore.

In practice, the use of matrix algebra facilitates the computations, especially when there are loops in the flow chart. Such loops correspond to recursive relations in the system, and they are quite normal. Production of electricity requires coal, and mining of coal requires electricity, and there are numerous examples of this kind. See Heijungs and Suh (2002) for a presentation of the algebraic structure of the inventory analysis.

17A.4 Inventory Analysis – Some Complications

So far the straightforward approach. There is often one important complication. For the incandescent bulb, there are seven unit processes, so we need to compute seven scaling factors. Furthermore, there are seven intersectoral commodities, so there are seven commodity balances. Each balance is an equation, so we have seven equations in seven unknowns. Under normal conditions, such a set of equations can be solved. However, it may happen that the number of equations is not equal to the number of unknowns. An important class of unit processes that give rise to this complication is those that provide two or more functions at the same time. There are numerous examples of such processes: cow breeding yields milk, meat, manure, skins, calves, sperm, and some other co-products, cracking of crude oil yields a whole range of fuels, and electrolysis of kitchen salt yields chlorine, caustic soda and hydrogen gas. When an LCA study is carried out for a PVC package, it means that production of chlorine is involved. This unit process is, however, intricately connected to the production of caustic soda and hydrogen gas. If we compute a scaling factor for chlorine production, we automatically include a certain amount of caustic soda and hydrogen gas in our product system.

Thinking in terms of equations that correspond to economic flows and unknowns that correspond to unit processes, the occurrence of multi-functional processes can be understood to create more equations than unknowns (Heijungs and Suh 2002). Such a set of equations is in general not solvable without additional assumptions. The extra equations are often not compatible with the other ones. In terms of equations and unknowns, there are several solutions to this problem:

- Add unknowns until there are as much as there are equations
- Delete equations until there are as much as there are unknowns
- Try to find an approximate solution

Remarkably, current thinking in LCA concentrates on the first solution. Procedures that effectively add unknowns to the set of equations are known as allocation procedures. The terminology may be somewhat confusing, as there is no relation with the economic mechanism with the same name, which deals with the choice of options under scarcity of budget or other resources. Adding unknowns to the equations corresponds to adding unit processes to the system.

There are three principally different ways to add unit processes to the system. The first way is to divide the multi-functional process into two or more mono-functional processes, simply by examining the process in more detail and disentangling the process into a number of constituent processes. For instance, it may happen that a plant produces two products with two different machines. With a global view, the plant can be described as one multi-functional process, but closer inspection reveals that there are in fact two mono-functional processes. It will be clear that this method is of limited value. It will not solve problems related to truly coupled production, like the case of chlorine and caustic soda.

The second way is known as the substitution method. The idea is that co-production of caustic soda and hydrogen gas makes that these come available to the market, and that less of these chemicals need to be produced by the traditional methods. Thus, co-producing a certain amount of caustic soda in chlorine production effectively substitutes (or avoids) other ways of producing this amount. It seems reasonable to subtract the fuel and material needs and the emissions of the substituted process from our PVC packaging system. In other words, we add a specification of a unit process 'production of caustic soda' to our flow chart of PVC packaging, and we will be prepared to find negative scaling factors for this substituted (or avoided) unit process. In practice, it may be difficult to find unit process specification for substituted processes. After all, these are activities that have not taken place, and providing data for things that are not there is of course cumbersome, both technically and philosophically speaking.

The third way may be called the partitioning method. The idea is that we can assign a 'fair' share of fuel and material needs and emissions to the different functional outputs of a multi-functional process. For instance, 50% to chlorine, 40% to caustic soda, and 10% to hydrogen gas. Or, more sophisticated, 50% of the electricity needs to be assigned to chlorine, but the entire 100% of the leaked chlorine. If we analytically decouple a multi-functional process into a number of mono-functional processes, and if we are allowed to choose independent scaling factors for these originally coupled processes, we will have introduced more scaling factors, some of which (those for caustic soda and hydrogen gas production) will be found to be zero. The large problem with the partitioning method is, of course, that the numerical values of the so-called allocation factors is difficult to decide upon. Main movements are towards reflecting physico-chemical causalities (based, for instance, on thermodynamics) and towards reflecting socio-economic causalities (based, for instance, on market prices).

We will not address the other possibilities for finding an answer to the inventory problem. However, it is clear that, whether we choose for allocation or for one of the other solutions, whether we choose within allocation for substitution or for partitioning, and how we choose within these approaches for one particular mode of obtaining the required data, the inventory analysis of an LCA is not based on purely mechanical manipulations with purely empirical data. Value judgments enter the analysis, most clearly at the allocation step, but also at several other points. This stresses the importance of a good procedural embedding, to prevent LCA working as a 'hired gun'.

Let us return to the example. We change the unit process data of Table 17A.1 by making the electricity production process (process 3) a multi-functional process. We assume that process heat now is collected to be delivered to users; see Table 17A.5.

Under normal conditions, the inventory can not be solved. We will carry out the allocation in two different ways, to demonstrate how results depend on such choices.

For the substitution method, we need to introduce data for an 'avoided' or 'substituted' process. This is a process that produces heat, without co-producing electricity. We will assume the process data to be as in Table 17A.6.

Notice that we now have a process producing a flow 'co-heat', and a process producing a flow 'heat'. The distinction in name has been made to emphasize the fact that there are two ways of production, hence there are two 'brands' so to say. It is like tomatoes cultivated with traditional and with organic culture. Although the tomatoes themselves may be indistinguishable, we must give them a distinct feature, and hence a different name. A crucial step in the substitution method is to explicitly announce that the heat that is produced in process 11 substitutes the heat co-produced in process 3'. Hence, we establish the (possibly debatable) equivalency between the flows heat and co-heat. Now, there are eight processes and eight flows, and the system of equations can be solved. We find new scaling factors and a new inventory table (see Table 17A.8 below).

Table 17A.5 Additional data for the unit processes in the flow chart that constitutes the life cycle of the fluorescent bulb

Process 3':	production of electricity and co-heat
Economic inflows:	500 kg fuel
Economic outflows:	1,000,000 MJ electricity
	1,000,000 MJ co-heat
Environmental emissions:	1,000 kg CO_2 to air
	100 kg SO_2 to air

Table 17A.6 Data for the unit process of heat production, which is assumed to be an 'avoided' process when the multi-functional process 3' is allocated with the substitution method

Process 11:	Production of heat
Economic inflows:	0.2 kg fuel
Economic outflows:	1,000 MJ heat
Environmental emissions:	0.3 kg CO_2 to air
	0.04 kg SO_2 to air

For the partitioning method, we proceed quite differently. Process 3' is separated into two independently variable processes: one (process 3a) which produces electricity without heat, and one (process 3b) which produces heat without electricity. For some reason (for instance, based on energetic value), we decide to allocate 80% of the burdens to electricity and 20% to heat. Such a partitioning key may be based on economic values, on energy content, or on a different physical or socio-economic parameter. Anyhow, given these allocation factors 0.8 and 0.2, this leads to the allocated process specifications as in Table 17A.7.

Again, we have eight flows and eight processes, and the system of equations can be solved.

We may observe that the results are different. The results with the substitution method are as in Table 17A.8, whereas Table 17A.9 shows the results with the partitioning method.

We will not discuss the advantages and disadvantages of the several approaches to allocation. There exists an extensive literature on this. For here, it is important to recognize the fact that the inventory analysis is not a mere mechanical processing of data, but that difficult and debatable choices must be made. These apply, as

Table 17A.7 Results of the application of the partitioning method with allocation factors of 0.8 and 0.2 to the data for the multi-functional process 3'.

Process 3'a:	Production of electricity
Economic inflows:	0.8 × 500 kg fuel
Economic outflows:	0.8 × 1,000,000 MJ electricity
Environmental emissions:	0.8 × 1,000 kg CO_2 to air
	0.8 × 100 kg SO_2 to air
Process 3'b:	Production of co-heat
Economic inflows:	0.2 × 500 kg fuel
Economic outflows:	0.2 × 1,000,000 MJ co-heat
Environmental emissions:	0.2 × 1,000 kg CO_2 to air
	0.2 × 100 kg SO_2 to air

Table 17A.8 Inventory table for the system, modified with a co-producing process, and allocated with the substitution method

Emission/resource	Incandescent bulb	Fluorescent bulb	Unit
Outputs			
CO_2 to air	0.017	0.0023	kg
SO_2 to air	0.0012	0.00012	kg
Copper to soil	4 E–5	1.6 E–5	kg
Inputs			
Sand	2 E–5	8 E–6	kg
Copper ore	0.0001	0.0006	kg
Crude oil	0.007	0.0007	kg

Table 17A.9 Inventory table for the same modified system as in Table 17A.8, but now allocated with the partitioning method

Emission/Resource	Incandescent bulb	Fluorescent bulb	Unit
Outputs			
CO_2 to air	0.02	0.0026	kg
SO_2 to air	0.0016	0.00016	kg
Copper to soil	4 E–5	1.6 E–5	kg
Inputs			
Sand	2 E–5	8 E–6	kg
Copper ore	0.0001	0.0006	kg
Crude oil	0.0096	0.00096	kg

we have seen, to the choice of the allocation method and to the choice of the 'avoided' process or allocation factor, but they apply at many more places, like setting of system boundaries, degree of region- and time-specificity of the processes, assumptions with respect to consumer behaviour, and so on. It is impossible to illustrate all these complications here, but we will bring forward one topic, related to consumer behaviour.

The functional unit was set at 10 h of light, with two different kinds of bulb. It can be argued that consumers will be less precise in switching off lights when they leave a room if they use economical, or 'green' for that sake, bulbs, like incandescent bulbs. One could carry out a survey, or just speculate, on the degree to which this behavioural difference exists. This then could lead to an adjustment in the basis for comparison, for instance comparing 10 h of light with an incandescent bulb against 12 h of light with a fluorescent bulb. One could, on the other hand, also argue that such a behavioural difference should not be taken for granted, that consumers should be informed on the adverse effects of lighting fluorescent bulbs when no one is in the room, and that a proper comparison should be based on how we expect that people use their bulbs in future.

Many similar questions can be posed. There is not one single answer to such questions, and it should be clear that a refined reflection on the exact goal and scope of the LCA may be needed. This may well involve the consultation of experts and stakeholders from different disciplines and pressure groups. It should make clear that LCA is not the objective process with reproducible results that was strived for some time ago. It is generally recognized by now that LCA is a tool for decision-support, and that this, amongst others, implies that a proper in-context-placing of the LCA process and its results is needed.

17A.5 Impact Assessment – Definition of Impact Categories, Selection of Characterization Models, Classification and Characterization

The inventory tables that were discussed above illustrate what life cycle inventory analysis is about. But they do not show the complexities that are involved in a typical case in real life. Then, the number of unit processes may grow to 500, with a more or less equal number of economic flows. Although this makes the analysis more difficult and time consuming, the use of dedicated LCA software will have no problems with systems of this size. The number of environmental flows is a different subject, however. It can rise up to 1,000. From a computational point of view, this is not especially problematic, but the interpretation of the results becomes unfeasible. How is one to make sense of a table with 1,000 different items, ranging from dichlorobenzene to sulfur dioxide and from bauxite to whales? It is most likely that one product alternative is better for one emission while the other is better for the other one. The impact assessment condenses for each product alternative the long inventory table into a small number of impact category indicators, or even into a single weighted index.

The first steps of impact assessment deal with the definition of impact categories and the selection of characterization models. In this example, we will define three hypothetical impact categories: resource scarcity, ecotoxicity and global warming, and we will associate these with three hypothetical characterization models. These express the impacts of a number of interventions in terms of resource depletion units (RDU), square meters of ecosystem affected, and kg of CO_2-equivalents with respect to global warming. Characterization factors are derived from these models, and their hypothetical values are listed in Table 17A.10.

Classification is a step that sometimes is misunderstood: it is the assignment of inventory results to the selected impact categories. Notice that no calculations are performed, and that the characterization factors provided by the characterization models are only used in determining if a certain item is assigned to a certain category or not. Table 17A.11 gives the table that result after classification of Table 17A.4.

The next step is characterization. This involves the use of the characterization factors for numerically converting the LCI results to so-called indicator results. For ecotoxicity for the incandescent bulb, for instance, we have contributions by SO_2 and by copper: $(20 \times 0.002) + (10,000 \times 4E-5) = 0.04 + 0.4 = 0.44\,m^2$ ecosystem (see Table 17A.12).

We see that characterization results in a table that is shorter (in a real-world LCA study, several hundreds of inventory results are condensed into a dozen of indicator results) and that is easier to understand. However, we still cannot formulate an unambiguous preference, as fluorescent bulbs are much better on ecotoxicity and global warming, but somewhat worse

Table 17A.10 Hypothetical characterization factors for three impact categories, derived from three hypothetical characterization models

Category 1:	Resource scarcity
Environmental resources:	0.1 RDU/kg sand
	10,000 RDU/kg copper ore
	250 RDU/kg crude oil
Category 2:	Ecotoxicity
Environmental emissions:	$20\,m^2$ ecosystem/kg SO_2 to air
	$10,000\,m^2$ ecosystem/kg copper to soil
Category 3:	Global warming
Environmental resources:	0.5 kg CO_2-equivalents/kg SO_2 to air
Environmental emissions:	1 kg CO_2-equivalents/kg CO_2 to air

Table 17A.11 Classification table, containing the inventory items assigned to the selected impact categories

Impact category	Incandescent bulb	Fluorescent bulb	Unit
Resource scarcity			
Sand	2 E–5	8 E–6	Kg
Copper ore	0.0001	0.0006	kg
Crude oil	0.012	0.0012	kg
Ecotoxicity			
SO_2 to air	0.002	0.0002	kg
Copper to soil	4 E–5	1.6 E–5	kg
Global warming			
CO_2 to air	0.024	0.003	kg
SO_2 to air	0.002	0.0002	kg

Table 17A.12 Characterization table, containing the converted and aggregated inventory items assigned to the selected impact categories

Impact category	Incandescent bulb	Fluorescent bulb	Unit
Resource scarcity	4.	6.3	RDU
Ecotoxicity	0.44	0.16	m^2 ecosystem
Global warming	0.024	0.003	kg CO_2-equivalent

on resource scarcity. A final judgement depends, amongst others, on the societal importance of resource scarcity in relation to the other impact categories. And this also depends on the meaning of the numbers.

17A.6 Impact Assessment – Normalization and Weighting

Normalization aims to add still more meaning to the indicator results. This can be done by expressing them relative to a reference value. A usual choice for this reference value is the annual global size of the impact categories. Given an inventory of annual world-wide emissions and resource extractions, the tables with characterization factors can be applied to compute the indicator results for the world over 1 year. This provides a reference value for each impact category. See Table 17A.13.

The normalization data can now be used, with Table 17A.14 as result.

Observe the peculiar unit 'year': the normalized indicator results all have the same unit, which expresses the period of time that the functional unit of product can be held responsible for the world problems. Of course, this time period is very small for 10 h of light.

Also observe that the decision problem has not been solved yet: the fluorescent bulb is still better in two respects and worse in one respect. But the removal of arbitrary choices of unit facilitates making the impact categories commensurable. Suppose that somehow, e.g., by shadow prices or questionnaires, it has been found that the inter-category weighting factors are as in Table 17A.15.

Table 17A.13 Hypothetical normalization data, based on world for 1 year, used as a reference value in normalization

Impact category	Value	Unit
Resource scarcity	1 E11	RDU/year
Ecotoxicity	5 E9	m² ecosystem/year
Global warming	1 E10	Kg CO_2-Equivalent/year

Table 17A.14 Characterization table, containing the converted and aggregated inventory items assigned to the selected impact categories

Impact category	Incandescent bulb	Fluorescent bulb	Unit
Resource scarcity	4.0 E–11	6.3 E–11	year
Ecotoxicity	8.8 E–11	3.3 E–11	year
Global warming	2.4 E–12	3.0 E–13	year

Table 17A.15 Hypothetical weighting data, used in weighting of normalized indicator results

Impact category	Value	Unit
Resource scarcity	5	–
Ecotoxicity	35	–
Global warming	60	–

Table 17A.16 Weighting table, containing the converted and aggregated normalized indicator results

Impact category	Incandescent bulb	Fluorescent bulb	Unit
Weighted index	3.4 E–9	1.5 E–9	year

Application of these weights to the normalized results yields Table 17A.16.

Now, the first 'unambiguous' judgement appears: the fluorescent bulb beats the incandescent bulb by a factor of more than 2.

17A.7 Interpretation

What has been presented above is of course a caricature. The final result (3.4 against 1.5) suggests that fluorescent bulbs are better, but we know that this ignores uncertainties, data gaps, assumptions, value judgements, and many other sources of error. A systematic analysis of the robustness of the results is an essential ingredient of a careful decision-making procedure. In LCA, the interpretation phase is meant to provide an explicit place for such analyses. Unfortunately, operational methods and clear rules are not yet standardized. In this example, we will show a small number of examples.

One example has already been shown. This was in Section 17A.4, where two different allocation methods were discussed and their results were compared. In general, one should make different choices for at least:

- Product performance characteristics
- System boundaries
- Process data
- Allocation methods and factors
- Choices of impact categories
- Characterization factors
- Normalization data
- Weighting principles and data

All types of results will be obtained. Monte Carlo techniques may be helpful here, but they cannot be used for all types of uncertainties. In the end, one will develop an idea of the bandwidth of the results, and consequently of the robustness of a judgement. For instance, when all reasonable choices lead to 3.4 with a standard deviation of 1 for the incandescent bulb and 1.5 with a standard deviation of 1 for the fluorescent bulb, one may still say that the latter beats the former. But when the standard deviations are much larger, and the confidence intervals largely overlap, decision-making will become problematic.

Apart from the parametric or probabilistic variation of data and the variation of assumptions and choices, there are interpretation-related issues that reside more at the procedural level. Are the product alternatives really comparable, also from the perspective of consumers? Have the right stakeholders been involved in the correct way? These questions may have a large impact on the usefulness and credibility of the LCA study, but they are not discussed in this more modeling-oriented example.

Another useful approach to interpretation is the contribution analysis, in which results are decomposed into contributing processes, flows or categories. One may, for instance, find out that 90% of the weighted index for the incandescent bulb is due to ecotoxicity, and that this in turn is largely determined by the emission of copper. Such results are useful for providing a check for counterintuitive results. They also help one to focus on key issues for process and product improvement. Thus one can step from using LCA for static assessment to using LCA for adaptive analysis.

17A.8 Concluding Remarks

In this chapter, the working details of LCA have been described. We sometimes touched upon issues relating to the possibilities and limitations of LCA. However, we did not explore that area. Chapter 12 discusses such issues at length.

Although we focused on the computational aspects of LCA, we still were able to only discuss the most important aspects in a not too rigid way. A full discussion, based on matrix algebra and including references to other material, is in Heijungs and Suh (2002).

Reference

Heijungs, R., & Suh, S. (2002). *The computational structure of life cycle assessment*. Dordrecht: Kluwer.

Chapter 18
Integrated Assessment

Jean-Paul Hettelingh, Bert J. M. de Vries, and Leen Hordijk

Contents

18.1 Introduction and Scope 385
18.2 Integrated Assessment as Multi Focal Methodology to Support Environmental Policies.. 387
  18.2.1 Definitions of Integrated Assessment 388
  18.2.2 Integrated Assessment Modelling and Participatory Integrated Assessment....................... 389
  18.2.3 The Structured Analysis of Causes and Effects 390
  18.2.4 Uncertainty.. 390
18.3 The RAINS Model and Its Use as an Example of Integrated Assessment....................................... 392
  18.3.1 Acidification.. 392
  18.3.2 Integrated Assessment Models for Acid Rain 392
  18.3.3 Description of the RAINS Model 392
  18.3.4 The Use of RAINS in the Negotiation Process........ 394
  18.3.5 Lessons Learned.. 400
18.4 The TARGETS Model: A World Model to Explore Sustainable Development 401
  18.4.1 The TARGETS Model Set-Up................................. 401
  18.4.2 Future Scenarios in TARGETS: Three Utopias....... 405
  18.4.3 Risks: Dystopical Tendencies 406
  18.4.4 Lessons Learned.. 407
18.5 The IMAGE Model's Role in the Construction of New Greenhouse Gas Emissions Scenarios..... 407
  18.5.1 The Special Report on Emissions Scenarios (SRES) 409
  18.5.2 The Framework: Four Narratives and Their Interpretation 410
  18.5.3 Model Implementation.. 412
  18.5.4 Some SRES Results 414
  18.5.5 SRES Evaluation... 415
18.6 Conclusions and Outlook 417
References.. 417

Jean-Paul Hettelingh(✉)
Coordination Centre for Effects at the Netherlands
Environmental Assessment Agency (MNP),
Bilthoven, The Netherlands
e-mail: jean-paul.Hettelingh@pbl.nl

18.1 Introduction and Scope

The development of environmental policy with respect to the current and future quality of air, water and soil cannot be considered in isolation. Alleviation of one environmental problem may lead to an increase in the importance of another problem. For example, the reduction of nitrogen oxide emissions may decrease the risk of acidification but can, under particular circumstances, lead to an increase in the formation of tropospheric ozone. Climate change may lead to changes in land cover that may affect biodiversity. Another example often mentioned is the relationship between sulphur dioxide emissions in which the risk of acidification is increased, while the enhanced greenhouse effect is, through the formation of sulphate aerosols, decreased. This is why relationships between environmental problems need to be recognised by policies aimed at reducing these problems. This also holds true for policies affecting socio-economic activities, since these can generate externalities in the environment. For example, both the fifth and the sixth Environmental Action Programmes of the European Commission identify five target sectors (industry, energy, transport, agriculture and forestry, and tourism), all having an important – but different – impact on the environment.

Environmental science – applied to support policy development – addresses different target sectors and a variety of environmental issues in facing the challenge of meeting both disciplinary and inter-disciplinary scientific requirements for formulating adequate hypotheses for verification and validation. Conflicts that may arise between the theories developed to understand environmental problems and their linkages, and the knowledge based on data, calls for a multi-disciplinary approach to relationships between environmental issues and socio-economic developments. This in turn puts high demands – possibly with significant variations – on data requirements. Scientific experiments to test hypotheses on climate change, for example, are not easy, considering the wide range of spatial and temporal scales involved. The availability of data is limited. This applies both to the impact of climate change, of which the risks can be simulated to occur over several decades, and the socio-economic causes, which may change dramatically during that time. Spatial and temporal scales of environmental problems will vary, depending on regional, national, continental or global characteristics. National environmental policy may be adequate to deal with environmental issues related to such factors as waste water, whereas formulating policies to influence transboundary environmental problems – such as air pollution – requires synchronization of national policies.

Thus the identification and analysis of the system of causes and impacts of environmental problems may depend on the way in which boundaries of the system are defined. The use of models (see Chapter 17) may help to unravel this complex piece of fabric, and ensure that model outputs with varying inputs are made compatible and mutually comparable. However, the complexity of causes and impacts of environmental issues requires more frequent use of one model to analyse the effectiveness of alternative policy instruments. Moreover, the output of one model may have to become the input for another model in order to describe the relevant relationships. This led two decades ago to the development of so-called *Integrated Environmental Models (IEM)*.

In Hafkamp (1984: 25) an Integrated Environmental Model is defined as a type of model in which a great diversity of real-world phenomena, e.g. phenomena related to the natural environment, and their interdependencies are represented simultaneously. Seen from this viewpoint, an IEM can include the modelling of:

- *Environmental impacts* due inter alia to air pollution (e.g. acid rain from sulphur and nitrogen compounds), solid and toxic waste
- *Ecological systems*: terrestrial and marine ecosystems, predator-prey systems etc.
- *Resources management*: exploitation of energy sources (i.e. coal, oil, gas), forests

- *Economic systems*: macro-, meso- and micro-scale systems
- *Social and political systems*: social layers, activity pattern, e.g. consumer organizations, unemployment rates
- *Transportation systems*: (rail)road networks, airport and harbour facilities
- *Demographic developments*: age structure, educational profiles

Many IEMs have been constructed over the past two decades, with detailed overviews to be found in the literature (e.g. Hafkamp 1984; Brouwer 1987; Braat and van Lierop 1987; Hettelingh 1990; Rotmans 1990; Morita et al. 1994; Hordijk and Kroeze 1997). More recent IEM models, with applications emphasising climate change scenario assessments, can be found in IPCC (2001c: chapter 2).

In the past few years it has become common practice to use the term 'assessment' to reflect the interaction between scientists and policy-makers in the process of the identification and analysis of causes and impacts of an environmental issue (GEA 2001). Nowadays, assessment does not necessarily imply the use of a model, but might also include other methods such as expert panels. The term *Integrated Assessment Model* (IAM) is increasingly used when models are implicated, although consensus is still lacking about the kind of processes to which this term may apply. Jäger points out that '*Integrated Assessment, Integrated Assessment Models*, and *Integrated Environmental Assessment* have been used in recent years to describe parts of the process of linking science and policy, in particular for global environmental issues' (Jäger 1998: 145). However the term Integrated Assessment Model has also been used for other issues. An overview of Integrated Assessment Models used in acidification, for example, can be found in Hordijk and Kroeze (1997). The discussion on the backgrounds and terminology related to Integrated Assessment and Integrated Assessment Modelling is still ongoing in the scientific literature. While we can provide the reader with a short overview of the different viewpoints and definitions in this chapter, we can only offer a snapshot of the state of the art.

For this reason we have chosen to present the information in the following order. Section 18.2 is devoted to a reflection on the (historical) context of – what is now termed – Integrated Assessment, including an overview of current definitions. In Section 18.3 a basic system-analytical framework for the description and understanding of linkages between cause, effect and policy evaluation of environmental issues is presented. This framework is then used in Sections 18.4–18.6 as an anchor point to describe well-known assessment tools and applications in the fields of acidification (the RAINS model) and climate change (IMAGE and TARGET). The chapter concludes, in Section 18.7, with an overview of lessons learned and an attempt to present an outlook for Integrated Assessment.

18.2 Integrated Assessment as Multi Focal Methodology to Support Environmental Policies

Effective environmental policy requires a comprehensive means of dealing with the definition, localization and assessment of damage to the natural environment from human activities. With the growing scale of the interactions between human development and the environment, the complexity of these interactions has also increased. Today's concern about the greenhouse effect and its environmental consequences is an example of an emerging consciousness that human activities, biochemical processes and climate are interrelated (see also Clark and Munn 1986). Environmental problems such as acid rain, soil erosion, ozone depletion, deforestation and surface water pollution are all associated with interrelated human activities. They occur all over the world but reflect multi-regional rather than global characteristics. Studies on these problems require a multi-disciplinary scientific collaboration and an integrative – or at least coherent – way of describing and formalising the system of causes and effects. Investigating global change of the environment requires an understanding of the structure of causes and environmental problems, as well as identification of their related temporal and spatial scales.

The temporal scale of environmental problems has to do with economic and social activities that are often undertaken for a benefit that is measurable within the time span of the current human generation. The use of pesticides, for example, may have increased the quantity of agricultural output over the past years but the persistence of these pollutants can jeopardise the quality of natural products for a long time to come. The related spatial scale may extend far beyond the regional scale within which the economic and social development occurred. Many studies – with the exception of climate change research – do not address interactions between human activities and environmental quality in a sufficiently general way. Before the recently emerging awareness of the cascade of sources and impacts of climate change and linkages with air pollution, the tendency was to focus on (a) environmental problems in isolation from one another, (b) immediate impacts of ameliorative measures above a sufficiently long-term horizon, and (c) local or national scales in favour of sufficiently broad spatial scales.

The Club of Rome study, which used the World2 and World3 models to address the limitations of natural resources in satisfying human development, should be considered the cradle of the later adopted broad and integrative view on the complex relationships between the environment and human activities (Meadows et al. 1972). This study used a modelling approach based on the system dynamics approach of Forrester (1961, 1971, 1980) to provide an analysis of linked temporal trends involving the population, the economy, the use of natural resources and the environment. Many complex models were developed in response to the Club of Rome report; these included the Mesarovic-Pestel World Model (focusing on the carrying capacity), the MOIRA model (focusing on international agricultural relationships), the Latin America World Model (optimization of life expectancy) and the UN World Model (focusing on multi-regional economic-environmental relationships). See Barney (1981: 603–681) for a comprehensive overview. The study of the Club of Rome attracted a wide audience of scientists, the general public and policy-makers. The 'business as usual' approach in the development of policies had until then not sufficiently emphasised the constraints imposed by our natural resources.

The increased awareness of growth constraints led – for example – to an Environmental Message, which President Carter of the United States presented to the congress in May 1977. The message 'directed the Council on Environmental Quality and the Department of State, working with other federal agencies, to study the "probable changes in the world's population, natural resources, and environment through the end of the century"… to serve as the foundation of our longer-term planning' (Barney 1981: iii). The resulting 'Global 2000 report to the President' does not mention the term 'integrated assessment' specifically.

However, it attempted to link a large variety of issues and models available at federal agencies and institutions to address issues of population, Gross National Product, climate, technology, food, fisheries, forestry, water, energy, fuel minerals, non-fuel minerals and the environment in many regions around the globe. It turned out to be a difficult task to establish these links: 'currently the principal limitation in the US Government's long-term global analytical capability is that the models for various sectors were not designed to be used together in a consistent and interactive manner'. The study was not restricted to the use of quantitative models but also included the opinion of a large number of expert advisers. Under the assumption that national policies regarding population stabilization, resource conservation and environmental protection would remain essentially the same, the study concluded that problems to preserve the carrying capacity of the earth and to sustain a decent life for the human beings are enormous (Barney 1981).

The interest in this kind of multi-focal analysis for the support of policy development materialised further after the 1970s in the development of methods to integrate policy views and alternatives with scientific and technical knowledge. The integrated (environmental) models already mentioned served as a tool; however, other methods to bring together different disciplines and stakeholders were also developed. Holling (1978) introduced the concept of *Adaptive Environmental Assessment (AEA)*, carried out by a team consisting of mathematical modellers, research scientists from various disciplines, resource managers with experience in monitoring and regulation, and finally policy analysts or decision-makers with a broad responsibility for defining management objectives and options. While these efforts in integrated analyses used to be anchored on the limitations of the *carrying capacity* of our resources, nowadays a new anchor-point has been found in a common (scientific and political) objective of developing adequate policies to alleviate the causes and effects of *climate change*. It is unclear to what extent the recent discussion on integrated assessment has taken sufficient notice of earlier work requiring integrated analysis. The concept has, in fact, been revived specifically in the context of climate change, leading to a broad discussion on 'integrated assessment' (IA). This is partly reflected in multiple efforts to find an appropriate definition for IA.

18.2.1 Definitions of Integrated Assessment

The National Acid Precipitation Assessment Program (NAPAP) established in 1980 in the USA, co-ordinated and administered the first study explicitly applying integrated assessment to generate information needed for policy and regulatory decisions on acidic deposition. The aim of the study was to 'examine the relationship among fossil fuel combustion, acids and other pollutants formed by emissions, and the effects of these pollutants on the environment, the economy and human health.' NAPAP defines integrated assessment as:

- *An interdisciplinary activity wherein findings from diverse disciplines are co-ordinated to produce a better understanding of the cumulative impacts of acidic deposition* (NAPAP 1990).

The background and definition of Integrated Assessment have been made subject to debate in a concerted action of the European Commission entitled, 'European Forum on Integrated Environmental Assessment', described in more detail in Tol and Vellinga (1998), and includes the following definitions:

- *Assessment is integrated when it draws on a broader set of knowledge domains than are represented in the research product of a single discipline. Assessment is distinguished from disciplinary research by its purpose: to inform policy and decision making, rather than to advance knowledge for its intrinsic value* (Weyant et al. 1996, quoted from Tol and Vellinga 1997, 1998).
- *In general, integrated assessment can be defined as an interdisciplinary process of combining, interpreting and communicating knowledge from diverse scientific disciplines in such a way that the whole cause–effect chain of a problem can be evaluated from a synoptic perspective with two characteristics: (1) IA's should have added value over a single disciplinary-oriented assessment; (2) IA's should provide useful information to decision-makers* (Rotmans et al. 1996; quoted from Tol and Vellinga 1997).
- *Integrated assessment is policy-motivated research to develop an understanding of the issue, not based on disciplinary boundaries, but on boundaries defined by the problem, (1) to offer insights to the research community for prioritization of their*

efforts; (2) to offer insights to the decision-making community on the design of their policies (Rotmans et al. 1996; quoted from Tol and Vellinga 1997).

- *The two defining characteristics (of integrated assessment) are: (1) that it seeks to provide information of use to some significant decision-maker rather than merely advancing understanding for its own sake, and (2) that it brings together a broader set of areas, methods, styles of study, or degrees of certainty, than would typically characterise a study of the same issue within the bounds of a single research discipline* (Parson 1995; quoted from Tol and Vellinga 1997).
- *Integrated Assessment can be described as a structured process dealing with complex issues, using knowledge from various scientific disciplines and/or stakeholders, such that integrated insights are made available to decision makers* (Rotmans 1998).

One might remark here that these definitions do not make any reference to the environment, probably because climate change happens to be the context in which these definitions were formulated. To apply these to the assessment of environmental issues other than climate change, an explicit reference to the environment is required. The National Institute for Public Health and the Environment for the European Environmental Agency formulated the following definition, including a reference to the environment (EEA 1995):

- *Integrated Environmental Assessment (IEA) is the interdisciplinary process of identification and analysed appraisal of all relevant natural and human processes and their interactions that determine both the current and future state of environmental quality, and resources, on appropriate spatial and temporal scales, thus facilitating the framing and implementation of policies and strategies.*

In the first five definitions, Tol and Vellinga (1998) point out 'in any of these definitions "integrated" conveys a message of "multi- or interdisciplinarity", and "assessment" a message of "policy relevance"'. This is also true for the last definition. Currently, it is doubtful whether the most appropriate definition can be identified (see also Hordijk 1995). The reference to policy relevance is not explicit in the definitions of Integrated Modelling, although the effort to link models has often been triggered by a policy question, as illustrated, for example, in the 'The Global 2000 Report' to the President mentioned earlier.

18.2.2 Integrated Assessment Modelling and Participatory Integrated Assessment

The communication between the experts and modellers on the one hand, and the policy analysts on the other, were not explicitly seen – in the days when models were linked – as part of the integration process. In the current discussion, both the hard quantitative and soft qualitative elements – including the involvement of policy analysts and other stakeholders – are seen to be part of integrated assessment. The distinction between Integrated Assessment and Integrated Assessment Modelling is not clear-cut, although they have at least the same requirement to collect knowledge from different areas, identify data needs, point out uncertainties and show linkages (Hordijk 1995). The involvement of policy in Integrated Assessment requires a comprehensive summary, presentation and packaging of modelling results, whereas Integrated Assessment Models are designed for the analysis of a broad range of policy options. It has become recent practice to make a distinction between Integrated Assessment Modelling and Participatory Integrated Assessment (Tol and Vellinga 1998). The first part, the modelling, refers to the policy-relevant use of mathematically formalised processes into computer models. The second Participatory Integrated Assessment, aims at the inclusion of more qualitative elements – such as value judgements – which can not be easily incorporated in a model. It also aims at the inclusion of decision-makers to obtain dialogue between the analysts and other stakeholders (see also Hisschemöller et al. 2001). In this sense one might argue that Participatory Integrated Assessment shows similarities to Holling's Adaptive Environmental Assessment mentioned earlier.

In practice, organising comprehensive communication between stakeholders is not straightforward. While one is seeking a process in informal workshops, another is seeking a process in conducting projects, in which timely interaction with policy-makers is formally organised. Examples of the latter include the support of the acidification and ozone strategy of the European Commission, the work of the Framework Convention on Climate Change (FCCC) or the activities under the Convention on Long-range Transboundery Air Pollution (LRTAP-Convention). A more detailed description of the FCCC and LRTAP-Convention is provided in the following sections on RAINS and IMAGE. Generally speaking, 'consensus' (between and within groups of scientists and policy analysis) here can be seen as being of great importance. Under the LRTAP-Convention, for example, consensus-based agreements are pursued to reduce European-wide emissions of sulphur and nitrogen oxides, of tropospheric ozone precursors, and of emissions of heavy metals, Polychlorinated Organic Pollutants and particulate matter. The scientific and policy-development part of the LRTAP- Convention is organised in Working Groups (policy) and Task Forces (technical). A Task Force on Integrated Assessment Modelling assesses policy alternatives formulated by the Working Group on Strategies in an iterative process until policy analysts have sufficient material to start negotiations. The material may consist of quantitative results on the cost of emission abatement relative to national energy structures, and the risk of environmental damage due to the long-range transport of emissions that are finally deposited in the natural environment. Thus insight into the chain of causes and effects of air pollution are obtained relative to alternative emission abatement strategies.

18.2.3 The Structured Analysis of Causes and Effects

The structured analysis – Integrated Assessment Modelling and Participatory Integrated Assessment – of the cause–effect chain and the interaction with policy analysts can be expressed in a number of entities reflecting 'Driving Forces', 'Pressures', 'State', 'Impact' and 'Response' (DPSIR), illustrated in Fig. 18.1. Driving forces behind environmental problems include a variety of human activity, such as production and consumption processes structured around economic sectors, and other underlying causes of environmental problems. Driving Forces are generally in consequence of meeting the needs of a society at individual and collective levels. Driving forces lead to 'Pressures' on the environment exerted by proximate causes (e.g. use of natural and biological resources, and emissions). Pressures affect the quality – the 'State' – of the various environmental compartments (air, water and soil) in relation to their functions. Changes in the 'State' may have 'Impacts' on ecosystems, humans, materials and amenities, and resources. This sequence of causes and effects is finally appraised through the assessment of different policy options as 'Response' to an undesired impact (see also EEA 1995: 14, 1999: 427). The DPSIR analysis of policy alternatives often involves the interpretation of a (set) of policy measures in terms of a change in driving forces and related pressures, and the way in which these changes propagate impacts, both in time and space. A sequence of iterations involving alternative policy measures has been made to allow for a comparison between iteration results. A package of policy measures and the simulated results in time or space are often termed scenarios. Scenarios generally describe possible future pathways of society (People), the environment (Planet) and the economy (Profit). Multiple scenarios can be identified, depending on the manner in which the three P's are assumed to interlink the subject to appropriate constraints. Broad constraints on processes affecting economic, institutional and cultural constellations allow a large variety of scenarios to be identified in climate change research under the FCCC, as we will see later in this chapter. Under the LRTAP-Convention the focus is on policy scenarios, generally describing future pathways of causes and effects due to policy measures. Policy scenarios form a sub-set of multiple scenarios, in which policy is only one of the elements. Policy scenarios are relatively simple to 'secure', compared to model variables or parameters. The 'translation' of policy measures into an altered magnitude of a relevant parameter or variable in the DPSIR modelling scheme can be applied in four different DPSIR levels (see Fig. 18.1). In general, measures are made operational at the level of causes, 'Driving Forces' or 'Pressures'. 'Driving Forces' can be adapted by imposing tax measures on polluting processes. Policies addressing 'Pressures' can either include emission limits or apply other sector-specific stimuli. Policy measures on the effect side can influence the 'State' through changed land-use practices, for example, whereas curative measures might include an operational practice in Scandinavian countries of liming soils and lakes to protect or reclaim soil, and water, from the 'Impact' of acidification.

18.2.4 Uncertainty

One of the vital components – and perhaps the current major setback – of the Integrated Assessment of policy options is the uncertainty of scenario results. Answers to questions about the *future* impact of climate change, for example, and the efficiency of policy measures *today* can not be verified and are, furthermore, likely to be error prone. The uncertainty of results obtained through Integrated Assessment is unknown. The only consolation is that the analysis of a variety of scenarios using one consistent DPSIR scheme (and underlying analytical tools) – as in policy scenarios – leads to results of which the error component is comparable from one scenario to another. The magnitude of error propagation in the DPSIR analysis does not necessarily vary *between* policy scenarios. Thus it can be assumed that differences between policy scenario results are primarily due to differences in policy alternatives defined for each individual scenario, which is exactly the purpose of Integrated Assessment. However, this also implies that results can only be interpreted *relatively* (in comparison to another scenario) and not *absolutely*. This reasoning also ignores uncertainties due to the use of a wrong model, lack of knowledge of actual processes (epistemological uncertainty), and perspective-dependent ('world view') uncertainties. Methods have been designed to take into account the various (cumulative) origins – the pedigree – of the uncertainty of input parameters and assessment results. This method is known as the NUmerical Spread Assessment Pedigree (NUSAP) approach of Funtowicz and Ravetz (1990). The extent to which the overall uncertainty can be affected through participatory processes – in addition to the processes inherent to the modelling process – is similarly a subject of current debate. The use of scenarios is also a way of depicting uncertainties, as will be discussed for the TARGETS model in the Section 18.4. An in-depth treatment of the uncertainty aspect of Integrated Assessment would go beyond the scope of this chapter, but see Hettelingh (1990); and van Asselt and Rotmans (1996, 2002).

Let us note here that further development of the Integrated Assessment concept will be a challenge, acknowledging that 'verification and validation of numerical models describing complex systems is impossible' (Oreskes et al. 1994). This chapter illustrates a limited treatment of uncertainty relative to applications of Integrated Assessment of acidification and climate change, which is the subject of the following sections.

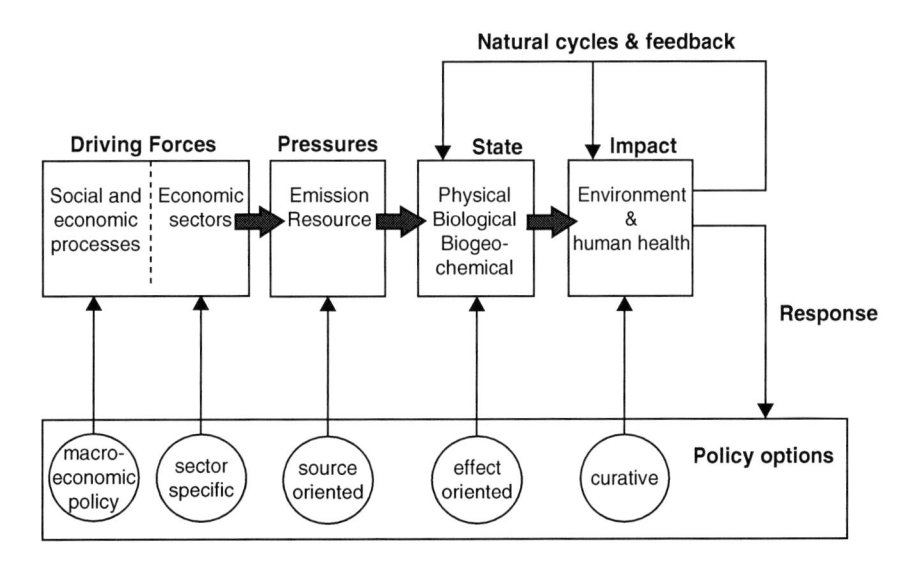

Fig. 18.1 Illustration of the cause-effect relationship including a characterisation of policy options. The scheme illustrates the iterative assessment of the influence of policy option alternatives on driving forces, pressures, state or impacts of environmental issues. Thus, an analysis based on Integrated Assessment Modelling of causes and effects, can interact with a variety of policy options which can be specified through a process of Participatory Integrated Assessment. This may ultimately lead to an identification of an appropriate policy package

18.3 The RAINS Model and Its Use as an Example of Integrated Assessment

18.3.1 Acidification

The problem of acidification has been recognised in Europe since the late 1960s and has been a political issue since the late 1970s. Harm to lakes and the fish population in Scandinavia and damage to forests in Germany have been triggering public concern ever since. It was soon understood that acidification is transboundary in nature, all the more reason for finding an international solution. In 1979 the LRTAP-Convention was signed, followed by a series of protocols on reducing relevant emissions, including sulphur dioxide, nitrogen oxides, volatile organic compounds and ammonia. From 1979 on, the LRTAP-Convention used results from science to back up its negotiations on further emission reduction. Working groups, task forces, integrated programmes and co-ordinating centres were instrumental in assessing the state of science and technology and informing the LRTAP-Convention negotiators. Since 1985 the negotiators started to consider the use of computer models (which we would now call Integrated Assessment Models) to base their protocols on. One of these models is the Regional Acidification INformation and Simulation (RAINS) model (Alcamo et al. 1990; Tuinstra et al. 1999; Schöpp et al. 1999). Further on in this section we will present a short description of RAINS and its use. After overviewing the Integrated Assessment models (IAMs) for acidification, we will shortly describe RAINS.

18.3.2 Integrated Assessment Models for Acid Rain

Hordijk and Kroeze (1997, further indicated by HK) presented the overview of IAMs for acid rain to be used in this section. In their review HK included the IAMs containing at least three of the five components that an IAM for acid rain might include. The five components are sources of emissions, atmospheric transport and transformation, environmental effects, abatement and mitigation options, and monetary evaluation of scenario results. HK found 20 IAMs that met the criteria: 6 for Europe, 6 for North America and 8 for other countries or continents. All 20 models make use of a source–receptor matrix assuming a simple linear relationship between emissions and deposition (in most cases gridcell-based). The European models hardly differ in spatial and temporal resolution. The North American models show much more variation than the European ones.

All models included in the review simulate SO_2 emissions and dispersion, whereas all but two include environmental effects. Only eight models take NO_x emissions into account, while emissions of NH_3 are only modelled in the Co-ordinated Abatement Strategy Model (CASM; see Cough et al. 1994) and RAINS. All but one include control costs, and almost all can be used for scenario analysis. Only eight of the models can be used for optimization analysis. Virtually all models considered are deterministic models that do not take stochastic variables or uncertainties into account.

The level of integration in the different models varies. In most models the sub-models for emissions, atmospheric transport and impact are fully integrated. The models included in the review differ with respect to their effect modules. Some include relatively detailed process-based models (NAPAP), but all present versions of the European models limit their impact modules to the critical load concept (an indicator for long-term environmental damage). HK concluded their review with the observation that most of the models are strongly policy-oriented, with several having actually been used in international negotiations.

18.3.3 Description of the RAINS Model

The RAINS model is modular, as can be seen in Fig. 18.2. The first module addresses energy use, agricultural and other productive activities, while related emissions and control costs are reflected in the following two modules. The fourth module deals with atmospheric dispersion, while the fifth, and last module addresses environmental effects.

The environmental effects module compares deposition of acidifying and eutrophying compounds to critical loads. A critical load is generally defined as 'a quantitative estimate of an exposure to one or more pollutants below which significant harmful effects on specified sensitive elements of the environment do not

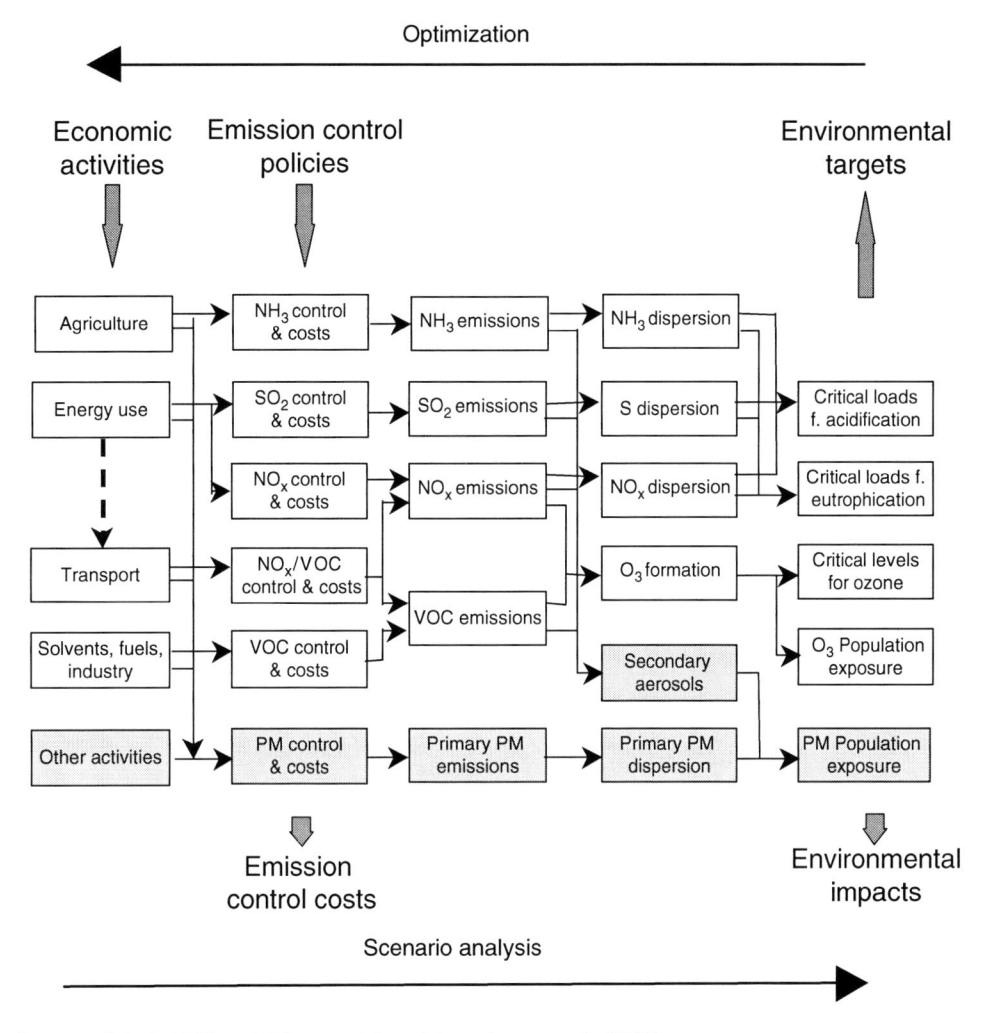

Fig. 18.2 Scheme of the RAINS model Source: Adapted from Amann et al. (2001)

occur according to present knowledge' (Nilsson and Grennfelt 1988). RAINS operates on a relatively large spatial scale: emissions are calculated at country levels and deposition determined at grid cell levels (150 × 150 km), while critical loads have been computed and mapped for about 1.5 million ecosystems. Cumulative distribution functions of critical loads for acidification and eutrophication are computed in all 150 × 150 km grid cells to allow comparison with deposition values for assessment of environmental impacts. The temporal scale of the RAINS assessments is 1 year, with computation results becoming available in tabular form every 5 years. Critical load data are constant over time to reflect levels of acidification (by sulphur and nitrogen deposition) and eutrophication (by nitrogen deposition) that do not damage ecosystems (e.g. forest soils, surface waters).

The first module uses the characteristics of different energy sources and agricultural activities to calculate emissions of three acidifying compounds (sulphur dioxide, nitrogen oxides, ammonia). It contains an extensive and detailed database on energy use and agricultural activities for 38 European countries and regions, 21 energy sources and 6 economic sectors. This degree of detail allows users to address fairly specific questions (e.g. the change in emissions if all power plants in Poland were to use natural gas instead of coal and oil) and more general issues, such as the total of ammonia emissions from all agricultural activities in all of Europe.

The module on abatement costs includes cost and efficiency of a range of abatement techniques for SO_2. These include such factors as substitution of high sulphur fuels with fuels containing less sulphur, limestone injection, wet flue-gas desulphurization, and regenerative processes for sulphur abatement. Control technologies for NO_x emissions in RAINS are combustion modification, selective catalytic reduction, catalytic converters on cars, and several ways to reduce process emissions. The following options can be used for ammonia: low nitrogen fodder, stable adaptation, bio filtration, covered manure storage, low ammonia application of manure, and stripping and adsorption techniques. This module produces emissions per sector, year and geographical grid cell that will be used in the atmospheric dispersion module.

The atmospheric dispersion module uses results from air pollution transport models developed by the Norwegian Meteorological institute for EMEP (the European Monitoring and Evaluation Programme). These source receptor matrices determine the deposition of acidifying and eutrophying compounds on so called EMEP-grid cells of $150 \times 150 \, km^2$ (see e.g. EMEP 1998) or more recently of $50 \times 50 \, km^2$. The module output can be shown both in the form of European deposition maps, either for separate compounds or for all compounds collectively, and in the form of a percentage distribution of the origins of the deposition in a particular grid cell anywhere in Europe.

Maps of deposition of acidifying and eutrophying compounds are compared with maps of critical loads in the environmental effect module. The Coordination Centre for Effects in the Netherlands develops methodologies and databases allowing the inclusion of critical loads in the RAINS model based on data from individual countries covering most of Europe (Hettelingh et al. 2001, 2007a; Posch et al. 2001, 2003). The RAINS model can be operated in two distinct modes: 'forward running' for the scenario analysis mode, where the regional pattern of acid deposition that will result from the use of particular combinations of fuels can be predicted, along with the costs and environmental benefits of alternative emission control strategies. In the optimization mode, RAINS can determine the least expensive way to arrive at a predetermined deposition level. The model user can specify environmental targets in various ways: realising a certain percentage reduction of excess over critical loads or determining the deposition pattern of a past year, etc. This RAINS mode has been used extensively in political negotiations in Europe.

18.3.4 The Use of RAINS in the Negotiation Process

In 1979, 30 countries adopted the LRTAP-Convention, establishing a general forum for international negotiations on emission reductions. Under the framework of the Convention a series of specific protocols lay down concrete agreements (see www.unece.org/env/lrtap/status/lrtap_s.htm):

1979 Geneva: The Convention on Long-range Transboundary Air Pollution

1984 Geneva: The Protocol on Long-term Financing of the Co-operative Program for Monitoring and Evaluation of the Long-range Transmission of Air Pollutants in Europe (EMEP)

1985 Helsinki: The Protocol on the Reduction of Sulphur Emissions (SO_2) or their Transboundary Fluxes by at least 30% (*technology based*)

1988 Sofia: The Protocol concerning the Control of Emissions of Nitrogen Oxides (NO_x) or their Transboundary Fluxes (*technology based*)

1991 Geneva: The Protocol concerning the Control of Emissions of Volatile Organic Compounds (VOC) or their Transboundary Fluxes (*technology based*)

1994 Oslo: The Protocol on Further Reduction of Sulphur Emissions (*effect-based*)

1998 Aarhus: The Protocol on Heavy Metals and Persistent Organic Pollutants (POP) (*technology based*)

1999 Gothenburg: Protocol to abate Acidification, Eutrophication and Ground-level Ozone (*multiple effects-based*)

In 1984, the parties signed a protocol for financing joint activities of the Co-operative Program for Monitoring and Evaluation of the Long-range Transmission of Air Pollutants in Europe (EMEP) in order to establish a framework for monitoring and for the scientific assessment of the acidification problem. The next three protocols addressed concrete obligations for reducing emissions of SO_2 (1985), NO_x (1988) and VOC (1991). These three protocols used simple 'flat rate' concepts for determining the international

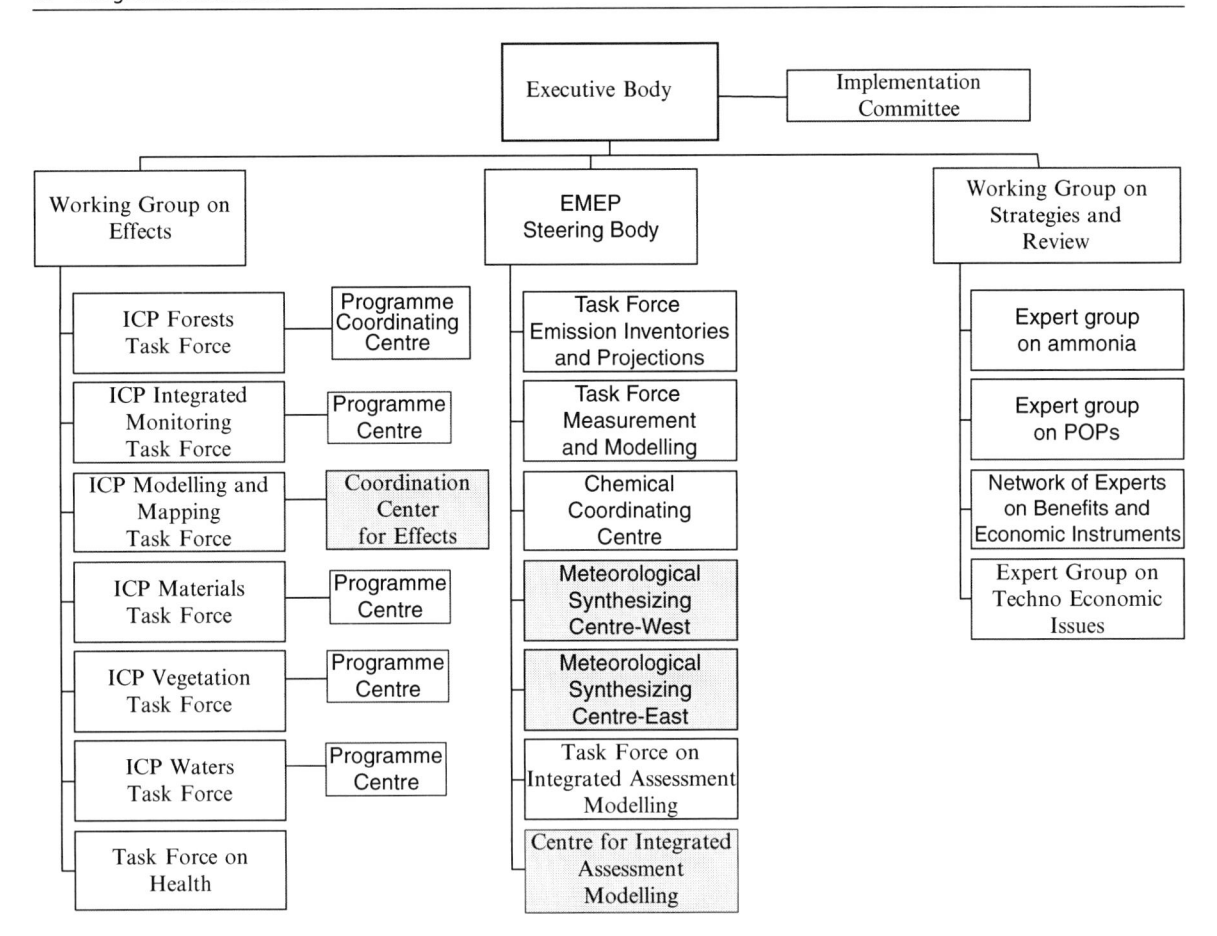

Fig. 18.3 The Structure of bodies under the LRTAP Convention, images the underlying scientific systems analysis. Grey shaded boxes indicate scientific groups involved in the Integrated Assessment of Air Pollution to support policies negotiated in the Working Group on Strategies and Review. The RAINS model is located at the Centre for Integrated Assessment Modelling at IIASA (Austria), while critical loads and atmospheric methods used in RAINS are located at the Coordination Centre for Effects at MNP (the Netherlands) and The Meteorological Synthesizing Centres West (Oslo) and East (Moscow). All centres collaborate with underlying scientific networks

distribution of reduction obligations, i.e. all countries agreed to cut back their emissions by an equal percentage (or stabilise them, as in the case of NO_x) relative to a base year.

Having reached agreement on initial emission reductions for the three pollutants, the negotiators under the Convention wanted to establish a refined approach to increase the cost-effectiveness of an advanced (and more expensive) second step on acidifying emission controls. This led to an important change in the character of the protocols: adoption of an effect-based approach in place of a flat-rate one. The required scientific support and the use of this knowledge for the support of this new approach necessitated an optimal use of the Convention's structure (see Fig. 18.3).[1]

Figure 18.3 shows how scientific bodies involved in integrated assessment (grey-shaded) are located within the current organization of the Convention. The Executive Body of the LRTAP Convention employed a special Task Force on Integrated Assessment Modelling

---

[1] Some of the bodies shown in Fig. 18.3 were either differently named at the time of the 1994 protocol or non-existent. Bodies relevant to integrated assessment interacted in the same way as at present. However, in 1994 IIASA, with its RAINS model, participated in the Convention without being formally established; it was known then as CIAM and participated under the EMEP Steering Body.

(TFIAM) to assist in preparing the next (1994) sulphur protocol and used the latest scientific understanding of the acid rain problem wherever possible. The TFIAM was given several important guidelines. First, environmental effects were to be evaluated in terms of critical loads rather than through detailed analysis of individual ecosystems. Second, TFIAM was to identify cost-effective solutions, i.e. the least expensive ways of meeting particular emission reduction goals derived from the comparison of critical loads to deposition of acidifying and eutrophying compounds. Third, environmental benefits were to be evaluated in physical rather than in monetary terms. The TFIAM decided to use the RAINS model as its principal tool to generate alternative policy scenarios for emission reduction. The input to alternative policy scenarios, generally based on cost minimization subject to varying constraints on exceedance of critical loads and emission reductions, are negotiated in the Working Group on Strategies and Review (WGSR). The TFIAM translates the inputs to RAINS model parameters while ensuring that outputs are transparent and easy to understand by policy negotiators. The review of the inputs entering the RAINS model is ensured by national representatives and stakeholders through formal reporting to appropriate bodies within the LRTAP-Convention (see Fig. 18.3).

Underlying the LRTAP-Convention structure shown in Fig. 18.3 are scientific networks in the field of environmental impacts, atmospheric chemistry, energy, emission and cost data. Scientific consensus on the knowledge to be incorporated within RAINS forms the core of reliability for the RAINS results and subsequent use in the negotiation process.

Development of methodologies and databases for the assessment of critical loads takes place at the Coordination Centre for Effects (CCE; www.mnp.nl/cce) located at the Environmental Assessment Agency in the Netherlands. The CCE collaborates with a network of more than 27 National Focal Centres that provide critical-load data and other scientific institutions. The development of atmospheric dispersion methodologies for these compounds, ozone and particulate matter takes place at the Meteorological Synthesizing Centre West (MSC-W) in Norway and via its network, while the Centre East in Russia (MSC-E) and collaborating institutions focus on heavy metals and polychlorinated organic compounds. Finally, RAINS data on energy use and costs are compiled at CIAM in collaboration with national experts, while emission data inputs include information from official reporting activities overseen by EMEP (e.g. Vestreng and Klein 2002). The RAINS model is located at the Centre for Integrated Assessment Modelling in IIASA (Austria).

For the scientific support of the 1994 protocol two other IAMs, the CASM and the Abatement Strategies Assessment Model (ASAM) (ApSimon et al. 1994), were used to check on consistency and for uncertainty analysis. The model runs were conducted in close collaboration with what is now called the Working Group on Strategies and Review (WGSR), the chief negotiation body of the Convention. Typically, WGSR requested a particular set of simulations, after which it studied the results and requested either different simulations or more refined ones. As a result, the analysis underlying the second sulphur protocol experienced a distinct evolution. Early simulations focused on a comparison between the emission reduction that would result from new regulations in the European Union and a simple 60% cut in emissions by all the countries involved. Where early simulations showed that attainment of the critical loads would be very costly and difficult, the WGSR developed a different approach, asking countries to determine 'target loads' as intermediate goals. These targets were to be based on critical loads, but were allowed to be higher, depending on technical, social, economic and political factors in each country. This approach proved to be very problematic, because no specific guidelines were provided for establishing the target loads. Therefore not many countries established target loads, and those who did, showed a large variety in values. For this reason, the negotiating countries decided not to proceed with the target load approach.

The final set of RAINS simulations looked at the effects of emission reduces sufficient to close the gap between 1990 deposition levels and the critical loads by at least 60%. This was the approach ultimately incorporated into the second sulphur protocol, which, after negotiations are completed, will aim at reducing overall emissions by 53%.

After the Oslo Protocol had been signed in 1994 the awareness that acidification was not only caused by sulphur increased. The understanding that nitrogen should be included as well dramatically increased the complexity of the linkages between sources and effects of air pollution, as illustrated in Fig. 18.4.

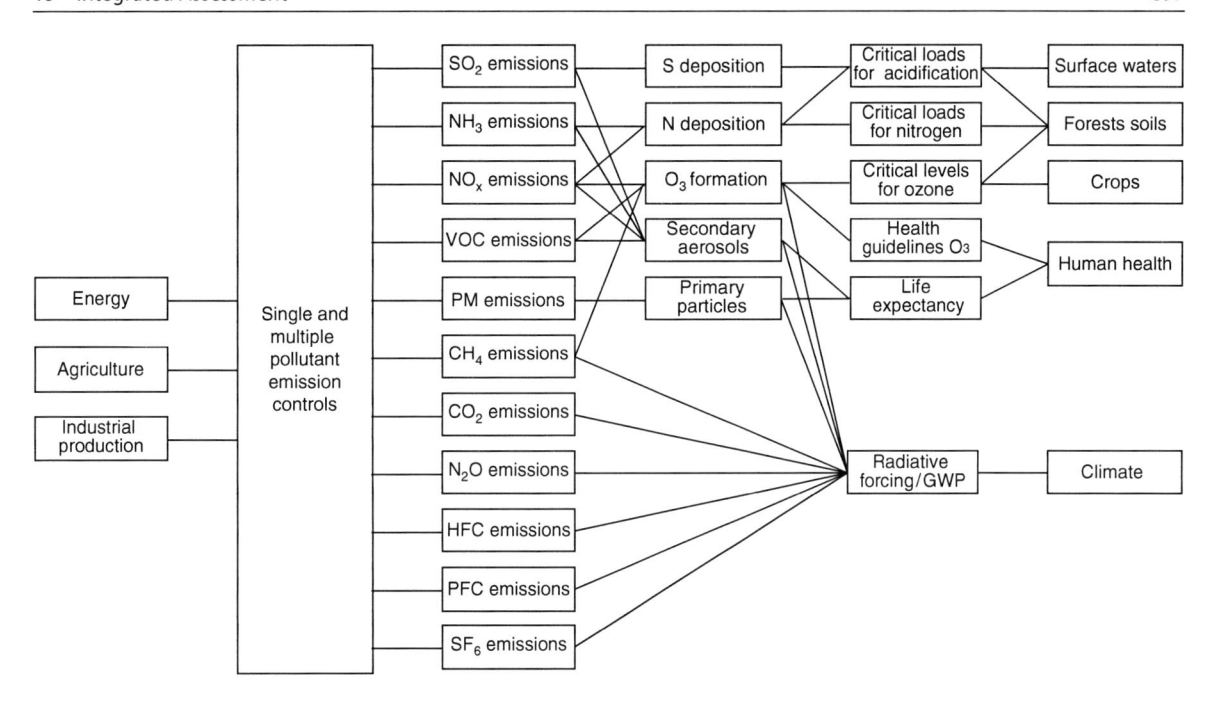

Fig. 18.4 The multiple source and multiple effect relationships between sources of emissions causing the excess of critical thresholds for acidification and eutrophication to forest soils and surface waters, ozone injury to crops, impacts to human health and the change of climate Source: Adapted from Amann (2003)

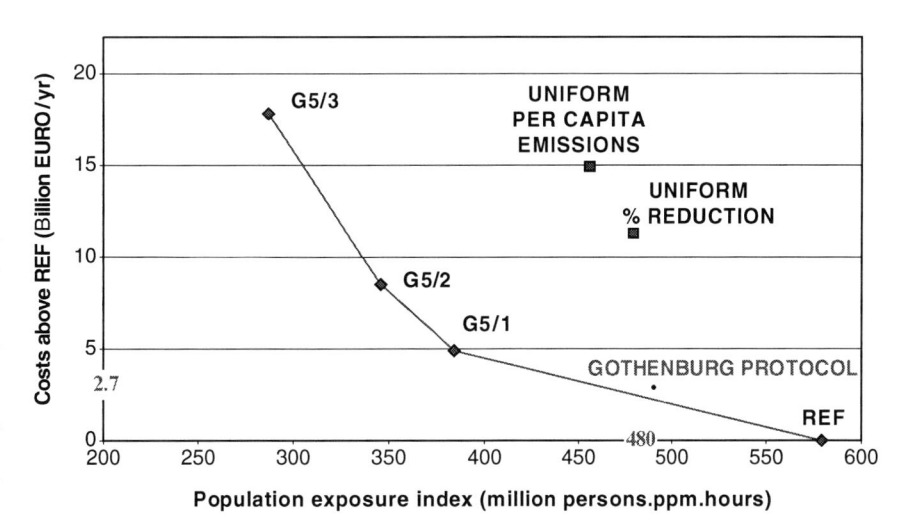

Fig. 18.5 The relationship between costs of the reduction of emissions of tropospheric ozone precursors and population exposure. Three scenarios are shown i.e. G5/1, G5/2, G5/3 compared to cost/exposure ratios obtained with the Gothenburg protocol, the reference case (REF) and two non-optimised results

A further protocol was to address emissions of sulphur dioxide, nitrogen oxides, ammonia and volatile organic compounds (VOCs) simultaneously, while considering multiple effects, i.e. acidification (by sulphur and nitrogen), eutrophication and population exposure to tropospheric ozone (by nitrogen oxides and VOCs). The inclusion of human health in addition to ecosystem effects implies the subjecting the respective RAINS optimization of emission reductions to health guidelines for the ambient concentration of ozone and critical loads. Scientific support of this protocol required the computation and mapping of multiple critical thresholds (above which damage may occur), i.e. the critical load for acidification and eutrophication and critical levels for ozone (Hettelingh et al. 2001, 2007a; Posch et al. 2001). Integrated assessment using RAINS (Schöpp et al. 1999) had to identify emission reduction alternatives which limit the exceedance of targeted pollutants over critical thresholds in a cost-effective manner. This led to the simulation of a large number of scenarios, generating various ratios between ecosystem protection, population exposure and emission reduction cost. The so-called G5/2 scenario was considered the most cost-effective (see Fig. 18.5) and became the basis for the 1999 Gothenburg Protocol.

Figure 18.5 illustrates the discrepancy between the Gothenburg Protocol result and the G5/2 starting point of the negotiations. The protocol leads to a population exposure similar to what would have been obtained with non-optimised emission control, i.e. a uniform percentage reduction over all countries or equal emissions per capita in Europe. The latter two options, however, would have been much more expensive than the Gothenburg Protocol.

Nonetheless, in terms of acidification, the Gothenburg Protocol proved to be successful. The reductions of both sulphur and nitrogen emissions resulted in a better protection of European ecosystems in 2020 as compared to 1990. Figure 18.6 illustrates

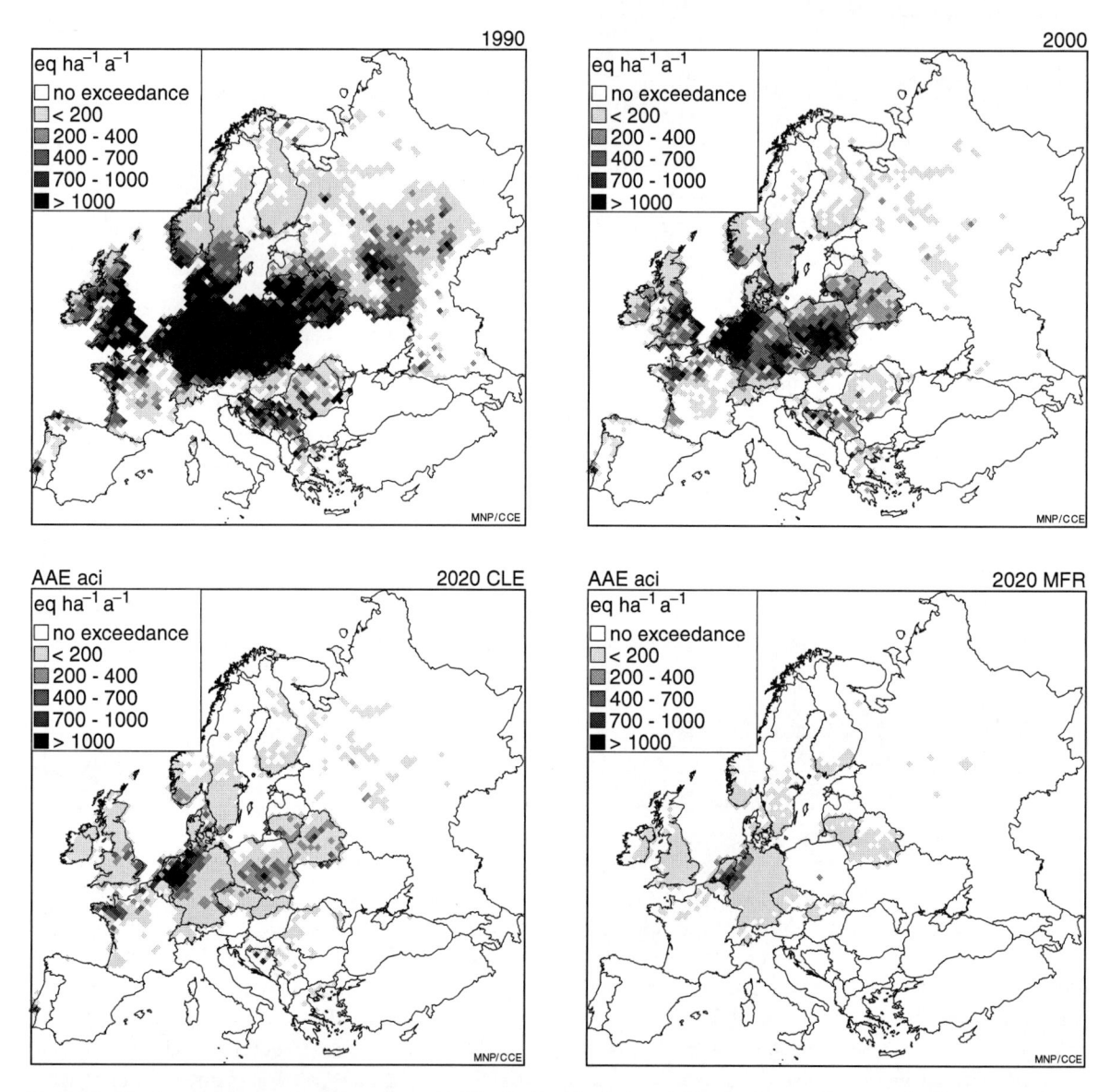

Fig. 18.6 The Average Accumulated Exceedance (AAE) of critical loads for acidification in each $50 \times 50\,\text{km}^2$ grid cell in 1990 (top left), 2000 (top right), in 2020 due to depositions that result from emission reductions under Current Legislation (CLE-2020; bottom left) and from the application of Maximum Feasible Reductions (MFR-2020). A grid cell is shaded when it contains one or more ecosystems that are at risk of acidification, i.e. that have a positive AAE
Source: Hettelingh et al. (2007b)

the success of the European policies in reducing acidification. It shows that both the magnitude and area coverage of the Average Accumulated Exceedances (AAE, see Posch et al. 2001) are reduced between 1990 and 2020. The AAE in an EMEP grid-cell is the ecosystem-area weighted sum of the exceedances of all critical loads computed in that grid cell. A dark shading in Fig. 18.6 indicates a high AAE of critical loads for acidity caused by excessive acid deposition.

Figure 18.6 illustrates where the AAE diminishes between 1990 (top-left map in Fig. 18.6) and 2020. For 2020 two maps are shown. One map shows the impact of emission reductions according to Current Legislation (CLE-2020; bottom-left map in Fig. 18.6). The other illustrates the impact of applying best available emission reduction technologies to obtain Maximum Feasible Reductions (MFR-2020; bottom-right map in Fig. 18.6). The percentage of the European ecosystem

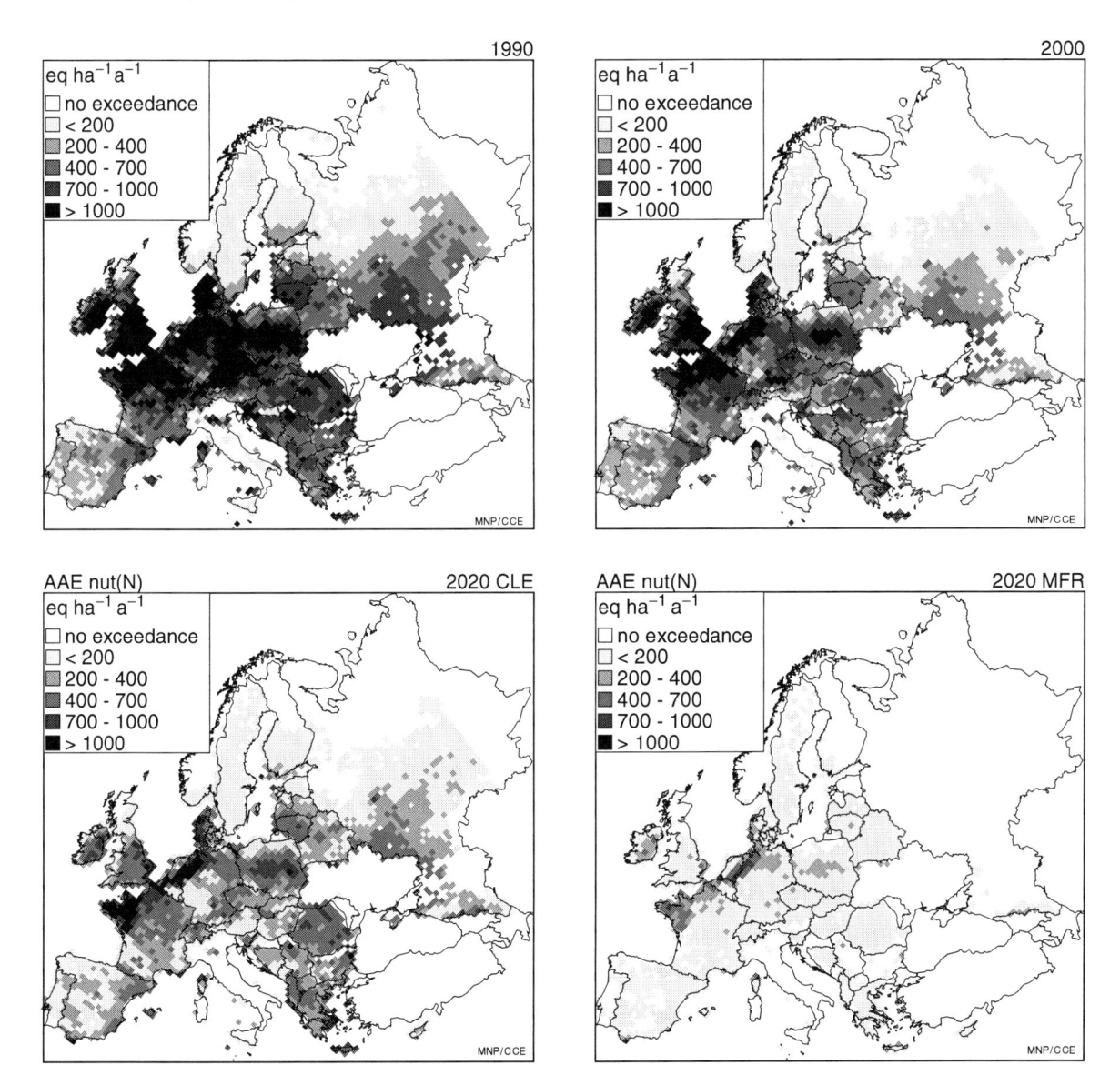

Fig. 18.7 The Average Accumulated Exceedance (AAE) of critical loads for eutrophication in each 50 × 50 km² grid cell in 1990 (top left), 2000 (top right), in 2020 due to depositions that result from emission reductions under Current Legislation (CLE-2020; bottom left) and from the application of Maximum Feasible Reductions (MFR-2020). A grid cell is shaded when it contains one or more ecosystems that are not protected from the risk of eutrophication, i.e. that have a positive AAE
Source: Hettelingh et al. (2007b)

area that is protected from acidification according to CLE-2020 and according to MFR-2020 is computed to be 94% and 99% respectively. The reduction of the risk of acidification in the recent past is also supported by findings on recovery of surface waters in Europe (Stoddard et al. 1999).

The protection of ecosystems from eutrophication, which results from the reduction of nitrogen oxide emissions is less striking, as can be seen from Fig. 18.7.

Figure 18.7 illustrates the magnitude and distribution of the AAE of critical loads for eutrophication. A dark shading in Fig. 18.7 indicates a high AAE of critical loads for eutrophication caused by excessive deposition of reduced as well as oxidized nitrogen. It turns out that the protection (meaning an AAE of zero or negative) of ecosystems from the risk of eutrophication covers about 58% of the ecosystems according to CLE-2020 (bottom-left map in Fig. 18.7). This leaves about 42% of the ecosystems at risk which are distributed in most EMEP-grid cells over Europe (note that a grid cell is entirely shaded if it contains one or more areas at risk). If MFR-2020 would be in place, a protection of 90% could be obtained. Even in this case (see Fig. 18.7 bottom right map) large areas in Europe would included ecosystems at risk. The implication is that even the application of currently known best available technologies would not yield full protection in 2020. However, a comprehensive integrated approach to reduce emissions (e.g. from energy combustion, agriculture) that cause climate change as well as air pollution could lead to further protection.

For the support of a possible revison after 2007 of the Gothenburg Protocol, the RAINS model is being extended to enable the assessment of sources and effects of particulate matter (Amann et al. 2001) and climate change. This is in line with the long-term strategy of the LRTAP-Convention. Its strategic view on the interaction between policy requirements and scientific support now includes the exploration of linkages between climate change and air pollution (see Fig. 18.4). This is embedded in the follow up of the RAINS model which is now called the Greenhouse gas and Air Pollution Interactions and Synergy (GAINS) model (www.iiasa.ac.at/rains/gains). The relationship between the impacts of climate change and of air pollution is now also a subject in the future modelling of critical loads. This includes the extension of critical loads to address both bio-geochemical as well as biological endpoints. Finally, dynamic modelling is increasingly becoming important in the effect based approach under the LRTAP-Convention to assess time trends of recovery (Hettelingh et al. 2007a).

18.3.5 Lessons Learned

RAINS and other IAMs were able to play such a large role in the negotiations because they enjoyed considerable credibility among both scientists and policy-makers. It is surprising that the scientific community accepted RAINS, considering the model is fairly crude compared with the analytical techniques of most scientific disciplines. One possible reason for this acceptance is the fact that teams of scientists from different disciplines developed the RAINS modules, with review by outside bodies and publication in peer-reviewed journals. Even more important, perhaps, RAINS components were developed as adjuncts to major international programmes for monitoring and research (EMEP, CCE, national emission inventories). A third reason for the models scientific credibility may lie in the fact that special studies were commissioned to address questions that could not be handled by RAINS. One such study looked at the relationship between economic restructuring in Eastern Europe and acid rain abatement strategies (Amann et al. 1992), while another explored options for international cost sharing (Klaassen and Sliggers 1993).

The model's credibility among policy-makers can be attributed to several factors. First, the negotiators had implicit faith in EMEP and CCE, and their scientific networks, having brought them into the policy process in the first place. Second, there was a fairly close relationship between the modellers and the policy advisors in WGS, so that the latter's numerous requests for simulations were met in a timely and relevant way. Third, the countries themselves provided the data used in RAINS. Fourth, the home base for the RAINS model, the International Institute for Applied Systems Analysis (IIASA), was respected as an independent institution involving scientists from different countries.

Another important aspect is the structure of the LRTAP-Convention, which images the interaction between policy and science, while providing a mechanism for making progress. Protocols evolved to become more complex as policy requirements were extended to include multiple pollutants and multiple effects. These requirements stimulate the scientific community to tailor available knowledge to the timetable of the LRTAP-Convention. In this way the structure of the LRTAP-Convention provides a traceable record of comprehensive policy–science interactions backed by reports, and so supplies the necessary memory to make progress. At the same time, publications in the scientific literature sources ensure that the science supporting LRTAP-Convention policy is reviewed and made public.

The LRTAP-Convention negotiations provide an example of participative integrated assessment that started before the first articles on this issue appeared in the scientific literature. It is important to note that the LRTAP-Convention negotiations have a long history (with roots in bilateral policy discussions in such countries as Finland and the former countries: USSR, West Germany and Czechoslovakia), which paved the way for successful multilateral negotiations. Moreover, the number of participants in the negotiations was and is quite small compared with those in climate negotiations. Further, it should be recognised that the implementation of LRTAP-Convention protocols is not as costly as the implementation of an envisaged climate protocol. In other words, the success of the LRTAP-Convention is partly due to the limited number of players and the limited economic impacts of the protocols. On the other hand, the interaction between scientists and policy-makers via a computer model and a generally accepted database has proven to provide a solid basis for fertile negotiations leading to an important improvement of environmental quality in Europe.

The support of negotiations under the LRTAP-Convention is based on a number of *ceteris paribus* assumptions including:

- Economic growth and the development of economic indicators
- Social and cultural perspectives
- Confounding factors such as the influence of world trade agreements, national governmental institutions
- Checks and balances between environmental, economic and social interest

The system and model domain of the RAINS model are tailored to the problem of air pollution causes and impacts. The evolving economy in Eastern European countries after 1989 were implicitly incorporated into emission reduction projections of the countries involved. The model structure was not subjected to (and perhaps did not need) any change to reflect the changing reality that occurred after the re-unification of Germany.

This may be different for the case of global climate change over the next century. The economic and social conditions vary widely among the global parties involved in the negotiations to curb climate change. The climate change issue is associated with essential activities of energy use, food production, and broad temporal and spatial scales with complex underlying dynamics involving non-linearities, and delays and feedback. There are large socio-economic and institutional impediments to effective response strategies. The next section explores integrated assessment methodologies designed to take the challenges up on climate change.

18.4 The TARGETS Model: A World Model to Explore Sustainable Development

Here we will discuss two Integrated Assessment Models (IAMs) and the way in which they have been constructed and used. The first one is the Tool to Assess the Regional and Global Environmental and health TargetS (TARGETS), developed as part of a Global Change research project at the Netherlands Environmental Assessment Agency at Dutch National Institute for Public Health and the Environment (RIVM). The second one is the Integrated Model to Assess the Greenhouse Effect (IMAGE), which has been developed and used since the early 1990s as a tool to explore climate change causes, effects and policies. Both models focus on global environmental change, more particularly climate change, and the quest for sustainable development.

18.4.1 The TARGETS Model Set-Up

'Limits to Growth', the report brought out by the Club of Rome in 1971, was the first attempt at surveying the long-term future of humankind on this planet through the use of a formal computer model (Meadows et al.

1974, 1991). In retrospect we can see two functions: (1) to awaken people to the particular problems of exponential population and economic growth and, secondly (2) to intensify the scientific debate waging on these problems within the various disciplines and to make [more] explicit many of the uncertainties and controversies. Subsequent years saw a series of other publications exploring – either with or without models – the future of 'the Earth's system' (see, for instance, OECD/Interfutures 1979) and Barney (1981). When integrated assessment of global climate change was initiated in the 1990s, the emerging realization was that because the uncertainties were so numerous and so large, linkages to (government) policies would be more challenging than the assessment of (regional) air pollution, as described in the previous section. This feeling was strengthened by the growing doubt about the 'makeability' of society.

On the other hand, it is increasingly recognised that population and economic growth poses such worldwide problems for us that integrated scientific insights and formal, model-based, investigations will be imperative – realising that the models are far from perfect. This undercurrent is gaining ground now that sustainability and sustainable development occupy an increasingly prominent place on the political agenda, although this still has to take form. To explore these issues in a model-based, innovative way did the Dutch National Institute for Public Health and the Environment (RIVM) carry out the objective of the project 'Global Dynamics and Sustainable Development'. The project resulted in the TARGETS model, which has been used to make the concept of sustainable development operational on the global scale and to investigate uncertainty and the role of value-laden assumptions in scenario construction (Rotmans and de Vries 1997; de Vries 2001a).

A number of working hypotheses emerging in the course of the project are listed below:

- The system was to be seen as a whole, i.e. a collection of well-defined interacting sub-systems.
- The so-called Driving Forces–Pressure–State–Impact–Response (DPSIR) chains and the concept of transitions were to be used as the ordering framework.
- Sub-models would be described on the basis of meta models, in other words, simplified 'expert models', providing a good representation of the main scientific insights and 'stylised facts' related to the long-term dynamics in the particular sub-system.
- Integration would not only be horizontal (between sub-systems), but also vertical (modelling human behaviour on the environmental dynamics 'substrate').
- Scientific uncertainties and the associated scientific and societal controversies would be dealt with explicitly and systematically.
- Computer simulation models were to be used to guarantee quantitative consistency and reproducibility, and make more explicit models.
- Insights, based on the simulation model, would be presented as far as possible interactively using good visualization techniques.

The result, a simulation model called TARGETS, comprises five sub-models: population and health; water; land and food; energy, and biochemical cycles. Each of the sub-models and the integrated model operate independently for the world as a whole, with the relevant model calibration and validation taking place in the 1900–1990 period. The method has been grafted to system dynamics, and a new simulation environment, M, for model development has been developed, fostering interactive communication (de Bruin et al. 1996). Scenarios for the future are being constructed using the cultural theory of Thompson et al. as the guiding principle in bringing uncertainties and controversies into focus, and assessing scenarios in terms of risks (see, for example, Thompson et al. 1990; Schwarz and Thompson 1990; Adams 1995).

Figure 18.8 sketches the TARGETS model in terms of the sub-models and their interaction. The economic component consists of a scenario generator, which reproduces GWP (Gross World Product) projections of such agencies as the World Bank and IPCC. The Population and Health model generates the size of the population, which, along with the GWP, is the main driving force behind the demand for water, food and energy. This derived demand serves as the input for the sub-models, Energy, AQUA and TERRA, after which the outputs for satisfying this demand are simulated. Important interactions also occur within the bio-geo-chemical cycles of the elements, carbon (C), nitrogen (N), sulphur (S) and phosphorus (P), simulated in the CYCLES model. Under the assumption that the demand for water, food and energy is always met, a number of indicators, like food intake per capita, access to clean water, temperature rise, etc., will result.

18 Integrated Assessment

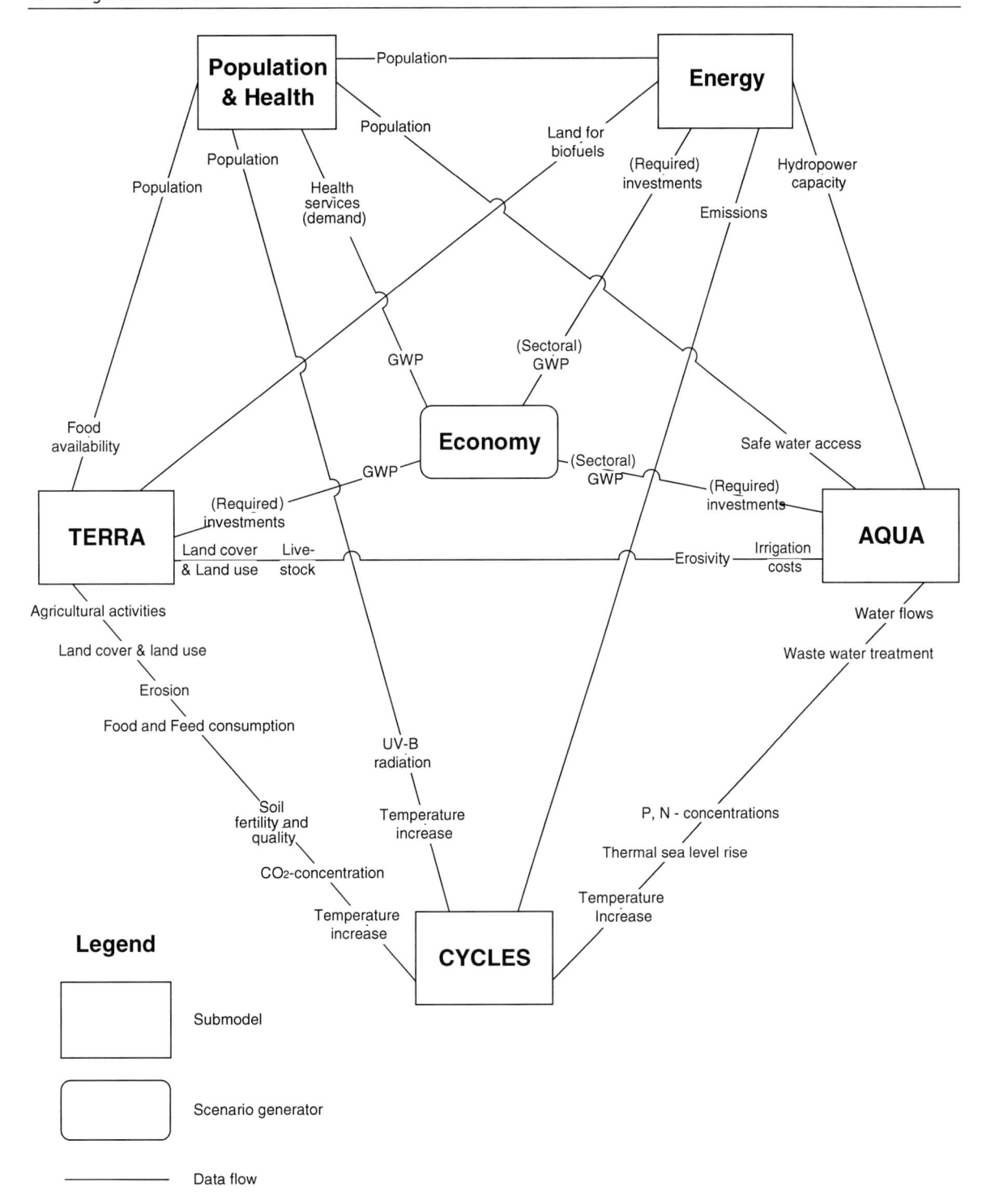

Fig. 18.8 Overview of the interconnections between the submodels in the TARGETS model

Just as the main feedback in the integrated model, these will in turn influence the fertility, and disease and mortality, of the (world) population. Another factor of influence is the GWP per capita and how far it is spent on health-care services. Changes in the development of the population influence in turn the demand for water, food and energy. The number of economic inputs for sustaining the water–food–energy system has been determined in a co-ordinate calculation. On this basis, we can obtain an idea of the extent to which degradation of the natural system, via reduced productivity, will affect freely disposable consumption, or, in reverse, to what extent technological breakthroughs will actually facilitate further growth of consumption.

The Population and Health sub-model describes the demographic development based on changes in fertility, morbidity and mortality. Birth determinants are related to the level of economic welfare (GWP/capita), literacy level of women and life expectancy (Hilderink 2000; Niessen 2002). Disease and death are a function of age, habits (e.g. smoking) and environmental factors (e.g. malnutrition and lack of clean drinking water) and split up into welfare and literacy classifications. In this way demographic and epidemiological transition are simulated. The model has been implemented for four countries and the world as a whole in the context of validation, with a number of sensitivity calculations made to allow interpretation of the developments between 1900 and 1990. The most important outcome of this sub-model, in relation to others, is the size of the population which, along with the economic activity levels, determines the demand for energy, water and food.

In the *Energy* sub-model the use of fossil and non-fossil energy carriers are simulated on the basis of the population and economic development (de Vries et al. 2001; van Vuuren and de Vries 2001). Energy demand is translated into the use of secondary energy-carriers, with autonomous and price-induced energy-saving, and relative fuel prices, being the most important determinants. Electricity demand is met by investing in fossil-based, non-fossil-based (nuclear, solar/wind) and hydropower plants. The resulting demand for fuel leads to investments in exploration, exploitation and transport of fossil fuels. Fuel prices are a function of depletion and technological learning behaviour. Rising fuel prices induce energy efficiency investments and the penetration of biomass-derived fuels, both important aspects of the transition from fossil to non-fossil energy sources. The world version is calibrated for the 1900–1990 period. The most important information flow to other sub-models is formed by the emissions of combustion gases (CO_2, SO_2 and NO_x) and the required investments.

The *AQUA water* sub-model consists mainly of the simulation of the hydrological cycle in interaction with human activity (Hoekstra 1998). The land, atmosphere, ocean and ice compartments distinguish between one or more water reservoirs linked via fluxes, like evaporation, precipitation and melting. Water quality is considered in the form of four water classes, differing in concentration, and originating from such compounds as nitrogen and phosphorus, which form part of the CYCLES sub-model. Water demand, influenced by water price and efficiency-increasing technologies, is met by applying costs which depend on the potential of available water of a certain quality. The resulting water transition has a quantity and a quality component. The model has been implemented in several basins to test its validity, with the world version calibrated for the 1900–1990 period in conformance with estimates from the literature. Information flows to other sub-models constitute the availability of clean drinking water, costs of irrigation and the required investments for water supply.

The *TERRA Land and Food* sub-model simulates inputs and outputs of food production to meet food demand. The most important soil characteristics and (changes in) natural matter cycles are taken into consideration by splitting the world into two regions according to the FAO categorization of developed and developing countries, with seven land classes, three climate zones and three soil fertility classes. The core of the model describes the land compartments and the fluxes between them. Food demands and demands for animal and timber products are calculated from the size of the population and the economic welfare (GWP/capita). To meet this demand, land is reclaimed and irrigated at rising marginal costs. The productivity of agricultural land depends on the land category and soil fertility (both of which depend on the element cycles in the CYCLES sub-model – see below), an exogenous technology factor and the amount of inputs, for example, of artificial fertilisers. The world version of the model is calibrated for 1900–1990 using data obtained mainly from FAO. Important data for other sub-models are the demand for irrigation water, fertiliser use and the required investments.

The biochemical CYCLES sub-model describes the cycles of the basic elements, carbon (C), nitrogen (N), phosphorus (P) and sulphur (S), on the global scale in comparison with human-induced emissions caused by fuel, water and land use (den Elzen et al. 1997). For this purpose, the model makes a distinction between compartment models for the atmosphere (troposphere/stratosphere), terrestrial biosphere (vegetation, two organic and inorganic soil layers, groundwater and surface water) and the oceans (one warm and cold column, each with a surface layer and four deeper layers). The exchange fluxes between the compartments – net primary production and soil respiration – and their main internal processes –biological, chemical and physical – are simulated. This also applies to the different biochemical feedback embodied in the CO_2 and the N-fertiliser effect. The resulting concentrations form the input for the climate model, consisting of a simple energy balance box. Model calibration has confirmed the strong link between N and C cycles and the need for studying their interaction. The change in CO_2 concentration (fertilization effect for agriculture) and temperature rise (consequences for agriculture and health effects) form an important link in the integrated TARGETS model.

18.4.2 Future Scenarios in TARGETS: Three Utopias

The TARGETS model has been used to explore possible futures differing according to our interpretation of them and the valuation of past and future. Following the cultural theory, a distinction is made between three 'active' attitudes on sustainability: hierarchist, egalitarian and individual perspectives, and between worldview and management/policy style. The first, worldview, indicates how people imagine the world to be and how they then interpret the past and speculate about the future. This interpretation describes the measures people take, given their adherence to a certain worldview. Of the nine combinations in existence, those in which the worldview and management/policy style coincide are worked out in detail. In these so-called utopias, the way the world functions and the policy pursued are in agreement, with this being reflected in a particular set of parameter values. The cultural perspectives are used here to implement multiple scenarios as a means to deal with uncertainties over and above perspective-independent uncertainties. In this way, room is made for divergent interpretations of the past and expectations for the future.[2] This way of dealing with uncertainty should be seen as a complement, not an alternative to other methods of dealing with uncertainty such as stochastic methods or the NUSAP approach, for example (van der Sluijs et al. 2001).

Table 18.1 provides a condensed overview of the model relationships and model parameters for which perspective-dependent values are used and, on a qualitative note, the values applied to the egalitarian and individualist perspective. The list shows mainly assumptions related to (the tangibility of) human behaviour, technological possibilities and the characteristics of the natural system. The hierarchist perspective takes a middle position, except for progress in efficiency technology, where there is less optimism than for the egalitarian and the individual.[3] Evidently, some of the assumptions may turn out to be scientifically unsound – but at present each of them can be and is defended in the scientific literature.

Table 18.1 The TARGETS model: implementation of two of the three worldviews/policy styles

Worldview/policy style	Egalitarian	Individualist
Population and health		
Effectiveness of birth control programmes	Large	Small
Speed of introducing anti-contraceptives	Medium	Steadily lower
Importance of economic welfare ($/capita GNP)	Low	High
Importance of welfare in a broad sense	High	Low
Effectiveness of health care/spending	Medium	High
Influence of health determinants	High	Medium
Progress in medical technology	Limited	High
Priority in health care	Preventive	Curative (private clinics)
Natural resources		
Growth of (water/food/energy) demand with rising income ($ per capita per GNP)	Small	Large
Saturation level for food (kcal per day per capita)	Low (3,470)	High (4,250)
Sensitivity of water demand to price	Low	High

(continued)

---

[2] Of course, this method can lead to a presentation of a certain worldview or policy which is sometimes caricatural. For a more extensive discussion on using cultural perspectives in scenario construction, see Rotmans and de Vries (1997).

[3] For the hierarchist perspective we take the Business-as-Usual scenario from the IPCC as being representative for the hierarchist outlook of many large organizations. It should be noted here that this scenario is partly meant as a political warning signal to take measures.

Table 18.1 (continued)

Worldview/policy style	Egalitarian	Individualist
Cost of energy-saving measures	Medium	Steadily lower
Progress in efficiency technology (water/energy)	High but limited	Medium(water)/high
Government interference with public supply efficiency technology	Large	None (i.e. market-directed)
Share of animal products in diet (% in 2100)	Low (5%)	High (40%)
Carbon tax/subsidies	Rapid and high	None
Use of groundwater stocks	Make lower	Market-driven
Importance of desalinization technology etc.	Slight (expensive)	Large
(Potential) agricultural land (Gha)	Low (2.8)	High (3.8)
(Potential) irrigated area (Gha)	Low (0.32)	High (0.46)
Effectiveness and use of artificial fertiliser	Medium/limited	Large/much
Change in food yields due to rising temperature	Highly negative	None
Yield change due to [net] CO_2 fertiliser effect	None	Positive
Extent of [cheap] oil/gas reserves	Small	Large
Costs of non-fossil alternatives	Falling somewhat	Falling sharply
Natural system		
Melting rate of glaciers	High	Low
Sensitivity of pole ice to temperature rise	High	Low
C cycle: overall effect of terrestrial feedback	Positive (temp.)	Strong CO_2 and N fertilization
N cycle: accumulation of 'missing nitrogen'	Esp. on land (soils, ground water)	Esp. in atmosphere (as harmless N_2)
Climate system: geographical feedback	Highly positive	Slightly positive
Climate system: radiation effects	Limited cooling	Strong cooling

18.4.3 Risks: Dystopical Tendencies

In the above scenarios it is assumed that those who are 'in power' have the 'correct' worldview – utopian futures. A discrepancy between the worldview and management/policy style is also possible, giving rise to dystopias. We have also explored such futures in which worldview and policy style belong to completely different perspectives, and both are pursued for 100 years without feedback.[4] An example of such an (extreme) scenario might be called the 'egalitarian nightmare', in which the growth in population and the economy continue, more efficient technology hardly gets off the ground and a (strong) climate policy is lacking. However, the natural system is just as sensitive as the egalitarians feared it would be. In this situation the climate change taking place causes the food harvests to drop; starvation leads to mass deaths and both size and life expectancy of the world population drops in the second half of the next century. In fact this is the variant described in the book, Limits to Growth, which has received the most media attention. The opposite is also possible: anthropogenic climate change is minor but, with the egalitarian perspective dominating, large efforts are made to reduce greenhouse gas emissions. In such a situation, individualists will see their fears become reality: large lost opportunities.[5]

To what extent do our model experiments provide information on the probability of the different utopias and dystopias? Part of the answer can be acquired by exploring the robustness for perturbations. To this end, important variables have been diversified to see what happens if there are discrepancies between worldview and policy style – with the dystopias as extremes. The hierarchist proves to be quite vulnerable to a situation in which environmental action groups are correct in their presupposition that the (climate) system is extremely sensitive. An important reason for this are the highest greenhouse gas emissions shown in the hierarchist utopia, which are due to the combination of large growth and slow technological progress. The Achilles heel of the individualist lies in his/her presupposition of a wide availability of high-quality raw materials and the continually increasing efficiency in the use of raw materials. If this turns out to be false, a dystopia unfolds in which the existent inequity between humans would probably be further aggravated by mass food and water shortages, migration, etc.[6] The egalitarian is particularly

---

[4] In reality, there will be a dynamic interaction between prevailing worldviews and actions. Such dynamic changes in the adherence to certain perspectives was explored in Janssen and de Vries (1998).

[5] Interestingly, no one may ever know because the reduction of greenhouse gas emissions will prevent some climate changes from happening, a factor impossible to prove.

[6] Compare this to scenarios such as Barbarisation in Gallopin et al. (1997) and the picture sketched by Kaplan (1996); www.gsg.org.

poorly off where there is no basis for measures proposed (by the government) when economic growth is low and if measures are skirted or held off using all sorts of manoeuvres. In this case a failing population and natural resources policy can result in overtaxing the, fragile, natural system with all the known consequences.

18.4.4 Lessons Learned

The TARGETS model is aimed at global change at a high level of aggregation. It is a meta model, not an expert model, in the sense that it attempts to frame issues and to provide a context for debate. The TARGETS research added at least three elements to the more speculative discussions on global change. First of all, there are the model results – numerically consistent and representing insights on how sub-systems of the world function. Secondly, a start has been made on consistently associating feedback between sub-systems in the analysis – the integrated approach. Thirdly, the use of perspective-based scenarios is an invitation to participative research at the science–policy interface. In practice, it has turned out to be difficult to generate such participative model-based involvement – there is still a long way to go (e.g. van Asselt 1997; van Asselt et al. 2001).

The simulation experiments carried out using the TARGETS model indicate a highly uncertain future even if expressed in its most basic, quantifiable characteristics: i.e. between 5.3 and 13 billion people in 2100 with an average life expectancy of 66–88 years, an annual income of between US$4,000 and US$40,000 per capita, a food intake of zero to a 70% increase and, finally, 40% less to about three times more water and energy per capita. The increasing pressure on the natural environment can be managed for the next 100 years if new and powerful technologies, geared to more efficient resource use, are put into practice and if natural ecosystems – flora and fauna – are not (too) vulnerable. This would lead to an individualist utopia – a highly managed, high-tech world with a materialist outlook. If these conditions cannot be met, the future will develop in a less attractive and dystopian way. The burden of environmental degradation may be shifted on to the weakest members of society, and could lead to an impoverished world of islands of extreme wealth in the middle of mass suffering and crime. It is also possible that the egalitarian ideal of frugality will gain ground so that material growth will be less or less energy- and material-intensive. This could happen through ecological disasters, 'grassroots' environmental initiatives, changes in consumer preferences and lifestyles, rebirth of spiritual/moral movements or a combination of all these.

It is quite probable that the future will contain a mixture of elements of these utopias and dystopias, with the mainstream unfolding along the lines of the hierarchist perspective. In this case, many of the present trends will continue until a collective change of course becomes unavoidable. If this happens on the basis of well-founded knowledge, and carrying capacity, a hierarchist ideal will unfold, with incremental policy measures taken on the basis of institutional expertise, conventions and control.

In our view the most interesting research lessons from this endeavour are that:

- The concept of sustainable development is more of a guiding principle than something to be defined scientifically, and will therefore always be value-laden.
- Integrated assessment demands the use of a variety of interdisciplinary observations, concepts and models, the last item representing one of the greater challenges.
- A modular approach on the basis of meta models, i.e. simplified disciplinary expert models, may be most attractive for this reason.
- More research should focus on the interfaces between different disciplines, as these are often the least well-researched and controversial, and therefore open to value-oriented interpretation and judgement.
- The most promising use of such tools as TARGETS may be in interactive, participatory sessions with stakeholders in the broadest sense of the word.

18.5 The IMAGE Model's Role in the Construction of New Greenhouse Gas Emissions Scenarios

The IMAGE model is rooted in a combination of early climate change models and system dynamics. Initially, IAMs were constructed as deterministic models, as were the early climate change models, consisting of sets of differential/difference equations. The IMAGE

1.0 model (Integrated Model to Assess the Greenhouse Effect) in Rotmans (1990) is an example of such an IAM, simulating the cause–effect chain responsible for climate change on a global aggregated scale using only physico-chemical relationships. In the 1990s IAMs gradually expanded to include socio-economic aspects and climate impacts. Climate change IAMs were also increasingly used – or at least claimed to be useful – to provide policy-makers with a coherent synthesis of all aspects of climate change (See Box 18.1). The development of the IMAGE model followed this trend along two development paths.t

> **Box 18.1** Climate change
>
> The threat of human-induced climate change has rapidly become one of the most prominent environmental issues on the global agenda (O'Riordan and Jager 1996). It is intricately related to the debate about sustainable development, which started in the 1980s with the publication of the report, 'Our common Future' of the World Commission on Environment and Development (WCED 1987). To assess the causes and consequences of human-induced climate change, the Intergovernmental Panel on Climate Change (IPCC) established three working groups in 1988. The first surveyed the scientific evidence on climate change. The second assessed environmental and socio-economic impacts of climate change, and the third explored response strategies. The three working groups published reports in 1990, 1996 and 2001 (www.ipcc.org).
>
> Climate change refers to the anticipated change in climate as a consequence of anthropogenic emissions of gases which enhance the natural greenhouse effect, often outlined in the previously discussed Pressure-State-Impact-Response (PSIR) framework. Observations show a steady increase in the atmospheric concentrations of gases known to trap heat, most notably carbon dioxide (CO_2), methane (CH_4) and dinitrogen oxide (N_2O). The resulting radiative forcing will, according to present insights, cause a rise in global average surface temperature. This will have an array of impacts on the global environment, among them, changes in regional climate regimes and a rise in sea level. Humankind may respond basically in two ways to these changes: adaptation in a variety of ways and mitigation through emission reductions.

The first path saw the introduction of a geographically explicit description of land use. This led to the IMAGE 2.0 model, with a more detailed terrestrial carbon cycle model, a natural vegetation model and a land-cover model using a grid scale of 0.5×0.5 degrees (Alcamo 1994). The second path endeavoured to take account of socio-economic aspects through linkages with energy, demographic and macro-economic models, and by making the model more open and transparent for use in interactive policy-support sessions. Here, the team made use of the experiences of the successful RAINS model, as well as the lessons from the TARGETS project, both discussed in the previous sections. The result was the IMAGE2.1 model (Alcamo et al. 1998) and, subsequently, the IMAGE 2.2 model (IMAGE-team 2001), including the integrated population and health model (Hilderink 2000) and the Targets-IMage Energy Regional model (TIMER; de Vries et al. 2001).

The IMAGE 2.1 model has been used for several, quite different, applications, including the Delft Dialogue workshops, greenhouse gas emission scenario construction for the IPCC SRES (Special Report on Emission Scenarios) analysis and the Dutch COOL project on climate change policy options (Berk et al. 2001). Furthermore, the model was applied in evaluating energy futures for China (van Vuuren et al. 2002) and investigating the synergy between acidification and climate change in Europe (Alcamo et al. 2002). The Delft workshops were set up to improve the dialogue between the scientific world and policy-makers with respect to the issue of climate change, for which a coherent policy and scientific structure such as the one underlying the LRTAP Convention (see Section 18.3.4) is under development. These workshops resulted in the construction of the Safe Landing Analysis to determine the short-term range of emissions that would still lead to desirable stabilization levels of CO_2 concentrations in the long term (Swart et al. 1998; van Daalen et al. 1998). The SRES analysis resulted in a set of emission pathways up to 2100, which can be expected to emanate from the different ways the world might evolve the coming years (Nakícenovíc et al. 2000; de Vries et al. 2000). In the next section we will discuss the SRES analysis in more depth, covering set-up, process, results and evaluation. An overview of the IMAGE model structure is given in Fig. 18.9; for more details please refer to the literature cited.

Fig. 18.9 Overview of the IMAGE2.2 model

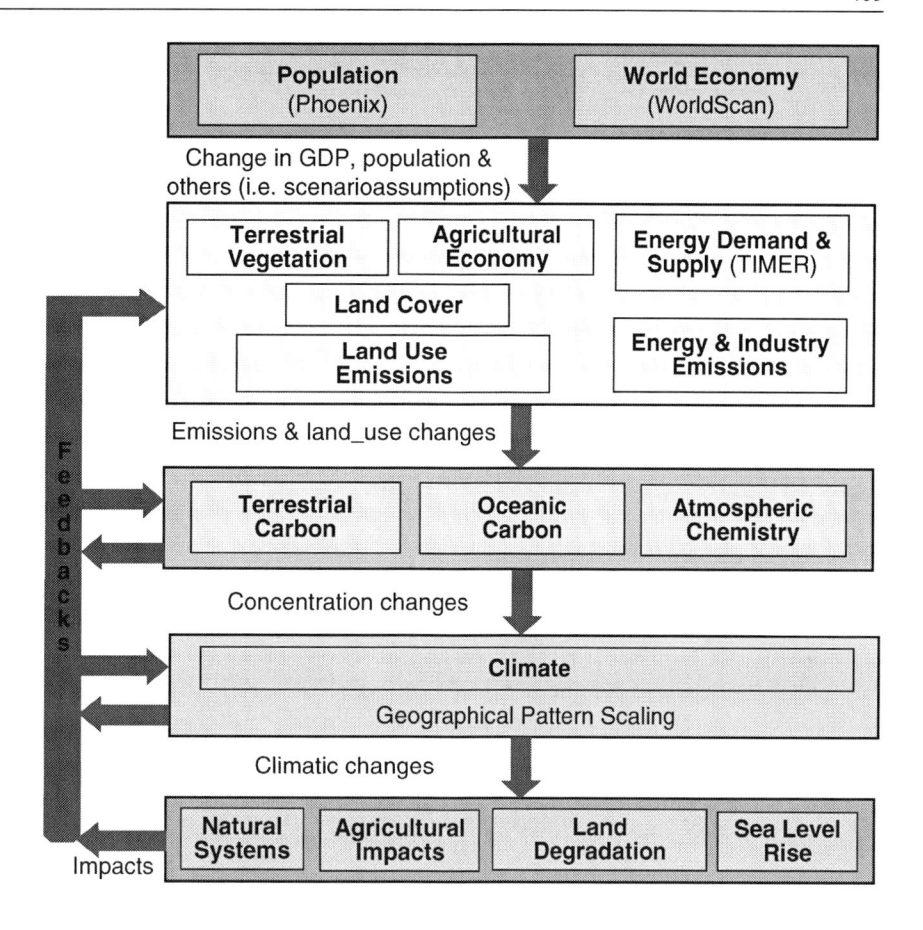

[Diagram: Population (Phoenix), World Economy (WorldScan) → Change in GDP, population & others (i.e. scenarioassumptions) → Terrestrial Vegetation, Agricultural Economy, Energy Demand & Supply (TIMER), Land Cover, Land Use Emissions, Energy & Industry Emissions → Emissions & land_use changes → Terrestrial Carbon, Oceanic Carbon, Atmospheric Chemistry → Concentration changes → Climate / Geographical Pattern Scaling → Climatic changes → Natural Systems, Agricultural Impacts, Land Degradation, Sea Level Rise → Impacts; with Feedbacks loop]

18.5.1 The Special Report on Emissions Scenarios (SRES)

In 1992 the Intergovernmental Panel on Climate Change (IPCC) published six non- intervention emission scenarios (IPCC 1991; Leggett et al. 1992), providing alternative emissions trajectories for the main direct and indirect greenhouse gases (CO_2, CH_4, N_2O, CFCs, SO_2, NO_x, VOCs, CO). These trajectories described how these emissions could – globally and regionally – evolve in the absence of explicit additional climate mitigation policies. The scenarios were intended for use by atmospheric, climate and impact scientists for the preparation of scenarios on atmospheric composition, climatic changes and impacts. The work updated and extended earlier research prepared for the IPCC's first Assessment Report. The IS92 scenarios, reviewed in the 1994 IPCC Supplementary Report (Alcamo et al. 1995), were found to be useful as a basis for climate analysis. However, they were less useful for purposes for which they were not developed, such as evaluating alternative policies.

Yet one felt that new scenarios were needed because the world had changed: the socialist economies of Eastern Europe and the former Soviet Union had opened up; population growth was lower than expected and economic growth, especially in Asia, was higher than expected. Thinking about the future in the context of climate change also resulted in much new scientific research. Hence, reason enough for IPCC to start planning on the development of a Special Report on Emissions Scenarios (SRES), in which new scenarios would not only be designed on the basis of the literature, but with a new approach. Some of the scenario requirements as formulated in the IPCC mandate, where the scenarios should be transparent, reproducible, and internally consistent, without any additional explicit climate policy (i.e. the Kyoto Protocol is *not* taken into account). They were to cover all greenhouse gases, including other radiatively active pollutants such as sulphates and particulates, involve various modelling frameworks and disciplines, allow regional disaggregation and be rich in the sense of spanning a wide spectrum of alternative futures, structural changes and salient uncertainties (Nakícenovíc et al. 2000).

In the first step the scenario literature was reviewed and alternative narratives (or storylines) and a rough quantification developed. In the second step, six modelling groups volunteered to analyse these scenarios in detail and to develop more detailed emissions trajectories for all GHGs.[7] The preliminary results from these modelling groups and the accompanying storylines were put into an 'open process' in the third step, in which all experts in this area could contribute to the scenarios. In the final step, the report was drafted and submitted for subsequent expert and government review according to IPCC rules and procedures. The resulting Special Report on Emissions Scenarios (SRES) was officially accepted by the IPCC in May 2000 (Nakícenovíc et al. 2000; for more details see also www.ipcc.org and www.sres.ciesin.). As intended, these SRES scenarios formed the input for the Third Assessment Report (TAR) of the IPCC and comprised an important ingredient in the construction of new mitigation scenarios (IPCC 2001c; Morita et al. 2000).

The scenario construction process has been an interesting, but at times, a confusing sequence of discussions, calculations, data exchanges and drafting sessions. It is next to impossible to give an objective account of such a complex, interactive process, which has often had the character of 'negotiated science'.[8] Only a brief sketch of the process and content of scenario construction is given here, with a discussion of some of the more controversial issues. A few observations in advance:

- The members of the SRES-team did not represent larger constituencies; the group more or less formed itself around the chairperson – who was asked by the IPCC – and a large majority came from OECD countries.
- The majority of the scientists involved had a background in natural or environmental science/engineering; interaction with other networks such as the economist-dominated Energy Modelling Forum were minor.
- Most if not all of the models used – all from modelling groups within the OECD – were only transparent for those who developed and used them. For this reason, the degree and nature of model diversity and input harmonization is hard to assess.

Although the process was an interactive and participatory one, many of the stakeholders – or would-be stakeholders who were either unaware or sceptical – were either not at all or only later involved. In this sense, the scenario process will remain part of the climate change debate for a long time to come, though with fluctuating emphasis on science, impacts and (costs of) mitigation/adaptation strategies.

18.5.2 The Framework: Four Narratives and Their Interpretation

Scenarios are descriptions of possible (alternative) images of the future. Most definitions imply the use of several – qualitative or quantitative – models to bring in some rationale or logic to th e scenario, also referred to as consistency and/or plausibility. Scenarios are neither predictions nor forecasts, nor are they 'objective'. They are, instead, a reflection of authors' interpretations and valuations of a limited number of past and present events, behaviours and structures. In the very first stages of the SRES process, it was decided to use two axes (or dimensions) to organise the scenarios: the degree of globalization vs. fragmentation and the degree of orientation on material vs. social/ecological values. The four 'scenario families' changed names several times; Table 18.2 lists the scenarios developed over the past few years, and indicates with which of the various SRES-families they may be associated (see also Box 18.2). The scenario family names are often suggestive enough – which is why the SRES team decided to choose the simple neutral names: A1, A2, B1 and B2.

---

[7] These models, or modelling groups, were the Asian Pacific Integrated Model (AIM), Japan; the Atmospheric Stabilization Framework (ASF) model, USA; the Integrated Model to Assess the Greenhouse Effect (IMAGE) model, The Netherlands; the Multiregional Approach for Resource and Industry Allocation (MARIA), Japan; the Model for Energy Supply Strategy Alternatives and their General Environmental Impact (MESSAGE) model, Austria and the Mini Climate Assessment Model (MiniCAM), USA. For a description and applications of the IMAGE model, see de Vries et al. (2000) and Alcamo et al. (1998). For the macro-economic model, WorldScan, see CPB (1999) and for an overview of all scenarios and models, see Kram et al. (2000).

[8] The authors have personal experience with these processes.

Table 18.2 SRES Scenarios: what's in a name?[a]

Type[a]	Scenario name	Author(s)
A1	Markets First	UNEP (2002)
	Tiger World	Nakićenović et al. (2000)[a]
	Just Do It/Da Wo; Business Class	Shell International (2000, 2002)
	Path A Competition	Bossel (1998)
	New Global Age	OECD (1997)
	Individualist/Hierarchist Utopia	Rotmans and de Vries (1997)
	Conventional Worlds: reference	Gallopin et al. (1997)
	Conventional Wisdom/ Hypermarket	Nakićenović (1995), EC (1996)
	Sustaining High Economic Growth/Driving into Tensions/High Prices; Capacity Constraints Case	EC (1990), IEA World Energy Outlook (1996)
A2	Security First	UNEP (2002)
	Divided World/Cultural Pluralism	Nakićenović et al. (2000)
	A Passive Mean World/ Rich Get Better	AC/UNU 1997/1998
	The Clash of Civilizations	Huntingdon (1997)
	Barbarization	Gallopin et al. (1997)
	Battlefield	EC (1996)
B1	Policy First	UNEP (2002)
	Sustainable Development	Nakićenović et al. (2000)
	Egalitarian Utopia	Rotmans and de Vries (1997)
	Policy Reform/Great Transitions	Gallopin et al. (1997)
	Forum: Energy Savings Case	EC (1999), IEA World Energy Outlook (1996)
	Ecologically Driven	Nakićenović (1995)
	Changing Course	Schmidheiny et al. (1992)
	Sustained Growth/ Dematerialization	Kassler (1994)
	Renewables-Intensive Global Energy Scenario (RIGES)	Johansson et al. (1993)
B2	Sustainability First	UNEP (2002)
	Regional stewardship/ Mixed Bag	Nakićenović et al. (2000)
	Great Transition	Raskin et al. (2002)
	Prism	Shell International (2002)
	Path B Partnership	Bossel (1998)

[a] The scenario names mentioned here were used during the SRES process but were finally all omitted in favour of the neutral names A1, B1, A2 and B2.

Box 18.2 Summary of the four narratives for the SRES scenarios

The *A1 scenario* assumes continuing globalisation and high economic growth, as well as a focus on the material aspects of life. This scenario represents a continuation of the 'modernization' process of the last 50 years in its Western form – 'Westernization' or 'Americanization'. Democratic governments and market capitalism are seen as the only viable way forward. The scenario includes such factors as further liberalization and privatization, secularization and emancipation of women. Dominant motives are worldly achievements, material wealth, competition and outer adventure, risk-proneness and optimism with regard to the blessings of technology.

In the *A2 scenario*, globalisation is slowing down or reversing as peoples and governments focus on cultural identity and traditional values. In large parts of the less industrialised regions, the Western form of the 'modernization' process of the last 50 years is actively resisted. Autocratic, nationalist or fundamentalist regimes provide one answer to rising social and political unrest in responding to feelings of exclusion and marginalization. Other answers to impending crises may lead to a successful revival of traditional lifestyles or high-tech niches. The 'cultural pluralism' of this world can take on the form of fragmentation, with protectionism and conflicts reigning, while for some regions it may be the road to rediscovery and strengthening of their own values. Dominant motives are nationalist/ethnic and cultural pride, observance of traditional values, authoritarian behaviour and associated obedience, fatalist attitudes with regard to the 'world problématique', and resource and technological opportunism.

The *B1 scenario* assumes continuing globalization and economic growth, and a focus on the environmental and social – immaterial – aspects of life. We interpret this as the continuation of a balanced, transformed 'modernization' process. Governance at all levels and regulated forms of market capitalism are seen as the way forward. The scenario includes the strengthening of NGOs concerned on the issues of sustainability and equity. A modest and decent world is bureaucratic, regulated, but also in search of fairness and sustainability.

(continued)

> **Box 18.2** (continued)
> Dominant motives are orientation on quality-of-life in a broad sense, co-operation, 'inner adventure', risk-averseness, judicious use of resources and technology and optimism with regard to human nature.
>
> In the *B2 scenario,* people increasingly perceive and seize the means to manage their local/regional situation in an equitable and sustainable manner. A culture-oriented anti-globalization movement may pave the way for new developments: decentralised settlements on the basis of bioregions, and centralised but high-energy and material-extensive infrastructure. Some regions can develop indigenous resources; others can focus on changing socio-political structures or turning consumption into less materialistic and more spiritually and psychologically rewarding activities. Ineffectiveness of global institutions makes it difficult to tackle global (environmental) problems – good intentions but no action. Dominant motives are orientation on local/regional quality-of-life, local co-operation, 'inner adventure', risk-averseness, an autarchic and judicious view on resources and technology, and optimism regarding human nature.
> Source: IMAGE-team (2001)

The two axes are by no means complete; such important aspects as governance and technology are either absent or implicit. Neither should any of these future worlds be interpreted as a blueprint of some probability or desirability. Some regions of the world will evolve in one direction, influencing the evolution of other regions. The industrialised regions of the world will search for manageable compromises between unrestrained capitalism and rapid high-tech developments on the one hand, and the search for meaningful, healthy and peaceful lives for their citizens on the other. Some of the less industrialised regions are following quickly in this process of 'modernization', and in doing so, are enriching and changing it. Others are not willing or able to follow, at least not yet, and cope by being marginalized or exploited, or by finding their own form of development. All these processes are taking place simultaneously. The scenarios tell stories about the possibility of certain regions, values and mechanisms becoming more dominant than others.

None of the four scenario families should be seen as a stable endpoint either. They interact dynamically: i.e. where forces characterizing one family become dominant, counterforces will emerge and push the family in other, new, directions. For instance, advocates of liberalization and privatization (in accordance with the free-market ideology) may be so successful that the resulting inequity and overexploitation strengthen environmentalist and fundamentalist forces. A considerable number of analyses explain the events on and after September 11th 2001 along these lines, e.g. Barber's Jihad vs. McDonald. Similarly, strong and successful implementation of a global (environmental/trade) governance regime may elicit strong forces to evade regulation and make it ineffective. A drift towards regional cultural identity and trade protectionism may, in combination with expanding global communication networks, induce such revivals as modernism and globalism. For a more detailed description of the scenarios in terms of narrative, key elements, characteristics, dominant motives and counteracting forces, please refer to IMAGE-team (2001).

18.5.3 Model Implementation

During the discussions it became clear that the constraints posed by available models, time and manpower would enormously reduce the (potential) richness of the qualitative narratives. For instance, the few macro-economists involved translated the difference between the 1 and 2 worlds as a difference in transport costs, while energy modellers introduced (energy) trade constraints. On the other hand, the difference between the A and B worlds was implemented by energy/land modellers as a difference in optimism with regard to the efforts put into and the success of 'green' innovation, and in saturation and dietary trends. The difficulty of dealing with diversity – even in the absence of socio-political and cultural dynamics in the models – showed up painfully in the implementation of the B2 scenario. Although the available models are all derived from OECD experiences in the last 2–3 decades, they are used to describe the complex processes in economies-in-transition and in economies of less developed regions where the agricultural and informal sectors are crucial elements of economic activity. Due to this 'model homogeneity' and lack of resources and involvement from modellers from less developed regions, the B2-scenario became a 'medium scenario' without even touching some of the crucial elements in the narrative.

There is another reason for the use of formal models to be regarded with some suspicion, notwithstanding their necessity and usefulness in introducing (numerical) consistency. The models use hypotheses – informal mental models or 'meta-models' if one likes – which guide the relevance and interpretation of data and assumptions. Figure 18.10 is an attempt to make such hypotheses explicit: it shows the four quadrants given by the two axes and, for each of these, a possible interpretative logic.

Several of the – sometimes implicit – mental maps about the particular 'world' behind one or more of the SRES scenarios, and/or models and arising during the discussions, are illustrated below and in Fig. 18.10:

- *Population and income*: faster growth of income (GDP/cap) accelerates the demographic transition by increasing the rate of fertility decline.
- *Technology*: more market-oriented economic growth in combination with free trade will accelerate economic growth through innovative entrepreneurial activities. Stability will be preserved as income will trickle down to lower-income segments of the population.
- *Nature of economic growth*: with rising income (GDP/cap), the fraction of agriculture in GDP declines and the fraction of industrial value added (IVA) rises. This is followed by a fall in IVA and a rise in Services Value Added (SVA), which, besides monetarization and formalization of the economy,

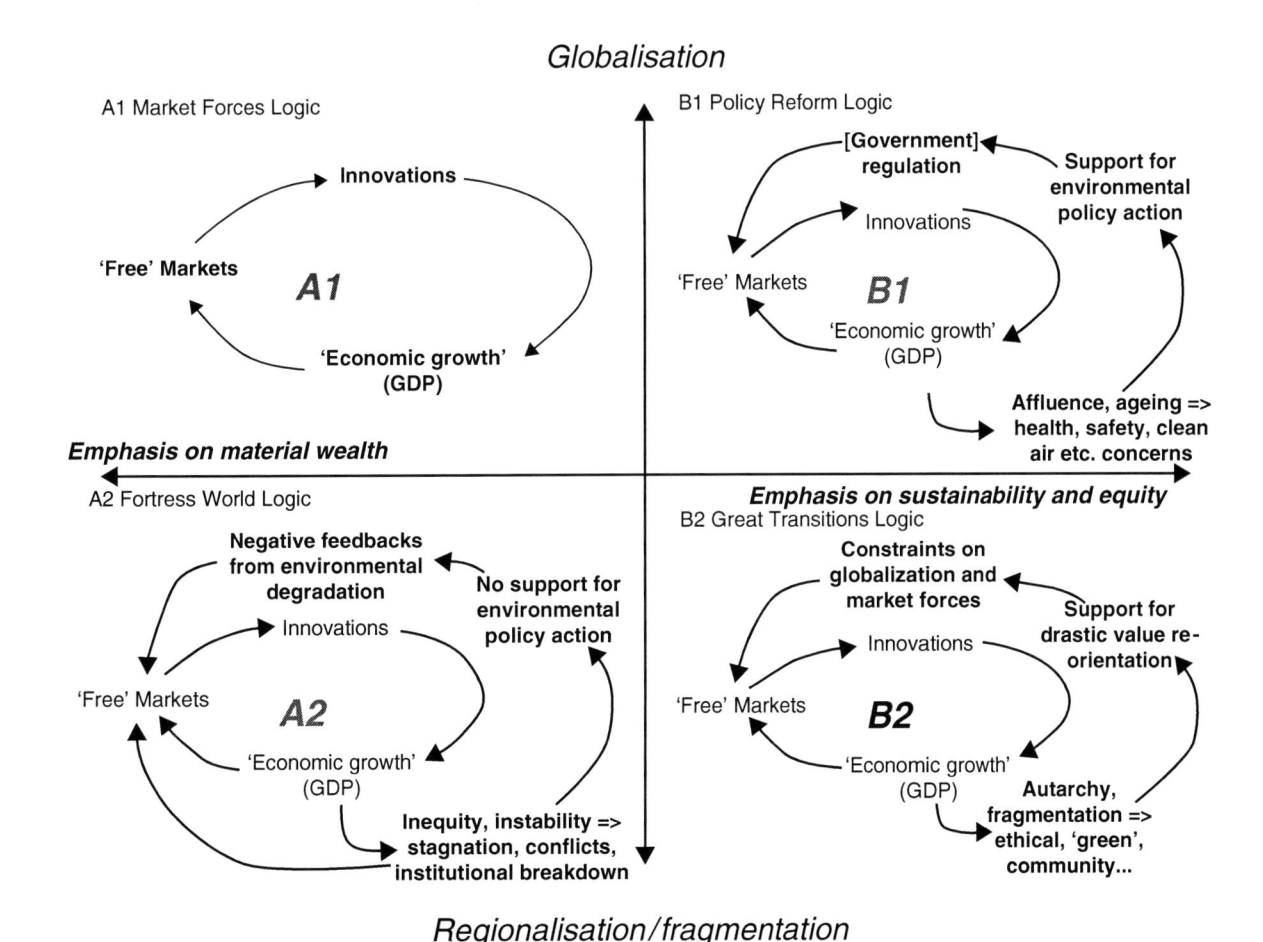

Fig. 18.10 Classification axes for the four SRES scenario families. For each scenario a representation of the mental map behind the dominant economy-environment logic is drawn

reflect all kinds of consumer and producer trends (service economy, 'information age', 'experience economy').
- *Consumer behaviour*: with rising income (GDP/cap), people will tend to become less material-oriented and focus more on immaterial values (health, environment...) – the so-called dematerialization.
- *Political organization*: the general trend towards 'modernization' will not only comprise income growth, fertility decline and urbanization, but also a transition from traditional, often authoritarian, values and governance structures, to (post-)modern, egalitarian ones.

The quantification process also made it necessary to harmonise at least some of the assumptions. It was decided that all modelling groups would use the same projections for world population and Gross World Product (GWP). Each scenario family was implemented in detail by one of the modelling groups, providing the so-called 'marker' scenario. To qualify for the status of 'fully harmonised scenario', population growth has to be within 5% of the respective marker scenario value and GWP and final energy demand within 10% (Nakícenovíc et al. 2000). A total of 40 scenarios have been considered, including one marker scenario for each family and two additional marker scenarios (A1F[ossil] and A1T[echnology], next to an A1B[alanced]) to illustrate the importance of energy technology choices.

18.5.4 Some SRES Results

The greenhouse gas emission trajectories, the emission pathways for sulphur – which postpone the warming effect – and the resulting atmospheric greenhouse gas concentrations have all been published and discussed extensively (see Nakícenovíc et al. 2000; IPCC 2001a–c). The 'Summary for Policymakers' lists some of the main conclusions, cited below.

Working Group I on *the scientific basis*:

- An increasing body of observations gives a collective picture of a warming world and other changes in the climate system.
- Emissions of greenhouse gases and aerosols due to human activities continue to alter the atmosphere in ways that are expected to affect the climate.
- Confidence in the ability of models to project future climate has increased.
- There is now stronger evidence that most of the warming over the last 50 years is attributable to human activities.
- Human influences will continue to change the atmospheric composition throughout the 21st century, and global average temperature and sea level are projected to rise under all IPCC SRES scenarios.

Working Group II on *impacts, adaptation, and vulnerability*:

- Emergent findings are that recent regional climate changes, particularly temperature increases, have already affected many physical and biological systems.
- Projected changes in climate extremes could have major consequences.
- Adaptation is a necessary strategy at all scales to complement climate change mitigation strategies.
- Those with the least resources have the least capacity to adapt and are the most vulnerable.
- Adaptation, sustainable development and enhancement of equity can be mutually reinforcing.

Working Group III on *mitigation*:

- Alternative development paths can result in very different greenhouse gas emissions.
- Significant technical progress relevant to greenhouse gas emissions reduction has been made since the Second Assessment Report in 1995 and has been faster than anticipated.
- Estimates of costs and benefits of mitigation actions differ (for a variety of reasons).
- As a result, estimated costs and benefits may not reflect the actual costs and benefits of implementing mitigation actions.
- Cost-effectiveness studies with a century time-scale estimate that the costs of stabilising CO_2 concentrations in the atmosphere increase as the concentration stabilization level declines and that different baselines can have a strong influence on absolute costs.
- The successful implementation of greenhouse gas mitigation options need to overcome many technical, economic, political, cultural, social, behavioural and/or institutional barriers which prevent the full exploitation of the technological, economic and social opportunities of these mitigation options.

Overall we can say that human-induced climate change is now with us, and will continue to be so in the decades to come. Impacts may be severe and unequal, with adaptation being necessary. And, finally, reduction of greenhouse gas emissions is and will continue to be an enormously complex task.

The IMAGE team has simulated the chain from demographic and economic drivers via emissions from energy use and land-use/cover changes to greenhouse gas concentrations and (aggregate) climate parameter changes (de Vries 2001b). Figure 18.11 shows the CO_2 concentrations and the CO_2 equivalent concentrations for the six SRES scenarios in the 1971–2100 period. Figure 18.12 shows the simulated change in global average surface temperature for two SRES scenarios: one with the highest (A1F) and the other with the lowest (B1) greenhouse gas emissions. The uncertainties caused by introducing an uncertainty in climate sensitivity are also indicated.

18.5.5 SRES Evaluation

It is probably too early to assess the process, content and impacts of the SRES. Besides, as concluded by the IPCC Working Group III (IPCC 2001c), decision-making in climate change is essentially a sequential process taking place under general uncertainty. Nevertheless, this does not prevent us delivering a few

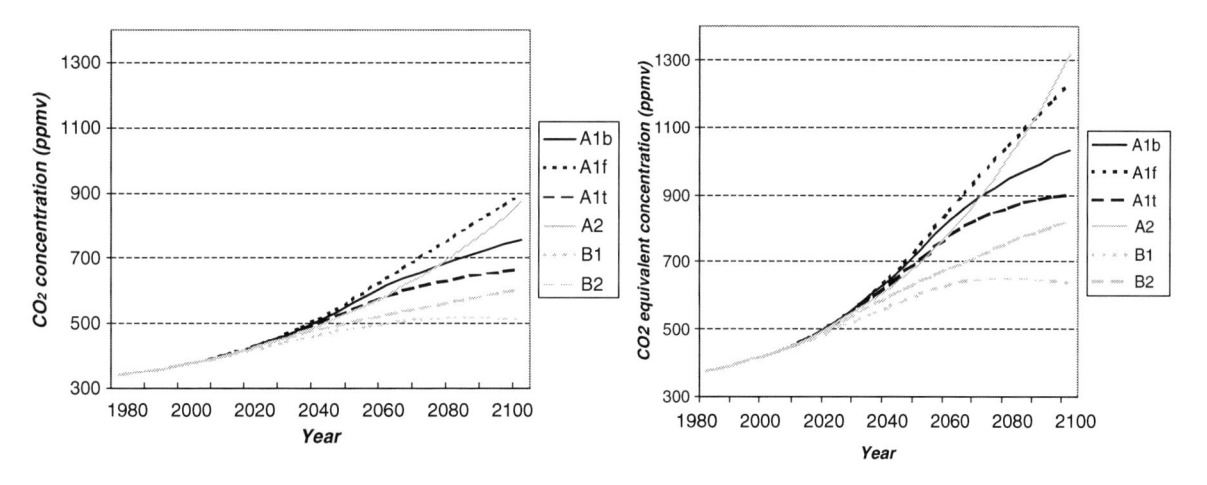

Fig. 18.11 Global CO_2 concentrations (left) and CO_2-equivalent concentrations (right) for the six SRES scenarios. CO_2 equivalent is a measure to present the impact on radiative forcing of all greenhouse gases in terms of CO_2. The CO_2 equivalent values are generally 20–30% higher than the CO_2 values. This figure sums the Kyoto gases: CO_2, CH_4, N_2O, HFC, PFC and SF_6

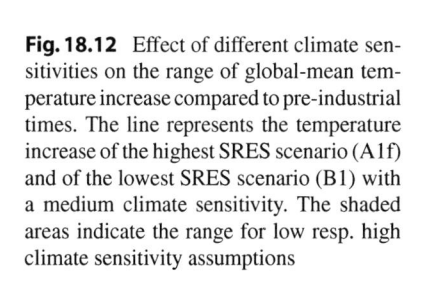

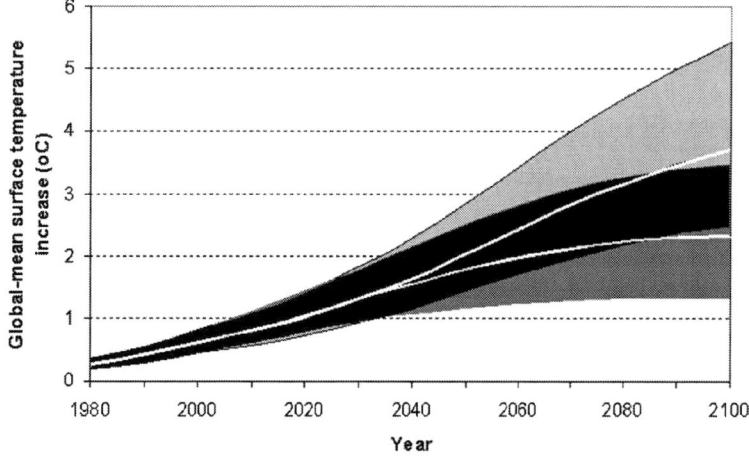

Fig. 18.12 Effect of different climate sensitivities on the range of global-mean temperature increase compared to pre-industrial times. The line represents the temperature increase of the highest SRES scenario (A1f) and of the lowest SRES scenario (B1) with a medium climate sensitivity. The shaded areas indicate the range for low resp. high climate sensitivity assumptions

points of constructive criticism, summarised under the points below.

1. With regard to the process of scenario construction using storylines, there has been much confusion about what scenarios mean or how they can be interpreted. For example, are they utopias? Personal convictions? More-or-less plausible or probable? The main reason for the confusion is that quite a few people/groups reject one or more of the scenarios as being totally implausible – often meaning undesirable – without looking at the set as a whole. For instance, the B1 scenario was initially dismissed by environmental NGOs as impossible and undesirable because it was thought to make policy-makers and some stakeholders believe that the climate problem would be solved 'automatically' by just having 'sustainable development'. More recently, however, the same B1 scenario has been dismissed as having implausibly high economic growth rates, in particular, for the less developed regions, and associated income convergence. Such critics confuse the attempt to cover a variety of possible futures with a partisan/strategic use of scenarios.

2. Such confusion is almost unavoidable given the large amount of ignorance, and hence value-laden interpretations of the various storylines that make any model-based quantification unavoidably a matter of guesses, values and taste. In particular, the absence of the often incomplete, implicit and crude inclusion of the role of governance, entrepreneurial and innovation dynamics and socio-cultural mechanisms will give rise to an ongoing debate. Of course, one may argue that this is exactly what the scenario exercise is meant to stimulate.

3. A serious flaw in the scenarios may be their one-way track approach to a future, in which certain values and motives prevail without any feedback, notably, from the particular climate change itself. For instance, an A2-world in which greenhouse gas concentrations rise to over 1,000 ppmv will most probably seriously thwart even the most modest economic activity and resource exploitation levels in this scenario. In fact, serious climate change impacts may be the foremost, if not the only, event that will convince the important stakeholders to take action – mitigative as well as adaptive. Table 18.3 suggests some of the socio-political complexities in the perception and associated attitude on climate change.

4. Despite these shortcomings, the SRES scenarios have provided, and continue to provide, a workable and intelligent framework to discuss the important issues to come: who will have to adapt to what and when, how to realise reduction in greenhouse gas emission [growth] and how to distribute the efforts and costs in terms of 'what gases?' 'how many?' 'how fast?' and 'where?'.

Table 18.3 Climate change (CC) in the four SRES scenarios

	A1	B1	A2	B2
Probability of CC	Medium to large A2 > A1 > B2	Smallest	Highest	Small to medium A1 > B2 > B1
Attitude towards CC	Economic growth has priority	Other environ. concerns have priority; no-regrets and side-benefits	Muddling through – not an issue/fatalist	Local environ. concerns; incapable of solving global environ. issues
Actions if CC is large	Barricades; adapt on your own; mitigate tensions by compensating poor	Formulate and implement global CC policy [too late but feasible], including adaptation	Not feasible: world community incapacitated; cope on your own	Some regions suffer; cope on your own, but [ineffective] attempts to share the burdens
Actions if CC is small	Pride: 'you see, ours is the best of all possible worlds'	Pride: 'you see: the high-tech and SD-orientation did it'	'Well, this time we were lucky…so far'	'Well, we were lucky… but our local concerns may have helped'
Attitude towards CC policy	Reluctant, although rich enough to make compensatory gestures	Reluctant if it slows down the narrowing of the N-S income gap	No expectations: impossible anyway	Low expectations: full of good intentions, but they fail

The debate on allocating emission allowances based on a variety of indicators around responsibility and capacity is gaining momentum. It is only now that more real stakeholders – such as insurance companies and fossil-fuel exporting companies and regions – are starting to feel the possible consequences of climate change, but at the same time there are attempts to adapt to or mitigate it. The game has just begun.

18.6 Conclusions and Outlook

RAINS and other IAMs were able to play a large role in the negotiations on air pollution because they enjoyed considerable credibility among both scientists and policy-makers. IMAGE is a widely acknowledged climate change model among scientists; TARGETS is less well known, possibly because of its global approach. However, the role of climate change models supporting climate change policies is still developing.

The use of IAMs to support policies in the field of air pollution has been successful in the sense that it supported various protocols with increasing complexity, i.e. extending from the assessment of a single pollutant and single effects to multiple pollutants and multiple effects. The interaction between science and policy regarding air pollution became successfully embedded in consensus-based procedures, more so than seems to be the case regarding climate change.

This is due to a number of reasons. One is the organizational structure of the bodies under the LRTAP Convention, which mirrors the underlying scientific networks. Another reason may be that climate change is a less structured problem area than air pollution, making climate change IAMs more complex than air pollution IAMs. The time horizon over which climate change develops is longer, while underlying system dynamics, feedback and uncertainties combine in many more ways than those underlying air pollution. This complicates options for consensus, let alone for a policy structure that mirrors consensus-based scientific networks. Finally, the issue of 'makeability' of society is a prominent element in the assessment of options to curb climate change, much more than is the case for air pollution.

In the near future the concept of multiple pollutants and multiple effects under the LRTAP Convention may be stretched to include climate change. The reason is to step up opportunities for cost-efficient policies that take full account of linkages between climate change and air pollution policies. In the field of climate change research, increasing attention is given to the exploration of co-benefits and co-damage of climate change policies. Thus CC policies become subject to analysis for identifying positive feedbacks (co-benefits) or negative feedbacks (co-damage) affecting other environmental issues.

We close with some thoughts regarding the modelling methodologies which may be expected to support future multiple actor and multi-impact assessments. The IAMs described in this chapter use 'traditional' methodologies embedded in the natural sciences: deterministic and based on (difference) equations describing, for example, material fluxes, biological processes and geo-chemical cycles. They will fall short if progress is to be made in dealing with multi-actor and multi-impact assessments. The 'traditional' methodologies have inherent limitations if applied to complex long-term spatial dynamics with non-linearities, thresholds, feedbacks and delays that can combine to unexpected phenomena. They are also poorly representative of socio-economic dynamics. Hence IAMs will have to gradually include methodologies – included in this chapter in relation to more complex systems – which may stretch the limits of the science premise regarding verification and validation. The focus will increasingly include spatial and behavioural heterogeneity, structural change processes, and communication in and openness to interactive use in policy exercises. New approaches such as multi-agent modelling, evolutionary and institutional economics and simulation gaming may well turn out to be the seeds for these developments.

References

Adams, J. (1995). *Risk*. London: UCL Press.

Alcamo, J. M. (Ed.). (1994). *IMAGE 2.0 Integrated Modeling of Global Climate Change*. Dordrecht: Kluwer.

Alcamo, J. M., Shaw, R. M., & Hordijk, L. (Eds.). (1990). *The RAINS Model of Acidification*. Dordrecht: Kluwer.

Alcamo, J. M., Bouwman, A., Edmonds, J., Grübler, A., Morita, T., & Sugandhy, A. (1995). An evaluation of the IPCC IS92 emission scenarios (Ch. 6). In J. T. Houghton et al. (Eds.), *Climate change 1994* (p. 338). Cambridge: Cambridge University Press.

Alcamo, J. M., Leemans, R., & Kreileman, E. (1998). *Global change scenarios of the 21st century – Results from the IMAGE 2.1 model*. London: Elsevier.

Alcamo, J. M., Mayerhofer, P., Guardans, R., van Harmelen, T., Onigkeit, J., Posch, M., et al. (2002). An integrated assessment of regional air pollution and climate change in Europe: Findings of the AIR-CLIM project. *Environmental Science and Policy, 5*(4), 257–272.

Amann, M. (2003). *Linkages and synergies of regional and global emission control*. Workshop of the Task Force on Integrated Assessment Modelling (TFIAM) of the UNECE-CLRTAP, 27–29 January 2003. Laxenburg: International Institute for Applied Systems Analysis.

Amann, M., Hordijk, L., Klaassen, G., Schoepp, W., & Sorensen, L. (1992). Economic restructuring in Eastern Europe and acid rain abatement strategies. *Energy Policy, 12*, 1186–1197.

Amann, M., Johansson, M., Lükewille, A., Schöpp, W., Apsimon, H., Warren, R., et al. (2001). An integrated assessment model for fine particulate matter in Europe. *Water, Air and Soil Pollution, 130*, 223–228.

ApSimon, H. M., Warren, R. F., & Wilson, J. J. N. (1994). The abatement strategies assessment model – ASAM: Applications to reduction of sulphur dioxide emissions across Europe. *Atmospheric Environment, 28*, 649–663.

Barney, G. O. (1981). *The global 2000 report to the president. Entering the twenty-first century* (Vols. 1 and 2). Charlottesville, VA: Blue Angel.

Berk, M., van Minnen, J., Metz, B., & Moomaw, W. (2001). *Climate OptiOns for the Long term (COOL) – Final report: Global dialogue*. Bilthoven: RIVM.

Bossel, H. (1998). *Earth at a crossroads – Paths to a sustainable future*. Cambridge: Cambridge University Press.

Braat, L. C., & van Lierop, W. F. G. (1987). Integrated economic-ecological modeling. In L. C. Braat & W. F. J. van Lierop (Eds.), *Economic-ecological modeling, studies in regional science and urban economics* (Vol. 16, pp. 49–68). Amsterdam/New York: North-Holland.

Brouwer, F. M. (1987). *Integrated environmental modelling: Design and tools*. Dordrecht/Boston: Kluwer.

Clark, W. C., & Munn, R. E. (Eds.). (1986). *Sustainable development of the biosphere*. Cambridge: Cambridge University Press.

CPB. (1999). *WorldScan, the core version*. The Hague: CPB.

Cough, C. A., Bailey, P. D., Biewald, B., Kuylenstierna, J. C. I., & Chadwick, M. J. (1994). Environmentally targeted objectives for reducing acidification in Europe. *Energy Policy, 22*, 1055–1066.

de Bruin, J., Vink, P., & van Wijk, J. (1996). M – A visual simulation tool. San Diego: The society for computer simulation *Simulation in the Medical Sciences*, 181–186.

de Vries, H. J. M. (2001a). Perceptions and risks in the search for a sustainable world – A model-based approach. *International Journal of Sustainable Development, 4*(4), 434–453.

de Vries, H. J. M. (2001b). Objective science? The case of climate change models. In P. Goujon & B. Heriard Dubreuil Bertrand (Eds.), *Technology and ethics, a European quest for responsible engineering*. Leuven: Peeters.

de Vries, H. J. M., Bollen, J., Bouwman, L., den Elzen, M., Janssen, M., & Kreileman, E. (2000). Greenhouse-gas emissions in an equity-, environment- and service-oriented world: An IMAGE-based scenario for the next century. *Technological Forecasting and Social Change, 63*(2–3), 137–174.

de Vries, H. J. M., van Vuuren D. P., den Elzen M. G. J., & Janssen, M. A. (2001). *The Targets Image Energy Regional (TIMER) model – Technical documentation*. Bilthoven: RIVM. (Report 461502024)

den Elzen, M., Beusen, A. H. W., & Rotmans, J. (1997). An integrated modeling approach to global carbon and nitrogen cycles: Balancing their budgets. *Global Biogeochemical Cycles, 11*(2), 191–215.

EC. (1990). *Energy for a new century – The European perspective*. Brussels: Report DGE.

EC. (1996). European energy to 2020 – A scenario approach. *Energy in Europe* (Spring-Special Issue). Brussels: DG XVII.

EC. (1999). European Union Energy Outlook to 2020. *Energy in Europe* (Special issue, November). Brussels: DG XVII.

EEA. (1995). *A general strategy for integrated environmental assessment at the European Environmental Agency*. Scoping study prepared by the National Institute for Public Health and the Environment (RIVM), Bilthoven and the European Environmental Agency, Copenhagen.

EEA. (1999). *Environment in the European Union at the turn of the century*. Copenhagen: EEA. (Environmental Assessment Report No. 2)

EMEP. (1998). *Transboundary acidifying air pollution in Europe, Parts 1 and 2*. Oslo: Norwegian Meteorological Institute. (EMEP/MSC-W Status Report 1/98)

Forrester, J. W. (1961). *Industrial dynamics*. Cambridge, MA: MIT Press.

Forrester, J. W. (1971). *World dynamics*. Cambridge, MA: Wright-Allen Press.

Forrester, J. W. (1980). System dynamics – future opportunities. In A. A. Legastro Jr., J. W. Forrester, & J. M. Lyneis (Eds.), *System dynamics, studies in the management science* (Vol. 14). Amsterdam/New York: North-Holland.

Funtowicz, S. O., & Ravetz, J. R. (1990). *Uncertainty and quality in science for policy*. Dordrecht: Kluwer.

Gallopin, G., Hammond, A., Raskin, P., & Swart, R. (1997). *Branch points: Toward global scenarios and human choice*. Stockholm: Stockholm Environment Institute (SEI). PoleStar Series Report No. 7.

GEA. (2001). Global environmental assessment project. Cambridge, MA: Kennedy School of Government/Harvard University Press. www.ksg.harvard.edu/gea/.

Hafkamp, W. A. (1984). *Economic-environmental modelling in a national-regional system*. Amsterdam: North-Holland.

Hettelingh, J.-P. (1990). *Uncertainty in modelling regional environmental systems: The generalization of a watershed acidification model for predicting broad scale effects*. Laxenburg: International Institute for Applied Systems Analysis. (Research Report RR-90-3)

Hettelingh, J.-P., Posch, M., & de Smet, P. A. M. (2001). Multi-effects of critical loads used in multi-pollutant reduction agreements in Europe. *Water, Air and Soil Pollution, 130*, 1133–1138.

Hettelingh, J-P., Posch, M., Slootweg, J., Reinds, G. J., Spranger, T., & Tarrason, L. (2007a). Critical loads and dynamic modelling to assess European areas at risk of acidification and eutrophication. *Water, Air and Soil Pollution: Focus, 7*, 379–384.

Hettelingh, J.-P., Posch, M., Slootweg, J., & van 't Zelfde, M. (2007b). Status of European critical loads with focus on nitrogen. In J. Slootweg, M. Posch, & J.-P. Hettelingh (Eds.), *Critical loads of nitrogen and dynamic modelling*. Coordination Centre of Effects Progress Report 2007, MNP Report 500090001/2007 (available from www.mnp.nl/cce).

Hilderink, H. (2000). *World population in transition – An integrated regional modelling framework. Faculteit der Ruimtelijke Wetenschappen*. Groningen: Groningen University.

Hisschemöller, M., Tol, R. S. J., & Vellinga, P. (2001). The relevance of participatory approaches in integrated environmental assessment. *Integrated Assessment, 2*, 57–72.

Hoekstra, A. Y. (1998). *Perspectives on water – An integrated model-based exploration of the future*. Delft: Technological University.

Holling, C. S. (Ed.). (1978). *Adaptive environmental assessment and management*. New York/London: Wiley.

Hordijk, L. (1995). Integrated assessment models as a basis for air pollution negotiations. *Water Air and Soil Pollution, 85*, 249–260.

Hordijk, L., & Kroeze, C. (1997). Integrated assessment models for acid rain. *European Journal of Operational Research, 102*, 405–417

Huntingdon, S. (1997). *The clash of civilizations and the remaking of the world order*. New York: Simon & Schuster.

IEA (1996). World Energy Outlook. Paris: International Energy Agency.

IMAGE-team. (2001). *The IMAGE 2.2 implementation of the SRES scenarios: A comprehensive analysis of emissions, climate change and impacts in the 21st century*. RIVM CD-ROM Publication 481508018.

IPCC. (1991). *Climate change: The IPCC response strategies*. Intergovernmental Panel on Climate Change (IPCC)/UNEP/WMO.

IPCC. (2001a). *Climate change 2001 – The scientific basis*. Summary for Policy Makers and Technical Summary of the Working Group I Report, UNEP/WMO.

IPCC. (2001b). *Climate change 2001 – Impacts, adaptation, and vulnerability*. Summary for Policy Makers and Technical Summary of the Working Group II Report, UNEP-WMO.

IPCC. (2001c). *Climate Change 2001 – Mitigation*. Summary for Policy Makers and Technical Summary of the Working Group III Report. UNEP/WMO.

Jäger, J. (1998). Current thinking on using scientific findings in environmental policy making. *Environmental Modelling and Assessment, 3*(3), 143–153.

Janssen, M., & de Vries, B. (1998). The battle of perspectives: A multi-agent model with adaptive responses to climate change. *Ecological Economics, 26*, 43–65.

Johansson, T., Kelly, H., Reddy, A., & Williams, R. (Eds.). (1993). *Renewable energy*. Washington, DC: Island Press.

Kaplan, R. (1996). *The ends of the earth*. New York: Random House.

Kassler, P. (1994). *Energy for development*. London/The Hague: Shell International Petroleum Company.

Klaassen, G., & Sliggers, J. (1993). *Cost sharing for the abatement of acidification in Europe: the missing link in the sulphur protocol*. Working Paper, Task Force on Integrated Assessment Modelling.

Kram, T., Morita, T., Riahi, K., Roehrl, R. A., van Rooien, S., Sankowski, A., et al. (2000). Global and regional greenhouse gas emissions scenarios. *Technological Forecasting and Social Change, 63*(2–3), 335–372.

Leggett, J., Pepper, W., & Swart, R. (1992). Emissions scenarios for the IPCC: An update. In J. T. Houghton, B. A. Callander, & S. K. Varney (Eds.), *Climate change*. Cambridge: Cambridge University Press.

Meadows, D. H., Meadows, D. L., Randers, J., & Behrens, W. W. (1972). *The limits to growth: A report for the club of Rome's project on the predicament of mankind*. New York: Universe Books.

Meadows, D. H., Behrens, W., Meadows, D. L., Naill, R., Randers, J., & Zahn, E. (1974). *Dynamics of growth in a finite world*. Cambridge, MA: Wright-Allen Press.

Meadows, D. H., Meadows, D. L., & Randers, J. (1991). *Beyond the limits. Global collapse or a sustainable future*. London: Earthscan.

Morita, T., Kaihuma, Y., Harasaw, H., Kai, K., Dong-Kumand, L., & Matsuoka, Y. (1994). *Asian-Pacific integrated model for evaluating policy options to reduce GHG emissions and global warming impact*. Report of the National Institute for Environmental Studies, Tsukuba, Japan.

Morita, T., Nakícenovíc, N. & Robinson, J. (2000). Long-term scenarios on socio-economic development and climatic policies [Special issue]. *Environmental Economics and Policy Studies, 3*(2,), 65–299.

Nakícenovíc, N. (1995). *Global energy perspectives to 2050 and beyond*. Laxenburg: IIASA/WEC.

Nakícenovíc, N., et al. (2000). *Special report on emissions scenarios (SRES) for the intergovernmental panel on climate change (IPCC)*. Cambridge: Cambridge University Press.

NAPAP. (1990). *1990 Integrated Assessment Report* (p. 520). Washington, DC: The NAPAP Office of the Director.

Niessen, L. (2002). *Roads to health – Multi-state modelling of population health and resource use*. Groningen: Groningen University.

Nilsson, J., & Grennfelt, P. (Eds.). (1988). *Critical loads for sulphur and nitrogen*. Copenhagen: Nordic Council of Ministers.

OECD. (1979). *Interfutures – Facing the future: Mastering the probable and managing the unpredictable*. Paris: OECD.

OECD. (1997). *Towards a new global age – Challenges and opportunities* (Policy report). Paris: OECD.

Oreskes, N., Shrader-Frechette, K., & Belitz, K. (1994). Verification, validation and confirmation of numerical models in the earth sciences. *Science, 263*, 641–646.

O'Riordan, T., & Jager, J. (Eds.). (1996). *Politics of climate change – A European perspective*. London/New York: Global Environmental Change Series/Routledge.

Parson, E. A. (1995). Integrated assessment and environmental policymaking. *Energy Policy, 23*(4/5), 463–475.

Posch, M., Hettelingh, J.-P., & de Smet, P. A. M. (2001). Characterization of critical load exceedances in Europe. *Water Air Soil Pollution, 130*, 1138–1139.

Posch, M., Hettelingh, J.-P., Slootweg, J., & Downing, R. J. (2003). *Modelling and mapping of critical thresholds in Europe*. Bilthoven: RIVM. (CCE Status Report 2003)

Raskin, P., Banuri, T., Gallopin, G., Gutman, P., Hammond, A., Kates. R., et al. (2002). *Great transition – The promise and lure of the times ahead*. Boston: Stockholm Environment Institute (SEI).

Rotmans, J. (1990). *IMAGE: An integrated model to assess the greenhouse effect*. Dordrecht/Boston: Kluwer.

Rotmans, J. (1998). Methods for IA: The challenges and opportunities ahead. *Environmental Modelling and Assessment, 3*(3), 155–179.

Rotmans, J., & de Vries, H. J. M. (Eds.). (1997). *Perspectives on global futures: The TARGETS approach*. Cambridge: Cambridge University Press.

Rotmans, J., Dowlatabadi, H., Filar, J. A., & Parson, E. A. (1996). Integrated assessment of climate change: Evaluation of methods and strategies. In S. Rayner & E. L. Malone (eds.), *Human choices and climate change: A state of the art report*. Washington, DC: Batelle Pacific Northwest Laboratories.

Schmidheiny, S. (1992). *Changing course – A global business perspective on development and the environment*. Cambridge, MA: MIT Press.

Schöpp, W., Amann, M., Cofala, J., Heyes, C., & Klimont, Z. (1999). *Environmental Modelling and Software, 14*(1), 1–9.

Schwarz, M., & Thompson, M. (1990). *Divided we stand – Redefining politics, technology and social choice*. New York: Harvester Wheatsheaf.

Shell International. (2000). *Global energy scenarios*. London: Global Business Environment.

Shell International. (2002). *People and connections – Global scenarios to 2020: Public summary*. London: Global Business Environment.

Stoddard, J. L., Jeffries, D. S., Lükewille, A., Clair, T. A., Dillon, P. J., Driscoll, C. T., et al. (1999). Regional trends in aquatic recovery from acidification in North America and Europe 1980–95. *Nature, 401*, 575–578.

Swart, R., Berk, M., Janssen, M., Kreileman, E., & Leemans, R. (1998). The safe landing approach: Risks and trade-offs in climate change. In J. M. Alcamo, R. Leemans, & E. Kreileman (Eds.), *Global change scenarios of the 21st century – Results from the IMAGE 2.1 model* (pp. 193–244). London: Elsevier.

Thompson, M., Ellis, R., & Wildawsky, A. (1990). *Cultural theory*. Boulder, CO: Westview Press.

Tol, R. S. J., & Vellinga, P. (1997). Approaches to integrated assessment. In A. Sors, A. Libertore, S. Funtowicz, J. C. Hourcade, & J. L. Fellous (Eds.), *Prospects for integrated environmental assessment: Lessons learnt from the case of climate change*. Proceedings of Conference, Toulouse, October 24–26, 1996. European Commission, DG XII. Report EUR 17639.

Tol, R. S. J., & Vellinga, P. (1998). The European forum on integrated environmental assessment. *Environmental Modelling and Assessment, 3*(3), 181–191.

Tuinstra, W., Hordijk L., & Amann, M. (1999). Using computer models in international negotiations; the case of acidification in Europe. *Environment, 41*(9), 32–42.

UNEP. (2002). *Global Environmental Outlook 3*. Nairobi: United Nations Environment Program (UNEP).

van Asselt, M. B. A. (1997). *TARGETS for ULYSSES – Work package*. Zurich/Bilthoven: EAWAG/RIVM.

van Asselt, M. B. A., & Rotmans, J. (1996). Uncertainty in perspective. *Global Environmental Change, 6*(2), 121–157.

van Asselt, M. B. A., & Rotmans, J. (2002). Uncertainty in integrated assessment modelling: From positivism to pluralism. *Climate Change, 54*(1–2), 75–105.

van Asselt, M. B. A., Langendonck, R., van Asten, F., van der Giessen, A., Janssen, P., Heuberger, P., et al. (2001). *Uncertainty and RIVM's environmental outlooks – Documenting a learning process*. Bilthoven/Maastricht: Report RIVM/ICIS.

van Daalen, C., Thissen, W., & Berk, M. (1998). The Delft process: Experiences with a dialogue between policy makers and global modellers. In J. Alcamo, R. Leemans, & E. Kreileman (Eds.), *Global change scenarios of the 21st century – Results from the IMAGE 2.1 model* (pp. 267–285). Oxford: Pergamon/Elsevier.

van der Sluijs, J. P., Potting, J., Risbey, J., van Vuuren, D., de Vries, H. J. M., & Beusen, A. (2001). *Uncertainty assessment of the IMAGE/TIMER B1 CO_2 emissions scenario using the NUSAP method*. Utrecht University, Utrecht. (NOP Report Report no: 410 200 104)

Vestreng, V., & Klein, H. (2002). *Emission data reported to UNECE/EMEP: Quality assurance and trend analysis and presentation of WebDab*. Oslo: Norwegian Meteorological Institute. (EMEP/MSC-W Note 1/2002)

van Vuuren, D., & de Vries, H. J. M. (2001). Mitigation scenarios in a world oriented at sustainable development: The role of technology, efficiency and timing. *Climate Policy, 1*, 189–210.

van Vuuren, D., Fengqi, Z., de Vries, H. J. M., Kejun, J., Graveland, C., & Yun, L. (2002). Energy and emission scenarios for China in the 21st century – Exploration of baseline developments and mitigation options. *Energy Policy, 31*, 369–387.

WCED. (1987). *Our common future*. Oxford: Oxford University Press.

Weyant, J., Davidson, O., Dowlatabadi, H., Edmonds, J., Grubb, M., Parson, E.A., et al. (1996). Integrated assessment of climate change: An overview and comparison of approaches and results. In J. P. Bruce, H. Lee, & E. F. Haites (Eds.), *Climate change 1995: Economic and social dimensions – Contribution of Working Group III to the Second Assessment, Report of the Intergovernmental Panel on Climate Change*. Cambridge: Cambridge University Press.

Part III
Context and Perspectives

Chapter 19
Environmental Policies in Their Cultural and Historical Contexts

Maurie J. Cohen

Contents

19.1	Introduction	423
19.2	Science, Romanticism, and the Environment	424
19.3	Studies of National Character	424
19.4	Culture, History, and the Environment: A Review of Research Perspectives	425
19.4.1	Comparative Environmental Politics and Policymaking	426
19.4.2	Cross-national Surveys of Public Opinion	426
19.4.3	Social Construction of Risk	427
19.4.4	National Innovation Styles and Technology Policy	427
19.4.5	Public Understanding of Science and the Production of Environmental Knowledge	428
19.4.6	Responses to Global Environmental Change in National Contexts	428
19.5	Conclusion	429
References		429

19.1 Introduction

Scholars working within the field of comparative environmental policy have regularly noted the disparity in how different countries react to ecological threats. The 1986 Chernobyl accident, a catastrophe that spread measurable amounts of radioactivity across a broad stretch of northern Europe, provides a particularly poignant illustration of the ways in which predominant public responses to environmental risk can vary. During the months following the incident, a number of commentators quipped that the ill-effects of atmospheric dispersal oddly seemed to stop at the German-French border. These remarks were motivated by the sardonic observation that while Germans typically refused to eat locally-grown vegetables during the months following the incident, their French neighbours evidenced no similar vigilance.

On a broader scale, we have witnessed over the past decade cross-national variation in the form of Dutch environmental advocates cooperating with industry in a way hardly imaginable in Germany, British eco-warriors burrowing themselves into underground bunkers to obstruct the construction of new roads, and American communities aggressively protesting the construction of hazardous waste incinerators. These multifarious forms of agitation around environmental concerns invariably lead to very different political responses and policy outcomes.

The traditional approaches used to explain these variations generally look at differences in the national institutions available to address environmental problems and in countrywide economic organisation. In other words, orthodox social scientists grounded in an understanding of ecological risk as a scientific object of study overwhelmingly focus their attention on cross-national disparities in regulatory structures and systems of governance and this becomes the limits of their analyses.

There can be little question that institutional organisation and economic interests have important bearing on the formulation of national environmental commitments. However, to constrain evaluations to these so-called structural considerations is to disregard the subtle ways that countries vary culturally and historically and how these differences shape the production of place-specific environmental knowledge.

M.J. Cohen (✉)
Environmental Policy and Sustainability at the New Jersey Institute of Technology, Newark, USA
e.mail: maurie.cohen@njit.edu

There has now emerged a school of thought, that devotes greater credence to the social construction of environmental issues (see, for example, Hannigan 1995; Irwin 1995). Adherents of this perspective are more likely to highlight the cultural and historical factors that influence individual countries' predominant tendencies to privilege certain ecological threats as paramount and to dismiss other risks as more trivial. It is to a description of this more culturally and historically informed mode of environmental policy analysis that this chapter devotes its attention.

19.2 Science, Romanticism, and the Environment

Within the ranks of mainstream environmental policy analysis it has become commonplace to treat ecological systems purely as objects of scientific scrutiny. Such an approach for conceptualising nature has given rise to a vast number of specialised fields. Regardless of the issue – for instance the cancer-inducing effects of industrial chemicals or the impacts of particular tree-harvesting practices on soil erosion – the response is to seek scientific explanations and scientifically-grounded solutions to environmental dilemmas.

This 'scientised' approach represents a distinct intellectual departure from previous modes of thought regarding the environment.[1] In earlier eras it was common for elements of the natural surround, indeed the biosphere itself, to be infused with religious, magical, and spiritual significance. As science has become more pervasive in contemporary life our more modern cast of mind has pressed these understandings to the periphery. Indeed, there has been active effort by governments, through education and other spheres of public policy, to completely repudiate such purportedly outmoded notions. After all, to be modern is generally thought to entail the abandonment of such anachronistic patterns of thought and the embrace of science and rationality.

Nonetheless, even in the most advanced nations of the world a variety of alternative epistemologies for producing environmental knowledge continues to persist and, in many cases, to coexist alongside more seemingly sophisticated ways of knowing. While some of these perspectives are informed by age-old religious tenets, today's most pervasive holistic and experiential knowledge systems have roots in the influential Romantic movements of the 18th and 19th centuries. It is necessary to speak of Romanticism in the plural sense here because these intellectual developments were situated in specific national contexts and drew their power from local cultural and historical influences (Porter and Teich 1988).

For instance, Teutonic Romanticism, as initially articulated by Goethe, was keenly influenced by German ecclesiastical and political debates of the 18th century (Cohen 1999; Harrington 1996). In contrast, English Romanticism took as its point of departure the rise of industrialism and a quest for refuge in an idyllic appreciation for rural landscapes (Bate 1991). The Romantic movement in the United States, was powerfully informed by frontier myths and ideals of a pristine and untrammelled nature (Nash 1967). Elsewhere, for example in the Netherlands, the relative stability of religious doctrine during the 18th and 19th centuries severely limited the emergence of a distinctly Dutch Romantic voice (Rupke 1988).[2]

The style and timbre of Romanticism, as well as the way these outlooks combined with more scientific forms of knowledge in the public domain, were thus strongly influenced by factors unique to the culture and history of individual places. There is also a remarkable immutability to these qualities as they recursively embed themselves in educational institutions, social norms, and public policies. Because of the persistence of these considerations, we can quite reasonably speak about countries having different styles for the interpretation of ecological information.

19.3 Studies of National Character

To contend that the production of environmental knowledge differs cross-nationally is to encroach quite closely on the domain of inquiry traditionally

---

[1] There is a vast literature on pre-modern conceptions of nature. Key sources offering an introduction to these perspectives include Glacken (1967) and Thomas (1983).

[2] Locally-inspired Romanticism in the Netherlands was largely limited to a 19th century Dutch school of Romantic painting centred around the writer Willem Bilderdijk.

known as the study of national character. This area of research suggests that people sharing geographical space and having certain institutional commonalities maintain similar dispositions and temperaments with respect to their natural surround. Moreover, the notion of national character suggests that it is possible to formulate defensible methodologies that enable us to claim, for example, that Germans are punctilious and that the Dutch have a propensity for cleanliness. Perhaps not surprisingly, this area of research can readily lend itself to the propagation of hackneyed stereotypes and it indeed does suffer from a chequered history (Schooler 1996). Because of these issues it is worth devoting attention to a brief discussion of how current initiatives to study national character differ from past endeavours.

The modern study of national character dates back to World War II when national governments recruited small groups of anthropologists and psychiatrists to formulate social-psychological profiles of the combatant countries. The United States was arguably the most aggressive in pursuing this line of research and the American military's Office of War Information commissioned Margaret Mead and Ruth Benedict to produce several pieces of work (Hoffman 1992). These prominent scholars sought to explain national tendencies such as Japanese tidiness or Russian tempestuousness in terms of nationally common child-rearing styles. Military planners in the United States had access to similar research pertaining to several occupied European countries and some of this information may have been used to prepare American liberation units during the final stages of the war.

It was not until the following decade that the investigation of national character began to shed its psychoanalytic commitments. The most important contribution in this regard was Almond and Verba's (1963) work on national predispositions for democracy. These authors looked at such issues as civic obligation, public trust, and interpersonal associativeness as factors responsible for the shaping of resolve for representative government. However, at the time, scholars of comparative politics deemed culture and history to be less important than cross-national institutional or economic differences and this work did not receive the attention that with hindsight it no doubt deserved. A related problem that diminished the significance of culture and history as relevant explanatory variables was the difficulty of conducting research on these issues in the absence of a normative framework that invariably privileged some national characteristics over others.[3] For example, critics treated Banfield's (1958) depiction of the clannishness of southern Italians quite harshly, claiming that such portrayals were deterministic and encouraged the perpetuation of negative stereotypes.

The work of social psychologist Inkeles (1997) has been central to efforts to move the study of national character onto a more rigorous footing within mainstream social scientific research. Through the formulation of the concept of 'modal personality'. Inkeles and his colleagues have sought to devise methodologies for linking sociocultural factors to specific behavioural responses in national groups.

Despite these developments there can be little question that throughout most of the post-World War II era the subject of national character has carried an incontrovertible stigma. While few researchers would deny the existence of country-specific traits, the elusiveness of these qualities to conventional research techniques has tended to place their investigation beyond the bounds of serious social science scholarship. More recently, with the end of the Cold War and the efforts of transnational policymaking bodies such as the European Union to expand their influence, it has become more difficult to side-step the issue of national character. These developments have led to widespread interest in how culture and history predetermines public policy outcomes (e.g. Fukuyama 1995; Huntington 1996; Lipset 1996; Putnam 1993).

19.4 Culture, History, and the Environment: A Review of Research Perspectives

The increase in attention that the social sciences are now devoting to the complex relationship among culture, history, and public policy extends into the area of environmental politics. Interestingly, scholars of comparative politics carried out some of the first studies of cross-national variability in environmental policymaking during the 1970s and 1980s as part of an effort to understand the divergent ways countries developed

---

[3] There can also be little question that the study of the cultural and historical predeterminants of public policies, variables that do not lend themselves quite so readily to quantitative and computer-driven methodologies, went against the grain of the times.

programs for ameliorating air and water pollution problems (Enloe 1975; Lundqvist 1980; Teich et al. 1997; Vogel 1986). These efforts were principally designed to explore how different political systems and institutional structures gave rise to distinct forms of regulatory intervention.

Current work in comparative environmental politics and policy builds on this foundation, but devotes greater attention to the cultural and historical determinants of the outcomes. There is now a growing body of evidence that these previously disregarded factors do indeed temper environmental decision making at the macropolitical level of the nation-state, as well as the production of situated environmental knowledge. For purposes of the current discussion we can organise into six categories the diverse methodologies that researchers are currently using to explore the connections between culture, history, and environmental policy making. Most of this work takes place at the conjunction of several interdisciplinary fields, namely comparative politics, environmental social science, and science and technology studies.

19.4.1 Comparative Environmental Politics and Policymaking

The study of comparative environmental politics and policymaking has begun to broaden its scope in recent years to look at how more conventional concerns such as the institutional and economic structures of nations combine with cultural and historical determinants to produce distinct policy outcomes. Research has focused on a variety of specific policy areas including acid rain (Boehmer-Christiansen and Skea 1993), vehicle emissions (Boehmer-Christiansen and Weidner 1995), wind energy (Breukers and Wolsink 2007), and ecological taxation (Andersen 1994), as well as broader themes such as environmental regulation (Gouldson and Murphy 1998), ecological modernisation (Weale 1992; Hajer 1995), and overall capacity for environmental reform (Wintle and Reeve 1994; Jänicke and Weidner 1997; Andersen and Liefferink 1997).

The bulk of this research has focused on Europe, largely as a result of the relatively intimate level of dialogue that takes place among proximate nations and the need to promote some degree of policy convergence within the context of the European Union (EU). Moreover, the EU has become over the past decade increasingly influential in the internal affairs of its constituent members at the same time that it is expanding its geographic range by drawing in countries from Eastern Europe. In a related sense, research has focused on the emergence of widening environmental policy cleavage between the EU and the United States and the serious implications that this pattern of divergence has on the global governance of critical environmental problems (Vig and Faure 2004).

Secondary policymaking institutions have played central roles in the wave of interest in comparative environmental politics and policy. For example, the Organisation for Economic Cooperation and Development (OECD), a Paris-based consortium of the world's wealthiest nations, has been an important forum for discussing new environmental policy concepts and tools. The OECD also conducts instructive peer reviews of the environmental performance of its member nations that have the potential to feed back into domestic political processes (see, for example, OECD 1996). Smaller bodies such as the Nordic Council of Ministers and the United Nations Economic Commission for Europe serve a similar role on a more geographically delimited basis (Christiansen 1996; Lafferty 1996).

19.4.2 Cross-national Surveys of Public Opinion

The polling of public opinions and attitudes regarding the environment has been for about three decades a fertile field of investigation. Most of this research has been informed either by Catton and Dunlap's (1980) pioneering notion of a New Ecological Paradigm or Inglehart's (1977) concept of a post-material revolution. Though there are important differences between these two theoretical approaches, they both suggest a Maslow-inspired social and political evolution of advanced societies away from a primary preoccupation with economic scarcity toward enhanced concern for issues of social equity and environmental responsibility. Surveys have been regularly administered to a cross-section of nations using compatible methodologies in an attempt to expose the extent to which the hypothesised processes of change might indeed be taking place (Mertig and Dunlap 1995, Inglehart 1995). Specific questioning typically focuses on the relative importance that members of the public confer upon the environment as a political objective (in comparison to other issues such as crime, housing, and social welfare), as well as public willingness to pay for the financial cost of environmental protection.

These surveys reveal that in virtually all countries the environment regularly rises and falls in its importance as a social concern. Considerable research has been carried out to determine the factors that are responsible for the episodic upticks and downticks in public interest and most of these assessments suggest that the state of the economy correlates quite closely with environmental commitment. Governmental efforts to seriously undermine environmental regulations, as occurred in the United States during both the 1980s and mid-1990s, frequently provoke a reaffirmation of environmental priorities. Despite these customary movements in public opinion concerning the environment, the relative standing of the dozen wealthiest countries has remained fairly consistent over the past three decades. The Scandinavian nations typically rank highest in public support for strong environmental standards (see, for example, Dunlap et al. 1993; Mertig and Dunlap 1995; Inglehart 1995).[4]

19.4.3 Social Construction of Risk

The social constructivist school of risk studies stands in stark contrast to the dominant policy perspective that suggests environmental and technological hazards are objectively determinable. Adherents of this perspective maintain that nature is comprised of any number of uncertainties and the threats to which we choose to devote our attention are invariably products of cultural and historical shaping. Scholars working in this field have drawn important attention to how understandings of risk are embedded in institutional forms, power relations, and other elements of social structure (see, for example, Douglas and Wildavsky 1982; Johnson and Covello 1987; Dake 1992; Cohen 2000b).

One important avenue for highlighting the socially constructed qualities of hazard identification has been the study of risk perception in different countries where

it becomes readily evident that public responses to danger are by no means cross-nationally consistent. Scholars working in this area have focused the bulk of their attention on the ways in which governments differently use science for policy purposes (Jasanoff 1986, 1990; Jasanoff and Wynne 1998) and the form of national protest movements that coalesce in response to particular issues (Nelkin and Pollak 1981; Joppke 1993; Halfmann 1999). Nuclear power, because of its centrality in environmental debates during the 1970s and 1980s, has provided the backdrop for much of this work, but as the dual issues of risk assessment and risk management have come to be important organising concepts in environmental affairs, attention has diffused to a much broader array of concerns such as biotechnology (Tucker et al. 2006) and global environmental change (Bickerstaff and Walker 2001).

19.4.4 National Innovation Styles and Technology Policy

The trend toward increasing globalisation and its consequences for the economic prospects of individual nations has facilitated the emergence of a field of scholarship interested in how countries innovate technologically. Attention has been directed toward both the cultural determinants underlying entrepreneurship and invention, as well as national styles for encouraging such creativity (Lundvall 1992; Nelson 1993). Related to this process is the emergence, particularly in western Europe, of the discourse of ecological modernisation, a policy programme designed to seek conciliatory interventions to harmonise economic growth and environmental protection (Hajer 1995; Mol 1995; Cohen 1996). Quite central to ecological modernisation is an emphasis on technological enhancements to improve production efficiencies through the use of alternative energy sources, the redesign of manufacturing processes, and the export of advanced environmental equipment. The outpouring of interest in eco-efficiency, industrial ecology, and transition management is a direct result of these developments (e.g. Schmidheiny 1992; Weizsäcker et al. 1997; Porter and van der Linde 1995; Elzen et al. 2005).

Andrew Jamison and his associates have examined the cultural and historical determinants that have shaped national technology policy and how govern-

---

[4] For a more encouraging portrayal of environmental attitudes in the United States see Kempton et al. 1995.

ment intervention in individual countries has helped to promote ecological modernisation (Jamison 1998; Jamison and Baark 1999). This work describes how seemingly similar countries have responded differently to the challenges of industrialisation, both during the 19th and early 20th centuries and in the post-World War II era, and how these experiences are now conditioning particular approaches to ecological modernisation. The divergent shaping of policy, and the fate of ongoing modernisation processes in different countries, has attracted substantial interest (e.g., Lundqvist 2000; Archambault 2004; Cohen 2006; Reveil 2007).

19.4.5 Public Understanding of Science and the Production of Environmental Knowledge

The comparative study of how the publics of different countries engage with scientific information has long been an area of inquiry (Miller 1996; Durant et al. 1989). Public trust in science differs markedly between countries (Procter 2006). National governments in particular have evinced considerable interest in the public uptake of science, owing to the persistence of a conviction that scientific literacy has powerful bearing on economic competitiveness. The perceived connection between public resolve for science and economic performance has promoted the regular use of social surveys to gauge lay attitudes, interest, and understanding of scientific themes. It has also become common for multinational organisations to administer these assessments, to publicise the resultant rankings of scientific competence, and to deliberate over the significance of departures from expectations.[5] As science has become increasingly essential in recent decades for the production of environmental knowledge, research has begun to examine how science as a mode of cultural communication combines with tacit and lay experiential evidence to produce situated environmental interpretations (Yearley 1991; Michael 1991; Beck 1992; Taylor and Buttel 1992; Harrison and Burgess 1994; Eden 1996, 1998; Irwin and Wynne 1996; Irwin 1997).

This line of investigation has taken an interest in the cross-national variation of environmental knowledge. For example, Carolyn Harrison and her colleagues have used household surveys, in-depth discussion groups, and ethnographic techniques to explore the assimilation of environmental themes among groups of Dutch and British respondents (Harrison et al. 1996; Burgess et al. 1998). In a similar vein, Cohen (1998, 2000a) has used several large-scale surveys of the public understanding of science to examine the differing levels of lay commitment to a 'scientific ethic' among a cross-section of European countries. This work illustrates how the production of environmental knowledge derives from a complex array of cultural and historical factors and then becomes translated into country-specific policy responses.

19.4.6 Responses to Global Environmental Change in National Contexts

As global environmental change has come to occupy an important position on policy-making agendas worldwide, the seemingly unbridgeable differences in national priorities have become particularly salient. International negotiations to achieve binding agreements to limit greenhouse gas emissions have been protracted. While the well-publicised discord between the developed and developing countries on this issue stems from deep-seated differences in economic organisation and political influence, it could be argued that the aims of the world's wealthiest nations should be more congruent. This lack of compatibility has led some analysts to suggest that the divergent negotiating positions between, for example, the United States and Germany, are grounded in the two countries different cultural values and historical traditions (Wynne 1993, 1994).

In response to concerns about global environmental change, the international community began to raise to prominent attention the importance of sustainable development and the 1992 Earth Summit held in Rio de Janeiro represented a watershed in this regard. One of the outcomes of this widely heralded conclave was the publication of an action plan known as Agenda 21 designed in part to encourage individ-

---

[5] Much of the government-sponsored work that takes place on the public understanding of science is informed by the so-called 'deficit model.' This interpretation contends that the lay public suffers from an insufficiency of scientific understanding and it is therefore necessary to augment this capacity. For a strong critique of this approach we refer to Wynne (1996).

ual countries to chart their own courses toward greater environmental responsibility. Some of the world's wealthiest nations, such as Sweden, Germany and the Netherlands, have started to move cautiously and ambivalently forward to articulate what sustainable development might mean in practice while other countries have remained steadfastly on the side lines. Despite a cascade of rhetoric from all directions, by virtually any rigorous assessment sustainable development has to date proven to be an elusive aspiration (Lafferty and Meadowcroft 2001).

Though this general assessment suggests a rather dim view of actual policy accomplishments consistent with sustainable development there are indeed some encouraging intimations emerging in certain places. Important precursors are beginning to suggest that some countries are demonstrating greater ingenuity and vigour in the pursuit of policies consistent with the goals of sustainability and a small handful of researchers have looked at the cultural and historical determinants conditioning this capacity (e.g., Jamison and Østby 1997).

19.5 Conclusion

On the basis of the review offered in this chapter, we can provisionally assert that the hegemony of the uniquely modern notion of nature as an exclusive object of scientific interpretation is coming to an end. It is not difficult to falsely caricature such an assertion as an attack on both science and rationality more generally. Critics might point out that the extraordinary power of science rests on its accumulated successes over the past several centuries and the lay public is foolish to deviate from a scientifically informed path. In response, we can simply note that sociologists and historians have long argued that it is a mistake to uncritically accept this naive view of science, however dominant it may be in certain circles.[6]

This chapter advances a far less ambitious claim and focuses singularly on the production of environmental knowledge among ordinary members of the public. As such, it suggests that people are likely to draw on any number of tacit and locally grounded epistemologies to formulate everyday, workable interpretations of their natural surround. These understandings will then be channelled into environmental politics where they will influence the design and implementation of public policies.

Such observations should not be taken to mean that ordinary people are fundamentally opposed to science, but rather that they are inclined to generate meanings appropriate for particular situated contexts. Alan Irwin has employed the useful term 'citizen science' to describe this process of lay environmental knowledge production. Social scientists have at present only a partial understanding of how the lay public generates citizen science, but there is little question that the specific cultural and historical context has great bearing on the public policy process.

References

Almond, G., & Verba, S. (1963). *The civic culture: Political attitudes and democracy in five nations*. Princeton, NJ: Princeton University Press.

Andersen, M. (1994). *Governance by green taxes: Making pollution prevention pay*. Manchester: Manchester University Press.

Andersen, M., & Liefferink, D. (1997). *European environmental policy: The pioneers*. Manchester: Manchester University Press.

Archambault, S. (2004). Ecological modernization of the agriculture industry in southern Sweden: Reducing emissions to the Baltic Sea. *Journal of Cleaner Production, 12*, 491–503.

Banfield, E. C. (1958). *The moral basis of a backward society*. New York: Free Press.

Bate, J. (1991). *Romantic ecology: Wordsworth and the environmental tradition*. London: Routledge.

Beck, U. (1992). *Risk society: Towards a new modernity*. London: Sage.

Bickerstaff, K., & Walker, G. (2001). Public understandings of air pollution: The 'localisation' of environmental risk. *Global Environmental Change, 11*, 133–145.

Boehmer-Christiansen, S., & Skea, J. (1993). *Acid politics: Energy and environmental policies in Britain and Germany*. London: Belhaven Press.

Boehmer-Christiansen, S., & Weidner, H. (1995). *The regulation of vehicle emissions in Britain and Germany: The catalytic conversion*. London: Pinter.

Breukers, S., & Wolsink, M. (2007). Wind power implementation in changing institutional landscapes: An international comparison. *Energy Policy, 33*, 2737–2750.

Burgess, J., Harrison, C., & Filius, P. (1998). Environmental communication and the cultural politics of environmental citizenship. *Environment and Planning A, 39*(8), 1445–1460.

---

[6] See for, example, Latour (1987) for a sociologist's perspective and Jacob (1997) for a view from historiography.

Catton, W., & Dunlap, R. (1980). A new ecological paradigm for post-exuberant sociology. *American Behavioral Scientist, 24*(1), 14–47.

Christiansen, P. (Ed.). (1996). *Governing the environment: Politics, policy, and organization in the Nordic countries*. Copenhagen: Nordic Council of Ministers.

Cohen, M. (1996). Risk society and ecological modernisation: Alternative visions for post-industrial nations. *Futures, 29*(2), 105–119.

Cohen, M. (1998). Science and the environment: Assessing cultural capacity for ecological modernization. *Public Understanding of Science, 7*(2), 149–167.

Cohen, M. (1999). Science and society in historical perspective: Implications for social theories of risk. *Environmental Values, 8*(2), 153–176.

Cohen, M. (2000a). Ecological modernisation, environmental knowledge, and national character: A preliminary analysis of the Netherlands. *Environmental Politics, 9*(1), 77–105.

Cohen, M. (Ed.). (2000b). *Risk in the modern age: Social theory, science, and environmental decision-making*. Basingstoke: Macmillan Press.

Cohen, M. (2006). Ecological modernization and its discontents: The American environmental movement's resistance to an innovation-driven future. *Futures, 38*, 528–547.

Dake, K. (1992). Myths of Nature: Culture and the social construction of risk. *Journal of Social Issues, 48*(1), 21–37.

Douglas, M., & Wildavsky, A. (1982). *Risk and culture: An essay on the selection of technological and environmental dangers*. Berkeley, CA: University of California Press.

Dunlap, R., Gallup, G., & Gallup, A. (1993). *Health of the planet*. Princeton, NJ: Gallup International Institute.

Durant, J., Evans, G., & Thomas, G. (1989). The public understanding of science. *Nature, 340*(6), 11–14.

Eden, S. (1996). Public participation in environmental policy: Considering scientific, counter-scientific, and non-scientific contributions. *Public Understanding of Science, 5*(3), 183–204.

Eden, S. (1998). Environmental issues: Knowledge, uncertainty, and the environment. *Progress in Human Geography, 22*(3), 425–432.

Elzen, B., Geels, F., & Green, K. (Eds.). (2005). *System innovation and the transition to sustainability: Theory, evidence, and policy*. Cheltenham: Edward Elgar.

Enloe, C. (1975). *The politics of pollution in comparative perspective: Ecology and power in four nations*. New York: David McKay.

Fukuyama, F. (1995). *Trust: The social virtues and the creation of prosperity*. New York: Free Press.

Glacken, C. (1967). *Traces on the rhodian shore: Nature and culture in western thought from ancient times to the nd of the eighteenth century*. Berkeley, CA: University of California Press.

Gouldson, A., & Murphy, J. (1998). *Regulatory realities: The implementation and impact of industrial environmental regulation*. London: Earthscan.

Hajer, M. (1995). *The politics of environmental discourse: Ecological modernization and the policy process*. Oxford: Clarendon Press.

Halfmann, J. (1999). Community and life chances: Risk movements in the United States and Germany. *Environmental Values, 8*(2), 177–197.

Hannigan, J. (1995). *Environmental sociology: A social constructivist perspective*. London: Routledge.

Harrington, A. (1996). *Re-enchanted science: Holism in German culture from Wilhelm II to Hitler*. Princeton, NJ: Princeton University Press.

Harrison, C., Burgess, J., & Filius, P. (1996). Rationalizing environmental responsibilities: A comparison of lay publics in the UK and the Netherlands. *Global Environmental Change, 6*(3), 215–234.

Harrison, C. M., & Burgess, J. (1994). Social constructions of nature: A case study of conflicts over the development of Rainham Marshes. *Transactions of the Institute of British Geographers, 19*(3), 291–310.

Hoffman, L. (1992). American psychologists and wartime research on Germany 1941–45. *American Psychologist, 47*(2), 264–273.

Huntington, S. (1996). *The clash of civilizations and the making of a new world order*. New York: Simon & Schuster.

Inglehart, R. (1977). *The silent revolution: Changing values and political style*. Princeton, NJ: Princeton University Press.

Inglehart, R. (1995). Public support for environmental protection: Objective problems and subjective values in 43 societies. *Political Science and Politics, 28*(1), 57–72.

Inkeles, A. (1997). *National character: A psycho-social perspective*. New Brunswick, NJ: Transaction.

Irwin, A. (1995). *Citizen science: A study of people, expertise, and sustainable development*. London: Routledge.

Irwin, A. (1997). Risk, the environment, and environmental knowledge. In M. Redclift & G. Woodgate (Eds.), *The international handbook of environmental sociology* (pp. 218–226). Cheltenham: Edward Elgar.

Irwin, A., & Wynne, B. (Eds.). (1996). *Misunderstanding science? The public reconstruction of science and technology*. Cambridge: Cambridge University Press.

Jacob, M. (1997). *Scientific culture and the making of industrial west*. Oxford: Oxford University Press.

Jamison, A. (Ed.). (1998). *Technology policy meets the public, pesto papers 2*. Aalborg: Aalborg University Press.

Jamison, A., & Baark, E. (1999). National shades of green: Comparing the Swedish and Danish styles in ecological modernization. *Environmental Values, 8*(2), 199–218.

Jamison, A., & Østby, P. (Eds.). (1997). *Public participation and sustainable development: Comparing European experiences, pesto papers 2*. Aalborg: Aalborg University Press.

Jänicke, M., & Weidner, H. (1997). *National environmental policies: A comparative study of capacity building*. Berlin: Springer.

Jasanoff, S. (1986). *Risk management and political culture: A comparative study of science in the policy context*. New York: Russell Sage Foundation.

Jasanoff, S. (1990). *The fifth branch: Science advisers as policymakers*. Cambridge, MA: Harvard University Press.

Jasanoff, S., & Wynne, B. (1998). Science and decision-making. In S. Rayner & E. L. Malone (Eds.), *Human choice and climate change* (pp. 1–87). Columbus, OH: Battelle Press.

Johnson, B., & Covello, V. (1987). *The social and cultural construction of risk*. Dordrecht: Reidel.

Joppke, C. (1993). *The social struggle over nuclear energy: A comparison of West Germany and the United States*. Berkeley, CA: University of California Press.

Kempton, W., Boster, J. S., & Hartley, J. A. (1995). *Environmental values in American culture*. Cambridge: MIT Press.

Lafferty, W. (1996). The politics of sustainable development: Global norms for national implementation. *Environmental Politics, 5*(2): 185–208.

Lafferty, W., & Meadowcroft, J. (Eds.). (2001). *Implementing sustainable development: Strategies and initiatives in high consumption societies*. Oxford: Oxford University Press.

Latour, B. (1987). *Science in action: How to follow scientists and engineers through society*. Cambridge, MA: Harvard University Press.

Lipset, S. (1996). *American exceptionalism: A doubled-edged sword*. New York: W.W. Norton.

Lundqvist, L. (1980). *The hare and the tortoise: Clean air policies in the United States and Sweden*. Ann Arbor, MI: University of Michigan Press.

Lundqvist, L. (2000). Capacity building or social construction? Explaining Sweden's shift towards ecological modernisation. *Geoforum, 31*, 21–32.

Lundvall, B. -Å. (1992). *National systems of innovation: A comparative analysis*. London: Pinter.

Mertig, A., & Dunlap, R. (1995). Public approval of environmental protection and other new social movement goals in Western Europe and the United States. *International Journal of Public Opinion Research, 72*(2), 145–156.

Michael, M. (1991). Discourses of danger and dangerous discourses: Patrolling the borders of science, nature, and society. *Discourse and Society, 2*(1), 5–28.

Miller, J. (1996). *Public understanding of science and technology in OECD countries: A comparative analysis*. OECD Symposium on Public Understanding of Science and Technology, Tokyo, 5–6 November.

Mol, A. (1995). *The refinement of production: Ecological modernization theory and the chemical industry*. Doctoral thesis, Amsterdam: University of Amsterdam.

Nash, R. (1967). *Wilderness and American mind*. New Haven, CT: Yale University Press.

Nelkin, D., & Pollak, M. (1981). *The atom besieged: Extraparliamentary dissent in France and Germany*. Cambridge, MA: MIT Press.

Nelson, R. (1993). *National innovation systems: A comparative analysis*. Oxford: Oxford University Press.

OECD. (1996). *Environmental performance in OECD countries: Progress in the 1990s*. Paris: OECD.

Porter, M., & van der Linde, C. (1995). Green and competitive: Ending the stalemate. *Harvard Business Review, 73*(5), 120–133.

Porter, R., & Teich, M. (Eds.). (1988). *Romanticism in national context*. Cambridge: Cambridge University Press.

Procter, J. (2006). Religion as trust in authority: Theocracy and ecology in the United States. *Annals of the Association of American Geographers, 96*, 188–196.

Putnam, R. (1993). *Making democracy work: Civic traditions in modern Italy*. Princeton, NJ: Princeton University Press.

Reveil, A. (2007). The ecological modernisation of SMEs in the UK's construction industry. *Geoforum, 38*, 114–126.

Rupke, N. (1988). Romanticism in the Netherlands. In R. Porter & M. Teich (Eds.), *Romanticism in national context* (pp. 191–216). Cambridge: Cambridge University Press.

Schmidheiny, S. (1992). *Changing course: A global business perspective on development and the environment*. Cambridge, MA: MIT Press.

Schooler, C. (1996). Cultural and social-structural explanations of cross-national psychological differences. *Annual Review of Sociology, 22*, 323–349.

Taylor, P., & Buttel, F. (1992). How do we know we have global environmental problems? Science and the globalization of environmental discourse. *Geoforum, 23*(3), 405–416.

Teich, M., Porter, R., & Gustafsson, B. (Eds.). (1997). *Nature and society in historical context*. Cambridge: Cambridge University Press.

Thomas, K. (1983). *Man and the natural world: Changing attitudes in England 1500–1800*. London: Allen Lane.

Tucker, M., Whaley, S., Sharp, J. (2006). Consumer perceptions of food related risks. *International Journal of Food Science and Technology, 41*, 135–146.

Vig, N., & Faure, M. (Eds.). (2004). *Green giants: Environmental policies in the United States and the European Union*. Cambridge, MA: MIT Press.

Vogel, D. (1986). *National styles of regulation: Environmental policy in Great Britain and the United States*. Ithaca, NY: Cornell University Press.

Weale, A. (1992). *The new politics of pollution*. Manchester: Manchester University Press.

Weizsäcker, E., Lovins, A., & Lovins, H. (1997). *Factor four: Doubling wealth, halving resource use*. London: Earthscan.

Wintle, M., & Reeve, R. (Eds.). (1994). *Rhetoric and reality in environmental policy: The case of the Netherlands in comparison with Britain*. Aldershot, Hants: Avebury.

Wynne, B. (1993). Implementation of greenhouse gas reductions in the European community: Institutional and cultural factors. *Global Environmental Change, 3*(1):101–128.

Wynne, B. (1994). Scientific knowledge and the global environment. In M. Redclift & T. Benton (Eds.), *Social theory and the global environment* (pp. 169–189). London: Routledge.

Wynne, B. (1996). May the sheep graze safely: A reflexive view of the expert-lay knowledge divide. In S. Lash, B. Szerszynski, & B. Wynne (Eds.), *Risk, environment, and modernity: Towards a new ecology* (pp. 44–83). London: Sage.

Yearley, S. (1991). *The green case: A sociology of environmental issues, arguments, and politics*. London: Harper Collins.

Chapter 20
National Policy Styles and Waste Management in The Netherlands and Bavaria

Jan Eberg

Contents

20.1 Introduction: Back to the Roots 433
20.2 Waste Policy in the Netherlands and Bavaria 435
20.3 Integrated Waste Management 435
20.4 National Policy Styles and Related
       Factors of Influence 436
20.4.1 Physico-Geographical Factors 437
20.4.2 Policy Style 437
20.4.3 Summary: The Netherlands
       versus Bavaria 441
20.5 Conclusion: Contextual Factors
       and Learning Opportunities 442

References 443

20.1 Introduction: Back to the Roots

Many people will remember the controversy that arose around the Brent Spar oil platform in the summer of 1995. Plans by oil company Shell to sink their drilling platform in the open sea were opposed in public actions by the environmental organisation Greenpeace, leading on to wider protests, especially in Germany. Eventually, Shell withdrew its plans.

The question arises why the Germans in particular were so concerned, something even Greenpeace was at a loss to explain. Some commentators pointed to 'a difference in national character'. One of them was Dutch journalist Willem Beusekamp, who related the Brent Spar affair to the aspect of the German 'national soul'

J. Eberg (✉)
Politics and Social Sciences at the Institute of Safety and Secutrity Management, Hogeschool of Utrecht, The Netherlands
e-mail: jan.eberg@casema.nl

embodied in the phrase *Den Wald im Kopf*, translatable as 'forest-mindedness', and equivalent to an almost mythical alliance with nature (Beusekamp 1995). It seems that nations may differ when it comes to the values accorded to nature and this in turn affects the processes of political decision-making. These national characteristics, including the socio-cultural roots of a country, are part of a broader set of institutional factors. In this chapter we examine the influence of these institutional factors on contemporary national waste policy.

Like other European countries, the Netherlands has a relatively advanced waste management system that has evolved over several decades. As elsewhere, Dutch waste policy has its own specific features. This case study describes how the features of national waste policy are influenced by certain relatively stable system parameters, including national policy style and characteristic modes of policy-making and policy implementation. The study clearly shows that culture, and more specifically the politico-cultural context of a given national policy system, is an important factor in explaining differences in the way national waste policies are developed and the achievements recorded to date.

This central argument is illustrated with data from a comparative study of waste policy in the Netherlands and the southern German state of Bavaria, which, although not an independent state, has a waste policy of its own. For ease of reference, we shall sometimes refer to both as 'countries'. The waste problems in the Netherlands are Bavaria are essentially the same, and in both countries conscious efforts are being made to resolve them effectively (see Box 20.1). However, a reconstruction of Dutch and Bavarian waste policy development reveals different policy trajectories and different policy styles, and indicates that of the two Bavarian waste policy has been the more successful (Eberg 1997). This can be shown to be related to

Box 20.1 Dutch and Bavarian waste in figures

Because of the heterogeneity and instability of the various different categories of waste, historical comparisons of waste volumes are problematical. Inter-country comparison of waste statistics is rendered even harder, moreover, by differences in data availability, processing and presentation. Rather than reproducing tables or graphics on waste production or Treatment and Disposal (T&D) methods, it seems more appropriate to discuss selected figures and trends, to give a general impression of the volumes of waste produced and processed in the Netherlands and Bavaria. The focus here will be on two familiar categories: hazardous waste and municipal waste. Hazardous wastes are produced in relatively large amounts at a limited number of specific locations, municipal wastes in relatively small amounts at numerous locations.

In the Netherlands the total amount of hazardous waste generated annually has roughly doubled over the last three decades, from about 1 million to 2 million tonnes. The overall composition of this waste has fluctuated, however. On the one hand, certain constituent streams no longer arise because of changes to production processes. On the other hand, new streams have been generated, while at the same time hazardous waste collection has improved, thanks both to a growing willingness, on the part of businesses and households, to hand in their hazardous (household) waste, and to better regulation, control and enforcement. In Bavaria the total volume of hazardous waste generated annually also doubled between 1970 and 2000, from 300,000 to 600,000 t.

Generation of municipal solid waste (MSW) is often expressed as an annual volume per head of the population. According to data complied by Statistics Netherlands (CBS), Dutch MSW production increased from 281 kg in 1970 to 566 kg in 2000. This is above the European average and about the same as in France and Germany. (Among OECD countries, the USA scores highest, with over 700 kg per capita.) At the end of the 1990s the total volume of Dutch MSW was about 10 million tonnes and still growing. The problem with these figures is the definition and classification of the constituent elements of municipal waste, which besides household waste may comprise bulky waste, small business waste and sometimes even construction and demolition waste. Certain fractions of the MSW stream may be collected separately and either recycled or subjected to specific forms of processing (in particular: kitchen and garden waste, glass, paper, and small chemical waste). The remaining fraction is called residual waste. The Bavarian Ministry of Land Development and Environment (StMLU) reports on 'recyclable waste' and 'residual waste' as the two components of MSW; these include all the commonly collected wastes cited above. In 2000 approximately 500 kg of MSW was collected per head of the Bavarian population. Between 1988 and 2000 the amount of residual waste fell from 450 to 206 kg per capita, implying a steady growth of recycling efforts. Since 1995 the share of recyclable waste has exceeded that of residual waste. The total volume of MSW generated annually remains roughly stable at around 6 million tonnes. In the Netherlands, too, the rate of municipal waste recycling has risen, standing at about 45% at the end of the 1990s. In Bavaria, this figure was then about 55%.

In most European countries, although a minor fraction of the MSW stream is composted, disposal in landfills is still the T&D option most commonly adopted. On average in the EU about 65% of MSW is processed in landfills. In the Netherlands, the MSW stream is split into fractions of which a large part is recycled. Residual waste is increasingly being incinerated (with growing efficiency of energy production). Incineration of residual waste has risen to roughly 55%. In Bavaria, where there has been rapid expansion of incineration capacity (in combined heat and power plants) since the beginning of the 1990s, over 85% of residual waste is processed this way.

These figures and trends indicate that the Netherlands and Bavaria both produce large volumes of waste. The substantive differences between the Netherlands and Bavaria cannot be represented by mere statistics, however, but only emerge from consideration of the overall quality of waste policy, as reflected in the organisation of waste management infrastructure, planning of waste processing capacity, perception of risks and hazards and use of waste reduction technologies. The overall standard of Bavarian waste policy, in its effects of preventing and handling hazardous wastes, as well as in recycling and reducing municipal wastes, is relatively high as compared to other countries.

institutional differences with respect to such factors as the specific role of government and cooperation between government and other actors, but also to particular natural and cultural circumstances in the respective countries. What, then, are the policy styles and other key factors that have shaped waste policy in the Netherlands and Bavaria? And to what extent can waste policy concepts that have proved effective be adopted at will by other countries? Answers to these questions can help improve our understanding of the waste policy development process, on the one hand, and assess the scope for internationalising waste policy, on the other.

20.2 Waste Policy in The Netherlands and Bavaria

One of the principal aims of Dutch waste policy is to work towards an environmentally sound system of integrated waste management. This means implementing the so-called *Lansink ladder*, as proposed by Dutch MPs Lansink et al. in a motion adopted in 1979. According to this concept, waste prevention – at the top of the ladder – should take priority wherever possible, with recycling as a second-best and processing by incineration or landfills as a strategy of last resort. In actual fact, though, waste management in the Netherlands for a long time resembled an 'inverted Lansink ladder', climbed rather than descended. Since the 1990s the situation has improved. There are now certain restrictions on disposal in landfills and a growing proportion of Dutch waste is now collected and processed as separate fractions. The average recycling rate has also risen to approximately 70%. Results on waste prevention lag behind, however, but this requires a greater degree of adaptation and more concerted efforts than the other strategies.

In Germany the constituent states or *Länder* must comply with federal 'framework' legislation on waste management (as well as certain other issues), with the federal republic laying down basic provisions and leaving the *Länder* to legislate on particular requirements and details according to their own needs. In comparison with other *Länder*, Bavaria established an environmental programme and regulatory system relatively early on. The Bavarian government was, for example, the first in Europe to establish an environment ministry (1970), set up an organisation for the treatment of hazardous wastes (1966) and build a hazardous waste incinerator (1971). Bavarian waste policy is characterised by central planning, and by development and use of innovative technology. Bavaria has retained its head start to this day and is still a forerunner in the field of environmental management.

One clear difference between Bavarian and Dutch waste policy is that whereas in the Netherlands policies on hazardous and non-hazardous waste for a long time evolved separately, in Bavaria they have been covered by a single set of policy arrangements from the beginning. It was not until 1994 that the Dutch Chemical Wastes Act and the Waste Substances Act were abrogated and incorporated into the Environmental Management Act. Subsequently, in 1996, the Commission for the Future Organisation of Waste Disposal (*Commissie Toekomstige Organisatie Afvalverwijdering*, CTOA) recommended establishing a single National Waste Management Plan for all types of waste, which was published 5 years later, in 2001.

20.3 Integrated Waste Management

It is the Bavarian model for managing hazardous wastes, in particular, that is relatively progressive and remains exemplary in the international context. The Bavarian government has identified itself with the role of risk manager and guarantees a closed processing system. The Bavarian organisations for the management of hazardous wastes are not final processors like AVR Chemicals in the Netherlands but run integrated facilities. This means that little waste is treated one-way. Following the principle of 'cascade-processing' or 'down-cycling', whenever possible, hazardous waste is treated chemically and/or physically, separated, processed biologically or recycled. Only residual fractions and certain intractable forms of hazardous waste are incinerated or processed in landfills. In addition, the government advises waste-generating companies on the scope and opportunities for prevention and recycling. State-of-the-art technology is applied at all Bavaria's integrated processing facilities.

Under this coherent, centralised waste policy system non-hazardous wastes are also processed according to the concept of integrated waste management: prevent where feasible, recycle as far as possible, incinerate as much as necessary, and landfill as little as possible. This has led to at least two striking results. First, a sophisticated recycling industry has developed that works even better than the federal German system for recycling (mainly) packaging waste. Second, a relatively large number of municipal waste incinerators have been built throughout Bavaria. Of the 50 or so incinerators operating in Germany today, 20 are in Bavaria. These differ from the municipal waste incinerators in operation in the Netherlands in several respects. The aggregate capacity of the 20 Bavarian incinerators is less than that of the 10 Dutch facilities. Furthermore, in Bavaria, where application of modern technology is generally seen as an important issue, incinerators were from an early date fitted with air pollution control systems and built as combined heat and power plant. In the Netherlands, in contrast, the first incinerator that operated as a cogeneration plant was not brought on stream until 1998. Again in contrast to the Netherlands, Bavaria is a major producer of waste-processing technology. Of the three most promising new incineration technologies for non-hazardous waste on the market today, two are Bavarian.

General waste policy developments show that, compared with the Netherlands, Bavaria got off to an early start and has retained its lead to the present day. While Bavarian waste policy was already quite mature at the beginning of the 1980s, Dutch waste policy was only just developing. Besides this difference in tempo, the two countries also differ in terms of *how* waste policy has been developed. Right from the outset Bavarian waste policy has been confidently structured and robustly elaborated, while in the Netherlands development has been fragmented, with several delays in the planning and implementation process. As mentioned, moreover, Bavaria has played an innovative role in the development of waste processing technologies. Comparing the two, then, we could label Bavaria a 'pioneer' and the Netherlands a 'follower'.

20.4 National Policy Styles and Related Factors of Influence

The evolution of Dutch and Bavarian waste policy has been strongly influenced by the institutional structure and politico-cultural climate in the respective countries. Together, these factors have helped create a characteristic mode of political decision making or, phrased differently, a specific policy style. Richardson (1982) defines a policy style as 'the interaction between (a) the government's approach to problem-solving, and (b) the relationship between government and other actors in the policy process'. This leads to a categorisation of societies in terms of four basic policy styles: anticipatory/active problem-solving vs. reactive problem-solving in combination with either a consensus or an impositional (i.e. 'command-and-control') relationship. From this perspective the Dutch policy style would be typified as reactive-consensual and the Bavarian style as active-impositional. However, this leaves little room for gradation, and it may therefore be more fruitful to define the concept of policy style more broadly, as the characteristic mode of policy-making and policy implementation made possible by the sum total of institutional factors underlying the policy process. Four categories of policy style factors can then be specified:

- Historico-cultural factors, e.g. state formation and the development of socio-cultural 'pillars'
- Legal-administrative factors, e.g. development of a constitutional system and bureaucratic structure
- Socio-economic factors, e.g. modernisation and the development of specific economic sectors
- Politico-organisational factors, e.g. democratisation and the development of decision-making structures.

All these factors contribute to the evolution of a specific politico-administrative tradition and their combined specification can provide a unique portrait of national (or, in the case of Bavaria, state) policy.

National policy style is an important explanatory factor for understanding the waste policy of a given country. As Richardson and Watts (1985) remark: '… countries tend to regulate, say, pollution, in much the same way (style) as they regulate other policy sectors'. In addition, specific policies on waste and environmental issues are also influenced by two other factors. First, the national image of nature, as expressed by historically moulded beliefs about the value of nature and beliefs about mankind's relationship with the natural environment. Second, policy style and image of nature are intimately linked to the physico-geographical features of a country, i.e. its location, ecological structure and so on, which define a variety of basic constraints and possibilities.

The following sections elaborate on these issues of physical geography, policy style and national image of nature. The chapter concludes with a summary in which Dutch and Bavarian institutional characteristics are taken as key explanatory factors for the evolution of their respective waste policies.

20.4.1 Physico-Geographical Factors

Without neglecting other key variables such as climate, state formation and socio-economic development, the geographical situation of a country can certainly be said to form an institutional mould. The fact that the Netherlands is a flat, coastal country accounts for the relative prominence of the petrochemical industry, a high population density and intensive agriculture, especially arable and livestock farming. One consequence of this has been described as follows by the Dutch economist Jan Pen: 'An unfortunate feature of the Netherlands is that we have a typical transit economy: import flows in, export flows out, and we make money from these streams. Our income, however, is just a fraction of what passes through, and it is this throughput that determines the scale of pollution'. In addition, contaminated river sediment is carried downstream from France, Belgium, Switzerland and Germany.

Bavaria is over one and a half times the size of the Netherlands, but has a smaller population of about 12 million, compared with some 16 million in the Netherlands. In comparison, Bavaria has more open space in absolute terms, while the Netherlands has a greater area under cultivation. While the Netherlands is flat and 'wet', Bavaria is partly covered with forests and mountains and is considerably more rural. It is an inland state which, because of its mountainous and wooded mid-European structure, has been decentrally cultivated. The difference in the number of municipalities is especially striking: in round figures, 500 in the Netherlands compared with 2,000 in Bavaria. These differences in physico-geographical structure have had an impact on the distribution of waste incinerators and other processing facilities in the respective countries, owing mainly to constraints on transportation.

20.4.2 Policy Style

20.4.2.1 Historico-Cultural Factors

The Dutch cultural historian Herman Pleij has written that the Dutch 'culture of compromises' emerged during the late medieval era when people were obliged to deliberate with their neighbours about the draining of water from private land in order to keep their feet dry. Tolerance and cooperation were also promoted by 'the struggle against the sea' (Zahn 1991: 39). The country's negotiating structure can be traced back to the Republic of the Seven Provinces (1579–1795), in which peaceful coexistence between the States General and the independent provincial states led to development of a political culture of 'settling and straightening'. In a climate of religious pluralism as well as foreign threats the Republic also became proficient in compromise (Kennedy 1995). A culture of pragmatic tolerance arose in which the politics of pacification and accommodation can also be situated (Lijphart 1968). This culture dominated the 20th century, as exemplified by the typically Dutch phenomenon of 'pillarisation', the coexistence of four social-cultural 'pillars': Protestants, Catholics, Socialists and Liberals. The other characteristic features of Dutch culture are Calvinism and commerce (cf. Weber 1905). Some cases of environmental damage that have occurred in the Netherlands can be traced to stark examples of miserly business acumen. When it comes to the ecological problematique, the Dutch seem to be especially concerned about *global* deterioration.

Bavaria is one of the oldest 'states' in Europe and has deep-rooted socio-cultural values. Its cultural cornerstones are tradition and progress. Cultural conservation is constitutionally prescribed and the Bavarian self-reproductive endeavour 'to be the best' has always spawned strong economic and technological development. To many Bavarians their *Freistaat* ('Free State') represents independence, identity and stability. Historically, Bavaria has been influenced by the development of both the French and German state (cf. Elias 1939). This has resulted in a socio-cultural blend of French pride and predilection for hierarchy and monarchic leadership with the German qualities of thoroughness, integrity and intellect. There is wide public involvement in social organisations in Bavaria. Religiously, the *Freistaat* is predominantly Catholic and Bavarian society still reflects certain basic Catholic principles of social order such as a hierarchic cosmology (authority comes from 'on high'; 'below' is a preparation for 'above'), a conception of society as an organic entity not a sum of individuals (mutual cooperation is necessary: 'solidarism') and subsidiarity ('higher' agencies and authorities should do only what 'lower' agencies and authorities cannot achieve by themselves). As far as environmental issues are concerned, Bavarians are personally involved and willing to take action, especially when it comes to conservation of their own land or *Heimat*.

20.4.2.2 Legal-Administrative Factors

The Dutch and German legal system are both based on the tradition of Roman law. They have developed into quite different forms, however, with repercussions on the administrative structures of the two countries.

'The orientation of Dutch law is reflected in its so-called 'opportuneness principle'. Dutch public prosecutors have some discretion to decide whether it is 'opportune' to prosecute a case or to drop it. This legal provision is both a reflection of, and has allowed for, the pragmatic, negotiating, and tolerant elements in Dutch regulatory style' (van Waarden 1995). The ultimate example of Dutch tolerance is 'condonation' (*gedogen*): legal acceptance of what is formally forbidden. This is something waste processors may experience in everyday practice, where illegal situations are often condoned after deliberations with controlling officials. Van Waarden explains the conduct of Dutch government officials as follows: 'Dutch civil servants are neither amateurs nor exclusively lawyers, nor are they trained in elite schools. As a result, Dutch administrators lack the strong identification with the state, as in France, but also with civil society, as in Britain, or with legalism, as in Germany'. And: 'Because there is no central civil service, there is practically no rotation between departments, a factor which has enhanced fragmentation in policies'.

'The bureaucrats of the German Federal Republic are the modern-day offspring of the Prussian officialdom of earlier centuries, which has been accurately described as 'a social elite recruited on the basis of competitive examination and dedication both to efficiency and to the principle of autocracy'' (Heady 1991). Top positions in the federal bureaucracy are still filled mostly by individuals with legal training (ibid), which, according to Hofstede (1991), correlates with Germany being a 'strong uncertainty avoidance country'. Implementation of policies before there is full scientific certainty to justify them is the quintessence of *Vorsorge* politics, or precautionary politics. *Vorsorge* is an integrative concept, combining acceptance of challenges and innovations, anticipation of new regulatory structures and aversion of new risks and their effects. This principle can be applied generally, but is pre-eminently suitable for environmental policy where feedback is problematic, uncertainties are manifest and risks are high. The precautionary principle is part and parcel of German environmental policy. According to Weale (1992), it is supported by a technocratic and legalistic administrative structure. It is a structure marked by caution and by experts and courts having a significant influence on the policy process. Consequently, opportunities abound for using 'best technical means' and identifying 'common interests'. All in all, this is to the good of environmental policy.

The origins of *Vorsorge* politics lie in early 19th-century Bavarian 'public service pragmatism', which demanded expertise and conscientious obedience from its officials. The system of civil law and bureaucracy developed by Count Montgelas has served as an example for a large part of the contemporary federal republic (Kraus 1992). During the 19th century sovereignty in Bavaria passed from the monarch to the state to the people. An important heritage from this period are Bavaria's constitutionalism, a strengthening of its resolve for self-government, its own Constitutional Court and its formal system of direct democracy with, among other things, local and state plebiscites. These plebiscites are sometimes used as sophisticated instruments of environmental policy.

20.4.2.3 Socio-Economic Factors

Of the many socio-economic factors relevant to policy style, we here single out two issues: land development and technological development.

The Netherlands had its 'era of glory' in the Golden Age of the 17th century. However, the country's economy was heavily dependent upon international trade and collapsed after agricultural prices fell throughout Europe, a situation aggravated by foreign protectionism. Attention shifted to the colonies, where trade was spoiled by French occupation, however, and during the 19th century the Dutch economy lagged behind that of neighbouring states. It had no choice but to adapt and convert and according to Lintsen et al. (1995) this was done by learning from foreign experience. In this sense the Netherlands was not an innovator but a critical follower. Only when a new technology became cheaper or at least as reliable as traditional methods did transfer take place and was the technology adopted or imported. Entrepreneurs acted rationally and cautiously. New impulses arose from the 'new' fight against the water: land reclamation for a growing population and the gigantic *Deltaplan* for flood protection. In these projects, development of land and hydraulic engineering skills went hand in hand.

While the Dutch tend to view technology as a means, in Bavaria it is seen as an end in itself. Technology equals progress, and progress (combined with tradition) means independence. For Bavaria, technological development is an integral part of maintaining the state's own culture in its federalist competition with the outside world. There has always been an emphasis on new, modern technology and this has had a strong effect on the social and economic structure of Bavaria. Around 1900 the Bavarian capital of Munich competed with the federal capital of Berlin for a leading role in scientific development. More than half a century later, Bavarian President Franz Josef Strauss forged ahead down this road, which some at the time even termed 'technology drunkenness'. He personally arranged for the first pilot nuclear reactor in Germany to be built near Munich and made a personal effort to promote the interests of companies like Messerschmitt and BMW (Strauss 1989). Technological development was for Strauss not just a private fascination, but also a prerequisite for sound socio-economic development: 'I am convinced that, if we want to achieve humane labour conditions, further our social standards and preserve our environment, we must apply modern science and technology' (ibid.). A well-known slogan reads 'Progress speaks Bavarian', a qualification with which Strauss was in full agreement.

20.4.2.4 Politico-Organisational Factors

Theodor Heuss, the first president of the Federal Republic of Germany, from 1949 to 1959, hit the nail on the head when he once said: 'One cannot create culture with politics; but perhaps one can make politics with culture' (Heuss 1964). Differences between Dutch and Bavarian culture have brought about quite different policy styles.

The Dutch policy style can be said to be pluralist and consensual. A central feature of Dutch social and administrative relationships is 'coexistence': between provinces, between religions, between socio-cultural pillars and between political parties. The pillarisation system outlined earlier represented a combination of political ideology and social organisation. The leaders of the respective pillars – more or less over the heads of their rank and file – were willing to negotiate and compromise. Lijphart (1968) and Daalder (1971) have defined the Dutch political system as a 'consociational democracy'. This leads on the one hand to social harmony but on the other to tortuous discussions and endless negotiations in which conflict avoidance is afforded the highest priority. Dutch political history may be described as a continuation of the 'ideological triangle' formed by the three main political traditions of liberalism, confessionalism and social democracy (de Beus et al. 1989). According to Middendorp (1991), 'the Dutch political system has been rather stable as regards its *structure* (a predominance of Christian-democratic/Liberal connecting minimal-winning coalitions), but since the late 1960s its *culture* has changed from predominantly traditionalist, deferent, politically apathetic to politicized, ideologized and partly deconfessionalized, thus having become ideological heterogeneous with ideologically competing elites'. Although neo-corporatism in environmental policy has brought certain comparative advantages (such as 'covenants': negotiated agreements between government and industry), it has also led to situations in which 'economic policy swept away environmental interests' (Wintle 1994).

In Bavaria policy and planning are rooted in a centralist administration. Count Montgelas' 'revolution from above' (März 1994) spawned structural politics and true 'Weberian' bureaucracy, based on rational-legal authority. A decisive factor in attaining the goals of government has been the willingness of the counties and local politicians (notwithstanding their powers of local self-administration) to faithfully comply with whatever is planned or recommended by the state. At the same time, though, a second type of authority distinguished by Max Weber was present in Bavaria: charismatic authority. From the late 1960s through to the 1980s, this was expressed by the leadership of Franz Josef Strauss, which accounted for a significant period of political stability. Bavarian politics has long been dominated by conservative forces. Because of the hegemony of the ruling Conservative Party (CSU) there were neither power switches nor political coalitions. This is quite remarkable, for Bavaria is a heterogeneous society and social movements have ample democratic opportunity to make their voices heard. Yet what unites Bavaria is its tradition and the common acknowledgement that its path of progress is the right one. The combination of tradition and progress has a 'catch-all' effect. Political stability provided a favourable backdrop for decision-making, socio-economic development and environmental management. Opposition forces striving for even better environmental and waste policies arose only after the Strauss era, from the early 1990s onwards, represented mainly by the citizens initiative group *das Bessere Müllkonzept*.

20.4.2.5 National Image of Nature

In a sense, each nation can be said to entertain certain beliefs and values concerning nature, as reflected in such terms as 'environmental consciousness' for a nation's general awareness of environmental problems (Rootes 1995), 'myth of nature' for its interpretation of ecosystem stability (Schwarz and Thompson 1990; Thompson et al. 1990), 'idea of nature' and 'conception of nature' for the meaning imbued in nature (Wissenberg 1993) and 'image of nature' as the result of the social construction of environmental knowledge (Hannigan 1995). Hannigan, in particular, demonstrated that concepts as 'nature', 'ecology' and 'environmentalism' are culturally grounded and both socially constructed and contested. Here, we shall use the term 'national image of nature' to express the cultural bias of a nation in this respect. Mind that the term is not fixed in meaning. Still, it is another key institutional factor shaping waste policy, alongside national policy style.

When it comes to conceptions of nature, the European countries share a common cultural heritage stemming from the Greek and Judeo-Christian traditions (cf. Dobson and Lucardie 1993; Boersema 2001). Roughly stated, the first tradition produced Reason, 'mastering by explanation', while the second was responsible for diffusion of the 'dominion' theme from the book of Genesis. Together, these traditions have influenced our culture directly through Christianity, the Renaissance and Humanism, as well as indirectly by way of the Enlightenment. The basic principles of this trajectory were an anthropocentric worldview, a techno-scientific paradigm and an 'economistic ideology'. More important, however, is that traditions broke with the 'pagan' cosmology of animism. Nature became desanctified and objectified, and man was no longer in a state of *unio mystica* ('one with nature'). Man gained the ability to intervene in the course of nature. This means that, within certain bounds, humankind can now decide to what extent such intervention shall take place. Ultimately, this depends on how man will value nature, which in turn varies across cultures.

The physico-geographical features of a country are an institutional variable close to nature. In terms of the amount of virgin land within their borders, the relationship between Bavaria and the Netherlands is that between a 'have' and a 'have-not'. Bavaria still has Alps and native forests, while in the Netherlands the last area of ancient woodland is presumed to have been cleared at the end of the 19th century. Although there is quite a diversity of – small-scale – ecosystems, especially heaths and moor land, the entire country is basically cultivated or has been otherwise altered by man. Of the natural elements, it is principally water with which the Dutch have had to live and struggle. For the Netherlands it has been an ever-present adversary, fought at first defensively and later offensively. Even today there is an array of water-related aphorisms to remind the Dutch of their struggles as well as successes in coping with the challenges it posed (Hampden-Turner and Trompenaars 1993).

What the water was to the Netherlands, the forest was to Bavaria (as well as much of the rest of Germany), albeit in a different sense. While the water was the adversary of the Dutch, the Bavarian forest served as an inspiration for patriotism and Romanticism (cf. Schama 1995). It was thanks to the forest, for example, that the Teutons were able to defeat the Romans. In patriotic art the oak tree figured prominently and in romanticist German literature, too, the forest played a major role. Many fairy tales by the brothers Grimm are tales of the forest, for example. It was Goethe, however, who truly personified Romanticism, and he was emulated by hundreds of other writers. As a cultural movement, Romanticism was a reaction to the rational and mechanistic worldview of the Enlightenment. It has had a strong impact on German society, with its well-developed traditions of philosophy and the humanities. In this way, ideological influences have contributed to a broad environmental awareness in Germany. What people strive for there is *Heimatschutz*, or 'homeland protection'. It is not only the values of *nature* that must be protected, but *cultural* elements too. What is to be preserved is no less than the total physiognomy of the country. As van Dieren (1995) explains: '*Umwelt*' means much more than 'environment'. *Umwelt* represents a combination of progressive and conservative concepts, in an often paradoxical mixture. The *Umwelt* movement encompasses almost everything that is critical and emancipatory, a cacophony of green feminism, pacifism, and alternative science, new life styles and critical consumerism, an alliance that cuts right through German society and from which no one escapes, and including – as a queer appendix – the old romantic conservationism'. The preoccupation of the Germans with their natural environment expresses itself as an 'alliance of consciousness' with the forest, the *Wald-im-Kopf*, or 'forest-mindedness', cited at the beginning of the chapter. During the 1980s, forest die-back as a result of air pollution (*Waldsterben*) gave a tremendous boost to environmental awareness in Germany (Richardson and Watts 1985). The theme formed the main motive for establishing the environmental organisation Robin *Wood*.

In the Netherlands, Romanticism arrived late and its impact was modest (cf. Zahn 1991). Its most popular exponents are probably the poet Herman Gorter, with his laudatory poem 'May', and the conservationist and teacher Jac. P. Thijsse, co-founder of the Association for the Preservation of Natural Monuments. Nature mysticism was not a source of inspiration for modern environmental awareness, however. It is more likely that Dutch efforts to combat contemporary waste problems derive from Calvinist moralisation of the domestic and from the rites of cleanliness (Schama 1991): 'The well-kept home was the place where the soiling world was subjected to tireless exercises in moral as well as physical ablution. ... To be filthy was to expose the population to the illicit entry of disease and the vagrant vermin that were said to be its carriers. ... Conversely, to be clean was to be patriotic, vigilant in the defence of one's homeland, hometown and home against invading polluters and polluted invaders.' Even today, the Dutch are highly motivated when it comes to pollution *clean-up*.

Another way a country's image of nature is often denoted is by characterising it in terms of the 'relative dominion of man over nature', according to the sixfold classification of Zweers (1994). In this typology the Netherlands would be a 'benevolent ruler', still ruling over nature but at the same time realising his dependence on it; striving to develop nature and place it as far as possible at the service of human society, but understanding that exploitation and repression are fundamentally wrong. In the same typology the closest characterisation of Bavaria would be the 'steward',

who does not rule over nature but manages it in the name of the 'owner' to whom he is responsible: in the Christian variant God, in the secular variant mankind. The setting is conservationist, with the emphasis on the maintenance of resources (natural capital, of which only the interest may be consumed), yet the purport is still anthropocentric.

20.4.3 Summary: The Netherlands versus Bavaria

A lot more could be said about each of the factors described here, but the data and examples presented should suffice for an understanding of the institutional backgrounds that have influenced waste policy in the Netherlands and Bavaria. Table 20.1 below summarises the main physico-geographical factors, policy style factors and features of the national image of nature in the respective countries.

Table 20.1 Synopsis of institutional factors influencing Dutch and Bavarian waste policy

The Netherlands	Bavaria
Physico-geographical factors	
• Flat, coastal	• Hilly, continental
• Water-land	• Wood-land
Policy style factors	
Historical-cultural factors	
• Calvinism	• Catholicism
• Pillarisation; pragmatic tolerance	• Federalism; 'national' identity
Legal-administrative factors	
• Opportuneness (government decides)	• Constitutionalism (power of law prevails over government)
• Policy of the lesser evil (*gedogen*)	• Precautionary politics (*Vorsorge*)
Socio-economic factors	
• Transit economy	• High-tech service economy
• Technology as a 'means'	• Technology as an 'end'
Politico-organisational factors	
• Pacification, deliberation, consensus	• Centralism, formalism, 'Weberian' bureaucracy
• Fragmented institutionalisation	• Political stability
• Middle-class liberalism	• Cultural conservatism
National image of nature	
• The environment is part of the economy	• The environment is part of culture
• Cleanliness (Calvinism)	• *Wald im Kopf* (Romanticism)
• 'Benevolent ruler'	• 'Steward'

The Netherlands is a flat, densely populated and intensively cultivated country. Historically, its sea coast and rivers formed the basis for a transit economy with heavy industry. Dutch culture is imbued with commerce and Calvinism, as well as a tradition of pragmatic tolerance. In the 20th century, this tradition found expression in pacification, pillarisation and a condoning 'policy of the lesser evil' (*gedoogbeleid*). Furthermore, officialdom has no clear professional identity, policy-making suffers from fragmented institutionalisation and decision-making is geared towards negotiation and consensus. The Dutch image of nature is bound up with the country's 'eternal struggle against the water'. There is no longer any real nature and the environment is considered part of the economy. Commitment to the environmental issue is rather shallow: 'clean-up' actions can be considered as handed-down Calvinist mores.

Dutch waste policy was first elaborated during the 1970s as a response to the problems of the day. There was no coordination and industrial mismanagement resulted in a series of environmental scandals. Although there has been growing government intervention from the 1980s onwards, this has had to compete with strong agricultural and industrial lobbies, bolstered up by traditional middle-class forces. In the field of waste management, much is still condoned and the ambitious goals set are still far from being attained. The policy actors involved tolerate one another but achieve only little progress. The result is a polite truce rather than constructive cooperation. Consequently, a situation prevails in which economic continuity takes priority over sustainable development.

Bavaria is a hilly, wooded state that is less densely populated than the Netherlands. It underwent tempestuous post-war transformation, evolving rapidly from a traditional, agricultural region to a modern, high-tech service economy. The Bavarian Model of tradition and progress is rooted in a culture of 'national' identity, federalism, constitutionalism, centralism, formalism and *Vorsorge* (precaution). The public administration system is characterised by a 'Weberian' bureaucracy, based on rational-legal authority. The Bavarian image of nature has been nourished by Romanticism and the widely felt empathy with the health of the forest. The natural environment is considered part of culture. This has led to a fairly sincere, 'internalised' commitment to resolving the environmental issue: environmental protection as cultural conservatism.

In Bavaria waste policy has been part of centralised structural politics since the 1970s. Thanks to the hegemony of the ruling CSU and the ensuing political stability, long-term programmes of structural politics could and can indeed be implemented. The government's conservationist policy enjoys broad public support and represents an integration of public interests and beliefs. In contrast to the Netherlands, these conditions have contributed to economic continuity *and* sustainable development. They have also enabled Bavaria to become a forerunner in the field of environmental management. And because of its political efforts to balance economic and environmental interests, Bavaria can even be regarded as the birthplace of the policy concept of ecological modernisation.

20.5 Conclusion: Contextual Factors and Learning Opportunities

Waste policy is the outstanding example of a policy domain that can always do better. The paradoxical situation is that we speak of successful waste management systems precisely in countries where pollution is very serious. This applies to both the Netherlands and Bavaria. It should therefore be stressed that Bavarian waste policy is only *relatively* successful, that is, with respect to the array of economic activities and the overall amount of waste generated. Besides, Bavarian waste policy has achieved the degree of success it has because waste problems have proved to be manageable and political-administrative solutions effective. Bavaria owes this success to a favourable combination of factors:

- Its physico-geographical situation, particularly its central European location, its devolved agricultural base and its expanses of unspoilt nature
- Its national image of nature, based on the attitude of the 'steward' who preserves and manages and
- Its state policy style, which brought to maturity the concept of *Vorsorge*

These three factors have influenced the manner in which the state's waste policy has developed. Successful waste policy requires more than policy programmes, though. In this field, particularly, achievement of policy goals is a matter of 'co-production', i.e. constructive interplay between the beliefs and actions of policy-makers, policy implementers and target groups. The success of any given policy instrument depends on the extent to which the envisaged policy changes are regarded as meaningful by the target groups involved (Grin and Van de Graaf 1996). This fourth factor, which may be termed the '*implementation climate*', is likewise favourable in Bavaria. The political engagement and environmental awareness of the Bavarian population is considerable, and in the Bavarian administrative system bureaucratic alertness and faithful obedience prevail.

It is the constellation of the above four factors which constitutes the political context for the evolution of waste policy in Bavaria and has determined the success of that policy. Within this setting the potential for steering refractory policies such as environmental policy and waste policy is limited, for three main reasons. First, because of the nature of the waste management policy domain itself, where the scope for rigid governance is constrained by the broad range of actors and respective decision levels as well as the various policy aspects described above. Second, because of the multitude of interdependencies among these actors. And third, because the co-evolution of waste policy with the social context also has unintended and unforeseen effects. In many countries, there is governance of environmental and waste issues nevertheless, in one country more hierarchically than in the other and in one perhaps more effectively than in the other. To a fairly reasonable extent these dynamics can be described and explained retrospectively. But how can we possibly steer these dynamics? One option is 'modulation': to account for an evolutionary logic of change and development (Arentsen and Eberg 2001). Modulation is a goal-setting rather than a goal-seeking strategy and therefore appropriate to deal with intractable policy issues such as climate change or waste management. 'The role of government in this respect is that of an alignment actor, a matchmaker, and a facilitator of change. … This leads to a different set of policy recommendations, aimed at shaping, facilitating and initiating interactions instead of achieving predefined outcomes. Modulation policies aim to exploit mechanisms of change and windows of opportunity' (ibid.). 'Modulation strategies should be focused on and designed for *learning*' (cf. Sabatier and Jenkins-Smith 1993; Eberg 1997).

Bavarian waste policy cannot simply be transferred to other countries with comparable waste problems. Nonetheless, the notion of *Vorsorge* or precautionary politics can serve as an example for national environmental and waste policy elsewhere. Today the concept has been embraced, in varying degrees, in a number of countries. As a policy approach it has helped shape Dutch environmental policy programmes and the Environmental Action Programmes of the European Union. As a true policy strategy, however, the precautionary principle has been applied mainly in Germany (and, outside Europe, in California). This is reflected in German efforts in the fields of wind power, waste prevention and waste processing technology, CO_2 reduction, environmental regulation and public environmental investments. Precautionary politics is a structural element of the German and Bavarian policy style and, therefore, very difficult to transfer. Still, other countries could *learn by adaptation* (cf. Rose 1993): "Adaptation occurs when a program in effect elsewhere is the starting point for the design of a new program allowing for differences in institutions, culture, and historical specifics. … Adaptation involves two governments in a one-to-one relationship of pacesetter and follower. Japan's late nineteenth-century adaptation of programs from the United States and European countries is a textbook example of this relationship".

Because it has been relatively more successful in implementing effective environmental policy, Bavaria can serve as an attractive example for other countries to learn from. In particular, the interdisciplinary research, interactive policy processes and efforts in the field of sustainable technology development that characterise the Bavarian policy programme might serve as a useful guide. Learning by adaptation is both desirable and necessary. It is desirable in order to harmonise and enhance environmental policy in Europe. And it is necessary in order to keep up with economic development in the process of Europeanisation and globalisation.

References

Arentsen, M. J., & Eberg, J. W. (2001). *Modulation of sociotechnical change as climate change challenge*. Dutch national research programme on global air pollution and climate change. Report no.: 410 200 100. Enschede: Twente University, 85, 63, 81.

Beus, J. W. de, van Doorn, J. A. A., & Lehning, P. B. (1989). *De ideologische driehoek. Nederlandse politiek in historisch perspectief*. Meppel/Amsterdam: Boom.

Beusekamp, W. (1995). 'Brent Spar', news paper article, *de Volkskrant*, June 11, 1995.

Boersema, J. J. (2001). *The torah and the stoics on humankind and nature. A contribution to the debate on sustainability and quality.* Leiden: Brill.

Daalder, H. (1971). On building consociational nations. The case of the Netherlands and Switzerland. *International Social Science Journal, 23,* 355–370.

Van Dieren, W. (1995). *Het groene universum. Reisverhalen uit de wereld van het milieu* (p. 227). Amsterdam: Van Lennep.

Dobson, A., & Lucardie, P. (Eds.). (1993). *The politics of nature. Explorations in green political theory.* London: Routledge.

Eberg, J. (1997). *Waste policy and learning. Policy dynamics of waste management and waste incineration in the Netherlands and Bavaria.* Delft: Eburon.

Elias, N. (1939, reprint 1978). *The civilising process. The history of manners, and state formation and civilisation.* Oxford: Basil Blackwell.

Grin, J., & van de Graaf, H. (1996). Implementation as communicative action. An interpretative understanding of interactions between policy actors and target groups. *Policy Sciences, 29*(4), 291–319.

Hampden-Turner, C., & Trompenaars, F. (1993). *The seven cultures of capitalism. Value systems for creating wealth in the United States, Britain, Japan, Germany, France, Sweden, and The Netherlands* (p. 267). New York: Piatkus.

Hannigan, J. A. (1995). *Environmental sociology. A social constructionist perspective.* London/New York: Routledge.

Heady, F. (1991). *Public administration. A comparative perspective* (4th ed., pp. 208 and 210). New York: Marcel Dekker.

Heuss, T. (1964). *Geist der Politik. Ausgewählte Reden* (p. 11). Frankfurt am Main: Fischer Bücherei.

Hofstede, G. (1991). *Cultures and organisations. Intercultural cooperation and its importance for survival* (p. 127). London: HarperCollins.

Kennedy, J. C. (1995). *Nieuw Babylon in aanbouw. Nederland in de jaren zestig* (p. 12). Amsterdam, Meppel: Boom.

Kraus, A. (1992). *Grundzüge der Geschichte Bayerns* (p. 146). Darmstadt: Wissenschaftliche Buchgesellschaft.

Lijphart, A. (1968). *The politics of accommodation: Pluralism in the Netherlands.* Berkeley: University of California Press.

Lintsen, H. W. (Red.) (1992–1995). *Geschiedenis van de Techniek in Nederland. De wording van een moderne samenleving, 1800–1890.* Walburg Pers.

März, P. (Ed.). (1994). *Geschichte des modernen Bayern. Königreich und Freistaat* (p. 31). München: Bayerische Landeszentrale für politische Bildungsarbeit (A95).

Middendorp, C. P. (1991). *Ideology in Dutch politics. The democratic system reconsidered, 1970–1985* (p. 282). Assen/Maastricht: Van Gorcum.

Richardson, J. J. (Ed.). (1982). *Policy styles in Western Europe* (p. 13). London: Allen & Unwin.

Richardson, J. J., & Watts, N. S. J. (1985). *National policy styles and the environment. Britain and West Germany compared* (p. 2 and 33). IIUG discussion paper 85–16. Berlin: WZB.

Rose, R. (1993). *Lesson-drawing in public policy. A guide to learning across time and space* (p. 31). Chatham, NJ: Chatham House.

Rootes, C. (1995). Environmental consciousness, institutional structures and political competition in the formation and development of Green parties. In D. Richardson & C. Rootes (Eds.), *The green challenge. The development of green parties in Europe* (pp. 232–252). London: Routledge.

Sabatier, P. A., & Jenkins-Smith, H. C. (1993). *Policy change and learning. An advocacy coalition approach.* Boulder, CO: Westview Press.

Schama, S. (1991). *The embarrassment of riches. An interpretation of Dutch culture in the golden age* (375 ff.). London: Fontana.

Schama, S. (1995). *Landscape and memory.* London: HarperCollins Publishers.

Schwarz, M., & Thompson, M. (1990). *Divided we stand. Redefining politics, technology and social choice.* New York: Harvester Wheatsheaf.

Strauss, F. J. (1989). *Die Erinnerungen* (p. 601, 603). Berlin: Siedler Verlag.

Thompson, M., Ellis, R., & Wildavsky, A. (1990). *Cultural theory.* Boulder, CO: Westview Press.

van Waarden, F. (1995). Persistence of national policy styles: A study of their institutional foundations. In B. Unger, & F. van Waarden (Eds.), *Convergence or diversity? Internationalisation and economic policy response* (pp. 333–372). Aldershot: Avebury, 357 and 358.

Weale, A. (1992). *The new politics of pollution.* Manchester, New York: Manchester University Press.

Weber, M. (1904–1905). *Die protestantische Ethik und der 'Geist' des Kapitalismus.* Archiv für Sozialwissenschaft und Sozialpolitik (20, 21). Also in: Gesammelte Aufsätze zur Religionssoziologie, Volume 1. Tübingen: Mohr (1920).

Wintle, M. (Ed.) (1994). *Rhetoric and reality in environmental policy. The case of the Netherlands in comparison with Britain* (p. 141). Aldershot: Avebury.

Wissenberg, M. (1993). The idea of nature and the nature of distributive justice. In A. Dobson, P. Lucardie (Eds.), *The politics of nature. Explorations in green political theory* (pp. 3–20). London: Routledge.

Zahn, E. (1991). *Regenten, rebellen en reformatoren. Een visie op Nederland en de Nederlanders.* Amsterdam: Contact.

Zweers, W. (1994). Radicalism or historical consciousness: On breaks and continuity in the discussion of basic attitudes. In W. Zweers, & J. J. Boersema (Eds.), *Ecology, technology and culture. Essays in environmental philosophy* (pp. 63–72). Cambridge: White Horse Press.

Chapter 21
Land Use in Zimbabwe and Neighbouring Southern African Countries

Ignas M.A. Heitkönig and Herbert H.T. Prins

Contents

21.1	Introduction	445
21.2	Rainfall	447
21.3	The Ecology of Savannahs	448
21.4	Current Land Use	450
21.4.1	Wildlife Outside Protected Areas: An Asset or a Liability?	450
21.4.2	Agricultural Policies and Game Laws	451
21.5	Future Land Use Changes	451
21.5.1	How the 'Tragedy' of the Commons Can Become a Celebration	453
21.5.2	CAMPFIRE Installs Pride and Conservation	454
21.5.3	Can Campfire Be Copied?	455
References		456

21.1 Introduction

In southern Africa some 54 million people occupy just over 3 million km^2 of land. The four southern African countries discussed in this chapter are Zimbabwe, Botswana, Namibia and South Africa. Most of the southern Africa land mass lies between 500 and 1,000 m above sea level, is fairly flat (Botswana) or slightly undulating, with interspersed Inselbergs. The Drakensberg in the south-east of South Africa (Lesotho) and the eastern Highlands of Zimbabwe are the only mountain ranges with peaks over 2,000 m.

I.M.A. Heitkönig (✉)
Resource Ecology Group of Wageningen University,
The Netherlands
e-mail: ignas.heitkonig@wur.nl

Southern Africa, as here defined, has a total area of about one-third that of the USA, or 3,100,000 km^2, and a human population of about one-fifth, or 54,000,000 inhabitants. In terms of core statistics like population density, Gross Domestic Product and environmental variables like rainfall the four countries differ substantially, sometimes by an order of magnitude (Table 21.1).

Their histories, too, are dissimilar, as are the indices of political stability. Amongst the features they share, however, are immense inequalities in income: about 70% of the Namibian population, for instance, is dependent on subsistence farming, which contributes only 1.5% to GDP (Barnard 1998), and which renders summary statistics like these fairly meaningless. The countries also share a high population growth rate of about 3% per annum and a high incidence of HIV infections. AIDS is now the top killer in each country, accounting for at least 10% of all deaths and creating increasingly severe shortages in mortuaries in urban and rural areas alike. Accordingly, human population growth rates are expected to decline considerably over the next few decades. Moreover, the countries share a rather low economic growth rate of about −1% to +3% per annum. Lastly, they share a common concern about their general environment and the natural resources soil, water and biodiversity in particular. All four countries are signatories to the 1992 Rio Convention on Biodiversity and have committed themselves to safeguard their environment as a whole and biodiversity in particular.

Summary statistics also mask another key feature that these countries share: gradual expansion of subsistence farming areas and of the associated poverty, at the expense of wilderness. This is particularly well visible in densely populated Zimbabwe and at several sites in South Africa, but it also occurs elsewhere. Almost by definition, wilderness occurs in rural areas, and successful conservation is therefore highly depen-

Table 21.1 Country profiles of the study region, comprising South Africa, Zimbabwe, Botswana and Namibia on the southern African subcontinent. (Note that actual values in this table may change considerably, depending on the source consulted.)

	South Africa	Zimbabwe	Botswana	Namibia
Land area[a] (km²)	1,270,871	390,620	581,730	823,988
Avg. precipitation range in 80% of the area[b] (mm/year)	250–1,200	400–1,250	300–600	50–550
Population[a]	39,357,000	11,377,000	1,570,000	1,660,000
Population/km²	31	29	3	2
GDP per capita[c] (US$)	3,215	508	3,000	3,600
Livestock[d] (head)	48,900,000	8,870,000	4,475,000	5,813,000
Permanent pasture[e] (km²)	839,280	172,000	256,000	380,000
Livestock/km² pasture[f]	58	52	18	15
Protected areas[g,h] (km²)	66,119	49,970	104,988	112,160
Protected areas[g] (% area)	5	13	18	14

[a] FAO (2001).
[b] Excluding the often isolated driest and wettest fragments of the land area.
[c] US State Department (2001), estimates for 1995, except South Africa for which 1999 estimates are provided.
[d] FAO (2001), estimates for 1998 and comprise herded or free-ranging cattle, sheep and goats, but not pigs, camels or poultry.
[e] FAO (2001), estimates for 1998. In South Africa about 66% of the land area is allocated to pasture, in the other countries about 45%.
[f] Inflated values due to non-differentiation between free grazing and stabled animals.
[g] State or lower government controlled protected areas only.
[h] World Conservation Monitoring Centre (2001); South Africa's prince Edward islands and Marion island are excluded.

dent on human activities in neighbouring communities. In their day the colonial systems of government operated centralised systems of management and extension in each country, including a single, unarticulated regime of conservation. That the South African National Parks Board also has provincial counterparts does not alter the basic premise that conservation was, and to a large extent still is, a centralised operation that bypasses neighbouring communities. The boundaries of protected areas were drawn on maps at provincial or national Parks Board headquarters, without consideration given to land use patterns of local inhabitants in such areas. Seeking more appropriate models of nature and resource conservation, Martin and Taylor (1983) built upon earlier work and proposed a system of regional land use planning in which nature conservation and nature protection is but one element, well integrated within the context of other land use activities, by allowing dispersal of animals onto neighbouring lands. The latter feature would permit community-based use of wildlife in the form of commercial hunting operations, for example.

This chapter discusses an example of community based management named with the acronym CAMPFIRE (Communal Areas Management Programme For Indigenous Resources) This programme evolved through grassroots initiatives and was successively introduced in 12 districts (Box 21.1).

To appreciate the nature of community-based resource management and its local successes and failures,

> **Box 21.1** How CAMPFIRE grew from WINDFALL and spread to 12 districts
>
> - A bird's-eye view of the development of management of indigenous resources by communal, peasant farmers, based on Child (1995) and Child (2000).
> - The Sebungwe region in NW Zimbabwe covers 40,000 km² of erodable soils and is sparsely inhabited. Introducing wildlife exploitation as a land use in communal areas in the 1980s was first attempted through WINDFALL (Wildlife Industry's New Development for All). Wildlife was still managed by the Wildlife Department, so it was still a public asset; income was passed back to various (not just the wildlife providing) District Councils to implement projects. But the system was not transparent, and it was not clear to these Councils where the money came from. Wildlife projects resembled projects funded from other government or donor sources. Farmers received very little financial benefits, and were left uninformed about the aims and procedures (Child 1995). The loop between wildlife use and small financial benefits was far too long and winded.
> - Where WINDFALL ended, CAMPFIRE started. District Councils were granted appropriate authority to manage wildlife, and a process
>
> (continued)

Box 21.1 (continued)

was introduced where Councils channelled part of wildlife revenues to communities. The communities subsequently decided how to allocate these revenues as cash to households, or as a contribution towards building infrastructure like schools or water pumps. The time span between the actual hunting of game and receiving the benefits thereof was short. Much emphasis was also placed on making clear where the money came from when another bag of it was literally handed over to the community gathering. Subsistence farmers now felt the benefits of wildlife in their pockets, or at least controlled the use of the money. Guided by a set of clear rules and agreements among several shareholders, revenue distribution is a process, rather than a simple action.

- CAMPFIRE is a process with about five distinctive steps – and a long list of detailed actions that go with it. *Step 1* asks for the installation of the legal framework and a supportive environment. District councils can then apply for proprietorship on a voluntary basis. *Step 2* increases the awareness of the value of wildlife, at District Council level. *Step 3* deals with marketing wildlife and tourism effectively, both nationally and internationally. *Step 4* ensures devolving wildlife-derived money from District Council level to community level, with full participation, accountability, and transparency. *Step 5* enables the management of wildlife and related resources at District and community level.
- The success of the first CAMPFIRE initiatives in Sebungwe in the North-West and Beitbridge in the South inspired increasingly more District Councils to follow suit. Each one of them agreed to abide by the CAMPFIRE guidelines concerning democracy, accountability, transparency, equity, and benefit (Child 1996). In 1989, 16 wards took part in a CAMPFIRE programme. This number grew and stabilised at about 70 wards in 1992 and 1993, then involving about 70,000 households. Equally impressive is that Councils retained progressively less revenue: from 44% in 1989 to 21% in 1993. Income devolved to communities rose from 7% in 1989 to 57% in 1993.

let us first consider in more detail the variable climate and ecology of the region, its system of protected areas, subsidies for commercial agriculture and the various means for utilising wildlife as a source of income.

21.2 Rainfall

In this ancient landscape, Namibia is the driest of the four countries (with precipitation ranging from less than 50 mm in the western Namib desert to max. 500 mm in the north-east) and Zimbabwe the wettest (from 400 mm to about 1,250 mm in the east); see Table 21.1. Most rainfall occurs in the warm and wet season (October–March) season with the rest of the year generally cool and dry. This gives rise to a typical savannah landscape away from the coastal belt, where rainfall exceeds 450 mm. Savannahs are characterised by a more or less contiguous layer of grasses, with a highly variable component of shrubs and trees. At the lower rainfall end of the spectrum, in eastern Namibia and in western to central Botswana, trees are sparse, and shrubs irregularly distributed. In the semi-arid and arid western part of Namibia and South Africa grasses are also sparse, and shrubs take over, while at the high rainfall end (eastern, mountainous parts of South Africa and Zimbabwe) dense forested stands occur. Moreover, rainfall patterns over the past century appear to have followed a rather clear cycle, 9 years of below-average rainfall being generally followed by 9 years of above-average rainfall (Tyson 1987). Most rain falls as showers, often as localised thundershowers. They typically release 10–40 mm of rainfall within 30–60 min. Part of this water runs off towards gullies, stained yellowish brown with soil, with the rest draining into the soil but evaporating soon afterwards, leaving relatively little for plant growth. On average, no more than ten heavy showers make up half annual rainfall. In the wet season, many rivers become swollen after such showers upstream, but water levels quickly recede. In the dry season these rivers are no more than broad tracts of sand, rocks and pebbles with an odd puddle of water.

Three main river systems drain the subcontinent: the Zambezi demarcates Zimbabwe to the west and north, the Limpopo separates Botswana and Zimbabwe from South Africa, and the Orange River drains the central part of South Africa and forms a border with southern Namibia. However, it is evaporation rather than surface water runoff that accounts for the bulk of annual water loss: from about 1,100 up to over 3,000 mm in South Africa (Huntley et al. 1989) and even to 3,700 mm in Namibia (Barnard 1998). In Namibia it is estimated that about 83% of rainfall evaporates, 14% is lost through plant evapotranspiration, 2% enters drainage systems, with a mere 1% recharging groundwater supplies (Barnard 1998). This high rate of loss through evaporation clearly leads to a permanent water shortage, even in the best of years in terms of rainfall.

21.3 The Ecology of Savannahs

The savannah vegetation is particularly well suited for supporting a wide range of herbivorous animal species, ranging in size from below 1 kg up to around 4,000 kg. In a particularly good rainy season, the grasses flourish and develop a high, green biomass in the wet season, followed by rapid desiccation and subsequent yellowing during the dry season. Fresh plant growth is invariably of high nutritional quality, which simply means a high nutrient content per gram of food. Further wet season growth basically means these nutrients become 'diluted' with fibre. While this is excellent for providing structure to the plant, fibre contains very few nutrients. Hence, towards the end of the wet season there is plenty of biomass and therefore food, but of rather low quality. This high, green biomass of grass in the wet season turns into a high, dry biomass in the dry season, which is highly inflammable. Bush fires are therefore a typical characteristic of savannahs, although both the frequency and the extent of such fires have increased by an order of magnitude over the past few hundred years. The current rate of CO_2 production by savannah fires on the African continent is higher than that due to the burning of fossil fuels in western Europe.

In Namibia and Botswana, where long-term average rainfall is low, annual fluctuations in total precipitation are proportionately large. As a result, there are almost no areas in these countries that are suitable for traditional dry-land agriculture. Both the annual fluctuations in total rainfall and its patchy spatial distribution have a great impact on vegetation. Animals such as wildebeest that cannot survive without surface water, manage to find grazing pastures and surface water even in arid areas (<300 mm annual rainfall), if they are not restricted in their movements. They are reported to cover up to 500 km a year to reach sites with fresh grass growth (Main 1996). In general, only extreme circumstances such as a prolonged drought will cause wildlife populations to decline.

The wildlife species of greatest importance in our context here include such prominent animals as buffalo, wildebeest, giraffe, warthog, hippopotamus, duiker, kudu, eland, oryx, waterbuck, hartebeest and impala, all so-called even-toed ungulates. The main livestock species, cattle, goat, sheep and pigs, also belong to this group. Zebra, rhinoceros and elephant complete the picture. It is useful to make a functional distinction among these species, based on their main food source. Although all eat plants and are therefore termed herbivores, the 'grazers' among them feed almost exclusively on grass, while the 'browsers' have a diet made up of forbs and the fruits and leaves of shrubs and trees. The grazers include waterbuck, hartebeest, wildebeest, buffalo, hippopotamus, zebra and white rhino. Cattle, introduced into southern Africa by humans about 1,700 years BP, are also typical grazers. Typical browsers include the small duiker species, giraffe, kudu and black rhino. Some species, like elephant, impala, springbok and eland, graze mainly during the warm wet season, but switch to browsing in the cool dry season when grass quality has declined while leaves from woody plants are still green. The third category of animals are often referred to as 'intermediate' or 'mixed' feeders and in this savannah environment include both goat and sheep, known to have been introduced into the area about 2,000 years BP.

There are two important differences between the array of wild animals on the one hand, and human livestock species on the other. The first is that livestock species are herded, in contrast to game. All livestock products, from milk to hides to dung, as well as draught power are therefore readily available to herders, and require minimal energy. The second difference is that wildlife is generally immune to indigenous disease, while livestock, by and large, is not. One important indigenous disease of major significance, *Trypanosomiasis* or Nagana, sleeping sickness, is spread by tsetse flies (see Box 21.2: Animal Diseases). These flies suck blood and in doing so may transmit the protozoa from one host to another. Cattle suffer terribly from the disease, and humans abhor the painful sting inflicted by the fly. This insect, in particular, is therefore a blessing in disguise, for it has protected parts of the savannah from invasion by livestock-herding pastoralists and settlers.

Box 21.2 Animal diseases

- Diseases in livestock play an important role in the shaping of human land use in savannas. Only three diseases are maintained solely by wildlife (African swine fever in warthog, malignant catarrhal fever in wildebeest, and bovine petechial fever in bushbuck), but none of them are a severe threat to livestock in the study region (Grootenhuis 2000). In contrast, many livestock diseases have a major impact on wildlife populations, either directly by infection, or through human intervention. Among the multitude of viral, bacterial, protozoal, and helminth- (worm-)based diseases prevalent in livestock, we selected four to illustrate this point.
- Rinderpest is a rapidly mutating, highly contagious viral disease introduced into Africa with cattle arriving in Somalia from India in 1889. It literally decimated both livestock and game populations throughout the eastern and southern part of Africa within a decade. Vaccines were commonly used in East Africa up to the 1960s. Wild animals quickly build up long-lasting resistance to the virus, which is also carried to the next generation (Plowright 1982). The virus can be passed from game to cattle, and *vice versa*, but such cross-species transport has no predictable consequences. (Examples and results from vaccination trials mentioned in Plowright 1982). Outbreaks still occur, but appear mainly limited to eastern Africa.
- Foot and Mouth Disease, a highly contagious viral disease which also affected the livestock industry again in north-western Europe in 2001, has caused large parts of South Africa, Zimbabwe and Botswana to be fenced-off. FMD occurs naturally in many game species, notably buffalo, who build up immunity. Livestock is far more susceptible, and incurs large losses. In the past century, FMD eradication campaigns were carried out in military style operations, culling almost all game in affected regions. Yet, it proved almost impossible to kill the last remaining wild animals. Later, thousands of kilometers of fences were erected to separate infected livestock and game in the East from non-infected animals further to the west in South Africa and Zimbabwe, and to achieve a similar separation in Botswana. Veterinary controls are carried out near the fences. Stringent export regulations stipulate that only FMD-free meat can be exported to Europe and North America. If FMD antibodies are present, either through infection or through vaccination, the export will end, crippling the economy. The spread of the disease is almost certainly limited by the fences, but is that due to limited contact between buffalo and cattle? Condy and Hedger (1974) herded susceptible non-infected cattle together with FMD carrying buffalo. They grazed and drank together but nothing happened: the virus was not transferred to cattle in the 2.5 years duration of the experiment. The exact mode of transmission remains still unresolved.
- In southern Africa, sleeping sickness (*Trypanosomiasis* or 'Nagana') is primarily a cattle disease, caused by the protozoa *Trypanosoma rhodesiense*, of which the tsetse fly (*Glossina morsitans*) is the carrier. Both male and female tsetse fly sting to suck blood from their victims. If the victim carries trypanosomes, the tsetse fly can carry them to its next victim, thus spreading the disease, similar to the way in which a female mosquito stings, draws blood, and can spread malaria. The drugs available to treat the disease have unpleasant and dangerous side effects, like renal and neurologic impairment. Also, they cannot be used as a prophylaxis, unlike anti-malaria drugs. Therefore, eradication of the disease has centered on getting rid of tsetse flies, rather than of the trypanosomes, through bush destruction, spraying of insecticides (including DDT), and the dramatic elimination of game through destruction (Zimbabwe: Tsetse Fly Act, 1930).
- Tuberculosis or bovine tuberculosis (TB), caused by infection with the bacterium *Mycobacterium bovis*, is easily spread among and between cattle and game, and can be transferred to humans through contact. There is no vector involved in the transmission. Once it is established in wildlife, it can become endemic without threatening the survival of the population. Tuberculosis is well established in buffalo populations in South Africa (Kruger NP, Umfolosi GR), and has also been diagnosed in giraffe, lechwe, kudu, and duiker (Huchzermeyer et al. 1994; *in* Grootenhuis 2000). Buffalo carries many cattle diseases, yet it is a much desired animal for hunting, particularly in southern Africa where it is one of the 'big five'. Disease-free buffalo fetched prices up to US$25,000 per head on game auctions in South Africa during 2001.

Ticks are also blood-suckers, are another important, unpleasant vector of diseases, and are similarly difficult to control in livestock (Grootenhuis 2000).

21.4 Current Land Use

Human-induced habitat changes, through livestock overgrazing and mismanagement of fire, have a complex impact on wildlife. In many cases, large grazers like roan and sable antelope, tsessebe and hartebeest will suffer a dramatic reduction in numbers, because their favoured grasses are no longer available. However, other species like impala and kudu, with a preference for forbs and the leaves of shrubs and trees, may proliferate thanks to the increase in bush density that is often associated with overgrazing and a deterioration of the grass layer (Child 1995). Once the vegetation has been altered to a significant extent, it is very difficult and sometimes even impossible to restore the habitat to its pre-overgrazing stage (Rietkerk et al. 1996). Hence, animal species that have been locally lost cannot be readily re-introduced, except at great effort and high cost, in a long process that starts with vegetation restoration.

Hunting and poaching are often posited as the main cause for the decline and disappearance of wild animal species. However, there is strong evidence that the disappearance of many grazer species in southern Zimbabwe, like hartebeest, wildebeest, zebra and warthog, was triggered by a series of droughts in the 1980s and following a deterioration of their grassy habitat. Browsing species such as kudu, impala and giraffe were much less affected by the droughts and tended to maintain their numbers. White rhino, a grazer, is probably the only species that has been hunted to extinction in Zimbabwe, the last individual being shot in 1912 (Child 1995). Black rhino almost followed suit in the 1980s, but this species could be saved by dramatically intensified law enforcement programmes at the time, including the use of helicopters to track and shoot poachers. Attempts to eliminate buffalo from numerous areas in Zimbabwe in a bid to eradicate Foot and Mouth Disease (FMD) proved only partially successful; some animals escaped being shot, despite a highly intensive campaign including culling teams and the use of helicopters. Outside protected areas, both lion and wild dog have been eliminated by man, because of their presumed adverse effects on livestock. In conclusion, where habitats remain favourable, most animal species can withstand even heavy hunting, provided there is no serious competition for land or other resources with local populations. Where there is competition with humans, the animals invariably lose out.

21.4.1 Wildlife Outside Protected Areas: An Asset or a Liability?

Protected areas are by no means the only sites where wildlife is to be found, for it also roams on private and communal land. No less than 90% of Namibia's large-mammal wildlife occurs outside state-controlled conservation areas. Privately owned commercial farms comprise 44% of the total land and shelter 80% of the larger game species. Communally managed areas cover 41% and harbour around 9% of the same species (Yaron et al. 1993). South Africa's large game species occur almost exclusively outside communal areas. Formal conservation areas account for 7% of the total land mass and harbour less than half the total number of large mammal species. The greatest numbers occur on privately owned commercial farms, the aggregate area of which grew exponentially during the last 20 years of the past century (Eloff 1996, in Hearne and McKenzie 2000). In Zimbabwe, too, more wildlife is conserved outside government-protected parks than in them (Child 2000). In South Africa and in Zimbabwe this is attributed mainly to the commercial success of private game management initiatives, and in Zimbabwe also in part to the communal management initiatives taken in the past 2 decades.

Wildlife is a renewable resource and landholders – whether private or communal – have, *de facto*, almost complete control over the future of wildlife on their land. Survival of the animals and the state of their habitat therefore depend on the decisions taken by these landholders, who will tolerate significant wildlife numbers only if there is sufficient incentive to do so. Private landowners with wildlife on their land which they thereby own, may derive substantial earnings from selling their game to hunters. This forms an incentive to maintain the wildlife on their property, as an asset. In communal areas the opposite is the case: landholders typically have no rights of ownership over wildlife, and any usufructuary rights over this resource are generally insecure or difficult to get upheld in practice. In these areas, wildlife is likely disappear entirely for lack of incentive to conserve it, despite the fact that it is forbidden to hunt wildlife for subsistence (Child and Chitsike 2000). Moreover, many wild animals such as elephants and baboons often cause severe damage to agricultural fields, leading to a substantial loss of income for which little or no compensation can be claimed (Deodatus 2000). This renders the animals even more vulnerable to local extinction by inducing local people to kill the raiders of their crops. Under these circumstances, wildlife is clearly a liability.

21.4.2 Agricultural Policies and Game Laws

Agricultural policies are probably the main reason why wildlife generates far less revenue than livestock in southern Africa. In the past, national governments have engaged in a variety of activities to destroy wildlife in the name of agricultural expansion and rural development (e.g. Plowright 1982). For much of the last century there was a strong focus on the technology of meat production, coupled with a tendency for governments to apply agricultural pricing controls rather than market forces. Game laws made commercial use of wildlife illegal, and ranchers considered wildlife to be valueless and even detrimental for their livestock earnings. Wildlife was replaced by livestock, which was heavily subsidised because of the political power of the cattle-ranchers in a government-dominated command-and-control economy. These economic distortions were even worse in communal areas: traditional hunting became poaching, traditional weapons were disallowed, and several communities were relocated to enable protected areas to be established. Traditional systems of resource control disappeared and landholders were expected to bear the costs of wildlife, while being denied its benefits. Authority over wildlife was put in the hands of administrators in a government department dealing with flora and fauna, making law enforcement outside protected areas virtually impossible. Centralised control over wildlife created an open-access property regime, and the absence of control mechanisms led to a scramble for the use of common resources. Consequently, a buffalo was now snared for meat, rather than sold at great profit to, say, an American hunter, who would leave the meat for the rural population (Child 2000).

Such institutional vacuums and especially open-access property regimes are considered to be a primary cause of the widespread ecological problems throughout Africa. Experience with wildlife in southern Africa in the latter part of the 20th century has shown that the root causes of many conservation problems are institutional and that solutions lie in developing effective property rights, pricing and markets (Child 2000).

21.5 Future Land Use Changes

The two most densely populated countries in our study region, Zimbabwe and South Africa, face popular demands for redressing the racial imbalance in land ownership. After Zimbabwe's war of independence (1980) the new government drew up plans to allocate previously 'white' farmland to black inhabitants, and the post-apartheid South African government (1994) followed suit. Moreover, the communities neighbouring many conservation areas claimed access to the protected sites for livestock grazing or agriculture ('Cattle at the gates of Kruger National Park'), despite the marginal value of such conversion and the poor returns expected. Engelbrecht and van der Walt (1993), for instance, have calculated that only about 1,200 families could derive a livelihood from maize crops and livestock if all 20,000 km^2 of the Kruger National Park were sacrificed to such subsistence agriculture and grazing. With 2,000 people on the Kruger NP payroll in the second half of 2001, the organisation was expected to increase its profits to ZAR 35,000,000 (US$ 3,500,000) that year (African Eye News Service 2001). Because of the protected status of conservation areas there was no large-scale conversion of conservation areas to farmland and pasture.

Keeping herded or fenced-in livestock provides a supply of milk, meat, blood and hides to the livestock keeper at little expenditure of energy. Obtaining the same products from wild animals is either impossible (milk, draught power) or requires high energy and financial expenditure to hunt and transport the animal products (carcasses, hides, horns) to sites for further processing. Tourists are willing to watch or hunt *and* pay for this activity, making wildlife-based land use a potentially viable option.

Wildlife-based land use (Box 21.3) is a far better provider of income and jobs than subsistence farming, although not necessarily so relative to commercial livestock farming. Mixed commercial livestock and wildlife farming in South Africa has provided higher returns on livestock than on wildlife (excluding safari hunting) in the case of sheep (Skinner et al. 1986), goats (Kerley et al. 1996) and cattle (Bigalke et al. 1992). The same holds for Namibia (Yaron et al. 1996) and Zimbabwe (Child 1995), but here this is due to market distortions, in the form of subsidies for livestock farming.

Box 21.3 Options for wildlife-based land use

Broad options for wildlife utilisation include consumptive and non-consumptive use. Consumptive use includes subsistence hunting, which refers to the procurement of wild animals in order to meet household needs, and commercial hunting, which is the exploitation of wild animals specifically for sale. It also includes farming wildlife for the production of meat, hide, horns, and other products. Non-consumptive, commercial use of wildlife includes game viewing and photographic safaris, and farming for the sale of live animals.

- Subsistence hunting is only allowed in Botswana and Namibia within, and in Tanzania outside of southern Africa. In all cases it is limited to peoples of hunting/gathering origin or established practice, such as the (!Kung) San, who are only allowed to hunt in the border region of northern Namibia/Botswana. Hitchcock (2000) examined the hunting activities of this group, and argued that the off take rates of the hunters appear to be sustainable. The (!Kung) San are highly sociable and communicative within the tribe, and typically make decisions based on group consensus. In the 1990s community-based Natural Resource Management programmes were introduced here, but it seems too early to say whether these programmes will enable the San to become economically self-sufficient.
- Game ranching is an activity where wildlife roams the farm rather than livestock, and where wildlife products may include meat, skin, horns, etc. Game is either slaughtered, like livestock, but more commonly it is hunted by local or foreign tourists. Game ranches typically stock more than one species of wildlife and usually have a combination of grazers and browsers. Financial profits are potentially high, but so is the amount of initial capital required. To surround an existing 20 km² ex-livestock farm with game-proof fencing, to build appropriate tourist accommodation and to install other infrastructure and equipment costs at least US$300,000. The acquisition of game costs another US$300,000, making it a land use activity for the more well-to-do section in society. In South Africa, game ranches now cover about 80,000 km² of the land area (Hearne and McKenzie 2000), well above the aggregate of protected areas.
- Culling is a hunting activity to limit population sizes in conservation areas where animal numbers tend to overshoot the sustainable resource base. It is a controlled management action, constrained within a national legal and administrative framework (Cumming 1983). The list of types of culling operations drawn up by Caughly (1983) includes economic, manipulative, pre-emptive, and adaptive culling. In all cases, animals are removed from the population of one or more species in an effort to 'balance' long-term food production with food consumption. Fenced conservation areas have a limited amount of space, which, even in southern Africa's largest Parks, appears never sufficient to allow undisturbed ecosystem processes to take place. Wildlife management often includes culling actions, although it appears seldom possible to be certain about the effect on the ecosystem of the seemingly most appropriate culling action. This is mainly due to the complicated feed-back mechanisms in multi-species ecosystems (Caughly 1983; Bell 1983). In the past, culling was mostly carried out by park staff, with help of local professional hunters. Current operations are often commer-

(continued)

> **Box 21.3** (continued)
>
> cialised to help obtain additional income for the conservation area. The animal products are often sold locally to inhabitants (meat) or locally processed and sold to tourists (meat, hides, craft).
> - Wildlife utilisation in communal areas is an activity where wildlife-based revenues through tourism complement (Namibia, Zimbabwe) or replace (Zimbabwe) subsistence livestock keeping. In Zimbabwe, communal wildlife utilisation gained much momentum through the CAMPFIRE initiative, which followed a WINDFALL initiative in 1980 (see Box 21.1). CAMPFIRE takes on different forms, depending a.o. on districts, locally available natural resources, and wishes of the community itself (Child 1995). Elephant are not omnipresent anymore after intensive campaigns to eradicate game in the first part of the 20th century, but where they are now present and commercially hunted, they fetch high prices. Buffalo also have a patchy distribution, for the same reason as elephant, but since they are hosts to many cattle diseases, severe veterinary restrictions are imposed on the transport of buffalo products. Warthog are almost omnipresent, but do not fetch high prices when hunted. Local situations therefore require local solutions. Fundamentally, CAMPFIRE is not a project of which a blueprint can be implemented, but it is a process to take place at different levels, from household to the District Council, based on voluntary participation.

On the other hand, Jansen et al. (1992; in Child 2000) analysed the financial returns of over 100 commercial operations in the drier 83% of Zimbabwe, with long-term average rainfall below 750 mm. Although beef producer farms had been subsidised for a long time, only 5% of these farms were profitable. In contrast, over 50% of the wildlife enterprises made a profit, despite the large capital investments required, subsidies to livestock farmers, restrictions on game meat markets, and the eradication of buffalo to protect the cattle industry. In the case of safari hunting, wildlife wins hands down over livestock farming, despite market distortions (e.g. Child 2000: 351). If market distortions like drought relief and livestock subsidies are removed, game viewing is expected to be even be more profitable than safari hunting in Namibia (Barnes and de Jager 1995).

21.5.1 How the 'Tragedy' of the Commons Can Become a Celebration

Livestock is a private asset in communal areas, but the grass and browse on which livestock feed is a common resource. Wildlife is a common resource, too. Resource use in communal areas is regulated by open access property regimes: there are no identified owners, no-one can be excluded from using the resource, and each person has a right to withdraw resources. This makes both wildlife and vegetation vulnerable to exploitation for maximum personal gain, at the expense of the community. Moreover, open access regimes do not encourage personal investment in the resource, because the benefits will be harvested by other resource users. This has become known as 'The tragedy of the commons' (Hardin 1968). Hence, both wildlife and the food supply of all animals is prone to unbridled exploitation as long as the open access property regime remains fully intact and law enforcement is absent.

What is needed is a change in the institution controlling open access areas. This might be achieved by changing from 'no ownership', 'free access' and 'free withdrawal' to a set of collective choice property rights (Schlager and Ostrom 1992) which include the right of management, but also the rights of exclusion and alienation. It means making the adults in the community members of a group that takes decisions on the management of common resources. In this way, besides being users, community members also become shareholders. Those who do not want to participate may be excluded and/or alienated from using the common resources.

In Zimbabwe such a change was implemented by a directive from the then-Prime Minister Robert Mugabe. The Communal areas were divided into a hierarchy of Wards and Villages, which were given control over the resources in their respective subdivisions. Communal areas are now divided into basic village units comprising approximately 100–150 households (1,000 people) and 'their' land. Each village is represented by a politically elected Village Development Committee, whose chairperson represents the village at ward level, which comprises up to ten villages. The chairperson of the Ward Development Committee serves on the Rural District Council, the grassroots or bottom-up representation of both commercial and peasant farmers at district level (Child 1995).

The Rural District Councils, with no financial resources (David), meet and deal with government sponsored, top-down central administration officials (Goliath) to plead their case for further development. Goliath has meanwhile weakened, owing to cash flow problems, and David gained strength, through earning income from CAMPFIRE-originated wildlife utilisation. This income-earning consolidated the household-village-ward-district relationship. If the revenue is then apportioned equitably, with a fair proportion going to landholders with wildlife, this system would entail sound democratic accountability. The returns to shareholders from a resource-based production system become linked to environmental inputs. 'Profits' are related directly to how people use or abuse their resources to influence their own future 'earnings'. It brings benefits from natural resources and accountability for their sustainable use close together (Child 1995). In this way, wildlife can become an asset in communal areas too.

What are also needed, besides the change in the institution controlling open access areas by implementing collective choice property rights, are changes to the restrictive game laws and to agricultural policies. For wildlife to be legally utilised in communal areas, its use should be decriminalised. For wildlife to be profitable, it needs to be valued, it needs to fetch a fair price, and it needs to be marketed. For wildlife utilisation to help reverse the negative spiral of land degradation, agricultural subsidies favouring livestock farming should be abolished.

Prior to 1960 Zimbabwe prohibited landholders from using wildlife commercially. Then, landholders were granted that right under a permit issued by the Wildlife Department. From 1975 onwards, the Parks and Wildlife Act enabled commercial farmers to use their wildlife without bureaucratic constraints, and without payment of fees. However, even by 1980 landholders were still strongly prejudiced against wildlife, and safari operations brought only limited financial returns owing to the low game prices. These were, in turn, kept low by the Wildlife Department. In the last decade, wildlife departments have been forced to substantially raise prices to meet cut-backs in state allocation of funds (typical for developing countries). This has improved the position of landholders, giving them an additional market for wildlife products through tourism. Zimbabwe's wildlife sector is environmentally, financially and economically sustainable and more productive than livestock in terms of profit, foreign exchange, and employment (Child 2000).

21.5.2 CAMPFIRE Installs Pride and Conservation

The amount of money generated by wildlife in communities that have adopted the CAMPFIRE programme is small, and can easily be replaced by a foreign donor. However, its impact in transforming the community has been large, because rural people have learned to do things for themselves, with their own resources. It is an achievement no donor project could realise (Child 2000). It has instilled an important sense of care for natural resources on communally owned lands in Zimbabwe, despite the still limited financial revenues per household. Nevertheless, wildlife use is the best strategy for maximising financial returns for arid and semi-arid lands, even under distorted market circumstances. Increasing land pressure from outside and steady population growth within allow no concurrent increase in livestock herds. Rather, they spur new strategies, based on a cash-economy, rather than on a cow-economy. Under these circumstances, in which financial returns are the motive for production, land use will be based on costs and benefits and new opportunities arise that benefit from the use of wildlife (Prins et al. 2000).

Child (2000) reckons that the profitable and efficient conservation of wildlife and wildlife habitat will have profound implications for future managers of government-protected areas, for there can be no justification for large subsidies to maintain commercially unproductive national parks in the midst of rural poverty. In the view of that author the role of these parks must be reconsidered. Current initiatives and experiments involving protected areas and their neighbouring communities in southern Africa appear to prove him right.

21.5.3 Can Campfire Be Copied?

The transformation of common property resource use and the implementation of wildlife utilisation is not readily copied in other areas, nor in other countries. Where wildlife habitat is in reasonably good shape, and wildlife is present in at least moderate numbers, utilisation of this resource is possible. But where the habitat is badly degraded, and animals are almost gone, it will likely be a futile exercise. The main problem is not to get the animals back, but to restore their habitat (Scheffer et al. 2001). Near lakes, fisheries could be an alternative land use option (Child 1995) and, in high rainfall areas, forest exploitation in the form of selective logging. Many districts in Zimbabwe currently appear to have opted for the CAMPFIRE approach on their communal lands, especially those who have not faced either political strife or badly degraded resources which are no longer worth being managed.

In South Africa, only few communal areas exist and these are all devoid of wildlife and wildlife habitat. It therefore appears futile to simply transfer the CAMPFIRE approach to Zimbabwe's southern neighbour. Instead, it seems more appropriate to try and reduce the major pressure on the land by developing means of partnerships between protected areas and neighbouring or inhabiting communities. Botswana's Wildlife Areas go some way in this respect. Another initiative, on the Madikwe Game Reserve, has recently been described by Davies (2000). Here, former cattle ranches were stocked with wildlife, and three stakeholders began to participate in the process of improving the socio-economic situation of poor communities. The government establishes and manages resource conservation. The private sector finances, develops and manages the tourism operations. The government charges concession fees, which are used for reserve management activities and community development programmes. The communities are responsible for identifying their own development priorities and for ensuring that they are carried out in a responsible, democratic and transparent manner. They also participate in developing policies for the Game Reserve. Appropriate institutions have been established in a participatory manner to accommodate these parties. The Reserve developed rapidly, and the success of each party individually and in combination shows that this model can be a profitable means of both development and conservation.

Livestock, and cattle in particular, have an important social and cultural significance in Namibia, where a man without cattle has no social status. Having wildlife instead of cattle does nothing to change this: one does not 'have' or 'own' wildlife since it cannot be herded and brought home, nor is it a single person's property. Wildlife utilisation through tourism in Namibian communal situations is therefore complementary to livestock herding. It broadens resource dependence, which is particularly important in low-rainfall areas with their unpredictable rainfall events. The wildlife-based benefits are increasingly distributed through 'conservancy' systems, in which defined communities are given custodial rights and responsibilities over their land (Richardson 1998), rather similar to the system of wards in CAMPFIRE. To be given resource management rights and associated responsibilities, conservancy committees must have their constitution approved.

All communal wildlife-based initiatives rely heavily on community-approved structures and processes, on democratic principles, and on treating resources as an asset. They also rely on full government backing, and on a steady or growing flux of tourists into southern Africa. If either of these fundaments crumble, the wise use of resources will likewise crumble. In addition, the most successful models of both communal and private wildlife-based land use stem from grassroots initiatives, which were invariably built by rowing against the economic or legal current. We know of no successful model which stems from the policy offices of central government. It takes considerable energy and dedication to convince provincial or national governments to issue directives or change legislation in order to remove the final legal obstacles to wise resource use. Only then, however, can these grassroots initiatives grow into more widely applied models that do even more than conserve resources. They can provide jobs, higher earnings, increased scope for children's education, etc. They provide development, thus becoming policy.

The political and social unrest in Zimbabwe since the end of 1999, with a focus on the re-allocation of land from whites to blacks, overtly backed by the country's president in spite of several court rulings, has led to the virtual elimination of tourist visits to the country. This has dealt both the communal and the private wildlife-based enterprises a very severe blow, the effects of which are still to be determined. South Africa also has a long history of internal political and social instability, and the increased crime rate has not done much to reassure the minds of potential tourists. Namibia and Botswana appear to fare a lot better in this respect, and a continuation of their internal stability may allow these countries to considerably increase their wildlife-based tourist revenues over the next decade.

References

African Eye News Service. (2001). News item on http://allafrica.com.
Barnard, P. (Ed.). (1998). *Biological diversity in Namibia: A country study*. Windhoek: Namibian National Biodiversity Taskforce.
Barnes, J., & De Jager, J. L. V. (1995). *Economic and financial incentives for wildlife use on private land in Namibia and the implications for policy*. Research Discussion Paper 8, 1–21. Directorate of Environmental Affairs, Windhoek.
Bell, R. H. V. (1983). Decision-making in wildlife management with reference to problems of overpopulation. In R. N. Owen-Smith (Ed.), *Management of large mammals in African conservation areas* (pp. 145–172). Pretoria: HAUM.
Bigalke, R. C., Berry, M. P. S., & Van Hensbergen, H. J. (1992). *Cattle versus game; an African case study*. Unpublished Report. Stellenbosch: Department of Nature Conservation, University of Stellenbosch.
Caughly, G. (1983). Dynamics of large mammals and their relevance to culling. In R. N. Owen-Smith (Ed.), *Management of large mammals in African conservation areas* (pp. 115–126). Pretoria: HAUM.
Child, B. A. (1996). The practice and principles of community-based wildlife management in Zimbabwe: The CAMPFIRE programme. *Biodiversity and Conservation, 5*, 369–398.
Child, B. A. (2000). Making wildlife pay: Converting wildlife's comparative advantage into real incentives for having wildlife in African savannas, case studies from Zimbabwe and Zambia. In H. T. Prins, J. G. Grootenhuis, & T. T. Dolan (Eds.), *Wildlife conservation by sustainable use* (pp. 335–387). Boston: Kluwer.
Child, G. (1995). *Wildlife and people: The Zimbabwean success*. Harare: Wisdom Foundation.
Child, G., & Chitsike, L. (2000). "Ownership" of wildlife. In H. H. T. Prins, J. G. Grootenhuis, & T. T. Dolan (Eds.), *Wildlife conservation by sustainable use* (pp. 247–266). Boston: Kluwer.
Condy, J. B., & Hedger, R. S. (1974). The survival of foot and mouth disease virus in African buffalo with non-transference of infection to domestic cattle. *Research and Veterinary Science, 16*, 182–185.
Cumming, D. H. M. (1983). The decision-making framework with regard to the culling of large mammals in Zimbabwe. In R. N. Owen-Smith (Ed.), *Management of large mammals in African conservation areas* (pp. 173–186). Pretoria: HAUM.
Davies, R. (2000). Madikwe game reserve: A partnership in conservation. In H. H. T. Prins, J. G. Grootenhuis, & T. T. Dolan (Eds.), *Wildlife conservation by sustainable use* (pp. 439–458). Boston: Kluwer.
Deodatus, F. (2000). Wildlife damage in rural areas with emphasis on Malawi. In H. H. T. Prins, J. G. Grootenhuis, & T. T. Dolan (Eds.), *Wildlife conservation by sustainable use* (pp. 115–140). Boston: Kluwer.
Engelbrecht, W. G., & Van der Walt, J. (1993). Notes on the economic uses of Kruger National Park. *Koedoe, 36*(2), 113–119.
FAO. (2001). www.fao.org.
Grootenhuis, J. G. (2000). Wildlife, livestock, and animal disease reservoirs. In H. H. T. Prins, J. G. Grootenhuis, & T. T. Dolan (Eds.), *Wildlife conservation by sustainable use* (pp. 81–113). Boston: Kluwer.
Grootenhuis, J. G., & Prins, H. H. T. (2000). Wildlife utilisation: A justified option for sustainable land use in African savannas. In H. H. T. Prins, J. G. Grootenhuis, & T. T. Dolan (Eds.), *Wildlife conservation by sustainable use* (pp. 469–482). Boston: Kluwer.
Hardin, G. (1968). The tragedy of the commons. *Science, 162*, 1243–1248.
Hearne, J., & McKenzie, M. (2000). Compelling reasons for game ranching in Maputaland. In H. H. T. Prins, J. G. Grootenhuis, & T. T. Dolan (Eds.), *Wildlife conservation by sustainable use* (pp. 417–438). Boston: Kluwer.
Hitchcock, R. K. (2000). Traditional African wildlife utilisation: Subsistence hunting, poaching, and sustainable use. In H. H. T. Prins, J. G. Grootenhuis, & T. T. Dolan (Eds.), *Wildlife conservation by sustainable use* (pp. 389–415). Boston: Kluwer.
Huchzermeyer, H. F. K. A., Brückner, G. K., Van Heerden, A., Kleberg, H. H., Van Rensburg, I. B. J., Koen, P., et al. (1994). Tuberculosis. In J. A. W. Coetzer, G. R. Thomson, & R. C. Tustin (Eds.), *Infectious diseases of livestock with special reference to southern Africa* (pp. 1425–1444). Oxford: Oxford University Press.
Huntley, B. J., Siegfried, R., & Sunter, C. (1989). *South African environments into the 21st century*. Cape Town: Human & Rousseau.
Jansen, D. J., Bond, I., & Child, B. A. (1992). *Cattle, wildlife, both or neither? – Results of a financial and economic survey of commercial ranches in southern Zimbabwe*. WWF Multispecies project, Project Paper 27.
Kerley, G. I. H., Knight, M. H., & De Kock, M. (1996). Wildlife and ecotourism as alternative landuse options for valley bushveld. *Grassland Society of South Africa, Special Publication*, 32–34.
Main, M. (1996). *Kalahari. Life's variety in dune and delta*. Cape Town: Southern Book.
Martin, R. B., & Taylor, R. D. (1983). Wildlife conservation in a regional land-use context: The Sebungwe region of Zimbabwe. In R. N. Owen-Smith (Ed.), *Management of large mammals in African conservation areas* (pp. 249–270). Pretoria: HAUM.
Plowright, W. (1982). The effects of rinderpest and rinderpest control on wildlife in Africa. *Symposia of the Zoological Society of London, 50*, 1–28.
Prins, H. H. T., Grootenhuis, J. G., & Dolan, T. T. (2000). *Wildlife conservation by sustainable use*. Boston: Kluwer.
Richardson, J. (1998). Economics of biodiversity conservation in Namibia. In P. Barnard (Ed.), *Biological diversity in Namibia: A country study* (pp. 227–278). Windhoek: Namibian National Biodiversity Taskforce.
Rietkerk, M., Ketner, P., Stroosnijder, L., & Prins, H. H. T. (1996). Sahelian rangeland development: A catastrophe. *Journal of Range Management, 49*(6), 512–519.

Scheffer, M., Carpenter, S., Foley, J. A., Folke, C., & Walker, B. (2001). Catastrophic shifts in ecosystems. *Nature, 413*, 591–596.

Schlager, E., & Ostrom, E. (1992). Property-rights regimes and natural resources: A conceptual analysis. *Land Economics, 68*(3), 249–262.

Skinner, J. D., Davies, R. A. G., Conroy, A. M., & Dott, H. M. (1986). Productivity of springbok Antidorcas marsupialis and Merino sheep Ovis aries during a Karoo drought. *Transactions of the Royal Society of South Africa, 46*(2), 149–164.

Tyson, P. D. (1987). *Climatic change and variability in southern Africa*. Cape Town: Oxford University Press.

US State Department. (2001). www.state.gov.

World Conservation Monitoring Centre. (2001). www.unep-wcmc.org.

Yaron, G., Healy, T., & Tapscott, C. (1993). *The economics of living with wildlife in Namibia*. Windhoek: Namibian Institute for Social and Economic Research.

Yaron, G., Healy, T., & Tapscott, C. (1996). The economics of living with wildlife in Namibia. In J. Bojö (Ed.), *The economics of wildlife: Case studies from Ghana, Kenya, Namibia, and Zimbabwe* (pp. 81–114). Washington, DC: AFTES Working Paper No. 19, World Bank.

Chapter 22
Climate Change Policy of Germany, UK and USA

Richard van der Wurff

Contents

22.1	Introduction	459
22.2	Climate Change	460
22.3	Climate Change Policy Negotiations	461
22.4	Climate Change Interests	462
22.5	Climate Change Perceptions	464
22.5.1	Climate Change Perceptions and Policies in Germany	465
22.5.2	Climate Change Perceptions in the United Kingdom	466
22.5.3	Climate Change Perceptions in the United States	467
22.5.4	Climate Change Perceptions and Policy Preferences	468
22.6	Integration of Results	469
22.7	Conclusions	470
References		470

Indeed, what is striking to many observers is that whilst 'hard' scientific proof of cause and effect (between given levels of CO_2 emission and measurable changes in climate) is still as elusive as ever, the social and cultural significance attached to the uncertainties involved has escalated rapidly since the mid 1980s. It is this, rather than new scientific evidence, which has helped to propel policy communities in many countries (under the aegis of the UN) towards the possibility of new global agreements on CO_2-reductions.

Grove-White et al. (1991: 6)

R. van der Wurff (✉)
Department of Communication and the Amsterdam School of Communications Research *ASCoR*, University of Amsterdam, The Netherlands
e-mail: R.J.W.vanderWurff@uva.nl

22.1 Introduction

International climate change politics provides a clear example of how cultural differences, conflicts of interest and scientific assessments interact to shape environmental policy-making. This section will explore these interrelationships by analysing the role of the United States, the United Kingdom and Germany in international climate change negotiations. All three countries are important and influential players on the international stage. From the start of the international climate change negotiation process, they have taken very different positions and favoured very different policy options – reflecting country specific approaches to dealing with the complexities and uncertainties involved in climate change. Given the global character of the climate system though, they also needed to reach a minimum of agreement amongst themselves as well as with other countries in order to develop global mitigation and adaptation strategies. This has resulted in a long and arduous process of negotiation, that will be briefly introduced in the following sections.

In the remaining part of this section, we shall argue that the origins of the diverging policy positions of the United States, the United Kingdom and Germany are as much of a cultural as of an economic nature. These origins encompass differences in national perceptions as well as differences in national interests. These differences shape scientifically based yet country-specific understandings of the causes, effects and solutions of climate change. Consequently, they predetermine country-specific preferences for particular solutions. By analysing these cultural and economic origins of country-specific policy preferences, we aim to clarify part of the cultural, scientific and political-economic dynamics that shape environmental policymaking in the modern, globalising yet culturally heterogeneous world.

Jan J. Boersema and Lucas Reijnders (eds.), *Principles of Environmental Sciences*,
© Springer Science+Business Media B.V. 2009

22.2 Climate Change

Climate change basically is a catch-all phrase that refers to possible climatological consequences of increasing greenhouse gas emissions. Greenhouse gases (GHGs) are gases such as carbon dioxide (CO_2), methane (CH_4), nitrous oxide (N_2O) and chlorofluorocarbons (CFCs). They are found in small quantities in the upper layers of the Earth's atmosphere, where they play a vital role in maintaining the Earth's energy balance. They give free way to incoming (short wave) solar radiation, and absorb part of the outgoing (longer wave) radiation that is reflected by Earth. That way, they increase the global average temperature by about 33°C to its present 15°C. Without this 'natural' greenhouse effect, the Earth would simply be too cold to sustain human life.

However, anthropogenic emissions of greenhouse gases have increased dramatically in the last 200 years. These emissions cause a gradual build up of greenhouse gas concentrations in the Earth's atmosphere. This enhances the 'natural' greenhouse effect. Of course, it is difficult to draw the line between a 'natural' and an 'enhanced' greenhouse effect, or between 'natural' (pre-industrial ?) and 'anthropogenic' emissions for that matter. Still, there is widespread concern that greenhouse gas emissions have increased just too much and too quickly, and therefore will cause irreversible climatological changes.

Actual climatological changes that will occur because of the enhanced greenhouse effect can only be predicted in general terms. These include the well-known predictions that the global mean surface temperature will most likely increase between 1°C and 3.5°C by 2100, that the average sea level will increase between 15 and 95 cm in the same period, and that precipitation and wind patterns will change (IPCC 1995). More detailed predictions for specific regions or countries can not (yet) be made. The complexity of the climatological system and the wide natural variability in our climate make such more detailed predictions unfeasible.

Our knowledge decreases even further when we attempt to predict social-economic and health effects of climate change. Then uncertainties about changes in regional climates are reinforced by the unpredictability of (long term) social-economic developments (Glantz 1995). Potentially, climate change will bring about changes in the availability of (clean) water, changes in agricultural productivity, increases in floodings of coastal areas, and a spread of tropical diseases. These consequences will be especially felt by people that live in coastal areas, that are dependent on agriculture, and that have limited resources to protect themselves; in sum, by people in developing countries and small island states. Yet, there is no definite proof that climate change might not have beneficial consequences for some or many of these people in these countries as well.

To prevent the potentially costly social-economic and health effects of climate change, governments can either aim to reduce greenhouse gas emissions (*mitigation* policies) or to prepare for coming changes in climate (*adaptation* policies). However, formulating and implementing these policies is an extremely complicated political activity, due to the long term character, the pervasiveness, the global scope and the imponderabilities of climate change. Firstly, policy makers have to co-ordinate their mitigation and adaptation policies across traditional political geographical and temporal boundaries. Isolated policy measures can have no significant impact. Secondly, the pervasiveness of sources and effects of climate change requires that climate change policies are tuned with many other sectoral policies (including energy policy, agricultural policy, transport policy and international trade policy) as well. Thirdly, (future) costs and benefits of climate change (policies) are as of yet uncertain, due to the uncertainties involved in climate change. Nevertheless, negotiators have to agree now on how present and future burdens of greenhouse gas emission reductions are to be distributed across past, present and future emitters (e.g., on a per capita basis, on economic/efficiency grounds, or on the basis of current emission levels).

In sum, climate change policymaking requires dealing with uncertainties in a policy game that transcends traditional spatial, temporal and issue boundaries and in which the stakes are very high. Numerous interests will be involved and affected in distinct and partly unknown ways. Given these characteristics, it should be no surprise that there exists widespread disagreement on how much and how rapidly emissions should be reduced, and by whom in particular. What is more remarkable is that indeed major attempts have been made to develop international climate change policies. These efforts culminated for the time being in the adoption of the United Nations Framework Convention for Climate Change (UNFCCC) in 1992, in which emission stabilisation targets for the period 1990–2000 for industrialised countries were included; and the adoption of the Kyoto Protocol in 1997, that includes emission reduction targets for the post-2000 period.

22.3 Climate Change Policy Negotiations

The negotiations that led to the adoption of the UNFCCC and, later, the Kyoto Protocol are characterised by contradictions both between industrialised and developing countries as well as within industrialised and within developing countries. Along the North-South dimension, the main issue of contention concerns policy efforts (in terms of emission reductions at home and North-South technology and financial transfers) that should be expected from industrialised and developing countries respectively, given the differences in their 'responsibilities' for (past) greenhouse gas emissions. Within the group of developing countries, the main battle line runs between the vulnerable small island states (organised in AOSIS) that support stringent policies, and other developing countries that are much more hesitant. Within the group of industrialised countries, finally, contradictions exist especially between the United States that favours flexible and cautious policies, and Germany that (in cooperation with some other continental European countries) supports the adoption of relatively stringent targets for industrialised countries in particular. Since industrialised countries still are the major emitters of greenhouse gas emissions, it are especially conflicts within the latter group of countries that (did) shape the outcome of climate change negotiations.

Negotiations in 1991–1992 on the UNFCCC focused on the adoption of greenhouse gas emission reduction targets for industrialised countries for the period until the year 2000. Germany was a proponent of vigorous targets. At home Germany aimed to reduce CO_2 emissions by 25% and expected to reduce emissions of other GHG by 50% in the period 1987–2005. During the negotiations, it supported the inclusion of binding targets in the UNFCCC. The United States, on the other hand, had the least vigorous domestic target resulting in an actual increase of domestic emissions over the 1990–2000 period. Consistent with its domestic approach, the United States opposed the adoption of binding targets during the UNFCCC negotiations and instead advocated its bottom-up, no-regret approach.[1] The position of the United Kingdom fitted neatly in between these two extremes. Its formal domestic target was more in line with US policies; yet its actual achievements are more in line with the German results. During the negotiations, it played an important mediating role between Germany and the United States. The eventual outcome was the adoption of 'quasi commitments', according to which industrialised countries were required to develop national policies 'with the *aim* of returning individually or jointly' GHG emissions to their 1,990 levels by the year 2000 (art. 4(2)(b) of the UNFCCC; italics added).

During these negotiations on the UNFCCC in the early 1990s, the United States appeared to become the 'outcast' of international climate change politics. However, since then it has regained a more central and leading role. During negotiations on emission reduction targets for the period beyond 2000 at several Conferences of the Parties to the UNFCCC, the United States gradually gave up its initial resistance to legally binding targets, but only in exchange for other countries' acceptance of the US preferences for flexibility in policy implementation. Thus, the way was cleared for the inclusion of quantified emission limitation and reduction objectives (QELROs) in the Kyoto Protocol, adopted in 1997. Whereas the UNFCCC commitments were 'only' political commitments to stabilise emissions at 1990 levels by 2000, these QELROs are legally binding commitments that should result in overall reduction of emissions by 5% for the period 2008–2012 compared to 1990 levels (Kyoto Protocol 1997, art 3(1)). However, departing from the UNFCCC quasi commitments, the QELROs vary from country to country,[2] are to be calculated on an aggregated rather than a gas-by-gas basis, include sinks, and may be (partly) realized through joint implementation and emission trading.[3] These latter elements have all been long-standing preferences of the United States.

---

[1] No-regret policies are climate change policies whose costs are (more than) compensated for by their benefits in other areas than climate change, and that thus contribute 'free of cost' to climate change mitigation.

[2] The United States committed itself to reduce emissions by 7%, Japan by 6% and the EU by 8%, whereas Australia was allowed to increase its emissions by 8% (Kyoto Protocol 1997, Annex B). Later the EU decided that in order to reach the EU target jointly, the United Kingdom would have to reduce emissions by 12% and Germany by 21% (Merkus and Ruyssenaars 1998: 12).

[3] Also in contrast with previous discussions, the phrase 'joint implementation' in the Kyoto protocol is reserved for joint implementation between Annex I countries (i.e. industrialized countries and countries in transition). Joint implementation between Annex I and non-Annex I countries is discussed in terms of the Clean Development Mechanism.

The adoption of binding targets in the Kyoto Protocol has been heralded by the environmental community as a victory, and attributed to what is considered a significant shift in the policy position of the United States. Yet, taken actual emission reduction obligations and preferences for policy instruments into account, climate change negotiations rather show a remarkable consistency in policy positions. Both in the beginning and the end of the 1990s, Germany aimed for the most vigorous emission reduction targets, whereas the United States continuously supports a much less vigorous target. Both in the beginning and the end of the 1990s, the United States advocated a flexible and 'efficient' approach, whereas Germany continuously favoured a target-oriented approach and emphasised the responsibility of industrialised countries. Taken this into account, the Kyoto Protocol reflects an increasing dominance of the US point of view, rather than a fundamental shift in the US policy position itself.

Our main question in the remaining part of this section will be how these differences in policy positions, that shaped international climate change policymaking throughout the 1990s, can be explained. Given the uncertainties in the prediction of regional climatological, social-economic and health effects of increasing greenhouse gas emissions, these differences in policy preferences can not be attributed to objective differences in the costs and benefits of climate change (policies) – which would be a relatively standard explanation of the variety of policy positions in international politics. Therefore, we need to look into more subjective perceptions of the risks, costs and benefits involved. Our main argument will be that these subjective perceptions are at least as much determined by country-specific traditions in environmental awareness (in short: culture) as by country-specific constellations of climate change related interests.

22.4 Climate Change Interests

Climate change interests are 'relative stable properties of individuals or collectivities [...] which provide them with actual or potential reasons' to support or oppose climate change policies (Hindess 1993: 293). In an interest-oriented explanation of climate change politics, these interests are considered to determine policy preferences (and even subjective perceptions).

As far as climate change and other environmental issues are concerned, we can distinguish between three categories of interests: polluter interests, victim interests and support interests (Von Prittwitz 1990). Polluter interests provide actors reasons to oppose emission reductions, because of the benefits they derive from the use of fossil fuels and other GHG emitting activities and because of the related costs they will have to bear when mitigation policies are developed. Victim interests provide actors with reasons to support emission reductions, because of the costs of climate change they (will) have to bear. Support interests, thirdly, provide actors with reasons to support emission reductions, too, because of the benefits these actors derive from climate change mitigation measures (e.g., by selling environmental equipment). These costs and benefits of climate change (policies) can be aggregated at the national level, resulting in estimates of 'national' climate change interests; or they can be estimated for major domestic economic actors, whose mutual power relations in turn determine which of their particular interests become the 'dominant' interests in a country.

Costs of climate change mitigation measures may include both declining benefits from GHG emitting activities because of a reduction in the scope of these activities, or they may concern the costs of introducing more energy efficient or less GHG emitting technologies itself. In the former case, no technological change is assumed and costs are considered to be proportional to emission reductions. In the latter case, no decline in benefits of GHG emitting activities is assumed, and costs are assumed to be inversely related to the environmental efficiency of previous technology. After all, the more efficient energy is already used or the less GHGs are already emitted, the more expensive further improvements will be.

When we equate polluter interests to the costs of reducing fossil fuel use or GHG emissions, without any change in technology, then aggregated national climate change interests in Germany and the United Kingdom turn out to be rather alike, and very dissimilar from the US constellation of interests (see left side of Fig. 22.1). The (dis-)similarities between the German, UK and US constellations of interests mainly reflect differences in (per capita) fossil fuel use and GHG emissions that make national polluter interests in the United States stronger than in both European countries; and differences in manufacturing energy efficiency and world market shares of environmental industries that make support interests

Polluter interests equal reduced benefits　　　　*Polluter interests equal technology costs*

```
                                    V
                                    S
                                    P
  Germany   United Kingdom  United States    Germany   United Kingdom  United States
```

Legend: horizontal lines represent interests

V: victim interests　　　　────────── strong interests

S: support interests　　　　────── medium interests

P: polluter interests　　　　─── weak interests

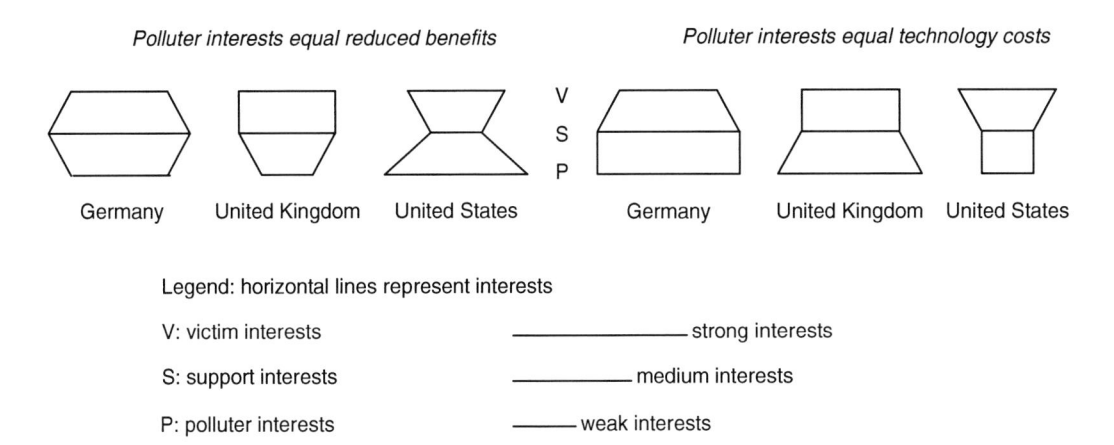

Fig. 22.1 National climate change interest profiles
Source: Adapted from Von Prittwitz (1990: 124)

stronger in Germany. Victim interests are assumed to be similar in any case, because there is no sufficient evidence that effects of climate change will be more costly in one compared to the other two countries.

The resulting US constellation of interest reflects a juxtaposition of polluter and victim interests, suggesting a political stale-mate in climate change politics. The relative strength of support interests in both European countries suggest a policy environment that is more favourable to climate change policymaking. The relative weakness of polluter interests in the United Kingdom, moreover, suggests that the UK constellation of interests is more conducive to climate change policymaking than either the German or US constellations. Of course, this outcome is not consistent with actual policy positions, in which Germany developed by far the most vigorous policies.

When polluter interests are interpreted as reflecting the costs of introducing more energy efficiency and less GHG emitting technologies, polluter interests in Germany and the United Kingdom are largest (see right part of Fig. 22.1). Since economic production in these countries is already less energy intensive and less polluting in terms of GHG emissions, it can be expected that further reductions in energy use and GHG emissions will be more costly than in the United States. In that case, most vigorous climate change policies could be expected to occur in the United States. Again, this outcome is not consistent with actual policies. National climate change interests therefore do not offer an adequate explanation for the observed differences in climate change policy positions of these three countries.

Constellations of *dominant* climate change interests in the three countries are much more alike than constellations of *national* interests. These dominant profiles moreover suggest that most vigorous policies could be expected in the United Kingdom, and least vigorous policies in Germany (see Fig. 22.2). These expectations mainly reflect differences in the size of (energy-intensive) industries with major polluter interests in the three countries. These industries are larger (in terms of their contribution to GDP and employment) in Germany, followed by the United States and then the United Kingdom. The resulting ordering of dominant profiles does clearly not correspond with the observed differences in national climate change policy positions either.

Still, the analysis of dominant interests does suggest a partial explanation for the observed differences in climate change policy positions. US industries are more energy-intensive than their international competitors. Although the international market is not so important for these industries, they still would loose when global emission reduction measures would be adopted. German industries, on the other hand, are on average less energy-intensive compared to their international competitors. They would therefore gain from the global adoption of emission reduction measures. This would explain why German industries would support the adoption of vigorous *international* policy targets. Yet, it would still not explain why German industries would support the adoption of vigorous policy targets at home – which is exactly what the German government did.

Fig. 22.2 Dominant climate change interest profiles
Source: Adapted from Von Prittwitz (1990: 124)

Germany   United Kingdom   United States

Legend: horizontal lines represent interests

V: victim interests ——————— strong interests

S: support interests ————— medium interests

P: polluter interests ——— weak interests

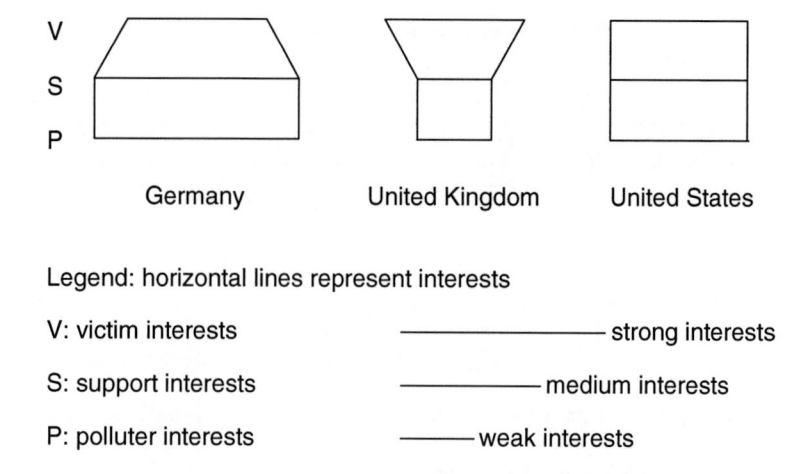

In addition to these assessments of current interests, we also assessed changes in interests over time. Reflecting the twofold interpretation of polluter interests that was discussed above, changes in interests over time can be assumed to influence policy positions in two ways: countries that in the past experienced relatively rapid emission reductions can be predicted to either support vigorous climate change policies because they would expect to experience a continued process of emission reductions in the future; or to oppose vigorous climate change policies, because these countries expect to experience slower reductions in the future, due to the law of diminishing returns. In any case, an analysis of changes in energy supply and economic production in the last 25 years does not support either interpretation.

Since the early 1970s, per capita GHG emissions decreased most rapidly in the United Kingdom. Likewise, per capita energy supply increased most slowly in this country compared with Germany and the United States. In addition, also emissions and energy supply per unit of GDP decreased somewhat more quickly in the United Kingdom than in Germany and considerably more quickly than in the United States. Underlying causes of these diverging developments include diverging trends in energy production (efficiency of energy generation, replacement of fossil fuels by non-fossil fuels), resource consumption (energy efficiency of manufacturing, resource-intensity of agriculture), and more general social-economic developments (changes in the branch structure of the domestic economy, and in shifts between different transportation modes). Essentially, these changes amount in Germany to an 'ecological modernisation' of manufacturing; i.e. a relative de-linking of energy consumption and GHG emissions from economic production through shifts in the branch structure from traditional energy-intensive to 'modern', less energy-intensive industries. In the United Kingdom, on the other hand, environmental stresses are reduced by a process of de-industrialisation, i.e. a de-linking of energy consumption and GHG emissions from economic production through a decline of manufacturing industries as such. In the United States, thirdly, both de-industrialisation and 'ecological modernisation' of manufacturing can be observed, though at a much slower speed than in the two European countries.

These estimates suggest that either the United Kingdom or the United States would be the most strongest supporter of climate change policies (when support would be either proportionally or inversely related to previous emission reductions achieved). Both outcomes are not in line with the observed climate change policy positions of these three countries.

In sum, we have to conclude that climate change interests do not offer an adequate explanation for the differences in policy positions. On the contrary, most of our assessments of climate change interests suggest that most vigorous climate change policies should be expected in the United Kingdom. We therefore have to look for an alternative explanation.

22.5 Climate Change Perceptions

World views are 'stable patterns in the observation and interpretation of the world (Jachtenfuchs 1993: 10; my translation).' They encompass assumptions about the basic characteristics of reality (cognitive aspect), a system of ethical values (normative aspect), and aesthetic criteria (symbolic aspect) (Jachtenfuchs 1993: 11–12; see also Buiks and Kwant 1981: 31–33). These world views can be considered to enable and structure rather than distort or bias perceptions of 'reality'; a reality that otherwise would just be made up of an incomprehensible multitude of sensations (see Koningsveld 1976: 126–153). From this perspective, world views rather than interests occur as the main determinants of subjective perceptions and policy preferences.

World views can be reconstructed along various dimensions. We can emphasise stable patterns in the perceptions of environmentalists as opposed to industrialists. We can emphasise stable patterns in the perceptions of left wing as opposed to right wing politicians. And we can emphasise country-specific patterns in climate change perceptions. In the next paragraphs we will argue that regardless of differences in climate change perceptions that can be observed within countries, indeed remarkable country-specific traits in climate change perceptions exist that characterise perceptions of both environmentalists and industrialists, and both left and right wing politicians. We will argue that these country-specific perceptions offer a better explanation for the differences in climate change policy positions than the interests that were investigated in the previous paragraphs.

22.5.1 Climate Change Perceptions and Policies in Germany

Our research[4] shows that in Germany climate change is generally perceived as a central issue, as a serious threat and a serious challenge. This view of climate change reflects several traditions in German culture, and is expressed in Germany's vigorous climate change policy position.

Various observers have argued that Germans in general take issues (or even life) seriously (Hampden-Turner and Trompenaars 1993: 207–236, 251). Perhaps related, Germans on average are rather aversive of taking risks. Other general characteristics identified in German culture include a focus on targets, a holistic approach and an orientation on technology (how to change) rather than science (how to explain) (see Boehmer-Christiansen 1992). These characteristics give rise to largely 'ideological' debates on climate change, in which a major gap exists between the thoroughness of the analysis and the inadequacy of concrete measures.

Especially the holistic approach and risk-aversiveness observable in German culture are reflected in the conceptualisation of nature that dominates German debates. On average, Germans consider nature a global and indivisible unity (Boehmer-Christiansen 1988). This unity is rather unstable. Nature should thus be left alone. German forests are considered the archetypical example of this nature.

Reflecting these general characteristics, environmental issues, and especially global environmental issues, are taken more seriously in Germany than in the United Kingdom or in the United States. It is therefore not surprising that Germans indicate in relatively large numbers to favour the environment above economic development; or that relatively many Germans report to behave themselves environment-friendly (e.g., by sorting waste). More striking is the finding that the 'willingness to pay' for environmental problems diminishes rapidly when concrete financial trade-offs are proposed. As far as solutions are concerned, Germans tend to expect much more from science and technology in the future – although at the same time *current* trends in science and technology are evaluated more negatively – than do the British and the Americans. (See Table 22.1.)

Against this cultural background, climate change is perceived as the 'ultimate' environmental issue, overarching and integrating a range of other environmental and non-environmental issues. Climate change raises 'fundamental' questions and requires global, long-term and integrated solutions. It presents an urgent and serious threat, but also a major challenge that must be

---

[4] The reconstruction of country-specific climate change perceptions is based upon approximately 20 interviews per country with representatives of actors that are involved in the domestic or international climate change policymaking process (government, political parties, industry, environmental organizations, experts), a secondary analysis of international surveys on environmental awareness that were held in these countries in the early 1980s and 1990s, and several more qualitative and historical studies of environmental awareness and policymaking in these countries.

Table 22.1 Climate change and environmental perceptions

Percentages of respondents…[a]	G	UK	USA
…considering that 'nature is sacred because it is created by God' (early 1990s)	18	24	41
…considering that 'nature is spiritual or sacred in itself' (early 1990s)	27	14	23
…considering it definitely/probably true that 'nature would be at peace and in harmony if only human beings would leave it alone' (early 1990s)	75	61	50
…agreeing that 'nature's balance is very delicate and easily upset' (early 1980s)	92	77	71

Percentages of respondents…[b]	G	UK	USA
…rating environmental problems 'very serious in our nation' (early 1990s)	67	36	51
…considering themselves 'a great deal or fair amount' concerned about environmental problems (early 1990s)	64	81	85
…considering the environment to be 'very or fairly bad' (early 1990s),			
–in their community	22	27	28
–in their nation	42	36	46
–and in the world	85	76	66
…considering global warming 'very serious' (early 1990s)	73	62	47

Percentages of respondents…[c]	G	UK	USA
…giving priority to environmental protection, 'even at a risk of slowing down economic growth' (early 1990s)	73	56	59
…preferring a society that limits instead of emphasizes economic growth (early 1980s)	42	25	18
…saying to be 'willing to pay higher prices so that industry could better protect the environment' (early 1990s)	59	70	65
…saying to be 'willing to pay more taxes (more than 250 DM/UR./US$ per year) to protect the environment' (early 1980s)	27	81	20
…saying to 'avoid using certain products that harm the environment' (early 1990s)	81	75	57
…always/often to make a special effort to sort materials for recycling (early 1990s)	83	44	60
…(strongly) agreeing that 'modern science will solve our environmental problems with little change to our way of life' (early 1990s)	34	20	20
…considering 'greater S&T development' more necessary to solve environmental problems than 'basic change in nature of society' (early 1980s)	48	25	31

[a] International Social Science Program (1995), items v18 and v23; Kessel and Tischler (1984), item v116.
[b] Dunlap et al. (1992), Tables 1, 2 and 6.
[c] Dunlap et al. (1992), Tables 14 and 16; International Social Science Program (1995), items V12 and V56; Kessel and Tischler (1984), items V201, 423 and V449.

dealt with immediately by far-reaching structural policies and technological innovation.

This 'principled' and technology-oriented view of climate change conforms to both the domestic and international climate change policies of the German government. The perceived seriousness of the issue, the observed faith in technological developments and the belief that economic and environmental considerations are not diametrically opposed give rise to and allow the adoption of relative vigorous targets (that then in practice may turn out to be less achievable). Additionally, these climate change perceptions results in strong support for international regulation; attempts to formulate policies that address climate change at relative structural levels of analysis; a belief that policies should be implemented as soon as possible; and a focus on North-North cooperation based on the principle that industrialized countries are primarily responsible for climate change.

22.5.2 Climate Change Perceptions in the United Kingdom

In the United Kingdom, climate change is much more perceived as a peripheral issue, not really meriting attention on its own. As a consequence, climate change policies mainly build upon side-effects of policies that aim to realize other objectives, including the deregulation of the energy sector. This approach to climate change is compatible with general trends in UK culture. However, correspondences between climate change policies, climate change perceptions and environmental traditions in the United Kingdom are less straightforward than in Germany.

UK culture is described by observers as 'analytical', 'individualistic' and expressing a preoccupation with oppositions and abstract debates. Scientific explanations are put before technological change, and money as abstract value is put before industrial activity (Boehmer-Christiansen 1992; Hampden-Turner and Trompenaars 1993). These characteristics are reflected in a relatively atomised perception of environmental issues, in which a clear distinction is made between emissions of pollutants and pollution as such; local environmental issues are emphasised; and 'the country-side' is perceived as a place to escape industrial society (Boehmer-Christiansen 1988; Short 1991: 75). The individualistic orientation, expressed in values attached to individual judgement and individual freedom, paradoxically seems to lead to feelings of helplessness vis-à-vis environmental pollution and to a remarkably strong willingness to pay higher prices and taxes for environmental protection. This latter finding suggests that British consider environmental and economic factors to be at right angles with each other. (See again Table 22.1.)

Because of the focus on local environmental issues and the country-side, climate change is not at the centre of environmental awareness in the United Kingdom. Moreover, the scientific and analytical tendencies in UK culture lead to an emphasis on uncertainties that legitimate a 'wait-and-see' approach rather than a focus on risks that would call for precautionary action. Related, costs of climate change policies are emphasised and 'industrial competitiveness' is identified as major concern that should be properly reflected in climate change policymaking. Likewise, international aspects of climate change are perceived through a 'realist' lens. Whereas in Germany perceived global characteristics of climate change and the necessity of global co-operation are emphasised, in the United Kingdom international aspects are mainly discussed in terms of 'getting other countries to do as much as possible'.

These characteristics are reflected in the cautiousness of UK domestic climate change policies, that build upon beneficial side-effects of other policies (liberalisation of the energy sector, fiscal measures, impact of European agricultural policies); as well as in the diplomatic, mediating role played by the UK during international climate change negotiations. Fortunately for the United Kingdom, emission reductions realised through other policies are large enough to allow the United Kingdom to comply with its internationally agreed commitments. That way, and by deliberately taking position between continental European countries and the United States, the United Kingdom could outflank foreign pressures, especially from 'Brussels', for more vigorous targets that – in the UK perspective – would have burdened UK industry with too excessive costs, given the perceived threat of climate change.

22.5.3 Climate Change Perceptions in the United States

Climate change in the United States is clearly perceived from a pragmatic and political perspective. It is not considered one of the most important environmental issues, but rather serves as battleground for a range of actors that want to realise other objectives. The need to balance costs against benefits and to maintain flexibility in policymaking are recurring themes in domestic and international climate change policy positions – two spheres that are not clearly distinguished in the US debate.

US culture is characterised by observers as 'extremely analytical', 'strongly individualist' and 'inner-directed'. These characteristics are assumed to underlie strong biases in US culture to focus on 'facts and figures', to 'get down to basics' and to consider 'self-interest' a basic social category (Hampden-Turner and Trompenaars 1993: 20–72; compare Rayner 1993: 26). Partly confirming these tendencies, nature is conceptualised in clear anthropocentric and religious terms. This conceptualisation of nature gives rise to a belief that nature is relatively stable and that human interference with nature is not too detrimental. Consistently, science and technology are, in general terms, evaluated relatively positively. Moreover, risks are considered to be an inevitable corollary of human life. Yet, on the other hand, nature is highly valued and considered to be important, too. These tendencies are reflected in a long-term preoccupation with 'wilderness' in US environmental awareness and politics.

These characteristics give rise to environmental concerns that are considered to be strong by US respondents themselves, but that in comparison with environmental concerns in Germany and the United Kingdom turn out to be relatively weak – especially as far as concern about global environmental issues is concerned. On the other hand, US citizens still indicate to be relatively willing to pay higher prices and taxes for environmental protection and to engage relatively frequently in environment friendly behaviour (sorting waste and the like). (See once more Table 22.1.)

Reflecting this cultural background, US climate change perceptions can best be described as pragmatic, politicised, and trans-national. The pragmatic bias is expressed in a preference for a 'hands-on approach', a focus on commercial opportunities and cost-effectiveness. The political bias is reflected in a tendency to consider climate change essentially an issue at stake in a political-economic struggle between interest groups, between Congress and the Administration, between isolationist and international-oriented forces, and between 'regulationalists' and 'free marketeers'. Uncertainties in climate change science are, in this respect, primarily interpreted in political terms, as 'assets' that are manipulated by actors; and US self-interest and self-interest of US companies are relatively openly discussed and advocated. Thirdly, domestic and international aspects of climate change are not clearly distinguished from each other, and the preferred role of the United States in global climate change politics ('leadership', or 'domestic issues first') is a recurring issue in the climate change debate.

These climate change perceptions underlie the emphasis in US domestic and international policies on bottom-up approaches and cost-effectiveness. Maintaining flexibility (by not adopting 'inflexible' binding targets) and realising emission reductions at the lowest costs possible (by allowing the reduction of emissions in developing countries instead of industrialised countries, and by using market instruments) have and still are key issues in the US input in international negotiations. However, the limited vigorousness of US climate change policies can not be directly related to US climate change perceptions. Of course, the US focus on quantifiable costs and benefits of environmental policies suggest that US climate change policy targets will not be too vigorous, because especially current quantifiable benefits are still negligible. Still, US climate change perceptions do not suggest *a priori* that US policy targets will be less vigorous than UK policy targets.

22.5.4 Climate Change Perceptions and Policy Preferences

The analysis of climate change perceptions results in the reconstruction of remarkably different and remarkably stable country-specific world views, that have been described above. These reconstructed country-specific world views are well compatible with the observed differences in national climate change policy positions (see Table 22.2). Especially the German case represents striking parallels in this respect. From general orientations observed in German culture to specific interpretations of climate change, a principled approach and a focus on global environmental issues are predominant. These characteristics are reflected in the vigorous domestic and international policy position of the German government.

Also in the UK case, clear correspondences arise between the perceived contradiction of environmental and economic factors, the emphasis on local environmental issues, and a general analytical bias, on the one hand, and the consideration of climate change as a not too threatening and peripheral issue, on the other hand. These characteristics give rise to relatively little domestic support for climate change policies. Still, climate change perceptions alone can not explain why the United Kingdom played a mediating role in global climate change negotiations. Here, international pressure (especially in the framework of the EU) and the 'free' emission reductions achieved must be integrated in the analysis.

Connections between the US climate change perceptions, trends in environmental awareness, and US climate change policies, thirdly, are least equivocal. On the one hand, US climate change policies and politics reflect the pragmatic and politicised tendencies

Table 22.2 Worldviews and policy positions

	Germany	United Kingdom	United States
Basic cultural characteristics	*Zielrationalismus*; rule-making; 'deep' thinking; technological orientation; aversion of risk	Individualistic and analytic; money instead of industry; principle against principle	Universalist-analytical; codification, basics and facts; focus on present; international ánd isolationalist outlook
...related to specific... conceptualization of nature	nature as a global unity; unstable; *Wald*	nature as local, divisible entity; environment as peripheral issue; *countryside*	anthropocentric orientation (utility and aesthetics); focus on species and natural parks; *wilderness*
...that underlies... national type of environmental awareness	'principled'; threat ánd challenge; global orientation; technological change	scientific-sceptical; local nature; willingness to pay; individual freedom	religious-anthropocentric; political; S&T 'as usual'
...that in turn presents... Climate change as....	global; ecological challenge; overarching	cost; uncertain; energy issue	political issue; not too threatening
...which results in the following... policy position	targets; structural approach; technological innovation	meeting international commitments; energy market liberalisation policies	soft 'targets'; joint Implementation; voluntary approach; cost-benefit analysis

that fit so well in US culture. Yet, on the basis of climate perceptions alone, there seems to be no adequate explanation why US policy objectives are less vigorous compared to UK objectives.

22.6 Integration of Results

Finally, we will therefore investigate whether interests and climate change views combined offer a better explanation.

As far as Germany is concerned, the analysis in terms of interests and in terms of perceptions reinforce each other. It was concluded that in Germany strong (support) interests with respect to international climate change policy exist. These strong support interests reflect a relatively high degree of energy efficiency of German manufacturing industries compared to UK and US manufacturing industries. These support interests express and reinforce strong environmental concerns in Germany, including German perceptions that vigorous climate change policies are needed. In that respect, climate change interests and climate change perceptions do go hand-in-hand in Germany.

Secondly, it was concluded that although climate change is considered to be not a very serious issue in the United Kingdom, the 'free' GHG emission reductions achieved through other policies allow the UK government to participate fully in international negotiations and to comply with international agreed climate change policy commitments. This assessment builds upon the consideration that the United Kingdom is only a moderate player at the international stage that cannot influence negotiations decisively. The analysis of changes in climate change interests over time, moreover, showed that changes in energy consumption and production patterns in the United Kingdom were indeed most conducive to climate change policymaking. The analysis of interests thus explains why UK climate change policies were more vigorous than could be expected on the basis of domestic climate change perceptions alone. Therefore, also in the UK case, interests and climate change views complement each other, and together offer a better explanation of UK climate change politics.

Finally, interests seem to tilt the balance in exactly the other direction in the United States. Also in this country it is believed that climate change is not too serious a threat. In combination with other cultural characteristics this leads to a perception of climate change as primarily a political issue, an issue of contention between actors that aim to use climate change to further their self-interests. Because of a focus on nature conservation and on quantifiable costs and benefits, costs of climate change policies will tend to be considered larger than benefits. In combination with a stronger position in international negotiations, and changes in emission ratios and energy consumption and production patterns that result in smaller 'free' GHG emission reduction opportunities than in the United Kingdom, it can be expected that support for vigorous domestic and/or international climate change policy targets will be smaller in the United States than either in Germany or the United Kingdom.

This analysis of climate change policies in terms of interests and perceptions combined can be summarised in terms of integrated 'subjective' and 'objective' interests. Country-specific constellations of these integrated interests can again be expressed in climate change policy profiles (see Fig. 22.3).

In Germany, the perception of climate change as a serious threat gives rise to large victim interests.

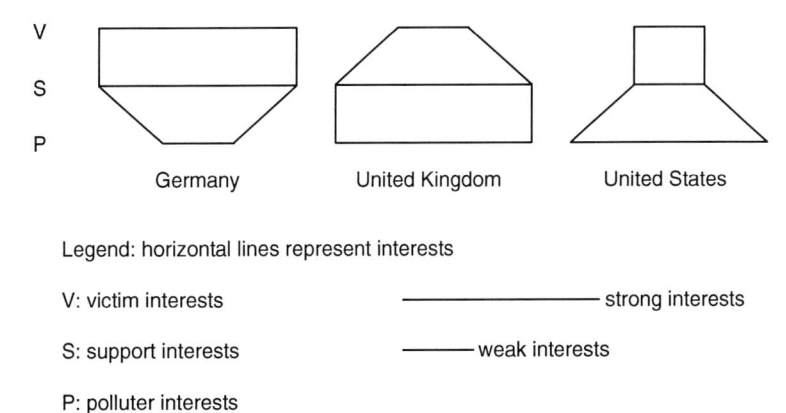

Fig. 22.3 Integrated climate change interest profiles
Source: Adapted from Von Prittwitz (1990: 124)

Legend: horizontal lines represent interests

V: victim interests
S: support interests
P: polluter interests

——————— strong interests
——— weak interests

Because of the faith in technological progress and energy efficiency of German manufacturing, support interests are large, too. Polluter interests, however, are smaller than in the other two countries, because environmental and economic concerns are not perceived to be diametrically opposed. In the United Kingdom and the United States, on the other hand, climate change views give rise to smaller victim interests and larger polluter interests than in Germany. Yet, as argued above, subjective support interests differ considerably in these two countries, resulting in very different climate change interest profiles. These subjective climate change interest profiles match very well with the observed differences in climate change policy positions. They are also consistent with the observation that Germany and the United States take relative extreme positions in international climate change policymaking, whereas the United Kingdom takes a more intermediate position.

22.7 Conclusions

The analysis of climate change policy positions of three major players showed that these positions can only be explained against the background of diverging country-specific assessments of the urgency, seriousness and other ecological attributes of climate change. Given the uncertainties involved in climate change policymaking, these different climate change perceptions must be considered equally valid from an environmental-scientific point of view. Moreover, on the basis of surveys and other studies, these climate change perceptions could be traced back to country-specific traditions in environmental awareness that predate climate change policymaking. Consequently, these climate change perceptions must be considered 'genuine' interpretations of climate change that shape domestic politics and international policy positions in country-specific ways. International climate change politics can only be explained when we take these country specific perceptions into account.

The underlying patterns in environmental awareness that shape country-specific climate change perceptions could be traced back to the early 1980s in quantitative terms (using existing surveys) and up to the beginning of this century in qualitative terms. This suggests that 'national' cultures are relatively robust in the face of increasing trans-national (media) exchanges and environmental interdependencies. It can therefore be expected that 'nations' will continue to play a major role in international negotiations, though increasingly more as cultural entities rather than as state actors.

References

Boehmer-Christiansen, S. (1988). Pollution control or Umweltschutz. *European Environment Review, 2*(1), 6–10.

Boehmer-Christiansen, S. (1992). *Environmental threat perception: Scientific or cultural? Anglo-German contrasts in environmental policy-making for acid rain abatement and the role of environmental threat perception*. Paper prepared at ISA symposium 'Current Developments in Environmental Sociology', The Netherlands.

Buiks, P. E. J., & Kwant, R. C. (1981). *Cultuursociologie; Een perspectief op cultuurvorming, cultuurbeweging en cultuurbeleid*. Alphen a/d Rijn: Samsom Uitgeverij. (In Dutch)

Dunlap, R. E., Gallup, G. H. Jr., & Gallup, A. M. (1992). *The health of the planet survey. A preliminary report on attitudes toward the environment and economic growth measured by surveys of citizens in 22 nations to date*. (Updated June 1992). New Jersey: George H. Gallup International Institute.

Glantz, M. H. (1995). Assessing the impacts of climate: The issue of winners and losers in a global climate change context. In S. Zwerver, R. S. A. R. Van Rompaey, M. T. J. Kok, & M. M. Berk (Eds.), *Climate change research. Evaluation and policy implications* (pp. 41–54). Amsterdam: Elsevier.

Grove-White, R., Kapitza, S., & Shiva, V. (1991). *Public awareness, science and the environment*. Paper for international conference on an Agenda of Science for Environment and Development into the 21st Century, Austria, November 1991.

Hampden-Turner, C., & Trompenaars, F. (1993). *The seven cultures of capitalism. Value systems for creating wealth in the United States, Britain, Japan, Germany, France, Sweden, and the Netherlands*. London: Piatkus.

Hindess, B. (1993). Interests. In W. Outhwaite & T. Bottomore (Eds.), *The Blackwell dictionary of twentieth-century social thought* (pp. 293–294). Oxford: Blackwell.

International Social Science Program. (1995). *Machine readable codebook. ZA study 2450. ISSP 1993. Environment*. Cologne: Zentralarchiv für empirische Sozialforschung.

IPCC (1995). *IPCC second assessment synthesis of scientific-technical information relevant to interpreting article 2 of the UN Framework Convention on Climate Change 1995*. Geneva: IPCC.

Jachtenfuchs, M. (1993). *Ideen und Interessen: Weltbilder als Kategorien der politischen Analyse*. Working Paper AB III/2. Mannheim Centre for European Social Research, Universität Mannheim (in German).

Kessel, H., & Tischler, W. (1984). *Umweltbewußtsein. Ökologische Wertvorstellungen in westlichen Industrienationen*. Berlin (in German): WZB/Sigma.

Koningsveld, H. (1976). *Het verschijnsel wetenschap; Een inleiding tot de wetenschapsfilosofie*. Amsterdam/Meppel (in Dutch): Boom.

Kyoto Protocol to the United Nations Framework Convention on Climate Change. (1997). http://unfccc.int/kyoto_protocol/items/2830.php

Merkus, H., & Ruyssenaars, P. (1998). Drafting of Dutch post-Kyoto climate policy starts in NEPP-3. *Change, 43*, 11–13.

Rayner, S. (1993). Prospects for CO_2 emissions reductions policy in the USA. *Global Environmental Change, 3*, 12–31.

Short, J. R. (1991). *Imagined country. Society, culture and environment*. London: Routledge.

United Nations Framework Convention for Climate Change (UNFCCC). (1992). http://unfccc.int/2860.php

Von Prittwitz, V. (1990). *Das katastrophen-paradox. Elemente einer theorie der umweltpolitik*. Opladen: Leske und Budrich.

Further Readings

On the Negotiations of the UNFCCC

Bodansky, D. (1993). The United Nations framework convention on climate change: A commentary. *The Yale Journal of International Law, 18*, 451–558.

Oberthür, S. (1993). *Politik im Treibhaus. Die Entstehung des internationalen Klimaschutzregimes*. Berlin: Sigma. (In German)

On the Ongoing International Climate Change Negotiations

Earth Negotiations Bulletin, volume 12. International Institute for Sustainable Development.

On Climate Change Politics in Germany, the United Kingdom and the United States

Boehmer-Christiansen, S. (1995a). Britain and the international panel on climate change: The impacts of scientific advice on global warming part I: Integrated policy analysis and the global dimension. *Environmental Politics, 4*, 1–18.

Boehmer-Christiansen, S. (1995b). Britain and the international panel on climate change: The impacts of scientific advice on global warming part II: The domestic story of the British response to climate change. *Environmental Politics, 4*, 175–196.

Hatch, M. T. (1993). Domestic politics and international negotiations: The politics of global warming in the United States. *Journal of Environment and Development, 2*(2), 1–39.

Hatch, M. T. (1995). The politics of global warming in Germany. *Environmental Politics, 4*, 415–440.

O'Riordan, T., & Jäger, J. (Eds.). (1996). *Politics of climate change. A European perspective*. London: Routledge.

On the Role of Ideas and Interests in Environmental Politics

Haas, P. M. (1989). Do regimes matter? Epistemic communities and Mediterranean pollution control. *International Organization, 43*, 377–403.

Schwarz, M., & Thompson, M. (1990). *Divided we stand. Redefining politics, technology and social choice*. New York: Harvester Wheatsheaf.

Sprinz, D., & Vaahtoranta, T. (1994). The interest-based explanation of international environmental policy. *International Organization, 48*, 77–105.

Weale, A. (1992). *The new politics of pollution*. Manchester: Manchester University Press.

Chapter 23
Technical Progress, Finite Resources and Intergenerational Justice[1]

Wilfred Beckerman

Contents

23.1 Introduction 473
23.2 The Illusion of Intergenerational Justice............. 473
23.3 How Much Richer Are Future Generations Likely To Be?.. 476
23.4 The Resources Constraint..................................... 478
23.5 Our Real Obligations to Future Generations...... 480
References 481

23.1 Introduction

The structure of my argument is as follows. First, I shall try to show that the widespread belief that society should adopt some policy of 'sustainable development' on account of considerations of justice or equity between generations and the 'rights' of future generations is false. I am not concerned here with criticisms of the concept of 'sustainable development' as such, and I have published my objections to this concept elsewhere (Beckerman 1995). In this chapter I try to show that its underlying philosophical premises concerning the rights of future generations are mistaken.

The fact that future generations may not have any rights does not mean, however, that we have no moral obligations to take account of the interests that they will have. Our problem then is to try to predict what their most important interests will be. The chapter goes on to argue that economic growth is likely to continue at a pace that will mean that future generations will be vastly richer than people are today. Furthermore, this growth will not be impeded by any resource constraints. By contrast with the improvement that this will bring in the relief of major problems, such as widespread poverty and environmental pollution, there will be no improvement in the various sources of suffering imposed on the vast majority of the world's population on account of violation of basic human rights or fear of such violation. The most important bequest we can make to future generations, therefore, is to bequeath to them a more decent society than the one in which most people live today, namely a society characterised by tolerance and respect for basic human rights.

23.2 The Illusion of Intergenerational Justice

One of the results of the growing concern during the last three decades with the impact of economic growth on the environment has been the growing popularity of the concept of 'sustainable development'. There are numerous interpretations of this concept and various ways in which the different versions of the concept have been defended. But a very common feature of most arguments in favour of sustainable development and one that is frequently encountered in environmental discourse is some appeal to notions of justice between generations and the corresponding need to respect the 'rights' of future generations to inherit a

W. Beckerman (✉)
Balliol College, Oxford, UK
e-mail: wildfred.beckerman@econs.ox.uk

[1] This chapter is reproduced with slight adaptations from: W. Beckerman (2001). Technical progress, finite resources and intergenerational justice (Chapter 7). In C. Ekko, van Ierland, J. van der Straaten, & H. Vollebergh (Eds.), *Economic growth and the valuation of the environment: A debate*. Aldershot, UK: Edward Elgar (with permission).

Jan J. Boersema and Lucas Reijnders (eds.), *Principles of Environmental Sciences*,
© Springer Science + Business Media B.V. 2009

certain state of the environment or to receive adequate compensation in some form or other. It is no exaggeration to say that the moral claim of 'sustainable development' is usually alleged to rest largely on its appeal to intergenerational 'equity'. Concern with equity is, in fact, often contrasted with the standard economist's concern with simply maximising the future stream of utility over some relevant time period. For example, one of the leading authorities on 'green' political philosophy, Robert Goodin, writes that the objective of sustainable development 'contrasts with the directive of ordinary expected-utility maximisation to go for the highest total payoff without regard to its distribution interpersonally or intertemporally' (Goodin 1983). Or, as an authority on the concept of sustainability, bluntly puts it, 'Sustainability is primarily an issue of intergenerational equity' (Norgaard 1992).

Whereas the problem of distributive justice within any given society at any point in time has occupied philosophers for over 2,000 years, its extension to intergenerational justice is relatively very new. But it is already well known that any attempt to construct a theory of justice between generations encounters special difficulties, such as those set out by Rawls, who believed that the problem of justice between generations subjected ethical theory to 'severe if not impossible tests' (Rawls 1972).

One might well ask, at the outset, therefore, as does Brian Barry (1978), whether there is really any need for a theory of intergenerational justice and whether, instead, one could not be satisfied with defining our obligations towards future generations on the basis of common humanity. Barry's view is that in spite of the difficulties such a theory is possible and also that it is necessary. The argument of this chapter is that it may well be impossible and is certainly unnecessary.

The structure of the argument is, first, the major premise that all common theories of justice imply conferring 'rights' on people that give a special status to certain of their interests. The minor premise is that unborn people cannot be said to have any rights. The conclusion is then that the interests of unborn people cannot be protected or promoted within the framework of any common theory of justice. The first of these propositions is the generally accepted (but not unanimous) view of the scope of theories of justice. The second premise is also not new and may be thought to be non-controversial, or even obvious, although some reputable philosophers dissent from it.[2] But taken together the two propositions lead to a conclusion that is less obvious.

As regards the major premise, there are, of course, many different theories of justice. But there is a wide consensus that any theory of justice necessarily implies attributing rights (and hence counterpart duties) to somebody or some group. For example, if it is claimed that people ought to be rewarded in accordance with their need or their merit, once their need or merit has been established they have a right to the corresponding reward. Anybody who claims that some theory of justice does not imply the attribution of rights needs to formulate the theory in such a way that it cannot be transposed in this manner into a proposition about rights.

As regards the minor premise, namely that future generations cannot be said to have any rights, it must be stated at the outset that this chapter is not concerned with overlapping generations. I am abstracting from the rights of small children to be looked after properly by their parents, for example.

Whether or not such categories of people have rights or whether their treatment is to be governed by principles of morality that lay outside the domain of justice is a difficult and debatable issue. But the distinguishing feature of such cases is the *physical* impossibility of the people in question to exercise their rights or to make claims that correspond to their rights. Here, however, I am concerned with the rights of unborn people. In their case it is *logically* impossible for them to exercise their rights or to make claims on behalf of their rights or to delegate those rights to anybody else.

The main reason why future generations cannot be said to have any rights is simply that – as pointed out many years ago by De George – this is implied by the present tense of the verb 'to have' (De George 1981).

---

[2] For example, 'Part Three' of Partridge (Ed.). (1981) is entitled 'Can Future Generations be Said to Have Rights?' and the contributions by three philosophers, Feinberg, Pletcher and Baier, to this volume maintain that they can. Kavka, in the same volume, is somewhat ambivalent on this question since although he nowhere claims that future generations have 'rights' he does claim that such of their interests as can be predicted with certainty have equal status with the interests of people alive today. In the same volume Warren is also somewhat ambivalent when she writes 'To say that merely potential people are not the sort of things which can possibly have moral rights is by no means to imply that we have no obligation toward people of future generations, *or that they (will) have no rights that can be violated by things which we do now*' (our italics) (Warren, in Partridge 1981). (See also Beckerman and Pasek 2001).

Unborn people cannot now be said to have anything, whether it is long hair, a taste for Mozart, or 'rights'. It makes no sense to attribute a property to a non-existent entity. To say that 'X has Y' or 'X is Z' when there is no X is just nonsense.

Similarly, one must also reject the notion that future generations have rights because they have interests. For even if the possession of certain vital interests are believed to carry with it the possession of corresponding rights, this still cannot apply to unborn people since unborn people cannot have interests either, or anything else for that matter. I presume that they *will* have certain interests, notably those that are common to virtually all human beings, such as an interest in material well-being, freedom from fear and persecution, self-respect and immunity from institutional humiliation, and other basic human rights enumerated in various international conventions. And I believe that this imposes certain moral obligations on us. But that is another matter, to which I shall return later on.

The notion that future generations can have rights is presumably based on the fallacy of assuming that unborn people constitute some class of people waiting in the wings to be born. But there is no such class of people. As Hillel Steiner puts it: 'In short, it seems mistaken to think of future persons as being already out there, anxiously awaiting either victimisation by our self-indulgent prodigality or salvation through present self-denial' (Steiner 1983).

This is not simply a linguistic – if decisive – point. It has substantive aspects. For however widely society wishes to draw the boundary around the rights that future generations *will* have, it cannot encompass rights to something that no longer exists, such as an extinct species. For example, it is surely absurd to say that we have a right to see live dodos. And it would be equally absurd to say that we ever *had* the right to see live dodos.

It may well be true that if somebody steals my property and burns it and then, in a fit of remorse, throws himself on the fire and dies, I still have some rights. But I can no longer say that I have a right to the property in question or that I have a claim against the villain who destroyed it. The right to the property that I had possessed was destroyed when the property was destroyed, and the person against whom I had a claim no longer exists. Any moral (or legal) right that I may well have would be a right to redress in the form of compensation from his family, or friends, or descendants. But this is no longer the same right as the one that was violated. It is a right to something different and is matched by a claim against different people.

This has important implications for our obligations towards future generations. For if we cannot claim that we have a right to see a live dodo it cannot be claimed that the inhabitants of the Mauritius islands three centuries ago violated our rights by failing to save the dodo from extinction. And it would be absurd to claim that we *had* a right to see a live dodo before it was exterminated in the same way that we had a right to some property before some villain burnt it. Before the dodo was exterminated we did not exist. In other words, the people alive at any point of time only have a right to what is available and only have claims against others to respect those rights.

Indeed, insofar as ordinary adults can be said to have a moral right to something or other, it must presumably mean that they have a moral right to choose whether to exercise the right, or claim to exercise it, or complain if they are denied the exercise of that right, or authorise somebody else to exercise the right in their place, or even waive the right. In practice, of course, some of these options may not be open to many people today, particularly in those countries of the world where people live under authoritarian regimes. *In practice* they are unable to exercise their moral rights. But it would not be impossible for them to do so if the regimes in which they live changed. There is no logical obstacle. But given the flow of time it is not *logically* possible for us to insist that inhabitants of Mauritius three centuries ago refrain from hunting the dodo or from taking action to preserve it, on the grounds that its extinction around the end of the seventeenth century deprived us of our right to see it. Nor could we – if we so wished – waive our right to see a live dodo by saying 'OK. Go ahead. Hunt it if you like. We think it is a rather silly bird anyway'.[3] Again, this is a logical impossibility, not a question of whether, in practice, one can exercise some right.

The logical flaw in the notion that future generations have 'rights' outside their relationships among contemporaries cannot be remedied by the assertion that their rights or their interests are being represented today by environmentalist pressure groups and the like. Of course, anybody can claim to represent the interests of future generations, in the same way that one can claim to represent the interests of any existing

---
[3] This point was developed by Hillel Steiner (1983).

person or persons, such as a client or a child or a handicapped person. But claiming or believing to represent the interests of any person or person does not mean that one is really doing so, let alone confer rights on them. As it happens I claim to represent the interests of future generations in the sense that I advocate policies that I believe are in their interest. But to do so I have no need to claim that I am representing the 'rights' of future generations.

To recapitulate, therefore, all theories of justice attribute rights (and counterpart obligations). Since future generations (of unborn people) cannot be said to have any rights the interests that they *will* have cannot be protected within the framework of any theory of justice. But the rejection of intergenerational justice does not mean that we have no moral obligations to take account of the interests that future generations will have. For justice does not exhaust the whole of morality.[4] It leaves room for benevolence, or common humanity or virtue. One does not save the life of a drowning child out of consideration of justice any more than one allows one's neighbour to use one's telephone when his or her own is out of order on account of any 'right' that he or she may have to do so.[5] A wider conception of morality goes beyond 'rights' and 'justice' and incorporates what are sometimes known as 'imperfect obligations'. These can be defended in terms of 'virtue', rather than in terms of duties arising out of justice.

It may be asked what difference it makes if we identify our moral obligations to future generations by appeal to some theory of justice rather than by appeal to the interests that they will have and our moral obligation to take account of these interests. The answer is that it makes a great difference to the relative importance we attach to different obligations. For 'rights' are generally regarded as having 'trumping' power over mere interests. The interests approach, which comes more naturally to economists, calls for a balancing of interests. The rights approach, by contrast, imposes a side constraint on any such weighting procedure. In the present context, for example, the rights/justice approach, as embodied in the 'sustainable development' objective, would imply that any weighting of the interests of present and future generations (using a suitable discount rate perhaps) should be subject to the constraint that future generations have a right to inherit the same environment as we have now ('strong sustainability') or corresponding compensation in the form, perhaps, of more man-made capital ('weak sustainability').

Thus switching from a perspective of justice to one of some wider concept of morality changes the relative importance that we should attach to our various moral obligations to future generations. For 'justice' between generations is inevitably a matter of distributive justice. And this must mean justice in the distribution of some scarce resource. Concern with distributive justice between generations, therefore, concentrates on rights to some scarce resource. By contrast, the 'interests' approach invites us to rank our 'imperfect obligations' to future generations according to the strength of the interests that we can safely predict will be experienced by future generations, without any predisposition that this must take the form of scarcity of resources. We have to start from a clean sheet and ask ourselves which of the future generations' most vital interests can be most confidently predicted. When we do this it appears, for reasons given below, that the supply of 'scarce' resources is probably a very minor problem as far as the vital interests of future generations are concerned and that the most important bequest we can make to future generations is a radically more just society, characterised by greater respect for basic human rights than those in which most of the world's population now live.

23.3 How Much Richer Are Future Generations Likely To Be?

Of course, all long-range predictions of economic growth rates are hazardous. How many people, for example, would have predicted 10 years ago that the Japanese economy was soon to enter into a period of prolonged economic stagnation? However, in predicting the likely world growth rate over the next century we do not think we are running any great risks. For the longer the period the more one can abstract from possible medium-term forces, such as the catching-up on wartime dislocation that characterised the 1950s in

---

[4] Lucid reminders of this include, notably, Rawls (1972), Raz (1984), and O'Neill (1996).

[5] By grounding our concern for future generations in a wider concept of morality we do not need to rely on Gauthier's solution to the problem of intergenerational justice, namely to rely on mutually advantageous contracts between overlapping generations (Gauthier 1986). In any case, Temkin seems to have refuted this solution (Temkin 1995.).

many Western countries, or the recuperation from the oil shocks that characterised the 1970s, or eccentric bursts of dogmatic monetarism that characterised the 1980s, or the speculative excesses and financial profligacy in some parts of the world that marked the later 1990s. In the very long run these forces can be seen to be relatively transient. To predict growth rates over the next century one must fall back on an appraisal of what are the really fundamental underlying forces determining economic growth potential in the modern world. In my view these are scientific and technological progress and the accumulation of human capital. I believe that these will lead to a growth rate in real world per capita incomes over the course of this new century of somewhere between 1% and 2% per annum.

My reasons for this are as follows. The average growth rate of real income per head in the world over the last 40 years has been 2.1% per annum.[6] And there are two reasons to believe that the future growth rate is likely to be at least as high as this, if not higher. First, in the very long run the main source of growth in incomes per head is technological and scientific progress. This is a function of variables which are all tending to increase, some at a phenomenal rate. In particular, the number of highly educated people in the world, especially those having technological and scientific qualifications, is increasing so rapidly that it far surpasses the corresponding number of people having similar qualifications only two or three decades ago, and is likely to go on expanding rapidly. The main source of current high levels of income and output in the modern world is not so much physical capital or material resources but human capital – that is knowledge, training, skills and attitudes. And there is no physical limitation on the growth of this human capital.

Second, the rate of international diffusion of innovation and technical progress – which many studies have shown to be decisive in determining growth rates – will continue to accelerate. This is partly on account of the increasing 'globalisation' of economic activity. This does, of course, bring with it certain problems, but it also means that the technical progress and innovations in one country will spread more quickly than in the past, as already seems to have been the case. This will be intensified by one of the more favourable aspects of the policy revolution of the 1980s, namely the widespread conversion to freer and more competitive markets (including the labour market) than had

been the case previously. Some countries – particularly those in the ex-Soviet bloc – are having great difficulty adjusting to a new competitive environment, and it may be decades before they really move into the modern world as far as the operation of their economies is concerned. But, in the longer run, they will no doubt do so. This could unleash vast potential growth rates in many major countries, such as India and Latin America – and even Africa – where, in fact, some signs of this are already visible.

These two underlying forces for *long-run* growth suggest that the average annual long-run growth of output per head over the next century should be above that of the last 40 years. Since this has been 2.1% my projection of between 1% and 2% per annum seems on the cautious side and might be giving excessive weight to the slowing-down during the last decade or so. But to simplify the argument we shall assume a single figure of 1.5% as the annual average growth rate of real incomes per head over the next 100 years. The power of compound interest being what it is, this means that world average real incomes per head in the year 2100 would be about 4.4 times as high as they are now!

And it should not be thought that the above guess at the annual average growth rate of gross world product (GWP) over the next 100 years is a fanciful figure. A recent draft report of the IPCC (the UN Intergovernmental Panel on Climate Change) adopted, for purposes of estimating possible levels of energy use and carbon emissions, four possible 'storylines' (to use their terminology describing possible scenarios of rates of growth of population and incomes). These put 'per capita GWP' at between 4.3 times as high as it is today and 20 times as high! In other words, my 'guesstimate' is at the bottom of the range adopted by the IPCC. It is thus a conservative and modest estimate.

And the IPCC scenarios are by no means fanciful. As one of the contributors to their work points out, even on the assumption that world incomes will rise about tenfold over the course of the century, namely to about $300 trillion at present prices, this would be consistent with per capita incomes in the rich countries rising at only 1% per annum and those of the developing countries rising at only 3% per annum.[7] The former figure is well below its long-run trend rate

---

[6] Calculations based on Table D 1a in Maddison (1995).

[7] Anderson (1998b: Table 1 and p. 8). Anderson's introductory note points out that the scenarios in question are still the subject of discussion in the relevant IPCC group and should not be interpreted as representing any final agreed consensus.

in the twentieth century and almost inconceivably low given the long-run influences on modern economic growth set out above. And given the scope for 'catching up' among developing countries the latter figure corresponds only to their having reached, by the year 2100, merely the average income level enjoyed in the rich countries today. Given the international transmission of technical knowledge and productive techniques it is equally virtually inconceivable that – taken as a whole – they will have failed to achieve this.

23.4 The Resources Constraint

Finally, one popularly perceived threat to the future growth of prosperity is the oft-alleged danger of using up so-called 'finite resources'. A full-scale exposition of the factual and theoretical reasons why one can ignore this possibility would be beyond the scope of this contribution. Some of them have been set out in detail elsewhere (Beckerman 1995; Cooper 1994; Schelling 1995). As I have often pointed out we have managed very well without any supplies at all of Beckermonium, the product named after my great grandfather who failed to discover it in the nineteenth century (Beckerman 1972, 1974).

The main reason why we will never run out of any resource, or even suffer seriously from any sudden reduction in its supply, is that whenever demand for any particular material begins to run up against supply limitations a wide variety of economic forces are set up to remedy the situation. These forces start with a rise in price, which, in turn leads to all sorts of favourable feedbacks, notably a shift to substitutes, an increase in exploration, and technical progress that brings down the costs of exploration and refining and processing as well as the costs of substitutes. In the end, of course, the relative prices of the final goods in which the materials in question may be embodied may still rise, which leads to a reduction in demand for them. If, for example, coal were ever to become a very scarce commodity its price would rise to the point where, like other scarce minerals such as diamonds, it would be used only for jewellery or certain very special industrial uses. We would never run out of it. And the process would take place very, very gradually, allowing time for economies to adapt. Key materials only disappear overnight in science fiction stories.

It was this total failure to allow for the way that markets work which led the Club of Rome in 1972 (Meadows et al. 1972) to issue its alarming prediction of the imminent exhaustion of many key minerals. In the event, during the following 20 years the consumption of these materials more or less matched, or exceeded, the levels of 'known reserves' that existed in that year. Yet the known reserves at the end of the period finished up being about as big as or, in some cases, much bigger than they were at the outset (Beckerman 1995: 53).

In fact the falsification of predictions of rapidly approaching shortages has a very distinguished pedigree, especially in the field of energy predictions. But this does not seem to deter later authoritative bodies from making equally mistaken falsified predictions. For example, the great economist Jevons predicted shortages of coal supply back in 1865 in a very sophisticated piece of applied economics which would still compare favourably with most contemporary applied economics (Jevons 1865). In spite of the decisive falsification of his predictions during the course of the subsequent decades, 90 years later, in 1955, the 1955 UN Atoms for Peace Conference made estimates of both 'proven' and 'ultimately recoverable' reserves of fossil fuels, which are now seen to be one-quarter and one-twelfth, respectively, of current estimates (Anderson 1998a). The following is a selection out of innumerable equally falsified predictions in later years.[8]

Countries with expanding industry, rapid population growth... will be especially hard hit by economic energy scarcities from now on (Amory Lovins 1974, mentioned in Mills 1999).

The supply of oil will fail to meet increasing demand before the year 2000, most probably between 1985 and 1995, even if energy prices are 50% above current levels in real terms (MIT Workshop 1977).

The diagnosis of the U.S. energy crisis is quite simple: demand for energy is increasing, while supplies of oil and natural gas are diminishing. Unless the U.S. makes a timely adjustment before world oil becomes very scarce and very expensive in the 1980s, the nation's economic security and the American way of life will be gravely endangered (Executive Office of the President, National Energy Program, 1977, mentioned in Mills 1999).

The oil-based societies of the industrial world cannot be sustained and cannot be replicated. The huge increases in oil prices since 1973 virtually guarantee that the Third World will never derive most of its energy from petroleum (Worldwatch Institute 1979, mentioned in Mills 1999).

What seems certain, at least for the foreseeable future, is that energy, once cheap and plentiful but now expensive and lim-

---

[8] Where not otherwise indicated the predictions below are included in many such predictions listed in Mills (1999).

ited, will continue to rise in cost (Union of Concerned Scientists 1980).

Conservative estimates project a price of $80 a barrel (in 1985) even if peace is restored to the Persian Gulf and an uncertain stability maintained (National Geographic 1981).

The falsification of past predictions of energy shortages is the result of three major forces, stimulated in most cases by the economic feedback mechanism outlined above. First, estimates of recoverable energy resources continually increase. Second, continuous technical progress is being made in the efficiency with which conventional energy is used. And, third, there is also substantial progress and innovation in the exploitation of renewable sources of energy.

As regards the first factor, namely supplies of conventional fossil fuels, current expert opinion is that 'In sum, the availability of fossil fuel resources can be measured in units of hundreds – perhaps thousands – of years. The availability of renewable energy resources (including geothermal resources), even if used on an immensely expanded scale has no known time limit (Anderson 1998b). As can be seen in the Table 23.1, the number of years' consumption at 1994 rates of consumption that are covered by total resources (including those not yet identified but that are likely to be insofar as prices provide the necessary incentive) varies from 240 years for oil to 1,570 years for coal. Combining oil, natural gas and coal in units of oil equivalent 'The total reserves of fossil fuels are currently thought to be around 5,000 Gtoe (gigatons of oil equivalent) or about 700 times the current annual rate of world consumption' (Anderson 1998a).

Of course, it may be argued that it is still too soon to claim that the pessimistic predictions will not be fulfilled even in the very long run. The above estimates of oil reserves are admitted to be subject to a significant margin of uncertainty. But some of the major uncertainties are on the favourable side. For example, there are also some unconventional sources of oil that are well known but that are not currently economically and technically viable on a significant scale, but which might well become so as a result of further major cost reductions in mining and processing of these sources.[9]

And we have not yet taken account of the scope for technological progress in renewable energy (especially solar and wind power), or the likely continuation in the trend towards greater efficiency in the use of energy and the declining energy intensity of output in developed countries. Nor have we taken account of geothermal energy.[10]

As for the technical progress in the use of renewable energy it should be born in mind that the total energy received from the sun is about 10,000 times total world energy consumption and if only a very small fraction of this can be harnessed in an economically viable manner the whole energy problem disappears. And, indeed, the technical progress in the harnessing of solar energy has been substantial. Already photovoltaic systems and solar-thermal power stations, such as those now operating in California,

Table 23.1 Consumption rates and aggregate global resources of fossil fuels (gigatons of oil equivalent)

	Consumption/ year (1994)	Discovered	Further	Total resources
Oil[a]	3.37	333	481	814
Natural gas	1.87	333	537	870
Coal	2.16	1,003	2,397	3,400
Total	7.40	1,669	3,415	5,084

[a] The figures for oil comprise what are known as conventional and unconventional sources, the latter including oil shales, tar sands and coal bed methane.
Source: Anderson (1998a,) referring to data in Rogner (1997)

---

[9] These include, in particular, oil shale in the western USA, heavy and extra heavy oil of the kind found in Venezuela, and bitumen (natural tar) such as found in Alberta, Canada. If, as seems quite possible, market conditions and their resulting incentives to improved extraction and processing made these sources economically viable, they could add some thing in the region of a further 15 trillion barrels of oil. And even without some major rise in the price of conventional oil supplies it is quite likely that they would become viable. For example, in Alberta there have been major cost reductions in recent years so that oil production costs from these sources have fallen to about $9 per barrel, which, while much higher of course than the very cheap oil in the Middle East, is below the cost of North Sea oil and much lower than oil costs in most of the USA. If technological and market development brought a large proportion of these 'unconventional' sources into the picture then, of course, at current rates of consumption there would be enough oil to last for several centuries (American Petroleum Institute 1995).

[10] According to a recent review by Mock et al. (1997), estimates for hot dry rock alone 'are orders of magnitude larger than the sum total of all fossil and fissionable resources'. While it is not yet known how much of these resources might be economically viable, the technological capacity for deep drilling already exists so in principle it would be physically feasible to tap this resource. Insofar as the price of conventional fuels rose significantly one can expect that there would be sufficient incentive for technical progress to be made in the exploitation of this source of energy. As Anderson puts it 'To sum up, the availability of fossil fuel resources can be measured in units of hundreds – perhaps thousands – of years. The availability of renewable energy resources, even if used on an immensely expanded scale, has no known time limit' (Anderson 1998a: 443).

manage to convert about 10% of the incident solar energy into electricity, and with further developments in the pipeline are expected to be converting about 20% of it.[11] The main constraints, therefore, are costs and storage.

It is true that, given the pace of technical progress especially in fuel cells that can store or produce electricity, there is good reason to believe that their costs will continue to fall. On the other hand, it is possible that some of the recent fast decline in the costs of certain renewable forms of energy will not be maintained into the future and, indeed, there are already signs of this happening. Furthermore, the fossil fuels start with an enormous initial advantage and do not seem inclined to allow it to be eliminated. The advances in gas-fired power generation, in particular, is one of the reasons why even wind power, which is far cheaper than solar, has not yet made much progress outside a few niche markets. And while many major oil producers and automobile manufacturers are spending large amounts on research into renewable fuels for various purposes, they are spending much bigger amounts on finding new resources of fossil fuels and in reducing the costs of exploiting them.

Thus, given the prospects for supplies of fossil fuels summarised above, any significant threat to the market for fossil fuels might simply be met by competitive reductions in their prices. So the future is much more likely to be one of competition between renewable and non-renewables, leading to further long-term declines in (real) energy prices, than one characterised by any shortages on account of an exhaustion of supplies of fossil fuels.

23.5 Our Real Obligations to Future Generations

I have argued that insofar as future generations cannot be said to have any particular *rights* to anything, the starting point for assessing our moral obligations to distant generations has to be some prediction of what their most important *interests* are likely to be. This should be followed up with some assessment of what effects our policies will have on their interests and how far they conflict with the interests of the present generation.[12] This does not mean, of course, that one can emerge with any 'lexicographic' ordering of priorities in general terms. For example, abuses of human rights can range from horrific behaviour to minor restrictions on peoples' freedom of movement or freedom to dispose of their property. Poverty can range from mass starvation to isolated instances of temporary poverty in generally affluent communities as a result of some transient bad luck or other exceptional circumstances. Environmental problems can range from the elimination of atrocious urban air conditions that were found in major cities of the industrialised countries until relatively recently or the absence of clean drinking water today in most parts of the developing countries at one extreme, to the reduction in noise levels from the occasional neighbourhood street party, at the other. But it is still possible to arrive at some judgement as to which of their interests, taken by and large, will be most important.

We have argued above that future generations will be vastly richer than we are today. So although there will always be pockets of acute poverty – as are found, for example, even in the richest countries today – one can expect that widespread poverty in the world will be eradicated. Furthermore, although there will always be local environmental problems, which may even get worse in the early stages of industrial development in many countries, a much richer world will both permit and stimulate most countries to improve their local environmental conditions and to find ways of adapting to such global environmental problems that may persist. Other problems may prove more persistent and difficult to deal with, such as those associated with the consequences of family breakdown or the problems of crime and drugs in contemporary societies.

But I believe that in the very long run the most important and permanent threat to human welfare will

---

[11] On conservative assumptions concerning the duration of sunlight and conversion efficiency, only about 0.25% of the area now under crops and permanent pastures would be needed to meet all of the world's primary energy needs. Even if these needs rise – as they may well do – fourfold over the course of the next century this still means that only 1% of this land area would be needed to supply the whole world demand for energy.

[12] To some extent this is a circular procedure, of course. One's prejudices, intuitions and predilections for certain priorities among obligations will naturally colour one's prediction of what one perceives to be the most salient and relevant features of the future evolution of human society. But there is no way of breaking out of the circle by appealing to some external objective formula for ranking the obligations.

continue to be the violation of basic human rights. For the safest prediction that can be made for the long-term future is that there will always be potential conflict between peoples for all sorts of different 'reasons' and that can all easily lead to horrific violations of basic human rights. At the same time one can predict with great confidence that people will always want life and security, and freedom from fear, discrimination and humiliation.

Except in some Utopian scenarios human wants will always expand more or less in line with what is available, so that, whatever we do now about the future availability of resources and however much technical progress expands our potential for producing goods and services, there will always be conflicting interests in the way that potential output is shared out. Furthermore, conflicts of interest over material possessions are by no means the only causes of conflict, any more than are cultural differences. There is no shortage of other causes. Even within any given culture or civilisation there are conflicts of various kinds between interests, objectives and values, which will divide members of any community. The past century has probably seen the most widespread unprecedented savagery on a mass scale of any time in human history (Glover 1999). There is no reason to believe that human nature will get any better. Human compassion does not seem to keep pace with technological progress.

The greatest contribution that can be made, therefore, to minimise human suffering and promote human welfare will be constant vigilance in the defence of human rights. Our greatest bequest to future generations, therefore, will be a much more 'decent' society than the one that exists in most countries of the world today, namely one in which there is greater respect for basic human rights than is the case today, and, although respect for basic human rights may be in short supply in most countries of the world today, this is not because it is a 'scarce resource' in the same sense as minerals, certain species and so on are often believed to be. Extending respect for basic human rights will also make a major contribution to well-being in most countries today. So there is no conflict of interest between generations as there is believed to be in the case of sharing out finite resources. Hence, on almost any conception of what purpose is served by theories of justice there is no need for a theory of intergenerational justice to guide us as to what an equitable share-out would be. It is indeed paradoxical that many environmental activists attack the materialist obsession to which the defenders of continued economic growth are alleged to be attached, yet seem to be much more concerned with our material legacy to future generations than with other far more important ingredients of human welfare. If the arguments presented here are valid, material obstacles will be among the least of the problems that future societies will have to face.

References

American Petroleum Institute (1995). *Are we running out of oil?* Discussion Paper no. 0181, December, Washington, DC.

Anderson, D. (1998a). On the effects of social and economic policies on future carbon emissions. In *Mitigation and adaptation strategies for global change* (pp. 437–438). Dordrecht: Kluwer.

Anderson, D. (1998b). Explaining why carbon emission scenarios differ, paper for IPCC Special Report on Emission Scenarios, May 1998, Geneva, 30.

Barry, B. (1978). Circumstances of justice and future generations. In R. Sikora, & B. Barry (Eds.), *Obligations to future generations* (p. 205). Philadelphia: Temple University Press.

Beckerman, W. (1972). Economists, scientists and environmental catastrophe. *Oxford Economic Papers, 24*(3), 327–344.

Beckerman, W. (1974). *In defence of economic growth*. London: Duckworth (US edition, *Two cheers for the affluent society*, St. Martin's Press, New York, 1975).

Beckerman, W. (1995). *Small is stupid* (Chapter 4). London: Duckworth (US edition, *Through green-colored glasses*, Cato Institution, Washington, DC, 1996).

Beckerman, W. & Pasek, J. (2001). *Justice, posterity and the environment*. Oxford: Oxford University Press.

Cooper, R. (1994). *Environmental and resource policies for the world economy* (Chapter 2). Washington, DC: Brookings Institution.

De George, R. T. (1981). The environment, rights, and future generations. In E. Partridge (Ed.), *Responsibilities to future generations*. New York: Prometheus Books.

Gauthier, D. (1986). *Morals by agreement* (299 ff.). Oxford: Clarendon Press.

Glover, J. (1999). *Humanity: A moral history of the twentieth century*. London: Cape.

Goodin, R. (1983). Ethical principles for environmental protection. In R. Elliot & A. Gare (Eds.), *Environmental philosophy* (p. 1). Brisbane: University of Queensland Press.

Jevons, W. S. (1865). *The coal question*. London: Macmillan.

Maddison, A. (1995). *Monitoring the world economy*. Paris: OECD.

Meadows, D. et al. (1972). *The limits to growth*, A Report to the Club of Rome. New York: Universe Books.

Mills, M. P., (1999). *Getting it wrong: Energy forecasts and the end-of-technology mindset*. Washington, DC: Competitive Enterprise Institute.

MIT Workshop on Alternative Energy Strategies (1977). *Energy: Global prospects 1985–2000*. Cambridge, MA: MIT Press.

Mock, J. E., Tester, J. W, & Wright, M. W. (1997). Geothermal energy from the Earth: Its potential impact as an environmentally sustainable resource. *Annual Review of Energy and the Environment, 22,* 305–356.

National Geographic (1981). A special report in the public interest, New York.

Norgaard, R. (1992). *Sustainability and the economics of assuring assets for future generations,* Policy Research Working Paper WPS 832. Washington, DC: World Bank.

O'Neill, O. (1996). *Towards justice and virtue.* Cambridge: Cambridge University Press.

Partridge, E. (Ed.). (1981). *Responsibilities to future generations.* New York: Prometheus Books.

Rawls, J. (1972). *A theory of justice* (p. 284). Oxford: Clarendon Press.

Raz, J. (1984). Right-based morality. In J. Waldron (Ed.), *Theories of rights.* Oxford: Oxford University Press.

Rogner, H-H. (1997) An assessment of world hydrocarbon resources, *Annual Review of Energy and the Environment, 22,* 217–262.

Schelling, T. (1995). Intergenerational discounting. *Energy Policy, 23*(4/5), 395–401.

Steiner, H. (1983). The rights of future generations. In D. Maclean & P. Brown (Eds.), *Energy and the future* (p. 159). Totowa, NJ: Rowman & Littlechild.

Temkin, L. (1995). Justice and equality: Some questions about scope. In E. F. Paul, F. D. Miller, & J. Paul. (Eds.), *Social philosophy and policy* (79ff.), 12, no. 2, Cambridge University Press.

Chapter 24
Sustainability is an Objective Value[1]

Herman E. Daly

Contents

24.1 A Technical Problem... 483
24.2 Some Conflicts in Economic Philosophy............. 484
24.3 Objective Value and Totalitarianism.................. 486
24.4 The Big Philosophical Issue 486
24.5 Whitehead's Lurking Inconsistency................... 487
24.6 Purpose and Value 488
24.7 More Neo-Darwinist Fallout 489
24.8 The Purposeful Perpendicular........................... 489
References... 490

24.1 A Technical Problem

The loss of natural functions has traditionally not been recognised in national income accounting (Hueting 1980). Loss of environmental function has been an unmeasured reduction in both productive capacity and direct welfare. To account for this loss in true national income it is necessary to value natural functions in order to subtract the loss. This requires prices for natural functions, which in turn requires supply and demand curves. Roefie Hueting has proposed a supply function that is the marginal cost curve of restoration of the natural function. His difficulty arises with the demand curve that is unknown because markets for many natural functions do not exist, and even if they did, most interested parties (for example future generations, other species) are not allowed to bid in the market. The logic of income accounting requires the subtraction of the value of sacrificed ecological functions. But sacrificed functions cannot be valued in the same way as other goods and services because the demand curve cannot be defined – that is, cannot be defined in the same way as other demand curves, namely in terms of individual preferences expressible in markets. Hueting's resolution is a perpendicular 'demand curve', an expression of objective value, not individual preferences (Hueting 1991). The objective value is sustainability. This entails a rejection of the dogma that individual subjective preferences are the sole source of value, and introduces collective objective value as an additional source.

Roefie Hueting has been led by the logic of practical problem solving into a rather basic conflict with the dogma that all value arises from private subjective preferences. He needs a perpendicular demand curve at a level of environmental exploitation that is sustainable in order to determine prices proper for the calculation of sustainable national income. Income is by definition (Hicks 1948) an amount of output such that its production during this year does not impair our capacity to produce the same amount next year. In other words, productive capacity must be maintained intact, there must be no net consumption of capital, either man-made or natural capital. Since we currently consume natural capital without deducting it, our calculation of national income is erroneous both by the quantity of natural capital consumed and by the price distortions caused by the drawdown of that natural

H.E. Daly (✉)
School of Public Policy of University of Maryland,
College Park, USA
e-mail: hdaly@umd.edu

---

[1] This chapter is reproduced with slight adaptations from: H. E. Daly (2001). Roefie Hueting's perpendicular 'demand curve' and the issue of objective value (chapter 5). In C. Ekko, van Ierland, J. van der Straaten, & H. Vollebergh (Eds.), *Economic growth and the valuation of the environment: A debate.* Aldershot: Edward Elgar (with permission).

capital. The condition of ecological sustainability has to be imposed in some way before one can calculate national income that is true to the very concept of income as sustainable production. So Roefie Hueting is led by standard economic logic to his imposition of a perpendicular demand curve at a sustainable level of exploitation of natural functions. It is not a gratuitous desire on his part to impose his personal preferences on the rest of the world. It is the honest national income accountant's professional duty to measure true income, not some amalgam of capital drawdown and production. Hueting appeals to the Brundtland Commission and other political bodies who have advocated sustainability as a social goal, in at least partial justification for his treating sustainability as an objective value.

He might also appeal more explicitly to the very concept of income. Strictly speaking the term 'sustainable income' is a redundancy because income is by definition sustainable. The very concept embodies the implicit purpose of prudential behaviour, of avoiding unplanned impoverishment by inadvertent consumption of productive capacity. National income is not a 'value-free' fact – it is a concept built around a prudential purpose. Its definition is not decided by aggregating individual preferences. If we did not have the purpose of avoiding capital consumption and consequent impoverishment (and, yes, of increasing wealth) there would be little reason to calculate national income in the first place. But practice has strayed far from theory, both because the purposive element in the concept has been forgotten, and the 'new scarcity' has been overlooked or denied. Consequently, what we currently call national income is decidedly unsustainable, necessitating the awkward pleonasm 'sustainable national income' for the corrected figure, in order to again convey the original meaning of income.

24.2 Some Conflicts in Economic Philosophy

To understand valuation we must pay some attention to the valuer – the valuing agent or self that is presupposed by valuation. Economists take this valuer to be the human individual. The market weights and aggregates individual valuations. Individual values, usually called preferences by economists, are taken to be subjective in the sense that, if they were objective, individuals could agree on them and enact them collectively. It is assumed that there is no objective value or standard for judging preferences. Values are by assumption reducible to subjective individual preferences. By motivating individual choices these subjective preferences become causative in the real world through the market. But they are not causative through collective action because they are assumed to lack the degree of objectivity necessary for the agreement presupposed by collective action.

Since the marginalist revolution economics has accepted a subjectivist theory of value: that value is rooted in utility conceived as the satisfaction of individual preferences. Diminishing marginal utility underlies the demand curve, which in combination with supply (based on increasing marginal costs) determines prices. Earlier in the history of economic thought value had been considered objective in the sense of being rooted in 'real costs', especially labour cost. Today some ecologists believe in an energy cost theory of value, so objective cost theories of value have not completely died out. Among economists, however, cost means 'opportunity cost', the best alternative benefit forgone. Thus the ultimate root of cost is the same as that of benefit – subjective preference, whether enjoyed or forgone. I am not objecting to opportunity cost, or to marginal utility. My use of the term 'objective value' in this chapter refers not to costs but to preferences – the notion of objectively good preferences. This notion has been rejected by economists who hold that preferences are purely personal and subjective. Moreover, any appeal to the concept of objective value in this sense is thought to be merely the veiled imposition of the speaker's personal preferences on everyone else. Indeed, this must follow from the dogmatic assumption that the only locus of value is subjective individual preference. If this is true then any claim on behalf of objective value can only be, at best confusion, and at worst an attempt to 'undemocratically' promote one's own preferences at the expense of others. Of course if the value of democracy, too, is simply a personal preference rather than an objective value, then it is hard to see why it should be privileged over nondemocratic pursuit of one's own preferences. But for now my point is simply that the non-existence of objective value is an assumption, not a conclusion of rational argument or empirical investigation. It may have started out as a methodological assumption, but today it functions as an ontological axiom.

Hueting's perpendicular demand curve represents an objective concept of sustainability in two senses: first, that a sustainable level of aggregate resource use is objectively definable ecologically and subject to at least crude measurement; second, that sustainability so defined is itself an objective value whose authority over private preferences should be accepted by individuals and expressed by their democratic representatives. Surprisingly, even the objectivity of sustainability in the first sense seems controversial, and Roefie Hueting and Lucas Reijnders (1998) had to write an article countering the notion that the very concept of ecological sustainability is subjective. My concern here is with objectivity in the second sense: that sustainability, in addition to being objectively definable (at least as definable as the concept of 'money'), is a good thing – an objective value worthy of being a goal of public policy. If the objective value of sustainability conflicts with private subjective preferences, then too bad for private subjective preferences. As offensive as this last statement is to economists, most would accept that preferences for murder and robbery should be ruled out, presumably because they conflict with objective value. But they remain reluctant to rule out 'revealed preferences' for unsustainable levels of consumption, because they do not recognise sustainability as an objective value.

Others, including myself, have been led to the same conflict as Hueting, but from a somewhat different starting point. If, instead of measuring national income in a way that reflects sustainability, one is trying to design a policy for actually attaining sustainability in a market economy, one encounters the same problem. For example, the policy of tradable permits for depletion or pollution requires as a first step that total quantity extracted or emitted be limited to an amount that is sustainable. In this case it is the supply function that becomes a perpendicular at the chosen total quantity. Supply is decided socially by reference to the objective value of sustainability, and demand is left to the market. But demand is only allowed to determine the price that rations the fixed total quantity that has been set socially. Demand, subjective individual preference, is not allowed to influence quantity – it only determines the rationing price, subject to the total quantity that is set by the objective value of sustainability.

It is interesting that one approach leads to a perpendicular demand curve, and the other to a perpendicular supply curve. This difference in analytical representation deserves further reflection and explanation, but for present purposes I want to emphasise that both analytical adaptations represent the introduction of objective value, and therefore both conflict with the fundamental dogma that all value arises from subjective individual preferences.

Is it possible to go too far in granting monopoly rights to subjective individual preferences in the determination of value? The force of the preference dogma was recently brought home to me in a conversation with a young professor of environmental economics. He confided to me that personally he had a strong preference for sustainability, but since he doubted that sustainability was derivable from the individual preferences of the population, he could not justify devoting time and effort as a professional economist to furthering the idea. I was shocked by what he said. However, if all value really is reducible to subjective preference then his position is not unreasonable – except one wonders why he is so diffident about asserting his own preferences, unless he just happens to have a personal preference for diffidence rather than assertiveness.

Of course the same logic applies to all social goals – full employment, a just distribution of income, avoidance of inflation, and indeed also the promotion of aggregate economic growth. Many social goals cannot be derived from individual preferences, so more and more they are thought by young economists, who reflect the current university teaching, not to exist. Goals like full employment and distributive justice were recognised in an era in which the dogma of individual preference was not so well established, and they continue to command respect thanks to historical inertia and to our enormous capacity to believe contradictory things. But more and more I suspect that they will fall into the same orthodox disrepute as sustainability as we continue to try to live by the dogma of subjective personal preference, and its corollary, the denial of objective value. I believe that sustainability is an objective public value whose legitimacy does not derive from private subjective preferences any more than does democracy or justice. We do not submit the institution of democracy itself to a popular vote, nor do we allow free market participants to sell themselves into slavery even if that is their preference.

24.3 Objective Value and Totalitarianism

A major question that arises in this context is whether departure from the dogma of individual preferences would lead to a slippery slope that could end in totalitarianism. One can certainly understand this concern. I want nevertheless to suggest that it is the preference dogma that is today the broader path to totalitarianism, and that only a commitment to objective value can save us from it.

The argument has already been made by Lewis (1944) and I need only try to summarise it. A good place to begin is Lewis's statement, so shocking to modern prejudices, that 'A dogmatic belief in objective value is necessary to the very idea of a rule which is not tyranny or an obedience which is not slavery.'

Nothing could be more contrary to the dogma of subjective preference. Yet Lewis's logic is both simple and compelling. If you and I disagree in our purposes or preferences, and neither of us believes in objective value, then there is nothing that either of us can appeal to in an effort to persuade the other. I can only restate and clarify my preferences, and you can only do the same. I hope that once I have made my preferences clear to you, you will agree with them, and you hope the same about me. But that usually does not happen. Our different preferences or purposes have no authority beyond the strength of personal conviction with which we hold them. Once our differences have been made unmistakably clear the only resolution is coercion, either by physical force or psychological manipulation. Only if we accept the reality of objective value whose authority trumps our personal preferences is there any possibility of reasoning together and of being genuinely open to persuasion.

We may not agree in our perceptions of objective value either, but as long as we are trying to discern more clearly a reality whose existence we both recognise, there is reason for at least a modicum of patience, tolerance, and goodwill. But why should I be tolerant of your subjective preferences which have no more authority than mine, indeed none at all to me since I am by assumption guided only by my own preferences. If I happen to be the stronger or the cleverer I will have my way. It makes no sense to appeal to my 'moral sensitivity' unless we believe that there is something real to which we should be sensitive. A 'moral compass' implies the existence of objective value, a true magnetic north to lure the sensitive needle towards itself. If our individual moral compass needles point in different directions we can try to sensitise them by reasoning together, but we must sensitise them to objective magnetic north, not to our own subjective preferences. If we believe there is no magnetic north, then we should find an alternative use for our compass – such as throwing it at the cat.

24.4 The Big Philosophical Issue

In what follows I want to reconsider critically a major feature of our culture that provides the larger context for the economist's preference dogma: the separation of the world of private subjective preferences – that make no truth claims and are therefore not cross-examined – from the world of purposeless efficient causation in which truth claims are both made and cross-examined (Newbigin 1986). To put it paradoxically, how can it be that the only things that are supposed to have value in public discourse are 'value-free' facts?

Why has this assumption that all value is rooted in individual preferences become a basic dogma in economics? I suspect because it is coherent with, and likely derivative from, a larger cultural assumption that excludes purpose from science. If objective value exists then its attainment obviously constitutes a purpose, and the hallmark of modern science is the exclusion of final causation and the focus only on efficient causation. Purpose and value have been confined to the private subjective world of individual experience in which one person's experience or preference is as good as another's. In the public world of facts upon which agreement is expected and truth claims are made, efficient causation reigns, and purpose is not allowed. Our age is often called 'pluralistic' – but we are pluralistic only in the private realm of values and purposes. In the public realm of fact, pluralism would be considered irresponsible indifference to the truth.

Of course, purpose continues to exist in the private subjective world of individuals and is presumed by economists to be causative in the public world, as noted earlier. Even this degree of causative efficacy via purpose as individual preference is an embarrassment to many scientists who, consistent with the overall banishment of teleology from science, dismiss our conscious experience of individual purpose as an epiphenomenon, an illusion. Economists usually do not go this far, but they do try to confine to the private subjective world of individuals and devote their efforts to explaining mechanistically how these individual actions, motivated by private preferences, give rise to public consequences, to definite prices and allocations under different market structures. This market allocation is usually taken as implicitly good, in the same way that democracy is considered good, although such goodness is impossible to demonstrate on the basis of personal subjective preferences with no appeal to objective value. Some people like markets and some do not, some people have a preference for democracy and others do not – just as some people like apples better than oranges. If one insists on deriving all value from the private sphere of individual preferences, and to deny any notion of publicly objective value, then one must accept the consequences, however nihilistic.

The problem is that economists, and with them modern culture, do not believe in objective value. Modern culture believes passionately in objective facts in the public world of efficient causation from which purpose has been expunged, and allows purpose to exist only in the private world of subjective experience in which no truth claims are allowed. One person's purposes are as good as another's; but one person's facts have to stand up to public scrutiny. An attitude that, as just noted, would be considered irresponsible indifference to truth in the public realm of facts is considered humble tolerance in the realm of purpose. This is because, in our plausibility structure, purpose is considered less real than fact. Therefore discussing purpose is less serious than discussing fact – we can afford to be 'tolerant' about subjective matters, like dreams, that don't really matter. Besides, if we start taking objective value seriously we may end up with religious wars like we had before the Enlightenment. If purpose is, as many scientists claim, just an illusion, or if, as economists claim, one person's purposes are as good as another's, then it would indeed make sense not to pay too much attention to purpose, and that benign neglect would contribute to peace and tranquillity. Furthermore, the elimination of final causation from the study of nature has been enormously fruitful in physics, chemistry, and even biology. Why not apply the same philosophy to the study of everything, of all that is, including economics? This indeed is the current programme. The problem is that this programme is leading us to conceptual absurdity, political paralysis and ecological catastrophe – because it is founded on an inconsistency.

24.5 Whitehead's Lurking Inconsistency

Alfred North Whitehead (1925) recognised this cultural contradiction back in 1925, and referred to it as the 'lurking inconsistency':

A scientific realism, based on mechanism, is conjoined with an unwavering belief in the world of men and of the higher animals as being composed of self-determining organisms. This radical inconsistency at the basis of modern thought accounts for much that is half-hearted and wavering in our civilisation. ... It enfeebles [thought], by reason of the inconsistency lurking in the background. ... For instance, the enterprises produced by the individualistic energy of the European peoples presuppose physical actions directed to final causes. But the science which is employed in their development is based on a philosophy which asserts that physical causation is supreme, and which disjoins the physical cause from the final end.

Whitehead went on to observe that, 'It is not popular to dwell on the absolute contradiction here involved'.

Biologist Charles Birch (1990), a keen student of Whitehead, has written an insightful book entitled *On Purpose* in which he begins to come to grips with Whitehead's lurking, radical inconsistency, which Birch restates as follows: 'The central symbol of ecological thinking in this book is purpose. It has become the central problem for contemporary thought because of the mismatch in modernism between how we think of ourselves and how we think and act in relation to the rest of the world'. Economics involves both thinking about ourselves and thinking and acting in relation to the rest of the world, the environment, thus neatly straddling the two poles of the lurking inconsistency.

In the emerging transdiscipline of ecological economics many of us have spent the last decade criticising economists for their neglect of the embeddedness of the economy in the larger ecosystem, and for their (our) ignorance of ecology in general. It has been the economist who needed correction and the ecologist who supplied it. I think this was, and still is, entirely necessary. However, as we try to develop policy on the basis of that theoretical understanding, it seems that ecologists are not only becoming less helpful, but also something of an obstacle. Why is this? I think Whitehead gives us the clue in the quotation above.

The enfeeblement of modern thought, noted by Whitehead, is evident today in the environmental movement, especially as it is promoted by biologists and ecologists, at least by their most visible representatives. Their science and philosophy is mechanistic. No final causes or purposes are permitted into their neo-Darwinian world of efficient causation by random mutation and natural selection. This mechanical process, over long time periods, is held to explain not only the evolution of all living things from a common ancestor, but also, in some versions, the emergence of the common ancestor itself from the 'primordial chemical soup'. For human beings in particular, random mutation and natural selection are thought to determine not only such characteristics as eye colour and height, but also intelligence, consciousness, morality and capacity for rational thought.

Powerful though it is, the neo-Darwinist explanation nevertheless faces severe difficulties even in the realm of mechanism. But leave those aside. My point is that it is obviously inconsistent to declare the world void of purpose, and then exempt one's self from that declaration by urging some policy in pursuit of – guess what? – a purpose! The manipulator (policy-maker) credits himself or herself with the very capacity for purposeful action in pursuit of objective value that he or she denies to the manipulated (the rest of the world). Herein lies the broad path to totalitarianism, alluded to earlier.

24.6 Purpose and Value

We, and perhaps higher animals in general, directly experience purpose, and within limits, act in a self-determining manner guided or lured by purpose. If we are part of nature then so is purpose; if purpose is not part of nature then neither, in large part, are we. However, the immediate reality of final cause and purpose that we all directly experience, must, in the mechanist's view, be an 'epiphenomenon' – an illusion that itself was selected because of the reproductive advantage that it chanced to confer on those under its influence. The policy implication of the dogma that purpose is not causative in the world is *laissez-faire* beyond the most libertarian economist's wildest model. The only 'policy' consistent with this view is, 'let it happen as it will anyway'. It is odd that the illusion of purpose should be thought to confer a selective advantage while purpose itself is considered non-causative – but that is the neo-Darwinist's problem. Economists do not go so far as to declare purposes illusory, but, as we have seen, they attain nearly the same result by confining them to the realm of private subjectivity exempt from the discipline to which public claims of objective truth are submitted. If one person's preferences are as good as another's, and preference is the ultimate source of value, then there is really nothing for us to talk about. There remains, however, a great deal for us to fight about.

But, economists do vigorously affirm at least one public purpose, one apparently objective value. That purpose is of course aggregate growth in GDP. It is thought to be derivable from individual preferences by the (invalid) argument that more and more preference satisfaction by more and more individuals must necessarily result from aggregate growth in GDP. But the 'New Scarcity' (remember the subtitle 'more welfare through less production') means that GDP growth has unmeasured costs that might be growing faster than the measured benefits (usually taken as GDP itself). So the inference is unwarranted. However, even an unworthy and unwarranted purpose, such as GDP growth for ever, will dominate the absence of purpose. Economists, for all their (our) shallowness and ignorance of the natural world, will continue to dominate ecologists in the policy forum simply because they affirm a purpose while the ecologists do not – and logically cannot as long as they remain faithful neo-Darwinists.

The relevance of the lurking inconsistency to the new subdiscipline of 'conservation biology' should be evident – conservation is, after all, a policy in the service of a purpose. What are we trying to conserve? Biodiversity? Habitat? Why are they valuable, and which parts of the biota are more valuable? Although economists do not know how to value biodiversity, it seems that biologists are even more clueless, having purged their science of the very concept of value because it is tainted with teleology. But the very existence of conservation biology means that some biologists want to affirm purpose at least implicitly. According to economists, preferences are the ultimate standard of value and expression of purpose. Witness economists' attempts to value species by asking consumers how much they would be willing to pay to save a threatened species, or how much they would accept in compensation for the species' disappearance. The fact that the two methods of this 'contingent valuation' give different answers only adds comic relief to the underlying tragedy which is the reduction of value to taste.

24.7 More Neo-Darwinist Fallout

Biologists have taken extreme pains for many years to rid their science of any trace of teleology. As Whitehead (1925) remarked:

Many a scientist has patiently designed experiments for the purpose of substantiating his belief that animal operations are motivated by no purposes. He has perhaps spent his spare time writing articles to prove that human beings are as other animals so that purpose is a category irrelevant for the explanation of their bodily activities, his own activities included. Scientists animated by the purpose of proving that they are purposeless constitute an interesting subject for study.

Teleology has its limits, of course, and it is evident that mechanism has constituted an enormously successful research paradigm for biology (if for the moment we allow the biologist the implicit purpose whose achievement defines their success). The temptation to elevate a successful research paradigm to the level of a complete worldview is perhaps irresistible. But mechanism, too, has its limits. To deny the reality of our most immediate and universal experience (purpose) because it does not fit the mechanistic paradigm is radically anti-empirical. To refuse to recognise the devastating logical contradictions that result from the denial of purpose is profoundly anti-rational. That people already unembarrassed by the fact that their major intellectual purpose is the denial of the reality of purpose itself should now want to concern themselves deeply with the relative valuation of accidental pieces of their random world is incoherence compounded. If there is objective value then its attainment becomes a public purpose. Even if value is subjective it remains causative for individuals who act on it as a private purpose. Neo-Darwinists who do not accept the reality of purpose in either sense owe it to the rest of us to remain silent about valuation – and conservation as well.

According to biologists the existence of any species is an accident, and its continued survival is always subject to cancellation by random mutation and natural selection anywhere in the interdependent ecosystem. For people who teach this doctrine to sophomores on Monday, Wednesday and Friday to devote their Tuesdays, Thursdays and Saturdays to pleading with Congress and the public to save this or that species is at least surprising. Naturally the public asks these biologists what purpose would be served by conserving certain threatened species? Since most leading biologists claim not to believe in purposes, ends, or final causes, this is not an easy question for them to answer. They reveal the inconsistency that Whitehead saw lurking in the background by the feebleness and wavering half-heartedness of their answers. They tell us about biodiversity, and ecosystem stability and resilience, and about a presumed instinct of biophilia that we who systematically drive other species to extinction are nevertheless supposed to have encoded in our genes.

But the biologists are too half-hearted to affirm any of these descriptive concepts as an abiding purpose, and thereby challenge the fundamental assumption of their science. For example, biophilia could be appealed to as a virtue, a persuasive value, a *telos* rather than a wishfully imagined part of the deterministic genetic code. But that would be to admit purpose. Instead the biologists try to find some overlooked mechanistic cause that will make us do what we suspect we ought to do, but cannot logically advocate it without acknowledging the reality of purpose. Yet, without purpose and value, the biologists' appeals to the public are both logically and emotionally feeble. Is it too much to ask the neo-Darwinist think about the 'lurking inconsistency' – to speculate about the possibility that the survival value of neo-Darwinism itself has become negative for the species that really believes it?

24.8 The Purposeful Perpendicular

Why the above excursion into the problems of neo-Darwinism in an essay on valuation and national accounts? The issue under consideration is the omission of purpose and objective value from our cultural worldview, and that omission is most evident in and influential through neo-Darwinism. But economics is not far behind in its denial of objective value. It hangs on to purpose only in the attenuated, but still causative, notion of personal preferences. One small but important barrier to this march of insane purposelessness is Hueting's perpendicular social demand curve. This humble perpendicular represents the assertion of objective value and final causation in a world whose plausibility structure recognises only efficient causation. It is the insertion of a public purpose, a *telos*, an objective value, into both the value-free world of biology and the subjectivistic world of economics. Neither world welcomes it. The perpendicular asserts that sustainability is indeed not derivable from individual preferences, unless individual preferences, like sensitive compass needles, respond to the pull of 'magnetic north', to the lure of objective value. In the latter case it is of course objective value that is luring and persuading preferences, and consequently subjective preferences would not be the ultimate source of value.

The Enlightenment, with its rejection of teleology, certainly illuminated some hidden recesses of superstition in the so-called Dark Ages. But the angle of its cold light has also cast a deep shadow forward into the modern world, obscuring the reality of purpose. To attain the purpose of using the biosphere sustainably we will first have to reclaim purpose itself from the dark shadows.

References

Birch, C. (1990). *On purpose*. Australia: New South Wales University Press.

Hicks, J. (1948). *Value and capital* (2nd ed.). Oxford: Clarendon Press.

Hueting, R. (1980). *The new scarcity. More welfare through less production?*. Amsterdam: North-Holland. (Dutch ed. 1974)

Hueting, R. (1991). Correcting national income for environmental losses: A practical solution for a theoretical dilemma. In R. Costanza (Ed.), *Ecological economics. The science and management of sustainability* (pp. 194–213). New York: Columbia University Press.

Hueting, R., & Reijnders, L. (1998). Sustainability is an objective concept. *Ecological Economics, 27*(2), 139–47.

Lewis, C. S. (1944). *The abolition of man*. London: Macmillan.

Newbigin, L. (1986). *Foolishness to the Greeks. The gospel and western culture*. Grand Rapids MI: Wm. B. Eerdmans.

Whitehead, A. N. (1925). *Science and the modern world* (p. 76). London: Macmillan.

Chapter 25
A Rejoinder to Wilfred Beckerman and Herman Daly[1]

Roefie Hueting

Contents

25.1 Introduction 491
25.2 Natural Resources and Life Support Systems ... 492
25.3 Extrapolation 493
25.4 A Rejoinder to Herman Daly 494
25.5 Hicks's Concept of National Income and Sustainable National Income 496
25.6 Vision 497
References 501

25.1 Introduction

According to Wilfred Beckerman, environmental considerations provide no grounds whatsoever for concern about the physical conditions under which future generations will have to live. In his opinion, a study of Sustainable National Income (SNI) designed to estimate the distance between actual and sustainable levels of production and consumption is therefore entirely superfluous. The value of such a study rests, moreover, so Beckerman holds, on the erroneous notion that rights can be conferred upon future generations. Any undertaking on behalf of such generations can at best be based on 'imperfect obligations' borne of moral considerations.

Let me start with the second point. Conferring rights has nothing whatsoever to do with studying Sustainable National Income. The SNI according to Hueting is not based on the rights of future generations, nor on intergenerational equity, but on the *preferences* of the present generation for handing down the vital functions of our physical surroundings (the environment) intact to generations to come. There are two grounds for assuming such preferences. First, the existence of 'blockages' preventing these preferences from being expressed (Hueting and de Boer 2001b). Second, the postulate that 'man derives part of the meaning of existence from the company of others'. These others include in any case his children and grandchildren. The prospect of a safer future is therefore a normal human need, and dimming of this prospect has a negative effect on welfare' (Hueting 1987).

Regardless of these considerations, though, why should one not speak of the rights of future generations? Unlike Wilfred Beckerman, I am a complete layman in the realm of philosophy (of law). Nonetheless, I do not see how his reasoning can stand up to scrutiny. Wilfred is obviously right in stating that if something no longer exists (an animal species, for example) there is no sense in speaking of rights and obligations. But why should we not be able to confer upon future generations the right to dispose over vital environmental functions that *do* still exist, among them the functions of life support systems, of which natural ecosystems (biodiversity) form an important part? If we assign such rights, surely these would be balanced by our obligation not to destroy those systems? Then surely Beckerman's logical condition is satisfied that rights should always have their counterpart in obligations? Surely the indeed unsatisfiable condition that future generations be able to exercise or waive such rights is not a *conditio sine qua non?*

I put these questions to C. W. Maris, specialized in philosophy of law, and W. Achterberg, specialized in environmental, moral and political philosophy, both at the University of Amsterdam. Their answer: 'yes, indeed'. They had more to say, though.

---

[1] This chapter is reproduced with adaptations from: R. Hueting (2001). Rejoinders to symposium authors (chapter 13). In C. Ekko, van Ierland, J. van der Straaten, & H. Vollebergh (Eds.), *Economic growth and the valuation of the environment: A debate*. Aldershot: Edward Elgar (with permission).

Jan J. Boersema and Lucas Reijnders (eds.), *Principles of Environmental Sciences*,
© Springer Science+Business Media B.V. 2009

One's answer depends on how 'rights' are defined. It is, in short, a question of semantics. If rights are defined such that the bearer thereof is to be able to choose whether or not to exercise them – if, in other words, they are defined in terms of the protection of individual autonomy – then there is an identity problem. 'Future persons' cannot, after all, be individually identified. No contract can therefore be concluded with them entailing specific rights and obligations. Beckerman's discourse proceeds from this definition of rights. However, if rights are defined in terms of the protection of interests, there is no problem of identity. Interests are often diffuse and not specifically attached to discrete persons but to groups of persons, such as future generations, for example. In the latter case we can proceed from the reasonable assumption that there will be people in the future and that these 'future persons' will have an interest in being able to dispose over those environmental functions that are indispensable for human life (see, for example, Feinberg 1974). It is, therefore, surely possible to confer rights of the disposal over those vital functions upon future generations, balancing them with an obligation to hand down these functions undamaged.

Beckerman argues the first point – that there is no need for concern about future generations – solely by extrapolating observed trends *vis-à-vis* technological progress. Two points of criticism can be raised here: (1) his extrapolation is only partial; (2) it is unduly simple.

25.2 Natural Resources and Life Support Systems

On the first point, Beckerman asserts that my exclusive focus is on stocks of natural resources and in his contesting the need for concern he refers solely to stocks of non-renewables, albeit that renewable energy sources are mentioned in a discussion of fossil fuels. Non-renewables formed in slow geological processes are clearly important natural resources. Beckerman's treatment of these resources is inconsistent and takes inadequate account of the many different functions of the resources. Thus he argues that 'we will never run out of any resource or even suffer seriously from any sudden reduction in its supply'. Further on, though, he approvingly quotes authors who hold that fossil carbon resources will run out in several hundred or thousand years.

He asserts, furthermore, that 'if, for example, coal were ever to become a very scarce commodity its price would rise to the point where, like other minerals, such as diamonds, it would be used only for jewellery or certain very special industrial uses'. In that case, however, its function as a major source of energy would clearly be lost. Because natural resources formed in geological time should for practical purposes be considered finite and because prices may be expected to soar as resources peter out, ongoing dissipation of these resources will lead ultimately to loss of most, if not all their current functions. With ongoing dissipation, the question is not whether functions will be lost, but when.

Beckerman moreover ignores the fact that burning all fossil carbon stocks would lead to dramatic climatic changes (Graedel and Crutzen 1993).

Non-renewables are not the only resources on which humankind depends, however. Contrary to Beckerman's assertion, in my work I have always stressed the importance of the life support systems of our planet. Life support systems are the processes that maintain the conditions necessary for current life on earth. This comes down to maintaining equilibria between narrow margins. The processes may be of biological or physico-chemical nature, or a combination thereof. Examples of biological processes include the carbon and nutrient cycles, involving the extraction of such substances as carbon dioxide, water and minerals from the abiotic environment during the creation of biomass and the return of these substances to the abiotic environment during decomposition of biomass. Examples of physico-chemical processes include the water cycle and regulation of thickness of the stratospheric ozone layer. These examples show that there is interaction between the processes, whereby equilibrium may be disturbed. The water cycle, for example, may be disturbed by large-scale deforestation.

Life support systems constitute a key element in the estimation of sustainable national income. Life support systems are not stocks but *processes*. These processes are irreplaceable and non-substitutable; and they are threatened by the ongoing growth of production and consumption by ever more people. Their equilibrium is vulnerable, extremely complex and as yet only very imperfectly understood. There are already instances in which meddling with those processes has had major deleterious consequences. Cases in point include the recent collapse of a number of fish populations in the North Atlantic and several other important fishing grounds (FAO 1990–1998; Tolba and El Kholy 1992) and, in the more distant past, the collapse of agriculture and a highly evolved culture on Easter Island (due to erosion and the clearcutting of forests) and in what is now southern Iraq (due to salinization) (Reijnders 1996).

Natural ecosystems are part of the life support systems on which humankind depends. The threat to such ecosystems goes far beyond the extinction of Beckerman's Dodo (Goodland 1995; Lovelock 1979; Odum 1971). Species are currently being lost at a rate exceeding new species formation by several orders of magnitude (Hueting and Reijnders 1998) and the area of land covered by natural ecosystems is dwindling rapidly. This is fast reducing the availability of resources for plant breeding and pharmaceuticals production and altering the composition of the atmosphere (Goodland 1995; Graedel and Crutzen 1993; Lovelock 1979; Tolba and El-Kholy 1992).

These processes, absolutely vital for future generations, have no part in Wilfred's extrapolations. They are not even mentioned, in fact.

25.3 Extrapolation

My second point of criticism is that Beckerman merely extrapolates historical trends. Attempts to shed light on future economic developments are unfortunately not that straightforward, however. I might well do the same as Wilfred. Consider the following: in estimating the distance between the standard world income and its sustainable counterpart Tinbergen and Hueting arrive at a rough figure of 50% of current world income (see Tinbergen and Hueting 1991). The provisional results of a modelling exercise show that this estimate is by no means extreme (Verbruggen et al. 2001). This study also demonstrates that the present 'sustainability gap' cannot be bridged with the technology currently available, implying an additional need for changes in consumption patterns. Not that long ago, on the relevant geological time scale, this gap was virtually non-existent, for environmental functions were then free goods. Despite gigantic technological strides being made, then, with regard to those functions of our physical surroundings that permit our very survival, there is an empirical trend of growing scarcity. Entirely in line with Wilfred's method, I might then extrapolate this trend and conclude: if we continue down the present path, humankind as a species is doomed to extinction, regardless of technological progress. Obviously more information can be gained from tracing the development of the gap between SNI and actual national income than from extending lines.

I share Wilfred's optimism about technological progress, although not *in extremo*, particularly when it comes to renewable energy resources to substitute for fossil fuels. However, I would point out a number of errors and omissions that render his extrapolations unduly simple.

1. In Beckerman's view, scientific and technological progress is the fundamental driving force determining production growth. I agree. However, intellect and our physical surroundings are entirely complementary. All produced goods, including human-made capital goods, are a combination of intellect and elements of our physical surroundings. Without the latter, we humans are left empty-handed. The issue is therefore whether the functions of our physical environment remain intact. A number of elements of our environment are irreplaceable and non-substitutable. These Beckerman ignores, as I have already noted. Exhaustible resources must be substitutable (in time) by other elements of our physical surroundings that can provide the same services. Wilfred's assertion that this will remain possible *ad infinitum*, regardless of the levels to which production and consumption soar, is pure speculation and thus at odds with the wisdom of the precautionary principle. He argues this position with almost exclusive reference to energy. But what about the resources required for food production and the world's finite stocks of freshwater? There is no agreement about the adequacy of natural resources to feed an ever-growing world population, particularly given the finite nature of freshwater resources and the ongoing deterioration of land suitable for agriculture due to soil compaction, erosion, salinization and other processes (Kendall and Pimentel 1994; Tolba and El-Kholy 1992). For Wilfred, however, all this is absolutely no cause for concern, for his extrapolations assume ever-rising per capita income, *ad infinitum*.

2. Beckerman asserts that the price mechanism will ensure that there are no disastrous shocks. There have been serious famines in the recent past, however, in some cases compounded by shortages of drinking water, and new famines are anticipated in the near future (Kendall and Pimentel 1994; Tolba and El-Kholy 1992).
3. Beckerman's extrapolations are unduly simple, for he overlooks at least six factors of relevance.

First, he makes no allowance for the fact that any price rise in real terms means a decline in the volume of national income and therefore a check on production growth.

Second, Wilfred overlooks the fact that, for a given technology, product costs will rise progressively as the yield (effect) of environmental measures is increased. Technological progress leads to higher yields, of course. As production increases further, however, so too must the yield of the measures in order to maintain the same state of the environment, while the fact of progressively rising costs with rising yield remains unaltered. There is thus a 'race' between environmental technology and production growth, the outcome of which cannot be predicted.

Third, Wilfred neglects the fact that the vast bulk of production growth is generated by the approximately 30% of industries that cause most pollution, utilize the greatest amount of land and deplete natural resources most rapidly (Hueting 1981; Hueting et al. 1992). A shift towards environmentally more benign activities therefore implies a substantial check on production growth. Such a shift is already unavoidable if sustainability is to be achieved.

Fourth, Wilfred neglects to consider the impact of increasing land use on, *inter alia*, species extinction, because he ignores the importance of life support systems for human survival.

Fifth, Wilfred does not take into account that in a long historical time series energy consumption (and CO_2 emissions) is found to run parallel to production and that there is negligible substitution by renewable energy sources, that is, energy derived from the sun (see Tinbergen and Hueting 1991). This trend he does *not* extrapolate. Although in the recent past there have been brief periods in which production growth, as measured in national income, was not accompanied by rising energy use, overall in the last 50 years this has been the case (Adriaanse et al. 1997; de Bruyn et al. 1998; Schipper and Meyers 1992). As a rule, moreover, renewable energy is currently much more expensive than energy generated using fossil fuels. In the case of photovoltaic power, the price may even be far higher than electricity from a coal-fired plant (Johansson et al. 1993). And although I share Wilfred's conviction that the cost of renewable energy will fall substantially in the future, implementation costs will do so far less. The latter will be extremely high, in my estimate, which will check production growth. For this check to be removed and continued production growth achieved, the renewables will have to become *cheaper* than current fossil energy. Whether this will ever be the case we cannot predict. The same holds for substitutes for any other resource.

Sixth, Wilfred makes no allowance for the fact that, for there to be growth of the total mass of production, the productivity of the approximately 30% most environmentally burdening sectors that generate most of that growth would have to rise by far more than the overall average. Consequently, the yield of environmental measures in these sectors would likewise have to increase by far more than the average across the national income, in order to maintain the same state of the environment – possibly a factor two or three more, with progressively rising elimination costs and a severe check on production growth as a result. A correct estimate of this factor requires more than a back of the envelope calculation, however.

If due allowance is made for all these factors (and their interrelationships), the growth and other figures predicted by Wilfred may well prove to be a factor 10 to 20 too high. But that requires a far more extensive analysis than Wilfred has presented and, as I have said, I have major doubts about the usefulness of such prognostications.

25.4 A Rejoinder to Herman Daly

In the debate on nature and the environment there are few people with whom I feel such affinity as Herman Daly. He too deems production growth, as measured in the national income, to be at the heart of the environmental problem. I had therefore expected my rejoinder to Herman to be easy. But Herman has a habit of making people pause for reflection and I indeed prove no exception. In his study he poses a key question that is scarcely ever asked: is there such a thing as objectively good preferences? If the answer is 'yes', then according to Daly this means that alongside subjective preferences there are also objective preferences, since the subjects subordinate their preferences to what is objectively good (or ought to do so). This leads logically to the thesis that objective values exist, alongside subjective.

One of the artefacts used by Daly in defence of this thesis is my perpendicular demand curve, that is to say, perpendicular in the relevant range. In doing so, he misinterprets this curve. This must have been a conscious move, because in January 1999 we corresponded about the background of the curve, on which occasion I plainly stated my view on the matter. This view has been reported (and clarified relative to earlier formulations) in my publications of 1992 and thereafter: the perpendicular curve is based on an *assumption* regarding *subjective* individual preferences, namely that subjects have an absolute preference for attaining standards of sustainability for vital environmental functions; the absolute character of these preferences is bound up with the nature of a standard; the assumption is inevitable because preferences can be deduced only partially and is legitimate because of the existence of what I have termed 'blockages' (Hueting and de Boer 2001b). Daly posits that my perpendicular 'demand curve' is intended to be an expression of objective value, not individual preferences. This is then compatible with Daly's own perspective, in which the attainment of sustainability is an end that is objectively good and one to which the subjects conform their individual ends or wants and thus their preferences (reflecting the relative weights of those wants); these individual preferences are thus objective, for they are derived from an objectively good, generally valid end. This, at least, is how I understand Daly.

With his interpretation, Daly apparently wished to initiate a debate on preferences and the underlying ends or wants (two words for the same notion), and not only those of human beings but also of other organisms. For this I am grateful, although I myself have a different outlook. For the same holds true for perspectives as for assumptions: rendering them explicit improves the quality of the information, because readers then know what they are up against.

Of somewhat less apparent interest are Herman's remarks about a perpendicular supply curve for environmental functions. At the end of the present section I hope to demonstrate that such a curve is an impossibility.

In treating Herman's study, I shall first expound my view on economics, which I shall then confront with his theses. However, these theses in fact imply an outlook on the world that co-determines Daly's perspective on economics: there is thus an underlying perspective or 'vision'. To this I shall turn after discussing Hick's concept of national income and Sustainable National Income. Those who dispute the scientific nature of the portion of my rejoinder ensuing from 'vision' and thus from personal experience and beliefs in the widest sense, I shall not counter.

My view on economics, by no means unique, can be summarized as follows. Economics rests on assumptions. The basic assumption is that in their dealings with scarce means human beings endeavour to achieve maximum satisfaction of their wants; 'endeavour', because mistakes are possible, and may sometimes be serious. This is a tautology, because the opposite is nonsensical, but it is nonetheless of the utmost importance and all too frequently forgotten. Wants or ends are meta-economic and are not for economists to judge. Ultimately, the economic problem boils down to the problem of choice arising from scarce means and conflicting ends (or 'purposes' in Daly's terminology). The bottom line – the satisfaction of wants, that is welfare – is not amenable to direct measurement. And so we make do with measurable factors which are assumed, by reasoned argument, to have an influence on welfare.

From this it follows that an assumption regarding wants must *always* be made. The assumption must be reasoned through, it must be plausible. In my work the plausibility and legitimacy of the assumption of preferences for sustainability are defended. It is not only the prices of environmental functions that are unknown because preferences are unknown; the correct market prices of goods that are produced and consumed at the expense of these functions are likewise unknown, and from that I draw the conclusion that the standard national income is also based on assumptions and is therefore just as hypothetical as the Sustainable National Income (SNI). There is also another argument, one that is scarcely ever mentioned in discussions, if at all. To hold market prices and the national income based on these prices for certainties is to assume implicitly that there is absolute consistency between wants and actions. This consistency has been disputed, *inter alia* by Marcuse (1964), who posits that we are caught up in the existing structure of production and consumption and in our actions are consequently hampered in satisfying our true wants. Others have expressed similar views (see footnote on page 88 of Hueting 1980). There is no way to decide this issue one way or the other: it is a question of plausibility, and I myself hold the opposite of Marcuse's thesis to be the more plausible. But I certainly do not reject the thesis as nonsense, when, for example, I see parents, *nolens volens,* driving their children to school or the playground because of the dangers of traffic (see, for example, Hueting 1974). Kuznets would probably characterize these expenditures as intermediate and thus as costs (see Hueting 1974). Summarizing: (1) people buy cars, meat and holiday flights because they like them; and (2) the assumption that the positive value of these products is higher than the negative value of the inherent loss of vital functions, which is implicitly made when constructing and using national income figures, is disputable; other assumptions are defensible.

Proceeding from this view, I shall now turn to what I consider to be the principal issues raised in Herman's study.

25.5 Hicks's Concept of National Income and Sustainable National Income

For Daly the only valid definition of (national) income is that of Hicks's *Value and Capital* and it is this that should, he holds, be my basic point of reference. Hicks's concept of national income, says Daly, is 'built around a prudential purpose' and consequently not based on individual preferences. That purpose can be formulated in two ways. The first is: net consumption of capital is to be avoided. If capital is consumed this counts as negative income, for which an appropriate correction should be introduced to arrive at the correct figure. National income, in Daly's view, is therefore by definition sustainable and the term Sustainable National Income a pleonasm.

But Tinbergen, in the 1930s a chief architect of the system of National Accounts and the concept of national income, and implicitly also Kuznets, for whom the same applies, albeit to a lesser degree, did proceed from individual preferences, with Tinbergen making clearly explicit the assumptions he thereby made (see Hueting 1974; Tinbergen and Hueting 1991). Given that official national incomes are based on this concept and that their growth enjoys top priority in economic politics the world over, to the detriment of the environment, there is an eminently practical reason for conforming to this concept in correcting for environmental loss: comparability. But there is also a theoretical reason for following Tinbergen rather than Hicks, and it is equally important. From the view on economics outlined above, Hicks's income concept rests on an assumption *vis-à-vis* individual preferences: that subjects are prepared to sacrifice a portion of their present consumption, namely up to the time-dependent and therefore variable level required to keep constant the capacity for producing consumption goods, expressed as the aggregate stock of man-made and natural capital. That level (the Hicksian income according to this first formulation) may thus be lower than the level actually realized, even at the start of the year. In Hicks's day the implicit assumption regarding individual preferences was anything but robust: structural, long-term production growth was (and for many still is) an eternal given and the environment was still conspicuously absent in the economic literature (see Tinbergen, in Hueting 1974; Tinbergen and Hueting 1991). Applied to the situation post-1970, however, the assumption is certainly robust.

This can best be clarified with reference to a second, oft-used formulation of Hicksian income: income is what one can spend (consume) in the course of a year without being worse off at year's end. But whether one is better or worse off depends on one's preferences for current consumption relative to future consumption options. If the latter weigh more heavily in a person's preference pattern, then in pursuing Hicks's 'prudential purpose' (Daly's phrase) he or she will be better off, but at a substantially lower level of consumption. If current consumption weighs more heavily, they will be worse off.

Stringing individual years into a long-term time series, it is only the second formulation that yields a constant national income. If preferences for sustainability are moreover assumed to predominate, this yields a national income that is the maximum achievable while securing sustainability; this is then the SNI according to Hueting. But, in contrast to the Hicksian income, this SNI does rest on a robust assumption regarding preferences: that subjects are 'better off' if, and only if, vital environmental functions remain available *ad infinitum*. Contrary to the situation in Hicks's day, those preferences involve a willingness to accept the sacrifice of a substantial reduction in consumption and a drastically changed consumption pattern. Such willingness and acceptance can be neither proven nor falsified and can therefore only be assumed. To dispute that securing sustainability requires sacrifice is to dispute the existence of the environmental problem.

In addition, the magnitude of this sacrifice, and with it the level of the Sustainable National Income, can in practice only be estimated with the aid of quantitative sustainability standards established with the greatest scientific rigour (see Hueting and Reijnders 1998). Again, this is not something to which Hicks attended, because in his day the environment was not an issue. Neither, certainly, did Hicks foresee a substantial sacrifice of consumption. In designing the concept of 'national income' Tinbergen likewise ignored the environment, because it was not then a factor of significance; as from the early 1970s, however, he was a firm advocate of correcting national income for environmental loss (see Hueting 1974; Tinbergen and Hueting 1991). Whether Hicks, too, adopted a similar position in his later life we do not know. Daly as well as Salah El Sarafy, whom he cites, ignore both the assumption regarding preferences and the practical necessity of estimating physical sustainability standards. My conclusion: in the novel situation of a new fundamental category of scarce goods, the established concepts of the economic literature cannot be applied gratuitously.

25.6 Vision

The second point, more than the first, is concerned with the deep, philosophical questions raised by Herman in our correspondence, which I have interpreted above as his underlying perspective or 'vision'. Herman makes it especially difficult for me here, for when he uses the word 'purpose' he appears to have two entirely different senses in mind, which he does not disentangle: (a) '(human) ends', in the established economic sense of the term, that is meaning wants or needs (fulfilling wants is the same as attaining ends); and (b) the general philosophical or religious meaning of 'object for which a thing exists, final cause' (Oxford Dictionary). What is concerned in the latter case, as I understand it, are such questions as: Is there a purpose to life, to existence, or is it without purpose? What is the meaning or purpose of existence? This impression is reinforced by Daly's use of the term to encompass all other organisms on the planet, too.

Daly answers the question whether human beings and other organisms have a purpose unshakingly in the affirmative. As an agnostic, my own answer must be: I do not know. On the sole basis of Daly's text I cannot deduce what exactly he holds to be the purpose, but my interpretation is that conservation of nature and the environment, and thus sustainability (or, more religiously coloured, 'stewardship'), are certainly part of it.

I use this interpretation to bridge the gap to the strictly economic sense of the term 'purpose'. Daly addresses my concerned question: If individual preferences are rejected, by whom or by what are our preferences determined? Not, evidently, by government. Neither is that indeed feasible, of course, certainly not in the case of sustainability. Although many governments have rallied behind 'sustainability', following the Brundtland Report, they still continue to give top priority to the conflicting goal of 'production growth': governments build roads, fill in estuaries to construct harbours, open up natural areas to development and neglect to make the market operate in a legislative framework that might lead to sustainability (internalization). The Brundtland Report is itself a matter of conflicting goals, it should be added (see Hueting 1990). Daly's answer to my question, now, is the following: 'purpose' is an objective value, a moral compass ('magnetic north'), a generally valid goal which is objectively good and to which both individual subjects and governments (should) subordinate their other objectives.

But this answer begs new questions, such as: (1) How do such 'objectively good' goals arise, and in particular that of 'sustainability'? (2) Alternatively, do they not arise but simply exist? (3) In that case must they be 'eternal' and therefore immutable? (4) How do individuals recognize the objectively good goals to which they (should) conform? From a close reading of Daly's text I distil the following answers to these questions. These objectively good goals are simply there. They are immutable over time. The subjects are well aware what these goals involve. They also know they should (really) show the courage or willingness to subordinate their other wants or goals accordingly – in the realm of sustainability, the goal of additional consumption or unprotected sex, for example. They should also refuse to allow themselves to be manipulated by governments that give priority to goals that conflict with the objectively good. Sustainability is, for sure, an objectively good goal.

Here we have a view on economics derived from a particular outlook on life, on human existence. Herman Daly is a believer. As an agnostic, I lack such faith. But before the reader rejects as eccentric the notion of a person's faith influencing his or her view on economics (including sustainability), let me recall Adam Smith, often cited as the founding father of economics. For was it not Smith who said: the baker bakes his bread as well as he can, not to serve his fellow man but in his own interest? Herein, too, lies embodied an outlook, on the human species that determines the perspective from which economics is practised.

Above, I have stated my own view on economics. In consciously misinterpreting my perpendicular demand curve, Herman is in fact implicitly asking me: What is your underlying view or 'vision'? Challenged thus, and for this reason alone, I shall answer that question. Compared with 50–65 years ago, when I held a firm belief, there remains today but one element: a belief in information. I proceed from the assumption that if individual subjects are in possession of appropriate factual information, they will make fewer and less serious errors in endeavouring to attain their goals or, in other words, maximize their welfare (see above). I make no pronouncement on the virtue of those goals (Hueting 1974). In religious terms: it is God (rather than man) who is the judge of motives. I thus assume that when subjects have made their choices on the basis of correct information the outcome thereof is 'good' in the sense just mentioned.

The hardest problems arise in the case of collective decisions concerning collective goods such as the environment (sustainability), democracy, and war and peace. As history shows, it is in these realms, above all, that agreement on the validity of information is often lacking, that information is often manipulated or, worse, supplanted by disinformation. History also shows that subjects experience bitter remorse, after the event, if they have acquiesced unprotestingly, due to ignorance, indifference, impotence (real or imagined) or lack of courage, to decisions which for them have proved disastrous and which were taken by individual subjects constituting government, even if those subjects were elected democratically.

Examples are rife, even restricting myself to what I myself have witnessed (sometimes at close hand). In the democratic elections of 1930, Hitler's NSDAP became Germany's second largest party; in 1933 he was elected Chancellor, by way of the democratic process. A small number of individual subjects informed the world about the lies on which that victory rested and about subsequent turns of event. I was seven when I first heard that information. With my own eyes I saw how the majority chose not to believe that information ('the easy way'?) and how already then, 4 years prior to the outbreak of war, resistance to an inevitable occupation was therefore mounted by a small group of individual subjects. Today, that information is universally believed, and remorse is bitter. Shortly after the war, the democratically elected individual subjects comprising the Dutch government chose to undertake two 'police actions' in Indonesia. The vast majority of the population acquiesced. From Dutch citizens in Indonesia information was available to the effect that those actions constituted a colonial war against freedom fighters. There were but few who gave credence, however. The vast majority believed the aim was to restore law and order, which had been disturbed by a *small* group of 'terrorists'. That majority also included the *small* group of individuals from whom I had heard what was going on in Germany 10 years prior and who had mounted resistance, regardless of personal consequence. Today, remorse is the predominant sentiment.

In India, marked on the constitutional atlas as a democracy, I have seen how (widespread) child labour is accepted by much of the population; in some of its manifestations the term slave labour is closer to the truth. This acceptance is likely to decline, however, when information on the consequences for these children sinks into people and the size of families decreases. In my experience, the latter will only be the case if the status of women is improved. For thousands of years slavery and the denial of women's rights, including passive and active voting rights, were accepted facts. In ancient Greece, where the word and notion of democracy were born, the very concept of suffrage for women was not even raised, even by women; slavery was an accepted fact of life. The same held true in classical Rome. Indeed, denial of women's suffrage was generally accepted in both Europe and the USA until the 20th century. In those countries, slavery was long accepted by the vast majority of the population and their churches. Today, shame predominates. At any particular time, some groups of individuals consider themselves *genuinely* superior. Herein lies part of the cause of these historical events, and of what is still happening today. But human perspectives and the preferences based upon them are continuously evolving.

All these phenomena, from personal experience or from literature, have led me to a different outlook from that of Herman Daly on a number of issues. Let me mention the principal of these. (1) My faith in such abstractions as 'government objectives' has declined. (2) Democracy, contrary to what Herman holds, certainly does find itself subjected to a popular vote – every day anew. (3) Individuals having different preferences which, contrary to what Herman holds, are not a priori either good or bad, may, again contrary to what Herman holds, most surely engage in dialogue. They may exchange information, validate that information and discuss interpretation thereof. Indeed, democracy requires that they do so as an intensive and ongoing process. (4) Perspectives on good and evil evolve over time and differ from culture to culture, often markedly so. But on one point we can wholeheartedly agree: individuals should not allow themselves to be manipulated by other individuals, whether embodying government or otherwise.

Above, most of what there is to say about preferences for sustainability has, *mutatis mutandis,* already been said. Individual subjects, be they citizens or politicians, are confounded in the extreme by the economic information with which they are confronted. Largely through the media and other coverage of economic affairs, there has evolved an almost universal belief in a conflict between environment and employment and in production growth being able to go hand in hand with environmental conservation – and, indeed, even in the necessity of such growth for conservation, whereas the opposite is true: environmental protection creates jobs and conflicts with production growth (Hueting 2001). There is absolutely no justification for proceeding from individual behaviour, because of the existence of what I have termed blockages (Hueting and de Boer 2001b). Scarcely, if ever, can subjects form a conception of the *physical* benefits of extremely complex processes, the physical costs (as negative benefits) of disturbance thereof and the technological (im)possibility of intervention in such disturbance. This is also one of the insurmountable problems of contingent valuation (see Hueting 1989, 1992, 1995). Although on the scientific side our information is better and more consistent, here too there are still major uncertainties. Let me mention the two examples that to me appear the most important: (1) the forced greenhouse effect and (2) the macro-effect of species extinction. The uncertainty regarding the first I presume to be universally familiar. The latter uncertainty is discussed in Hueting (1974), based on the work of Eugene Odum and also personally validated by him. In short, the continuing burden on the environment may lead to an ecological crisis in which man's continued existence is at stake, but there is no way of indicating at what level of burden such a crisis will occur. In the estimation of SNI according to Hueting, species conservation is nonetheless taken as a sustainability standard. That norm is based on the precautionary principle and on the fact that the rate at which species are becoming extinct today is at least a factor 10,000 higher than the rate at which new species are evolving.

My conclusion: there is considerable scope for improving and debating information *vis-à-vis* sustainability, as well as for debating what 'preferences for sustainability' precisely are, before judgement can be made about those preferences. I conclude, and here I differ with Herman, that an assumption about preferences is inevitable as well as legitimate.

Herman posits that estimation of a Sustainable National Income is a different matter from designing a policy for attaining sustainability in a market economy. But the opposite is true: estimates of SNI are based precisely on simulated internalization of sustainability costs in market prices (Verbruggen et al. 2001). He also posits that introduction of tradable permits leads to a perpendicular supply curve (and thus to an objective value). However, supply curves for environmental functions are made up of increasing costs per unit reduction of environmental burden, and thus per unit function restored or supplied. They are empirically grounded and have been estimated since the early 1970s (see Hueting, 1974 for several examples). Obviously, introduction of tradable permits does not change the characteristic, progressively rising shape of the curves. The sole objective is to achieve a certain level of environmental burden (for example emissions) as efficiently as possible. Any standard of environmental burden, sustainable or otherwise, can be represented as a limit and therefore as a perpendicular. Subsequently preferences, and therefore a *demand*, can be assumed which correspond to the (sustainability) standard. The degree of availability of environmental functions is determined by the standard (and thus by demand). The cost of supply depends on the shape of the specific elimination cost or supply curve determining the point of intersection with the standard or demand curve. This is shown *inter alia* in Fig. 2.7 of Hueting and de Boer (2001b) for a sustainability standard D.

A second objection is that there is no such thing as a perpendicular supply curve for environmental functions. Herman here appears to be making the same error as Costanza et al. (1997). A perpendicular supply curve intersects the abscissa. This means that the function (and its services) can be supplied at zero cost up to the perpendicular, which boils down to the statement that functions (including the vital functions of the world's ecosystems) are free goods, because they are 'supplied' free of charge. This is an inappropriate representation of the opportunity costs associated with provision of a typical environmental function. Consider the graphs A and B in Fig. 25.1.

The horizontal axis is some index of quantity and quality (that is: availability) of an environmental function, and the vertical axis is money units. In graph A, a certain amount X_0 of the function is available as a free good, but the cost of increasing the function's availability beyond this level is infinite. This corresponds to the 'indestructible land' of the classical economists, such as John Locke, Jean Baptiste Say and David Ricardo. But it is certainly *not* the typical situation when it comes to environmental functions, including those of the planet's forests, marine ecosystems, wetlands or arable lands, all of which are clearly *destructible* (more or less irreversibly).

Now consider the question of how to value this indestructible asset. Say that it is productive land and that the product (say corn) has a money price. If, after every man has taken as much land as he wants, there is still some to spare, then the resource will have a zero price (free good). If, on the contrary, the amount is less than what people would utilize if it were freely available, the land will command a 'rent' as a factor of production, in accordance with established economic theory. However, this rent does *not* measure the cost of producing land, nor is it an opportunity cost in the sense of production having to be sacrificed to obtain more land. There is no question, in this situation, of loss of functions due to depletion of land, nor of marginal cost of an additional unit of land: its supply is fixed. So 'rent' is a value category that arises through imputation back to land as a factor of production, reflecting the usefulness of the land as a source of income through the market value of the goods produced with that land.

Consider by contrast environmental functions from, say, aquatic of forest ecosystems. These are not indestructible, neither is their quantity fixed. The amount of environmental functions available in any period of time becomes an endogenous variable. Today, the functions of ecosystems are no longer free goods and their supply curves are typically of the form shown in Fig. 25.1 (graph B), as demonstrated by research (see for example de Boer 1996; Hueting 1974). Measures must – and can – be taken in order to restore and safeguard the functions that one wishes to maintain, to the extent that irreversible losses have not yet occurred. These measures form (opportunity) costs. So sacrifices in terms of consumption (or family planning: lower national income, more environment) are evidently unavoidable. The costs of the measures form the basis for constructing a supply curve in the normal economic sense of the concept: the supply of a desired economic good through human activity involving the sacrifice of an alternative. The necessary sacrifices are obviously not reflected in the perpendicular that Daly and Costanza et al. (1997) have in mind. Therefore this curve is not a supply curve in the economic sense and cannot consequently be used for economic valuation of environmental functions and their services: all it tells us is that the cost of any quantity of function exceeding that indicated by the perpendicular would approach infinity.

Ecosystems as we find them, the result of hundreds of millions of years of evolution, cannot be quantitatively extended by human action. But they certainly can be – and in fact are – cut back by human (productive and consumptive) action. At the same time, however, these systems can be maintained and restored by human action. It looks as if what Daly and Costanza et al. (1997) have in mind is that irreplaceable goods cannot be supplied. However, it is possible to formulate

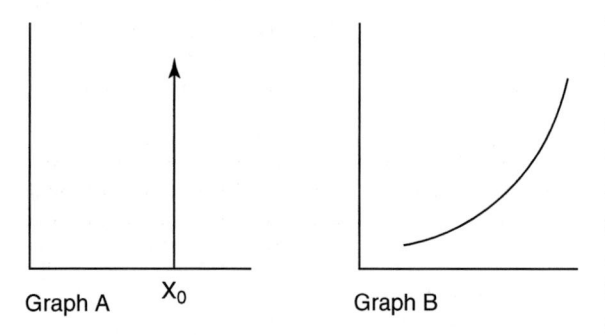

Fig. 25.1 Cost functions for environmental services

measures (what has to be done and, above all, from which activities must people refrain?) in order to safeguard (the functions of) irreplaceable ecosystems, and to estimate the (opportunity) costs thereof. This yields an elimination or supply curve: the less risks one wishes to take, the higher the costs (see Hueting and de Boer 2001a; Hueting et al. 1998).

References

Adriaanse, A., Bringezu, S., Hammond, A., Moriguchi, Y., Rodenburg, E., Rogich, D., & Schutz, H. (1997). *Resource flows: The material basis of industrial economies*. Washington, DC: World Resources Institute.

de Boer, B. (1996). Calculation of sustainable national income in the Netherlands: some results, paper prepared for the workshop *Valuation Methods for Green Accounting. A Practical Guide,* organized by The World Bank, UN Statistical Office and Ecological Economics, Washington, DC, March 20–22.

de Bruyn, S. M., van den Bergh, J. C. J. M., & Opschoor, J. B. (1998). Economic growth and emissions: Reconsidering the empirical basis of environmental Kuznets curves. *Ecological Economics, 25,* 161–175.

Costanza, R., d'Arge, R., de Groot, R., Farber, S., Grasso, M., Hannon, B., Limburg, K., Naeem, S., O'Neill, R. V., Paruelo, J., Raskin, R. G., Sutton, P., & van den Belt, M. (1997). The value of the world's ecosystem services and natural capital. *Nature, 387,* May 15, 253–260.

FAO (1990–1998). *Yearbooks of fishery statistics, catches and landings, 1990–1998*. Rome: FAO.

Feinberg, J. (1974). Non-Comparative Justice, in *Philosophical Review, 83*(3), 297–358.

Goodland, R. (1995). The concept of environmental sustainability. *Annual Review of Ecology and Systematics, 26,* 1–24.

Graedel, T. E., & Crutzen, P. J. (1993). *Atmospheric change*. W.H. Freeman: New York.

Hueting, R. (1974). *New scarcity and economic growth: More welfare through less production?* Amsterdam: Agon Elsevier (Dutch ed.), Amsterdam: North-Holland (English ed., 1980).

Hueting, R. (1980). Environment and growth, expectations and scenarios. In S. K. Kuipers & G. J. Lanjouw (Eds.), *Prospects of economic growth*. Amsterdam: North-Holland.

Hueting, R. (1981). Comments on the report A low energy strategy for the United Kingdom. In G. Leach et al. (Eds.), *Working party on integral energy scenarios*. The Hague: The International Institute for the Environment and Development (IIED), 20 May.

Hueting, R. (1987). An economic scenario that gives top priority to saving the environment. *Ecological Modelling, 38*(1/2), 123–140.

Hueting, R. (1989). Correcting national income for environmental losses: towards a practical solution. In Y. Ahmad, S. El Serafy, & E. Lutz (Eds.), *Environmental accounting for sustainable development* Washington, DC: The World Bank.

Hueting, R. (1990). The Brundtland report: A matter of conflicting goals. *Ecological Economics, 2,* 109–117.

Hueting, R. (1992). The economic functions of the environment. In P. Ekins & M. Max-Neef (Eds.), *Real-life economics: Understanding wealth creation* (pp. 61–69). London: Routledge.

Hueting, R. (2001). Three persistent myths in the environmental debate. In E. C. van Ierland, J. van der Straaten, H. R. J. Volleberg (Eds.), *Economic growth and valuation of the environment* (pp. 78–89). *A Debate*. Cheltenham, UK: Edward Elgar.

Hueting, R., & de Boer, B. (2001a). The parable of the carpenter. *International Journal of Environment and Pollution, 15*(1), 42–50.

Hueting, R. & de Boer, B. (2001b). Environmental valuation and sustainable national income according to Hueting. In E. C. van Ierland, J. van der Straaten, H. R. J. Volleberg (Eds.), *Economic growth and valuation of the environment. A Debate* (pp. 17–77). Cheltenham, UK: Edward Elgar.

Hueting, R., Reijnders, L. (1998). Sustainability is an objective concept. *Ecological Economics, 27*(2), 139–147.

Hueting, R., Bosch, P., de Boer, B. (1992). *Methodology for the calculation of sustainable national income, statistics Netherlands*, Statistical Essays, M44, SDU/Publishers, 's-Gravenhage. Also Published as WWF International Report, Gland, Switzerland, June 1992.

Hueting, R., Reijnders, L., de Boer, B., Lambooy, J., Jansen, H. M. A. (1998). The concept of environmental function and its valuation. *Ecological Economics, 25*(1), 31–35.

Johansson, T. B., Kelly, H., Reddy, A. K. N., & Williams, R. H. (1993). *Renewable energy*. Washington, DC: Island Press.

Kendall, H.-W., & Pimentel, D. (1994). Constraints on the expansion of the global food supply. *Ambio, 23,* 198–205.

Lovelock, J. E. (1979). *Gaia, a new look at life on earth*. Oxford: Oxford University Press.

Marcuse, R. (1964). *One dimensional man, the ideology of industrial society*. Boston: Beacon.

Odum, E. G. (1971). *Fundamentals of ecology* (3rd ed.). Philadelphia, Penn: W.B. Saunders.

Reijnders, L. (1996). *Environmentally improved production processes and products*. Dordrecht: Kluwer.

Schipper, L., & Meyers, S. (1992). *Energy efficiency and human activity*. Cambridge: Cambridge University Press.

Tinbergen, J. & Hueting, R. (1991). GNP and market prices: wrong signals for sustainable economic success that mask environmental destruction. In R. Goodland, H. Daly, S. El Serafy, & B. von Droste zu Huishoff (Eds.), *Environmentally sustainable economic development: Building on Brundtland* (pp. 39–52). Paris: United Nations Educational, Scientific and Cultural Organization.

Tolba, M. K., & El-Kholy, O. A. (Ed.). (1992). *The world environment 1972–1992*. London: Chapman & Hall.

Verbruggen, H., Dellink, R., Gerlagh, R., Hofkes, M., & Jansen, H. M. A. (2001). Alternative calculations of a sustainable national income for the Netherlands according to Hueting. In E. C. van Ierland, J. van der Straaten, H. R. J. Volleberg (Eds.), *Economic growth and valuation of the environment, A Debate* (pp. 275–312). Cheltenham, UK: Edward Elgar.

Chapter 26
Transitions to Sustainability as Societal Innovations

Anna J. Wieczorek and Frans Berkhout

Contents

26.1	Introduction	503
26.2	Sustainability Transitions as Systems Innovations	504
26.2.1	Multi-causality	505
26.2.2	Structural Change	505
26.2.3	Uncertainty	505
26.2.4	Multiple Actors	506
26.3	Multilevel Perspectives on Technological Transitions	506
26.4	Historical Examples	508
26.5	Inducing Transitions	510
26.5.1	Transition Management	510
26.5.2	Transitions in a Global Context	510
26.6	Conclusion	511
References		511

26.1 Introduction

One of the great themes of the social debate about environmental protection has been the question whether environmental quality can be safeguarded without major economic or social change. With the advent of the notion of 'sustainable development' in the late 1980s a new consensus emerged which suggested that the economy and the environment could be complementary, so long as the economy internalised the costs of damage to the environment and technological innovation provided for smarter and cleaner ways of doing things. Sustainable development rested on the argument that it would be possible, through the adjustment of incentives and the application of knowledge, to reconcile increases in welfare with a healthy environment. This conviction grew out of the great successes achieved through environmental regulation, beginning in the 1960s, which had brought radical improvements in environmental quality – air, water and soil – in richer, industrialised countries. New technologies – less toxic products, more efficient production processes and a panoply of abatement techniques – modified the environmental impact of social and economic activities, while also enabling growing welfare. By adjusting the economic incentives of innovators in such a way that socially-desired trade-offs were made between economic and social welfare and environmental quality, growth and sustainability could be reconciled – so the argument ran (cf. Elkington 1994).

But great themes don't disappear; they lie dormant for a while, only to return in a different guise. Through the 1980s and 1990s it became increasingly clear that while many local and tangible environmental problems were being solved, new forms of environmental problems were emerging. These were being caused by the cumulative and global effects of social and economic activities, which – measured purely in terms of their energetic and material scale – were coming to have a major impact on biophysical systems. It had long been recognised that human activity has transformed land use and vegetation at the surface of the Earth (Clark et al. 1990), but it became apparent that the great biogeochemical cycles (such as the carbon, nitrogen, phosphorous and sulphur cycles) were also being profoundly modified by industrial and agricultural activity (Schlesinger 1991; Ayres et al. 1994), with often poorly-understood consequences for the functioning, dynamics and stability of atmospheric, hydrological and ecological systems.

A.J. Wieczorek (✉)
Institute for Environmental Studies (IVM), VU University
Amsterdam, The Netherlands
e-mail: anna.wieczorek@ivm.vu.nl

It appeared therefore, that although at the local scale environmental quality was improving (at least in richer countries), at the global level of earth systems – climate, oceans, biomes, watersheds – economic and social activities were having damaging, long-term and possibly irreversible impacts. A series of new global environmental problems – climate change, biodiversity loss, soil fertility loss – came to be framed as social and political problems, re-opening the debate about whether economic growth and environmental sustainability can be reconciled. If earth systems were being modified by the sheer scale of human intervention in global nutrient cycles and ecosystems, then these systems' sustainability might be secured only through the more or less radical reorganisation of key economic *systems of provision* related to energy, food, mobility and shelter. Social and biophysical systems therefore came to be seen as being embedded in each other; interacting and co-evolving. More recently the notion of 'coupled socio-ecological systems' has emerged to try to capture this interdependence and inter-linkage (Young et al. 2006), with the social realm and the biophysical realm each conditioning and influencing the other, but with the primary impetus for change and reorganisation coming from the anthropogenic 'forcing' of socio-ecological systems.

To engage in this analysis means changing the terms of analysis (Berkhout 2002). For economists, sociologists and historians of technology, it means thinking in terms of *systems*, rather than strictly in terms of technologies, sectors, industries and markets. It also means a renewed attention to long-term transformations in, or transitions of, industrial, technological and economic systems, and understanding how these are related to social, political and cultural change. This new-found interest in transitions has generated new concepts which are briefly outlined in this chapter. In addition there is new empirical research into historical transitions of all kinds, development of methods for envisaging and discussing technological and social futures, and more exploratory forms of policy-making. The key questions being asked are about appropriate models of change in systems, the directionality of change and about its governability.

In principle, research would like to be able to uncover the conditions under which transformations occur, and to be able to suggest ways of inducing and influencing the direction of such change towards sustainability. A starting point is that finding answers will not be easy. We know that 'sustainability transitions' will affect many areas of human activity – including nutrition, health, shelter, mobility – in which needs are met through complex, global and interlocking systems of provision (Elzen & Wieczorek 2005). Governance of these systems of provision is deeply problematic because of the range and variety of actors involved in configuring and enacting them and the unstructured character of problems, making objective-setting and coordination difficult.

26.2 Sustainability Transitions as Systems Innovations

In innovation studies the notion of systems change in the context of sustainability has gained broad attention in both academic (Kemp et al. 1998; Elzen et al. 2004; Geels 2005b; Hekkert et al. 2007) and policy arenas (The Netherlands National Environmental Plan NMP4: VROM 2001). *Innovation* describes the bringing into use of a new idea and is typically seen as occurring in response to a problem being solved in a new way. This may be through a change in technology, practice or in the organisation of activities. *Systems innovation* involves the bringing into use of many new ideas, leading over time to the transformation of the character of a system that includes many different sorts of components (techniques, inputs, skills, regulation, habits) and their relationships. To give a simple example, a complex of relationships have developed since the 1940s between the cooling and cooking of food in most households in developed countries. A range of technologies exist to provide cooling and cooking; food producers and retailers have developed products that respond to consumer tastes in respect of convenience, quality, nutrition and health; and architects and design companies have determined the arrangement of these functions in the modern kitchen. A systems innovation perspective would include the fridge, freezer, cooker, microwave oven, the cook, the recipe book and the supermarket within an analysis of the provision of a meal in the household.

Systems innovations are typically seen as changes that occur at various levels of analysis, which become aligned and connected – people, organisations, societal functions, society. At the beginning of a process of systems innovation there may be many costly and dif-

Table 26.1 Multiple levels in transitions[a]

Organisational hierarchy	Examples of transitions
Society	From hunter gathering society to sedentary, urban society
	From rural to industrial society
Societal functions	From sailing ships to steamships
	From telegraph to telephone
	From gas light to electric light
Organisations	From single to multiple divisions
	From production to service provider
	From hierarchical to horizontal coordination
People	From child to adult
	From student to worker

[a] Based on Geels (2005).

ficult misalignments, which over time are reduced as the socio-technical system tends increasingly towards a new order and stability. A sustainability transition is defined as a set of long-term, radical and mutually-reinforcing changes in the economic, technological, institutional and socio-cultural domains of a system that serves a societal function (Elzen & Wieczorek 2005).

26.2.1 Multi-causality

Transitions are not caused by change in a single factor – such as the introduction of a new technology – but are the result of the interplay of many factors that influence each other over time, leading to the emergence of new structures and processes. Technology does not function autonomously, but is an outcome of economic and technical choices and expectations. Only in association with human agency and institutions ('rules of the game') do artefacts fulfil functions (Geels 2004). The historian Thomas Hughes (1986) coined the idea of a 'seamless web' of technological, institutional and behavioural features of socio-technical systems. Societal functions are fulfilled by clusters of elements including technology, regulations, user practices, supply networks etc., which through their interaction create the opportunities and incentives for innovation and transformation (Rohracher 2001; Geels 2004). Terms like techno-institutional complex (Unruh 2000) or techno-economic network (Callon 1991) are also used to capture these multiple dimensions of socio-technical systems. System innovations are then changes from one kind of socio-technical system to another (Geels 2004).

26.2.2 Structural Change

If we discuss transitions as innovation processes, a distinction between *radical* and *incremental* forms of innovation is useful. Elzen et al. (2004) modify Abernathy and Clark's (1985) typology of innovations to explain system innovations as radical changes that disrupt the structure of socio-technical systems. This contrasts with incremental change in which components of a system may become different, but where the overall structure and relationships of technologies and actors remain unchanged. Incremental modifications of existing ways of doing things that can be used by customers with little or no extra instruction or cost, and which are fitted to established practices and infrastructures are normal and pervasive in capitalist economies. As they accumulate, successive generations of technologies form a pattern of path-dependent change (a technological trajectory) that tends to entrench particular ways of doing things. Accumulated advantages – including learning and scale advantages which affect relative prices – for established systems mean that radical changes tend to be more difficult and risky for entrepreneurs and consumers. Even though the dominant trajectories may face long-standing technical, environmental or social problems, responses will tend to be incremental adjustments, rather than more radical transformations (Freeman 1994).

26.2.3 Uncertainty

Socio-technical transitions are long-term processes (Vellinga and Herb 1999), but they include both fast and slow changes, and are never complete. In one way or another, even during periods of relative stasis, all systems are in a process of change and adjustment that is a precursor to a transformation of structure. System innovation is no different. It includes small, relatively fast changes in technology (the replacement of one vintage model with a new vintage, for instance) that would be counted as incremental, as well as the structural change at the level of the socio-technical system (reconfiguration of electricity supply networks to permit the penetration of intermittent, renewable supplies, for instance).

The main point is that the transition can only be envisioned imperfectly at the beginning. Technical, economic and other uncertainties always play a fundamental role in innovation. The inventor and the adopter of an innovation will usually not be able to judge fully what the benefits and the costs will be of a change in the way of doing things. For instance, although a system for hydrogen as fuel in automobiles can be designed and engineered, and many issues related to cost, safety and reliability might be at least partly answered, there will remain residual uncertainties over the rate of learning in alternative technologies, such as batteries, or the role of hydrogen-fuelled cars relative to other forms of transport. Questions like these prove to be extremely difficult to resolve *ex ante*.

The more radical the system innovation envisaged, the greater will be the number of these problems of alignment, many of them caused as a result of economic and social conflicts: over the benefits and costs of the alternatives, over the distribution of costs and benefits, over new economic and technical uncertainties about the consequences of the alternative, and so on. Resolving each of these problems takes time and some problems may turn out to be persistent. One of the reasons why visions of hydrogen as a widely-used energy carrier have not been translated into reality is that the technical and economic problems at the heart of these visions have been difficult to resolve (McDowell and Eames 2006). Another example of a long-standing alternative to the internal combustion engine has been the electric vehicle (Hirsch 2000). In recent times, the latter too has consistently failed to out-compete the incumbent system.

26.2.4 Multiple Actors

Socio-technical systems are composed of a wide range of actors, including innovating firms, consumers, regulators and many others. Systems innovations frequently involve the emergence of new actors and networks that tend to be disruptive to established actor networks (Christensen 1997). The broadening of actor networks expands the problem of coordination in system innovation. Stakeholders have their own visions and perceptions and are not free from vested interests. New governance of innovation is also called for, involving participative approaches that can bring the voice of diverse actors to bear on framing and motivating transition processes.

Summarizing, socio-technical transitions may be necessary to achieve sustainability, but designing and shaping transitions is difficult. A key challenge and the first step for research is to better understand the dynamics of transitions.

26.3 Multilevel Perspectives on Technological Transitions

A stylised model of transition processes has been given by Geels (2002, 2004, 2005b) who proposes a 'multi level perspective on system innovations' to analyse transitions as system innovations (see Fig. 26.1). This three-level model includes micro-, meso- and macro-levels.

The *meso-level,* or a *socio-technical regime*, describes a specific system, such as a system of private mobility based primarily on the automobile. This regime includes a multitude of components, including cars, fuels, roads, signalling systems, garages, regulations, police forces, drivers, information services, lifestyles and so on. The regime co-exists with other forms of personal mobility, both public and private. This concept of the socio-technical regime builds upon Nelson and Winter's (1982) notion of a technological regime, but is significantly widened to also include actors and institutions (Rip and Kemp 1998). The socio-technical regime accounts for stability of the ordering, practices and expectations of actors. Innovations that take place result in a chain of associated adjustments across the regime. Often these adjustments will be relatively local and have no implications for the structure of the system. Only infrequently does an irresolvable mis-alignment between different components of the regime occur, leading to instability that may induce structural change through larger adjustments in the relationships and organisation of regime components, including the introduction of new components (actors, technologies). The emergence of hybrid battery-internal combustion engine cars (such as the Toyota Prius) may be an example of an innovation that generates such instability with regime-changing consequences.

The *macro-level* of the model is called the *landscape* and describes broad, slow-changing factors that influence a variety of regimes, including cultural and

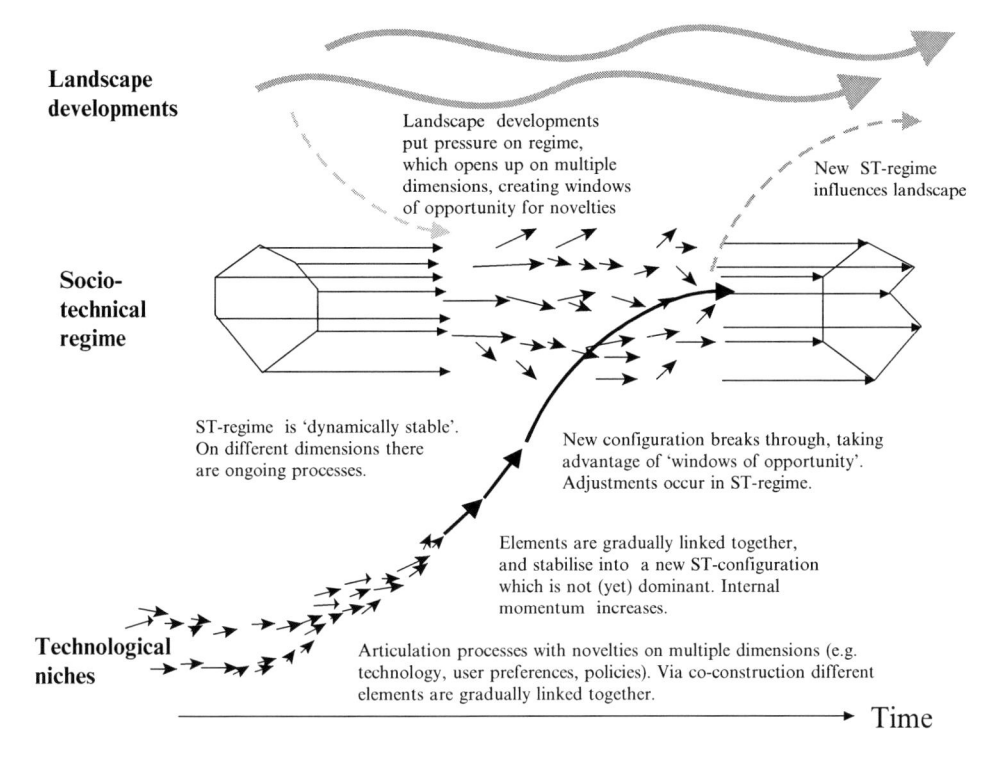

Fig. 26.1 A dynamic multi-level perspective on system innovations Source: Geels (2002: 1263)

normative values, changes in political ideology, demographic change, as well as exogenous factors with large impacts (such as recessions and wars).

The *micro-level* of the model is characterised as being composed of *niches*, within which innovations emerge as 'experiments' in relatively protected contexts. Theories of innovation and diffusion have made extensive use of the idea of niches (Windrum and Birchenall 1998). Niches play a role in enabling learning and as space to build social networks in support of innovations. Fostering niches for innovations has become a major theme in much recent scholarship on transitions which emphasizes the mixed role of policy, entrepreneurship and civil society movements in promoting and developing niches (Raven 2006; Smith 2007).

This three-level model of transitions stresses the interplay between dynamics at these multiple levels. Geels (2002) argues for a framework, which can account for the stability in a regime while showing how misalignment can cause instabilities which lead to transition. He argues that innovations that emerge in technological niches may, under conditions of misalignment and instability, link with and break through into an incumbent regime. Stresses within the regime and developments at the landscape level create windows of opportunity for alternative solutions to be sought by system actors and to become adopted.

Geels makes a distinction between two causes of change: internal drivers originating from the regime itself, when existing technologies can no longer deal with regime problems; and external pressures originating from the landscape such as a widespread fear of the consequences of climate change.

Once a new proto-system becomes embedded in the regime, it may become a substitute or a complement (think of microwave ovens and cookers in the modern kitchen), leading to the creation of a new regime. Geels identifies four phases of transitions:

1. Emergence of novelty in existing context
2. Technical specialisation in market niches and exploration of new functionalities
3. Wide diffusion, breakthrough of new technology and competition with established regime
4. Gradual replacement of established regime, wider transformations

While accepting this general framework, Smith et al. (2005) argue that the causes and dynamics of transitions may vary significantly. Specifically, they show how selection pressures on a socio-technical regime account for historically-situated transformation processes, depending on whether the resources leading to transition are located *within* the regime or are *external* to it; and whether there is a high *degree of coordination* between actors during the transition process or not. Using this framing, four different transition contexts are generated: (1) reorientation of trajectories, (2) endogenous renewal, (3) emergent transformation, and (4) purposive transition, depending on the degree to which response to selection pressures on the regime are dependent on availability of resources within the regime (y-axis) and the level of coordination (x-axis) (Fig. 26.2). One of the main points here is that regime transformation emerges not only as a result of the accumulation of novelty in niches, but also through changing conditions at the landscape level.

A closely-related literature has emerged on how innovation systems (as distinct from systems innovation) provide the resources which determine technological innovation, including radical systems innovation (Jacobsson and Johnson 2000; Hekkert et al. 2007). The innovation systems approach seeks to explain patterns of innovation by interpreting the complex interplay between firms, industrial networks and public institutions that operate in sectors, regions and countries. The emerging literature stresses the importance of a number of *functions* provided by these innovation systems, including the creation of technological knowledge, information exchange and the creation and regulation of markets.

26.4 Historical Examples

Analysis of past transitions give clues on patterns of change and may be useful for drawing conclusions that are relevant for inducing or steering. There exists a growing literature that applies the multi-level framework to the analysis of historical cases such as sail to steam ship (Geels 2005a), organic farming (Smith 2006) and bio-energy (Raven 2007). In Box 26.1 we shortly describe the Dutch transition from coal to gas from the socio-technical, multi-level perspective.

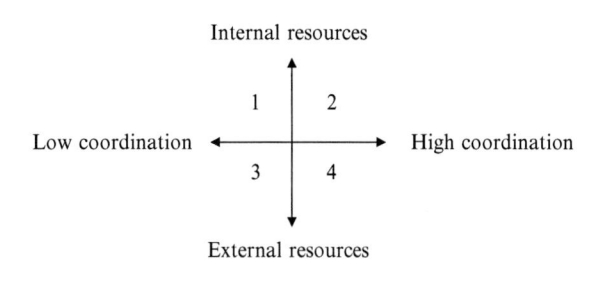

Fig. 26.2 Four transition contexts Source: From Berkhout et al. (2004: 67)

Box 26.1 Coal to gas change in The Netherlands as a system innovation

After substantial amounts of natural gas had been found in Groningen, in the north of The Netherlands, in 1959, a new system of gas supply was rapidly introduced in the country. Gas was available to consumers as early as in 1963, and by 1968 all municipalities were connected to the grid. Export of gas to other Western European countries also began. Natural gas as compared to coal was low cost, reliable and clean, and brought the state enormous tax revenues.

Originally, before the Groningen discovery, gas was a by-product of oil production or produced through gasification of coal, occupying a market niche with a maximum share of about 10% in The Netherlands. In view of rapidly-developing sales of oil products to households and industry, gas exploration did not have high priority for NAM (Nederlandse Aardolie Maatschappij) – established as a join venture for oil and gas exploration by the Shell subsidiary BPM (Bataafse Petroleum Maatschappij) and the Standard Oil Company of New Jersey Esso (later Exxon). The Dutch government however saw advantages in stimulating gas exploration for the national economy. It was seen as an opportunity to replace small gas factories with a national system of gas supply. An agreement was concluded between NAM and

(continued)

Box 26.1 (continued)

SGB (newly founded State Gas Company) according to which NAM was the producer and SGB was in charge of gas transportation. The high price of natural gas (33 Dutch cents per cubic meter) initially made gas uncompetitive compared to oil and coal. Attempts to market gas also failed due to conflicting interests of actors involved in its production and transportation.

The unexpected discovery of substantial amounts of gas in July 1959 opened new opportunities for several of these actors. It also coincided with a period of post-war growth in energy demand and a transition away from dirty and expensive coal to lower cost oil, with nuclear energy considered as another alternative. After initial difficulties with estimating the size of the Slochteren field, it turned out to be one of the world's largest known gas deposits, generating conflicting claims to the rights to production and transportation of gas. A period of tumultuous negotiations ensued. First plans to sell gas to big industrial users failed for technical reasons (primarily due to the need to adjust existing infrastructure to a new energy carrier), so the alternative – to sell gas to households and small business – was chosen. The opportunities of using gas at a national scale were so large that they presented serious competition to the existing energy supply system in The Netherlands, with coal and mining the major losers.

After difficult negotiations, a new institutional framework was created with the new gas policy announced in 1962. NAM continued to produce gas, while a new company, the Gas Unie, was charged with selling it to municipal distribution companies. In this arrangement, the state and the private sector took on new roles. Despite many difficulties and the size of the undertaking, the new gas network was constructed very quickly. The construction process was a highly managed task with a number of agreements easing the construction process. Standardisation of appliances sped consumer adoption of gas in homes and businesses. A major information campaign familiarised consumers with the new energy carrier, stressing its convenience, safety and cleanliness. To make gas attractive, its price was made equal to the price of oil. Rebates and subsidies were available for customers to stimulate consumers to exchange old boilers for new, more efficient ones. The introduction of gas also played into a widespread cultural commitment to modernisation and the renewal of the housing stock. Post-war there had been a considerable shortage of good-quality housing. Gas provided incentives and state revenues to introduce central heating into new houses. With time also industry converted to usage of gas, so that oil use declined significantly after 1969.

What makes this change a transition? Several aspects seem relevant. Gas use in homes had been a niche development. The transformation initiated by the Slochteren discovery built on accumulated experiences and existing infrastructure, technology and users. But the transition to natural gas was not smooth. Instead it was marked by active participation, struggle and negotiation. The Government orchestrated the creation of a new regime in which the oil industry continued to have an important stake, making choices that had important, long-term impacts. A central vision, laid out in the 1962 Gas Law, played a crucial role, including changes in the regulatory framework, a new pricing strategy, gas and equipment supply schemes and an information campaign. The state played an important function in network management, involving companies with expertise in energy markets and gradual change of expectations and interests. Cooption of new actors and keeping balance between public and private entrepreneurs was a cornerstone of the new regime. The changes that took place during the gas transition were of a co-evolutionary character: multi-dimensional (changes in regulatory framework, market mechanisms, technological system etc.), multi-actor and multi-level. The emergence of new consumer tastes, in a social and cultural context open to modernisation, further reinforced the transition.

Source: Elzen et al. (2004)

26.5 Inducing Transitions

Against this background, can we draw policy-relevant conclusions? Can transitions for sustainability be induced? Existing policy often constitutes a barrier to transition since the regulations are tuned to an existing regime with its own characteristics. As a result it may be problematic to promote innovations in the public domain (electric vehicles in the 1970s and 1980s for instance, faced various legal obstacles). But there is also another side. Public policy plays a vital role within innovation systems that stimulate and shape the rate and direction of innovation. Through support for scientific and engineering research they contribute to the production of knowledge, through regulation they contribute to the creation of new markets and through the provision of information they build awareness and skills.

26.5.1 Transition Management

One approach that focuses on inducing and influencing system change towards sustainability is 'transition management' (Rotmans et al. 2001, Loorbach 2007). Transition management has been developed as a participative process to steer learning (learning-by-doing and doing-by-learning) by actors cooperating on inducing specific systems changes. In this it draws on traditions of constructive technology analysis (Schot 1992) and on a literature on technology governance, which argues that the state increasingly plays a coordinating function in innovation systems.

Transition management builds on the multi-level perspective on system innovation. In that context, a transition management process begins with the organisation of a multi-actor network (transition arena) and the development of shared visions of futures (Tuinstra et al. 2003). It then builds upon bottom-up initiatives and innovations (micro level) and influences the existing socio-technical regime (meso level) causing its reconfiguration. Only then the new regime can influence the landscape of worldviews and trends (macro level).

The transition management approach focuses on anticipation and adaptation, as opposed to command and control. A number of proposals have been made about how to organise transition management exercises. Rotmans (2003) lists ten steps, whereas the Dutch government (VROM 2001) implements transitions according to six steps. In this paper we present the most generic four phases (Loorbach and Rotmans 2006):

1. Organisation of a multi-actor network (transition arena), which concerns problem definition; identification of stakeholders, establishment of preconditions for operation of the arena, and definition of transition themes.
2. Development of sustainability visions, which concerns establishment of a common, long-term view and perception of the problem involved as well as discussion of the differences in perception of the problem.
3. Exploration of transition pathways (scenarios) through experiments and joint actions; development and implementation of effective instruments for the specific transition theme.
4. Evaluation and learning, which concerns monitoring of the progress, intermediate goals and learning effects. On the basis of the evaluation, the agenda and visions are adjusted and a next transition round is being prepared.

26.5.2 Transitions in a Global Context

While most attention in the development of ideas about systems innovation and sustainability has been centred on industrialised countries, it is clear that these debates are highly relevant to processes of industrialisation in newly-industrialising countries as well. Taking a global perspective, the current and anticipated rate of economic and social changes in developing Asia make the region central to sustainable development at the global scale. Developing Asia is in the midst of an urban-industrial transition that in absolute terms of urban population growth and scale of economic activity is historically unprecedented (Rock and Angel 2005). Much of the existing literature on economic growth suggests that national economies pass through a fixed set of phases as they converge with the economies of OECD countries. This implies the development of energy- and materials-intensive production and consumption systems, based on technologies, infrastructures and lifestyles, imitating those of richer countries. A systems innovation approach applied in rapidly-developing economies would focus on creation of new socio-technical regimes, on the emergence of innovation systems and on the links between sustainability transitions in OECD and newly-industrialising countries (Berkhout et al. 2008). Not much work has been done on such an approach, but there is great scope for the future.

26.6 Conclusion

Technological innovation is widely-recognised as playing a major role in economic growth and improvements in general welfare. It also plays a vital role in determining the environmental impact of economies and societies.

Over the past decade or more, analysis of the role played by technology has changed profoundly. First, there has been increasing recognition that technologies and technological change need to be situated in its social and institutional context. This sociological and historical turn in innovation studies has greatly expanded the range of actors and factors which are now included to explain the rate and direction of technological change. Second, there has been a recognition that technological change happens at many institutional and temporal scales. This has generated new conceptual developments such as the multi-level perspective and a much greater concern for slow but radical change in the way things are done. By applying these system innovation perspectives, we may get closer to understanding of – and participating in – transitions towards sustainability.

Acknowledgements The authors would like to thank Patricia Timmerman and Beatrice Alders for their invaluable support in editing this chapter.

References

Abernathy, W. J., & Clark, K. B. (1985). Innovation: Mapping the winds of creative destruction. *Research Policy, 14*, 3–22.

Ayres, R. U., Schlesinger, W. H., & Socolow, R. H. (1994). Human impacts on carbon and nitrogen cycles. In R. Socolow, C. Andrews, F. Berkhout, & V. Thomas (Eds.), *Industrial ecology and global change*. Cambridge: Cambridge University Press.

Berkhout, F. (2002). Technological regimes, path dependency and the environment. *Global Environmental Change, 12*(1), 1–4.

Berkhout, F., Smith, A., & Stirling, A. (2004). Socio-technological regimes and transition contexts. In B. Elzen, F. W. Geels, & K. Green (Eds.), *System innovation and the transition to sustainability* (pp. 48–75). Cheltenham: Edward Elgar.

Berkhout, F., Angel D, and A. J. Wieczorek (eds) (2008) sustainability Transitions in Developing Asia: are alternative development pathways likely ? Technological Forecasting & social change, doi:10.1016/j.techfore. 2008.0003.

Callon, M. (1991). Techno-economic networks and irreversibility. In J. Law (Ed.), *A sociology of monsters, essays on power, technology and domination* (pp. 132–161). London: Routledge.

Clark, W. C., Turner, B. L., Kates, R. W., Richards, J., Mathews, J. T., & Meyer, W. (Eds.). (1990). *The earth as transformed by human action. Global and regional changes in the biosphere over the past 300 years*. Cambridge: Cambridge University Press.

Christensen, C. M. (1997). *The innovators' dilemma*. Cambridge, MA: Harvard Business School Press.

Elkington, J. (1994). Towards the sustainable corporation: Win-win-win business strategies for sustainable development. *California Management Review, 36*(2), 90–100.

Elzen, B., Geels, F. W., & Green, K. (Eds.). (2004). *System innovation and the transition to sustainability: Theory, evidence and policy*. Cheltenham: Edgar Elgar. pp 114–137.

Elzen, B., & Wieczorek, A. J. (2005). Introduction: Transitions towards sustainability through system innovation. *Technological Forecasting and Social Change Journal, 72*(6), 651–662.

Freeman, C. (1994). The economics of technical change. *Cambridge Journal of Economics, 18*(5), 463–514.

Geels, F. W. (2002). Technological transitions as evolutionary reconfiguration processes: A multi-level perspective and a case study. *Research Policy, 31*(8/9), 1257–1274.

Geels, F. W. (2004). Understanding system innovations: A critical literature review and a conceptual synthesis. In B. Elzen, F. W. Geels, & K. Green (Eds.). *System innovation and the transition to sustainability: Theory, evidence and policy*. Cheltenham: Edgar Elgar.

Geels, F. W. (2005a). Technological transitions as system innovations: A co-evolutionary and socio-technical analysis. Cheltenham: Edgar Elgar.

Geels, F. W. (2005b). Co-evolution of technology and society: The multi-level perspective and a case study, the transition in water supply and personal hygiene in the Netherlands (1850–1930). *Technology in Society: An International Journal, 27*(3), 363–397.

Hekkert, M. P., Suurs, R. A. A., Negro, S. O., Kuhlmann, S., & Smits, R. E. H. M. (2007). Functions of innovation systems: A new approach for analysing technological change. *Technological Forecasting and Social Change, 74*(4), 413–432.

Hirsch, D. A. (2000). *The electric vehicle and the burden of history*. New Brunswick: Rutgers University Press.

Hughes, T. P. (1986). Seamless web: Technology, science, etcetera, etcetera. *Social Studies of Science, 16*, 281–292.

Jacobsson, S., & Johnson, A. (2000). The diffusion of renewable energy technology: An analytical framework and key issues for research. *Energy Policy, 28*(9), 625–640.

Kemp, R., Schot, J., & Hoogma, R. (1998). Regime shifts to sustainability through processes of niche formation: The approach of Strategic Niche Management. *Technology Analysis and Strategic Management, 10*(2), 175–195.

Loorbach, D. A. (2007). Transition management: New mode of governance for sustainable development. Utrecht, The Netherlands: International Books.

Loorbach, D., & Rotmans, J. (2006). Managing transitions for sustainable development. In X. Olsthoorn, & A. Wieczorek (Eds.), *Understanding industrial transformation: Views from different disciplines*. Dordrecht, The Netherlands: Springer.

McDowall, W., & Eames, M. (2006). Forecasts, scenarios, visions, backcasts and roadmaps to the hydrogen economy: A review of the hydrogen futures literature. *Energy Policy, 34*(11), 1236–1250.

Nelson, R., & Winter, S. G. (1982). *An evolutionary theory of economic change*. Cambridge, MA: Belknap Press.

Raven, R. P. J. M. (2006). Niche accumulation and hybridisation strategies in transition processes toward a sustainable energy system: An assessment of differences and pitfalls. *Energy Policy, 35*(4), 2390–2400.

Rip, A., & Kemp, R. (1998). Technological change. In S. Rayner & E. L. Malone (Eds.), *Human choice and climate change* (pp. 327–399). Columbus, OH: Battelle Press.

Rock, M. T., & Angel, D. (2005). *Industrial transformation in the developing world*. Oxford: Oxford University Press.

Rohracher, H. (2001). Managing the technological transition to sustainable construction of buildings: A socio-technical perspective. *Technology Analysis & Strategic Management, 13*(1), 137–150.

Rotmans, J. (2003). Transitiemanagement: sleutel voor een duurzame samenleving. Assen: Van Gorcum.

Rotmans, J., Kemp, R., & van Asselt, M. (2001). More evolution than revolution: Transition management in public policy. *Foresight, 3*(1), 15–31.

Schlesinger, W. H. (1991). *Biogeochemistry: An analysis of global change*. San Diego, CA: Academic.

Schot, J. W. (1992). Constructive technology assessment and technology dynamics: The case of clean technologies. *Science, Technology and Human Values, 17*(1), 36–56.

Smith A. (2006). Green niches in sustainable development: The case of organic food in the United Kingdom. *Environment and Planning C: Government and Policy, 24*(3), 439–458.

Smith A. (2007). Translating sustainabilities between green niches and socio-technical regimes. *Technology Analysis and Strategic Management, 19*(4), 427–450.

Smith, A., Stirling, A., & Berkhout, F. (2005). The governance of sustainable socio-technical transitions. *Research Policy, 34*, 1491–1510.

Tuinstra, W., van de Kerkhof, M., Hisschemöller, M., & Mol, A. (2003). COOL: Exploring options for carbon dioxide reduction in a participatory mode. In B. Kasimir, J. Jäger, C. Jaeger, & M. Gardner (Eds.), *Public participation in sustainability science* (pp. 176–186). *A handbook*. Cambridge: Cambridge University Press.

Unruh, G. C. (2000). Understanding carbon lock-in. *Energy Policy, 28*, 817–830.

Vellinga, P., & Herb, N. (Eds.). (1999). Industrial transformation science plan, International human dimensions programme (IHDP) Report No.12. www.ihdp.uni-bonn.de/

VROM (2001). Where there is a will, there is a world. Fourth National Environmental Policy Plan (NEPP 4). The Hague, The Netherlands.

Windrum, P., & Birchenall, C. (1998). Is product life cycle theory a special case? Dominant designs and the emergence of market niches through co-evolutionary thinking. *Structural Change and Economic Dynamics, 9*(1), 109–134.

Young, O. R., Berkhout, F., Gallopin, G. C., Janssen, M. A., Ostrom, E., & van der Leeuw, S. (2006). The globalization of socio-ecological systems: An agenda for scientific research. *Global Environmental Change, 16*(3), 304–316.

Chapter 27
Agriculture and Food Problems

David Pimentel

Contents

27.1 Introduction 513
27.2 Soil and Land Resources 513
27.3 Water Resources 514
27.4 Energy Resources 515
27.5 Biotechnology 515
27.6 Societal Obstacles to Solution to the Agriculture and Food Problem 516
Bibliography 516

27.1 Introduction

Agriculture provides more than 99.7% of the world food supply; the oceans and aquatic ecosystems contribute less than 0.3%. With the human population projected to grow from its 2005 level of 6.5 billion to 9–11 billion by 2050, it will be increasingly difficult to meet future basic human food needs given the finite resources of the earth.

The status of the food supply has already become critical in many areas of the world. Based on data of the Food and Agricultural Organization and the World Health Organization, it is estimated that roughly 3.7 billion people are currently malnourished whereas about 800,000 suffer from hunger. Not only are hunger and malnutrition significant problems in and of themselves, but they also predispose people to infectious diseases. This relationship is evidenced by the growing number of people dying from infectious diseases and illness associated with such environmental problems as air pollution and chemical pollutants. Diseases in humans worldwide have increased during the past decade.

There is a growing imbalance between the quantity of food crops produced and the number of people who depend on these food resources. Some per capita crop yields have been decreasing; the most significant decrease is the per capita grain production, which started to decline in 1984 and continues to decline today. This decline is especially important because grains provide about 80% of the world's food. In addition, fish production has declined by about 10% per capita during the past decade.

The development of a productive and sustainable agriculture system is essential to stem the growing imbalance between food availability and number of people. Such a system depends significantly on solar energy and a favorable climate, but fertile soils, adequate fresh water, and fossil energy, and biodiversity of plant, animal, and microbe species are also essential. These natural resources, however, do not exist in infinite supply on the earth. An understanding of the relationship between agricultural production and natural resources, described in Sections 27.2 to 27.4, is vital to the future development of a sustainable agriculture system.

27.2 Soil and Land Resources

Worldwide, food and fiber crops are grown on 11% (~1,500 million hectares) of the earth's total land area. In 1960, when the world population numbered about 3 billion, approximately 0.5 ha of cropland was available per capita for crop production. Half a hectare of cropland per capita is needed to provide a diverse, nutritious diet of plant and animal products – similar to the

---

D. Pimentel (✉)
Ecology and Agriculture at Cornell University, Ithaca, USA
e-mail: dp18@cornell.edu

typical diet in the United States and Europe. Since the 1960s, as the global human population has increased, the average per capita availability of cropland has decreased to 0.23 ha, or about half the amount needed.

Essential fertile soil is degraded by water and wind erosion, plus the salinization and water logging of irrigated soils. Cropland erosion, a slow, insidious process, results in a range of soil losses, from about 10 t per hectare per year (t/ha/year) in the United States and Europe to 40 t/ha/year in China. Wind erosion is so serious that soil eroded in Africa is blown across the Atlantic and can be detected in Florida and Brazil. Worldwide, soil erosion losses average approximately 30 t/ha/year, or about 20-times faster than the rate of soil reformation. Note, approximately 500 years are required for 2.5 cm (1 in.) of topsoil to form under agricultural conditions.

Erosion adversely affects crop productivity by reducing water availability for plants and the water-holding capacity of the soil. It also diminishes nutrient levels and organic matter in the soil, as well as topsoil depth. All these soil components are vital to maximizing agricultural production. Escalating soil and land degradation is destroying crop and pasture land throughout the world, with more than 10 million hectares of productive cropland so severely degraded they have to be abandoned each year. In addition, salinization is causing the loss of about 10 million hectares per year. Because of these losses, approximately 30% of the world's cropland has been abandoned during the past 50 years. This loss of cropland, simultaneously with the expansion of the human population, has resulted in an urgent need for more agricultural land. In response to this need, more than 60% of the deforestation now occurring worldwide is directly related to the spread of agriculture and the shortage of productive cropland.

Available per capita cropland has declined 20% during the past decade, and is projected to decline at least 60% more by 2050, when the world population is projected to grow to between 9 and 11 billion. In addition, agricultural land degradation alone is projected to depress world food production by between 10% and 20% by the year 2020. These estimates emphasize the importance of implementing known and available soil conservation techniques, including the use of biomass mulches, no-till, ridge-till, terracing, grass strips, crop rotations, and combinations of all of these. The current high erosion rate throughout the world is becoming critical because of the slow rate of topsoil renewal.

Irrigation practices have to be improved to reduce the serious salinization problem that exists worldwide.

27.3 Water Resources

All vegetation requires and transpires massive amounts of water during the growing season. Agriculture consumes about 70% of the water removed from lakes, rivers, and aquifers. A corn crop that produces about 9,000 kg/ha of grain uses more than 5 million liters/ha of water during its growing season. To supply this much water to the corn, approximately 1,000 mm of rainfall, or 10 million liters of irrigation water, per hectare are consumed by the plants during the growing season. This means that about 1,000 l of water are required to produce 1 kg of grain. When grain and forage are fed to cattle, a total of 45,000 l of irrigation water are required to produce just 1 kg of beef.

The minimum amount of water required for food production is about 1 million liters per capita per year. As both local and global water shortages become more and more serious, supplying this much water for food production in the future will be difficult, if not impossible.

Rapid population growth combined with increasing total water consumption is depleting the available supply of fresh water. Even ground water resources are being rapidly mined and consumed. Between 1960 and 2000, the per capita availability of fresh water worldwide declined by about 60%. Another 50% decrease in per capita water supply is projected by the year 2025. A further decline in the water supply will cause a serious decline in the per capita food supply in the future.

The availability of water is complicated further by the uneven distribution of rainfall throughout the world. The impact of global warming on rainfall is a concern. Many areas and nations that depend on shared rivers and lakes for their water may face political conflicts as populations escalate and human water needs increase.

In addition to the unsustainable use of water, water pollution is another serious problem. Although considerable water pollution has been documented in the United States, it is more prevalent in countries where water regulations are not rigorously enforced or do not exist at all. Developing countries discharge approximately 95% of their untreated urban sewage directly into surface waters. Downstream, the polluted waters are then used for drinking, bathing, and washing. A startling example of this situation is evident in India, where, of the 3,119 towns and cities, only 8 have full wastewater treatment facilities. With an increasing world population, water pollution is projected to increase even more in the future.

27.4 Energy Resources

Although about 50% of all solar energy captured by photosynthesis is used by humans (see also Chapter 6.3), it is still insufficient to meet all the needs of modern human society. Over time, humans have enhanced energy with manpower and draft animals. Since the 1800s, when coal, oil, and natural gas were developed as fuels, ample supplies of these finite energy sources have been supporting the productivity of food production with fertilizers, pesticides, machinery, and irrigation. Fossil energy use generally improves the quality of human life, including providing water protection from diseases.

In industrialized countries typically between 15% and 20% of all fossil energy use serves the provision of food. Some developing nations with high rates of population growth are increasing fossil fuel use to augment their agricultural production to feed more people. For example, in China, fossil energy use in agriculture for fertilizers, pesticides, and irrigation has increased more than 100 fold since 1955.

Problems will arise from the finiteness of conventional natural gas and mineral oil resources that currently cover about 60% of current global energy demand. These have been formed over a period of 400 million years and are depleted at a high rate. In 2005 consumption of conventional mineral oil equaled the amount formed over a period of about one million years and consumption of natural gas the amount formed over a period of about 3 million years. Production of conventional mineral oil and natural gas is expected to peak in before 2010 and will decline thereafter. This will be associated with higher prices and will negatively affect the prospects of adequately feeding everyone, especially the poor.

If available renewable energy technologies, such as biomass and wind power, are developed, an estimated 200 quads of renewable energy could be produced worldwide (1 quad = 383×10^{18} J). However, this would require the cost of using about 20% of the world land area. And this estimated 200 quads of renewable energy are still less than half of the present energy consumption by the 2005 global population of 6.5 billion people.

27.5 Biotechnology

Biotechnology has been proposed as the savior for future agricultural production. However, biotechnology has been put to use in agriculture for nearly 25 years, but has resulted in only minimal increases in crop production. The reason for this minimal increase in crop production is that biotechnology has focused primarily on developing herbicide tolerance in crops; more than 75% of the agricultural land planted to GMO crops are planted to herbicide tolerant crops. Herbicide tolerance in crops does not improve weed control nor does it increase crop yields. It does, however, generally increase herbicide use, costs of weed control, and environmental pollution.

The biotechnology investigations on non-chemical insect control methods have focused primarily on the inserting genes from *Bacillus thuringiensis* (BT) into various crops, such as corn and cotton. The insect control results have been minimally successful with corn, for several reasons. With corn the focus of BT use has been aimed at the corn borer, a pest that actually causes minimal losses, rather than focusing on the corn rootworm complex, which causes the major pest losses in corn production. The use of BT in cotton has provided effective control of several serious cotton caterpillar pests.

In the future, biotechnology will be able to help improve crop yields, but the technology will not provide a second Green Revolution. Instead of focusing on herbicide tolerance in crops that increase the use of chemical pesticides, greater attention should be focused on increasing host plant resistance to insect pests and plant pathogens. Another important focus should be on the development of perennial grains. This development would save enormous amounts of fossil energy but also conserve soil and water resources.

27.6 Societal Obstacles to Solution to the Agriculture and Food Problem

A rapidly increasing human population will be the prime obstacle to seeking solutions to the global food problem. As the global population grows in the future, the per capita availability of finite resources of cropland, water, and energy resources will decrease. Already, there are roughly 3.7 billion people who are malnourished. The number of malnourished is projected to increase as per capita food production continues to decline because of shortages of the cropland, water, and energy resources that are essential for food production.

Food production per capita has been declining since 1983 and is expected to continue to decline in the future. Crop yields per hectare have increased only slightly since 1983, but a rapidly growing world population has more than offset the gains on a per capita basis. The worldwide per capita food situation is expected to further decline in the future, leading to increasing malnutrition.

Bibliography

Food and Agricultural Organization (2000). *The state of food insecurity in the world*. Rome: FAO (available at www.fao.org).

Gleick, P. H. (1993). *Water in Crisis*. Oxford/New York: Oxford University Press.

Pimentel, D., & Pimentel, M. (1996). *Food, energy and society*. Boulder, CO: Colorado University Press.

Patzek, T. W., & Pimentel, D. (2005). Thermodynamics of energy production from biomass. *Critical Reviews in Plant Sciences*, 14(5/6), 263–279.

Pimentel, D. (2001). The limitations of biomass energy. *Encyclopedia on Physical Science and Technology* (pp. 159–171) San Diego, CA: Academic.

Pimentel, D., Wilson, C., McCullum, C., Huang, R., Dwen, P., Flack, J., et al. (1997). Economic and environmental benefits of biodiversity. *BioScience*, 47(11), 747–758.

Troeh, F. R., Hobbs, J. A., & Donahue, R. L. (1993). *Soil and water conservation* (2nd ed.). Englewood Cliffs, NJ: Prentice-Hall.

World Health Organization (2000). *Turning the tide of malnutrition: Responding to the challenge of the 21st century*. Geneva (WHO/NHD/00.7)

World Population Prospects www.esa.un.org/unpp.

Youngquist, W. (1997). *Geodestinies: The inevitable control of earth resources over nations and individuals*. Portland, OR: National Book Company.

Chapter 28
Tracing the Sustainable Development of Nations with Integrated Indicators

Bastiaan Zoeteman

Contents

28.1 Why Look for an Integrated Framework? 517
28.2 Proposals for an Integrated Framework 518
28.3 Environmental Sustainability 519
28.4 Social Sustainability 521
28.5 Economic Sustainability 521
28.6 The Five Levels Applied in a Sustainability Scale 522
28.7 Calculating a Sustainability Index 522
28.8 Rating Sustainability of Countries 524
28.9 Applying the Sustainability Scale 524
   28.9.1 Taking Other Sustainability Interests into Account 524
   28.9.2 Eco-efficiency Reconciliates Environment and Economy 526
   28.9.3 Next Steps in Development 526
   28.9.4 Balance Between the Three Sustainability Aspects 527
   28.9.5 Reduction of Damage in Fall-Back Situations 527
References 528

28.1 Why Look for an Integrated Framework?

Managing sustainable development as a policy issue is a complex task. The European Union Sustainable Development Strategy (CEC 2001) has stimulated EU member states to develop national sustainable development strategies. However policy makers are mostly lacking reliable and practical tools to measure progress and to indicate where adaptation to new or unbalanced developments is needed. In this chapter a new approach to develop an integrated sustainability index for nations is proposed and illustrated.

Since the UN World Commission on Environment and Development (1987) published its report 'Our Common Future', sustainable development has been accepted widely by governments, businesses and civil society as a common goal. Sustainable development was defined in this report, I paraphrase, as our obligation to increase and not reduce overall development possibilities for the next generations. We should leave our Earth behind in a better shape than we found her at the moment we were born, not only for moral reasons, but also to promote future prosperity. Many attempts have been made to define the concept of sustainable development in a way that makes it possible to link operational targets to it. (Biesiot 1997; Bossel 1997; Daly and Cobb 1990; Hueting and Bosch 1990; Meadows 1998; Wackernagel and Rees 1996). As yet no definition obtained general support and I have not the pretense to provide the ultimat e solution to the operationalization of sustainability. But the proposals presented here may serve as a contribution from the viewpoint of a former policy-maker.

Sustainable development is a multidimensional concept. Many concentrate on its three dimensions of economy, ecology and social cohesion. Bossel (1999) made a valuable overview of sustainable development theories and indicators from a systems analysis point of view. He defined sustainable development as co-evolution of human and natural systems and identified six essential subsystems of the societal system. Three subsystems, including individual development, the

B. Zoeteman(✉)
Sustainable Policies in International Perspective at the Faculty for Economics and Business Administration at Tilburg University Telos, Brabants Centre for Sustainability Isues, Tilburg, The Netherlands
e-mail: zoeteman@uvt.nl

social system and government comprise the human system. Two subsystems, infrastructure and the economic system, form the support system. Natural resources and environment constitute the last subsystem: the natural environment.

I will use the terms social, economic and ecological for the subsystems Bossel defined as the human, the support and the natural system. I will integrate these three dimensions in a tool that allows for a broad characterization of levels of sustainability.

28.2 Proposals for an Integrated Framework

How do we know developments are moving into the right or wrong direction? Looking only at growth of GDP is not enough. Improving the eco-efficiency of the economy can still be insufficient when population grows and overall environmental degradation continues. Cleaning the environment can be counterproductive when the result is that large groups of employees lose their jobs. Solutions have to be optimized from different angles at the same time, using an overall indicator for sustainability. On the other hand, one indicator for such a complex issue as sustainability has the disadvantage that important aspects are obscured and decision-makers are tempted to neglect these. Yet I think the advantages outweigh the possible risks, as the more detailed information can always been shown where needed. Generally policy-makers do not lack detail but overview.

Many suggestions have been made which can be used to derive an overall indicator for sustainability. Especially useful here are the proposals of O'Conner (1995) of the World Bank, the 'Balaton Group' (see De Kruijf and Van Vuuren 1998), Bossel (1999), Winsemius and Guntram (2002) and Swarttouw (1999).

Also relevant is a hierarchy of human needs developed by Maslow (Maslow 1970; Maslow et al. 1998). He suggested five layers in the needs of human beings (survival, safety, social belonging, self-esteem and self-actualization) and the principle that a higher need can only be fulfilled if the lower needs are satisfied. Max-Neef (1991) proposed a somewhat similar set of seven psychological and social human needs (namely subsistence, protection, understanding, participation, identity, freedom and creation).

Against the background of these proposals I suggest five levels in the evolution of human consciousness and related attitudes relevant for our present society.

Our awareness of the need to develop in a sustainable way is closely linked to our awareness of the physical limits of the planet we are inhabiting. The look at our blue Earth from the Moon, the spread of deserts over the continents, the signals that we are for the first time in history changing our atmosphere by disrupting the ozone layer and changing the climate system, this and many more has made us conscious of the fact that we cannot neglect possible negative effects of economic growth. Industrialization has brought us not only a rapid growth in our ability to exploit nature and create wealth, it also forced us to become more aware of the potential detrimental consequences of our newly conquered powers. In the course of the 19th century employers in industrialized countries not only took into account social security of their employees; they recently also started to recognize their own long term interest in respecting nature as the source of health for their consumers and employees and as the stock of raw materials for their production processes. Kaku (1996), chief executive officer of Canon Inc. stated this aptly: 'the essence of a successful company is to strive to contribute on three dimensions: to its customers, to its employees, and to society.' No economic activity can in the long term exist on a war footing with its social and natural environment. This is a truth known and lived up to by many indigenous societies thousands of years ago. Modern society is learning to adapt its behavior to this old wisdom in order to restore sustainability at the present high level of economic activities. Corporate Social Responsibility therefore has gained considerable momentum, not because of governmental pressures alone, but also as a need to fulfill demands from the globalized market (Zoeteman et al. 2007).

Here I suggest five levels that describe the evolution of our consciousness ranging from being concerned with our own immediate physical interests to our concern with the welfare of all members of our society and the stability of our planet. These five levels are steps in the development of consciousness by all stakeholders, including government representatives, business leaders, members of non-governmental organizations such as labor unions and environment pressure groups, as well as of the public in general.

On the first level of consciousness our focus is on the well-being of our body and its immediate surroundings.

Only what we see and feel is of concern. On the second level our responsibility expands to the borders of the village or city where we live. On the third level our concerns comprise our country, while on the fourth it further expands to our continent and the globe. Finally our 'home', the environment we feel responsible for, equals the three-dimensional planet as a whole, as it moves through space in our solar system. With the expansion of the spatial limits of our consciousness also the time horizon of our concerns enlarges from a daily or weekly scope, to a monthly and yearly horizon. On the fourth level we reach for the first time a global consciousness and we become aware of possible consequences of our actions during the next decades. On the fifth level our impact and responsibility may achieve a scope of a century or more. A similar scope for the change in horizons of influence and responsibility has been given by Meadows et al. (1972). The levels of consciousness outlined here are summarized in Table 28.1.

These five levels of consciousness can be illustrated, as shown in Table 28.1, by the metaphor of war and peace for the attitude with which we approach our surroundings. The more aggressive we are, the narrower our horizon is. Wars are the magnified expression of the feelings of anger, frustration and hatred between two groups of people. Our external concern becomes in this case very narrow minded, short term oriented and destructive. This is typical for consciousness level 1 behavior. The other or the environment are seen as objects to be used and exploited for self-expression.

Level 2 consciousness is like a guerrilla, aiming at disrupting the enemy where possible. Government commands rules and businesses try to avoid or sabotage such rules as much as possible. Labor conflicts are often typical examples of this level.

At level 3, the case of a cease-fire and armistice, we accept the other party's continued existence and influence, but not voluntarily. It is by force or after careful calculation and negotiation that we sign the armistice agreement. Covenants between government, businesses and NGO's are good examples of this level.

At level 4 a peace agreement with the enemy signed. Such agreements can not only be found between two nations or municipalities, but also between a business and a labor union or a nature conservation group. This is the level of sustainable development, where we agree to fully accept the other party and its concerns and are willing to invest in consensus building and transparency in reporting. It is also at this stage that businesses eliminate barriers with civil society through voluntary implementation of corporate social responsibility.

At level 5 it is no longer a matter of calculation or consensus building, but it is trust and creative enthusiasm out of which new conditions to work and live are emerging. We start to interact in a synergistic way, bringing each other new possibilities that were unthinkable before. Business here commits, quoting again Kaku (1996) 'to serving humankind as a whole through its philosophy and activities.'

We can characterize these five levels as local, municipal, national, global and planetary awareness. In the next three sections a more detailed description will be given of the five levels of sustainability for the environmental, social and economic domain.

Table 28.1 Five levels of consciousness as a framework for sustainable development categories

Level of consciousness (Aggression metaphors or attitude towards the surroundings)	Consciousness horizon	
	Space	Time
1. Local (War/annihilation)	Direct surroundings	Hours/days
2. Municipal (Guerrilla/resistance)	Village/city	Weeks/month
3. National (Armistice/negotiation)	State	Years
4. Global (Peace/acceptance)	Continent/globe	Decades
5. Planetary (New possibilities/abundance)	Planet	Centuries

28.3 Environmental Sustainability

Environmental concern is aimed at providing adequate conditions for the growth and reproduction of all organisms including man, and contributing to the quality of ecosystems and thus to the community of living organisms as a whole. These conditions are initially often affected negatively when technology is introduced to support the expansion of economic and social activities. Degradation continues until improvement of environmental quality is recognized as a prerequisite for further economic and social development. The latter attitude is currently mostly absent when large groups of people fight for survival, which narrows their horizon to their

Table 28.2 Environmental sustainability levels

Level of environmental development	Illustration
1. Unlimited exhaustion and pollution	Toxic waste dumps
2. Continued exploitation	Metals in sewage
3. Protection of exploited nature	Acid rain
4. Precautions regarding long term effects	Climate change
5. Creation of new ecosystems	CO_2 sinks in deserts

momentary needs. However the need to maintain a healthy environment becomes often recognized when the primary needs of life are fulfilled and the damage caused by environmental degradation becomes evident. With this in mind the five environmental levels can be defined as follows (for a summary see Table 28.2):

Level 1: Wide ranging exhaustion of resources, unrestricted disposal of wastes and limited power of the government to counteract. The only way for a government to intervene is by strict command and control. Corruption is often involved. The start of industrialization corresponds often with this level of environmental sustainability.

Level 2: Exhaustion and pollution meet increasing legal restrictions from the government. Waste is discharged at locations where it has no immediate effect on local society e.g. outside the city limits or by means of high chimneys. Business is resisting as much as possible environmental measures as propagated by the green NGO's. The national government develops a set of legal measures forcing polluters to meet emission standards, which are enforced by local authorities with moderate effect.

Level 3: Nature starts to be protected. Management of ground water reservoirs and river basins is started. Businesses take responsibility to limit environmental damage through legal requirements e.g. to abate acid rain and smog. Government and business may negotiate agreements to reduce cost and obtain more commitment to meet long term emission reduction goals. Green planning can be introduced in addition to the existing legal framework.

Level 4: Co-existence of man and nature. Resources are used with a view to the needs of future generations. Waste is handled as a resource. Major actors take responsibility for the protection of nature and of the global commons. The precautionary principle is applied. The environmental issue is broadened to sustainable development.

Level 5: From co-existence to co-creation. Ultimately the challenge becomes not only to protect the environment but to develop and improve its qualities beyond the existing characteristics.

A proposal for the quantitative characteristics of these levels is given in Table 28.3.

Table 28.3 Quantitative characteristics of proposed sustainability levels

Characteristic	Sustainability level				
	1	2	3	4	5
Environmental resources indicators					
Natural capital (1990) (US$1,000/capita)	<1	1–3	3–10	10–30	≥30
Annual withdrawal of water resources (%) (1970–1998)	≥100	30–100	10–30	1–10	<1
Forest as % of original forest (1996)	<20	20–40	40–60	60–80	≥80
Environmental pollution indicators					
Carbon dioxide emissions per capita (t) (1995)	≥12	4–12	1.2–4	0.4–1.2	<0.4
Carbon dioxide emission per dollar of GDP (t/US$1,000) (1995)	≥3.2	0.8–3.2	0.2–0.8	0.05–0.2	<0.05
Maximum concentration of lead in gasoline (g/l) (1991–1996)	≥0.75	0.5–0.75	0.25–0.5	0.1–0.25	<0.1
Social security indicators					
Life expectancy at birth (years) (1995–2000)	<50	50–65	65–75	75–80	≥80
Urban population connected to sewer (%) (1993)	<10	10–80	80–98	98–99.5	≥99.5
Murders in urban environment per 100,000 population (1993)	≥27	9–27	3–9	1–3	<1

(continued)

Table 28.3 (continued)

Social security benefits expenditure (%GDP) (1985–1990)	<0.6	0.6–2	2–7	7–25	≥25
Combined first, second and third level enrollment ratios (%) (1993)	<40	40–60	60–80	80–95	≥95
Economic indicators					
Percentage of labor force in services (1990)	<20	20–40	40–60	60–80	≥80
Number of cars per 1,000 population (1993)	<58	58–115	115–230	230–460	≥460
Number of main telephone lines per 100 population (1991)	<1	1–20	20–40	40–60	≥60
Produced assets (1990) (US$1,000/capita)	<1.6	1.6–5.7	5.7–20	20–70	≥70

28.4 Social Sustainability

The essence of our social concern is meeting increasing levels of human needs, such as the need for a healthy life, income, education, safety, social functioning and self-expression. The five levels for sustainability in the social field are defined as follows (for a summary see Table 28.4):

Level 1: Survival is central. This level includes basic provisions for the short term such as clothes, a place to sleep, water, basic food and defense against acute life threatening circumstances, infirmities and diseases.

Level 2: This level implies a more structural provision in basic needs, including a home, easy access to food, safe drinking water and sewerage in the home, basic education, protection against theft and violence and the availability of care in case of chronic disease or infirmity.

Level 3: Social belonging and respect are central. Government measures promote adequate employment and care for all members of society. The role of social unions is well established.

Level 4: Actions reflect respect for individuality and self-esteem. Working conditions are flexible. Governmental guarantees for income and employment conditions are less needed. Time and money become interchangeable for the employee.

Level 5: Self-realization is becoming a main driver. Personal potentialities are expressed. New forms of employment, income, health care and social security are introduced. A growing group of people shifts from being employed to becoming employer. People start to function in different roles in networks. Basic social security on a global scale emerges as a human right.

A proposal for the quantitative characteristics of these levels is in Table 28.3.

28.5 Economic Sustainability

Our economic concern is to create wealth and ultimately abundance of material goods and services. A general characteristic of higher levels of economic sustainability is increasing knowledge intensity. The following five levels (summarized in Table 28.5) of economic development are suggested.

Table 28.4 Social sustainability levels

Level of social development	Illustration
1. Survival	Shelter, daily food
2. Security guarantees	Protection against crime, medical care
3. Social belonging	Job, sport
4. Self-esteem	Volunteering
5. Self-realization	Inner growth

Table 28.5 Economic sustainability levels

Level of economic development	Illustration
1. Autarctic, subsistence oriented	Food production
2. Crafts, basic commodity trade	Mining
3. Specialized industry, national protection	Ship building
4. Liberalized trade, mass individualization	Car production, recreation
5. Virtual trade, services dominated	Private banking

Level 1: The economy is undifferentiated and focused on food production in agriculture and fisheries. More than two-third of the labor force is working in agriculture.

Level 2: The start up of industrialized production of cloths, buildings and medicines. Crafts play an important role. The income of craftsman depends on skilled use of tools. Specialization and local trade. Half of the labor force of a country works in agriculture.

Level 3: Transport of people and goods to promote growth. Governments and businesses invest in infrastructure. Product differentiation. The economy is more capital intensive. Technical skills determine income. Industry provides work for about a quarter of the labor force while some 50% of the labor force works in services.

Level 4: Trade and recreation expand. Increasing role of media. Customer tailored design and trade lead to mass individualization of production and services. Income determined by logistic skills to handle complex networks. Social and environmental costs are internalized in prices by increased eco-taxes and reducing subsidies which undermine sustainability. Business recognizes the importance of increasing the level of social and environmental development. Multinationals become vulnerable to the 'world consciousness' that makes consumers world-wide change their purchasing preferences on the basis of the corporate public image. Dematerialization of the economy. Government policy geared to increase innovative power. Civil society takes over the initiating role of government in safeguarding social and environmental long term values in national and international forums, including WTO. Two-third of the labor force works in services, about one-third in industry and less than approximately 5% in agriculture.

Level 5: Information technology enhances activities. Major breakthroughs e.g. in the exploration of outer space, virtual reality, and in new realms of the human spirit. Income is dependent on cognitive skills. New financial, entertainment, spiritual and educational services emerge. Employment in agriculture and industry falls back to 20% of total. Every adult person has a PC and a zero-emission personal mode of transport.

A proposal for quantitative characteristics of these levels is in Table 28.3. It should be noted that the examples given are time dependent and will have to be periodically actualized.

28.6 The Five Levels Applied in a Sustainability Scale

On the basis of these five levels of social, environmental and economic sustainability, five triads describing levels of sustainable development have been defined. These have similarities in their spatial and time dimensions and in the type of policy instruments and enforcement practises.

A summary of the policy instruments characteristic for each level and domain is given in Table 28.6.

28.7 Calculating a Sustainability Index

Table 28.3 gives a quantitative characterization of the indicators for an Environmental Index, a Social Index and an Economic Index, which are the building blocks for the overall sustainability index.

Table 28.6 Sustainability levels of societies, characterized by policy measures for the three domains

Level of concern	Triads of levels of sustainable development (characterized by type of policy measures applied)		
	Environmental level	Social level	Economic level
1. Local	Prohibitions, black lists	Provision of free food and lodging	Property rights, subsidies
2. Municipal	Emission standards, licenses	Minimum income, insurance	Trade tariffs, import licenses
3. National	Agreements	Labor market interventions	Competition rules
4. Global	Emission trading, eco-taxes	Employability stimulation	Innovation stimulation
5. Planetary	Guarantees, codification	Personal growth facilitation	Education, information

1. The Environmental Index

For estimation of the Natural Resources Sub-index the Natural Capital in US dollars per capita (as calculated by the World Bank 1995), the Annual Withdrawal of Internal Renewable Water Resources and the Forest as percentage of the Original Forest Coverage are used. The index is rated by scoring the values in one of the five levels of sustainability as indicated in Table 28.3 and calculating the average position on the sustainability scale by weighing all aspects equally. The advantage of this method is that in case data for one aspect are lacking and have to be deleted, the overall score on the sustainability scale can still be calculated, be it less reliably. Also new criteria for the index can be added later without affecting the design of the system.

The Environmental Quality Sub-index is based on the per capita Fossil Energy Use as an indicator for overall emissions of pollutants, the Carbon Dioxide Emission per dollar of GDP as a measure for the eco-efficiency of the economy and the Maximum Content of Lead in Gasoline as an indicator for the legislative power of the environmental authorities and for urban pollution. In the future lead in gasoline will no longer be a useful parameter as lead will everywhere be substituted.

The overall Environmental Index is calculated by weighing both Sub-indices equally.

2. The Social Index

To compose a Social Index, Life Expectancy, percentage Urban Population connected to a Sewer, the Number of Murders per 100,000 population in the urban environment, the Social Security Benefits Expenditure and the Combined First, Second and Third Level Enrollment Ratio, indicating the level of education (UNDP 1996) are used.

3. The Economic Index

For the Economic Index four indicators were compiled: percentage Labor Force in Services, the Number of Cars per 1,000 population, the Number of Main Telephone

Table 28.7 The sustainability index of 24 countries in the year 1998/99

Nation	Environmental Index — Resource index[a]	Quality index	Total index	Social Index	Economic Index	Sustain-ability Index
1. Sweden	4.7	3.7	4.2	4.4	4.5	4.37
2. Canada	4.7	3.0	3.9	4.2	4.0	4.03
3. USA	4.0	3.0	3.5	3.8	4.3	3.87
4. Japan	3.3	3.3	3.3	4.0	4.3	3.87
5. Australia	4.3	2.0	3.2	3.6	4.3	3.70
6. Netherlands	2.0	3.0	2.5	4.2	4.0	3.57
7. Germany	2.7	3.0	2.9	3.6	3.8	3.47
8. Argentina	4.0	3.7	3.9	3.2	3.0	3.37
9. Israel	1.7	3.0	2.4	3.4	3.8	3.20
10. Spain	2.0	2.7	2.4	3.6	3.5	3.16
11. Brazil	4.0	3.7	3.9	2.4	2.8	3.03
12. Costa Rica	3.0	3.3	3.2	3.0	2.8	3.00
13. Poland	3.0	2.7	2.9	3.2	2.8	2.97
14. Saudi Arabia	2.3	2.0	2.2	2.8	3.3	2.77
15. Russian Fed.	4.3	1.3	2.8	2.6	2.5	2.63
16. Mexico	3.0	2.7	2.9	2.6	2.3	2.60
17. Indonesia	3.7	2.7	3.2	2.4	1.5	2.37
18. Iraq	2.0	2.0	2.0	2.0	2.8	2.27
19. Egypt	1.6	2.0	1.8	2.8	1.8	2.13
20. India	2.0	3.0	2.5	2.2	1.5	2.07
21. Nigeria	2.1	2.6	2.4	1.8	1.8	2.00
22. China	2.0	2.3	2.2	2.4	1.0	1.87
23. Afghanistan	1.7	3.7	2.7	1.2	1.3	1.73
24. Sierra Leone	2.7	3.3	3.0	1.0	1.0	1.67

[a] Remarkable are some countries with little natural resources like the Netherlands, Israel and Spain, which nevertheless manage to obtain a high economic and social performance.

Lines per 100 population (UNDP 1996) and the Value of Produced Assets per capita (World Bank 1995). The Produced Assets per capita corresponds closely with GNP per capita. Unless indicated otherwise all data were retrieved from the World Resources 1998–1999 report (World Resources Institute et al. 1998).

A Sustainability Index can now be calculated weighing the Environmental Index, Social Index and Economic Index equally. The integration is possible because all three sustainability domains have a common reference in the five sustainability levels. The Sustainability Index ranges from 1.0 till 5.0.

The quantitative characteristics of the five sustainability levels proposed in Table 28.3 have a somewhat arbitrary character and can be further improved. However, the method proposed here is flexible in the sense that new indicators can be added without changing the system, and new sub-indicators can easily be added.

Future inclusion of a wider range of indicators is recommended to obtain a more complete picture of the situation and to further improve the methodology.

Table 28.8 Triads of sustainability levels, characterized by index values

Level of concern	Triads of levels of sustainable development (characterized by average index values for 1995)		
	Environmental Index	Social Index	Economic Index
1. Local	3.3	1.3	1.3
2. Municipal	2.8	2.4	2.4
3. National	3.2	3.6	3.6
4. Global	4.3	4.4	4.4
5. Planetary	4.9	4.9	4.9

28.8 Rating Sustainability of Countries

For a selection of 24 countries, representing different regions in the world, the Sustainability Index for the year 1998/1999 has been calculated as shown in Table 28.7. Similar ratings can be made for cities and companies to analyze their performance and to design next development steps (Zoeteman and Harkink 2005).

Table 28.7 shows that among the nations studied Sweden is for all aspects the most sustainable one. Nations with a high rating on natural resources, such as Canada, the USA and Australia also tend to have a high Sustainability Index, with the Russian Federation as an exception, due to its very low environmental quality score and moderate economic development. For instance Brazil is doing better in this respect. Low sustainability scores are found in Africa and Asia. In Sierra Leone and similar countries such as Ethiopia the very low scores for the social and economic indices strongly reduce the sustainability performance.

Table 28.8 indicates a fundamental difference between the Social Index and the Environmental Index in their relation to the Economic Index. The Environmental Index rating first goes down with economic development and picks up again from level 3 onwards, while the relationship between the Economic Index en the Social Index is more linear. This clearly shows why the environment issues are at sustainability levels 1 and 2 opposing the interests of economic and social development and that a synergy between the three aspects evolves from level 3 onwards.

How the individual nations are positioned to each-other is shown in Figs. 28.1 and 28.2. Nations at sustainability level 1 are Sierra Leone, Afghanistan and China. Typical examples of level 2 are Mexico and the Russian Federation, while Spain and Israel are representatives of level 3. Although a number of nations are bordering level 4, such as Australia, Japan and the Netherlands, only few such as Sweden have crossed this threshold.

28.9 Applying the Sustainability Scale

The concept of a sustainability scale can help to indicate imbalances in a society or company that may be detrimental to further development. It can also help to identify the need for safety nets in case a community has to face a step backward.

28.9.1 Taking Other Sustainability Interests into Account

The sustainability concept used emphasizes the need for each of the three sustainability pillars to take into account the developments in the two other pillars. Gladwin (1999) pointed out that it is in the interest of environmentalists to help improve social conditions.

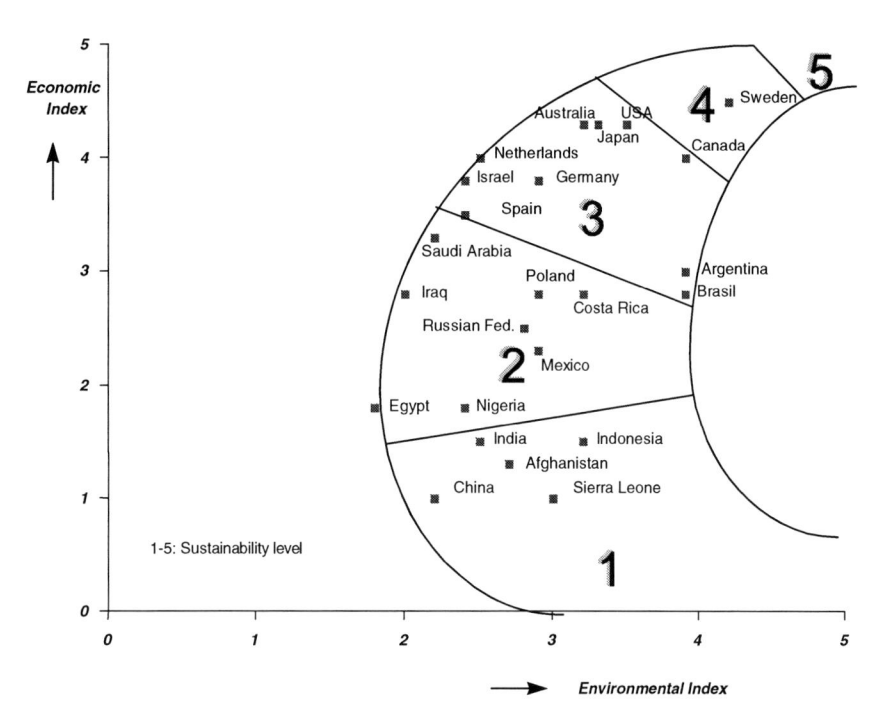

Fig. 28.1 The relationship between the Economic Index and the Environmental Index for 24 countries in the year 1998/99

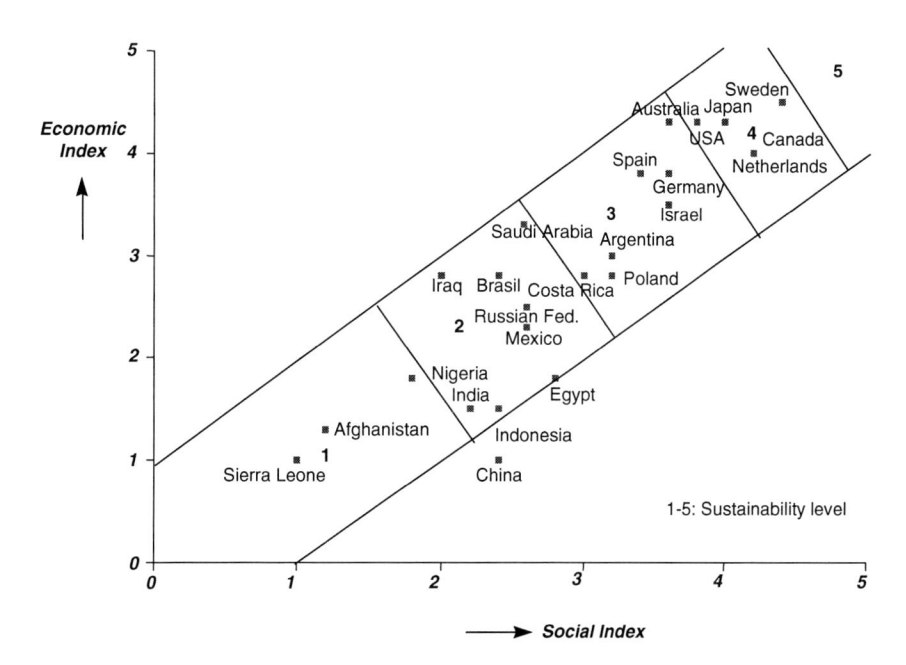

Fig. 28.2 The relationship between the Economic Index and the Natural Resources Index for 24 countries in the year 1998/99

Therefore companies and nations embarking upon level 4 have to aspire the ability to actively collaborate across the three pillars and make each intervention contribute to all three aspects. This approach has been called 'super-optimality' by Susskind (1999), inspiring participants to supersede the conventional aspiration to

Table 28.9 Data for GNP per capita, eco-efficiency (CO_2-emission per $1,000 of GDP), fossil energy use per capita, natural capital and life-expectancy for 24 countries in 1998/99

Nation	GNP/capita (US$)	CO_2 emission per $1,000 of GDP (t)	Fossil energy use/ capita (t)	Natural capital/ capita (US$1,000)	Life-expectancy at birth (years)
Japan	39,600	0.22	9.0	11	80.0
Germany	27,500	0.34	10.2	20	76.7
USA	27,000	0.78	20.5	103	76.7
Netherlands	24,000	0.34	8.8	7.6	77.9
Sweden	23,750	0.19	5.1	143	78.5
Canada	19,400	0.76	14.8	510	78.9
Australia	18,700	0.80	16.2	592	78.3
Israel	16,000	0.50	8.4	5.1	77.7
Spain	13,600	0.41	5.8	2.4	78.0
Argentina	8,000	0.45	3.7	61	73.2
Saudi Arabia	7,000	1.8	13.9	101	71.4
Brazil	3,640	0.36	1.6	21	67.1
Mexico	3,300	1.4	3.9	2.1	72.5
Poland	2,800	2.8	8.8	16	71.1
Costa Rica	2,610	0.56	1.5	5.0	74.7
Russian Fed.	2,240	5.2	12.2	6.8	64.4
Iraq	1,760	2.1	4.9	10	62.4
Indonesia	980	1.4	1.5	3.0	65.1
Egypt	790	1.9	1.5	3.3	66.0
China	620	4.5	2.7	0.37	69.9
Afghanistan	350	0.12	0.1	1.6	45.5
India	340	1.0	1.0	0.47	62.4
Nigeria	260	2.2	0.8	1.1	52.4
Sierra Leone	200	0.54	0.1	1.9	37.5

obtain an optimal outcome for one sector at the expense of the others.

28.9.2 Eco-efficiency Reconciliates Environment and Economy

How nations make the 'eco-efficiency U-turn' at level 2 is documented in Table 28.9. From a GDP per capita level of US$1,000 onwards carbon dioxide emission per US$1,000 of GDP diminish with a factor of about 2.2 for each doubling of GDP per capita. Due to an increase in the population this can still not be sufficient to achieve a decoupling between the economic growth and the environmental pressure.

Societies leaving level 3, such as Germany and the Netherlands, showed a difference in eco-efficiency by a factor 100 while GDP per capita increased by a factor of 10, compared to nations entering level 3, such as Indonesia and Iraq. Through increased eco-efficiency, countries may move from an antagonistic into a synergistic relation of environment and economy.

As Fig. 28.3 suggests, overall environmental quality seems, within wide margins, to be rather constant at different levels of the Economic Index as a result of the increasing eco-efficiency of the economy. An exception is the Russian Federation where pollution in 1998/1999 was larger than in comparable countries.

As the data in Table 28.9 indicate, nations with a large natural capital per capita such as Canada, the USA and Australia, seem to feel less need to focus on an efficient use of resources.

28.9.3 Next Steps in Development

Moving to higher levels of sustainability is a challenge. The overview given identifies major steps to improve sustainability, essentially moving from short term actions, looking solely at economic profit and ignoring external effects, towards longer term actions including social and ecological security at a global scale and a high degree of transparency. Sharing experiences will help to move upward faster. Although there are chances to leapfrog

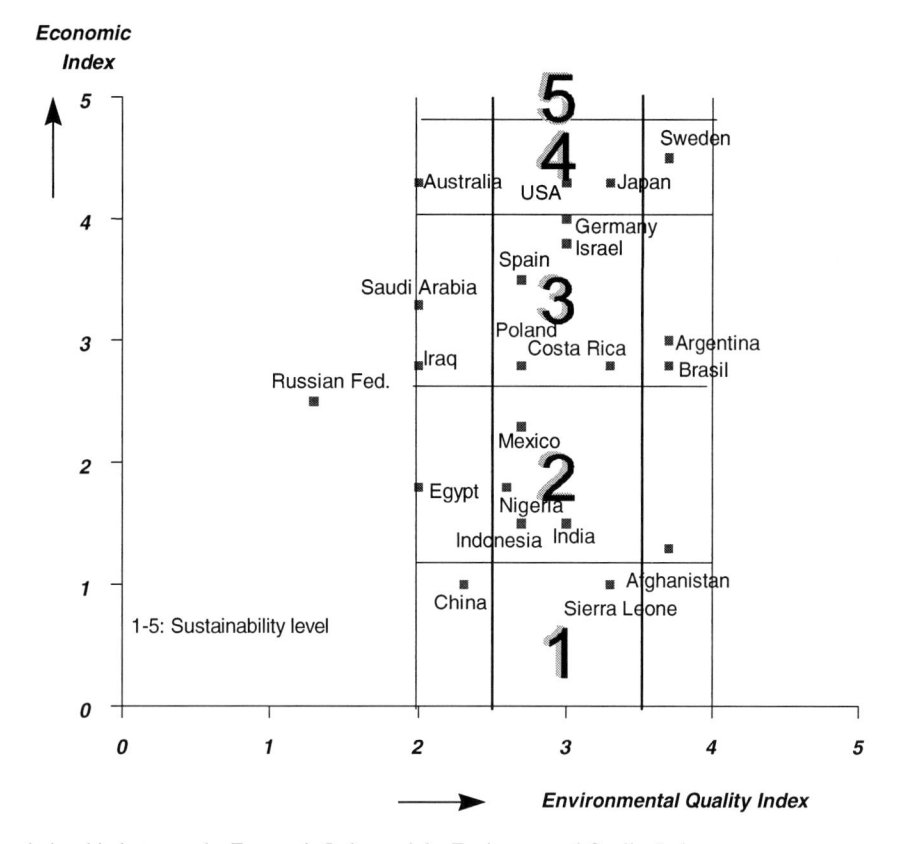

Fig. 28.3 The relationship between the Economic Index and the Environmental Quality Index

ahead there is little evidence that levels in the sustainability scale presented in Table 28.6 can really be skipped. But the framework presented may encourage pioneers to move forward by using it to self-assess the performance.

28.9.4 Balance Between the Three Sustainability Aspects

When the economy is overdeveloped compared with the social or environmental aspects a shift in priorities is advisable to avoid a forced correction later. Transparency of the performance on all three aspects and sensitivity for the signals given by the public are helping to prevent mistakes. Provisions such as social requirements for foreign direct investments and a practical form of environmental impact assessment, applied by internationally operating banks and insurance companies, will help to keep the three aspects in balance in dynamic growth situations.

28.9.5 Reduction of Damage in Fall-Back Situations

The level of sustainability of a country (or company) can fall back for many reasons, such as economic recession, social conflict or war. The behavioral attitudes dominant at the lower sustainability levels tend to be re-activated during fall-back. The interconnectedness with neighbours and the global economy make solutions often extra complicated. Yet as a general rule it is in all these situations desirable to build institutions that promote stability and trust. For similar reasons Soros (1998) emphasized the need to stabilize the flow of capital from the global center of the capitalist system to the periphery by establishing e.g. international credit insurance corporations or reduce rivalry between currency systems by better policy coordination. Nurturing collaboration mechanisms, which is the new quality emerging at sustainability level 3, will be of great value to prevent a fall-back and to restore the situation as soon as possible. In the case of economic restructuring, emissions often decline initially due to reduced economic activity. In such cases the creation of externally financed investment funds and other forms of external help such as local capacity building for environmental management during times of economic contraction, will promote a higher level of sustainability. However, as was shown for the case of the airlines industry, global competition and consumer expectation have an upward effect on sustainability attitudes, even during periods of economic crisis in the sector (Zoeteman et al. 2007).

References

Biesiot, W. (1997). Respite time and response time. *Balaton Bulletin, Fall 1997,* 12–13.
Bossel, H. (1997). Deriving indicators of sustainable development. *Environmental Modeling and Assessment, 1*(4), 193–218.
Bossel, H. (1999). *Indicators for sustainable development: Theory, method, applications.* A Report to the Balaton Group, International Institute for Sustainable Development, Winnipeg, Manitoba, Canada.
CEC (Commission of the European Communities). (2001). *A Sustainable Europe for a Better World, A European Union Strategy for Sustainable Development.* COM 264, 15-5-2001, Brussels.
Daly, H. E., & Cobb, J. (1990). *For the common good, redirecting the economy towards community, environment and a sustainable future.* London: Green Point.
De Kruijf, H. A. M., & Van Vuuren, D. P. (1998). Following sustainable development in relation to the north-south dialogue: Ecosystem health and sustainability indicators. *Ecotoxicology and Environmental Safety, 40,* 4–14.
Gladwin, T. N. (1999). *Global sustainability: Challenges for environment and social equity.* Paper presented at seminar on Sustainable Development, Ministry of Housing, Spatial Planning and Environment, The Netherlands.
Hueting, R., & Bosch, P. (1990). Note on the correction of national income for environmental losses. *Statistical Journal of the United Nations Economic Commission for Europe, 7*(2), 75–83.
Kaku, R. (1996). Cited in J. Jaworski, *Synchronicity, the inner path of leadership* (p. 164). San Francisco: Berrett-Koehler.
Maslow, A. H. (1970). *Motivation and personality.* New York: Harper & Row.
Maslow, A. H., Stephens, D. C., & Heil, G. (1998). *Maslow on management.* New York: Wiley.
Max-Neef, M. A. (1991). *Human scale development, conception, application and further reflections.* New York: The Apex Press.
Meadows, D. H. (1998). *Indicators and information systems for sustainable development.* A Report to the Balaton Group, The Sustainability Institute, Hartland Four Corners.
Meadows, D. H., Meadows, D. L., Randers, J., & Behrens, W. W. (1972). *The limits to growth.* New York: Potomac (Signet).
O'Conner, J. (1995). *Real wealth of nations.* Washington, DC: World Bank.
Soros, G. (1998). *The crisis of global capitalism* (pp. 176–187). New York: Public Affairs.
Susskind, L. (1999). *Super-optimization: A new approach to national environmental policy-making.* Paper presented at seminar on Sustainable Development, Ministry of Housing, Spatial Planning and Environment, The Netherlands.
Swarttouw, K. (1999) personal communication.
UNDP. (1996). *Human development report.* New York: Oxford University Press.
UN World Commission on Environment and Development. (1987). *Our common future.* Oxford: Oxford University Press.
Wackernagel, M., & Rees, W. (1996). *Our ecological footprint.* Gabriola Island, Canada: New Society.
Winsemius, P., & Guntram, U. (2002). *Thousand shades of green.* London: Earthscan.
World Bank. (1995). *Monitoring environmental progress: A report on work in progress.* Washington, DC: World Bank.
World Resources Institute, et al. (1998). *World resources 1998–99. A guide to the global environment, World Resources Institute.* Oxford/New York: United Nations Environment Programme/United Nations Development Programme/World Bank/Oxford University Press.
Zoeteman, K., & Harkink, E. (2005). Collaboration of national governments and global corporations in environmental management. In F. Wijen, K. Zoeteman, & J. Pieters (Eds.), *A handbook of globalisation and environmental policy* (pp. 179–210). Cheltenham: Edward Elgar.
Zoeteman, B., van Reisen, O., & Kaashoek, B. (2007). Do global critical events impact organizational sustainability attitudes? The case of the airline industry. *Aerlines Magazine, e-zine edition,* 35, 1–7.

Author Index

A
Abernathy, W.J., 505
Abrams, N.E., 297
Adams, J., 199, 402
Adriaanse, A., 9, 494
Ahmad, Y., 186
Akesson, A., 65
Alcamo, J.M., 392, 408, 409
Allenby, B.R., 92
Allen, P., 348, 359
Almond, G., 425
Amann, M., 397, 400
Anawar, H.M., 54
Andersen, M., 426
Anderson, D., 477–479
Anderson, T.L., 91
Aneja,V.P., 64
Angel, D., 510
ApSimon, H.M., 396
Archambault, S., 428
Archer, D., 66
Arentsen, M.J., 443
Arrigo, K.R., 61
Arrow, K., 10, 92
Arthur, W.B., 360
Atkinson, G., 189
Auer, M.T., 157, 159
Aurobindo, S., 349
Ayres, L.W., 91, 208
Ayres, R.U., 86, 91, 92, 208, 223, 503
Azar, C., 356

B
Baark, E., 428
Baird, C., 153
Bakker, J., 60
Bakker, K., 155
Baldwin, L.A., 64
Balmann, A., 366, 369
Balter, M., 24
Bandura, A., 136
Banfield, E.C., 425
Baptist, H.J.M., 172, 173
Barber, B., 308
Bardach, E., 257

Barnard, P., 445, 448
Barnes, J., 453
Barney, G.O., 387, 388, 402
Barry, B., 474
Bate, J., 424
Baumol, H., 262, 263
Bayerl, G., 28, 30
Bayne, B.L., 170
Beckerman, W., 473, 474, 478
Beck-Gernsheim, E., 245
Beck, U., 245, 284, 428
Beinart, W., 28
Bell, R.H.V., 452
Bell, S., 9, 343
Bennett, M., 191, 192
Berdowski, J.J.M., 66
Berger, T., 366, 369
Berkhout, F., 504, 508, 511
Berk, M., 408
Bermejo, M., 58
Berz, G., 68
Beusekamp, W., 433
Beus, J.W., 439
Bickerstaff, K., 427
Bigalke, R.C., 452
Birch, C., 487
Birchenall, C., 507
Blair, P.D., 178–185
Boehmer-Christiansen, S., 426, 465, 466
Boeker, E., 147
Boersema, J.J., 94, 143, 167, 339, 440
Bohm, P., 262
Bolin, B., 16
Bonus, H., 257
Boschma, R., 348
Bossel, H., 340, 352, 355, 360, 411, 517
Botts, L., 296
Boulding, K., 355
Boulton, A.J., 171
Bouwman, A.F., 61, 62
Braat, L.C., 386
Bramwell, A., 26
Braungart, M., 11
Brauni, E., 306
Bressers, J.T.A., 257, 260, 266
Breukers, S., 426

Brimblecombe, P., 7
Bromley, D.W., 91
Brouwer, F.M., 386
Budiansky, S, 26
Buiks, P.E.J., 465
Burgess, J., 428
Buttel, F., 428
Byrd, B.F., 34

C
Calabrese, E.J., 64
Callon, M., 505
Cann, M., 153
Cardinale, B.J., 58
Carmichael, J., 298
Carson, R., 7
Castells Cabré, N., 250
Castells, M., 264
Catton, W., 426
Caughly, G., 452
Cerin, P., 91
Chapman, J.L., 21
Chapman, P.M., 168
Chertow, M. R, 10
Child, B.A., 446, 447, 450, 451, 453, 454
Child, G., 446, 450–453, 455
Chitsike, L., 451
Christensen, C.M., 506
Christensen, V., 56
Christiansen, P., 426
Clapp, B. W, 26
Clark, J., 257
Clark, K.B., 505
Clark, N., 365, 369
Clark, W.C., 8, 353, 387, 503
Claussen, M., 19
Coase, R.H., 252
Cobb, J., 187
Cohen, B., 215
Cohen, D.K., 295
Cohen, M., 424, 427, 428
Cohen, S.J., 296
Cohn, N., 86, 94
Colborn, T., 5
Cole, D.H., 257
Commoner, B., 9, 10
Condy, J.B., 449
Connell, D.W., 152, 153
Connell, J.H., 155
Cook, R.B., 16
Cooper, R., 478
Corbain, A., 25
Costanza, R., 168, 169, 171, 172, 199, 500
Couclelis, H., 365
Cough, C.A., 392
Covello, V., 427
CPB., 410
Crosby, A.W., 28
Cross, C.F., 290

Crutzen, P.J., 63, 492, 493
Cumming, D.H.M., 452

D
Daalder, H., 439
Dahl, R.A., 282
Daily, G., 195
Dake, K., 427
Daly, H., 88, 89
Daly, H.E., 13, 187
Darmstadter, J., 13, 51
Dasgupta, P., 358
Davies, R., 455
De Boer, B., 499–501
De Bruin, J., 402
De Bruyn, S.M., 494
Deffuant, G., 366, 369
De George, R.T., 474
De Groot, R.S., 193, 195
De Haan, M., 189
De Jager, J.L.V., 453
DeMers, M.N., 164
De Moor, M., L.
Den Elzen, M., 351, 404
Deodatus, F., 451
Devall, B., 94
Devinny, J.S., 149
De Vries, H.J.M., (Bert), 347, 351–353, 356, 358, 360, 364, 370, 402, 404, 406, 408, 411, 415
Dietz, Th., 10
Difi glio, C., 198
Dirkx, G.P.H., 28, 29
Disse, M., 68
Ditz, D., 191
Dively, D.D., 193
Dobson, A., 440
Doran, J., 368
Dorfman, M., 192
Douglas, M., 86–87, 93, 427
Downs, A., 5, 6
Dreborg, K.H., 298
Drury, S.A., 21
Dryzek, J.S., 86, 89
Duchin, F., 354
Dunlap, R., 426, 427
Dunlap, R.E., 5, 129, 466
Dunn, W.N., 298
Durant, J., 298, 428

E
Easton, D., 276
Eberg, J., 288, 295, 433, 443
Eberg, J.W., 443
Edelstein-Keshet, L., 355, 360
Eden, S., 428
Edney, J.J., 118
Ehrlich, A., 9
Ehrlich, P.R., 9, 76, 133
Ekins, P., 185, 187, 189, 190, 197, 340

Elias, N., 437
El-Kholy, O.A., 493, 494
Elkington, J., 13, 503
Elliot, M., 297
Elzen, B., 427, 504, 505, 509
Endres, A., 257
Engelbrecht, W.G., 451
Engelen, G., 360
Engel, H., 68
Engels, A., 299
Enloe, C., 426
Etemad, B., 51
Etienne, M., 370
Ezrahi, Y., 282, 283

F

Faist Emmenegger, M., 231
Falkenmark, M., 65, 68
Farber, S., 199
Farthing, K., 166
Faure, M., 426
Feinberg, J., 492
Fenger, J., 59, 64, 65
Ferber, J., 366
Fiddaman, T.S., 352, 364
Fischer, F., 282
Fischer-Kowalski, M., 248, 354
Forrester, J.W., 387
Freeman, C., 505
Frewer, L., 93
Frissell, C.A., 161, 172
Fritts, T.H., 57
Frosch, R.A., 91
Fukuyama, F., 425
Funtowicz, S.O., 212, 283, 348, 391

G

Galanter, H., 256
Gallopin, G., 411
Gallopoulos, N.E., 91
Galloway, J.N., 61
Gardner, G.T., 107
Gardner, R., 93
Gauthier, D., 476
Geels, F., 353
Geels, F.W., 504–508
Gerstenfeld, M., 86, 94
Gibbons, M., 283
Giddens, A., 263, 264
Gilbert, N., 368
Glacken, C., 424
Glantz, M.H., 460
Glover, J., 481
Golley, F.B., 155
Goodin, R., 474
Goodland, R., 493
Gordon, D.C. Jr., 160, 161
Gordon, R.B., 52
Goudsblom, J., 85, 93, 352, 360

Gouldson, A., 426
Graedel, T.E., 52, 63, 492, 493
Gray, K., 91
Greenblat, C., 370
Greene, D.L., 198
Greening, L., 198
Grennfelt, P., 393
Grime, J.P., 155
Grin, J., 443
Groenewold, H., 347
Groombridge, B., 56, 57
Grootenhuis, J.G., 449, 450
Groot, R.S. de, 4
Grove-White, R., 459
Grubler, A., 356
Guinée, J.B., 269
Gumbricht, T., 164
Gunderson, L., 352, 359
Gunningham, N., 278
Gupta, J., 351

H

Haas, P., 291, 295
Haberl, H., 56, 354
Hafkamp, W.A., 386
Haftendorn, H., 91
Hahn, S., 28
Hajer, M., 295, 426, 427
Hajer, M.A., 295, 426–427
Halffman, W., 283
Halfmann, J., 427
Hamilton, K., 189
Hampden-Turner, C., 440, 465–467
Han, F.X., 65
Hannigan, J., 424
Hannigan, J.A., 440
Hannon, B., 360, 364
Han, X., 65
Hardin, G., 91, 92, 116, 453
Harman, W., 349
Harremoës, P., 158
Harrington, A., 424
Harrison, C., 428
Harrison, C.M., 428
Harrison, K., 277
Havelaar, A., 49
Havel, V., 6
Hawken, P., 328
Hawkins, K., 257
Heady, F., 438
Heal, G., 358
Hearne, J., 450, 452
Heaton, G., 10
Heclo, H., 285
Hedger, R.S., 449
Hekkert, M.P., 504, 508
Hemmelskamp, J., 263
Herb, N., 505
Herendeen, R.A., 338
Herman, R., 90

Hertin, J., 340
Hettelingh, J-P., 386, 391, 394, 397–400
Heuss, T., 439
Heyes, A.G., 184
Hicks, J., 186, 187, 483
Hilderink, H., 404, 408
Hindess, B., 462
Hirsch, D.A., 506
Hirsch, F., 121
Hisschemöller, M., 282, 283, 292, 294–296, 298, 351, 389
Hitchcock, R.K., 452
Hittelman, A.M., 68
Hoekema, A.J., 258
Hoekstra, A.Y., 404
Hoesel, G., 85
Hoffman, A.J., 254
Hoffman, L., 425
Hoffmann, R.C., 30
Hofstede, G., 438
Holdren, J.P., 9, 133
Holland, E.A., 58, 62
Holland, J., 348
Holling, C.S., 160, 170, 388
Holmberg, J., 89
Honrath, R.E., 144, 145
Hoppe, R., 292, 296, 298
Hordijk, L., 386, 389, 392
Houghton, J.J., 147
Howes, R., 192
Hubbert, M.K., 52, 53
Huber, J., 90
Huchzermeyer, H.F.K.A., 44
Hueting, R., 89, 483, 491, 493, 494, 496–501, 517
Hughes, J.D., 85
Hughes, T.P., 505
Huijbregts, M.A.J., 162
Huntingdon, S., 411
Huntington, S., 425
Huntley, B.J., 448
Huppes, G., 227, 239, 240, 257, 258, 269, 270, 275
Huston, M., 155
Hutter, K., 146

I
Imhoff, M.L., 56, 248
Inglehart, R., 426, 427
Ingold, T., 248
Inkeles, A., 425
Irwin, A., 424, 428

J
Jachtenfuchs, M., 464, 465
Jackson, A.R.W., 159
Jackson, J.B.C., 45, 58
Jacob, M., 429
Jacobsson, S., 508
Jäger, H., 26
Jäger, J., 386
Jager, W., 104, 361, 367–369
Jagtman, E., 172, 173
James, P., 191, 192
Jamison, A., 428, 429
Jänicke, M., 276, 426
Jansen, D.J., 453
Janssen, M., 348, 360, 366, 367, 406
Jantsch, E., 349
Jasanoff, S., 427
Jasch, C., 191
Jelsma, J., 298
Jenkins, M.D., 56
Jenkins-Smith, H.C., 285–288, 295, 443
Jevons, W.S., 88, 478
Johansson, P.-O., 198
Johansson, T.B., 411, 494
Johnson, A., 508
Johnson, B., 427
Jones, A., 144, 163
Joppke, C., 427
Jorgens, H., 276
Joss, S., 298

K
Kagan, B., 240, 257
Kagan, R.A., 257
Kahneman, D., 112
Karlson, L., 91
Karr, J.R., 167
Kasemir, B., 296, 297
Kassler, P., 411
Kauppi, P.E., 37
Kaye, G.W.C., 145, 150
Kelly, M.G., 165
Kempe, M., 29
Kemp, R., 504, 506
Kempton, W., 427
Kendall, H.-W., 493, 494
Kennedy, J.C., 437
Kerley, G.I.H., 452
Kessel, H., 466
Khalid, F., 94
Kirby, J., 91
Klaassen, G., 400
Kleijn, R., 224
Klein, H., 396
Klok, P.J., 260, 266
Komarov, B., 7
Koningsveld, H., 465
Kooistra, L., 165, 166
Kram, T., 410
Kraus, A., 438
Kroeze, C., 386, 392
Krug, E.C., 61
Kuik, O., 9
Kuntz-Duriseti, K., 269
Kwant, R.C., 465

L

Laby, T.H., 145, 150
Lafferty, W., 426, 429
Laidlaw, M.A.S., 65
Lakatos, I., 285
Lamphear, R.P., 65
Lange, G.-M., 354
Lastovicka, J., 65
Latour, B., 429
Leal, D.R., 91
Lee, I., 215
Leggett, J., 409
Leinders, J.J.M., 165, 166
Leopold, A., 90
Leroy, P., 241
Leuven, R.S.E.W., 166, 167
Lewin, R., 348
Lewis, C.S., 486
Lide, D.R., 150
Liebrand, W.B.G., 118
Liefferink, D., 426
Lijphart, A., 437, 439
Lindblom, C.E., 295
Linstone, H.A., 298
Lintsen, H.W., 438
Lipset, S., 425
Liston-Heyes, C., 184
Lofstedt, R., 93
Lomolino, M.V., 57
Loorbach, D.A., 510
Lopez, A.D., 49, 215, 216, 338
Lovelock, J.E., 16, 58, 493
Lovins, L.H., 328
Lowe, J.J., 21, 22
Lucardie, P., 440
Lucretius., 85
Luhmann, N., 240, 241
Lundqvist, L., 426, 428
Lundvall, B.-Ä., 427
Lutz, E., 186

M

MacKay, D., 151, 217
Maddison, A., 477
Mageau, M., 169, 171
Main, M., 448
Malthus, R.T., 9, 80–82
Mannion, A.M., 20–22
Manson, S.M., 350
Marchetti, C., 356
Marcuse, R., 496
Marsh, G.P., 6
Martin, R.B., 446
März, P., 439
Maslow, A.H., 6, 131
Mason, R.O., 298
Max-Neef, M., 367
Max-Neef, M.A., 518
Mayer, I., 297
Mayer, S., 202

Mayntz, R., 240
Mayo, D.G., 213, 215
McDonough, W., 11
McKenzie, M., 450, 452
McNeill, J.R., 26–28
Meadowcroft, J., 429
Meadows, D.H., 7, 76, 370, 387, 401, 402, 478, 519
Meadows, D.L., 360
Merkus, H., 461
Mertig, A., 426, 427
Meyers, S., 494
Michael, M., 428
Middelkoop, H., 166
Midden, C., 134
Middendorp, C.P., 439
Middleton, K., 28
Mihelcic, J.R., 143–145, 151, 153
Miller, J., 428
Miller, R.E., 178–185
Mill, J.S., 87, 88, 282
Mills, M.P., 478
Milton, K., 93
Mintzberg, H., 246
Mishan, E.J., 89
Mitroff, I.I., 298
Mock, J.E., 479
Mol, A., 427
Monson, R.K., 58
Moore, P.D., 22
Morecroft, J., 348, 349
Morgenstern, O., 269
Morita, T., 386, 410
Morse, S., 9
Mosler, H.J., 104
Müller, A., 348
Mulligan, M., 353
Munasinghe, M., 137
Munn, R.E., 8, 353, 387
Muradian, R., 57
Murphy, J., 426
Murray, C.J., 49
Murray, C.J.L., 215, 216, 338

N

Naeher, L.P., 65
Naess, A., 94
Nakícenovíc, N., 356, 408, 410, 411, 414
Nash, R., 424
Nelkin, D., 427
Nelson, R., 427, 506
Neutel, A.-M., 364
Newbigin, L., 486
Newell, P., 291
Niessen, L., 404
Nijkamp, P., 201–204
Nilsson, J., 292, 393
Noonan, J.P., 23
Noorman, K.-J., 354
Norgaard, R., 474
Nowotny, H., 283

O

Oates, W.E., 262, 263
O'Brien, J., 94
Odum, E.G., 493
Odum, E.P., 169, 170
Ogle S.M., 17
Olivier, J.G.J., 60, 66
O'Neill, O., 476
Opschoor, J.B., 89, 263
Oreskes, N., 391
O'Riordan, T., & Jager, J., 408
Ormel, J., 131
Osborn, A., 297
Østby, P., 429
Ostrom, E., 93, 118, 369, 453

P

Parson, E.A., 297, 389
Partridge, E., 474
Pasek, J., 474
Pauly, D., 54, 56
Pearce, D., 20, 189, 198, 351
Perman, R., 358
Perrault, D.R., 57
Pesaran, M.H., 52, 53
Pfister, C., 7, 27, 29
Pigou, A.C., 88, 89
Pimentel, D., 56, 493, 494
Platt, J., 58
Plowright, W., 449, 451
Pollak, M., 427
Poortinga, W., 132
Porteous, A., 4
Porter, E., 92
Porter, M., 197, 427
Porter, R., 424
Poulussen, P., 26
Primack, J.R., 297
Prins, H.H.T., 454
Procter, J., 428
Proops, J., 180, 184
Prüss, A., 49
Putnam, R., 425

R

Radkau, J, 29
Raga, G.B., 64
Rahmstorf, S., 67
Randers, J., 360
Rapport, D.J., 168
Raskin, P., 53, 411
Rauberger, R., 191
Raven, R.P.J.M., 507
Ravetz, J.R., 212, 283, 348, 391
Rawls, J., 263, 472, 474, 476
Rayner, S., 467
Raz, J., 476
Rees, W., 11, 517
Reeve, R., 426

Reijnders, L., 11, 34, 85, 90, 95, 485, 493, 497
Renn, O., 298, 299
Reveil, A., 428
Reynolds, D.R., 53
Richards, D.J., 92
Richardson, G.P., 355
Richardson, J., 455
Richardson, J.J., 436, 441
Ridken, R.G., 327
Riesman, D., 244
Rietkerk, M., 450
Rikhardsson, P., 192
Rip, A., 506
Ritov, I., 112
Roberts, N., 85, 358, 360
Rock, M.T., 510
Rodda, G.H., 57
Rodrigues, A.S.L.
Rogner, H-H., 479
Rohracher, H., 505
Rohr, Chr., 29
Rootes, C., 440
Rosa, E.A., 10
Rosen, R., 346, 360
Rose, R., 443
Rotmans, J., 353, 386, 388, 389, 391,
    402, 408, 411, 510
Ruddiman, W.F., 19
Ruiz, G.M., 57
Rupke, N., 424
Russell, C.S., 262
Ruth, M., 360, 364
Ruyssenaars, P., 461

S

Sabatier, P.A., 285–288, 295, 443
Sachs, W., 137
Sala, O.E., 57, 58, 69, 171
Samet, J.M., 65
Samier, H., 52, 53
Sandbrook, R., 89
Sands, P., 91
Schama, S., 440, 441
Scharpf, F.W., 256
Scheffer, M., 359, 455
Schelling, T., 194, 478
Schipper, L., 494
Schlager, E., 453
Schlesinger, W.H., 503
Schmidheiny, S., 427
Schmidt-Bleek, F., 11, 12, 90, 328, 338
Schneider, S.H., 269
Schön, D.A., 296
Schooler, C., 425
Schoot Uiterkamp, T., 354
Schopenhauer, A., 4
Schöpp, W., 392, 397
Schot, J.W., 510
Schumacher, E.F., 76
Schumpeter, J., 245

Schumpeter, J.A., 245
Schwartz, S.H., 128
Schwarzenbach, R.P., 152
Schwarz, M., 93, 402, 440
Seley, J.F., 298
Sen, A.K., 260, 261, 269
Sessions, G., 94
Shiva, V., 87, 94
Short, J.R., 466
Sieferle, R.-P., 27
Silveira, L., 58
Simberloff, D., 57
Simmons, I.G., 6, 7, 30
Simon, H.A., 348
Simonis, E.U., 248
Simon, J.L., 10, 88, 190
Simon, S., 190
Simons, S., 340
Sinclair, D., 278
Skea, J., 426
Skinner, J.D., 452
Slicher van Bath, B.H., 28
Sliggers, J., 400
Slovic, P., 213
Smeets, E., 211
Smil, V., 61
Smith, A., 507, 508
Smith, K.R., 58
Smith, M.A., 87
Smythe, K.D., 60
Söderbaum, P., 348
Sol, V.M., 340
Sonnlechner, C., 29
Spretnak, C., 94
Stagl, S., 202
Steadman, D.W., 94
Steg, L., 131
Steiner, H., 475
Sterman, J., 348, 349, 360
Sternberg, L., 58
Stern, N., 198, 199
Stern, P.C., 107, 129
Stirling, A., 200–202
Stoddard, J.L., 400
Strauss, F.J., 439
Suh S., 227
Susskind, L., 297
Swart, R., 67, 408

T
Tainter, J., 353
Taylor, P., 428
Taylor, R.D., 446
Teeuw, R.M., 166
Teich, M., 424, 426
Tellegen, E., 93
Temkin, L., 476
Ten Brink, B.J.E., 172, 338
Tesfatsion, L., 368, 369
Thomas, C.D., 57

Thomas, K., 29, 76, 424
Thomas, W.L. Jr., 6
Thompson, M., 93, 348, 402, 440
Tietenberg, T., 125
Tilman, D., 171
Tinbergen, J., 278, 493, 494, 496, 497
Tischler, W., 466
Tisdell, C., 252
Tolba, M.K., 493, 494
Tol, R.S.J., 388, 389
Tönnies, F., 245
Toth, F., 370
Toulmin, S., 348
Treweek, J., 164, 165
Trompenaars, F., 440, 465–467
Tucker, M., 427
Tuinstra, W., 392, 510
Tukker, A., 232
Turoff, M., 298
Tyson, P.D., 447

U
Udo de Haes, H.A., 210, 230
Ulanowicz, R.E., 171
Uljee, I., 369
Ulrich, R.S., 110
Unruh, G.C., 505

V
Valiela, I., 156, 157, 159
Van Asselt, M.B.A., 391, 401, 407
Van Daalen, C., 408
Van Dam, P.J.E.M., 26, 27
Van de Graaf, H., 443
Van de Kerkhof, M.F., 296, 298, 299
Van de Meent, D., 162, 218
Van der Linde, C., 197, 427
Van der Sluijs, J.P., 296, 348, 405
Van der Voet, E., 221
Van der Walt, J., 451
Van der Werf, W., 191
Van Dieren, W., 440
Van Dongen, H.J., 93
Van Grondelle, R., 147
Van Tatenhove, J., 241
Van Vuuren, D., 61, 62, 404, 408
Van Vuuren, P.P., 52
Van Waarden, F., 438
Van Wilgen, B.W., 57
Van Zanden, J.L., 26
Van Zon, H., 25, 26
Vavrousek, J., 91
Velders, G.J.M., 60
Vellinga, P., 388, 389, 505
Vellinga, P.M., 67
Verba, S., 425
Verbruggen, C., 28
Verbruggen, H., 9, 493, 499
Verburg, P., 365, 369

Verstegen, S.W., 26
Vestreng, V., 396
Vig, N., 426
Vitousek, P.M., 56, 61
Vlek, C., 133, 134
Vlek, C., & Keren, G.B., 119
Vogel, D., 257, 276, 426
Vogler, J., 88, 91, 93
Von Benda-Beckmann, K., 258
Von Gleich, A., 224
Von Holle, B., 57
Von Neumann, J., 269
Von Prittwitz, V., 462–464, 469
Von Weizsäcker, E.U., 11, 328, 427
Voogd, H., 201
Vos, H.B., 263

W

Wackernagel, M., 11, 338, 354, 517
Wadsworth, R., 164, 165
Wagner, B., 191
Walker, G., 427
Walker, J., 93
Walker, M.J. C, 21, 22
Walsh, P.D., 58
Wania, F., 151
Ward, J.V., 155, 156
Ward, P., 29
Wartenberg, D., 65
Watts, N.S.J., 436, 441
Weale, A., 285, 426, 438
Weaver, P., 298, 328
Weber, M., 240, 289, 437
Weidner, H., 426
Weiss, C., 285
Weterings, R.A.P.M., 89, 211
Wettestad, J., 291

Weyant, J., 388
White, G.F., 68
Whitehead, A.N., 487–489
White, L., 94
White, R., 365
Wieczorek, A.J., 504, 505
Wilber, K., 345
Wildavsky, A., 86–88, 93, 427
Williams, M., 22
Wilson, R., 214, 215
Wilson, R.C.L., 21
Wilting, H., 354
Windrum, P., 507
Winiwarter, W., 25
Winstanley, D., 61
Winter, S.G., 506
Wintle, M., 426, 439
Wissenberg, M., 440
Witte, F., 57, 58
Wolf, A.T., 91
Wolff, W.J., 57
Wolsink, M., 93, 426
Worm, B., 52
Worster, D., 25, 26
Wrisberg, N., 210
Wynne, B., 93, 427, 428

Y

Yaron, G., 450, 452
Yearley, S., 428
Young, O.R., 504

Z

Zahn, E., 437, 441
Zerbe, R.O., 193
Zhang, J., 64
Zock, J.P., 58

Subject Index

A

Absorption, 49, 146, 152, 195, 319
Acceptable risk, 87, 88, 93, 94, 114, 130, 213
Accounting national income, 483, 484
Accounting sustainable national income, 190, 483, 484, 491, 492, 495–497, 499
Adaptive Environmental Assessment (AEA), 388, 389
Advocacy coalition, 285–288, 292, 300
AEA. *See* Adaptive Environmental Assessment
Aerobic biological treatment, 320–321
Africa, 21, 33, 36, 37, 53, 55–57, 61, 62, 68, 136, 369, 445–456, 477, 514, 524
African countries, 445–456
Age estimation, 23, 24, 203
Agricultural policy, 451, 454, 460, 467
Agriculture, 7, 19, 21, 27, 33–47, 52, 56, 57, 69, 85, 134, 136, 138, 166, 178, 179, 183–185, 251, 265, 290, 293, 307, 356, 366, 369, 385, 393, 397, 400, 405, 413, 437, 447, 448, 451, 460, 464, 493, 513–516, 522
AMOEBA method, 12, 172, 174
Anaerobic biological treatment, 320, 321
Analysis, 5, 7, 8, 11, 13, 22, 29, 54, 76, 86, 90, 97, 99, 100, 102–106, 109, 111–114, 117, 132, 133, 135
Asbestos, 58, 324

B

Bacterial soil treatment, 323
Behaviour, 26, 30, 59, 63, 79, 80, 93, 94, 97–140, 156, 162, 166, 169, 170, 195, 239–245, 257, 259–261, 264–267, 269, 277, 279
   empirical study of, 99, 104–106
   measurement of, 99–102
   study design of, 102–104
   change, 122–124
   data analysis, 102–106, 112
   intervention, 126–127
Biodiversity, 5, 12, 19, 55–57, 64, 65, 84, 90, 106, 111, 125, 132, 136, 137, 155, 156, 162–163, 171, 174, 200, 230, 233, 241, 247, 249–251, 322, 340, 352, 385, 445, 488, 489, 491, 504, 513
Biogeochemical cycle, 15–17, 19, 58, 61, 63, 86, 157, 159
Bio-invasive species, 324
Biological soil treatment, 323
Bio monitoring, 162
Biotechnology, 427, 515

C

CAMPFIRE. *See* Communal Areas Management Programme For Indigenous Resources
Carbon-14, 24
Carbon cycle, 15, 16, 19, 408
Carrying capacity, 81, 87, 154, 312, 355–358, 387, 388, 408
Cellular automata, 351, 359, 365–366
CFC, 5, 42–44, 59, 60, 234, 322, 327, 409, 460
Chain Management by Life Cycle Assessment (CMLCA), 375
Characterization, 381–382, 518, 522
Chemical water treatment, 66, 150, 153, 158, 316, 318–321, 403, 515
Classification, 113, 160, 163, 178, 189, 213, 230, 242, 267, 277, 312, 345, 354, 381–382, 404, 413, 434, 441
Climate change, 8, 10, 29, 44, 53, 56, 57, 62, 66–69, 91, 125, 147, 170, 175, 197, 198, 200, 224, 226, 229, 230, 235, 248, 249, 268, 269, 288, 289, 291, 292, 295–297, 316, 322, 351, 365, 369, 386–391, 400–402, 406–410, 414–417, 443, 459–470, 477, 504, 507, 520
Climate change perceptions, 464–470
Climate change policy, 408, 459–470
Club of Rome, 7, 13, 208, 387, 401, 478
CMLCA. *See* Chain Management by Life Cycle Assessment, 375
Collective good, 98, 111, 132, 135, 242, 257, 260, 261, 498
Common goods, 111–113, 116, 118, 125, 140
Communal Areas Management Programme For Indigenous Resources (CAMPFIRE), 446, 447, 453–455
Comparative politics, 425, 426, 428
Complementarity, 347
Complexity, 33, 104, 110, 155, 156, 160, 162, 172–174, 196, 198, 203, 240, 245, 249, 255, 257, 258, 270, 276, 291, 308, 340, 345, 348–351, 386, 387, 396, 417, 460
Condensation, 145, 150, 151, 318, 319
Conservation of energy, 76, 77, 145
Constructive technology assessment, 311
Constructivist, 5, 87, 93, 287, 294, 310, 350, 427
Cost benefit analysis, 198, 210
Cost effectiveness analysis, 195–198
Critics of technology, 306, 309
Cross-national surveys, 426–427
Culling of species, 322, 449, 450, 452
Cultural bias, 86, 440
Culture, 30, 69, 86, 93, 113, 133, 134, 136, 137, 212, 240, 245, 246, 247, 249, 261, 263, 264, 284, 289, 296, 309, 317, 379, 412, 424–426, 433, 437, 439–443, 462, 465–469, 481, 486, 487, 493, 499
CYCLES model, 402–404

D

DALE. *See* Disability Adjusted Life Expectancy
DALY. *See* Disability Adjusted Lost Years
Data analysis, 102–106, 112
Defensive expenditure, 190–191
Demand, 12, 35, 45, 52, 53, 106–108, 110, 117, 121, 123, 126, 135, 138, 153, 157, 158, 164, 171, 178–185, 196, 197, 222, 226, 261, 281, 306, 310, 311, 326, 352, 355, 361, 363, 365, 367
Dematerialisation, 87, 90
Differential equations, 346, 351, 353, 354–356, 358, 359, 361, 365, 366, 367, 369
Dilemma, 8, 87, 93, 98, 106, 109, 111, 115–123, 126–128, 132, 133, 159, 174, 219, 243, 260–262, 265, 281, 282, 295, 300, 332, 339, 351, 367, 424
Disability Adjusted Life Expectancy (DALE), 338
Disability Adjusted Lost Years (DALY), 49, 215, 338, 340
Distributive justice, 241, 257, 262, 474, 476, 485
Dose-effect relation, 63, 64
Driving force, Pressure, State, Impact, Response (DPSIR), 209, 211, 224, 233, 352, 390, 391, 402
Dynamic simulation, 171, 354, 360–369

E

Eco-efficiency, 210, 526
Ecological footprint, 9, 11–13, 338, 354
Ecological history, 25, 30
Ecological modernization
Economic index, 522–527
Economic philosophy, 484–485
Economic sustainability, 521–522
Economic valuation, 193, 269, 500
Economy, 7, 8, 10, 11, 27, 29, 40, 43, 44, 46, 47, 87–92, 94, 96, 134–136, 138, 177–204, 207–232, 244–246, 249, 251, 256, 261–263, 270, 292, 312, 328, 332
Ecosystem
    health, 87, 92, 167, 170, 172, 174, 233, 364
    organisation, 155
    resilience, 155, 170, 171, 359, 366, 489
    services, 59, 89, 171, 195
    stability, 91, 155, 169, 171, 364, 440, 489
    sustainability
    vigour, 170, 171
Effects of pollution, 168
Electrostatic precipitators, 318
E=MC^2, 76
EMEP. *See* European Monitoring and Evaluation Programme
Emissions, 6, 7, 12, 17, 23, 27, 42–44, 50, 56, 59–61, 63, 65–67, 89, 98, 99, 123, 125, 135, 138, 144, 147, 154, 177, 183, 184, 188, 189
Empirical study, 99, 104–106
End-of-pipe technologies, 312, 313, 317–324, 328
Energy, 10–13, 15, 27, 33, 38–41, 43–46, 50–53, 61, 66, 67, 76–83, 92, 98–100, 104, 110, 113–115, 121, 123, 131, 135, 138, 139, 144–147, 150, 151, 156, 157, 169–171
    accounting, 182
    balance, 66, 144–147, 174, 405, 416
    budget, 156
    efficiency, 33, 44, 114, 123, 138, 197, 198, 242, 324, 325, 329, 332, 404, 462, 463, 469, 470
    flow, 156–157
    resources, 27, 46, 479, 493, 515, 516
    storage, 77, 325, 331
    technology, 330–332
Enthalpy, 150, 151
Entropy, 77–81, 150, 169, 208
Environment, 1–13, 15, 16, 19–30, 33–47, 49–52, 54, 58–61, 64, 65, 67–70, 75
    accounting, 111, 177, 182, 185–192
    behaviour, 97–139
    change, 19–24, 26, 30, 139, 158, 188–189, 200, 401, 427–429
Environmental flow, 184–185 (it is like environment commodity flow)
Environmental function, 4, 188–190, 195, 483, 491–493, 495, 497, 499, 500
Environmental history
    approaches, 27
    methodology, 25–30
    object, 25–30
    of deforestation, 25, 37
    of energy production and consumption, 27
    of global warming, 43–44, 46
    of industrialisation, 25, 39–44
    of inequality, 40–41
    of ozone depletion, 42, 46
    periodisation, 26–28
    of pollution, 25–28, 35, 41–44, 46
    of population, 28, 30, 34–38, 40, 41, 43, 45, 46
    of resource consumption, 41, 46
    of soil degradation, 46
    sources, 28–30
Environmental impact, 6, 7, 10, 11, 19, 35, 41, 50, 65, 90, 94, 98, 99, 106–115, 122–124, 133, 135–137, 157, 177, 183, 186, 190–192, 196, 208–210, 224, 225, 229, 231–234, 253, 273, 306, 316, 327, 328, 338, 340–342, 386, 393, 396, 410, 503, 511, 527
Environmental impact assessment, 208–210, 269, 273, 341, 342, 527
Environmental index, 522–525
Environmental knowledge, 423, 424, 426, 428, 429, 440
Environmental liability, 87–90, 252
Environmental management systems, 126, 210, 317
Environmental modelling, 21, 345–371
Environmental problems, 3–9, 26, 30, 33, 40, 46, 50, 51, 61, 70, 76, 85, 86–88, 93–95, 98, 99, 103, 106, 117, 120, 123, 125, 126, 130, 132, 134, 136–139, 143, 144, 151–153, 157, 158, 160, 174, 175, 191
Environmental qualities, 4, 9, 65, 98, 106, 112–116, 127, 132, 133, 134, 136–139, 143, 144, 159, 165, 167–174, 177, 185, 186, 188, 193, 247, 249, 252, 256, 257, 259, 260, 278, 286, 338, 339, 340, 387, 389, 401, 503, 504, 519, 523, 524, 526, 527
Environmental restoration, 188, 322–324
Environmental risk analysis, 211–219
Environmental risk assessment, 209–219
Environmental risk perception, 109–110, 140
Environmental sciences, 3–13, 19, 140, 163, 167, 172, 360, 370
Environmental scientists, 3, 8–9, 59, 76, 143, 144, 149, 151, 164, 165, 170, 174

Environmental security, 87, 91
Environmental stress, 102, 106–110, 120, 464
Environmental sustainability, 13, 98, 135–137, 189, 190, 192, 193, 195, 196, 210, 504, 519–521
Environmental technologies, 263, 305, 306, 312–316, 354, 494
Environmental utilization space, 92
Environmental valuation, 188–189, 192, 193, 199, 500
Equilibrium, 77–83, 87–89, 126, 129, 130, 150, 151, 155, 174, 188, 197, 210, 217, 218, 347, 355, 356, 358, 360, 361, 369, 492
European Monitoring and Evaluation Programme (EMEP), 394–396, 399, 400
Evolution, 7, 19, 21, 23, 29, 34, 78, 80–83, 283, 285, 291, 311, 346, 348, 349, 350, 351, 357, 366, 369, 371, 396, 412, 426, 436, 443, 488, 500, 518
Evolution theory, 87, 311
Exponential growth, 7, 9, 309, 354, 355
Extinction, 6, 8, 45, 56–58, 79, 82, 85, 94, 154, 269, 307, 322, 339, 450, 451, 475, 489, 493, 494, 499
Extrapolations, 217, 218, 234, 492–494

F
Factor X, 11, 328
Filters, 319
Filtration, 321
Financial flow, 191–192
Finite resources, 473–481
Fisheries, 50, 173–174, 363
Flaring, 319
Flotation processes, 320
Flow, 143, 145, 149, 156–157, 184–185, 191–192, 207, 209–210, 219–224, 230, 361, 376, 403
Flow accounting, 209, 210
Fluorescent bulb, 377–378, 380–382
Food, 6–9, 11–12, 15, 33–38, 41–45, 55–56, 60–61, 81–82, 91–92, 113–115, 134, 144, 154, 156–158, 165–166, 169, 171, 173, 179, 191, 212, 216–217, 251, 266, 319, 321, 326, 331, 352, 356–358, 361–365, 367, 388, 401–407, 448, 452–453, 493, 504, 513–516, 521–522
Formal system, 345–347, 438
Future generations, 8, 13, 64, 88–89, 93, 127, 195, 199–200, 322, 332, 339, 473–476, 480–481, 483, 491–493

G
Gaia theory, 16
Game law, 445, 451, 454
Gas separation, 305, 318–319
GCD. *See* Greatest Common Divisor), 336–337
Geographic Information Systems (GIS), 143, 160, 163–166, 365, 369
GHG. *See* Green House Gases
GIS. *See* Geographic Information Systems
Global Warming Potential (GWP), 147, 230, 250, 265, 268, 272, 278, 336, 338
Good Housekeeping, 273, 305, 306, 312, 313, 316–317
Governance, 28, 53, 239, 242, 259, 282, 292, 301, 411–412, 414, 416, 423, 426, 443, 504, 506, 510
Greatest Common Divisor (GCD), 336–337

Green House Gases (GHG), 6, 15, 17, 23, 44, 50, 60, 65–66, 138, 147, 150, 198, 230, 233, 235, 246, 250–251, 293, 298, 316, 351, 354, 385, 400, 406–410, 414–416, 428–429, 460–464, 469
GWP. *See* Global Warming Potential

H
Hazard, 50, 68, 109, 123, 160, 162, 211–212, 214–217, 224, 271, 294, 298, 312, 319, 321, 324, 423, 427, 434–436, 476–477
Human needs, 45–46, 98, 127, 130–132, 136, 327, 349, 491, 518, 521

I
IA. *See* Integrated Assessment
IAM. *See* Integrated Assessment ModelIBI. *See* Index of Biotic Integrity
Ice Ages, 17, 21–22, 34
IEM. *See* Integrated Environmental Models
IMAGE. *See* Integrated Model to Assess the Greenhouse Effect
Impact assessment, 210, 226, 229–230, 269, 273, 300, 341–342, 376, 381–382, 417, 527
Impact-oriented measures, 115
Incandescent bulb, 376–378, 380–383
Incineration, 221, 226, 254–255, 313, 318–319, 321–322, 325, 376–377, 434–436
Index of Biotic Integrity (IBI), 167, 171, 174
Indicators, 9, 12–13, 21–23, 46, 53, 56, 79, 131, 155, 159–160, 165, 167–175, 187–191, 197, 209, 224, 229–235, 258, 268–269, 271, 287, 338, 340, 352, 354, 358, 364, 370, 381–382, 392, 401–402, 417, 517–528
Individual preferences, 193, 198, 483–488, 490, 495–497
Industrial ecology, 87, 88, 91–92, 208–210, 232, 235–236, 313, 325, 427
Industrial metabolism, 87–88, 92
Innovation, 10, 27, 82, 86, 123, 134, 197, 223, 245–247, 249, 252, 256, 286, 293, 301, 307–309, 311, 329, 359, 367–368, 412–413, 416, 423, 427–428, 438
Innovation style, 427–428
Input-output analysis, 177–185, 191, 209, 232, 270
Institutional background, 441
Institutional characteristic, 436
Institutional factor, 204, 285, 433, 436, 440–441
Institutionalisation, 288–292, 300, 441–442
Institutions, 7, 29, 47, 91–92, 113, 133–134, 136–137, 193, 204, 209, 212, 224, 240–244, 247, 251, 254, 257, 259, 261–262, 264–266, 273, 275, 276
Instrument
    binding -, 246–247, 249–250, 254, 267, 269–270, 272–273, 277
    cultural -, 250, 266, 271–273
    economic -, 7, 86, 88, 263, 266, 271–273, 395
    option creating -, 251, 266, 271–273
    procedural -, 266, 271–273
    regulatory -, 247–248, 266, 270–272, 275, 277
    structural -, 246, 251, 266–267, 271–272, 273, 275
Intergovernmental Panel on Climate Change (IPCC), 65–67, 147, 235, 291, 295, 386, 402, 405, 408–410, 414–415, 460, 477Integrated Assessment (IA), 158, 255, 299, 345, 385–417

540   Subject Index

Integrated Assessment Model (IAM), 345, 386, 389–392, 395–396, 401, 408
Integrated Environmental Models (IEM), 386
Integrated framework, 517–519
Integrated Model to Assess the Greenhouse Effect (IMAGE), 365, 387, 389–390, 401, 407–410, 412, 415, 417
Integrated waste management, 435–436
Integrating concepts, 337
Integration, 9, 11–13, 153, 161, 166, 168, 172, 209, 240, 244–245, 247, 249–254, 268–270, 278, 296, 298, 335–343, 345, 354, 364, 371, 389
Intent-oriented measures, 114–115
Interactive simulation, 354, 370
Interactive technology assessment, 311–312
Intergenerational justice, 87–90, 94, 263, 473–481
Inventory analysis, 226–229, 376–381
IPAT equation, 9–11, 50, 327
IPCC. *See* Intergovernmental Panel on Climate ChangeIrreplaceable goods, 500–501
ISO 14001, 266–267, 273, 317

J
Justice between the generations, 90

K
Kuznets curve, 10
Kyoto protocol, 147, 246, 268, 271–272, 292, 409–410, 460–462

L
Landfill, 66, 227–228, 252, 254, 315–316, 321, 325, 434–435
Land use, 11–12, 50, 56–57, 134, 138, 154, 163, 166, 211, 221, 229–230, 233, 241, 250, 272, 331, 351, 365–366, 369, 390, 403–404, 408–409, 415, 445–456, 494, 503
Law, 28, 76–78, 80–82, 85–88, 90–91, 123–124, 126, 129, 137, 143, 145, 150–151, 156–157, 174, 208, 234, 240–241, 243, 246, 256–259, 265, 272–273, 276–277, 281, 340–341, 345–349, 368, 438, 450–454, 464, 491, 498,
LCA. *See* Life Cycle Assessment
LCM. *See*Least Common Denominator, 336–337, 340–341
Learning by adaptation, 443
Least Common Denominator (LCM), 336–337, 340–341
Liability, 87–91, 94, 208, 239, 241, 243, 249–250, 252, 256, 266–267, 270–271, 273, 275–276, 279, 450–451
Life cycle analysis, 135
Life Cycle Assessment (LCA), 135, 209–210, 220–221, 224–234, 250, 258, 268–270, 339, 354, 375–383
Life support systems, 491–494
Long Range Transboundery Air Pollution (LRTAP) Convention, 389–390, 392, 394–396, 400–401, 408, 417
Lurking inconsistency, 487–489

M
MAS. *See* Multi-agent Simulation
Mass balance, 144–145, 208–210, 232
Mass transport, 28, 148–149, 159, 174
Material flow analysis, 145, 219–224

MCA. *See* Multi Criteria Analysis
Measurement scale, 99–100, 104
Measuring impact, 9–13
Measuring sustainable development, 13
Milankovitch cycle, 17, 19, 20, 22
MIPS, 9, 11–13, 210, 338
Model Implementation, 405, 412–414
Multi-agent Simulation (MAS), 351, 360, 366–370
Multi-attribute evaluation, 112–113
Multi Criteria Analysis (MCA), 177, 201–204, 210, 337–338
Multilevel perspective, 506–508
Myths of Nature, 128–130, 137–138

N
National character, 424–425, 433
National income, 186, 190, 240–241, 483–485, 491–497, 499–500
National policy style, 433–443
Natural resources, 6, 9, 11, 50, 52–55, 68–69, 86, 88–89, 92, 129–130, 137, 185–186, 189, 285, 309, 340, 367, 377, 387–388, 405, 407, 445, 453–454, 492–494, 513, 518, 523–525
Nature, 4–5, 7–8, 10, 12, 19, 21, 26, 29–30, 44, 49–50, 52–58, 63, 69, 75–76, 78, 85–92, 94–95, 98, 105–107, 110–112, 122–123, 128–130, 137–138, 143, 152, 154–155, 163, 167, 170, 172, 174, 193, 203, 207
   views on, 128, 283
   appreciation of, 106, 110–111
   compensation, 87, 92
   conservation, 26, 57, 69, 86, 91, 172, 370, 446, 469, 519
Needs, Opportunities and Abilities (NOA), 120, 125, 127
Negotiation, 44, 50, 59–60, 87, 93–94, 129, 147, 252, 256, 258–259, 281, 289–290, 292–295, 341–343, 390, 392, 394–401, 417, 428, 439, 442, 459, 461–462, 467–470, 509, 519
Neo-Darwinist, 488–489
Nitrogen cycle, 16, 61, 157–158
NOA. *See* Needs, Opportunities and Abilities
Noise, 58–59, 64–65, 85, 98–99, 102, 106–107, 134, 136–137, 154, 239, 259, 268, 271, 273, 313, 318, 322, 331, 480
Non-renewable resources, 7, 52–53, 82–83, 188, 195–196, 313, 325, 329, 358
Normalization, 382
Nuclear power, 109, 214–215, 232, 288, 297, 306, 328, 427
Nutrients, 56–58, 61, 63–64, 152–153, 155–158, 160, 169–170, 174, 208, 275, 359, 448, 492, 504, 514

O
Objective value, 483–490, 495, 497, 499
Obligations to future generations, 476, 480–481
Ocean storage of CO^2, 323
Organizing concepts, 352–353
Ozone, 13, 15, 19, 42–43, 46, 60, 63, 65, 89, 152, 224, 230, 235, 248, 250, 254, 272, 289, 292, 321, 346, 385, 387, 389, 390, 393–394, 396–397
Ozone layer, 5, 15, 42–43, 59–60, 65, 152, 266, 316, 322, 346, 492, 518

Sustainable development, 6, 8, 9, 75–78, 82, 83, 89, 90, 98, 101, 112, 127, 128, 131–134, 137–139, 185, 189, 193, 195, 207, 305, 306, 309, 313, 330, 331, 352, 353, 355, 360–365, 367, 370, 371, 401–407, 411, 414, 416, 428, 429, 442, 473, 474, 476, 503, 510, 517–528

Sustainable development of nations, 517–528

Sustainable national income (SNI), 190, 483, 484, 491, 492, 495–497, 499

Sustainable production and consumption, 313, 327–330

Sustainable technologies, 135, 262, 306, 312, 313, 327, 329, 332, 443

System, 9, 11–13, 26, 29, 38, 41, 46, 49, 59, 60, 64, 77–83, 87, 91, 94, 114, 123, 125, 126, 130, 133, 135, 137, 144, 145

Systems innovations, 329, 504–508, 510

T

TARGETS. *See* Tool to Assess the Regional and Global Environmental and Health TargeTs

Tax, 117, 183, 242, 248, 251, 252, 262, 266, 268, 276, 363, 390, 406, 508

Technical progress, 346, 414, 473–481

Technological change, 85, 306, 307, 310–312, 328, 329, 332, 462, 466, 468, 504, 505, 511

Technological choice, 309

Technological determinism, 310, 311

Technological niche, 507

Technology
  assessment, 210, 296, 298, 301, 311, 332
  policy, 427–428
  and society, 306–310
  and Science, 242, 297, 298, 308, 310, 348, 392, 426, 439, 465, 467, 516

Teleology, 487–490

Thermodynamics, 76, 77, 145, 149–152, 156, 157, 174, 208, 217, 308, 346–348, 354, 379

Threshold Limit Value (TLV), 338

Tool to Assess the Regional and Global Environmental and Health TargeTs (TARGETS), 391, 401–408, 417

Totalitarianism, 486, 488

Tradable right, 249

Tragedy of the commons, 93, 243, 257, 260, 261, 265, 453–454

Transformation, 27, 36, 40, 61, 76, 77, 100, 151–153, 159, 163, 164, 209, 216–220, 307, 308, 313, 332, 365, 392, 442, 455, 504, 505, 507–509

Transition, 22, 25, 27, 34, 37, 39, 170, 352, 353, 365, 367, 402, 404, 411–413, 427, 503–511

Transport, 6, 7, 12, 19, 21, 27, 28, 38, 39, 45, 46, 57, 59, 61, 68, 69, 85, 98, 99, 104, 109, 112, 115, 117, 123, 125, 134, 135, 138

TRIAD method, 167–168, 174

Trickling filters, 320

Triple D technologies, 312, 315, 316

Triple M technologies, 313, 316–317

U

Unborn people, 474–476

Uncertainty, 88, 94, 118, 119, 158, 169, 172, 200, 202, 210, 217, 218, 232, 234–235, 250, 295, 339, 342, 348, 351, 367, 390–391, 396, 402, 405, 415, 438, 479, 499, 505–506

United Nations Framework Convention for Climate Change (UNFCCC), 460, 461

V

Value and capital, 496

Value orientation, 112, 118, 119, 128–131, 137, 369

Values, 11, 13, 56, 82, 98, 108, 117, 123, 127–137, 140, 148, 151, 163, 167–169, 171, 172, 174

Vicious circle, 82–83

Vinyl chloride, 326

Vorsorge, 438, 441–443

W

Waste
  heat, 150, 316
  incineration, 221, 322
  management, 138, 191, 192, 222, 223, 225, 227, 228, 229, 252, 273, 288, 433–443
  policy, 433–436, 440–443

Wastewater treatment, 66, 150, 153, 158, 318–320, 515

Water
  disinfection, 321
  resources, 52–54, 91, 106, 136, 493, 514–515, 520, 523

Weighting, 112, 113, 169, 199–204, 211, 212, 215, 230, 231, 234, 268, 269, 337, 339, 382, 476

Welfare, 9, 10, 11, 13, 99, 112, 185, 187, 188, 190, 191, 197–200, 241, 263, 264, 269, 353, 404, 405, 427, 480, 481, 483, 488, 491, 495, 498, 503

Wet scrubbing, 318, 319

Wildlife, 105, 149, 152, 316, 324, 327, 331, 338, 339, 446–456

Z

Zoo breeding programmes, 322

Subject Index

P

Papermaking, 307, 314
Perceived risk, 106, 108–110, 136, 212, 232
Performance measures, 100–102, 273
Permit, 91–92, 103, 105, 109, 125–126, 145, 181, 191–192, 197, 223, 231–232, 239–240, 244, 248, 250, 252–254, 256–258, 261, 265–268, 270–279, 326, 340, 345, 446, 454, 480, 493, 505
Perpendicular demand curve, 483–485, 495, 498
Persistent organo-chlorine compounds, 316
Philosophical issue, 486–487
Physical flow, 191
Physical space, 50, 68–70, 69, 163
Phytoremediation, 323
Plastics, 11, 113–114, 179, 215, 313, 321, 325–326, 329–330
Policy, 5–8, 10, 28–29, 40, 42, 47, 67, 69, 85, 88, 90, 92–93, 98–102, 104, 106, 110, 111
Policy
    accommodation, 294–295
    assessment, 390
    design, 139, 243–244, 273–274, 276
    development, 99, 208, 239–242, 251–253, 257, 276, 284, 385, 388, 390, 433–436
    efficiency in -, 210, 253
    equity in-, 249, 252–253, 413
    evaluation, 203–204, 387
    goal, 8, 124, 127, 240, 242, 254, 269–270, 284, 442
    implementation, 239–241, 251, 270, 274–276, 433, 436, 461
    integration, 252–254, 268–269, 278
    learning, 295–296, 299, 301
    negotiation, 461–462
    participatory -, 283
    preferences, 100, 459, 462, 465, 468–469
    process, 239–242, 276, 284–285, 287, 289, 293–294, 299, 300, 400, 429, 436, 438, 443
    style, 285, 294, 405–406, 433–443
Political culture, 284–285, 437
Polluted soil, 313, 323–324
Polluter pays principle, 7, 87–89, 248, 251–253, 256, 262, 275–276, 279
Pollution, 5–7, 9–10, 25–28, 33, 35, 41–44, 46, 50, 52, 54, 58–67, 85–86, 88, 90–91, 99, 106–107, 116–117, 119, 134
Population, 7, 9–11, 28, 30, 33–38, 40–41, 43, 45–46, 50–51, 53–57, 61, 64–65, 68–69, 76, 80–82
Population dynamics, 154–155, 174
Precautionary politics, 438, 441, 443
Precautionary principle, 87–88, 94, 158, 200, 213, 292, 301, 438, 443, 493, 499, 520
Preferences, 100–101, 104, 110, 112–113, 122, 131, 135–136, 137, 193, 198–199, 201, 203–204, 230, 233, 253, 261, 264, 269, 278, 289, 306, 342
Prisoner's dilemma, 243, 257, 260, 261, 265
Process adaptation, 306, 312, 324–327
Public opinion, 285, 426–427
Purposeful perpendicular, 489–490

Q

Quality, 3–13, 35, 52, 53, 57, 64, 65, 77, 92, 98, 102, 106, 111, 112, 114, 116, 117, 125, 126, 127, 128, 130, 134, 137, 139, 143
Quality of life, 41, 98, 101, 127–133, 135, 138, 139, 140, 412

R

Radiation, 15, 16, 59, 64–66, 145–147, 165, 215, 230, 313, 315, 318, 322, 328, 403, 406, 460
Raw materials, 98, 103, 134, 138, 171, 208, 225, 227, 314, 324–326, 340, 406, 518
Rebound effect, 197, 198, 329, 332
Reconstructing environmental change, 19–24
Regional Acidification INformation and Simulation (RAINS)-model, 343, 387, 392–401, 408
Reintroduction of extinct species, 324
Remote sensing, 165–167, 174
Renewable energy, 202, 324, 330, 331, 351, 479, 492–494, 515
Renewable resources, 52, 53, 82, 195
Resource accounting, 351, 354
Resource accounting concepts, 354
Resource productivity, 87, 90
Resources, 6, 7, 9, 11, 12, 25, 27, 29, 34, 35, 38–41, 44, 45, 46, 50
Restoration technology, 312, 313, 318
Risk, 8, 53, 58, 62, 64, 65, 87, 88, 89, 93, 94, 99, 106–110, 116–120, 122–124, 126, 129, 130, 136, 139, 140
Romanticism, 26, 424, 440–442

S

Savannah, 447–450
Savings
    genuine, 189
    net, 189
Scenario analysis, 223, 235, 342, 351, 391–394
Screens, 320
Sedimentation, 166, 320, 356
Sewerage system, 41, 49, 153, 314, 315
Smokestack, 3, 6, 41, 315
SNI. *See* Sustainable National Income
Social construction, 87, 109, 275, 424, 427, 440
Social index, 522–525
Social sustainability, 137, 521
Societal innovation, 503–511
Societal obstacles, 516
Socio-technical regime, 506, 508, 510, 511
Soil and land resources, 513–514
Solar energy, 15, 77, 80, 81, 146, 208, 230, 330, 331, 479–480, 513, 515
Solid waste treatment, 321–322
Solutions in context, 5–9
Southern Africa, 445–456
Space debris, 313, 322, 323
Special Report on Emissions Scenarios (SRES), 408–411, 413–416
Stakeholder approach, 342
Steady state economy, 87–90, 92, 94
Study design, 102–105
Subjective preferences, 483–487, 490, 495
Substance flow analysis, 209, 210, 220
Sulphur cycle, 63, 503
Supply curve, 485, 495, 499, 500, 501
Sustainability index, 517, 522–524
Sustainability scale, 522–524, 527